Potato and Sweetpotato in Africa

Transforming the Value Chains for Food and Nutrition Security

FSC
www.fsc.org
MIX
Paper from responsible sources
FSC® C013604

Potato and Sweetpotato in Africa

Transforming the Value Chains for Food and Nutrition Security

Edited by

Jan Low

International Potato Center (CIP), Nairobi, Kenya (CGIAR)

Moses Nyongesa

Kenya Agricultural and Livestock Research Organization

Sara Quinn

International Potato Center (CIP), Nairobi, Kenya (CGIAR)

Monica Parker

International Potato Center (CIP), Nairobi, Kenya (CGIAR)

www.cabi.org

CABI is a trading name of CAB International

CABI
Nosworthy Way
Wallingford
Oxfordshire OX10 8DE
UK

Tel: +44 (0)1491 832111
Fax: +44 (0)1491 833508
E-mail: info@cabi.org
Website: www.cabi.org

CABI
745 Atlantic Avenue
8th Floor
Boston, MA 02111
USA

Tel: +1 (617) 682-9015
E-mail: cabi-nao@cabi.org

A catalogue record for this book is available from the British Library, London, UK.

Library of Congress Cataloging-in-Publication Data

Potato and sweetpotato in Africa : transforming the value chains for food and nutrition security / edited by Dr Jan Low, International Potato Center (CIP), Nairobi, Kenya (CGIAR); Moses Nyongesa, Kenya Agricultural and Livestock Research Organization; Sara Quinn, International Potato Center (CIP), Nairobi, Kenya (CGIAR); and Monica Parker, International Potato Center (CIP), Nairobi, Kenya (CGIAR).
pages cm
Includes bibliographical references and index.
ISBN 978-1-78064-420-2 (hbk : alk. paper)
1. Potatoes--Breeding--Africa, Sub-Saharan. 2. Sweet potatoes--Breeding--Africa, Sub-Saharan. 3. Potatoes--Varieties--Africa, Sub-Saharan. 4. Sweet potatoes--Varieties--Africa, Sub-Saharan. 5. Potato industry--Africa, Sub-Saharan. 6. Sweet potato industry--Africa, Sub-Saharan. I. Low, Jan Wayland, editor.

SB211.P8P772 2015
338.1'75210967625--dc23

2015020015

ISBN-13: 978 1 78064 420 2

Commissioning editor: Joris Roulleau
Editorial assistant: Emma McCann
Production editor: Lauren Povey

Typeset by SPi, Pondicherry, India.
Printed and bound in the UK by CPI Group (UK) Ltd, Croydon, CR0 4YY.

Contents

Contributors

Abidin, P.E., International Potato Center (CIP)-Ghana c/o CSIR-SARI, PO Box 52 Nyankpala, Tamale, Ghana. E-mail: p.abidin@cgiar.org (Previous address: International Potato Center (CIP)-Malawi, PO Box 31600, Lilongwe, Malawi.)

Abong', G., Department of Food Science, Nutrition and Technology, University of Nairobi, PO Box 290053-00625, Nairobi, Kenya. E-mail: georkoyo@yahoo.com

Adofo, K., Council for Scientific and Industrial Research – Crops Research Institute (CSIR-CRI), PO Box 3785, Kumasi, Ghana. E-mail: kinfodda@yahoo.com

Afolabi, M.S., Department of Crop and Soil Science, Landmark University, Km 4 Ipetu,Omu Aran Road, PMB 1001, Omu Aran, Kwara State, Nigeria. E-mail: afolabimike97@yahoo.com

Afuape, S.O., National Root Crops Research Institute (NRCRI), Umudike, Ikot Ekpene-Umuahia Road, Umuahia, Abia State/southeastern PMB 7006, Nigeria. E-mail: solomonafuape@yahoo.com

Aggrey, B.N., Department of Horticulture, Jomo Kenyatta University of Agriculture and Technology, PO Box 62000-00200 Nairobi, Kenya. E-mail: abnyende@hotmail.com

Agili, S., International Potato Center Sub-Saharan Africa (CIP-SSA), PO Box 25171, Nairobi 00603, Kenya. E-mail: s.agili@cgiar.org

Aiyelaagbe, I.O., Department of Horticulture, Federal University of Agriculture, PMB 2240, Abeokuta, Nigeria. E-mail: ola_olu57@yahoo.com

Akoroda, M.O., Executive Director, Cocoa Research Institute, Idi-Ishin, Ibadan, Oyo State, Nigeria. E-mail: akoroda@yahoo.com (Previous address: Department of Agronomy, University of Ibadan, Ibadan, Nigeria.)

Alao, B.I., Agricultural and Rural Management Training Institute (ARMTI), Km 18, Ilorin Ajase-Ipo Highway, PMB 1343, Ilorin, Kwara State, Nigeria. E-mail: alaobashir@yahoo.co.uk

Almekinders, C., Wageningen University, Hollandseweg 1, Wageningen 6706 KN, The Netherlands. E-mail: conny.almekinders@wur.nl

Amadi, C.O., National Root Crops Research Institute, Umudike, Ikot Ekpene-Umuahia Road, Umuahia, Abia State/southeastern PMB 7006, Nigeria. E-mail: okeyamadi2003@yahoo.com

Amagloh, F.K., Food Processing Technology Unit, Faculty of Agriculture, University for Development Studies, PO Box TL 1882, Nyankpala Campus, Tamale, Ghana. E-mail: fkamagloh@uds.edu.gh

Andrade, M., International Potato Center (CIP), c/o Instituto de Investigacão Agraria de Moçambique (IIAM), Avenida das FPLM 2698, PO Box 2100, Maputo, Mozambique. E-mail: m.andrade@cgiar.org

Angsten, A., Johns Hopkins University, Baltimore, MD 21218, USA. E-mail: anne.angsten@gmail.com

Anil, S.R., Central Tuber Crops Research Institute (CTCRI), Thiruvananthapuram, Kerala, India. E-mail: shirlyanil@yahoo.com

Ano, A.O., National Root Crops Research Institute, Umudike, Ikot Ekpene-Umuahia Road, Umuahia, Abia State/southeastern PMB 7006, Nigeria. E-mail: aoanoh@yahoo.com

Anthony, N.M., PO Box 1983, Sapon, Abeokuta, Ogun State, Nigeria. E-mail: nyahabehanthony@yahoo.co.uk (Previous address: Sierra Leone Agricultural Research Institute, Njala Agricultural Research Centre (NARC), Njala, PMB 540, Freetown, Sierra Leone.)

Anyaegbunam, H.N., Umuahia Extension Services Programme, National Root Crops Research Institute, Umudike, Ikot Ekpene-Umuahia Road, Umuahia, Abia State/southeastern PMB 7006, Nigeria. E-mail: helenkol8@yahoo.com

Asante, K.I., West Africa Centre for Crop Improvement (WACCI), University of Ghana, Legon, PMB 30, Accra, Ghana. E-mail: asanteisaack@yahoo.com

Asfaw, A., International Potato Center (CIP), Sub-Saharan Africa Region, Nairobi, Kenya. E-mail: a.amele@cgiar.org

Asumugha, G.N., Umuahia Extension Services Programme, National Root Crops Research Institute, Umudike, Ikot Ekpene-Umuahia Road, Umuahia, Abia State/southeastern PMB 7006, Nigeria. E-mail: gnasumugha@yahoo.com

Atayese, M.O., Department of Plant Physiology and Crop Production, Federal University of Agriculture, PMB 2240, Abeokuta, Nigeria. E-mail: atayesem2000@yahoo.com

Attaluri, S.I., International Potato Center South, West and Central Asia (CIP-SWCA), Liaison Office, Regional Center of Central Tuber Crops Research Institute (RC CTCRI) Dumduma, Bhubaneswar-751019, Odisha, India. E-mail: s.attaluri@cgiar.org

Aura, J., PO Box 30197, GPO, University of Nairobi, Nairobi, Kenya. E-mail: aurajoel@yahoo.com

Bakare, H.A., Department of Food Service and Tourism, Federal University of Agriculture, PMB 2240, Abeokuta, Nigeria. E-mail: bakareha@funaab.edu.ng

Ball, A., HarvestPlus, Plot 15, East Naguru Road, PO Box 28565, Kampala, Uganda. E-mail: a.ball@cgiar.org

Bararyenya, A., Institut des Sciences Agronomiques du Burundi, PO Box 795, Bujumbura, Burundi. E-mail: barastere@yahoo.fr

Barker, I., Syngenta Foundation for Sustainable Agriculture, Schwarzwaldalle, 215, CH-4002, Basel, Switzerland. E-mail: ian.barker@syngenta.com

Belem, J., Institut de l'Environnement et de Recherches Agricoles (INERA), Burkina Faso, 01 BP 476, Ouagadougou, Burkina Faso. E-mail: jebelem2000@yahoo.fr

Bellstedt, D.U., Department of Biochemistry, University of Stellenbosch, Private Bag X1, 7602 Matieland, South Africa. E-mail: dub@sun.ac.za

Biryomumaisho, B., Uganda National Seed Potato Producers' Association (UNSPPA), PO Box 968, Kabale, Uganda. E-mail: bbirybenia@yahoo.co.uk

Bokosi, J.M., Department of Crops and Soil Science, Lilongwe University of Agriculture and Natural Resources, Bunda College Campus, PO Box 219, Lilongwe, Malawi. E-mail: jmbokosi@yahoo.com

Bonierbale, M., Genetics and Crop Improvement Global Program, International Potato Center (CIP), Apartado Postal 1558, Lima, Peru. E-mail: m.bonierbale@cgiar.org

Borus, D., University of Tasmania, Private Bag 45, Hobart 7001, Australia. E-mail: Dinah.Borus@utas.edu.au (Previous address: International Potato Center (CIP), Sub-Saharan Africa Region, Nairobi, Kenya.)

Botha, B., Irish Aid, PO Box 30735, Lilongwe, Malawi. E-mail: blessingsbotha@yahoo.co.uk

Boukhris-Bouhachem, S., Plant Protection Laboratory, National Agricultural Research Institute of Tunisia, Hédi Karray Avenue, 2049, Ariana, Tunisia. E-mail: bouhachem.sonia@iresa.agrinet.tn

Bwembya, S., Zambia Agriculture Research Institute, Mutanda Research Station, PO Box 110312, Solwezi, Zambia. E-mail: sbwembyac@yahoo.com

Carey, E.E., International Potato Center (CIP)-Ghana, c/o Council for Scientific and Industrial Research, Crops Research Institute (CSIR-CRI), PO Box 3785, Fumesua, Kumasi, Ghana. E-mail: e.carey@cgiar.org

Carhuapoma, P., Crop Management and Production Systems Division, International Potato Center (CIP), PO Box 1558, Lima 12, Peru. E-mail: p.carhuapoma@cgiar.org

Chaieb, I., Plant Protection Laboratory, National Agricultural Research Institute of Tunisia, Hédi Karray Avenue, 2049, Ariana, Tunisia. E-mail: ikbal_c@yahoo.fr

Chalwe, A., Zambia Agriculture Research Institute, Mutanda Research Station, PO Box 110312, Solwezi, Zambia. E-mail: ablechalwe@yahoo.com

Chilanga, T., Department of Agricultural Research Services (DARS), PO Box 2054, Blantyre, Malawi. E-mail: tchilanga@yahoo.co.uk

Chiona, M., Zambia Agriculture Research Institute (ZARI), PO Box 710129, Mansa, Luapula, Zambia. E-mail: martinchiona@yahoo.com

Chipungu, F., International Potato Center (CIP), 38 Jonateri Road, PO Box 511197, Chipata, Zambia. E-mail: f.chipungu@cgiar.org

Coad, J., Institute of Food, Nutrition and Human Health, College of Health, Te Kura Hauora Tangata, Massey University, Palmerston North, New Zealand. E-mail: j.coad@massey.ac.nz

Custers, R., Vlaams Instituut voor Biotechnologie (VIB), Rijvisschestraat 120, BE9052 Ghent, Belgium. E-mail: rene.custers@vib.be

Dalyop, T.Y., National Root Crops Research Institute, Potato Programme, Kuru, Jos, Plateau State 930101, Nigeria. E-mail: dalyopty@yahoo.com

Danquah, Y.E., West Africa Centre for Crop Improvement (WACCI), University of Ghana, Legon, PMB 30, Accra, Ghana. E-mail: edanquah@wacci.edu.gh

Demo, P., International Potato Center (CIP), Malawi Office, Area 11 Plot 36, Lilongwe, PO Box 31600, Lilongwe 3, Malawi. E-mail: p.demo@cgiar.org

Diaz, F., International Potato Center (CIP), PO Box 1558, Lima 12, Peru. E-mail: f.diaz@cgiar.org

Diaz, L., International Potato Center (CIP), PO Box 1558, Lima 12, Peru. E-mail: l.diaz@cgiar.org

du Plooy, C.P., Agricultural Research Council-Roodeplaat Vegetable and Ornamental Plant Institute (ARC-VOPI), Private Bag X293, East Lynne, Pretoria 0001, South Africa. E-mail: iduplooy@arc.agric.za

du Toit, E., Department of Plant Production and Soil Science, University of Pretoria, Pretoria 0002, South Africa. E-mail: nicodt@gwk.co.za

Dung, E.A., National Root Crops Research Institute, Potato Programme, Kuru, Jos, Plateau State 930101, Nigeria. E-mail: dungea2003@yahoo.com

Ebeniro, C.N., National Root Crops Research Institute, Umudike, Ikot Ekpene-Umuahia Road, Umuahia, Abia State/southeastern PMB 7006, Nigeria. E-mail: christyebeniro@yahoo.com

Ekwe, C.C., National Root Crops Research Institute, Umudike, Ikot Ekpene-Umuahia Road, Umuahia, Abia State/southeastern PMB 7006, Nigeria. E-mail: chiomaekwe@yahoo.com

Ekwe, K.C., National Root Crops Research Institute, Umudike, Ikot Ekpene-Umuahia Road, Umuahia, Abia State/southeastern PMB 7006, Nigeria. E-mail: kcekwe@yahoo.com

El Fahem, M., Groupement Interprofessionnel des Légumes (GIL), Rue Maouia Ibn Abou Sofiène, Le Belvédère, BP 178, 1002 Tunis, Tunisia. E-mail: info@agripack-sa.com

Esan, T.A., Department of Food Science and Technology, Federal University of Agriculture, PMB 2240, Abeokuta, Nigeria. E-mail: esantitilayo@gmail.com

Eyzaguirre, R., International Potato Center (CIP), Apartado 1558, Lima, Peru. E-mail: r.eyzaguirre@cgiar.org

Faber, M., South African Medical Research Council, PO Box 19070, Tygerberg 7505, South Africa. E-mail: mieke.faber@mrc.ac.za

Farayola, C.O., Agricultural Development and Management Department, Agricultural and Rural Management Training Institute (ARMTI), Km 18, Ilorin Ajase-Ipo Highway, PMB 1343, Ilorin, Kwara State, Nigeria. E-mail: walecharless@yahoo.com

Gachuiri, C., University of Nairobi, PO Box 29053-00100, Nairobi, Kenya. E-mail: gachuiri@uonbi.ac.ke

Gastelo, M., Genetics and Crop Improvement Global Program, International Potato Center (CIP), Apartado Postal 1558, Lima, Peru. E-mail: m.gastelo@cgiar.org

Gathaara, V.N., Kenya Agricultural and Livestock Research Organization (KALRO)-Kabete, PO Box 14733-00800, Nairobi, Kenya. E-mail: violet.gathaara@kalro.org

Gheysen, G., Department of Molecular Biotechnology, Ghent University, Coupure links 653, BE9000 Ghent, Belgium. E-mail: godelieve.gheysen@ugent.be

Gibson, R.W., Agriculture, Health & Environment Group, Natural Resources Institute, University of Greenwich, Chatham Maritime, ME4 4TB, UK. E-mail: nosbigrw@yahoo.com

Gichangi, A., Kenya Agricultural and Livestock Research Organization (KALRO)-Njoro, Private Bag Njoro-20107, Njoro, Kenya. E-mail: wanjugu67@gmail.com

Gildemacher, P.R., Royal Tropical Institute (KIT), PO Box 95001, 1090 HA Amsterdam, The Netherlands. E-mail: p.gildemacher@kit.nl

Grant, F.K., International Potato Center Sub-Saharan Africa (CIP-SSA), PO Box 3441, Dar es Salaam, Tarzania. E-mail: f.grant@cgiar.org

Greyling, R., Agricultural Research Council-Roodeplaat Vegetable and Ornamental Plant Institute (ARC-VOPI), Private Bag X293, Pretoria 0001, South Africa. E-mail: rgreyling@arc.agric.za

Griffin, D., Crops Research Centre, Teagasc, Oak Park, Carlow, Ireland. E-mail: denis.griffin@teagasc.ie

Grüneberg, W.J., International Potato Center (CIP), Apartado 1558, Lima, Peru. E-mail: w.gruneberg@cgiar.org

Guaf, E., National Agricultural Research Institute (NARI), Kana Aburu Haus Sir Alkan Tololo Research Centre, PO Box 4415, LAE 411, Morobe Province, Papua New Guinea. E-mail: elick.guaf@nari.org.pg

Hailemariam, G., International Potato Center (CIP), c/o International Livestock Research Institute (ILRI), PO Box 5689, Addis Ababa, Ethiopia. E-mail: gbiruzooo@yahoo.com

Hakizimana, S., Great Lakes Cassava Initiative (GLCI), Catholic Relief Services, PO Box 65, Kigali, Rwanda. E-mail: shakizimana@rw.caro.crs.org

Harahagazwe, D., International Potato Center (CIP), International Livestock Research Institute (ILRI) Campus, PO Box 25171, Nairobi 00603, Kenya. E-mail: d.harahagazwe@cgiar.org

Hareau, G., Social and Health Sciences Global Program, International Potato Center (CIP), Apartado Postal 1558, Lima, Peru. E-mail: g.hareau@cgiar.org

Haverkort, A.J., Plant Research International (PRI), Wageningen University and Research Centre, Droevendaalsesteeg 1, Wageningen, The Netherlands. E-mail: anton.haverkort@wur.nl (Also affiliated to Department of Crop Production and Soil Science, University of Pretoria, South Africa.)

Heremans, B., Department of Applied Biosciences, Ghent University, Valentin Vaerwyckweg1, BE9000 Ghent, Belgium. E-mail: betty.heremans@telenet.be

Hossain, M., Tuber Crops Research Center (TCRC), Bangladesh Agricultural Research Institute (BARI), Bangladesh House 74, 4th Floor, Road 07, Block H, Banani, Dhaka 1213, Bangladesh. E-mail: mohammad.hossain@cgiar.org

Huamani, K., International Potato Center (CIP), Apartado 1558, Lima, Peru. E-mail: supkhp@gmail.com

Hundayehu, M.C., International Potato Center (CIP), Hawassa, Ethiopia. E-mail: m.cherinet@cgiar.org (Previous address: Department of Plant Production and Soil Science, University of Pretoria, Pretoria 0002, South Africa.)

Hutten, R., Wageningen UR Plant Breeding, Wageningen University and Research Centre, Droevendaalsesteeg 1, Wageningen, The Netherlands. E-mail: ronald.hutten@wur.nl

Idowu, O.T.H., Institute of Food Security, Environmental Resources and Agricultural Research, Federal University of Agriculture, PMB 2240, Abeokuta, Nigeria. E-mail: hamzatoth@gmail.com

Jacobsen, E., Wageningen UR Plant Breeding, Wageningen University and Research Centre, Droevendaalsesteeg 1, Wageningen, The Netherlands. E-mail: evert.jacobsen@wur.nl

Juarez, H., Crop Management and Production Systems Division, International Potato Center (CIP), PO Box 1558, Lima 12, Peru. E-mail: h.juarez@cgiar.org

Jusuf, M., Indonesian Legumes and Tuber Crops Research Institute (ILETRI), Jl. Raya Kendalpayak Km 8, PO Box 66, Malang, East Java 65101, Indonesia. E-mail: m.jusuf751@gmail.com

Kabira, J., Kenya Agricultural and Livestock Research Organization (KALRO), National Potato Research Centre, PO Box 338-00217, Limuru, Kenya. E-mail: kari.tigoni@yahoo.com

Kaguongo, W., National Potato Council of Kenya (NPCK), NARL-KARI Campus, Waiyaki Way, PO Box 29982-00100, Nairobi, Kenya. E-mail: nkaguongo@npck.org; nkaguongo@yahoo.com

Kakuhenzire, R., International Potato Center (CIP)-Tanzania, PO Box 400, Mbeya, Tanzania. E-mail: r.kakuhenzire@cgiar.org

Kamundia, J., Kenya Agricultural and Livestock Research Organization (KALRO)-Njoro, Private Bag Njoro-20107, Njoro, Kenya. E-mail: johnkamundia@yahoo.com

Kanema, H., Barrick Lumwana Mining Company, PO Box 110199, Solwezi, Zambia. E-mail: hkanema@barrick.com

Kankundiye, L., Rwanda Agriculture Board (RAB), PO Box 5016, Kigali, Rwanda. E-mail: lydie2013@gmail.com

Karanja, L., Kenya Agricultural and Livestock Research Organization (KALRO)-Njoro, Private Bag Njoro-20107, Njoro, Kenya. E-mail: laurashali@yahoo.com

Karuniawan, A., Laboratory of Plant Breeding, Faculty of Agriculture, Padjadjaran University (UNPAD), Bandung 40600, Java, Indonesia. E-mail: akaruni1@unpad.ac.id

Kashaija, I.N., National Agricultural Research Organization (NARO), PO Box 295, Entebbe, Uganda. E-mail: ikashaija@naro.go.ug; inkashaija@gmail.com

Kazembe, J., International Potato Center (CIP)-Malawi, PO Box 31600, Lilongwe, Malawi. E-mail: kazembejohn@yahoo.co.uk

Kering, J., District Agricultural Office, Ministry of Agriculture, Bomet Sub County, Kenya. E-mail: joskering@yahoo.com

Khamassy, N., Horticulture Laboratory, National Agricultural Research Institute of Tunisia, Hédi Karray Avenue, 2049, Ariana, Tunisia. E-mail: khamassy.nouri@iresa.agrinet.tn

Khan, M.A., International Potato Center (CIP), Apartado 1558, Lima, Peru. E-mail: a.khan@cgiar.org

Kilambya, D.W., Socio-economic and Applied Statistic Programme, Kenya Agricultural and Livestock Research Organization (KALRO)-Kabete, PO Box 14733-00800, Waiyaki Way, Nairobi, Kenya. E-mail: daniel.wambua@gmail.com

Kimani, E., Kenya Plant Health Inspectorate Service (KEPHIS), PO Box 49592-00100, Nairobi, Kenya. E-mail: ekimani@kephis.org

Kinyua, J., Kenya Agricultural and Livestock Research Organization (KALRO) – Lanet, PO Box 3840-20100, Nakuru, Kenya. E-mail: jimmy.kinyua@yahoo.com

Kinyua, Z.M., Kenya Agricultural and Livestock Research Organization (KALRO)-Kabete, PO Box 14733-00800, Nairobi, Kenya. E-mail: kinyuazm@gmail.com

Kipkoech, D., Kenya Agricultural and Livestock Research Organization (KALRO), Tigoni, PO Box 338-00217, Limuru, Kenya. E-mail: davidnyamasia@gmail.com

Kleinwechter, U., Ferdinanstr, 22A, 12209 Berlin, Germany. E-mail: ukleinw@posteo.net

Kreuze, J.F., International Potato Center, PO Box 1558, Lima 12, Peru. E-mail: j.kreuze@cgiar.org

Kroschel, J., Crop Management and Production Systems Division, International Potato Center (CIP), PO Box 1558, Lima 12, Peru. E-mail: j.kroschel@cgiar.org

Kumar, A., Department of Plant Physiology, College of Basic Sciences and Humanities, GB Pant University of Agriculture and Technology, PO Pantnagar, US Nagar, Uttarakhand, India. E-mail: atulvkumar@gmail.com

Kumar, V.A., Department of Biochemistry, College of Basic Sciences and Humanities, GB Pant University of Agriculture and Technology, PO Pantnagar, US Nagar, Uttarakhand, India. E-mail: vandanaakumar@gmail.com

Kwapata, M.B., Horticulture Department, Lilongwe University of Agriculture and Natural Resources, Bunda College Campus, PO Box 219, Lilongwe, Malawi. E-mail: mkwapata@yahoo.com

Kwigizile, O.H., Agricultural Research Institute (ARI)-Uyole, PO Box 400, Mbeya, Tanzania. E-mail: wekishakwigizile@gmail.com

Labonte, D.R., Department of Horticulture, Louisiana State University (LSU), Baton Rouge, LA 70803, USA. E-mail: dlabonte@agcenter.lsu.edu

Landeo, J.A., Crop Improvement and Genetic Resources, International Potato Center (CIP), Apartado Postal 1558, Lima, Peru. E-mail: j.landeo@cgiar.com

Lang, A.J., National Root Crops Research Institute Potato Sub-Station, Kuru, Plateau State 930101, Nigeria. E-mail: potato@nrcri.org

Laurie, S.M., Agricultural Research Council-Roodeplaat Vegetable and Ornamental Plant Institute (ARC-VOPI), Private Bag X293, East Lynne, Pretoria 0001, South Africa. E-mail: slaurie@arc.agric.za

Lawal, O.I., Department of Plant Physiology and Crop Production, Federal University of Agriculture, PMB 2240, Abeokuta, Nigeria. E-mail: lawalisiaq@gmail.com

Lelgut, D., Kenya Agricultural and Livestock Research Organization (KALRO), PO Box 2787-20100, Njoro, Kenya. E-mail: lelgut@yahoo.com

Lemaga, B., Ethiopian Agricultural Transformation Agency, PO Box 708, Addis Ababa, Ethiopia. E-mail: berga.lemaga@ata.gov.et (Previous address: International Potato Center (CIP), Sub-Saharan Africa Region, Nairobi, Kenya.)

Lenka, D.M., National Root Crops Research Institute, Potato Programme, Kuru, Jos, Plateau State 930101, Nigeria. E-mail: lenkadaniel@yahoo.com

Low, J., International Potato Center Sub-Saharan Africa (CIP-SSA), PO Box 25171, Nairobi 00603, Kenya. E-mail: j.low@cgiar.org

Lukonge, E., Lake Zone Agriculture Research Institute Ukiriguru, PO Box 1433, Mwanza, Tanzania. E-mail: elukonge@yahoo.com

Lukuyu, B., International Livestock Research Institute (ILRI), PO Box 30709-00100, Nairobi, Kenya. E-mail: b.lukuyu@cgiar.org

Lung'aho, C., International Potato Center Sub-Saharan Africa (CIP-SSA), PO Box 75, Chimoio, Mozambique. E-mail: c.lungaho@cgiar.org

Ma, D., Xuzhou Sweetpotato Research Center (XSPRC), Xuzhou 221121, PR China. E-mail: daifuma@163.com

Macharia, I., Kenya Plant Health Inspectorate Service (KEPHIS), PO Box 49592-00100, Nairobi, Kenya. E-mail: macharia.isaac@gmail.com

Malinga, J., Kenya Agricultural and Livestock Research Organization (KALRO), PO Box 450-30200, Kitale, Kenya. E-mail: joyce.malinga@kalro.org

Maliro, M.F., Department of Crops and Soil Science, Lilongwe University of Agriculture and Natural Resources, Bunda College Campus, PO Box 219, Lilongwe, Malawi. E-mail: mmaliro@yahoo.com

Maraganedzha, T., Agricultural Research Council-Roodeplaat Vegetable and Ornamental Plant Institute (ARC-VOPI), Private Bag X 293, East Lynne, Pretoria 0001, South Africa. E-mail: tmaraganedzha@ufh.ac.za

Masinde, W.P., Department of Horticulture, Jomo Kenyatta University of Agriculture and Technology, PO Box 62000-00200, Nairobi, Kenya. E-mail: masinde_peter@yahoo.com

Mayanja, S., International Potato Center Sub-Saharan Africa (CIP-SSA), Kampala Liaison Office, PO Box 22274, Kampala, Uganda. E-mail: s.mayanja@cgiar.org

Mazengia, W., Centre for Crop Systems Analysis, Wageningen University, Droevendaalsesteeg 1, 6708 PB Wageningen, The Netherlands. E-mail: waga966@yahoo.com

Mbanasor, J.A., Department of Agribusiness and Management, Michael Okpara University of Agriculture, Umudike, PMB 7267, Umuahia, Abia State, Nigeria. E-mail: jambanasor@yahoo.co.uk

Mbiri, D., International Potato Center Sub-Saharan Africa (CIP-SSA), PO Box 25171, Nairobi 00603, Kenya. E-mail: d.mbiri@cgiar.org

Mbiyu, M., Kenya Agricultural and Livestock Research Organization (KALRO), Tigoni, PO Box 338-0217, Limuru, Kenya. E-mail: mirrywm@yahoo.com

McEwan, M., International Potato Center Sub-Saharan Africa (CIP-SSA), c/o International Livestock Research Institute (ILRI), PO Box 25171, Nairobi 00603, Kenya. E-mail: m.mcewan@cgiar.org

Mello, A.S., EMBRAPA Hortaliças, Final Av. W3 Norte Caixa, Postal 040315, Brasilia, Brazil. E-mail: alexandre.mello@embrapa.br

Miano, D.W., Department of Plant Science and Crop Protection, University of Nairobi, PO Box 29053-00625, Nairobi, Kenya. E-mail: dwatuku@yahoo.com

Migwi, P., Egerton University, PO Box 536-20115, Njoro, Kenya. E-mail: pemigwi@yahoo.co

Moleleki, L.N., Department of Microbiology and Plant Pathology, Forestry and Agricultural Biotechnology Institute (FABI), University of Pretoria, Lunnon Road, Pretoria, South Africa 0002. E-mail: lucy.moleleki@up.ac.za

Mphela, W.M., Agricultural Research Council-Roodeplaat Vegetable and Ornamental Plant Institute (ARC-VOPI), Private Bag X 293, East Lynne, Pretoria 0001, South Africa. E-mail: mmphela@arc.agric.za

Mtileni, M.M., Agricultural Research Council-Roodeplaat Vegetable and Ornamental Plant Institute (ARC-VOPI), Private Bag X 293, East Lynne, Pretoria 0001, South Africa. E-mail: mmtileni@arc.agric.za

Muriithi, G., Kenya Agricultural and Livestock Research Organization (KALRO) – Lanet, PO Box 3840-20100, Nakuru, Kenya. E-mail: gideon.muriithi@kalro.org

Muriuki, E.K., Kenya Agricultural and Livestock Research Organization (KALRO)-Kabete, PO Box 14733-00800, Waiyaki Way, Nairobi, Kenya. E-mail: muriukie96@gmail.com

Musabyemungu, A., Rwanda Agriculture Board (RAB), PO Box 5016, Kigali, Rwanda. E-mail: musabe2005@yahoo.fr

Musalia, L., Chuka University, PO Box 109-60400, Chuka, Kenya. E-mail: musalialm@gmail.com

Mwanga, R.O.M., International Potato Center Sub-Saharan Africa (CIP-SSA), Ntinda II Road, Plot 47, PO Box 22274, Kampala, Uganda. E-mail: r.mwanga@cgiar.org

Mwase, W.F., Forestry Department, Lilongwe University of Agriculture and Natural Resources, Bunda College Campus, PO Box 219, Lilongwe, Malawi. E-mail: westmwase@yahoo.co.uk

Mwenye, O., Department of Agricultural Research Services (DARS), PO Box 2054, Blantyre, Malawi. E-mail: omwenye@yahoo.co.uk

Myeza, N., Agricultural Research Council-Roodeplaat Vegetable and Ornamental Plant Institute (ARC-VOPI), Private Bag X293, Pretoria 0001, South Africa. E-mail: myezan@arc.agric.za

Nahayo, P., Kachwekano Agricultural Research and Development Research Institute, PO Box 421, Kabale, Uganda. E-mail: namuggak@gmail.com

Naico, A., International Potato Center (CIP), c/o Instituto de Investigacão Agraria de Moçambique (IIAM), Avenida das FPLM 2698, PO Box 2100, Maputo, Mozambique. E-mail: a.naico@cgiar.org

Namanda, S., International Potato Center (CIP), PO Box 22274, Kampala, Uganda. E-mail: s.namanda@cgiar.org

Namugga, P., Kachwekano Agricultural Research and Development Research Institute, PO Box 421, Kabale, Uganda. E-mail: namuggak@gmail.com

Namutebi, A., Department of Food Technology and Nutrition, Makerere University, PO Box 7062, Kampala, Uganda. E-mail: asmutebi@yahoo.co.uk

Ndirigwe, J., Rwanda Agriculture Board (RAB), PO Box 7231, Kigali, Rwanda. E-mail: ndrick3@gmail.com

Ndung'u, J., Kenya Agricultural and Livestock Research Organization (KALRO)-Njoro, Private Bag Njoro-20107, Njoro, Kenya. E-mail: enjohn75@yahoo.co.uk

Ndunguru, J., Mikocheni Agricultural Research Institute, Sam Nujoma Road, PO Box 6226, Dar es Salaam, Tanzania. E-mail: jndunguru2003@yahoo.co.uk

Ngamau, K., Department of Horticulture, Jomo Kenyatta University of Agriculture and Technology, PO Box 62000-00200, Nairobi, Kenya. E-mail: k.ngamau@yahoo.com

Ng'ang'a, N., Kenya Agricultural and Livestock Research Organization (KALRO), Tigoni, PO Box 338-00217, Limuru, Kenya. E-mail: nancy.nganga@kalro.org

Ngundo, G., Kenya Plant Health Inspectorate Service (KEPHIS), PO Box 49592-00100, Nairobi, Kenya. E-mail: gngundo@kephis.org

Njoloma, J.P., Horticulture Department, Lilongwe University of Agriculture and Natural Resources, Bunda College Campus, PO Box 219, Lilongwe, Malawi. E-mail: jnjoloma@yahoo.com

Nolte, P., University of Idaho, 1776 Science Center Dr., Idaho Falls, ID 83402, USA. E-mail: pnolte@uidaho.edu

Nshimiyimana, J.C., International Potato Center (CIP), PO Box 1269, Kigali, Rwanda. E-mail: njclaude5@yahoo.fr

Ntizo, S., Potato Research and Extension Program, Rwanda Agriculture Board (RAB), PO Box 5016, Kigali, Rwanda. E-mail: senkesha.ntizo@rab.gov.rw

Nwachukwu, S.C., Rural Development and Gender Department, Agricultural and Rural Management Training Institute (ARMTI), Km 18, Ilorin Ajase-Ipo Highway, PMB 1343, Ilorin, Kwara State, Nigeria. E-mail: stanley4nwachukwu@gmail.com

Nwosu, A.C., Department of Agriculture Economics, Michael Okpara University of Agriculture, Umudike, PMB 7267, Umuahia, Abia State, Nigeria. E-mail: aloysius_nwosu@yahoo.com

Nyaika, J.A., Horticulture Department, Lilongwe University of Agriculture and Natural Resources, Bunda College Campus, PO Box 219, Lilongwe, Malawi. E-mail: jnyaika@gmail.com

Nyekanyeka, T., International Potato Center (CIP)-Malawi, PO Box 31600, Lilongwe, Malawi. E-mail: tednyek@yahoo.com

Nyongesa, M., Kenya Agricultural and Livestock Research Organization (KALRO), Tigoni, PO Box 338-0217, Limuru, Kenya. E-mail: moses.nyongesa@kalro.org

Obi, J.I., National Root Crops Research Institute, Umudike, Ikot Ekpene-Umuahia Road, Umuahia, Abia State/southeastern PMB 7006, Nigeria. E-mail: obiomanrcri@nrcri.org

Ochieng-Obura, B., International Potato Center Sub-Saharan Africa (CIP SSA), PO Box 25171, Nairobi 00603, Kenya. E-mail: b.ochieng@cgiar.org

Oggema, J.N., Agricultural Development Corporation, PO Box 366, Molo-20107, Molo, Kenya. E-mail: judynata@gmail.com

Ogolla, J., Department of Food Science, Nutrition and Technology, University of Nairobi, PO Box 29053-00625, Nairobi (Kangemi), Kenya. E-mail: ogollajackline@gmail.com

Okoth, M.W., Department of Food Science, Nutrition and Technology, University of Nairobi, PO Box 29053-00625, Nairobi, Kenya. E-mail: mwokoth@uonbi.ac.ke

Olaiya, A.O., Department of Plant Physiology and Crop Production, Federal University of Agriculture, PMB 2240, Abeokuta, Nigeria. E-mail: olaiyaabideen@yahoo.com

Olubayo, F.M., Department of Food Science, Nutrition and Technology, University of Nairobi, PO Box 29053-00625, Nairobi, Kenya. E-mail: fmmogi@yahoo.com

Onditi, J., Kenya Agricultural and Livestock Research Organization (KALRO), Tigoni, PO Box 338-0217, Limuru, Kenya. E-mail: john.onditi@kalro.org

Opio, F., Association for Strengthening Agricultural Research in Eastern and Central Africa (ASARECA), Plot 5, Mpigi Road, PO Box 765, Entebbe, Uganda. E-mail: f.opio@asareca.org

Otazu, V., International Potato Center (CIP), Apartado Postal 1558, Lima, Peru. E-mail: v.otazu@cgiar.org

Otieno, S., Kenya Agricultural and Livestock Research Organization (KALRO), Tigoni, PO Box 338-0217, Limuru, Kenya. E-mail: susan.otieno@yahoo.co.uk

Ouedraogo, T.J., Institut de l'Environnement et de Recherches Agricoles (INERA), Burkina Faso, 01 BP 476 Ouagadougou, Burkina Faso. E-mail: jouedraogo5@hotmail.com

Ouma, J.O., Department of Food Science, Nutrition and Technology, University of Nairobi, PO Box 29053-00625, Nairobi (Kangemi), Kenya. E-mail: oumajames91@yahoo.com

Oyekanmi, A.A., Department of Plant Physiology and Crop Production, Federal University of Agriculture, PMB 2240, Abeokuta, Nigeria. E-mail: akeemoyekanmi@yahoo.com

Palohuhta, J.P., The Finnish Seed Potato Company Ltd, Leppiojantile 31, 91800, Tyrnava, Oulu, Finland. E-mail: jp.palohuhta@dnainternet.net; jp.palohuhta@spk.fi

Parker, M., International Potato Center Sub-Saharan Africa (CIP-SSA), PO Box 25171, Nairobi 00603, Kenya. E-mail: m.parker@cgiar.org

Peters, D., 2143 Woodbox Lane, #C Baltimore, MD 21209, USA. E-mail: daipeters@gmail.com

Quinn, S., International Potato Center (CIP), PO Box 25171, Nairobi 00603, Kenya. E-mail: s.quinn@cgiar.org

Rahaman, E., International Potato Center South, West and Central Asia (CIP-SWCA), Bangladesh House 74, 4th Floor, Road 07, Block H, Banani, Dhaka 1213, Bangladesh. E-mail: e.rahaman@cgiar.org

Ramathavhana, T., Agricultural Research Council-Roodeplaat Vegetable and Ornamental Plant Institute (ARC-VOPI), Private Bag X 293, East Lynne, Pretoria 0001, South Africa. E-mail: tramathavhana@arc.agric.za

Randrianaivoarivony, J.M., FIFAMANOR – Fiompiana Fambolena Malagasy Norveziana, BP 198, Antsirabe 110, Madagascar. E-mail: jmarcrandrianaivoarivony@yahoo.fr; it.fifamanor@moov.mg

Rausch Fernandes, F., EMBRAPA Hortaliças, Final Av. W3 Norte Caixa, Postal 040315, Brasilia, Brazil. E-mail: fernanda.rausch@embrapa.br

Riadh, I., Horticulture Laboratory, National Agricultural Research Institute of Tunisia, Hédi Karray Avenue, 2049, Ariana, Tunisia. E-mail: riadh.ilahy@iresa.agrinet.tn

Ricardo, J., Mozambique Institute of Agricultural Research (IIAM), Avenida das FPLM 2698, PO Box 2100, Maputo, Mozambique. E-mail: j.ricardo1999@yahoo.com.br

Rwomushana, I., Association for Strengthening Agricultural Research in Eastern and Central Africa (ASARECA), Plot 5, Mpigi Road, PO Box 765, Entebbe, Uganda. E-mail: i.rwomushana@asareca.org

Sakariyawo, O.S., Department of Plant Physiology and Crop Production, Federal University of Agriculture, PMB 2240, Abeokuta, Nigeria. E-mail: adentanwa@yahoo.co.uk

Sanni, L.O., Department of Food Science and Technology, Federal University of Agriculture, PMB 2240, Abeokuta, Nigeria. E-mail: sannilateef5@gmail.com

Sawi, M.K., Njala University, Private Mail Bag, Njala, Freetown, Sierra Leone. E-mail: memunamanyeh@yahoo.com

Schulte, R., Environment, Soils and Land Use Research Department, Teagasc, Johnstown Castle, Wexford, Ireland. E-mail: rogier.schulte@teagasc.ie

Schulte-Geldermann, E., International Potato Center Sub-Saharan Africa (CIP-SSA), PO Box 25171, Nairobi 00603, Kenya. E-mail: e.schulte-geldermann@cgiar.org

Schulz, S., German Technical Cooperation (GIZ), PO Box 10009, Addis Ababa, Ethiopia. E-mail: steffen.schulz@giz.de (Previous address: International Potato Center (CIP)-Ethiopia, PO Box 5689, Addis Ababa, Ethiopia.)

Sediane, L., Agricultural Research Council-Roodeplaat Vegetable and Ornamental Plant Institute (ARC-VOPI), Private Bag X 293, East Lynne, Pretoria 0001, South Africa. E-mail: lsediane@arc.agric.za

Sellami, M.H., Plant Protection Laboratory, National Agricultural Research Institute of Tunisia, Hédi Karray Avenue, 2049, Ariana, Tunisia. E-mail: hsellami@yahoo.fr

Shingiro, J.B., Rwanda Agriculture Board (RAB), PO Box 138, Huye, Rwanda. E-mail: bshingiro2000@yahoo.fr

Shofiur Rahaman, E.H.M., CIP, AVRDC Bangladesh House 74, 4th Floor, Road 07, Block H, Banani, Dhaka 1213, Bangladesh. E-mail: e.rahaman@cgiar.org

Shumbusha, D., Rwanda Agriculture Board (RAB), PO Box 138, Huye, Rwanda. E-mail: shudam@yahoo.fr

Simiyu, P.O., Department of Agricultural Economics, University of Nairobi, PO Box 29053-00625, Nairobi, Kenya. E-mail: philipsimiyu95@yahoo.com

Sindi, K., International Potato Center Sub-Saharan Africa (CIP-SSA), PO Box 1269, Kigali, Rwanda. E-mail: k.sindi@cgiar.org

Sobukola, O.P., Department of Food Science and Technology, Federal University of Agriculture, PMB 2240, Abeokuta, Nigeria. E-mail: olajidephilip@yahoo.com

Somé, K., Institut de l'Environnement et de Recherches Agricoles (INERA), Kamboinse, 01 BP 476, Ouagadougou 01, Burkina Faso. E-mail: koussao@hotmail.com

Song, Y.-S., Rural Development Administration (RDA), 293-5 Cheongcheon-ri, Cheonggye-meon, Muan-gun, Jeonranam-do, Republic of Korea. E-mail: leehu79@korea.kr

Souissi, R., Plant Protection Laboratory, National Agricultural Research Institute of Tunisia, Hédi Karray Avenue, 2049, Ariana, Tunisia. E-mail: sr_souissi@yahoo.fr

Sowe, S., Sierra Leone Agricultural Research Institute – Njala Agricultural Research Centre (NARC), Njala, PMB 540, Freetown, Sierra Leone. E-mail: sulaisowe@yahoo.com

Sporleder, M., Crop Management and Production Systems Division, International Potato Center (CIP), PO Box 1558, Lima 12, Peru. E-mail: m.sporleder@cgiar.org

Ssemakula, G., National Crops Resources Research Institute (NaCRRI), PO Box 7084, Kampala, Uganda. E-mail: nankingag@yahoo.com

Steyn, M., Department of Plant Production and Soil Science, University of Pretoria, Pretoria 0002, South Africa. E-mail: martin.steyn@up.ac.za

Struik, P.C., Centre for Crop Systems Analysis, Wageningen University, Droevendaalsesteeg 1, 6708PB, Wageningen, The Netherlands. E-mail: paul.struik@wur.nl

Subakanya, D., Barrick Lumwana Mining Company, PO Box 110199, Solwezi, Zambia. E-mail: dsubakanya@barrick.com

Tadese, Y., Centre for Crop Systems Analysis, Wageningen University, Droevendaalsesteeg 1, 6708PB, Wageningen, The Netherlands. E-mail: yenenesh.gebresilase@wur.nl

Thuranira, E.G., Kenya Agricultural and Livestock Research Organization (KALRO)-Kabete, PO Box 14733-00800, Waiyaki Way, Nairobi, Kenya. E-mail: thuranira2003@yahoo.com

Tindimubona, S., Uganda National Seed Potato Producers' Association (UNSPPA), PO Box 329, Kabale, Uganda. E-mail: sbtindi@gmail.com

Tjintokohadi, K., International Potato Center, East and Southeast Asia and the Pacific (CIP-ESEAP), BALITSA, Jl. Tangkuban Perahu no. 517, PO Box 8404, Lembang-Bandung 40391, Indonesia. E-mail: k.tjintokohadi@cgiar.org

Tonnang, H.E.Z., African Insect Science for Food and Health (ICIPE), PO Box 30772-00100, Nairobi, Kenya. E-mail: htonnang@gmail.com

Torrance, L., The James Hutton Institute, Invergowrie, Dundee, DD2 5DA, UK. E-mail: lesley.torrance@hutton.ac.uk

Udealor, A., National Root Crops Research Institute, Umudike, Ikot Ekpene-Umuahia Road, Umuahia, Abia State/southeastern PMB 7006, Nigeria. E-mail: ansudealor@yahoo.com

Valkonen, J.P.T., Department of Agricultural Sciences, University of Helsinki, PO Box 27, Helsinki, FI-00014, Finland. E-mail: jari.valkonen@helsinki.fi

Van den Berg, A.A., Agricultural Research Council-Roodeplaat Vegetable and Ornamental Plant Institute (ARC-VOPI), Private Bag X 293, East Lynne, Pretoria 0001, South Africa. E-mail: avdberg@arc.agric.za

Van Droogenbroeck, B., Institute for Agricultural and Fisheries Research, Burg. Van Gansberghelaan 115, BE9820 Merelbeke, Belgium. E-mail: bart.vandroogenbroeck@ilvo.vlaanderen.be

van Jaarsveld, P.J., South African Medical Research Council, PO Box 19070, Tygerberg 7505, South Africa. E-mail: paul.van.jaarsveld@mrc.ac.za

van Koesveld, M.J., Applied Plant Research (PPO), Wageningen University and Research Centre, Droevendaalsesteeg 1, Wageningen, The Netherlands. E-mail: flip.vankoesveld@wur.nl

Vernon, G., Campus Information and Visitor Relations, Day Hall Lobby, Cornell University, Ithaca, NY 14853, USA. E-mail: vg45@cornell.edu

Visser, J.C., Department of Biochemistry, University of Stellenbosch, Private Bag X1, 7602 Matieland, South Africa. E-mail: johanchrisvisser@gmail.com

Visser, R.G.F., Wageningen UR Plant Breeding, Wageningen University and Research Centre, Droevendaalsesteeg 1, Wageningen, The Netherlands. E-mail: richard.visser@wur.nl

Vossen, J.H., Wageningen UR Plant Breeding, Wageningen University and Research Centre, Droevendaalsesteeg 1, Wageningen, The Netherlands. E-mail: jack.vossen@wur.nl

Wasilwa, L., Kenya Agricultural and Livestock Research Organization (KALRO), Head Quarters, PO Box 57811-00200, Nairobi, Kenya. E-mail: lusike.wasilwa@kalro.org

Were, H.K., Masinde Muliro University of Science and Technology, Kakamega, Kenya. E-mail: werehkde@yahoo.com

Whitworth, J.L., United States Department of Agriculture Agricultural Research Service (USDA-ARS), 1693 South 2700 West, Aberdeen, ID 83210, USA. E-mail: jonathan.whitworth@ars.usda.gov

Williams, M.S.E., PO Box 1983, Sapon, Abeokuta, Ogun State, Nigeria. E-mail: mrthswilliams@yahoo.com (Previous address: Sierra Leone Agricultural Research Institute, Njala Agricultural Research Centre (NARC), Njala, PMB 540, Freetown, Sierra Leone.)

Woldegiorgis, G., Ethiopian Institute of Agricultural Research (EIAR), PO Box 2003, Addis Ababa, Ethiopia. E-mail: gebregiorgis2003@yahoo.com

Wustman, R., Applied Plant Research (PPO), Wageningen University and Research Centre, Droevendaalsesteeg 1, Wageningen, The Netherlands. E-mail: romke.wustman@wur.nl

Xie, J., Institute of Agro-products Processing Science and Technology of Sichuan Academy of Agricultural Sciences, Shi Zi Shan Road, Chengdu, Sichuan Province 610066, PR China. E-mail: xiejiang2010@gmail.com

Xie, K., CIP China Center for Asia Pacific (CCCAP), #709 Pan Pacific Plaza, A12 Zhongguancun South Street, Beijing 100081, PR China. E-mail: k.xie@cgiar.org

Yencho, G.C., Department of Horticultural Science, North Carolina State University (NCSU), PO Box 7609, Raleigh, NC 27695, USA. E-mail: yencho@ncsu.edu

Zhang, X., Institute of Agricultural Economics (LEI), Wageningen University and Research Centre, Droevendaalsesteeg 1, Wageningen, The Netherlands. E-mail: xiaoyong.zhang@wur.nl

Zhu, Y., Institute of Agro-products Processing Science and Technology of Sichuan Academy of Agricultural Sciences, Shi Zi Shan Road, Chengdu, Sichuan Province 610066, PR China. E-mail: zhuyongqing68@sina.com

Zimba, S.C., Horticulture Department, Lilongwe University of Agriculture and Natural Resources, Bunda College Campus, PO BOX 219, Lilongwe, Malawi. E-mail: sibozim@yahoo.co.uk

Preface

Potato (*Solanum tuberosum* L.) and sweetpotato (*Ipomoea batatas* Lam.) are the fourth and seventh most important crop plants in the world and of increasing importance in sub-Saharan Africa (SSA). During the past decade, the area under sweetpotato and potato expanded faster than other major food crops in Africa. Clearly, sweetpotato is emerging from its traditional role of the classic food security crop to a more commercialized crop, and in the case of the orange-fleshed sweetpotato, a healthy food for all. With urbanizing populations, the demand for potato has never been higher and its value as a nutritious, short-duration crop to fill gaps in cereal supply is increasingly recognized.

Yet research done on these two crops in Africa frequently gets lost in the grey literature. This is especially true of applied research presented at conferences. This book is a concerted effort by the African Potato Association (APA) (www.africanpotatoassociation.org), founded in 1983, to reverse that trend. It captures recent advances made in tackling the bottlenecks facing potato and sweetpotato value chains in Africa.

The book encompasses papers prepared for the Ninth Triennial African Potato Association Conference held in Naivasha, Kenya from 30 June to 4 July 2013. The conference was organized by the APA, who joined hands with three co-hosting institutions: (i) the Ministry of Agriculture, Livestock and Fisheries – Republic of Kenya; (ii) the National Potato Council of Kenya; and (iii) the International Potato Center (CIP).

The conference at the Great Rift Valley Lodge was attended by 287 participants (25% women) from 23 African countries and 15 non-African countries – the largest APA conference held to date. Attracting 56 participants from outside of Africa reflects the growing interest in understanding what is happening in African agriculture and a concerted effort by the APA to promote greater interaction between African and non-African potato and sweetpotato scientists and practitioners. Participants came from various research institutions, universities, development agencies, government institutions and private sector companies.

The theme of the conference was 'Transforming Potato and Sweetpotato Value Chains for Food and Nutrition Security'. Abstracts were solicited in five sub-themes. Each was then assigned to be presented as an individual oral presentation, part of a panel discussion or a poster presentation. The number of papers submitted by theme varied enormously, with themes on seed systems and breeding having the most submissions. Dr Moses Nyongesa of the Kenya Agricultural and Livestock Research Organization (KALRO) coordinated a committee of 31 scientists to review the abstracts and produce a book of

172 abstracts in time for the meeting. Luka Wanjohi, CIP's data manager, managed the information flow concerning both abstracts and subsequent papers.

In total, there were 56 oral presentations and ten participants in panel discussions. Authors involved in individual oral or panel presentations were invited to submit papers for publication. In total, 59 papers were received and edited by the Ninth Triennial APA Book Editing Team:

- Dr Jan Low, APA President 2011–2013 and CIP scientist;
- Dr Moses Nyongesa, the APA East Africa Representative 2014–2016 and a KALRO potato scientist;
- Ms Sara Quinn, CIP's regional communications officer; and
- Dr Monica Parker, a CIP potato scientist.

The team also acknowledges contributions made by Joseph Kigamwa to the editing effort, and the tremendous job of copyediting done by Priscilla Sharland.

Reflecting the papers received, the book is organized into six parts:

- Part I Advances in Sweetpotato and Potato Breeding;
- Part II Getting Seed Potato Systems Moving;
- Part III Addressing Bottlenecks in the Sweetpotato Seed Systems;
- Part IV Tackling Disease and Crop Management in Potato and Sweetpotato;
- Part V Innovations in Marketing, Processing Technologies and Value Chain Management for Potato and Sweetpotato; and
- Part VI Evidence on Nutritional Value, Product Acceptance and Changing Behaviours Concerning Sweetpotato.

Each part begins with a review paper (chapter) concerning a major subject within the field, written by well-known scientists in that field. Each review paper was the basis for a keynote presentation at the conference and is longer than the remaining papers in the part. Part I begins with an in-depth review of breeding progress in sweetpotato from 1992 through to 2012. Part II commences with a chapter that draws on recent experience in five SSA countries that informs on how to improve seed potato quality and supply. Part III starts with an analysis of the challenges encountered moving local sweetpotato seed systems to scale in SSA. Part IV begins with a review of the state of knowledge concerning viruses affecting sweetpotato and potato, and progress made in developing more sensitive techniques to detect them. Part V starts with a paper looking at what it would take to get potato processing moving, especially for fries and crisps, in African highland countries with high potato production but low investment in processing to date. Part VI begins with a comprehensive review of the progress made in understanding the nutritional value and impact related to the introduction and promotion of improved vitamin-A-rich orange-fleshed sweetpotato during the past 15 years in SSA, focusing on what kinds of behaviour changes are needed to assure uptake by producers and consumers.

In reading this book, we think you will agree with us that the quality and quantity of research being undertaken in Africa on potato and sweetpotato is relevant and exciting, reflecting the increased recognition of the potential of these crops to address food and nutrition security on the continent.

Jan Low
African Potato Association President 2011–2013

Acronyms

ABS, accelerated breeding scheme
ADC, Agricultural Development Cooperation
AEZ, agroecological zones
AFLP, amplified fragment length polymorphism
AGRA, Alliance for a Green Revolution in Africa
AgR4D, agricultural research for development
AI, activity index
AIDs, acquired immune deficiency syndrome
AIS, Agricultural Innovation Systems
AMMI, Additive main effects and multiplicative interaction
ANOVA, analysis of variance
AOAC, Association of Official Analytical Chemists
AP, after planting
APA, African Potato Association
APF, AgriProFocus
APLV, *Andean potato latent virus*
ARC, Agricultural Research Council
ARC-VOPI, Agricultural Research Council-Roodeplaat Vegetable and Ornamental Plant Institute
ARI, Agricultural Research Institute
ASARECA, Association for Strengthening Agricultural Research in Eastern and Central Africa
AUDPC, area under the disease progress curve
AVRDC, Asian Vegetable Research and Development Center
BMGF, Bill & Melinda Gates Foundation
BoA, Bureau of Agriculture
BSPM, boiled sweetpotato meal
Bt, toxin derived from *Bacillus thuringiensis*
CAADP, Comprehensive African Agriculture Development Programme
CARE, Carry American Relief Everywhere
CARE EEEGL, Carry American Relief Everywhere (CARE) Enterprise, Environment and Equity in the Great Lakes (a CARE project)
CBO, community-based organization
CBSPM, composite boiled sweetpotato meal

CCAFS, Climate Change, Agriculture and Food Security
CCD, composted cow dung
C-D, Cobb-Douglas (model)
CEC, cation exchange capacity
CES, constant elasticity of substitution
CFC, Common Fund for Commodities
CIDA, Canadian International Development Agency
CIE, Commission Internationale de l'Eclairage
CIP, International Potato Center
CMV, cucumber mosaic virus
COII, cytochrome oxidase II
COoA, Chencha Office of Agriculture
CP, coat protein
CP, crude protein
CPRI, Central Potato Research Institute
CRS, Catholic Relief Services
CSIR-CRI, Council for Scientific and Industrial Research, Crops Research Institute
CSRW, commercial storage root weight
CSSPM, composite sun-dried sweetpotato meal
CV, coefficient of variation
CVP, crystal violet pectate (medium)
DAP, diammonium phosphate
DARS, Department for Agricultural Research Services
DAS-ELISA, double antibody sandwich ELISA
DFID, Department for International Development (UK)
DIIVA, Diffusion and Impacts of Improved Varieties in Africa
DLS, diffused light storage/store
DM, dry matter
DMC, dry matter content
DONATA, Dissemination of New Agricultural Technologies in Africa
DRC, Democratic Republic of the Congo
DSPs, Decentralized Seed Producers
DTI, drought tolerance index
dwb, dry weight basis
EAC, East African Community
EC, electrical conductivity
ECA, East/Eastern and Central Africa
EFSA, European Food Safety Authority
EI, establishment risk index
EIAR, Ethiopian Institute of Agricultural Research
ELISA, enzyme-linked immunosorbent assay
ESA, East and Southern Africa
EST, expressed sequence tag
ETDF, equivalent thermal driving force
ETP, evapotranspiration
EU, European Union
FAO, Food and Agriculture Organization of the United Nations
FARA, Forum for Agricultural Research in Africa
FCR, feed conversion ratio
FFS, farmer field school
FP, farmers' practice (seed)
FPU, food production unit

FRG, farmers' research group
FS, farmer selection (seed)
FSE, farmer seed enterprise
FUNAAB, Federal University of Agriculture, Abeokuta
fwb, fresh weight basis
GAIN, Global Alliance for Improved Nutrition
GAP, gender action plan
GAPs, Good Agricultural Practices
GBS, genotyping by sequencing
GCA, general combining ability
GCM, general circulation model
GDP, gross domestic product
G × E, genotype by environment interaction
GEBVS, genomic estimated breeding values
GI, generation index
GIL, Groupement Interprofessionnel des Légumes
GIS, geographic information system
GLM, general linear model
GMO, genetically modified organism
GMP, good manufacturing practices
GPB, Golden Power Biscuit
GPS, global positioning system
GS, genomic selection
GTSPI, Genomic Tools for Sweetpotato Improvement
HARC, Holetta Agricultural Research Centre
HEBS, heterosis-exploiting breeding scheme
HI, harvest index
HIV, human immunodeficiency virus
HMNEH, Horticulture Mission for North East and Himalayan States
HPLC, high performance liquid chromatography
IA, investment area
IAA, indole-3-acetic acid
IARC, International Agricultural Research Centres
ICP-OES, inductively coupled plasma argon optical emission spectrometry
ICT, information and communication technology/technologies
IGS, intergenic spacer
IIAM, Mozambique Institute of Agricultural Research
ILCYM, Insect Life Cycle Modelling software
IMPACT, International Model for Policy Analysis of Agricultural Commodities and Trade
INERA, Institut de l'Environnement et de Recherches Agricoles
INES, Institut d'Enseignement Supérieur de Ruhengeri
INGO, international non-governmental organization
INRAT, Institut National de la Recherche Agronomique de Tunisie
IPGRI, International Plant Genetic Resources Institute
IPM, integrated pest management
IPPC, International Plant Protection Convention
IRR, internal rate of return
ISAR, Rwandan Institute for Agricultural Sciences
ISO, International Organization for Standardization
ISPMs, International Standards for Phytosanitary Measures
ISTA, International Seed Testing Association
ISTRC, International Society for Tropical Root Crops

IVOMD, *in vitro* organic matter digestibility
KALRO, Kenya Agricultural and Livestock Research Organization
KAPAP, Kenya Agricultural and Agribusiness Project
KARI, Kenya Agricultural Research Institute
KENAPOFA, Kenya National Potato Farmers' Association
KEPHIS, Kenya Plant Health Inspectorate Service
KIST, Kigali Institute of Food Science
LB, late blight
LBHT, late blight resistant heat tolerant
LED, light-emitting diode
LER, land equivalent ratio
LGA, local government area
LR, likelihood ratio
LSD, least significance difference
lsmean, least-squares mean
LSU, Louisiana State University
masl, metres above sea level
MI, Micronutrient Initiative
MLE, maximum likelihood estimate
MoA, Ministry of Agriculture
mos, microorganisms
MOU, memorandum of understanding
MP, microplant
MRDR, modified relative-dose–response test
MS, Murashige and Skoog (medium)
MT, microtuber
mtDNA, mitochondrial DNA
MTP, microtuber raised plant
NAADS, National Agricultural Advisory Services
NaCRRI, National Crops Resources Research Institute
NARC, Njala Agricultural Research Centre
NARI, National Agricultural Research Institute
NARO, National Agricultural Research Organisation
NARS, National Agricultural Research System
NASPOT, Namulonge Sweetpotato
NCM-ELISA, nitrocellulose membrane ELISA
NCSR, number of commercial storage roots
NCSU, North Carolina State University
NFL, normal fluorescent light
NGO, non-governmental organization
NIRS, near-infrared reflectance spectroscopy
NPCK, National Potato Council of Kenya
NPT, National Performance Trial
NPV, net present value
NRCRI, National Root Crops Research Institute
OECD, Organization for Economic Cooperation and Development
OFSP, orange-fleshed sweetpotato
OLS, ordinary least squares
PAR, photosynthetically active radiation
PBRSV, *Potato black ring-spot virus*
Pcb, *Pectobacterium carotovorum brasiliense*
PCR, polymerase chain reaction

PDR, pathogen-derived resistance
PEG, polyethylene glycol
PLRV, potato leaf roll virus
PMCA, Participatory Market Chain Approach
PNAP, Programme National pour l'Amélioration de la Pomme de terre
PPP, public–private partnership
PQBS, Plant Quarantine and Biosecurity Station
PRA, participatory rural appraisal
PRA, pest risk analysis
PRAPACE, Programme Régional d'Amélioration de la Pomme de Terre et de la Patate Douce en Afrique Centrale et de l'Est
PS, positive-selection (seed)
PT, preliminary trials
PTNRD, potato tuber necrotic ringspot disease
PVA, potato virus A
PVM, potato virus M
PVS, potato virus S
PVX, potato virus X
PVY, potato virus Y
PYT, preliminary yield trial
QDPM, Quality Declared Planting Material
QDS, Quality Declared Seed
qRT-PCR, quantitative real-time reverse transcriptase polymerase chain reaction
QTL, quantitative trait loci
RAA, reduced ascorbic acid
RAB, Rwanda Agriculture Board
R&D, research and development
RAE, retinol activity equivalents
RAPD, random amplified polymorphic DNA
RBDO, refined, bleached and deodorized oil
RBP, retinol-binding protein
RBS, Rwanda Bureau of Standards
RCBD, randomized complete block design
RDA, recommended daily allowance
REU, Reaching End Users (project)
RKN, root knot nematodes
RMT, rapid multiplication techniques
RSA, response surface analysis
R/V, ratio of total dry matter yield of roots to vines
SARI, South Agricultural Research Institute
SASA, Science and Advice for Scottish Agriculture
SASHA, Sweetpotato Action for Security and Health in Africa
SBCC, social and behaviour change communication
SCCI, Seed Control and Certification Institute
SE, standard error of the mean
SED, standard error of the difference of means
SFLCs, sweetpotato flour from limed chips
SFNLCs, sweetpotato flour from non-limed chips
siRNA, small interfering RNA
SNHP, State National Harmonization Program
SNNPR, Southern Nations Nationalities and Peoples' Region
SNPs, single nucleotide polymorphisms

SPCSV, sweet potato chlorotic stunt virus
SPFMV, sweet potato feathery mottle virus
SPHD, State Plant Health Director
SPHI, Sweet potato for Profit and Health Initiative
SPLCV, Sweet potato leaf curl virus
SPMMV, sweet potato mild mottle virus
SPPV, Sweet potato pakakuy virus
SPRO, State Plant Regulatory Official
SPS, sanitary and phytosanitary (measures)
SPSMV, sweet potato symptomless virus 1
SPSS, Statistical Program for Social Scientists
SPVC, Sweet potato virus C
SPVD, sweet potato virus disease
SPVG, Sweet potato virus G
SPVZ, sweet potato virus Z
SPV2, Sweet potato virus 2
SRE, soft rot *Enterobacteriaceae*
sRSA, small RNA sequencing and assembly of viral genomes
SSA, sub-Saharan Africa
SSPM, sun-dried sweetpotato meal
SSR, simple sequence repeat
SUN, Scaling Up Nutrition (initiative)
3G (a seed production model), '3 seed potato generation revolution'
TAS-ELISA, triple antibody sandwich
TATI, Tinarha Agri Tourism Initiative
TIPs, Trials for Improved Practices
T_{max}, daily maximum temperature
T_{min}, daily minimum temperature
TOT, Training of Trainers
TSNI, Towards Sustainable Nutrition Improvement (project)
UNICEF, United Nations Children's Fund
UNSPPA, Uganda National Seed Potato Producers' Association
UPOV, Union for Protection of Plant Varieties
USAID, United States Agency for International Development
USDA-APHIS, United States Department of Agriculture Animal Plant Health Inspection Service
VAD, vitamin A deficiency
VAS, vitamin A supplementation
VFAs, volatile fatty acids
VITAA, Vitamin A for Africa
VPCs, vegetatively propagated crops
WACCI, West Africa Centre for Crop Improvement
WATSAN, water and sanitation sector
WHO, World Health Organization
WTO, World Trade Organization
XSPRC, Xuzhou Sweetpotato Research Center
YWCA, Young Women's Christian Association
YWT, yellow water trap
ZARI, Zambia Agriculture Research Institute

Part I

Advances in Sweetpotato and Potato Breeding

1 Advances in Sweetpotato Breeding from 1992 to 2012

W.J. Grüneberg,[1]* D. Ma,[2] R.O.M. Mwanga,[3] E.E. Carey,[4] K. Huamani,[1] F. Diaz,[1] R. Eyzaguirre,[1] E. Guaf,[5] M. Jusuf,[6] A. Karuniawan,[7] K. Tjintokohadi,[8] Y.-S. Song,[9] S.R. Anil,[10] M. Hossain,[11] E. Rahaman,[12] S.I. Attaluri,[13] K. Somé,[14] S.O. Afuape,[15] K. Adofo,[16] E. Lukonge,[17] L. Karanja,[18] J. Ndirigwe,[19] G. Ssemakula,[20] S. Agili,[21] J.M. Randrianaivoarivony,[22] M. Chiona,[23] F. Chipungu,[24] S.M. Laurie,[25] J. Ricardo,[26] M. Andrade,[27] F. Rausch Fernandes,[28] A.S. Mello,[28] M.A. Khan,[1] D.R. Labonte[29] and G.C. Yencho[30]

[1]International Potato Center (CIP), Lima, Peru; [2]Xuzhou Sweetpotato Research Center (XSPRC), Xuzhou, China; [3]CIP Sub-Saharan Africa (CIP-SSA), Kampala, Uganda; [4]CIP-SSA, Kumasi, Ghana; [5]National Agricultural Research Institute (NARI), Papua New Guinea; [6]Indonesian Legumes and Tuber Crops Research Institute (ILETRI), Java, Indonesia; [7]Padjadjaran University (UNPAD), Java, Indonesia; [8]CIP East and Southeast Asia and the Pacific (CIP-ESEAP), Lembang-Bandung, Indonesia; [9]Rural Development Administration (RDA), Republic of Korea; [10]Central Tuber Crops Research Institute (CTCRI), Kerala, India; [11]Tuber Crops Research Center (TCRC), Bangladesh Agricultural Research Institute (BARI), Bangladesh; [12]CIP-South, West and Central Asia (CIP-SWCA), Bangladesh; [13]CIP-SWCA, Odisha, India; [14]Institut de l'Environnement et de Recherches Agricoles (INERA), Ouagadougou, Burkina Faso; [15]National Root Crops Research Institute (NRCRI), Umudike, Nigeria; [16]Council for Scientific and Industrial Research – Crops Research Institute (CSIR-CRI), Kumasi, Ghana; [17]Lake Zone Agriculture Research Institute (LZARDI), Ukiriguru, Tanzania; [18]Kenya Agricultural and Livestock Research Organization (KALRO), Njoro, Kenya; [19]Rwanda Agriculture Board (RAB), Kigali, Rwanda; [20]National Crops Resources Research Institute (NaCRRI), Kampala, Uganda; [21]CIP-SSA, Nairobi, Kenya; [22]Fiompiana Fambolena Malagasy Norveziana (FIFAMANOR), Madagascar; [23]Zambia Agriculture Research Institute (ZARI), Mansa, Zambia; [24]Department of Agricultural Research Services (DARS), Blantyre, Malawi; [25]Agricultural Research Council-Roodeplaat Vegetable and Ornamental Plant Institute (ARC-VOPI), Pretoria, South Africa; [26]Mozambique Institute of Agricultural Research (IIAM), Maputo, Mozambique; [27]CIP-SSA, Maputo, Mozambique; [28]EMBRAPA Hortaliças, Brasilia, Brazil; [29]Louisiana State University (LSU), Baton Rouge, Louisiana, USA; [30]North Carolina State University (NCSU), Raleigh, North Carolina, USA

*w.gruneberg@cgiar.org

Abstract

Sweetpotato, with a global annual planting area of approximately 9 million ha, is the second most important tropical root crop. It is widely adapted, being grown in more than 110 countries. Early maturing varieties grow in 3–4 months. It is hardy and has multiple uses. Both roots and foliage are edible and provide energy and nutrients in diets. Distinct quality types have different uses, with orange-fleshed sweetpotato being valued for its extremely high provitamin A content, and other types used in varied fresh and processed forms. Sweetpotato is easily bred, as true seed is easily obtained and generation cycles are short. There are five objectives of this review. The first objective is to briefly describe recent production and utilization trends by region; the second is to review knowledge about the origin and genetic nature of sweetpotato; the third is to review selected breeding objectives. The fourth objective is to review advances in understanding of breeding methods, including: (i) generation of seed through polycross nurseries and controlled cross breeding; (ii) a description of a new accelerated breeding approach; (iii) recent efforts to systematically exploit heterosis; and (iv) new approaches of genomic selection. The fifth objective is to provide information about variety releases during the past 20 years in West, East and Southern Africa, South Asia, East and South-east Asia, China and the Pacific.

Keywords: abiotic, accelerated breeding scheme, autopolyploidy, beta-carotene (β-carotene), biotic, breeding, controlled cross, genomic selection, heterosis, heterozygous, hybrid, molecular markers, orange-fleshed sweetpotato, origin, polycross, sweetpotato, traits

1.1 Introduction

Sweetpotato breeding was reviewed by Jones (1985) and Martin and Jones (1986), mainly against the background of breeding in the USA. Sweetpotato breeding was more recently reviewed by Grüneberg *et al.* (2009a,b) and by Lebot (2010). Carpena (2009) provides an overview of important varieties across different regions of the world. This review updates these previous reviews, highlighting recent advances in sweetpotato breeding methods. There are five objectives: (i) to briefly describe recent production and utilization trends by region; (ii) to review knowledge about the origin, centres of diversity and the genetic nature of sweetpotato; (iii) to review selected breeding objectives; (iv) to review recent advances in understanding of breeding methods; and (v) to provide information about variety releases during the past 20 years in the Americas, West, East and Southern Africa, South Asia, China, East and South-east Asia and the Pacific.

Distribution and importance

Sweetpotato was domesticated in tropical America about 6000 BC and reached Polynesia, Hawaii and New Zealand naturally or by early seafarers in pre-Columbian times. The Spanish introduced the crop to the Philippines in the 16th century, from whence it spread to other islands and the Asian mainland. By 1594, the crop was recorded in south China, where it was promoted to mitigate drought during the Qing Dynasty (ruling from 1644 to 1912). Portuguese seafarers introduced the crop into western Mediterranean Europe, Africa, India and parts of South-east Asia (O'Brien, 1972; Yen, 1976, 1982; Jia, 2013). According to the Food and Agriculture Organization of the United Nations (FAO), sweetpotato is currently cultivated in 117 countries in all tropical and subtropical regions of the world, with 104 million t of production in 2011. Asia is the world's largest sweetpotato producing region, with about 80% of annual production, followed by Africa, the Americas and Oceania with approximately 16%, 3% and 1% of annual production, respectively (FAOSTAT, 2011).

Trends in area cultivated from 1992 to 2011 by region (Fig. 1.1), notably show declines in Asia (from 6.4 to 3.6 million ha) and increases in Africa (from 1.2 to 3.2 million ha). Storage root yield trends for the same period show increases for all regions (Fig. 1.2). Yields in sub-Saharan Africa (SSA) are the lowest overall, while those of the West Pacific (China, Japan and Korea) are about four times higher (FAOSTAT, 2011)

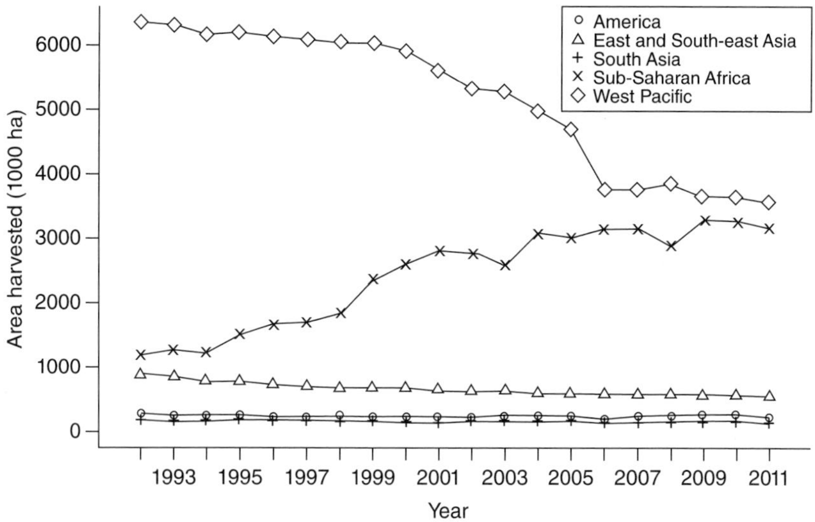

Fig. 1.1. Annual sweetpotato planting area by region. America is comprised of Argentina, Brazil, Cuba, Haiti, Peru and the USA. Sub-Saharan Africa includes East Africa with Burundi, Ethiopia, Kenya, Rwanda, Uganda and the United Republic of Tanzania; Southern Africa with Angola, Madagascar, Malawi, Mozambique and Zambia; and West Africa with Nigeria, Ghana and Mali. South Asia is comprised of Bangladesh and India. East and South-east Asia includes Indonesia, Papua New Guinea, the Philippines and Vietnam. West Pacific is comprised of China, Korea and Japan. (From FAOSTAT, 2011.)

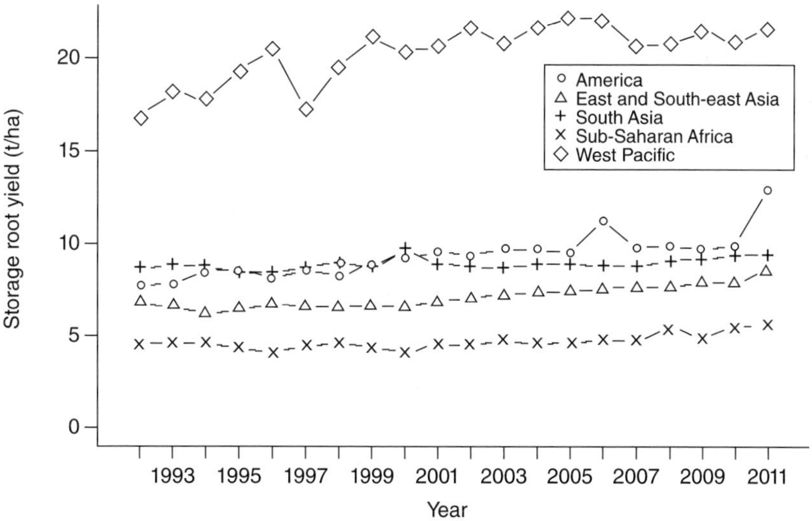

Fig. 1.2. Annual sweetpotato storage root yields by region. The composition of each region is the same as in Fig. 1.1. (From FAOSTAT, 2011.)

than global yields. Thus, there is significant potential to increase global yields through the use of improved cultural practices and varieties (Oswald *et al.*, 2009). Recent major increases in area in countries such as Nigeria and Tanzania reflect the crop's comparative advantage as populations increase and demands on production systems intensify.

An overview of the storage root yields from 2002 to 2011 of the 30 countries contributing to more than 99% of worldwide annual production is given in Table 1.1. Yield increases in the West Pacific (China, Japan and Korea), the USA, SSA and South Asia (India and Bangladesh) were about 1.4%, 2.1%, 1.2% and 0.5% per year, respectively, across the past two decades. Some countries in SSA reported yield decreases (Angola, Ghana and Nigeria), whereas the annual yield increases of around 14% across the past two decades in Mali and Tanzania are probably overestimates. Four countries in SSA (Kenya, Mali, Rwanda and Tanzania) reported yield increases larger than 3% per year and four additional countries (Madagascar, Malawi, Mozambique and Zambia) yield increases of 1–3% per year over the past two decades. We consider that the yield estimates for Ghana and Nigeria from FAO (Table 1.1) are highly inaccurate, most likely due to overestimation of the harvested area. National scientists estimate that in both countries yields per hectare are around 8 t/ha. Moreover, the yield estimates for Uganda are likely underestimates.

Progress in yield can be achieved by breeding (replacing old varieties by new)

Table 1.1. Storage root yields (t/ha) in 30 countries which contribute greater than 99% of annual global sweetpotato production.

	Year									
Country	2002	2003	2004	2005	2006	2007	2008	2009	2010	2011
Angola	4.2	3.8	4.5	4.6	4.4	5.9	6.5	6.7	6.3	6.6
Argentina	14.6	16.1	15.5	15.5	14.4	14.2	14.0	13.8	14.4	15.1
Bangladesh	9.1	9.0	9.0	8.9	9.0	9.1	9.7	9.6	9.9	9.8
Brazil	11.4	11.5	11.5	11.3	11.7	12.1	12.0	11.3	11.9	12.4
Burundi	6.7	6.5	6.7	6.5	6.7	6.7	6.6	6.7	6.9	7.0
China	21.7	20.8	21.7	22.2	22.1	20.7	20.8	21.6	20.9	21.7
Cuba	6.1	7.1	6.7	6.06	6.4	6.4	6.4	5.6	4.8	6.8
Ethiopia	10.0	10.6	9.9	8.1	7.3	8.4	8.0	8.4	9.0	9.0
Ghana	1.4	1.4	1.4	1.4	1.4	1.5	1.6	1.6	1.6	1.8
Haiti	3.0	2.8	2.7	3.0	4.7	3.0	3.0	3.6	3.3	4.6
India	8.6	8.6	8.9	8.9	8.7	8.7	8.9	9.0	9.2	9.3
Indonesia	10.0	10.1	10.3	10.4	10.5	10.7	10.8	11.2	11.3	12.3
Japan	25.4	23.7	25.0	25.8	24.2	23.8	24.8	25.3	21.8	22.8
Kenya	8.5	10.0	9.4	–	9.7	13.3	14.3	12.0	10.0	12.3
Madagascar	5.2	4.7	5.1	7.1	7.2	7.2	7.2	7.2	7.1	7.8
Malawi[a]	12.7	13.1	12.1	9.0	13.4	15.2	14.6	16.2	16.1	17.1
Mali	15.8	14.8	16.5	16.0	18.7	19.1	20.2	19.1	18.8	18.8
Mozambique	5.7	7.3	7.2	7.08	7.3	7.3	7.1	7.1	7.1	7.7
North Korea	12.6	13.7	13.1	13.2	13.1	13.5	13.6	13.5	13.8	13.6
Nigeria	3.0	3.1	3.1	3.2	3.4	2.2	3.0	2.9	2.9	2.9
Papua New Guinea	4.8	5.0	5.0	4.7	5.0	5.0	5.0	5.0	4.8	4.8
Peru	16.1	15.5	15.7	16.4	17.7	16.5	15.7	16.4	16.8	18.1
Philippines	4.5	4.5	4.5	4.8	4.8	4.9	4.9	4.9	5.0	5.0
Rwanda	6.6	5.9	5.6	6.0	5.7	5.7	5.5	6.5	7.5	8.1
South Korea	21.5	19.0	20.8	16.5	17.2	16.7	16.9	16.8	15.6	14.2
Tanzania	3.5	1.5	2.9	3.0	2.8	3.3	6.7	2.2	4.2	5.1
Uganda	4.4	4.4	4.4	4.4	4.5	4.5	4.5	4.5	4.6	4.8
USA	17.4	19.2	19.46	19.9	21.1	20.8	21.3	22.5	22.9	23.3
Vietnam	7.2	7.2	7.5	7.8	8.1	8.2	8.2	8.3	8.7	9.4
Zambia	17.0	14.7	14.9	14.9	14.5	14.6	14.6	15.4	16.9	18.4

[a]FAOSTAT (2011) statistics for Malawi are confusing: potato and sweetpotato production appear to be reported under potato – for this reason we used the statistics of FEWSNET (2009), the early warning system data system, which separates the two crops.

and by cultivation techniques (e.g. weed control, crop rotation and fertilizer input). For developed countries, about 50% of yield progress across crops is usually attributed to breeding progress (Wricke and Weber, 1986). Reported yield increases by FAOSTAT do not allow the separation of total yield progress into these two categories. Genetic gain studies for sweetpotato (i.e. by comparing old and new varieties on-farm or a new breeding population with a previous population on-station) have so far not been reported – a clear gap in sweetpotato research. Such studies would be useful to calibrate genomic selection (GS) models to predict trait performance. Based on extensive on-farm observations, we hypothesize that storage root yields of 15 t/ha for sweetpotato on poor soils can be obtained through combining three factors: (i) 'good' varieties; (ii) weeding; and (iii) disease-free or 'clean' planting material.

Uses, markets and varieties

Sweetpotato is used in a variety of ways for food, feed and processed products, with the principal uses varying by region. The literature on nutritional value of cooked and fried sweetpotatoes – as well as processing sweetpotato into food products such as bread, ready-to-eat breakfast, French fries, syrup, starch and beverages – was comprehensively reviewed by Woolfe (1992), Bovel-Benjamin (2007) and Padmaja (2009). In developing countries, the crop is mainly grown for homestead food and feed use and to sell to local markets for fresh consumption. Use of both vines and roots for pig feeding is important in China, Vietnam and Papua New Guinea (Peters, 2004). Padmaja (2009) provides details on use of the crop for cattle, poultry and fish feed.

All sweetpotatoes used both as human food and as animal feed are called 'dual-purpose' sweetpotatoes. Dual-purpose sweetpotatoes should have high foliage yields, because these are mainly used for sweetpotato-based silage and high-protein supplements (fodder) for livestock (Scott, 1991; Zhang *et al.*, 1993; León-Velarde and de Mendiburu, 2007). However, there may be a contradiction between the nutritional value for human food and the demand for extremely high digestibility by the feed industry (Zhang *et al.*, 1993), so that consideration should be given to breeding varieties exclusively for animal feed for areas where that is its dominant use. In China, much sweetpotato is also used in starch noodle production, and use for production of distilled spirits is common in East Asia. Purple-fleshed types, high in anthocyanin, are increasingly popular in China and Japan, used fresh or in a variety of processed snacks and as a source of natural food colouring (Timberlake and Henry, 1988; Gilbert, 2005; Liu, 2008; Ma, 2010).

Awareness of the high nutritional value of sweetpotato is driving increasing consumer demand for the crop among health-conscious consumers in the USA and Europe (USDA, 2015). Orange-fleshed sweetpotato (OFSP) can be used effectively to combat vitamin A deficiency (VAD) among vulnerable populations (Low *et al.*, 2007; Hotz *et al.*, 2012). The leaves of sweetpotato have nutritive values comparable to common dark-green leafy vegetables (Ishida *et al.*, 2000; Bovel-Benjamin, 2007) and leaves, including shoot tips and petioles, are an increasingly popular green vegetable in some regions of China and important in parts of Africa. Ornamental sweetpotatoes with strikingly varied foliage are commercially popular in the USA (Barnes and Sanders, 2012) and South Korea (Yeong-Sang Song, Korea, 2013, personal communication). To our knowledge, there is no significant use of sweetpotato starch in textile, paper, plywood and pharmaceuticals. The crop was traditionally a food security crop (Jia, 2013). It retains this role in many parts of the world, because it: (i) is high yielding; (ii) needs low amounts of water per unit of food and energy (see section 'Drought and other abiotic stresses'); (iii) provides relatively good yields under poor input and marginal soil conditions; and (iv) exhibits wide adaptability to climates, farming systems and uses (Diop, 1998; Hijmans *et al.*, 2002; Jiang *et al.*, 2004). All parts of the plant (roots, leaves and shoots) are edible. Moreover, the crop produces more edible energy per unit area and time (194 MJ/ha/day) than any other major food and it can support more people per hectare than any other crop (Norman *et al.*, 1984; Woolfe, 1992). There are efforts investigating the use of sweetpotato in bioethanol production in the USA (Estes, 2006, 2009) and China (Liu *et al.*, 2010; Wang *et al.*, 2013). On the basis of current technology, 1 t of bioethanol can be produced from approximately 8 t of fresh sweetpotatoes (Qiu *et al.*, 2010).

Two major quality classes of sweetpotato for fresh consumption are generally recognized (Martin and Jones, 1986; Kays *et al.*, 2005). The so-called 'dessert types' are high in β-carotene, have relatively low dry matter content (< 30%) and moist texture, with a high flavour impact due to sweetness and aroma. 'Staple types' typically lack β-carotene, have relatively high dry matter content (> 30%) with drier texture, and have lower flavour impact due to lower sweetness and aroma. A third quality class was recently coined by Tumwegamire *et al* (2011a), namely 'OFSP dry and starchy' also called 'sabor simple' in Latin America. These are OFSP varieties, high in β-carotene, but with staple attributes such as high dry matter. Nearly all new OFSP varieties bred in SSA are 'OFSP dry and starchy' to meet adult taste preference in SSA. This new OFSP type might also be attractive for markets in South America and South Asia. Sweetpotato breeding and seed programmes are largely supported through the public sector, driven to a varying extent by policies and to a minor extent by the needs of industry. Currently significant investment in sweetpotato breeding is directed towards the development of adapted, high-yielding OFSP varieties to be used for combatting VAD among vulnerable populations in SSA. These investments are additionally supported by 'going-to-scale' disseminations of OFSP varieties in SSA. We assume that the OFSP fraction of the total sweetpotato harvested area in Uganda is still low (around 5%), whereas the OFSP in Mozambique is 22% (TIA, 2012) of total sweetpotato production, so that in the medium term Mozambique could be the first country in SSA with significantly lowered VAD prevalence due to consumption of OFSPs. The general perception of sweetpotato as a 'poor person's crop' is changing in SSA towards a 'food security and health crop'. So far, there are no comparable investments in sweetpotato breeding in South and South-east Asia, in spite of very high VAD prevalence in these regions (UN-SCN, 2004). An important factor underlying increased investment in sweetpotato breeding in SSA was the biofortification programme of HarvestPlus (Pfeiffer and McClafferty, 2007), which is linked to the AgroSalud and Biofort programmes in Latin America. However, sweetpotato is now of minor importance as a food crop in the Americas.

What is biofortification? Biofortification refers to quality breeding aiming at the enhancement of provitamin A, iron and zinc contents in major food crops so that they reach about 50% of their respective recommended daily allowances (RDAs). The micronutrients provitamin A, iron and zinc are critically deficient in our food supply (UN-SCN, 2004) and billions of people are micronutrient deficient without being hungry (so-called 'hidden hunger'). In all countries in which VAD is a serious public health problem, OFSP breeding is a cost-efficient and sustainable vehicle to alleviate VAD and to improve public health. This holds true even if only small quantities of OFSPs are eaten. OFSP, biofortified with provitamin A, is considered by HarvestPlus (Bouis and Islam, 2012; Hotz *et al.*, 2012) to be the first biofortified crop ready to go to scale. Sweetpotatoes are not biofortified for iron and zinc, but OFSPs can contribute about 20%, 20%, 25% and 50% to the RDA of iron, zinc, calcium and magnesium, respectively, where the crop is used as a staple (e.g. Uganda; Tumwegamire *et al.*, 2011a). The target levels to reach 50% RDA, to be able to label sweetpotato as biofortified, for iron and zinc are 60 ppm and 40 ppm, respectively (Wolfgang Pfeiffer, Colombia, 2009, personal communication). Theoretically it is possible to double iron and zinc contents in sweetpotato storage roots, but this will require several breeding cycles (see sections on 'Quality' and 'Breeding Methods'). Fewer cycles may be needed if the bioavailability of iron and zinc is found to be much higher in OFSP roots than currently assumed. Leaves also contain iron and zinc (Woolfe, 1992; Ishida *et al.*, 2000; Bovel-Benjamin, 2007), whose bioavailability is also unknown. In addition, it is not clear to what extent iron levels in leaves are due to non-plant iron contamination of the samples.

For further details on uses and markets by regions, consult Loebenstein and Thottappilly (2009).

1.2 Origin of Sweetpotato, Wild Species and Centres of Genetic Diversity

Sweetpotato (*Ipomoea batatas*) is a polyploid, and is the only hexaploid species (6x = 90, x = 15) in section *Batatas* of the family *Convolvulaceae* (Table 1.2). How and where it originated have not been fully resolved. There are two hypotheses concerning the evolution of the sweetpotato ancestor. The most widely held hypothesis is that *I. batatas* evolved from interspecific hybridization between *Ipomoea trifida* and *Ipomoea triloba* (Austin, 1988). The second is that *I. batatas* developed by polyploidization in *I. trifida* (Kobayashi, 1984). Recent studies based on evaluation of chloroplast haplotypes and nuclear DNA indicate that it was domesticated separately in Central and South America through autopolyploidization of distinct populations of *I. trifida* or a close relative (Roullier *et al.*, 2011, 2013a). In Roullier's studies, tetraploid accessions previously classified as *I. trifida*, but later classified as *I. batatas* (Bohac *et al.*, 1993), shared haplotypes with cultivated sweetpotato in both the northern and the southern regions of domestication. Cytological, molecular and conventional genetic studies provide evidence for some differentiation of the genomes making up the hexaploid sweetpotato, based on pairing at meiosis and tetradisomic segregation ratios (Magoon *et al.*, 1970; Kumagai *et al.*, 1990; Buteler *et al.*, 1999; Kriegner, 2001).

South and Central America have long been recognized as the primary centre of genetic diversity of sweetpotato (Austin, 1978; Austin and Huamán, 1996; Zhang *et al.*, 2000). Secondary centres of diversity exist, however, on the island of New Guinea (Yen, 1974; Austin, 1988) and in East Africa (Zhang, D. *et al.*, 2004; Montenegro *et al.*, 2008). Evidence indicates that sweetpotato could have reached the New Guinea highlands around AD 1200 (Golson, 1976), but the penetration of the crop into Melanesia remains unclear. However, by the 19th century, sweetpotato was the most important staple food crop in New Guinea, and notably is

Table 1.2. Species, ploidy level, origin and accession availability at the International Potato Center (CIP) of *Ipomoea* section *Batatas.*

Species	Polyploidy	Origin	Accessions in CIP genebank
Ipomoea batatas	4x, 6x	New World	4616
Ipomoea cordatotriloba	2x	New World	100
Ipomoea cynanchifolia	2x	New World	3
Ipomoea grandifolia	2x	New World	123
Ipomoea lacunosa	2x	New World	5
Ipomoea littoralis	2x	Australia	–
Ipomoea × leucantha	2x	New World	13
Ipomoea ramosissima	2x	New World	32
Ipomoea tabascana	4x	New World	1
Ipomoea tenuissima	2x	New World	–
Ipomoea tiliacea	4x	New World	54
Ipomoea trifida	2x, 4x	New World	183
Ipomoea triloba	2x	New World	60
Ipomoea umbraticola	2x	New World	6

adapted to very different environments in New Guinea compared with China, Korea and Japan, where it became important nearly simultaneously. Without doubt, the sweetpotato has a secondary diversity centre in and around New Guinea (Yen, 1974; Austin, 1988). Although the genetic diversity in this secondary centre of diversity is considerable, this is probably not based on a large number of introduced clones, but due to isolated environments where the crop flowers and sets seed readily, giving rise to new varieties (Roullier *et al.*, 2013b). This ability of sweetpotato to rapidly develop genetic diversity – even on the basis of a relatively small number of clones – has also been driven by its genetic nature as a highly heterozygous hexaploid hybrid (see section 'Sexual Reproduction, Autopolyploidy and Population Genetics'). A further secondary centre of diversity of sweetpotato has been proposed in East Africa with the discovery of dry and starchy farmer varieties of OFSP (Gichuki *et al.*, 2003; Tumwegamire *et al.*, 2011b).

A recent molecular marker study with both chloroplast and nuclear microsatellite markers supports the existence of two geographically restricted gene pools for *I. batatas* in Central and South America (Roullier *et al.*, 2011) and the authors argued that sweetpotato could have evolved by independent domestications in Central America (including the Caribbean) and South America. Venezuela, Colombia, Ecuador and Peru are represented by 2930 *I. batatas* accessions in the International Potato Center (CIP) genebank (only 10% of these accessions are breeding lines or improved varieties). To date, there are not many *I. batatas* accessions from Central America in CIP's genebank, with 259 of 4616 accessions originating from Central America. Future germplasm collections and acquisitions should prioritize this region.

Crosses among wild species in the section *Batatas*

It is possible to re-synthesize new *Ipomoea* hexaploids (i.e. diploid *Ipomoea leucantha* × tetraploid *Ipomoea littoralis*; Nishiyami *et al.*, 1975). Most cross combinations among species in the *Batatas* section result in interspecific hybrids (Iwanaga, 1988; Freyre *et al.*, 1991; Orjeda *et al.*, 1991; Cao *et al.*, 2009). With the exception of *Ipomoea nil* (for grafting to induce flowering) and *Ipomoea setosa* (for grafting to induce flowering and to screen for viruses), wild *Ipomoea* species have not been used in applied sweetpotato breeding, probably because breeders so far have found sufficient genetic variation in *I. batatas* for most breeding needs by screening their own or foreign germplasm, gene-pool separation or moderate inbreeding. However, other species in the *Batatas* section are a potential resource for unforeseen biotic and abiotic resistance needs. The Global Trust (Dempewolf *et al.*, 2014) programme started an initiative to use wild relatives of major food crops and plans to evaluate the *Batatas* section in heat-stress environments. This gene pool could become a source of heat-stress tolerant genes useful for more intensive sweetpotato breeding for climatic change. Moreover, wild species in the section *Batatas* could be a new source of additional resistances to sweetpotato weevils and sweetpotato virus disease (SPVD). The number of accessions of wild species in the *Batatas* section held in trust at CIP is not large (Table 1.2). However, these wild accessions are maintained as true-seed populations and each accession is formed by a large number of heterozygous genotypes. In contrast to wild *Ipomoea* species, *I. batatas* accessions are nearly exclusively maintained at CIP as *in vitro* clones.

Finally, we note that close wild relatives of sweetpotato are very interesting for genomic studies of sweetpotato. The sweetpotato genome is extremely large (the haploid DNA content is 1.55–2.25 pg/C nuclei or 1515–2200 Mbp; Ozias-Akins and Jarret, 1994; Kriegner, 2001) and highly heterozygous, which makes sequencing the *I. batatas* genome as well as mapping studies for sweetpotato extremely cumbersome. For this reason, many argue that the diploid *I. trifida* be used for genome sequencing to obtain information about the *I. batatas* genome, as well as diploid *I. trifida* maps to anchor the sweetpotato genome (Awais Khan, Peru, 2013, personal communication). CIP is currently incorporating an *I. trifida* mapping population, comprising about 200 genotypes, into its genebank.

1.3 Sexual Reproduction, Autopolyploidy and Population Genetics

The evolutionary forces driving sweetpotato are hexaploidy (6x), high heterozygosity, easy true-seed set by out-crossing and rapid clonal propagation. The crop is an autopolyploid highly heterozygous clone hybrid. The term clone hybrid reflects its genetic nature and presents the opportunity of applying heterosis-exploiting breeding schemes (HEBS). The genetic response of sweetpotato is often surprising – some breeders refer to it as a 'genetic monster'. Due to polyploidy with an even number of chromosome sets, more or less regular meiosis makes sexual seed production possible. Many genotypes very easily develop true seeds in nature (escapes and in farmer fields). The plant has a relatively strong sporophytic self-incompatibility system (Martin and Cabanillas, 1966; Martin, 1968) so that self-pollination usually occurs at low frequency. New genotypes are developed by recombining one highly heterozygous hexaploid hybrid with another highly heterozygous hexaploid hybrid. Incompatibility alleles result in specific cross combinations being difficult to achieve, and seeds from controlled sweetpotato crossings have especially high value (only one to three seeds are obtained from a successful pollination).

Flowering is a prerequisite for sexual reproduction, but sweetpotato genotypes differ greatly in this respect. We have observed that nature selects for prolific flowering among escaped clones (Fig. 1.3). Sweetpotato flowers can be very attractive and the plant has become an ornamental in the USA (Craig Yencho, USA, 2013, personal communication) and Korea (Yeong-Sang Song, Korea, 2013, personal communication). Some genotypes flower easily during any season, others are day-length sensitive and some have problems flowering – for example at the Xuzhou Sweetpotato Research Center (XSPRC) in China, parental material is generally treated with short day lengths during summer. Day-length flowering can be stimulated by grafting on *I. nil* or *I. setosa* (Lam *et al.*, 1959; Wang, 1975; Jones, 1980). Readily and balanced flowering among genotypes is important to recombine genotypes in polycross and controlled cross breeding nurseries. In cases where rare genotypes with special attributes can be selfed, a rare recessive inherited trait becomes fixed in offspring comprising several clones. The frequency of self-incompatibility/compatibility in populations is material dependent.

Fig. 1.3. Feral sweetpotato at San Ramon, Peru: natural selection favoured abundant flowering.

In populations undergoing intensive breeding, the frequency of successful cross combinations, the frequency of successful crossings per genotype and the frequency of self-compatibility probably changes over time. For example, during the summer season of 2012/13 in Peru, 23 selected parents of the population Jewel (one of the first OFSP populations at CIP) were recombined in a complete diallel crossing scheme (529 cross combinations) resulting in 460 cross combinations with seed set (383 cross combinations with ≥ 10 seeds) and eight parents were clearly self-compatible (with ≥ 10 seeds from auto-fertilization). This contrasted with 16 selected parents of the population Zapallo (a population created in 2005) and the same crossing scheme (256 cross combinations) in the same summer season – the results were 179 cross combinations with seed set (174 cross combinations with ≥ 10 seeds) and five parents were clearly self-compatible (with ≥ 10 seeds from auto-fertilization). This may indicate that sweetpotato is becoming more compatible with breeding.

The autopolyploid segregation ratios of sweetpotato are usually complex (Jones, 1967). Sweetpotato has some advantages as a model crop for breeding clonally propagated crops, especially its extremely short recombination cycles. In the case of a single dominant allele, the segregation ratios are simple (Poole, 1955) and the same is true for self-compatible clones and recessively inherited traits. Self-compatibility in sweetpotato presents a huge opportunity to increase the number of genotypes for a desired rare and recessively inherited trait – a new unique population is formed in which the desired trait is fixed. Crossing rare clones with a recessive inherited trait to 'normal' parents most often results in failure – the recessive trait disappears as genetic load in the population. Double reduction is a phenomenon that leads to discrepancies from expected segregation ratios in autopolyploids (note: this problem does not exist in diploids). The two segregation extremes in an autopolyploid are random chromosome segregation and random chromatid segregation (Wricke and Weber, 1986). With the latter, double reduction is possible – that is, sister chromatids of a chromosome sort into the same gamete (alleles are identical and derived from the same chromosome). Chromosome segregation is more frequent for loci close to the centromere, whereas the probability of chromatid segregation increases with the distance of loci to the centromere.

Gallais (2003) describes segregation ratios in the presence of double reduction for hexaploids. Single-locus segregation ratios become more complicated due to dosage effects of dominant alleles (discrete ratios are not seen and single-locus segregation ratios become continuous). The complexity of segregation in a hexaploid makes it extremely difficult to develop sweetpotato genetic maps. Moreover, homozygous sweetpotato parents are not available to develop mapping populations. The development of homozygous genotypes by selfing is illusory for hexaploid sweetpotato. Even if plants are self-compatible it would require seven generations of selfing to reach an inbreeding coefficient of $F = 0.5$ (for the calculations, readers are referred to p. 124 of Gallais, 2003), whereas $F = 0.5$ is reached in diploids after one generation of selfing. For this reason, attempts to develop double-triploids for sweetpotato are underway.

For decades, theoretical descriptions of autopolyploid genetics were limited (usually restricted to tetraploids) until the book by Gallais (2003) was published. For a hexaploid crop, more genotypes are possible and heterozygosity is much larger compared with diploid crops. Even in the simple case of one locus and biallelism, a hexaploid already allows the formation of seven different genotypes, compared with three for a diploid. With multi-allelism at a single locus the number of possible genotypes greatly increases in a hexaploid as a function of the number of alleles. Genotypes can carry a large load of alleles (i.e. five hexaploid genotypes can carry up to 30 alleles, whereas at least 15 diploid genotypes are needed to carry the same amount of alleles). Most loci across the hexaploid genome are heterozygous. In the case of biallelism, equal allele frequencies ($p = q = 0.5$), and random mating (and absence of double reduction) results in nearly all loci being heterozygous (Fig. 1.4). Within the allele frequency range of about $q = 0.2$ to $q = 0.8$, the frequency of heterozygosity is still > 0.75 in a hexaploid.

The heterozygosity in sweetpotato genomes has certain consequences for the ability of the crop to change and adapt in nature and breeding. This can be observed for simple inherited traits, but is perhaps much more important for complex inherited traits controlled by many loci. Several surprising observations in sweetpotato populations can be explained by multiple alleles at one locus and/or extreme heterozygosity across many loci. The first observation is that sweetpotato is capable of developing a large genetic diversity with few introductions (e.g. the diversity observed today in Papua New Guinea or East Africa). In other words, sweetpotato has a larger effective population size and is less affected by genetic drift compared with diploids. The second observation is the extreme large genetic diversity for quality traits

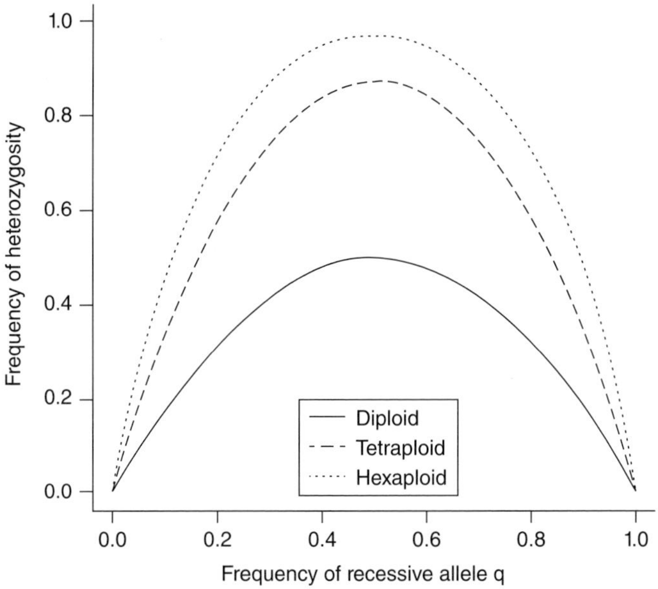

Fig. 1.4. Effect of ploidy level on the frequency of heterozygosity in a random mating biallelic population at equilibrium as a function of the frequency q of the recessive allele (p + q = 1), in the absence of double reduction. (From Gallais (2003), modified by inserting the hexaploid curve.)

(i.e. storage root shape/form, skin colour, flesh colour, stem and leaf form and colour, starch properties and micronutrient contents). On the other hand, it also has a larger 'genetic load' in the negative sense due to defective alleles compared with crops with low ploidy level. With moderate inbreeding (crossing relatives) and gene-pool separation this genetic load can be made more visible for selection. The third observation is that some attributes are very rarely found in sweetpotato germplasm and breeding populations (i.e. SPVD resistance or non-sweetness after boiling) – much worse is that they 'disappear' rapidly after recombination. Typically, less than 0.2% out of 1000 clones is resistant to SPVD in breeding populations at Namulonge in Uganda (Mwanga *et al.*, 2002a,b).

Frequency of recessive homozygosity (Fig. 1.5) and frequency of heterozygosity (Fig. 1.4) are obviously related. Recessively inherited traits are rarely expressed in a diploid open-pollinated crop in a wide range of allele frequency, but in autopolyploid crops (especially a hexaploid) the expression of a recessively inherited attribute is extremely rare, even if the recessive allele has medium frequency (q of 0.3–0.6). Only at high frequencies of the recessive allele (q > 0.7) can the desired recessive inherited attribute be observed with elevated frequencies (> 10%). This results in the paradox that a recessively inherited attribute is very rarely observed, although the recessive allele is present in the population with medium frequency. Breeding for recessive inherited attributes in sweetpotato is much more difficult than in diploids and the same is true for purging negative genetic loads in quantitatively inherited traits – it can be improved by crossing with relatives, controlled crossing by the 'best with the rest' (top clones are crossed with remaining parents) and gene-pool separation.

The extremely high frequency of heterozygosity (Fig. 1.4) in hexaploid populations indicates that the 'stimulus of heterozygosity' or heterosis might be very high in sweetpotato. During the past 5 years, a more intensive discussion has developed on HEBS for clonally propagated crops (Miles, 2007; Grüneberg *et al.*, 2009a). Actually, HEBS was proposed earlier for breeding clonaly propagated crops (Hull, 1945; Melchinger and Gumber, 1998), but the recommendations

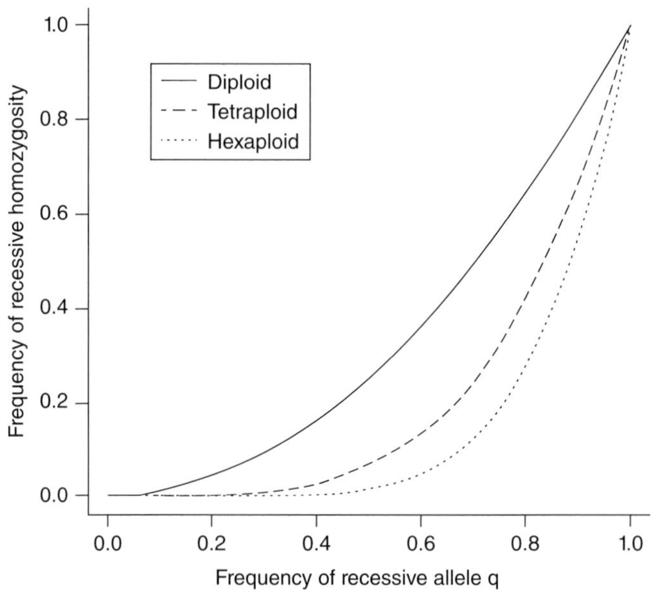

Fig. 1.5. Effect of ploidy level on the frequency of recessive homozygous genotypes in a random mating biallelic population at equilibrium as a function of the frequency q of the recessive allele (p + q = 1), in the absence of double reduction.

were buried in reports concerning heterosis in traditional hybrid crops. Arguments supporting applying HEBS in clonally propagated crops are: (i) all important clone crops are hybrids (clone hybrids); (ii) in cases where sexual reproduction is possible all clonally propagated crops are out-crossing species; and (iii) most clonally propagated crops are autopolyploids with considerably higher heterozygosity compared with the diploids in which HEBS have been applied. In theory, without large investments (simply by gene-pool separation and controlled recombination), large genetic gains might be realized. This holds true for quantitatively inherited traits (controlled enhancement of heterozygosity by inter gene-pool recombination) as well as qualitative inherited traits (controlled inbreeding by intra gene-pool recombination – see also section 'Breeding Methods').

1.4 Breeding Objectives and Genetic Variation

The multitude of potential breeding objectives in sweetpotato can be confusing. Owing to the large segregation potential and diversity and cultivation across a wide range of agroecological zones (Hijmans *et al.*, 2002) many different variety types can be developed. For clarity, we group all breeding objectives into those related to yield, quality and resistance. In reality, there is only one breeding objective – a better variety.

Variety types

Variety types are groups of varieties discriminated on the basis of their use or purpose and adaptation. Usually these are shaped on the basis of demands of agroclimatic zones and use (human consumption, animal feed, non-food industries). Often these groups are made more specific on the basis of colour, cooking quality, processing characteristics and adaptation to cropping systems as well as early or late maturity. A variety may belong to two or more groups (e.g. dual-purpose use for human food and animal feed). Breeders usually select for variety types in separate gene pools.

Formally, four variety types are distinguished in sweetpotato according to flesh colour, dry matter, total sugar and taste of storage roots. Twenty years ago, there were only two variety types: #1: the white, yellow or cream, dry, low-sweet or staple type (also called '*bonitos*' or '*ricos*' in the Caribbean; Baynes, 1972) and #2: the orange, moist, sweet or dessert type (Martin and Rodriguez-Sosa, 1985). A new variety type #3, 'OFSP dry and starchy' (Tumwegamire *et al.*, 2011a), is an OFSP that in the mouth feels and tastes rather bland, like 'OFSP sabor simple' in Latin America. Nearly all OFSP variety releases in SSA are categorized as OFSP dry and starchy (Appendix 1, at the end of the chapter). Varieties of this new type are also in the pipeline for the Amazon Basin (Appendix 3). Variety type #4 is the purple-fleshed type, usually dry and low in sweetness. Additional variety types may emerge due to specific suitabilities for boiling/microwaving (e.g. the variety Quick Sweet; Katayama *et al.*, 2006) or processing into chips, purée, juice, baby food and bakery products (Woolfe, 1992; Liu, 2008; Ma, 2010).

Informally, three more variety types are recognized (Appendix 1). The first is the 'dual-purpose' type for food and animal feed; the second is the 'good for industrial use' type – for both of these, there are no clear classification criteria. A variety classified as 'dual purpose' is usually a clone with acceptable storage root yield and abundant upper biomass production, sufficient to provide considerable fodder. A variety classified as 'good for industrial use' is most often a clone with high storage root yield and high starch content – sometimes associated with undesired form and size of storage roots. Within varieties classified as 'good for industrial use' screening is conducted for biofuel production (Estes, 2009; Liu *et al.*, 2010; Wang *et al.*, 2013). The third informal classification criterion is 'maturity period'. Yanfu *et al.* (1989) classified sweetpotato into short-duration or early-maturing (12–17 weeks after planting), medium-duration (17–21 weeks) and long-duration or late-maturing (> 21 weeks) types. In contrast to potato, this classification system is not much used in sweetpotato (Tarn *et al.*, 1992). The reason might be that Yanfu *et al.*'s threshold levels are not appropriate for farming systems. An improved formal maturity classification would be very useful for tropical areas where sweetpotato is used for piecemeal harvest (East Africa) and where sweetpotato needs to fit several other crops per year in a rotation system (South Asia and South-east Asia). The same holds true for subtropical areas with short rainfall seasons and temperate areas with short summers. We propose here a different classification system for maturity time: (i) 'early bulking' with < 100 days duration after planting; (ii) 'normal bulking' with 100–130 days duration; and (iii) 'late bulking' with > 130 days duration. Among new breeding materials in the pipeline at CIP in Peru, there are many clones that can be labelled as 'early bulking' (90-day sweetpotatoes are possible) and most come from hybrid populations (Federico Diaz, Peru, 2013, personal communication), indicating that earliness and hybrid vigour are associated in sweetpotato.

Storage root yield

Improvement of storage root yield is high priority in all countries where average yields are low (< 12 t/ha, see Table 1.1). However, many breeders rank yield and quality equally, because clones that do not meet consumer quality preferences are simply not permanently adopted. Without a doubt, breeders in high SPVD-pressure zones rank resistance breeding as the most important breeding objective. Susceptible varieties cannot realize their yield potential in farmers' fields where seed systems are not economically viable. Breeders in drought-prone areas rank resistance breeding to this abiotic stress as most important to realize the yield potential of new varieties and minimize the risk of adopting these varieties. Even in the USA, Martin and Jones (1986) emphasized that the yield trait was not the highest priority. With respect to the 'dessert type' in Asia and the Pacific, yield was ranked number five after: (i) eating qualities; (ii) nutritional value; (iii) appearance and uniformity; and (iv) early maturity (Lin *et al.*, 1983).

Storage root yield can be disassembled into components at two levels. The first level comprises those components forming the biological yield or total biomass production. These are net assimilation rate per leaf area (gross photosynthesis minus respiration), leaf area, leaf area duration, water and nutrient uptake, and water and nutrient utilization. The second level comprises the allocation of biological yield into above-ground biomass and root biomass (with storage and non-storage roots). Harvest index (HI) captures this biomass allocation. Measuring the amount of non-storage roots is extremely difficult, so HI is usually calculated by storage root yield divided by above-ground biomass and storage root production. Storage root yield components consist of storage root weight and number of storage roots. In the case of commercial storage roots, yield has two components: (i) commercial storage root weight; and (ii) number of commercial storage roots. Among all the yield components, applied breeding uses HI and commercial yield the most. This is because many yield components are either very difficult to measure or are correlated and to a certain extent complement each other.

Biological yield and HI also help inform the current storage root yield potentials of sweetpotato. This can be illustrated with an evaluation of germplasm held in trust at CIP (Tables 1.3 and 1.4 for yield traits, and later in the chapter Tables 1.7 and 1.8 for quality traits). To the best of our knowledge, this evaluation of 1174 clones from

Table 1.3. Mean ($\overline{x}$ by least-squares mean (lsmean) estimates), minimum (min) and maximum (max) genotypic values and variance components[a] estimates for sweetpotato yield traits (N = 1174 clones) evaluated in diverse environments (five environments in Peru).

Trait	$\overline{x}$	Min	Max	σ^2_G	σ^2_E	$\sigma^2_{G\times E}$	σ^2_ε
Storage root yield (t/ha)	19.0	0.0	55.5	19.8	27.2	115.9	48.3
Foliage yield (t/ha)	22.6	0.0	67.8	26.2	110.6	161.6	79.3
Biomass (t/ha)	41.5	1.8	97.5	52.8	106.5	313.1	154.1
Harvest index (%)	47.8	0[b]	100[c]	65.2	207.6	230.7	93.3
Storage root dry matter (% FM[d])	34.9	18.3	47.2	14.8	4.2	5.7	3.0

[a]Variance components: σ^2_G, variance component due to genotypes; σ^2_E, variance component due to environments; $\sigma^2_{G\times E}$, variance component due to genotype-by-environment interaction; and σ^2_ε, variance component due to plot error.
[b]lsmean estimate –10.5% set to 0.
[c]lsmean estimate 109.2 set to 100.
[d]FM, fresh matter.

Table 1.4. Pearson's correlation coefficients among yield and quality traits of sweetpotato (N = 1174 clones) evaluated in diverse environments (five environments in Peru) – correlations calculated as means across phenotypic correlations for each environment and replication to obtain approximations of genetic correlations.

	Storage root yield[a]	Foliage yield[a]	Biomass[ab]	Harvest index[c]
Foliage yield[a]	0.197			
Biomass[ab]	0.735	0.790		
Harvest index[c]	0.508	–0.582	–0.075	
Storage root dry matter[a]	0.168	0.095	–0.035	–0.204

[a]FM, fresh matter.
[b]Biomass = storage root yield + foliage yield.
[c]Harvest index = (storage root yield/biomass) × 100.

different regions of the world is the largest study ever undertaken for yield and quality in sweetpotato. The study was conducted in 2006 and 2007 in Peru across varying eco-geographic conditions – four locations and five environments, respectively: La Molina, San Ramon with fertilization and without fertilization, Chiclayo with two and four irrigation treatments, and Oxapampa (no quality traits were determined at Oxapampa). At each environment, the experiment was conducted in a complete randomized block design with two plot replications. Each plot comprised two rows with five plants per row. Planting distance was 0.25 m within rows and 0.9 m between rows. An extreme range for biological yield or biomass production, respectively, was observed with a genotypic minimum of 2 t/ha up to a genotypic maximum of nearly 100 t/ha (Table 1.3). The population mean was around 40 t/ha. On average about 48% of the biological fresh matter yield was allocated to storage root fresh matter yield. Assuming an average of 20.7% dry matter in the upper biomass (Federico Diaz, unpublished, $n = 6874$ breeding clones) and an average of 34.9% dry matter in storage roots (Table 1.3) it can be estimated that sweetpotato allocates 58% of the biological dry matter yield (11.3 t/ha) into storage root dry matter yield (6.6 t/ha). However, sweetpotato exhibits extreme variation in HI ranging from close to zero to nearly 100%.

Obviously HI is a key yield component for storage root yield, with a huge variation in sweetpotato. There are two ways to breed for higher storage root yield: the first is to increase biological yield and the second is to increase HI. Which strategy is expected to have larger genetic gains in the short and/or long term? During the past decade, variance component estimates have been increasingly used in sweetpotato to determine if a breeding objective merits investment (Grüneberg *et al.*, 2004, 2005; Tumwegamire, 2011; Tumwegamire *et al.*, 2011a). Variance components are the appropriate parameters to judge investments in breeding. Although there are many heritability estimates available for sweetpotato (Martin and Jones, 1986), this parameter already depends on the test capacity (number of environments and replications), which varies among studies and experiments, respectively.

The variance component due to genotypes (σ^2_G) provides information on the genetic variability, and instability of measurements of genotypes in different environments is captured by the variance component due to genotype-by-environment interactions ($\sigma^2_{G\times E}$), whereas biological and technical errors are captured by the variance component due to the plot error (σ^2_ε). With these three parameters, it is possible to calculate expected genetic gains and determine whether to invest in breeding. In our example (Table 1.3) comprising diverse sweetpotato germplasm in contrasting environments, the ratio of $\sigma^2_{G\times E}$ and σ^2_ε relative to σ^2_G for HI was estimated to be 1:3.54:1.43. Hence, for various test capacity scenarios the expected genetic gain for HI is larger than those expected for biomass (1:5.93:2.92) and storage root yield (1:5.85:2.44). On the basis of genetic correlations or approximations of genetic correlations (Table 1.4), it is possible to obtain information indirectly for selection for storage root yield by selecting for HI. The latter is more efficient than a direct selection on storage root yield. This leads to model calculations and simulation studies to optimize breeding strategies (for complex studies, refer to Longin (2007); for a simpler study, Grüneberg *et al.* (2004)).

CIP is working on appropriate weighting factors for HI in breeding programmes utilizing index selection. Usually the $\sigma^2_{G\times E}$ for HI is lower in studies with less diverse material and/or less diverse environments (Grüneberg *et al.*, 2004, 2005; Tumwegamire, 2011). For example a $\sigma^2_{G\times E}$ and σ^2_ε to σ^2_G ratio for HI of 1:0.46:1.24 (recalculated from data of Tumwegamire *et al.*, 2011a) indicates that during the selection process the HI has progressively lower $\sigma^2_{G\times E}$ and that HI is not the only important factor for high storage root yields. It could also be that HI stability is a key factor in selection of storage root yield stability. Overall, HI is a simple measureable trait and when selection in early breeding stages is conducted at two contrasting environments, the $\sigma^2_{G\times E}$ of HI can be captured early in the breeding process (see also section 'Breeding Methods') and it may enable the selection for storage root yield and storage root yield stability during early breeding stages. HI has, in diverse material and contrasting environments, high associations with storage root yields ($r = 0.508$, Table 1.4). In other words, more than 50% of storage root yields appear to be determined by HI.

Certainly there are limits to achieving genetic gains by augmenting HI, but in the short term HI has large potential to increase storage root yields in sweetpotato. However, breeders must take into account that varieties with very high HI are not desired by farmers, because above-ground biomass is needed as planting material (also see section 'Drought and other abiotic stresses'). This leads to a question – what is the optimal HI for sweetpotato? Medium- to high-yielding varieties such as Jewel and Xushu 18 have HI of 53.1% and 66.7%, respectively, in contrasting environments (Grüneberg *et al.*, 2005). This is perhaps too high for areas where planting material is a bottleneck. Grüneberg *et al* (2005) observed an HI of 42.4% for the popular African variety Tanzania, which is certainly medium to low, but not out of range for a 'good' HI. The variety CEMSA-74-228, with HI of 55.6% across 12 East African environments (Grüneberg *et al.*, 2004), is perhaps very close to optimal. In conclusion HI – especially HI stability and its association with storage root yield stability – continues to merit further investigation.

Commercial storage root weight (CSRW) and number of commercial storage roots (NCSR) are also considered valuable information by many breeders. Each plant in a sweetpotato field should have a high NCSR (four to six/plant) of medium size and good uniformity (8–23 cm in length and 5–9 cm in diameter) (Firon *et al.*, 2009) and fields should have 35,000–45,000 plants/ha (i.e. the target in Peru sweetpotato growing areas). A limitation of our study (Table 1.3) is that CSRW and NCSR were only determined in the environment of San Ramon with fertilization; $\sigma^2_{G\times E}$ cannot be calculated for CSRW and NCSR. However, the least-squares mean (lsmean) estimates at San Ramon (Table 1.5) show that: (i) on average 78% of the storage root yield was considered commercially marketable; (ii) on average a plant had about 0.5 kg of commercial storage roots; and (iii) an average of 2.3 storage roots per plant. The maximum genotypic value was 3.3 kg of commercial storage roots per plant. The 'environment specific variance component due to genotypes' was overestimated compared with σ^2_G by factors of 5.1, 6.6, 5.5 and 5.2 for storage root yield, foliage yield, biomass yield and HI, respectively (compare with Table 1.3), because environment specific σ^2_G estimates are inflated by $\sigma^2_{G\times E}$.

For NCSR, corresponding broad-sense heritabilities of 0.73, 0.40 and 0.83 were reported by Martin and Jones (1986). In our germplasm study, CSRW was strongly correlated with total storage root yield ($r = 0.940$) and breeders should ask themselves if determining non-commercial roots is necessary. All clones with high CSRW per plant (> 2.5 kg per plant) appear to exhibit high

Table 1.5. Mean ($\bar{x}$ by lsmean estimates), minimum (min) and maximum (max) genotypic values and variance components[a] estimates for sweetpotato yield traits evaluated at San Ramon with fertilization in 2006.

Trait	*N* clones	$\bar{x}$	Min	Max	σ^2_G	σ^2_ε
Storage root yield (t/ha)	1160	13.7	–0.5	74.2	100.8	42.5
Commercial root yield (t/ha)	1110	10.7	–1.9	61.1	63.8	46.3
Foliage yield (t/ha)	1200	26.1	–2.1	130.4	173.5	85.1
Biomass (t/ha)	1200	37.6	0.4	135.1	290.8	132.9
Harvest index (%)	1160	34.8	–3.6	104.4	340.9	142.8
Commercial root yield per plant (kg per plant)	1110	0.51	–0.1	3.3	0.14	0.19
Number of commercial roots (number per plant)	1110	2.3	–0.4	37.2	2.75	4.68

[a]Variance components: σ^2_G, variance component due to genotypes; and σ^2_ε, variance component due to plot error.

NCSR per plant (6.1–11.5 per plant, i.e. CIP clones 441341, 440652, 441608, 440157, 490065.25 and 400375, results not presented). CSRW and NCSR appear to be similarly important key traits for sweetpotato yields as HI and should be considered in all HI and HI stability studies. Nowadays, genes that are differentially expressed in non-storage and storage roots (e.g. 22 genes were found by You *et al.*, 2003) can be identified and these studies were recently reviewed by Firon *et al.* (2009). Certainly NCSR per plant is determined by fewer genes than storage root or biomass yields and it might be an interesting trait to include in studies on genomic selection (GS) for sweetpotato (see section 'Breeding Methods').

To breed for improved storage root yield, one must understand storage root initiation in sweetpotato and its interaction with the environment. Storage root initiation has been reviewed by Kays (1985), Ravi and Indira (1999) and Firon *et al* (2009). Storage roots only derive from adventitious roots arising from the underground stem portions of a vine cutting. Lateral roots (those roots arising from existing roots) do not form storage roots. Adventitious roots can be separated into 'thick' or 'thin' roots (Kays, 1985; Ravi and Indira, 1999). The former nearly always develop from the nodal area of the underground stem, whereas the latter arise primarily from internodal regions of the underground stem. Only thick roots can develop into storage roots (> 15 mm in diameter); however, a larger proportion of thick adventitious roots develop into pencil roots (< 15 mm in diameter). Thin adventitious roots nearly always develop into fibrous roots (< 5 mm diameter). The number of storage roots is determined early in sweetpotato, usually within less than 8 weeks after planting (Lowe and Wilson, 1975). For example, the number of storage roots in the variety Beauregard is determined within 3–6 weeks after planting (Arthur Villordon, USA, 2013, personal communication). Lignification of steles in thick adventitious roots causes irreversible storage root formation and is a result of unfavourable environmental soil conditions in early growing stages (Togari, 1950; Wilson and Lowe, 1973; Lowe and Wilson, 1975; Belehu *et al.*, 2004). The realization of the potential to become storage roots to a large degree determines the final storage root yield ($r = 0.412$, Table 1.6). We hypothesize that this could be developed into early screening methods for storage root yield. Moreover, the large $\sigma^2_{G\times E}$ for storage root initiation presents opportunities to select for storage root initiation stability (e.g. in Peru we observed that the check clone Tanzania is very sensitive to abundant water supply, whereas this does not affect check clone Resisto).

Breeders do not usually pay much attention to yield physiological traits and the overall assimilation potential. However, assimilation is not a simple function of net assimilation rate per leaf area, leaf area and leaf area duration. A very important factor

Table 1.6. Pearson's correlation coefficients among yield traits[a] of sweetpotato (N = 1110 clones) evaluated at San Ramon with fertilization in 2006 – correlations calculated as means across phenotypic correlations for each replication to obtain approximations of genetic correlations.

	RYLD[b]	CRYLD[b]	FYLD[b]	BIOM	HI	CRWP
CRYLD	0.940					
FYLD	0.065	0.094				
BIOM	0.638	0.627	0.805			
HI	0.681	0.595	0.513	0.004		
CRWP	0.672	0.717	0.327	0.146	0.792	
CRNP	0.423	0.412	–0.372	–0.011	0.689	0.743

[a]RYLD, storage root yield; CRYLD, commercial root yield; FYLD, foliage yield; BIOM, biomass = RYLD + FYLD; HI, harvest index = (RYLD/BIOM) × 100; CRWP, commercial root weight per plant; CRNP, commercial root numbers per plant.
[b]FM, fresh matter.

for assimilation is how efficient assimilates are incorporated from the leaf source into the sinks, and among these the storage root is a very dominant sink (Kays, 1985; Ravi and Indira, 1999). The sucrose concentration is high at the source and is moved in water via the phloem to sinks where the sucrose concentration is low. With the conversion to starch by hydrolysis in the storage roots, the sucrose concentration remains low in the storage root sink. Reciprocal graft experiments between sweetpotato and *I. trifida*, as well as among sweetpotato genotypes with poor or strong sink capacity, show how important this factor might be in sweetpotato yield formation. Carbohydrate accumulates in the leaves of shoots grafted onto genotypes with low sink capacity (Hozyo and Park, 1971; Ko *et al.*, 1993) and the source potential of low-yielding cultivars is increased when grafted onto genotypes with high sink capacity (Hahn, 1977; Zhong, 1991). Net photosynthetic rate drastically declines when root enlargement is restrained (Tsuno and Fujise, 1965). Note that the top five biomass-yielding clones in our study presented in Table 1.3 (biomass yield: > 90 t/ha in 199076.1, 401549, 420886, 401031 and 187016.2 (for details see lsmean values uploaded as 'sp_germ_2005-2006.pdf' on 'A sweetpotato breeding repository' available at http://sweetpotatobreeder.com)) were all clones with a strong storage sink capacity (high storage root yields of 35.6–55.5 t/ha). An active source appears to need a high sink capacity (Ravi and Indira, 1996a,b).

Certainly the sink is not the only driving force to assimilate carbohydrates. In photosynthesis (the source), it is needed to distinguish between light utilization and light uptake. Light utilization is determined by the net assimilation rate per leaf area. There are opinions that light utilization has already been well optimized during plant evolution (green plants have long existed in evolutionary history), whereas light uptake still offers opportunities. Light uptake is determined by leaf area, leaf area duration and leaf orientation to the incoming radiation. The leaf area relative to the soil surface is estimated by the leaf area index. Sweetpotato appears to exhibit a great magnitude of genetic variation for leaf area. Most sweetpotatoes rapidly cover the ground, but lack of canopy depth due to horizontal development of the canopy and poor leaf orientation, result in shading of leaves within the canopy. The optimum leaf area index of sweetpotato appears to be 3–4 (Tsuno and Fujise, 1965). Cultivars adapted to elevated altitudes in Africa are reported to be more erect and have lower leaf area indexes (Hahn and Hozyo, 1984). There is a pronounced period during the growing season in which the leaf area index of sweetpotato is larger than 3–4 (Kotama *et al.*, 1970). Compared with rice, sweetpotato has higher crop growth rates during the first 4–6 weeks after planting and later again at 10–15 weeks after planting (Tsuno, 1971); however, between these periods rice is superior to sweetpotato and this is the period during which sweetpotatoes usually have a leaf area index greater than 3–4. Most yield physiology studies trace back to findings of Tsunoda (1959), who observed that the highest yielding varieties produced relatively thick and small leaves in response to high light intensity, which allowed good light penetration. To our knowledge such aspects have not been further investigated during the past two decades, except in a study by Kelm *et al* (2000) with the two clones Jewel and Tanzania. Significant options for genetic improvement probably exist, as the optimal assimilating surface of a densely planted sweetpotato monocrop should be very different from that of a single wild sweetpotato plant. Certainly clones with many branches, exhibiting long extended internodes and long vines and a horizontal leaf orientation (thereby allocating a major proportion of assimilates into the canopy) are not optimized when planted densely as a monocrop. We further examine the performance and efficiency of underground roots to supply water and nutrients for assimilation in the section 'Drought and other abiotic stresses'.

Quality

Quality demands are driven by how sweetpotato is used. Most important are the needs for direct human consumption. Second are needs associated with use as animal feed. Quality required for the food industry is determined by the product. Traits needed for sweetpotato processed into chips are different from those needed for sweetpotato processed into Chinese noodles. This discussion focuses on quality for direct human consumption in the developing world. Demands for direct human consumption (boiling, roasting and mashing into purée) vary among societies and countries. Different taste preferences depend mainly on how people have been socialized and income. In this discussion, a distinction will be made between directly noticeable quality and *not* directly noticeable quality traits.

The first group of directly noticeable quality traits is storage root shape and form, flesh colour and skin colour. These three traits have medium to high heritabilities and therefore are also used as morphological descriptors (Huamán, 1991). Drawing again on the data for 1174 clones in Peru, the variation for storage root shape and form ranges from round (resembling large-size potato tubers) to very long (nearly resembling small cassava storage roots) (Fig. 1.6a). Many breeders, growers and consumers have an ideal for how a sweetpotato storage root should look, that is uniform shape 8–23 cm in length and 5–9 cm in diameter (Firon *et al.*, 2009). However, in most developing countries, a commercial storage root is simply defined on a weight basis, for example ≥ 100 g in the case of Malawi (Felistus Chipungu, Malawi, 2013, personal communication). The range in storage root flesh colour includes white, yellow, orange and purple (Fig. 1.6b). Yellow and orange colour in sweetpotato storage roots is determined by carotenoids. Fortunately, the proportion of β-carotene as dominant provitamin A is greater than 80% among the total carotenoid content in OFSP (Woolfe, 1992). For this reason, flesh colour alone can be used to predict β-carotene content of storage roots using colour charts (G. Burgos, R. Carpio, C. Sanches, P. Sosa, E. Porras, J. Espionza and W.J. Grüneberg, unpublished data). During the past 5 years, these colour charts have become widely used by the National Agricultural Research System (NARS) breeding programmes in SSA to estimate β-carotene contents of new selections. The purple flesh colour is determined by anthocyanins. Owing to the health-promoting effects of antioxidant anthocyanin substances such sweetpotatoes are also attractive for quality breeding. Moreover, such purple varieties can be used to obtain food colourants, which is a relatively new market for sweetpotato (Timberlake and Henry, 1988; Gilbert, 2005; Konczak, 2006). The storage root skin colour ranges from white, yellow, orange and brownish orange, red to dark purple (Fig. 1.6c). Consumers in most regions still tolerate a wide range of storage skin colour (white, brown, red and purple).

The second group of directly noticeable quality traits is mouthfeel and taste. Many believe that it is not possible to define the compound(s) which determines the 'sweetpotato taste'. Certainly, in breeding OFSPs local taste preferences are critical. Consumers like the orange-fleshed coloured clones as long as they are not associated with undesirable mouthfeel and taste. Adult consumers do not make many compromises with respect to this trait. For example, the first introduction of OFSPs into Africa – where the white, dry, low-sweet and bland type was nearly exclusively consumed – was hampered by the moist and sweet mouthfeel and taste of traditional OFSPs. The problem was solved by breeding for orange, dry and starchy varieties in SSA (Tumwegamire *et al.*, 2011a,b). As a consequence, there are now over 40 variety releases and new breeding materials for orange, dry and starchy sweetpotatoes (Appendices 1 and 3). Mouthfeel and taste depend much on dry matter, starch and sugar contents of storage roots. Laurie *et al.* (2012) observed significant correlation of maltose content with sensory sweet and sweetpotato-like flavour, which might serve as a tool for selection in early breeding stages. However, dry matter, starch content and sugars do not exclusively control taste and flavour. Hence, storage roots must be assessed by eating for taste and flavour quality breeding. While thousands of genotypes can by screened by microwaving, taste panels need to be conducted by experienced persons.

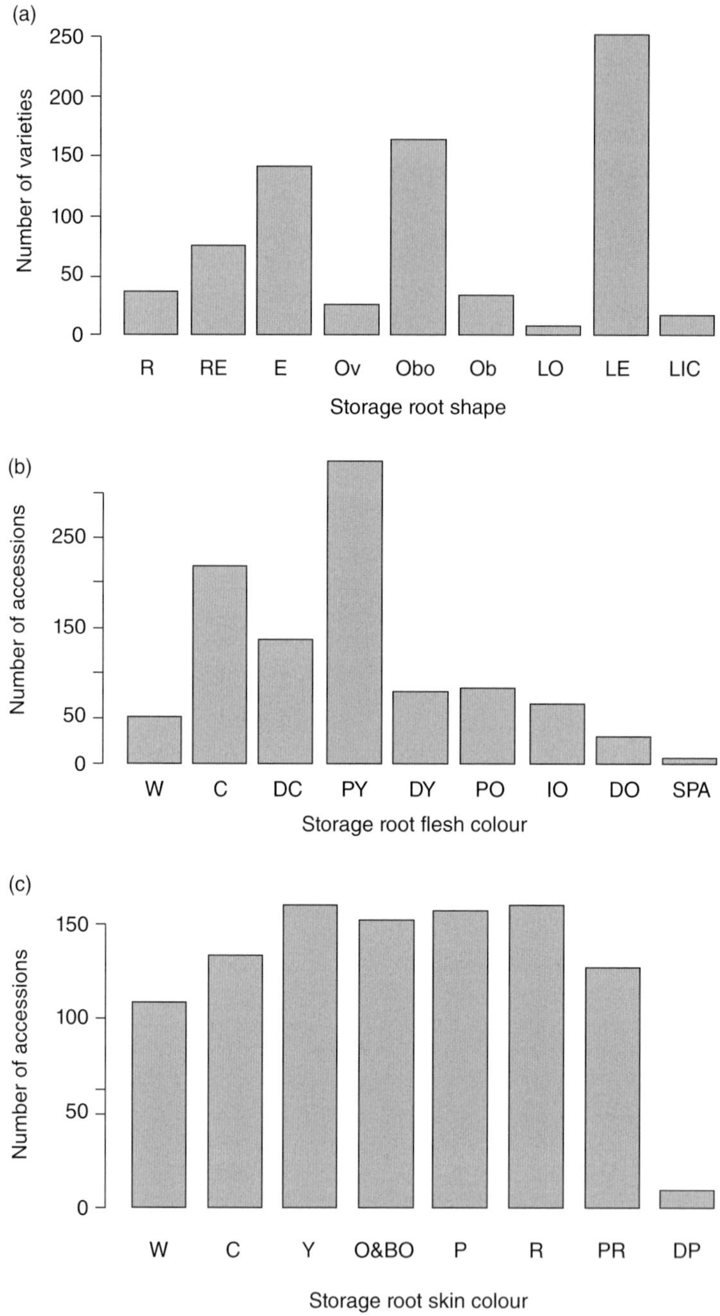

Fig. 1.6. Data bank information for storage root shape (a), flesh (b) and skin colour (c) for 1174 health status II clones held in trust at CIP and evaluated during 2006–2007 (see also Table 1.3). Root shape: R, round; RE, round elliptic; Ov, ovate; Obo, obovate inversely ovate outline; Ob, oblong; LO, long oblong; LE, long elliptic; LIC, long irregular or curved. Flesh colour: W, white; C, cream; DC, dark cream; PY, pale yellow; DY, dark yellow; PO, pale orange; IO, intermediate orange; DO, dark orange; SPA, strongly pigmented with anthocyanins. Skin colour: W, white; C, cream; Y, yellow; O&BO, orange and brownish orange; P, pink; R, red; PR, purple red; DP, dark purple. (From Huamán, 1991.)

Not directly noticeable quality traits are those associated with the nutritional value of sweetpotato and need to be determined by analytical methods. The obvious exception is β-carotene content of storage roots that turn storage roots yellow to dark orange. In the past, it was only possible to evaluate protein, starch, individual sugars, vitamins and micronutrients by complex analytical laboratory procedures. Owing to the cost and time required only a few clones were screened in a breeding programme. Such studies provided information about the range of the chemical composition in sweetpotato as described in the textbook by Woolfe (1992) or genetic variance component estimates of storage root dry matter, starch and β-carotene content on the basis of a few clones (Grüneberg *et al.*, 2005).

Studies indicate that most nutritional traits can be efficiently changed by breeding due to the large σ^2_G and low $\sigma^2_{G\times E}$ of these quality traits relative to σ^2_G. However, at the end of the breeding cycle, when only a few clones remain and most have been discarded, there is not much genetic variation left to enable finding genotypes which combine desired nutritional traits with desired yields. In other words, breeders who want to change quality in the entire crop need to evaluate quality in the early breeding stages. This requires fast throughput methods such as colour charts to predict provitamin A content. During the past 8 years it became possible to calibrate near-infrared reflectance spectroscopy (NIRS) with reference values from chemical analytic methods such as spectrophotometry for total carotenoids, high performance liquid chromatography (HPLC) for different carotenoids and inductively coupled plasma argon optical emission spectrometry (ICP-OES) for minerals (Lu *et al.*, 2006; zum Felde *et al.*, 2007; Lebot *et al.*, 2011). This technology is now in use for early breeding stages at CIP headquarters and in SSA (Grüneberg *et al.*, 2009a,b) and for germplasm evaluation (Tumwegamire *et al.*, 2011a; Tables 1.7 and 1.8). The aim of this study was to determine maximum genotypic values, variance components, heritabilities and approximations of genetic correlations for β-carotene, iron and zinc to obtain information on how quality could efficiently be improved in sweetpotato.

What has been learned during the past decade on how efficiently yield and quality can be improved in sweetpotato? The average commercial storage root contains 34.9% dry matter with 4.3% protein, 66.0% starch, 10.3% sucrose, 143.7 ppm β-carotene, 15.6 ppm iron and 9.3 ppm zinc (Table 1.7). However, this average is quite artificial because its estimation is across very different variety types. Dry matter, starch, individual sugars

Table 1.7. Mean ($\bar{x}$ by lsmean estimates), minimum (min) and maximum (max) genotypic values and variance components[b] estimates for sweetpotato quality traits (N = 1174 clones) evaluated in diverse environments (five environments in Peru).

Trait[a]	$\bar{x}$	Min	Max	σ^2_G	σ^2_E	$\sigma^2_{G\times E}$	σ^2_ε
Root yield (t/ha)	19.0	0.0	55.5	19.8	27.2	115.9	48.3
Root dry matter (% FM)	34.9	18.3	47.2	14.8	4.2	5.7	3.0
Protein (% DM)	4.3	2.7	8.9	0.3	6.2	0.7	0.5
Starch (% DM)	66.0	36.5	76.0	28.9	6.0	7.2	3.8
Sucrose (% DM)	10.3	2.0	33.1	12.2	0.7	5.5	3.0
Fructose (% DM)	1.7	0.0	11.1	1.6	0.0	0.6	0.3
Glucose (% DM)	2.2	0.0	16.0	3.0	0.1	1.0	0.5
β-Carotene (ppm DM)	143.7	1.8	1,220	14,751	2,262	4,640	1,817
Iron (ppm DM)	15.6	10.5	28.6	2.7	15.1	3.2	2.5
Zinc (ppm DM)	9.3	6.2	17.1	0.9	9.4	1.5	1.1

[a]FM, fresh matter; DM, dry matter.
[b]Variance components: σ^2_G, variance component due to genotypes; σ^2_E, variance component due to environments; $\sigma^2_{G\times E}$, variance component due to genotype-by-environment interaction; and σ^2_ε, variance component due to plot error.

and β-carotene can be positively or negatively associated (Table 1.8). For example, varieties with very high dry matter and starch content are low in sugars and usually have no or very low β-carotene content; or varieties with very high β-carotene content are usually high in individual sugar content and low in dry matter and starch content. Obviously, quality attributes vary widely in sweetpotato (Table 1.7) with respect to dry matter (18.3–47.2% fresh weight basis (fwb)), protein (2.7–8.9% dry weight basis (dwb)), starch (36.5–76.0% dwb), sucrose (2.0–33.1% dwb), fructose (0–11.1% dwb), glucose (0–16.0% dwb) and β-carotene (1.8–1220 ppm dwb). This results in extreme differences in storage root colour, texture and taste. Differences among sweetpotato varieties with respect to storage root iron (10.5–28.6 ppm dwb) and zinc content (6.2–17.1 ppm dwb) are less pronounced than differences in dry matter, protein, starch, sucrose, fructose, glucose and β-carotene content (data for individual clones are posted on the sweetpotato knowledge portal: http://sweetpotatoknowledge.org). Many sweetpotato varieties have greatly exceeded the minimum β-carotene target level required to be labelled as 'biofortified for provitamin A', but the best varieties as of 2014, only reach 50% of the iron and zinc targets and biofortified sweetpotato for iron and/or zinc are longer term objectives. However, as mentioned previously, the label 'biofortified' depends also on intakes and bioavailability. If iron in sweetpotato has a much higher bioavailability than currently assumed, for instance, this would have tremendous impact on breeding progress towards iron biofortified sweetpotatoes.

The σ_G^2 variance components for storage root dry matter, starch, individual sugars and β-carotene are large compared to σ_E^2 and $\sigma_{G\times E}^2$ (Table 1.7). In other words, these traits have a large genetic variation in sweetpotato and are not greatly affected by the environment and genotype-by-environment interactions. Notable negative genetic correlations exist between storage root β-carotene and dry matter content, and between storage root β-carotene and starch content, whereas positive correlations exist between storage root β-carotene and sugar content (Table 1.8). However, the magnitudes of these genetic associations are not sufficiently large to greatly slow breeding progress towards dry and starchy OFSP varieties that are rich in provitamin A. The positive genetic association between both trace minerals (iron and zinc) and β-carotene supports breeding,

Table 1.8. Pearson's correlation coefficients among yield and quality traits[a] of sweetpotato (N = 1174 clones) evaluated in diverse environments (five environments in Peru) – correlations calculated as means across phenotypic correlations for each environment and replication to obtain approximations of genetic correlations.

	RYLD[b]	FYLD[b]	DM[b]	PROT[c]	STA[c]	SUC[c]	FRUC[c]	GLUC[c]	BC[c]	FE[c]
FYLD	0.197									
DM	–0.168	0.096								
PROT	–0.114	–0.078	–0.071							
STA	–0.066	0.108	0.767	–0.232						
SUC	0.007	–0.068	–0.437	0.168	–0.788					
FRU	0.174	–0.060	–0.715	–0.164	–0.586	0.095				
GLU	0.170	–0.067	–0.718	–0.128	–0.608	0.117	0.982			
BC	–0.048	–0.086	–0.189	0.177	–0.425	0.462	0.059	0.072		
FE	–0.086	–0.065	–0.245	0.760	–0.458	0.362	0.004	0.054	0.264	
ZN[c]	–0.154	–0.087	–0.133	0.801	–0.310	0.291	–0.204	–0.145	0.213	0.822

[a]RYLD, storage root yield; FYLD, foliage yield; DM, dry matter content of storage roots; PROT, protein content of storage roots; STA, starch content of storage roots; SUC, sucrose content of storage roots; FRUC, fructose content of storage roots; GLUC, glucose content of storage roots; BC, β-carotene content of storage roots; FE, iron content of storage roots; ZN, zinc content of storage roots.
[b]FM, fresh matter.
[c]DM, dry matter.

but the magnitude of σ_G^2 for iron and zinc does not favour rapid genetic improvement. Finally, variance components and genetic associations favour the efficient breeding of varieties with low sugar content. Our findings are consistent with results of Woolfe (1992) and Courtney *et al.* (2008), who used US breeding material, and Tumwegamire *et al.* (2011a), who used East African germplasm. In summary, during the past decade, we have learned that: (i) extremely high provitamin A content can be combined with many other quality and yield attributes; (ii) breeding high iron and zinc varieties is a difficult and time-consuming process most likely achieved by several cycles of selection; and (iii) breeding for non-sweet sweetpotato varieties should be efficient and rapid.

Quality breeding is not as straighforward as theorized and we give three examples. The first is that raw non-sweet sweetpotato varieties do not necessarily stay non-sweet after boiling due to hydrolysis of starch to maltose in the cooking process by β-amylase. Without sufficient β-amylase, the normal hydrolysis of starch to maltose does not occur during cooking. This attribute is controlled by one recessive allele (Kumagai *et al.*, 1990). Breeding for non-sweet sweetpotato varieties requires both screening for low individual sugar content and low β-amylase activity in storage roots. Owing to the recessive inheritance of the attribute 'non-sweet after cooking', this breeding effort becomes a quite difficult task in hexaploid sweetpotato. The second example is an effort to change the two starch components amylose and amylopectin in sweetpotato (Lii and Tsai, 1996; Richardson *et al.*, 2000). The amylose content in sweetpotato starch is low (10–25%) (Noda *et al.*, 1998). High amylose starches as well as amylose-free starches are of interest in food and other industries (Richardson *et al.*, 2000; Ocloo *et al.*, 2011). Amylose-free genotypes have been generated using a transgenic approach (Kimura *et al.*, 2001), as have genotypes with increased amylose content (Shimada *et al.*, 2006). The third example is an effort to improve the baking quality of sweetpotato – so far sweetpotato flour is only used in mixtures with wheat flour in bread making since sweetpotato has no gluten. To improve the baking quality of sweetpotato flour, a glutenin gene of wheat was introduced into sweetpotato (variety Huachano). Among the 13 transformed events obtained, three expressed glutenin in high amounts (CIP, unpublished). For details on transgenic approaches to change sweetpotato quality the reader is referred to Kreuze *et al.* (2009).

Biotic and abiotic stress

Sweetpotato is affected by about 35 bacterial and fungal diseases, more than 20 viruses or virus-like agents, 20 nematodes and 20 insect species (Martin and Jones, 1986). Recently, fungal and bacterial diseases (Clark *et al.*, 2009), viruses (Loebenstein *et al.*, 2009), nematodes (Overstreet, 2009) and insects (Sorensen, 2009) affecting sweetpotato were reviewed in the textbook by Loebenstein and Thottappilly (2009). Only five pests and diseases are known to be economically important: (i) SPVD; (ii) sweetpotato weevils; (iii) nematodes; (iv) *Alternaria*; and (v) *Fusarium*. With respect to abiotic stresses the crop is affected by drought, heat, cold and salinity.

Worldwide, the greatest biotic constraint is SPVD across all regions. SPVD often causes serious yield losses, especially in high virus-pressure zones of SSA. Nearly all OFSP varieties bred outside of SSA that were introduced to East Africa failed because of SPVD. The critical component within SPVD is whitefly-transmitted *Sweetpotato chlorotic stunt virus* (SPCSV). This virus usually occurs in co-infection with other sweetpotato viruses in which SPCSV breaks the resistance of sweetpotato against other viruses (Ian Barker, Switzerland, 2009, personal communication). SPCSV often occurs in co-infection with aphid-transmitted *Sweetpotato feathery mottle virus* (SPFMV). Clear synergistic disease effects are observed by co-infection of SPCSV and SPFMV resulting in extreme yield losses (Milgram *et al.*, 1996; Gutiérrez *et al.*, 2003) – the so-called SPVD virus complex. Viruses can be grouped into gene pools and phylogenetic groups (strains). Four strains have been reported for SPFVM: (i) East African (EA); (ii) russet crack (RC); (iii) ordinary (O); and (iv) common (C). Virus coat-protein gene sequences have shown that EA SPFMV strains clearly form a separate cluster (Kreuze *et al.*, 2000). Two strains have been reported for SPCSV: (i) East African (EA); and (ii) West African (WA) (Loebenstein *et al.*, 2009). The EA strain was first identified in East Africa and is also present in Peru, while the WA strain was first identified in West Africa and is also found in the Americas and the Mediterranean, but so far not in East Africa (Hoyer *et al.*, 1996; Tairo *et al.*, 2005). There are two serotypes (SEA1 and SEA2) in the EA strain (Loebenstein *et al.*, 2009). A resistance working against one virus strain may not necessarily work against another strain of the same virus. This complicates virus resistance breeding and can result in unexpected resistance breakdown. Moreover, recombination between strains of SPFMV can be expected (Untiveros *et al.*, 2006). A substantial number of farmer varieties in SSA appear to have resistance to SPFMV (Carey *et al.*, 1998; Clark *et al.*, 2012), whereas so far no resistance to SPCSV has been reported.

All sweetpotato varieties need a certain degree of resistance to SPVD and there is considerable genetic variation for this attribute (Mwanga *et al.*, 2002b). A high resistance level to SPVD is especially needed in high virus-pressure zones of East Africa; however, this resistance level turns up in breeding populations at very low frequencies of ≤ 0.2%. The resistance or immunity to SPFMV is often not clearly pronounced, although inheritance of SPFMV resistance is supposedly recessive, but dosage effects might occur (Mwanga *et al.*, 2002b). All clones reported resistant to SPCSV (e.g. CIP-420269; Luis Salazar, Peru, 2005, personal communication) turned out not to be resistant. Under the recent Generation Challenge Programme, a high level of resistance to SPCSV was thought to be found in at least one clone, VJ08.330 (CIP-107729.9). Fortunately, this clone is self-compatible and all offspring clones evaluated so far showed the same level of resistance to SPCSV as the parent VJ08.330 (Kelvin Huamani, Peru, 2013, personal communication). This SPCSV resistance might be recessively inherited. A major problem in SPCSV resistance screening (triple antibody sandwich (TAS)-ELISA) is false negative results. The screening should be conducted by grafting on infected material in two subsequent years; the same holds if real-time PCR is used to screen for resistance to SPCSV instead of TAS-ELISA. In the field, sweetpotato virus pressure varies widely among environments and the final decision over whether a clone is resistant to SPVD takes many years. Moreover, virus symptoms observed in field evaluation are usually due to co-infections of several viruses and a SPCSV-resistant clone does not need to be necessarily free of all virus symptoms. Our experience has shown that use of nitrocellulose membrane (NCM)-ELISA testing for resistance screening is not very reliable. Assuming a recessive inheritance of SPFMV and SPCSV resistance, it appears that this inheritance is modified by quantitative variation. As mentioned previously, it is quite difficult to find recessive inherited traits in autopolyploids as long as the allele frequency q in breeding populations is < 0.7. To fix recessive resistance alleles without marker assisted selection is difficult but not impossible. It requires crossing 'the best with the rest', that is crossing confirmed SPVD-resistant clones with acceptable performance as male parents with the remaining parental material as female parents. It would be extremely helpful to find a molecular marker associated with SPCSV resistance, even if it is only a dominant marker. This would save 2–3 years of evaluation trials. Such a molecular marker would be much more attractive if it was codominant, to enable distinguishing of different heterozygous genotypes. In the absence of double reduction, a recessive genotype can only segregate out from the recessive homozygous male parent (rrrrrr) if female parents carry at least three copies of the recessive resistance allele (RRRrrr). At CIP, groups were formed comprising 'truly' SPCSV-resistant and 'truly' SPCSV-susceptible clones, aiming in a first step at molecular markers that could distinguish homozygous SPCSV-resistant from heterozygous and homozygous SPCSV-susceptible genotypes.

Another approach to achieve resistance to SPVD is transgenic. The most often used approach against viruses is the so-called pathogen-derived resistance (PDR) (Latham and Wilson, 2008). Genes or parts of genes are introduced into the plant genome encoding for virus protein. The produced virus protein activates an antiviral defence system in plants and the plant is ready for defence before the virus infects the plant. Sweetpotato has been transformed with this approach for resistance to SPFMV (Okada *et al.*, 2001), but the resistance broke down under field conditions in East Africa (Anon., 2004). Two reasons for this resistance breakdown are proposed (Kreuze *et al.*, 2009). The first is that the plants were not modified with a gene from the locally prevalent SPFMV strain in East Africa. The second reason could be a co-infection under field conditions with SPCSV. At least to a certain extent transgenic approaches are facing the same problems as classical breeding.

In regions with a pronounced dry season, the greatest constraints are sweetpotato weevils (*Cylas formicarius elegantulus*) in all parts of the tropics, *Cylas puncticollis* and *Cylas brunneus* in Africa, and *Euscepes postfasciatus* in the Caribbean. The weevils have a very large host plant range within the plant family *Convolvulaceae* (Austin, 1991). However, co-evolution between sweetpotato and *Cylas* weevils is quite short, because *Cylas* evolved in the Old World, with a common ancestor in Africa, and arrived with *C. formicarius* in the Americas after the discovery of the New World (Wolfe, 1991). Production losses can reach 60–100% across different regions of the world and even slightly to moderately infested storage roots are often not palatable because of bitter tastes (due to terpenoids and phenols) produced by the plant following weevil infestation (Uritaini *et al.*, 1975; Chalfant *et al.*, 1990; Dinh *et al.*, 1995; Stathers *et al.*, 2003).

In breeding for weevil resistance it is important to understand the biology as well as genetic and environmental preferences of weevils. Adult weevils start to feed on leaves shortly after planting, but this normally causes little damage. The main damage is caused by the larvae, which tunnel inside the storage root and stem (Bohlen, 1973). The biology is well documented for *C. formicarius* and to a certain extent also for *E. postfasciatus* (Sorensen, 2009). As plant stems enlarge, females start making holes in stems and fleshy roots near the soil surface to deposit eggs (these holes are covered by a faecal plug). On average, *C. formicarius* females lay about 120 eggs; whereas *E. postfasciatus* females are reported to deposit on average 106 eggs/month for 4–6 months and for this they prefer to use roots within 2 cm of the soil surface. The *C. formicarius* larvae hatch within less than 1 week, and burrow deep into stems and fleshy roots for about 2–3 weeks. After this period the larvae return to the plant surface at the soil line to pupate. The *C. formicarius* pupae transform into adults within 7–10 days, and these live about 2.5–3 months at higher temperatures (up to 8 months at lower temperatures). The *E. postfasciatus* adults live up to 6 months (laboratory observations) and do not fly. The weevils continue to feed and breed in storage roots remaining in the soil after harvest, as well as on other host plants and stored sweetpotatoes. Weevil populations increase with higher temperature, exposed storage roots, soil cracks in the dry season and length of the growing season. Sorensen (2009) lists plant attributes affecting the weevil population: (i) time needed for harvest; (ii) storage root density, dry matter and starch content; (iii) storage root depth; (iv) vine thickness; and (v) storage root chemistry. These traits could be targets for conventional breeding efforts.

Finding weevil resistance has been an objective for more than 50 years. With reference to variety releases (Appendix 1), conventional breeding has been successful in selecting weevil-resistant varieties only to a certain extent (and often this success is disputed). However, farmers in drought-prone areas clearly distinguish varieties on the basis of weevil susceptibility. For example, in Malawi it is believed that dense storage roots developed deep below the soil surface are less susceptible than less dense, moist-fleshed storage roots. There are several reports of varieties with the attribute of being less affected by weevils: TIB-2532 and TIS-70357 (Lema, 1992), Tamburin Putih (Jusuf, 2002), Porto Rico (Sorensen, 2009), New Kawogo (Stevenson *et al.*, 2009), Brazlandia Roxa (Fuentes and Chujoy, 2009), Santo Amaro

(CIP-400011; José Bienvenido Núñez, Dominican Republic, 2008, personal communication) and CIP-PZ06.120 and CIP-105058.2 (Appendix 3); for additional varieties see Appendices 1 and 2. Moreover, a number of varieties from Papua New Guinea are supposed to be less vulnerable to sweetpotato weevil (Carpena, 2009). The reduced weevil infestation of Santo Amaro is associated with the latex in the storage root skin of this variety (José Bienvenido Núñez, Dominican Republic, 2008, personal communication). Recent findings of compounds in the latex of the Ugandan variety, New Kawogo and the effect of these compounds on weevils may be of interest for breeding investment (Stevenson *et al.*, 2009).

Weevil resistance appears to be built up by a complex of traits. It would be useful to obtain more information about each trait supposedly related to weevil resistance and its association with overall weevil resistance. Certainly the inheritance of weevil resistance is quantitative, but heritabilities of each trait related to weevil resistance may be very different. The apparent inconsistency in weevil resistance among cultivars in different seasons and locations is not surprising, given that strong genotype-by-environment interaction has been observed for weevil damage in Malawi (Felistus Chipungu, unpublished) and it must be expected that environments differ in their suitability for weevil resistance selection. Moreover, a systematic error cannot be excluded due to feeding preferences of weevils among genotypes grown in small plots.

It remains unclear if conventional breeding for weevil resistance can result in 100% resistant varieties. For decades, considerable research has focused on breeding for resistance to *Cylas* and *Euscepes* weevils (Hahn and Leuschner, 1982) and many reports indicated that these efforts have so far shown little progress (Rolston *et al.*, 1979; Mullen *et al.*, 1985; Story *et al.*, 1996; Mao *et al.*, 2001).

Transgenic approaches were suggested to achieve weevil resistance in sweetpotato because solutions to the problem using conventional breeding were not visible. Initial work focused on transformation with proteins (i.e. trypsin and cysteine proteinase inhibitors) that decrease the digestibility of sweetpotato for insects (Cipriani *et al.*, 1999, 2001). This strategy was abandoned due to concerns regarding nutritional impact of such compounds on the human diet. Today transgenic approaches focus on toxins from *Bacillus thuringiensis* (Bt). Different Bt proteins have been tested on *C. puncticollis* and *C. brunneus* (Moar *et al.*, 2007). Several gene constructs have been developed and varieties have been successfully transformed by *Agrobacterium* (the toxic Bt protein is expressed in the plant) (Sefasi *et al.*, 2013). However, it appears so far that transgenic genotypes with *Bt* genes do not show the required resistance levels (Rukarwa *et al.*, 2013).

The major fungal disease of sweetpotato is Fusarium wilt caused by *Fusarium oxysporum* f. sp. *batatas*, but it is only a problem under temperate or cool subtropical climates (Armstrong and Armstrong, 1958; Jones, 1969; Collins, 1977). The disease was once important in the southern parts of the USA (Clark *et al.*, 2009); however, it is still a problem in South Africa (Thompson *et al.*, 2011) and Korea. The pathogen can persist in soil for many years, but appropriate crop rotations reduce disease pressure. Breeding resistant varieties has been very effective in the USA (Dukes *et al.*, 1975; Clark *et al.*, 2009) and China (Appendix 1). Very high heritabilities have been reported for resistance to this disease (Jones, 1969; Collins, 1977). Some strains of the tobacco pathogen, *F. oxysporum* f. sp. *nicotianae* can also cause wilt in susceptible sweetpotato and a new race of *F. oxysporum* f. sp. *batatas* was reported in California (Clark *et al.*, 1998).

Alternaria stem and petiole blight caused by *Alternaria* spp. may be found on sweetpotato in many parts of the world, and several species of *Alternaria* can infect sweetpotato (Lenné, 1991). The disease is only a problem in the African highlands where a more aggressive blight was first reported by Bruggen (1984). Both *Alternaria bataticola* and *Alternaria alternata* have been isolated from infected plants (Anginyah *et al.*, 2001; Osiru *et al.*, 2007). Disease severity varies greatly within the African highlands from minimal levels in less humid areas up to 25–50% of the plant infected elsewhere (Anginyah *et al.*, 2001).

In each reported location, cultivars differ in susceptibility. High levels of tolerance or resistance are frequently found (Sseruwu, 2012). Potential varieties in Uganda are routinely screened for Alternaria blight resistance (Mwanga *et al.*, 2003, 2009; Narayanin *et al.*, 2010).

Plant-parasitic nematodes can seriously damage sweetpotato. Many nematodes have a wide host range and nematodes can spread easily among infected sweetpotato storage roots. Among the most important genera of nematodes many feed on sweetpotato (i.e. *Meloidogyne, Pratylenchus, Ditylenchus, Xiphinema, Radopholus, Rotylenchulus* and *Helictotylenchus*). Only a few nematode species cause significant damage to sweetpotato (Overstreet, 2009). However, production losses of sweetpotato due to these nematodes are significant and in 1987 it was estimated that these losses were about 10% globally (Sasser and Freckman, 1987). Certainly these production losses can be much higher in areas where sweetpotato is grown frequently. One country with the great diversity of nematodes on sweetpotato is Uganda (Coyne *et al.*, 2003). The Peruvian coast is also a hotspot area of different nematodes (Mario Tenuta, Peru, 2013, personal communication). In the French Caribbean, 13 genera of nematodes were found on sweetpotatoes and wild relatives (Massese, 1969), with *Rotylenchulus reniformis* being the most common. In the Philippines, 13 genera and 22 species of nematodes were associated with sweetpotato (Gapasin, 1979) and the genera *Rotylenchulus* and *Meloidogyne* were the most common (present in 80% and 15% of the samples, respectively). In Korea *Meloidogyne* sp. were found in 30–60% of the sweetpotato fields (Dongro *et al.*, 2006); similar results were found in Kyushu (Japan) – 94% were *Meloidogyne incognita* (Iwahori and Sano, 2003). The major nematodes of sweetpotato found in Papua New Guinea and the Pacific were *Meloidogyne arenaria, M. incognita, Meloidogyne javanica* and *R. reniformis* (Vilsoni and Heinlein, 1982; Bridge, 1988). In the past *M. incognita* was considered the most important pest nematode of sweetpotato but *R. reniformis* appears to be increasing in importance (Overstreet and McGawley, 2000; Koenning *et al.*, 2004) at least in the USA. Generally root-knot nematodes of the genus *Meloidogyne* are the most important nematodes for sweetpotato production worldwide (*M. incognita* is extremely destructive to the root system) followed by the reniform nematodes in the genus *Rotylenchulus* (Overstreet, 2009).

Species of *Meloidogyne* are found throughout the tropics, subtropics and in temperate zones with a short winter. The number of species in the genus *Meloidogyne* is very large, but the primary species damaging sweetpotato are *M. incognita* and *M. javanica* (both have a very wide host range). For *M. incognita*, there are clear host × parasite interactions (variable virulence on genotypes; Lawrence and Clark, 1986) and these have been repeatedly used to breed new resistant varieties (Martin and Jones, 1986). Many pathotypes or races have been identified in *M. incognita* (Sano and Iwahori, 2005). The frequencies of *M. incognita* races can differ greatly (Sasser and Carter, 1982). It is expected that the frequency of each race is changing over time so that old and new races can break resistances in sweetpotato.

There has been recurrent success in breeding for root-knot resistance against new races of *Meloidogyne* spp. (Martin and Jones, 1986). A molecular marker linked to a dominant inherited resistance gene was identified using *M. incognita* race 3 (Ukoskit *et al.*, 1997). However, the resistance to root-knot nematode appears to be qualitatively (by one or few major genes) as well as quantitatively controlled (Mcharo *et al.*, 2005) and/or durable (Cervantes *et al.*, 2002). An *I. trifida* resistance against *M. incognita* may be controlled by two dominant genes (Komiyama *et al.*, 2006). The inheritance of *M. incognita* resistance by a single dominant gene might explain why root-knot nematode resistance can be easily found in hexaploid sweetpotatoes (see Overstreet (2009) for details) and incorporated into different breeding populations.

There are ten named species of *Rotylenchulus* but only two have been found associated with sweetpotato (Robinson *et al.*, 1997): *R. reniformis* and *Rotylenchulus borealis*. The first has a wide host range and occurs throughout the Americas, Africa, South Asia, South-east Asia and the Pacific; and *R. borealis* has a limited host range and has been found only in Europe and Africa. Differences in population development of the reniform nematode have also been reported among sweetpotato genotypes (Clark *et al.*, 1980) but tolerance or resistance appears to occur at lower frequencies than for root-knot nematodes. Two races of *R. reniformis* have been described on cotton in India (Dasgupta and Seshadri, 1971a,b). Differences in *R. reniformis* populations have been observed in the USA (McGawley and Overstreet, 1995; Agudelo *et al.*, 2005). Other nematodes that can cause local or regional problems in sweetpotato are: (i) the lesion nematode genus, *Pratylenchus* (*Pratylenchus coffeae* in Japan and China; Yoshida, 1985; Kukimura *et al.*, 1992; Feng *et al.*, 2000), but for Brazil, Anguiz and Canto-Sáenz (1991) reported that sweetpotato supported very little reproduction of this nematode; and (ii) stem nematodes, *Ditylenchus dipsaci* and *Ditylenchus destructor* (in China; Lin *et al.*, 1993; Zhang *et al.*, 2006) causing brown ring disease which is primarily a storage problem.

Drought and other abiotic stresses

Sweetpotato originates from the humid tropics in an ecosystem experiencing high average temperatures and significant rainfall. The crop disseminated into the semi-arid tropics and warm-temperate zones (Hijmans *et al.*, 2002). For sweetpotato in Peru, where all three agroecological zones are found within a close distance, top selections in arid costal lowlands (i.e. Adriano (CIP-105228.1), Alexander (CIP 105240.1), Arne (CIP-105086.1) and Benjamin (CIP 105085.2)) clearly differ from top selections in humid tropical lowlands (i.e. Abigail (CIP-194540.5), Isabel (CIP-189153.18) and Sumy (CIP-105523.1)). However, clones adapted across agroecological zones can be found, such as Xuzhou 18, as well as clones with striking yield advantages in low-yielding environments such as SR92.499-23 (Grüneberg *et al.*, 2005). For the underlying physiological mechanisms associated with adaptation to low-yield environments the reader is referred to Sattelmacher *et al.* (1994). For sweetpotato in Africa it is recognized that it is not possible to breed for adaptation across agroecological zones and for this reason CIP recommends decentralized sweetpotato breeding (Grüneberg *et al.*, 2009b). Genotype-by-environment studies are very limited and to conclusively show that it is not possible to breed for adaptation across the humid and semi-arid tropics strong cross-over interactions should be observed. Within the semi-arid tropics and warm-temperate zones the crop is affected by drought, flood, heat, cold and salinity. The effects of drought, flood, shade and salinity on sweetpotato were reviewed by Ravi and Indira (1999). In this contribution we will focus on the abiotic stress of drought.

A large fraction of the sweetpotato germplasm appears to be adapted to drought and exhibits adequate harvest in critical drought years (Anselmo *et al.*, 1988; Ding *et al.*, 1997; Xie *et al.*, 1998; Hou *et al.*, 1999; Chávez *et al.*, 2000; Wang *et al.*, 2003; Agili, 2012). Sweetpotato clearly needs an adequate water supply at planting and for several weeks thereafter (Indira and Kabeerathumma, 1988; Nair *et al.*, 1996; Ravi and Indira, 1996a). In the second and third months of growth, sweetpotato can tolerate moderate drought (mid-season drought) and in the fourth or fifth month can cope well with severe drought (terminal drought). Early season drought affects storage root initiation and the number of storage roots. Under typical semi-arid growing conditions, the crop requires 500 mm of water for a 4 month period (King, 1985; Onyekwere and Nwinyi, 1989; Chukwu, 1995). Assuming a storage root yield of up to 30 t/ha (33% storage root dry matter) this corresponds to a water requirement of 500 l/kg 'sweetpotato (dry)', which is clearly less compared with soybeans (2000 l/kg), rice (1600 l/kg), sorghum (1300 l/kg) and wheat (900 kg/l) (Pimentel *et al.*, 2004). The crop yields best when irrigated at 25% available soil moisture (Hernandez and Barry, 1966; Hammett *et al.*, 1982) – but at high soil moisture content (flood) suffers extreme yield decreases. The critical soil moisture for storage root yield decreases is around 20% available soil moisture (Hernandez and Hernandez, 1967; Chowdhury and Ravi, 1988; Indira and Kabeerathumma, 1988; Nair *et al.*, 1996).

Adaptation to drought-prone environments is achieved by drought escape and/or drought avoidance and/or drought tolerance (Blum, 1988). A proper timing of growth cycle (completion of the most sensitive developmental stages while water is abundant) is considered to be drought escape (earliness). Avoiding water-deficit stress by reducing evapotranspiration without affecting yields or with a root system capable of extracting water from deep soil layers is considered drought avoidance (reduction of water loss and/or maintenance of water uptake). Mechanisms which result in maintaining assimilation under reduced leaf relative water content are categorized as drought tolerance. Most of the supposed drought-tolerant plants cannot tolerate true drought – they escape and/or avoid drought. It appears that sweetpotato uses all three strategies to adapt to drought. There is a large genetic variability for earliness in sweetpotato (Yanfu *et al.*, 1989). The crop appears to have an astonishingly wide root distribution and architecture and can penetrate about 2 m deep into the soil to absorb water/nutrients from deep soil layers (Weaver and Bruner, 1927; Yoshida *et al.*, 1970). There are striking differences in rooting depth among genotypes and these appear to be correlated with the response of sweetpotato to drought (Yen *et al.*, 1964; Noel Pallais, unpublished). Deep rooting is an attribute difficult to investigate and it can only be hypothesized that sweetpotato is as different underground as above ground. Certainly associations of canopy characteristics, water use efficiency and storage root yields under water-deficit stress merit investigation. Drought-adapted germplasm may have distinct leaf morphology (narrow leaves) compared with regular types (broader leaves) (Francisco Vilaró, unpublished observations). Drought-adapted germplasm often has narrow leaves, erect to semi-erect growth types and reduced foliar area which could result in reduced loss of water. The two varieties Jewel (broader leaves) and Tanzania (narrow leaves) clearly differ in canopy attributes and water use efficiency (Kelm *et al.*, 2000). Tanzania is considered a clone well adapted to drought-prone areas and is used by CIP as a check across regions. Under water stress conditions the plant leaf water potential or leaf relative water content decreases (Sung, 1985a,b; Indira and Kabeerathumma, 1988; Chowdhury and Naskar, 1993; Ravi and Indira, 1995). Leaves permanently wilt when their water potential decreases to –1.3 MPa, and between –1.6 and –2.0 MPa the leaves senesce (Sung, 1985b; Ravi and Indira, 1995). However, at different growing stages (even early stages) the crop can recover from wilting and there are striking differences in this attribute among genotypes (Robert Laurie, South Africa, 2013, personal communication). The relative content of free amino acids, soluble sugars, ATP and chlorophyll a/b ratio appear to correlate with drought tolerance (Zhang, M.S. *et al.*, 2004, 2005) indicating an association of these compounds with drought tolerance and overall sweetpotato drought adaptation.

Van Heerden and Laurie (2008) investigated four sweetpotato varieties (Resisto, Excel, W-119 and A15) under long-term restricted water supply and found two contrasting responses to drought. Although restricted water supply decreased leaf relative water content similarly in Resisto and A15, the negative effects on stomatal conductance disappeared with time in A15 (indicating high drought acclimation in A15). The suppression of aboveground biomass accumulation during restricted water supply was considerably lower in A15 than in Resisto – photosynthesis on a leaf area basis in A15 was not inhibited, whereas CO_2 assimilation in Resisto was inhibited and A15 yielded much better than Resisto under conditions of restricted water supply. Other clones with similar responses to restricted water supply as A15 might be Chissicuana-2, Nhacutse-5, ADMARC, Xiadlaxakau, Nwanaqtsjo, 199062.1 and TIS-2534 (Maria Andrade, Mozambique, 2013, personal communication). Most drought-tolerance related parameters are very cumbersome to determine, even for a few clones. However, a fast throughput method, *in vitro* screening using polyethylene glycol salt, was found to be efficient and simple enough to select for drought tolerance in sweetpotato (Agili, 2012). Agili (2012) demonstrated that salt tolerance is associated with drought tolerance in sweetpotato. Chávez *et al.* (2000) selected varieties with tolerance to drought, salinity, and boron for the arid Pacific coast of Peru and northern Chile. Sweetpotato is considered to be semi-tolerant to salinity (tolerant to an electrical conductivity (EC) of 4.0 dS/m in irrigation water or an EC of soil saturated extract at 6–11.0 dS/m with yield reductions of 50%; Bernstein, 1974) and also semi-tolerant to boron (saturation extract of 2 mg/l; Wilcox, 1960). However, such studies have so far only considered a small fraction of sweetpotato germplasm (i.e. anecdotal clones were observed which survived EC ≤ 25.0 dS/m after some rain on salt-prone soils in northern Peru; Roberto Quiroz, Peru, 2013, personal communication).

A further attribute required by farmers in drought-prone areas is 'vine survival' (Yanggen and Nagujja, 2005; Lebot, 2010). For example, the variety Resisto was disappearing on farms after drought years in Mozambique due to inadequate vine survival under drought stress. Genotypes with strong and thick vines (often lignified) and medium to high upper biomass production provide sufficient planting material, which has a long storability and withstands short dry spells after planting. Vine survival became a key attribute for new variety releases in Mozambique (Maria Andrade, Mozambique, 2010, personal communication). However, yield under water-limited conditions is determined by yield potential and/or drought avoidance and/or drought tolerance – yield potential is defined as the maximum yield realized under non-stress conditions (Blum, 2005). Especially for sweetpotato it appears that many genotypes are simply not affected in the 'statistical sense' by drought because they have low yield potential (low storage root yield under non-stress conditions). Within this fraction of clones we observed two types of clones: the first not responding to water, but the second responding to non-stress water supply by increasing biomass production allocated nearly completely to foliage production (a typical clone with such a response to water is Tanzania). In breeding for areas with irregular rainfalls it may have been underestimated that drought-adapted clones need to adequately respond to rain (biomass production increases in association with HI stability such as for clones Chissicuana-2, ADMARC, Xiadlaxakau, Taca and TIS-2534; Maria Andrade, Mozambique, 2013, personal communication). The statement 'what is good under high-yielding environments is also good under stress environments' is nearly accepted as an axiom in breeding (Blum, 2005). However, how can a crop become adapted in evolution to stress if it is not grown under stress? We think that for sweetpotato and early breeding stages (see accelerated breeding scheme in section 'Breeding Methods') there is merit in investigating populations in a first selection step under stress (discarding all genotypes below the lowest acceptable value) followed by a second selection step under non-stress conditions. This can also be conducted as simultaneous selection under stress and no stress conditions so that very elegant index selection procedures (Pesek and Baker, 1969) could be applied for desired gains under abiotic stress conditions.

1.5 Breeding Methods

The breeding methods for a crop are not set in concrete. Depending on the pollination and propagation biology various options exist on how to breed a crop (Schnell, 1982). What is the pollination biology of sweetpotato? It is an open-pollinated crop propagated by cloning. For population improvement, sweetpotato should be treated as an open-pollinated crop and for variety development as a clonally propagated crop.

The general principle of breeding clonally propagated crops is to break normal clonal propagation by generating true seeds, which results in a new population and genetic variation. All subsequent propagation steps are asexual by clonal propagation in which selection is carried out (Grüneberg *et al.*, 2009a). This selection aims at a set of individuals superior to previous sets. Finally, superior clones are used to generate true seeds.

This process leads to recurrent cycles of recombination and selection and results in a combination of good attributes in genotypes which appeared in different genotypes before selection. However, in the medium and long term, recurrent selection also results in generation of new genotypes with trait performance outside of the distribution range of previous populations. For the extreme forces of this process the reader is referred to the fundamental long-term breeding experiments using maize as a model crop (Dudley, 1977).

The efficiency of a breeding method is determined by the genetic gain and the time needed to achieve the genetic gain. Across several recurrent selection cycles, high genetic gains across traits can only be achieved by structuring plant breeding into two components: (i) variety development; and (ii) population improvement. Gallais (2003) proposed a new way of thinking in breeding autopolyploid crops (in addition to new information about their population genetics), a comprehensive breeding scheme comprising variety development and population improvement. **Variety development** aims at the selection of the best or very few best clones (maximum response to selection and complete or nearly complete exploitation of the genetic variation). **Population improvement** aims at the selection of the 'best' parents to generate new genetic variation around an improved population mean (in practice the population mean across all traits for which the breeder desires improvement). Variety development and selection for the 'best' clone for the current needs of clients is relatively straightforward and what is the 'best' is usually best known locally on the ground. However, population improvement or identifying the best parents to create a new and better population for future selections is a challenge in sweetpotato, as it is for all other clonally propagated crops. Population improvement is indeed complex and should be carried out by an interconnection of breeders for an agrogeographic zone. It often requires more resources and capacities than small- to medium-sized breeding programmes can usually afford. Note that commercial breeding companies, especially smaller ones, also form alliances or crossing unions for strategic population improvement. For a better understanding, details and illustrations of the importance of population improvement, consult Gallais (2003) part III: 'Population improvement and varietal development'.

In 2009, sweetpotato breeding in Africa had the opportunity through the Sweetpotato Action for Security and Health in Africa (SASHA) project, funded by the Bill & Melinda Gates Foundation (BMGF), and Alliance for a Green Revolution in Africa (AGRA) projects, funded by the Rockefeller Foundation and BMGF, to organize sweetpotato breeding programmes on the basis of comprehensive breeding (for details, see Grüneberg *et al.*, 2009b). Sweetpotato breeding platforms were established with emphasis on population improvement providing NARS breeding programmes with improved true-seed populations (SASHA), whereas NARS breeding programmes emphasize variety development using these improved true-seed populations to select new varieties as well as better parents for their own breeding programmes (AGRA). In this way, it was possible for CIP to implement its concept of decentralized breeding in which each NARS partner maintains its independence and autonomy.

This breeding network for sweetpotato comprises four breeding platforms (at the National Crops Resources Research Institute (NaCRRI) in Uganda, the Mozambique Institute of Agricultural Research (IIAM), the Council for Scientific and Industrial Research, Crops Research Institute (CSIR-CRI) in Ghana and CIP in Peru) and 12 NARS breeding programmes (at NaCRRI, IIAM, CSIR-CRI, the Agricultural Research Council (ARC) in South Africa, the Kenya Agricultural Research Institute (KARI), the Agricultural Research Institute (ARI) in Tanzania, the Zambia Agriculture Research Institute (ZARI), the Department for Agricultural Research Services (DARS) in Malawi, the Rwanda Agriculture Board (RAB), the National Root Crops Research Institute (NRCRI) in Nigeria, the Ethiopian Institute of Agricultural Research (EIAR) and the Environment and Agricultural Research Institute (INERA) in Burkina Faso). CIP and NARS breeding programmes together aim at four strategic objectives with respect to efficiency of sweetpotato breeding methods: (i) more recombination and parents; (ii) accelerated breeding and improved allocation of breeding resources; (iii) more controlled cross breeding in addition to stepwise reduction of polycross breeding; and (iv) in the future, most likely heterosis-exploiting breeding schemes and molecular tools for sweetpotato breeding. However, by 2003, NARS in SSA had started alliances and exchanged seeds obtained in crossing blocks for strategic OFSP population improvement in the frame of the HarvestPlus programme, which were taken up and extended by SASHA in 2009. This first step towards decentralized OFSP breeding might serve today as a blueprint for South and South-east Asia.

We do not want to give the impression that sweetpotato breeding requires huge investments. Exactly the opposite is true, as can be seen from the history of sweetpotato breeding (Martin and Jones, 1986). All successful sweetpotato breeding programmes initiated in the past century such as those at Louisiana State University (LSU), North Carolina State University (NCSU), the XSPRC and NaCRRI had one characteristic in common – that they intensified recombination and conducted gene-pool separation (recombination of parents adapted to local needs). The intensification of recombination is nearly an axiom for success in sweetpotato breeding. In crop evolution of sweetpotato, farmers did and still use and test sweetpotatoes derived from true seeds (e.g. by gathering planting material in the fields for the next growing season).

To use and/or care for true-seed-derived plants is regarded as one of the driving factors of the formation of sweetpotato variation in the diversity centres, such as in and around Papua New Guinea (Yen, 1974; Schneider, 1995; Fajardo *et al.*, 2002). How efficient this process is can be seen from the fact that before DARS in Malawi had no crossing blocks the breeder Felistus Chipungu collected true seeds from clones in selections from local and introduced germplasm trials and in this way selected several new varieties: Nyamoyo, Sungani, Anaakwanire, Mathuthu, Kaphulira, Chipika and Kadyaubwerere (the first two are cream fleshed and the others are OFSP). This procedure is a ‘precursor’ of a polycross seed nursery. In the second half of the last century, a major sweetpotato breeding advance was to establish polycrosses for clones adapted to local needs (i.e. at LSU, NCSU and NaCRRI). Polycross recombination became the standard recombination technique (Martin and Jones, 1986), except in China where due to climatic conditions quite early controlled crosses were conducted (Daifu Ma, China, 2004, personal communication). The NCSU breeding programme became the blueprint of many NARS breeding programmes in SSA.

For theoretical reasons, controlled cross breeding should be superior to polycross breeding. From the practical point of view, polycrosses must not be necessarily inferior to controlled crosses. The reason is that controlled cross breeding requires more resources (especially skilled technicians) so that usually much more true seed can be generated in polycross than controlled cross breeding, which results in higher selection intensities. The SASHA breeding network continues strong support for polycross breeding, but encourages all breeders to do more controlled cross breeding. At the current stage it is a major mistake in sweetpotato breeding to conduct no recombination or to conduct recombination with a small set of parents (< 15). It appears that the major bottleneck in sweetpotato breeding is not creating large variability for selection but improvement of the population mean from one recurrent selection cycle to the next. Thus, the number of parents and the choice of parents is the most difficult task in sweetpotato breeding.

There are two strategies to raise more good crosses/families in sweetpotato population improvement: (i) increase the respective number of parents and cross combinations; and (ii) change from polycross to controlled cross breeding. A third strategy using offspring information to select for better parents is being tested at CIP in Peru. Most programmes work with 20–30 parents in polycross seed nurseries, and thus the programme of NCSU and the recommendations of Martin and Jones (1986) are used as examples. During the past 10 years the breeding platform in Uganda, which serves East and Central African NARS, increased the number of parents in polycross breeding from 25 to 150. This large number of parents was divided into two gene pools on the basis of simple sequence repeat (SSR) markers (David, 2012). CIP's breeding programme in Peru completely changed from polycrosses to controlled cross breeding in 2004, and two populations were formed: Jewel and Zapallo. This programme used about 200 parents per population recombined by crossing the 'best with the rest' (crossing five to eight top clones as male parents with the remaining clones as female parents) and completed in 2009 two recurrent selection cycles for each population (note not all cross combinations result in seed set). All NARS breeding programmes in Africa funded by AGRA established polycrosses and allocated at least some crossing capacity into controlled crosses. We assume that across all these sweetpotato breeding programmes the respective number of parents and cross combinations used have increased by a factor of about 15 compared with before 2003. The rationale underlying this change is solidly based on the findings of selection theory. There are two theoretical approaches to optimize number and size of crosses (Wricke and Weber, 1986): (i) 'risk minimization of raising no good crosses' (mathematical proof by Liapounoff inequality in Kendal and Stuart, 1958); and (ii) prediction of responses to selection on the basis of variance components estimations among and within families (Weber, 1979; Wricke and Weber, 1986). Both approaches result in the following recommendation: *as long as there is no prior knowledge of the value of the cross (no offspring information) the number of crosses needs to be maximized and the size of a cross needs to be minimized.* In other words, breeders should make as many cross combinations as possible when they do not have prior knowledge of the value of a cross. This is exactly what we did in breeding in Africa for Africa under SASHA. In the case of prior knowledge concerning a cross (e.g. offspring information due to test crosses), the breeder discards all parents from population improvement which are not good 'family makers'. This third strategy to raise more good families in population improvement could be the most efficient one and is being tested at CIP in Peru, Uganda and Mozambique (see 'Heterosis-exploiting breeding scheme (HEBS)' below).

In addition to raising more good crosses, a key factor in breeding is the time needed from the cross to variety release and the time required for one recurrent selection cycle in population improvement (selection of a new set of parents). The breeder Gerhard Röbbelen wisely said to his students: 'There is only one breeding objective: a better variety and to come with this at least one year before the competitor.' Traditional breeding schemes for clonally propagated crops take too long; consult Grüneberg *et al.* (2009a) for illustrations of a traditional breeding scheme. Donors are very reluctant to invest when it takes a decade to materialize concrete outputs and to reach clients (i.e. varieties in farmers' fields). In SASHA- and AGRA-funded breeding programmes only 2 years are used for later breeding stages before entering the variety release process. The recommended allocation of the test capacity is to enter about 150–300 clones into later breeding stages and to test these in two to three environments in a first stage, and to select 20–30 clones and test these in five to six environments in a second stage using no more than two plot replications. The rationale underlying this resource allocation in later breeding stages are the findings of intensive research of the resource allocation problem by selection theory (e.g. Cochran, 1951; Hanson and Brim, 1963; Finney, 1966; Utz, 1969, 1984; Mi *et al.*, 2014) including with parameters obtained from sweetpotato (Grüneberg *et al.*, 2004). A three-stage selection is only about 5–10% superior to a two-stage selection (at the optimum resources allocation) if $\sigma^2_{G\times Y}$ is large, which appears not to be the case in sweetpotato, at least in East Africa (Grüneberg *et al.*, 2004). Moreover, we consider variety release testing as the third and additional selection stage. It should be noted that the optimum around the maximum response to selection is flat (for yield or an index which includes yield as a component) so that the breeder is not moving out of the flat area as long as he/she allocates around one-third of the test capacity to the number of tested genotypes at the first selection stage and selects 'aggressively' (8–15% of clones). The value of a variety is not only determined by yield. We have had good experiences in later breeding stages with aggregating several traits using the Elston index (Elston, 1963); for examples with sweetpotato see Grüneberg *et al.* (2004). The weakness of the index – which is also an advantage – is that it does not attach weight to traits, but this can be relatively easily achieved (i.e. by including yield components such as HI or number of commercial storage roots into the index). However, in later breeding stages appropriate multi-trait selection procedures do not appear to be a major problem – in contrast with selection in early breeding stages and selection of new parents where multi-trait selection is very important. What remains interesting in research on selection in later breeding stages is the suitability of selection sites, for example by slope of regression lines and/or location-specific heritabilities (Mechelke, 1986), which may vary tremendously in sweetpotato breeding programmes.

Accelerated breeding scheme (ABS): The ABS targets the early stages of breeding clonally propagated crops to increase breeding efficiency. Where $\sigma^2_{G\times Y}$ is not very important, temporal variation of test environments can be replaced by spatial variation of test environments (Patterson, 1997). Thus, more locations can compensate for reduction in test years. ABS uses the simple fact that in breeding clonally propagated crops each true-seed plant is already a potential variety. An additional advantage of sweetpotato is the very short crop duration (3–5 months) and high propagation coefficient (up to 90 cuttings per plant within 3–4 months). Directly after the multiplication step of true-seed-derived plants the genotypes are tested in small 1 m row plots (three plants) in two to three environments without plot replications. All that is measured in early breeding stages in subsequent years is measured in 1 year at several environments. Different traits can be measured simultaneously and aggregated into an index or sequentially in the environments (so-called independent culling). About 150–300 clones are selected and the breeder enters these clones directly into the later breeding stages. ABS was proposed by Grüneberg *et al.* (2009a) and it is also discussed by Lebot (2010). The human and financial resources required are manageable if controlled cross breeding is applied with about 10–20 genotypes per family. However, in the case of polycross breeding with a generation of 50,000–100,000 seeds, ABS requires selection among single seed plants in the multiplication step to enter with 5000–10,000 clones into ABS.

ABS originated in 2005 by breeders under pressure by donors and by farmers calling for more adapted OFSP varieties to deliver new OFSP varieties within a few years. On the basis of the variance component estimates in later breeding stages (Grüneberg *et al.*, 2004) it was assumed that $\sigma^2_{G\times Y}$ was also not of much importance in early breeding stages. Instead of planting A clones (1 m row plots) in only one environment, three environments were used. The results of this A clone evaluation with several environments or ABS (Table 1.9) supported an ABS as theory predicted. The ratios $\sigma^2_G/\sigma^2_{G\times E}$ were 1:2.05, 1:1.47, 1:0.45, 1:0.28, 1:1.03 and 1:0.95 for storage root yield, foliage yield and dry matter, total carotenoid, iron and zinc contents of storage roots, respectively. It was decided to select 200 clones on the basis of the Pesek–Baker index (Pesek and Baker, 1969) using the square root of σ^2_G as the desired genetic gain and to enter these directly into later breeding stages. Similar results were obtained with the population Jewel during 2007 and the population Zapallo in 2006 and 2008 (results not presented). In April 2010 it was possible to launch four varieties together with INIA in Peru: Adriano (CIP-105228.1), Alexander (CIP 105240.1), Arne (CIP-105086.1) and Benjamin (CIP 105085.2). All these clones traced back to the population Zapallo 2006, which was crossed in 2005 and entered the field for the first time in 2006.

Using ABS in Mozambique enabled the release of 15 varieties in 2011: Amelia, Tio, Joe, Irene, Bela, Delvia, Cecilia, Ininda, Lourdes, Esther, Melinda, Erica, Jane, Namanga and Sumaia. Since 2009 several NARS breeding programmes in Africa have taken up ABS. Rapid uptake was no doubt driven by the donor, AGRA, as their 3 year grant required breeders to have advanced breeding

Table 1.9. Variance components[a] and operative heritability for observed traits in early breeding stages of the population 'Jewel 2005' planted at three locations (Loc) in Peru (San Ramon, La Molina and Cañete) without replications in 1 m row plots.

Traits[b]	σ^2_G	σ^2_E	$\sigma^2_{G\times E}$	*N* clones	*N* Loc	h^2
Storage root yield (t²/ha)	47.7	23.2	98.0	4,175	3	0.59
Foliage yield (t²/ha)	237.0	52.1	349.0	4,167	2	0.58
Dry matter content of roots (% FM)	13.94	8.18	6.22	2,709	2	0.82
Carotene content of roots (ppm DM)	33,651	3,453	9,539	2,709	2	0.88
Iron content of roots (ppm DM)	7.41	5.79	7.61	2,709	2	0.66
Zinc content of roots (ppm DM)	3.10	4.63	2.92	2,709	2	0.68

[a]Variance components: σ^2_G, genotypes; σ^2_E, environments; $\sigma^2_{G\times E}$, genotype-by-environment interactions; h^2, operational broad-sense heritability.

[b]FM, fresh matter; DM, dry matter.

clones by the end of the funding period to be eligible for further funding. All East African countries that received AGRA support for sweetpotato breeding are using ABS: Kenya, Rwanda, Tanzania and Uganda. However, more experiments are still needed to strengthen this new breeding scheme approach for sweetpotato and other clonally propagated crops.

Three types of studies are used to investigate the efficiency of ABS. The first is to estimate the variance components (σ^2_G and $\sigma^2_{G\times L}$) and corresponding heritabilities when ABS is applied in early breeding stages: $h^2 = \sigma^2_G / (\sigma^2_G + \sigma^2_{G\times L} / \mathrm{L})$, where h^2 denotes operative broad-sense heritability and L the number of locations and environments (see Table 1.9). Since it is an efficiency study at least with respect to yield traits, data should be recorded for all clones (discarding clones by visual selection results in bad estimates of σ^2_G and $\sigma^2_{G\times L}$). The second type of study on ABS efficiency is to estimate variance components (σ^2_G and $\sigma^2_{G\times L}$) and corresponding heritabilities when ABS is applied with check clone(s) and plant the selected fraction of clones again with the same check(s) for one further breeding stage to estimate the observed response to selection relative to checks (R_{obs} = the mean across the selected fraction relative to check(s) in year 2 minus the mean across all clones in ABS relative to check(s) in year 1). The third type of study on ABS efficiency is to estimate the variance components (σ^2_G, $\sigma^2_{G\times L}$, $\sigma^2_{G\times Y}$, $\sigma^2_{G\times L\times Y}$ and σ^2_ε) and corresponding heritabilities when ABS is applied in early breeding stages with plot replications (at least two plots per location) and replant all clones in year 2 without selection at the same locations and same plot replication numbers. The first two types of studies can be managed alongside ongoing efforts in applied breeding even when data are recorded for all clones and no visual selection is applied (breeders usually do not like to measure what obviously does not merit efforts, but for efficiency studies this needs to be made). The third type of efficiency study requires considerable resources in addition to ongoing efforts – all breeders are very reluctant to replant material once discarded and often consider such work a waste of time and funds. However, the third type of study separates all variance components of genotype-by-environment interactions ($\sigma^2_{G\times L}$, $\sigma^2_{G\times Y}$ and $\sigma^2_{G\times L\times Y}$) from σ^2_G and allows estimation of R_{obs} and prediction of response to selection for different breeding scenarios by model calculations. Moreover, estimates for plot error in early breeding stages are obtained, which differ from those in later breeding stages. For the time being only one third-type study is ongoing within the SASHA project, and soon we expect to have information concerning $\sigma^2_{G\times Y}$ and plot error in ABS.

What makes this new breeding scheme approach efficient in sweetpotato is still speculative. We assume it is associated with early capturing of the σ_G^2 and $\sigma_{G\times E}^2$ of yield-related traits, especially HI and HI stability in contrasting environments. ABS is certainly very efficient for the quality traits dry matter, starch, sugars and carotene content of storage roots. Currently, CIP scientists are working on appropriate weighting factors for yield-related traits in ABS in the context of index selection. Finally, we note that by ABS very short breeding cycles can be achieved in population improvement – selected clones enter later breeding stages, but are also used as parental material for the next cycle of recombination and selection – and by working with two populations the technical workload can be balanced with respect to crossing and field capacity each year – when one population is recombined the other is in the field and vice versa. An additional advantage of ABS is that farmer participatory selection approaches (Gibson *et al.*, 2008), which are critical for assessing client needs, can be very elegantly implemented in the ABS by conducting at one of the two to three environments the selection in cooperation with farmers. Consult Grüneberg *et al* (2009a) for further details.

Heterosis-exploiting breeding schemes (HEBS): The phenomenon of heterosis is well known in breeding as the increase in yield or other traits in hybrid offspring, which are significantly superior to those of the parents. In the case of sweetpotato, the frequency of heterozygosity indicates that the 'stimulus of heterozygosity' or heterosis might be very large (Fig. 1.4). It is hypothesized that the performance of quantitative traits in autopolyploid crops, such as sweetpotato, are largely determined by heterosis (Gallais, 2003). HEBS for clonally propagated crops have been proposed by Hull (1945), Melchinger and Gumber (1998), Miles (2007) and Grüneberg *et al* (2009a). There are several possible reasons why HEBS have so far found no place in applied clonal breeding. The major reason is that it is difficult to estimate heterosis in clonally propagated crops, because the total magnitude of heterosis is defined by the difference between the mean of two homozygous parents and their offspring (H = F1 – (P1 + P2)/2; where F1 denotes the offspring and P1 and P2 the homozygous parents). It is illusory to develop homozygous genotypes for sweetpotato and nearly all other clonally propagated crops. Nevertheless, it is possible to determine a fraction of the full amount of heterosis in clonally propagated crops, which is the 'heterosis increment' or 'heterotic gain' by crossing two heterozygous parents and use the **mid-parent–mid-offspring heterosis increment** as a parameter to obtain information about the exploitable amount of heterosis. The 'heterosis increment' or 'heterotic gain' has often been used to determine the magnitude of exploitable heterosis in traditional hybrid crops when homozygous inbred lines were not available or simply too weak to be used as parents (Moll *et al.*, 1965; Becker, 2011). The heterosis increment can also be determined by comparing intra gene-pool cross combinations (A and B) with inter gene-pool cross combinations (A × B hybrid population) – this corresponds to the classical heterosis experiment in maize by Moll *et al.* (1965).

There is no reason why the phenomenon of heterosis cannot be investigated in clonally propagated crops. For storage root yield we present an experimental cross population evaluated together with their parents and provide estimates of heterosis increments in sweetpotato (Table 1.10). The population was generated with 16 clones crossed in a factorial design using four varieties as male parents and 12 breeding clones as female parents. The field experiment was carried out at San Ramon (planting date: 15 April 2007; harvest date: 1 October 2007) and at La Molina in Peru (planting date: 15 December 2007; harvest date: 30 June 2008). Each cross combination was represented by 15–20 offspring clones. Each clone was planted in 1.5 m row plots with six plants and two plot replications per location. For many offspring the storage root yield family means were larger than the mid-parent means. Parents clearly differed in their combining ability. Heterosis increments of up to 58.7% (Wagabolige × SR02.174) were observed, and for high-yielding parental combinations we also found storage root yield offspring means larger than mid-parent means (i.e. Zapallo × SR02.174). To our knowledge this is the first detailed

Table 1.10. Storage root yield (t/ha) of four male and 12 female sweetpotato parents (underlined), their offspring means and heterosis increments of offspring on basis of mid-parent–mid-offspring estimates[a] (italics) evaluated at two locations, San Ramon and La Molina, in Peru.

		Male parent							
Female parent		INIA100 (25.2)		Zapallo (22.0)		Wagabolige (10.9)		Tanzania (23.3)	
SR02.132	(33.5)	26.8	*(−8.5%)*	21.5	*(−22.5%)*	17.3	*(−21.9%)*	28.4	*(−0.1%)*
SR01.024	(11.7)	19.5	*(5.6%)*	20.8	*(23.3%)*	16.8	*(48.9%)*	22.5	*(28.5%)*
SR01.022	(12.7)	16.6	*(−12.4%)*	19.1	*(9.9%)*	14.2	*(20.6%)*	22.7	*(26.0%)*
LM02.082	(18.4)	19.4	*(−11.2%)*	23.9	*(18.3%)*	16.6	*(13.4%)*	23.3	*(11.5%)*
SR02.174	(22.7)	27.4	*(14.7%)*	28.8	*(28.9%)*	26.6	*(58.7%)*	28.2	*(22.6%)*
SR02.177	(41.3)	23.2	*(−30.3%)*	22.9	*(−27.8%)*	17.3	*(−33.7%)*	25.2	*(−22.0%)*
LM02.032	(23.1)	20.3	*(−16.1%)*	19.2	*(−15.1%)*	15.6	*(−8.0%)*	21.5	*(−7.4%)*
LM02.035	(13.7)	18.2	*(−6.4%)*	18.9	*(5.8%)*	15.1	*(23.2%)*	17.9	*(−3.0%)*
SR90.021	(4.6)	14.6	*(−1.8%)*	11.5	*(−13.9%)*	11.1	*(43.5%)*	13.1	*(−6.6%)*
SR01.029	(8.6)	15.0	*(−11.3%)*	13.8	*(−10.1%)*	10.9	*(12.1%)*	14.6	*(−8.5%)*
SR01.005	(11.5)	15.1	*(−17.7%)*	12.9	*(−23.0%)*	8.0	*(−28.7%)*	12.7	*(−27.0%)*
SR01.002	(32.1)	24.5	*(−14.5%)*	19.1	*(−29.6%)*	18.3	*(−15.1%)*	20.3	*(−26.7%)*

[a]Mid-parent to mid-offspring correlation $r = 0.705$, Pearson's correlation coefficient, $N = 48$.

study of heterosis in sweetpotato and clonally propagated crops. Breeders should certainly be interested in doing more crosses of the type Zapallo × SR02.174.

To systematically increase the frequency of heterotic cross combinations, the breeder needs to work with separate gene pools and more precise mutually heterotic gene pools. CIP tested applied breeding populations, Jewel (PJ) and 'Zapallo' (PZ), in Peru to determine if they were mutually heterotic. PJ and PZ have different genetic backgrounds (origin of parental material) and have been developed independently since 2005 (no PJ clone is used as a parent in PZ and vice versa). Selected parents were tested by SSR markers and results indicated that PJ and PZ formed clearly segregated clusters and gene pools. In total, 6898 offspring clones were developed that traced back to 231 offspring derived from PJ × PZ crosses (49 PJ05 and 31 PZ06 clones). The hybrid population (PJ × PZ) exhibited on average a mid-parent–mid-offspring heterosis increment of 14% for storage root yield (dwb). We observed storage root yield (dwb) heterosis increments in > 70% of all offspring, about 25% of all offspring exhibited a heterosis increment of ≥ 26%, and two offspring had heterosis increments of close to 80% (Federico Diaz, Peru, unpublished). CIP considered this as a 'go decision' to start heterosis studies in the breeding platforms of Uganda and Mozambique. Moreover, CIP is continuing with the heterosis study in Peru after discarding parents found to be poor 'family makers' and/or which developed SPVD problems over time. After one reciprocal recurrent selection cycle we expect a yield jump in storage root yield (dwb) of about 30% in the next hybrid population. Such a HEBS cannot only increase the efficiency of population improvement for yield-related traits, but can also increase the efficiency of breeding for recessive inherited traits (such as resistance to SPVD and/or non-sweet sweetpotatoes) by moderate inbreeding through intra gene-pool recombination. However, the adoption of HEBS will clearly depend on whether breeders clearly see an advantage for their own breeding programmes.

Molecular tools for sweetpotato: Molecular tools have greatly improved our understanding about origin and centres of diversity of sweetpotato, but to date in sweetpotato applied breeding, molecular tools have not been used much. An exception is SSR markers for gene-pool subdivision, especially among parental material for breeding. Up to now expressed sequence tag (EST) sequencing has resulted in identification of about 1600 gene-based SSR markers for sweetpotato

(Schaftleitner *et al.*, 2010). Few SSR primers have been published for applied characterization of breeding material (Tumwegamire *et al.*, 2011b), but there are > 200 SSR primers available for sweetpotato and about 75 SSR primers are routinely used at CIP for gene-pool subdivision and marker association studies. This set of SSRs were used to confirm gene-pool subdivision of parental material at CIP in Peru (Federico Diaz, Peru, unpublished), to characterize parental material in the Uganda breeding platform (David, 2012) and to search for potential heterotic gene pools among accessions from China, Korea and Japan held in trust at CIP (Maria David, unpublished).

With respect to experimental breeding material, several studies have used random amplified polymorphic DNA (RAPD), amplified fragment length polymorphism (AFLP) and SSR markers for genetic mapping or marker-trait associations. There are two AFLP-based sweetpotato linkage maps available for Beauregard × Tanzania crosses (Cervantes, 2006; Cervantes *et al.*, 2008; Solis and Grüneberg, 2008) and a set of about 250 Beauregard × Tanzania clones are being processed to be available for international distribution by CIP's genebank. Marker associations have been reported for several yield, quality and resistant traits (Table 1.11), but so far none of these molecular markers have been validated and investigated for their efficiency in applied breeding material. Finally, we want to note that in early 2015, a new 4 year project called 'Genomic Tools for Sweetpotato Improvement (GTSPI)' will begin. This project will primarily focus on sequencing the *I. trifida* genome, developing high-throughput molecular marker systems such as genotyping by sequencing (GBS), developing statistical tools to process the huge amount of raw GBS data and testing genomic selection (GS) for sweetpotato. GS proposes the prediction of the performance of genotypes based on genomic data using the genomic estimated breeding values (GEBVs) approach. GS is a further extension of association mapping, but in contrast to the initial idea of association mapping by comparing different alleles of candidate genes, the basis of GS will be to associate a large number of single nucleotide polymorphisms (SNPs) and finally all sequence differences with differences in quantitative trait performance. With respect to applied breeding, GS is abandoning the idea of dividing quantitative variation into values of single quantitative trait loci (QTLs) followed by identifying candidate genes within QTLs. It is a statistical approach somewhat similar to Fisher's approach of considering quantitative variation as a result of small contributions of an infinite number of genes. This means it does not need information about genes contributing to quantitative trait performance. Theoretically, all historical information on phenotypic performance of genotypes can be used to generate GS prediction models, provided that DNA of these individuals is still available. It appears that among all the arguments in favour of GS one is missing, and that is the advantage that GS has the potential to predict many traits simultaneously. Nevertheless, the efficiency of GS has to be tested in applied breeding populations similar to that outlined above for ABS.

1.6 Released or Launched Varieties and New Breeding Material

This short but critical section provides a very condensed overview on released and launched sweetpotato varieties in the world and we focus on SSA. During 1994–2003 a total of 56 sweetpotato varieties were released in SSA (12 countries). Fifteen of these releases were OFSP. During this period, eight OFSP variety releases were made in Mozambique, but nearly exclusively with introduced varieties. During 2004–2013 a total of 89 sweetpotato varieties were released in these 12 countries in SSA and 62 of these variety releases were OFSP (Fig. 1.7). For variety releases in other regions in the world, consult Appendix 1. Predominantly grown varieties across regions and breeding material in the pipeline are listed in Appendices 2 and 3, respectively. The lists are updated on the web (http://sweetpotatobreeder.com) and on the Sweetpotato Knowledge Portal (www.sweetpotatoknowledge.org).

Table 1.11. Overview of molecular markers and trait association studies in sweetpotato.

Trait	Population	Marker	Reference
Sweetpotato virus disease (SPVD) resistance	Tanzania × Wagabolige mapping population	Two markers, E41M33.a and E38M36.u located on linkage groups 22 and 35, respectively, were highly significant ($P < 0.0001$) for resistance to sweetpotato chlorotic stunt virus (SPCSV), and marker S13.1130 located on linkage group 6 was highly significant ($P < 0.0001$) for resistance to sweetpotato feathery mottle virus (SPFMV). The markers explained 72% (SPCSV) and 71% (SPFMV) of variation	Mwanga (2001)
SPVD resistance	47 diverse clones in a training group (15 susceptible and 15 resistant) and a validation group (14 susceptible and three resistant)	Four amplified fragment length polymorphism (AFLP) markers resulted in 100% correct classification: E33M49.202 (cag202), E33M59.168 (cta168), E33M59.110 (cta110) and E33M59.334 (cta334)	Miano *et al.* (2008)
Root-knot nematode resistance	71 progenies of the F1 single-cross population produced from parent Regal (resistant) and and Vardaman (susceptible)	One random amplified polymorphic DNA (RAPD) marker was selected: $OP15_{1500}$; estimated recombination fraction of (0.2421 ± 0.057) between the marker and the root-knot-nematode-resistance gene	Ukoskit *et al.* (1997)
Southern root-knot nematode resistance	48 half-sibs developed at Louisiana State University (LSU) and 54 full-sibs developed by International Potato Center (CIP) in East Africa	Five and four AFLP markers with strong associations to resistance selected in LSU and CIP populations, respectively. The markers E33M61.218 (ctg218), E33M61.227 (ctg227), E33M59.098 (cta098), E33M49.267 (cag267), E33M61.232 (ctg232) – LSU population, and E33M61.228 (ctg228), E33M49.118 (cag118), E33M49.108 (cag108) and E33M59.148 (cta148) – CIP population, resulted in 88.78% and 88.04% classification efficiency, respectively	Mcharo *et al.* (2005)
Root-knot nematode resistance	Beauregard × Tanzania mapping population of North Carolina State University (NCSU) 240 individuals	Seven significant quantitative trait loci (QTLs) in Tanzania and two in Beauregard: E32M4920, E42M6022, E46M3201, E35M4414, E32M3722, E38M4512 and E36M3811 in Tanzania and E40M6008 and E42M3525 in Beauregard; each explained 20% of the observed variation	Cervantes (2006)

Continued

Table 1.11. Continued.

Trait	Population	Marker	Reference
Storage root dry-matter content (SRDM)	Beauregard × Tanzania mapping population of NCSU 240 individuals	Eight AFLP markers in Beauregard: four regions had positive effect on SRDM associated with E35M4511 (LG B05.26, *P* = 0.0247), E32M3202 (LG B07.40, *P* = 0.0098), E40M4010 (LG B11.61, *P* = 0.0138) and E36M5408 (LG B89, *P* = 0.049); four loci with negative effect – E42M3421 (LG B01.03, *P* = 0.0056), E43M5403 (LG B04.23, *P* = 0.0007), E36M5103 (LG B11.62, *P* = 0.0055) and E34M4906 (LG B12.70, *P* = 0.0006). In Tanzania four markers with a positive effect – E35M3603 (LG T01.05, *P* = 0.0064), E36M3808 (LG T05.25, *P* = 0.0224), E31M3208 (LG T06.32, *P* = 0.0021) and E46M6011 (LG T07.37, *P* = 0.0486); and one with negative effect – E43M3524 (LG T02.07, *P* = 0.025)	Cervantes (2006)
SRDM	Beauregard × Tanzania mapping population of CIP 135 individuals	E40M32_309.5 (8% variation, negative effect); E41M42_449 (8.4% variation, positive effect); E39M60_204 (9.2% and 8.3% variation, positive effect); E33M60_66 (7.6% negative effect); E31M37_71 and E40M34_191 (7.7% positive effect)	Solis and Grüneberg (2008)
β-Carotene content	Beauregard × Tanzania mapping population of NCSU 240 individuals	In Beauregard, four loci linked to E43M5403 (LG B04.23), E38M3725 (LG B08.48), E36M5103 (LG B11.62) and E44M4902 (LG B12.69) were significant. In Tanzania, four loci located near E45M3611 (LG T13.74), E40M3105 (LG T13.76), E46M3901 (LG T78) and E36M4015 (LG T82) were significant	Cervantes (2006)
Total carotenoid content	Beauregard × Tanzania mapping population of CIP 135 individuals	E31M36_446 (9.7% negative), E45M37_127 (7.5% negative), E41M42_449 (7.5% negative), E42M35_70 (7.8% negative), E31M37_345 (7.5% negative), E44M36_184.5 (10.5% negative), E43M49_119 (8.8% positive)	Solis and Grüneberg (2008)
β-Carotene content	Two contrasting groups (38 clones with high and 17 clones with low β-carotene)	Nine AFLPs achieved 100% correct classification: E33M62.240 (ctt240), E33M62.347 (ctt347), E33M49.224 (cag224), E33M59.067 (cta067), E33M61.186 (ctg186), E33M61.149 (ctg149), E33M62.092 (ctt092), E33M.099 (ctg099) and E33M62.167 (ctt167)	Mcharo and LaBonte (2010)

Continued

Table 1.11. Continued.

Trait	Population	Marker	Reference
Starch content	Beauregard × Tanzania mapping population of CIP 135 individuals	E41M42_449 (11.7% positive), E31M36_446 (9.6% positive), E34M51_194 (8.5% positive), E32M54_328 (7.2% positive), E40M34_191 (8.3% positive)	Solis and Grüneberg (2008)
Sucrose content	Beauregard × Tanzania mapping population of CIP 135 individuals	E31M36_446 (7.3% negative), E45M60_234 and E34M51_194 (6.7% negative), E32M54_328 (9% negative), E40M34_191 (6.7% negative)	Solis and Grüneberg (2008)
Maltose content	Beauregard × Tanzania mapping population of CIP 135 individuals	E33M54_292 (7.7% negative), E40M34_303 and E42M40_138 (13.4% positive), E42M45_148 (8.7% positive), E42M35_74 (9.7% positive)	Solis and Grüneberg (2008)
Storage root yield (SRYLD)	Beauregard × Tanzania mapping population of NCSU 240 individuals	In Beauregard, four markers (E33M4213, E35M3317, E33M6104 and E41M5911) located on linkage groups B02.10, B03.14, B09.53 and B11.65, respectively, with positive effect on SRYLD. Three other regions near markers E36M3610, E36M3414 and E42M3606 on linkage groups B05.29, B07.37 and B09.54, respectively, with negative effect on SRYLD (explained approx. 12% of total variation of SRYLD). In Tanzania, four regions showed positive effect: E37M3109, E41M3217, E34M3501 and E40M3309, on linkage groups T02.10, T06.35, T07.40 and T07.41, respectively (explained approx. 10% of total variation of SRYLD). Seven markers showed negative effect: E41M3215, E42M5002, E43M5018, E37M4301, E44M4513, E32M3722 and E38M3718, on linkage groups T01.05, T02.08, T02.12, T03.18, T06.32, T07.39 and T72, respectively (explained approx. 20% of variation of SRYLD)	Cervantes (2006)
SRYLD	Beauregard × Tanzania mapping population of CIP 135 individuals	E43M60_337 and E32M54_137 (13.6% variation, positive effect); E32M54_88 (11.4% variation, negative effect)	Solis and Grüneberg (2008)

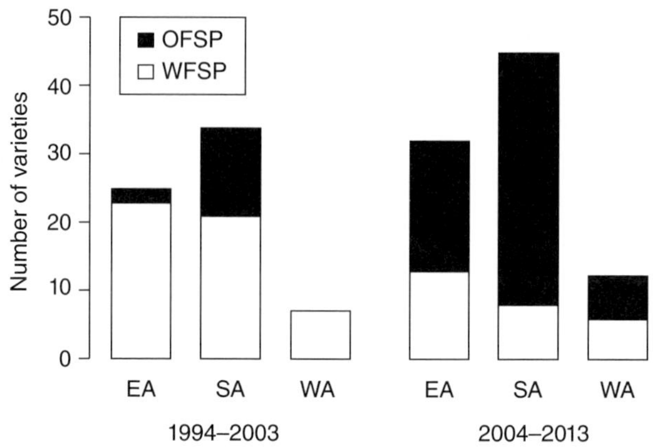

Fig. 1.7. Number of sweetpotato variety releases in SSA during 1994–2013 by subregion and flesh colour. EA, East Africa (Kenya, Rwanda, Tanzania and Uganda); SA, Southern Africa (Madagascar, Malawi, Mozambique, Republic of South Africa and Zambia); WA, West Africa (Burkina Faso, Ghana and Nigeria); OFSP, orange-fleshed sweetpotato; WFSP, white-fleshed or yellow-fleshed sweetpotato.

References

Agili, S. (2012) Yield evaluation, selection and drought tolerance indices of orange-fleshed sweetpotato (*Ipomoea batatas* Lam.) under water stress conditions. PhD thesis, Jomo Kenyatta University of Agriculture and Technology, Nairobi.

Agudelo, P., Robbins, R.T., Stewart, J.M. and Szalanski, A.L. (2005) Intraspecific variability of *Rotylenchulus reniformis* from cotton-growing regions in the United States. *Journal of Nematology* 37, 105–114.

Anginyah, T.J., Narla, R.D., Carey, E.E. and Njeru, R. (2001) Etiology, effect of soil pH and sweetpotato varietal reaction to Alternaria leaf petiole and stem blight in Kenya. *African Crop Science Journal* 9, 287–292.

Anguiz, R. and Canto-Sáenz, M. (1991) Reacción de cultivares de camote (*Ipomoea batatas* (L.) Lam.) a *Pratylenchus flakkensis* Sienhorst. *Nematrópica* 21, 197–201.

Anon. (2004) Monsanto failure. *New Scientist* 181(2433), 7 February, p. 7.

Anselmo, B.A., Ganga, Z.N., Badol, E.O., Heimer, Y.M. and Nejidat, A. (1988) Screening sweet potato for drought tolerance in the Philippine highlands and genetic diversity among selected genotypes. *Tropical Agriculture* 75, 189–196.

Armstrong, G.M. and Armstrong, J.K. (1958) The Fusarium wilt complex as related to the sweet potato. *Plant Disease Reporter* 42, 1319–1329.

Austin, D.F. (1978) The *Ipomoea batatas* complex. I. Taxonomy. *Bulletin of the Torrey Botanical Club* 105, 114–129.

Austin, D.F. (1988) *The Taxonomy, Evolution and Genetic Diversity of Sweet Potato and Related Wild Species*. International Potato Center (CIP), Lima, pp. 27–58.

Austin, D.F. (1991) Associations between the family Convolvulaceae and *Cyclas* weevils. In: Jansson, R.J. and Raman, K.V. (eds) *Sweet Potato Pest Management – a Global Perspective*. Westview Press, Boulder, Colorado, pp. 45–57.

Austin, D.F. and Huamán, Z. (1996) A synopsis of *Ipomoea* (Convolvulaceae) in the Americas. *Taxon* 45(1), 3–38.

Barnes, S.L. and Sanders, S.A. (2012) Advances in functional use of sweet potato, *Ipomoea batatas* (L.) Lam. *Recent Patents on Food, Nutrition and Agriculture* 4, 148–154.

Baynes, R.A. (1972) Sweet potato varieties in the Eastern Caribbean. *Caribbean Farming* 3(4), 20–21.

Becker, H.C. (2011) *Pflanzenzüchtung*. Eugen Ulmer KG, Stuttgart, Germany. (in German)

Belehu, T., Hammes, P.S. and Robbertse, P.J. (2004) The origin and structure of adventitious roots in sweet potato (*Ipomoea batatas*). *Australian Journal of Botany* 52, 551–558.

Bernstein, L. (1974) Crop growth and salinity. In: van Schilfgaarde, J. (ed.) *Drainage for Agriculture. Agronomy Monograph* 17, 39–54.

Blum, A. (1988) *Plant Breeding for Stress Environments*. CRC Press, Boca Raton, Florida.

Blum, A. (2005) Drought resistance, water-use efficiency, and yield potential – are they compatible, dissonant, or mutually exclusive? *Australian Journal of Agricultural Research* 56, 1159–1168.

Bohac, J.R., Austin, D.F. and Jones, A. (1993) Discovery of wild tetraploid sweetpotatoes. *Economic Botany* 47, 193–201.

Bohlen, E. (1973) *Crops Pests in Tanzania and Their Control*. Paul Parey, Berlin, 108 pp.

Bouis, H. and Islam, Y. (2012) *Delivering Nutrients Widely through Biofortification: Building on Orange Sweet Potato – Scaling up in Agriculture, Rural Development and Nutrition*. Focus 19, Brief 11. International Food Policy and Research Institute (IFPRI), Washington, DC. Available at: http://www.ifpri.org/sites/default/files/publications/focus19_11.pdf (accessed 4 January 2014).

Bovel-Benjamin, A.C. (2007) Sweet potato: a review of its past, present, and future role in human nutrition. *Advances in Food and Nutrition Research* 52, 1–59.

Bridge, J. (1988) Plant-parasitic nematode problems in the Pacific islands. *Journal of Nematology* 20, 173–183.

Bruggen, A.H.C. van (1984) Sweet potato stem blight caused by *Alternaria* sp.: a new disease in Ethiopia. *Netherlands Journal of Plant Pathology* 90, 155–164.

Buteler, M.I., Jarret, R.L. and LaBonte, D.L. (1999) Sequence characterization of microsatellites in diploid and polyploidy *Ipomoea*. *Theoretical and Applied Genetics* 99, 123–132.

Cao, Q., Zhang, A., Ma, D., Li, H., Li, Q. and Li, P. (2009) Novel interspecific hybridization between sweetpotato (*Ipomoea batatas* (L.) Lam.) and its two diploid wild relatives. *Euphytica* 169, 345–352.

Carey, E.E., Gichuki, S.T., Mwanga, R.O.M., Kasule, S., Fuentes, S., Macharia, C., and Gibson, R.W. (1998) Sweetpotato viruses in Uganda and Kenya: results of a survey. In: Akoroda, M.O. and Ekanayake, I.J. (eds) *Proceedings of the Sixth Triennial Symposium of the International Society of Tropical Crops – Africa Branch* (ISTRC-AB), 22–28 October 1995, Lilongwe (Malawi). ISTRC-AB, Ibadan, Nigeria, pp. 457–461.

Carpena, A.L. (2009) Important cultivars, varieties, and hybrids. In: Loebenstein, G. and Thottappilly, G. (eds) *The Sweetpotato*. Springer Science + Business Media BV, Houten, The Netherlands, pp. 27–40.

Cervantes, J.C. (2006) Development of a genetic linkage map and QTL analysis in sweetpotato. PhD thesis, North Carolina State University, Raleigh, North Carolina.

Cervantes, J.C., Yencho, G.C. and Davis, E.L. (2002) Host reactions of sweetpotato genotypes to root-knot nematodes and variation in virulence of *Meloidogyne incognita* populations. *HortScience* 37(7), 1112–1116.

Cervantes, J.C., Yencho, G.C., Kriegner, A., Pecota, K.V., Faulk, M.A., Mwanga, R.O.M. and Sosinski, B.R. (2008) Development of a genetic linkage map and identification of homologous linkage groups in sweetpotato using multiple-dose AFLP markers. *Molecular Breeding* 21, 511–532.

Chalfant, R.B., Richard, K.J., Dakshina, R.S. and James, M. (1990) Ecology and management of sweet-potato insects. *Annual Review of Entomology* 35, 157–180.

Chávez, R., Mendoza, H., Upadhya, M., Espinosa, J., Cabello, R., Arévalo, N., Wijntje, A., Scoffield, J., Zúñiga, P., Guevara, E. and Siles, P. (2000) *Mejoramiento genético y adaptación de camote (Ipomoea batatas) a las condiciones áridas y salinas. Idesia (Arica) Volume 18*. Facultad de Agronomía, Universidad de Tarapacá, Chile, p. 97.

Chowdhury, S.R. and Naskar, S. (1993) Screening of drought tolerance traits in sweet potato: role of relative water content. *Orissa Journal of Horticulture* 21, 1–4.

Chowdhury, S.R. and Ravi, V. (1988) Physiology of tuberization in sweetpotato with reference to moisture stress and seasonal influence. In: *Annual Report*. Central Tuber Crops Research Institute, Trivandrum, India, pp. 89–90.

Chukwu, G.O. (1995) Crop irrigation water needs of sweetpotato (*Ipomoea batatas*). *African Journal of Root and Tuber Crops* 1, 35–38.

Cipriani, G., Michaud, D., Brunelle, F., Golmirzaie, A. and Zhang, D.F. (1999) Expression of soybean proteinase inhibitor in sweetpotato. In: *CIP Program Report 1997–1998*. International Potato Center, Lima, pp. 271–277.

Cipriani, G., Fuentes, S., Bello, V., Salazar, L.F., Ghislain, M. and Zhang, D.P. (2001) Transgene expression of rice cysteine proteinase inhibitors for the development of resistance against sweet potato feathery mottle virus. In: *CIP Program Report 1999–2000*. International Potato Center (CIP), Lima, pp. 267–271.

Clark, C.A., Wright, V.L. and Miller, R.L. (1980) Reaction of some sweet potato selections to the reniform nematode, *Rotylenchulus reniformis*. *Journal of Nematology* 12, 218.

Clark, C.A., Hyun, J.-W. and Hoy, M.W. (1998) Relationships among wilt-inducing isolates of *Fusarium oxysporum* from sweetpotato and tobacco. *Plant Disease* 82, 530–536.

Clark, C.A., Holmes, G.J. and Ferrin, D.M. (2009) Major fungal and bacterial diseases. In: Loebenstein, G. and Thottappilly, G. (eds) *The Sweetpotato*. Springer Science + Business Media BV, Houten, The Netherlands, pp. 81–103.

Clark, C.A., Davis, J.A., Abad, J.A., Cuellar, W.J., Fuentes, S., Kreuze, J.F., Gibson, R.W., Mukasa, S.B., Tugume, A.K., Tairo, F.D. and Valkonen, J.P.T. (2012) Sweetpotato viruses: 15 years of progress on understanding and managing complex diseases. *Plant Disease* 96, 168–185.

Cochran, W.G. (1951) Improvement by means of selection. In: *Proceedings of the Second Berkeley Symposium on Mathematical Statistics and Probability*. University of California Press, Berkeley, California, pp. 449–470.

Collins, W.W. (1977) Diallel analysis of sweet potatoes for resistance to Fusarium wilt. *Journal of the American Society of Horticultural Science* 102, 109–111.

Courtney, M., Mcharo, M., La Bonte, D. and Grüneberg, W. (2008) Heritability estimates for micronutrient composition of sweetpotato storage root. *HortScience* 43(5), 1382–1384.

Coyne, D.L., Talwana, H.A.L. and Maslen, N.R. (2003) Plant-parasitic nematodes associated with root and tuber crops in Uganda. *African Plant Protection* 9, 87–98.

Dasgupta, D.R. and Seshadri, A.R. (1971a) Races of the reniform nematode, *Rotylenchulus reniformis* Linford and Oliveira, 1940. *Indian Journal of Nematology* 1, 21–24.

Dasgupta, D.R. and Seshadri, A.R. (1971b) Reproduction, hybridization and host adaptation in physiological races of the reniform nematode, *Rotylenchulus reniformis. Indian Journal of Nematology* 1, 128–144.

David, M. (2012) Molecular characterization of *Ipomoea batatas* L. 'sweetpotato' African landraces and modern varieties using microsatellite markers. MSc thesis, Universidad Nacional Agraria La Molina, Lima.

Dempewolf, H., Eastwood, R.J., Guarino, L., Khoury, C.K., Müller, J.V. and Toll, J. (2014) Adapting agriculture to climate change: a global initiative to collect, conserve, and use crop wild relatives. *Agroecology and Sustainable Food Systems* 38(4), 369–377.

Ding, C.W., Niu, F.X., Guo, X.D. and Hua, X.X. (1997) Identification on the drought resistance in sweetpotato genetic resource. *Journal of Henan Agricultural Sciences* 10, 3–5.

Dinh, N.V., Khang, H.L., The, N.V., Hung, H., Oanh, N.K., Ho, T.V. and Braun, A. (1995) Preliminary results on IPM application for sweetpotato at Thanhbinh, Hatay. In: *Research Results of the Faculty of Crop Science*. Hanoi Agricultural University, Hanoi, pp. 83–89.

Diop, A. (1998) *Storage and Processing of Roots and Tubers in the Tropics*. Food and Agriculture Organization of the United Nations (FAO), Agro-Industries and Post-Harvest Management Service. Agricultural Support Systems Division. FAO, Rome, pp. 38–50.

Dongro, C., Jaekook, L., Byeongyong, P. and Minam, C. (2006) Occurrence of root-knot nematodes in sweet potato fields and resistance screening of sweet potato cultivars. *Korean Journal of Applied Entomology* 45, 211–216.

Dudley, J.W. (1977) 76 generations of selection for oil and protein percentage in maize. In: Pollanck, E., Kempthorne, O. and Bailey, T.B. (eds) *Proceedings of the International Conference on Quantitative Genetics*. Iowa State University Press, Ames, Iowa, pp. 459–473.

Dukes, P.D., Jones, A. and Cuthbert, F.P. (1975) Mass evaluation of sweet potato seedlings for reaction to Fusarium wilt and root knot nematodes. *Proceedings of the American Phytopathological Society* 2, 133–134.

Elston, R.C. (1963) A weight-free index for the purpose of ranking or selection with respect to several traits at a time. *Biometrics* 19, 669–680.

Estes, E. (2006) *Economics and Trends in Sweet Potato Industry: Sweet Potato Speaks.* News for North Carolina Processors and Growers. North Carolina Sweet Potato Commission, Selma, North Carolina, March.

Estes, E.A. (2009) Marketing sweetpotatoes in the United States: a serious challenge for small-to-moderate volume growers. In: Loebenstein, G. and Thottappilly, G. (eds) *The Sweetpotato*. Springer Science + Business Media BV, Houten, The Netherlands, pp. 269–283.

Fajardo, D.D., La Bonte, D.R. and Jarret, R.L. (2002) Identifying and selecting for genetic diversity in Papua New Guinea sweet potato *Ipomoea batatas* (L.) Lam. germplasm collected as botanical seed. *Genetic Resources and Crop Evolution* 49, 463–470.

Famine Early Warning System Network (FEWSNET) (2009) Malawi Food Security Update. United States Agency for International Development (USAID). Available at: http://www.fews.net/sites/default/files/documents/reports/Malawi_FSU_November_2009_final.pdf (accessed 26 March 2015).

FAOSTAT (2011) Available at: http://faostat.fao.org/ (accessed 15 September 2014).

Feng, G., Yifu, G. and Pinbo, Z. (2000) Production and deployment of virus-free sweetpotato in China. *Crop Protection* 19, 105–111.

Finney, D.J. (1966) An experimental study of certain screening processes. *Journal of the Royal Statistical Society, Series B* 28, 88–109.

Firon, N., LaBonte, D., Villordon, A., McGregor, C., Kfir, Y. and Pressman, E. (2009) Botany and physiology: storage root formation and development. In: Loebenstein, G. and Thottappilly, G. (eds) *The Sweetpotato*. Springer Science + Business Media BV, Houten, The Netherlands, pp. 13–26.

Freyre, R., Iwanga, M. and Orjeda, G. (1991) Use of *Ipomoea trifida* (HBK.) G. Don germplasm for sweetpotato improvement. 2. Fertility of synthetic hexaploids and triploids with 2n gametes of *I. trifida*, and their interspecific crossability with sweet potato. *Genome* 34, 209–214.

Fuentes, S. and Chujoy, E. (2009) Sweetpotato in South America. In: Loebenstein, G. and Thottappilly, G. (eds) *The Sweetpotato*. Springer Science + Business Media BV, Houten, The Netherlands, pp. 415–440.

Gallais, A. (2003) *Quantitative Genetics and Selection Theory in Autopolyploid Plants*. Institut National de la Recherche Agronomique (INRA), Paris.

Gapasin, R.M. (1979) Survey and identification of plant parasitic nematodes associated with sweet potato and cassava. *Annals of Tropical Research* 1, 120–134.

Gibson, R.W., Byamukama, E., Mpembe, J. and Mwanga, R. (2008) Working with farmer groups in Uganda to develop new sweet potato cultivars: decentralization and building on traditional approaches. *Euphytica* 159, 217–228.

Gichuki, S.T., Barenyi, M., Zhang, D., Hermann, M., Schmidt, J., Glossl, J. and Burg, K. (2003) Genetic diversity in sweetpotato [*Ipomoea batatas* (L.) Lam.] in relationship to geographic sources as assessed with RAPD markers. *Genetic Resoures and Crop Evolution* 50, 429–437.

Gilbert, J.C. (2005) Coloring foods and beverages. *Food Technology* 59, 38–44.

Golson, J. (1976) Archaeology and agricultural history in the New Guinea Highlands. In: Sieveking, G., Longworth, I.H. and Wilson, K.E. (eds) *Problems in Economic and Social Archaeology.* Duckworth, London, pp. 201–220.

Grüneberg, W.J., Abidin, E., Ndolo, P., Pereira, C.A. and Hermann, M. (2004) Variance component estimations and allocation of resources for breeding sweet potato under East African conditions. *Plant Breeding* 123, 311–315.

Grüneberg, W.J., Manrique, K., Dapeng, Z. and Hermann, M. (2005) Genotype × environment interactions for a diverse set of sweet potato clones evaluated across varying ecogeographic conditions in Peru. *Crop Science* 45, 2160–2171.

Grüneberg, W.J., Mwanga, R., Andrade, M. and Espinoza, J. (2009a) Selection methods part 5: breeding clonally propagated crops. In: Ceccarelli, S., Guimarães, E.P. and Weltzien, E. (eds) *Plant Breeding and Farmer Participation*. Food and Agriculture Organization of the United Nations (FAO), Rome, pp. 275–322.

Grüneberg, W.J., Mwanga, R., Andrade, M. and Dapaah, H. (2009b) Sweetpotato breeding. In: Andrade, M., Barker, I., Cole, D., Dapaah, H., Elliott, H., Fuentes, S., Grüneberg, W.J., Kapinga, K., Kroschel, J., Labarta, R., Lemaga, B., Loechl, C., Low, J., Lynam, J., Mwanga, R., Ortiz, O., Oswald, A. and Thiele, G. *Unleashing the Potential of Sweetpotato in Sub-Saharan Africa: Current Challenges and Way Forward. Working Paper* 2009-1. International Potato Center (CIP), Lima, pp. 1–42.

Gutiérrez, D.L., Fuentes, S. and Salazar, L.F. (2003) Sweetpotato virus disease (SPVD): distribution, incidence, and effect on sweetpotato yield in Peru. *Plant Disease* 87, 297–302.

Hahn, S.K. (1977) A quantitative approach to source potentials and sink capacities among reciprocal grafts of sweetpotato varieties. *Crop Science* 17, 559–562.

Hahn, S.K. and Hozyo, Y. (1984) Sweetpotato. In: Goldsworthy, P.R. and Fisher, N.M. (eds) *The Physiology of Tropical Field Crops*. Wiley, New York, pp. 551–567.

Hahn, S.K. and Leuschner, K. (1982) Breeding sweetpotato for weevil resistance. In: Villareal, R.L. and Griggs, T.D. (eds) *Sweet Potato: Proceedings of the First International Symposium*. Asian Vegetable Research and Development Center, Shanhua, Tainan, Taiwan, pp. 331–336.

Hammett, H.L., Constantin, R.J. and Hernandez, T.P. (1982) The effect of phosphorous and soil moisture levels on yield and processing quality of 'Centennial' sweetpotatoes. *Journal of the American Society of Horticultural Science* 107, 119–122.

Hanson, W.D. and Brim, C.A. (1963) Optimum allocation of test material for two-stage testing with an application to evaluation of soybean lines. *Crop Science* 3, 43–49.

Hernandez, T.P. and Barry, J.R. (1966) The effect of different soil moisture levels and rates of nitrogen on production and quality of sweetpotatoes. *Proceedings of the International Horticultural Congress* 17, 327–335.

Hernandez, T.P. and Hernandez, T. (1967) Irrigation to increase sweetpotato production. In: Tai, E.A., Charles, W.B., Haynes, P.H., Hon, E.F.and Leslie, K.A. (eds) *Proceedings of the International Symposium on Tropical Root Crops*. University of the West Indies, St Augustine, Trinidad, pp. 31–38.

Hijmans, R., Huaccho, L. and Zhang, D. (2002) Description and analysis of a geo-referenced database of the global distribution of sweetpotato area. Proceedings, First International Conference on Sweet Potato. Food and Health for the Future, Lima, Peru. *Acta Horticulturae* 583, 41–49.

Hotz, C., Loechl, C., Lubowa, A., Tumwine, J.K., Ndeezi, G., Nandutu Masawi, A., Baingana, R., Carriquiry, A., de Brauw, A., Meenakshi, J.V. and Gilligan, D.O. (2012) Introduction of β-carotene-rich orange sweet potato in rural Uganda resulted in increased vitamin A intakes among children and women and improved vitamin A status among children. *Journal of Nutrition* 142, 1871–1880.

Hou, L.X., Xiao, L.Z., Kang, Z.H., Yang, X.P., Gu, J.H., Tang, B.J. and Liu, J.B. (1999) Identification on drought resistance of sweetpotato varieties. *Journal of Henan Agricultural Sciences* 2, 5–6.

Hoyer, U., Maiss, E., Jelkmann, W., Lesemann, D.E. and Vetten, H.J. (1996) Identification of the coat protein gene of a sweetpotato sunken vein closterovirus isolate from Kenya and evidence for a serological relationship among geographically diverse closterovirus isolates from sweetpotato. *Phytopathology* 86, 744–750.

Hozyo, Y. and Park, C.Y. (1971) Plant production in grafting plants between wild type and improved variety in *Ipomoea. Bulletin of the National Institute of Agricultural Science, Japan Series D* 22, 145–164.

Huamán, Z. (1991) *Descriptors for Sweet Potato*. International Potato Center (CIP), Asian Vegetable Research and Development Center (AVRDC) and International Board for Plant Genetic Resources (IBPGR), Lima.

Hull, F.H. (1945) Recurrent selection for specific combining ability in corn. *Journal of the American Society of Agronomy* 37, 134–145.

Indira, P. and Kabeerathumma, S. (1988) Physiological response of sweet potato under water stress. I. Effect of water stress during the different phases of tuberization. *Journal of Root Crops* 14, 37–40.

Ishida, H., Suzuno, H., Sugiyama, N., Innami, S., Tadokoro, T. and Maekawa, A. (2000) Nutritive evaluation on chemical components of leaves, stalks and stems of sweetpotatoes (*Ipomoea batatas* poir). *Food Chemistry* 68, 359–367.

Iwahori, H. and Sano, Z. (2003) Distribution of main plant-parasitic nematodes in sweet potato and taro fields in Kyushu and Okinawa, Japan. 3. Survey in the northern part in Kyushu Island (Fukuoka, Saga, Nagasaki and Ohita Prefs.). *Kyushu Plant Protection Research* 49, 83–87.

Iwanaga, M. (1988) Use of wild germplasm for sweet potato breeding. In: Gregory, P. (ed.) *Exploration, Maintenance, and Utilization of Sweet Potato Genetic Resources*. International Potato Center (CIP), Lima, pp. 199–210.

Jia, R. (2013) Weather shocks, sweet potatoes and peasant revolts in historical China. *The Economic Journal* 124(575), 92–118.

Jiang, X., Jianjun, H. and Wang, Y. (2004) Sweetpotato processing and product research and development at the Sichuan Academy of Agricultural Sciences. In: Fuglie, K.O. and Hermann, M. (eds) *Sweetpotato Post-Harvest Research and Development in China*. Proceedings of an International Workshop, 7–8 November 2001, Chengdu, Sichuan, PR China. International Potato Center (CIP), Bogor, Indonesia.

Jones, A. (1967) Theoretical segregation ratios of qualitatively inherited characters for hexaploid sweet potato (*Ipomoea batatas* L.). *Technical Bulletin* 1368. United States Department of Agriculture (USDA), Georgia Agricultural Experiment Station, Washington, DC.

Jones, A. (1969) Quantitative inheritance of Fusarium wilt resistance in sweet potatoes. *Journal of the American Society of Horticultural Science* 94, 207–209.

Jones, A. (1980) Sweet potato. In: Fehr, W.R. and Radley, H.H. (eds) *Hybridization of Crop Plants*. American Society of Agronomy and Crop Science Society of America, Madison, Wisconsin, pp. 645–655.

Jones, A. (1985) Sweetpotato. In: Bassett, M.J. (ed.) *Breeding Vegetable Crops*. AVI Publishing, Westport, Connecticut.

Jusuf, M. (2002) Adoption of sweetpotato varieties in major producing area. Appendix Tables b2i, b2ii, b2v. In: Rao, R. and Campilan, D. (eds) *Exploring the Complementarities of* in situ *and* ex situ *Conservation Strategies for Asian Sweetpotato Genetic Resources*. Proceedings of the Third International Workshop of the Asian Network for Sweetpotato Genetic Resources (ANSWER), 2–4 October 2001,

Denpasar, Bali, Indonesia. International Plant Genetic Resources Institute (IPGRI) – Asia, the Pacific and Oceania (APO), Serdang, Malaysia, pp. 98, 99 and 102.

Katayama, K., Komae, K., Kotyama, K., Kato, T., Tamiya, S., Kuranouchi, T., Komaki, K. and Nakatani, M. (2006) New sweet potato cultivar 'Quick Sweet' having low gelatinization temperature and altered starch structure. Paper presented at the Second International Symposium on Sweet Potato and Cassava – Innovative Technologies for Commercialization, 14–17 June 2005, Kuala Lumpur, Malaysia.

Kays, S.J. (1985) The physiology of yield in the sweet potato. In: Bouwkamp, J.C. (ed.) *Sweet Potato Products: A Natural Resource for the Tropics*. CRC Press, Boca Raton, Florida, pp. 79–132.

Kays, S.J., Wang, Y. and McLaurin, W.J. (2005) Chemical and geographical assessment of the sweetness of the cultivated sweetpotato clones of the world. *Journal of the American Society of Horticultural Science* 130(4), 591–597.

Kelm, M., Brück, H., Hermann, M. and Sattelmacher, B. (2000) Plant productivity and water use efficiency of sweetpotato (*Ipomoea batatas*) as affected by nitrogen supply. In: *CIP Program Report 1999–2000*. International Potato Center (CIP), Lima, pp. 273–279.

Kendal, M.A. and Stuart, A. (1958) *The Advanced Theory of Statistics*, Vol. 1. Griffin and Co., London.

Kimura, T., Otani, M., Noda, T., Ideta, O., Shimada, T. and Saito, A. (2001) Absence of amylose in sweet potato [*Ipomoea batatas* (L.) Lam.] following the introduction of granule-bound starch synthase I cDNA. *Plant Cell Reports* 20, 663–666.

King, G.A. (1985) The effect of time of planting on yield of six varieties of sweetpotato (*Ipomoea batatas* (L.) Lam.) in the southern coastal lowlands of Papua New Guinea. *Tropical Agriculture* 62, 225–228.

Ko, J.Y., Chen, C.Y. and Kuo, G. (1993) Activity of anomalous cambium and sink capacity in self and reciprocally grafted leaf cuttings of sweetpotato. *Journal of the Agricultural Association, China. New Series* 161, 1–10.

Kobayashi, M. (1984) The *Ipomoea trifida* complex closely related to sweet potato. In: Shideler, S.F. and Rincon, H. (eds) *Proceedings of the Sixth Symposium of the International Society of Tropical Root Crops*. International Potato Center (CIP), Lima, pp. 561–568.

Koenning, S.R., Kirkpatrick, T.L., Starr, J.L., Wrather, J.A., Walker, N.R. and Mueller, J.D. (2004) Plant-parasitic nematodes attacking cotton in the United States: old and emerging production challenges. *Plant Disease* 88, 100–113.

Komiyama, A., Sano, Z., Murata, T., Matsuda, Y., Yoshida, M., Saito, A. and Okada, Y. (2006) Resistance to two races of *Meloidogyne incognita* and resistance mechanism in diploid *Ipomoea trifida*. *Breeding Science* 56, 81–83.

Konczak, I. (2006) Anthocyanin-rich polyphenolic complex with enhanced physiological activity from a sweet potato cell line. Paper presented at the Second International Symposium on Sweet Potato and Cassava – Innovative Technologies for Commercialization, 14–17 June 2005, Kuala Lumpur, Malaysia.

Kotama, S., Chuman, K. and Tanoue, M. (1970) On the growth differentials of sweet potato to different soil fertility. *Bulletin of the Kyushu Agricultural Experimental Station* 15, 493.

Kreuze, J.F., Karyeija, R.F., Gibson, R.W. and Valkonen, J.P.T. (2000) Comparisons of coat protein gene sequences show that East African isolates of sweetpotato feathery mottle virus form a genetically distinct group. *Archives of Virology* 145, 567–574.

Kreuze, J.F., Valkonen, J.P.T. and Ghislain, M. (2009) Genetic engineering. In: Loebenstein, G. and Thottappilly, G. (eds) *The Sweetpotato*. Springer Science + Business Media BV, Houten, The Netherlands, pp. 41–63.

Kriegner, A. (2001) Genetic linkage mapping, determination of ploidy type, and genetic variability of hexaploid sweetpotato (*Ipomoea batatas* (L.) Lam.). PhD thesis, Universität für Bodenkultur Wien, Vienna.

Kukimura, H., Kimaki, K., Yoshinaga, M., Hidaka, M., Sakamoto, S., Yoshida, T., Tabuchi, S. and Ide, Y. (1992) The new sweet potato cultivar 'Beniotome'. *Bulletin of the Kyushu Agricultural Experimental Station* 27, 249–267.

Kumagai, T., Umemura, Y., Baba, T. and Iwanaga, M. (1990) The inheritance of β-amylase null in storage roots *Ipomoea batatas* (L.) Lam. *Theoretical and Applied Genetics* 79, 369–376.

Lam, S.L., Thompson, A.E. and McCollum, J.P. (1959) Induction of flowering in the sweet potato. *Proceedings of the American Society of Horticultural Science* 73, 454–462.

Latham, J.R. and Wilson, A.K. (2008) Transcomplementation and synergism in plants: implications for viral transgenes? *Molecular Plant Pathology* 9, 85–103.

Laurie, S.M., Faber, M., Calitz, F.J., Moelich, E.I., Muller, N. and Labuschagne, M.T. (2012) The use of sensory attributes, sugar content, instrumental data and consumer acceptability in selection of sweet potato varieties. *Journal of the Science of Food and Agriculture* 93(7), 1610–1619.

Lawrence, G.W. and Clark, C.A. (1986) Identification, race determination, and pathogenicity of root knot nematodes to resistant and susceptible sweet potato. *Journal of Nematology* 18, 617.

Lebot, V. (2010) Sweet potato. In: Bradshaw, J.E. (ed.) *Root and Tuber Crops. Handbook of Plant Breeding 7*. Springer Science + Business Media BV, Houten, The Netherlands, pp. 97–125.

Lebot, V., Ndiaye, A. and Malapa, R. (2011) Phenotypic characterization of sweet potato [*Ipomoea batatas* (L.) Lam.] genotypes in relation to prediction of chemical quality constituents by NIRS equations. *Plant Breeding* 130(4), 457–463.

Lema, K.M. (1992) Reducing weevil damage in sweetpotato using host-plant resistance and early planting and harvesting. In: *Fourth International Society for Tropical Root Crops – Africa Branch (ISTRC-AB) Symposium Proceedings*. ISTRC-AB, Kinshasa, pp. 345–346.

Lenné, J.M. (1991) *Diseases and Pests of Sweet Potato*. Natural Resources Institute, Bulletin 46. Natural Resources Institute, Chatham Maritime, UK.

León-Velarde, C. and de Mendiburu, F. (2007) Variedades de batata de doble propósito. In: León-Velarde, C. and Amable, R. (eds) *Producción y uso de la batata (*Ipomoea batatas *(L.) Lam.): estrategias de alimentación animal*. Universidad ISA, Santiago de los Caballeros, Dominican Republic, pp. 36–44.

Lii, C.Y. and Tsai, K.H. (1996) Effect of amylose content on the rheological properties of rice starch. *Cereal Chemistry* 73, 415–420.

Lin, M.S., Fang, Z.D. and Xie, Y.P. (1993) Responses of sweet potato to exudate of potato rot nematodes (*Ditylenchus destructor*). *Acta Phytopathologica Sinica* 23, 157–162.

Lin, S.S.M., Peet, C.R., Chen, D.M. and Lo, H.-F. (1983) Breeding goals for sweet potato in Asia and the Pacific – a survey of sweet potato production and utilization. In: Martin, F.W. (ed.) *Breeding New Sweet Potatoes for the Tropics. Proceedings of the American Society of Horticultural Science* 27(B), 42–60.

Liu, M., Gao, W., Wang, Z., Fan, X. and Dou, L. (2010) Technical analysis for the development of fuel ethanol production from sweetpotato in Northwest China. In: Ma, D.F. (ed.) *Sweetpotato in Food and Energy Security*. Proceedings of China Xuzhou Fourth International Sweetpotato Symposium and Fourth China-Japan-Korea Sweetpotato Workshop. China Agricultural University Press, Beijing, pp. 29–33.

Liu, Q. (2008) *Sustainable Sweetpotato Production Technology for Food, Energy, Health and Environment*. Proceedings of the Third China-Japan-Korea Workshop on Sweetpotato. China Agricultural University Press, Beijing.

Loebenstein, G. and Thottappilly, G. (eds) (2009) *The Sweetpotato*. Springer Science + Business Media BV, Houten, The Netherlands.

Loebenstein, G., Thottappilly, G., Fuentes, S. and Cohen, J. (2009) Virus and phytoplasma diseases. In: Loebenstein, G. and Thottappilly, G. (eds) *The Sweetpotato*. Springer Science + Business Media BV, Houten, The Netherlands, pp. 105–134.

Longin, C.F.H. (2007) Optimum allocation of test resources and comparison of alternative breeding schemes for hybrid maize breeding with doubled haploids. PhD thesis, University of Hohenheim, Stuttgart, Germany. Available at: http://opus.ub.uni-hohenheim.de/volltexte/2008/206/pdf/Diss_longin_2007.pdf (accessed 4 January 2014).

Low, J.W., Arimond, M., Osman, N., Cunguara, B., Zano, F. and Tschirley, D. (2007) A food-based approach: introducing orange-fleshed sweet potatoes increased vitamin A intake and serum retinol concentrations in young children in rural Mozambique. *Journal of Nutrition* 137, 1320–1327.

Lowe, S.B. and Wilson, L.A. (1975) Yield and yield components of six sweetpotato (*Ipomoea batatas*) cultivars. II. Variability and possible sources of variation. *Experimental Agriculture* 11, 49–58.

Lu, G., Huang, H. and Zhang, D. (2006) Application of near-infrared spectroscopy to predict sweet potato starch thermal properties and noodle quality. *Journal of Zhejiang University, Science B* 7(6), 475–481.

Ma, D.F. (ed.) (2010) *Sweetpotato in Food and Energy Security*. Proceedings of China Xuzhou Fourth International Sweetpotato Symposium and Fourth China-Japan-Korea Sweetpotato Workshop. China Agricultural University Press, Beijing.

Magoon, M.L., Krishnan, R. and Vijaya Bai, K. (1970) Cytological evidence on the origin of sweet potato. *Theoretical and Applied Genetics* 40, 360–366.

Mao, L., Story, R.N., Hammond, A.M. and Labonte, D. (2001) Effect of sweet potato genotype, storage time and production site on feeding and oviposition behavior of the sweetpotato weevil, *Cylas formicarius* (Coleoptera: Apionidae). *Florida Entomologist* 84, 259–264.

Martin, F.W. (1968) The system of self-incompatibility in Ipomoea. *Journal of Heredity* 59, 263–267.

Martin, F.W. and Cabanillas, E. (1966) Post pollen germination barriers to seed set in sweet potato. *Euphytica* 15, 404–411.
Martin, F.W. and Jones, A. (1986) Breeding sweet potatoes. *Plant Breeding Reviews* 4, 313–345.
Martin, F.W. and Rodriguez-Sosa, E.J. (1985) Preference for color, sweetness, and mouth feel of sweet potato in Puerto Rico. *Journal of Agriculture of the University of Puerto Rico* 69, 99–106.
Massese, C.S.L. (1969) The principal plant nematodes of crops in the French West Indies. In: Peachy, J.E. (ed.) *Nematodes of Tropical Crops*. Commonwealth Bureau of Helminthology, St Albans, UK, pp. 164–183.
McGawley, E.C. and Overstreet, C. (1995) Reproduction and pathological variation in populations of *Rotylenchulus reniformis*. *Journal of Nematology* 27, 508.
Mcharo, M. and LaBonte, D.R. (2010) Multivariate selection of AFLP markers associated with β-carotene in sweetpotatoes. *Euphytica* 175, 123–172.
Mcharo, M., LaBonte, D.R., Clark, C., Hoy, M. and Oard, J.H. (2005) Molecular marker variability for southern root-knot nematode resistance in sweetpotato. *Euphytica* 144, 125–132.
Mechelke, W. (1986) *Wertung von Umwelten für die Selektion in der Zuckerrübenzüchtung*. Arbeitstagung AG der Saatzuchtleiter, Gumpenstein 191–210. Bundesanstalt für alpenländische Landwirtschaft, Liezen, Austria. (in German)
Melchinger, A.E. and Gumber, R.K. (1998) Overview of heterosis and heterotic groups in agronomic crops. In: *Concepts and Breeding of Heterosis in Crop Plants*. Crop Science Society of America (CSSA) Special Publication no. 25. CSSA, Madison, Wisconsin.
Mi, X., Utz, H.F., Technow, F. and Melchinger, A.E. (2014) Optimizing resource allocation for multistage selection in plant breeding with R Package Selectiongain. *Crop Science* 54, 1413–1418.
Miano, D.W., LaBonte, D.R. and Clark, C. (2008) Identification of molecular markers associated with sweet potato resistance to sweet potato virus disease in Kenya. *Euphytica* 160, 15–24.
Miles, J.W. (2007) Apomixis for cultivar development in tropical forage grasses. *Crop Science* 47(S3), S238–S249.
Milgram, M., Cohen, J. and Loebenstein, G. (1996) Effects of sweetpotato feathery mottle virus and sweetpotato sunken vein virus on sweetpotato yields and rate of reinfection on virus-free planting material in Israel. *Phytoparasitica* 24, 189–193.
Moar, W.J., Mwanga, R.O.M., Odongo, B., Ekobu, M., Solera, M. and Ghislain, M. (2007) Progress towards engineering resistance to weevil in sweetpotato using Bt gene technology. In: *Biotechnology, Breeding and Seed Systems for African Crops*. Proceedings of a conference held on 26–29 March, Maputo, Mozambique. The Rockefeller Foundation, New York, p. 162.
Moll, R.H., Lonnquist, J.H., Vélez Fortuno, J. and Johnson, E.C. (1965) The relationship of heterosis and genetic divergence in maize. *Genetics* 52, 139–144.
Montenegro, Á., Avis, C. and Weaver, A. (2008) Modeling the prehistoric arrival of the sweet potato in Polynesia. *Journal of Archaeological Science* 35, 355–367.
Mullen, M.A., Jones, A., Paterson, D.R. and Boswell, T.E. (1985) Resistance in sweet potatoes to the sweetpotato weevil, *Cylas formicarius elegantulus* (Summers). *Journal of Entomological Science* 20(83), 345–350.
Mwanga, R.O.M. (2001) Nature of resistance and response of sweetpotato to sweetpotato virus disease. PhD thesis, North Carolina State University, Raleigh, North Carolina.
Mwanga, R.O.M., Yencho, C.G.C. and Moyer, J.W. (2002a) Diallel analysis of sweetpotatoes for resistance to sweetpotato virus disease. *Euphytica* 128, 237–248.
Mwanga, R.O.M., Moyer, J., Zhang, D., Carey, E.E. and Yencho, G.C. (2002b) Nature of resistance to sweetpotato virus diseases. *Acta Horticulturae* 583, 113–119.
Mwanga, R.O.M., Odongo, B., Turyamureeba, G., Alajo, A., Yencho, G.C., Gibson, R.W., Smit, N.E.J.M. and Carey, E.E. (2003) Release of six sweet potato cultivars ('NASPOT1' to 'NASPOT6') in Uganda. *HortScience* 38(3), 475–476.
Mwanga, R.O.M., Odongo, B., Niringiye, C.N., Alajo, A., Kigozi, B., Makumbi, R., Lugwana, E., Namakula, J., Mpembe, I., Kapinga, R., Lemaga, B., Nsumba, J., Tumwegamire, S. and Yencho, C.G. (2009) 'NASPOT 7', 'NASPOT 8', 'NASPOT 9 O', 'NASPOT 10 O', and 'Dimbuka-Bukulula' Sweetpotato. *HortScience* 44, 828–832.
Nair, G.M., Nair, V.M. and Sreedharan, C. (1996) Response of sweet potato to phasic stress irrigation in summer rice fallows. *Journal of Root Crops* 22, 45–49.
Narayanin, C.D., Thompson, A.H. and Slabbert, M.M. (2010) Greenhouse screening of South African sweet potato cultivars and breeding lines for tolerance to Alternaria blight caused by *Alternaria bataticola*. *African Plant Protection* 16, 10–13.

Nishiyami, I., Miyazaki, T. and Sakamoto, S. (1975) Evolutionary autoploidy in sweet potato (*Ipomoea batatas* (L.) Lam.) and its progenitors. *Euphytica* 24, 197–208.

Noda, T., Takahata, Y., Sato, T., Suda, I., Morishita, T., Ishiguro, K. and Yamakawa, O. (1998) Relationships between chain length distribution of amylopectin and gelatinization properties within the same botanical origin for sweet potato and buckwheat. *Carbohydrate Polymers* 37, 153–158.

Norman, M.J.T., Pearson, C.J. and Searle, P.G.E. (1984) *The Ecology of Tropical Food Crops*. Cambridge University Press, Cambridge.

O'Brien, P.J. (1972) The sweet potato: its origin and dispersal. *American Anthropologist* 74, 343–365.

Ocloo, F.C.K., Otoo, G., Nkansah, R.M., Mahami, T., Odonkor, S. and Quayson, E.T. (2011) Functional and physicochemical characteristics of starch obtained from gamma-irradiated sweet potato (*Ipomea batatas* L.). *Journal of Agriculture and Food Technology* 1(7), 116–122.

Okada, Y., Saito, A., Nishiguchi, M., Kimura, T., Mori, M., Hanada, K., Sakai, J., Miyazaki, C., Matsuda, Y. and Murata, T. (2001) Virus resistance in transgenic sweetpotato [*Ipomoea batatas* L. (Lam.)] expressing the coat protein gene of sweet potato feathery mottle virus. *Theoretical and Applied Genetics* 103, 743–751.

Onyekwere, P.S.N. and Nwinyi, S.C.O. (1989) Water requirements of sweetpotato (*Ipomoea batatas*). In: *Annual Report*. National Root Crops Research Institute, Umudike, Nigeria, pp. 48–51.

Orjeda, G., Freyre, R. and Iwanaga, M. (1991) Use of *Ipomoea trifida* germ plasm for sweet potato improvement. 3. Development of 4x interspecific hybrids between *Ipomoea batatas* (L.) Lam. (2n = 6x = 90) and *I. trifida* (H.B.K) G. Don. (2n = 2x = 30) as storage-root initiators for wild species. *Theoretical and Applied Genetics* 83, 159–163.

Osiru, M., Adipala, E., Olanya, O.M., Lemaga, B. and Kapinga, R. (2007) Occurrence and distribution of Alternaria leaf petiole and stem blight on sweetpotato in Uganda. *Plant Pathology Journal (Faisalabad)* 6, 112–119.

Oswald, A., Kapinga, R., Lemaga, B., Ortiz, O., Kroschel, J. and Lynam, J. (2009) Integrated crop management. In: Andrade, M., Barker, I., Cole, D., Dapaah, H., Elliott, H., Fuentes, S., Grüneberg, W.J., Kapinga, K., Kroschel, J., Labarta, R., Lemaga, B., Loechl, C., Low, J., Lynam, J., Mwanga, R., Ortiz, O., Oswald, A. and Thiele, G. *Unleashing the Potential of Sweetpotato in Sub-Saharan Africa: Current Challenges and Way Forward. Working Paper* 2009-1. International Potato Center (CIP), Lima, pp. 130–153.

Overstreet, C. (2009) Nematodes. In: Loebenstein, G. and Thottappilly, G. (eds) *The Sweetpotato*. Springer Science + Business Media BV, Houten, The Netherlands, pp. 135–159.

Overstreet, C. and McGawley, E.C. (2000) Geographical distribution of reniform nematode in Louisiana. In: *Proceedings of the Beltwide Cotton Conference*, January 2000, San Antonio, Texas. Beltwide Cotton, San Antonio, Texas, pp.168–171.

Ozias-Akins, P. and Jarret, R.L. (1994) Nuclear DNA content and ploidy levels in the genus *Ipomoea*. *Journal of the American Society of Horticultural Science* 119, 110–115.

Padmaja, G. (2009) Uses and nutritional data of sweetpotato. In: Loebenstein, G. and Thottappilly, G. (eds) *The Sweetpotato*. Springer Science + Business Media BV, Houten, The Netherlands, pp. 189–234.

Patterson, H.D. (1997) Analysis of series of variety trials. In: Kempton, R.A. and Fox, P.N. (eds) *Statistical Methods for Plant Variety Evaluation*. Chapman & Hall, London, pp 139–161.

Pesek, J. and Baker, R.J. (1969) Desired improvement in relation to selection indices. *Canadian Journal of Plant Science* 49, 803–804.

Peters, D. (2004) Use of sweet potato in pig production in Asia: agricultural and socio-economic aspects. animalscience.com Reviews, No. 4. *Pig News and Information* 25(1), 25N–34N.

Pfeiffer, W.H. and McClafferty, B. (2007) HarvestPlus: breeding crops for better nutrition. *Crop Science* 47, 88–105.

Pimentel, D., Berger, B., Filiberto, D., Newton, M., Wolfe, B., Karabinakis, E., Clark, S., Poon, E., Abbett, E. and Nandagpal, S. (2004) Water resources: agricultural and environmental issues. *BioScience* 54(10), 909–918.

Poole, C.F. (1955) Sweet potato genetic studies. *Technical Bulletin* 27. Hawaii Agricultural Experiment Station, Honolulu, Hawaii.

Qiu, H.G., Huang, J.K. and Yang, J. (2010) Bioethanol development in China and the potential impacts on its agricultural economy. *Applied Energy* 87(1), 76–83.

Ravi, V. and Indira, P. (1995) Investigation on the physiological factors limiting yield potential in sweetpotato under drought stress. In: *Annual Report*. Central Tuber Crops Research Institute, Trivandrum, India, pp. 37–38.

Ravi, V. and Indira, P. (1996a) Investigation on the physiological factors limiting yield potential in sweetpotato under drought stress. In: *Annual Report*. Central Tuber Crops Research Institute, Trivandrum, India, pp. 23–24.

Ravi, V. and Indira, P. (1996b) Anatomical studies on tuberization in sweetpotato under water deficit stress and stress free conditions. *Journal of Root Crops* 22, 105–111.

Ravi, V. and Indira, P. (1999) Crop physiology of sweetpotato. *Horticultural Reviews* 23, 277–338.

Richardson, P.H., Jeffcoat, R. and Shi, Y.-C. (2000) High-amylose starches: from biosynthesis to their use as food ingredients. *MRS Bulletin* 25, 20–24.

Robinson, A.F., Inserra, R.N., Caswell-Chen, E.P., Vovlas, N. and Troccoli, A. (1997) *Rotylenchulus* species: identification, distribution, host ranges, and crop plant resistance. *Nematropica* 27, 127–180.

Rolston, L.H., Barlow, T., Hernandez, T. and Nilakhe, S.S. (1979) Field evaluation of breeding lines and cultivars of sweet potato for resistance to the sweetpotato weevil. *HortScience* 14, 634–635.

Roullier, C., Rossel, G., Tay, D., McKey, D. and Lebot, V. (2011) Combining chloroplast and nuclear microsatellites to investigate origin and dispersal of New World sweet potato landraces. *Molecular Ecology* 20, 3963–3977.

Roullier, C.A., Duputié, A., Wennekes, P., Benoit, L., Bringas, V.M.F., Rossel, G., Tay, D., McKey, D. and Lebot, V. (2013a) Disentangling the origins of cultivated sweet potato (*Ipomoea batatas* (L.) Lam.). *PLoS One* 8(5), e62707.

Roullier, C., Kambouo, R., Paofa, J., McKey, D. and Lebot, V. (2013b) On the origin of sweet potato (*Ipomoea batatas* (L.) Lam.) genetic diversity in New Guinea, a secondary centre of diversity. *Heredity* 110, 594–604.

Rukarwa, R.J., Mukasa, S.B., Sefasi, A., Ssemakula, G., Mwanga, R.O.M. and Ghislain, M. (2013) Segregation analysis of cry7Aa1 gene in F1 progenies of transgenic and non-transgenic sweetpotato crosses. *Journal of Plant Breeding and Crop Science* 5, 209–213.

Sano, Z. and Iwahori, H. (2005) Regional variation in pathogenicity of *Meloidogyne incognita* populations on sweetpotato in Kyushu Okinawa, Japan. *Journal of Nematology* 35, 1–12.

Sasser, J.N. and Carter, C.C. (1982) Root-knot nematodes (*Meloidogyne* spp.): identification, morphological and physiological variation, host range, ecology, and control. In: Riggs, R.D. and Editorial Committee (eds) *Nematology in the Southern Region of the United States. Southern Cooperative Series Bulletin* 276, 21–32.

Sasser, J.N. and Freckman, D.W. (1987) A world perspective on nematology: the role of society. In: Veech, J.A. and Dickson, D.W. (eds) *Vistas on Nematology: A Commemoration of the Twenty-fifth Anniversary of the Society of Nematologists*. Society of Nematologists, Inc., Hyattsville, Maryland, pp. 7–14.

Sattelmacher, B., Horst, W.J. and Becker, H.C. (1994) Factors that contribute to genetic variation for nutrient efficiency of crop plants. *Z. Pflanzenernähr Bodenk* 157, 215–224.

Schafleitner, R., Tincopa, L.R., Palomino, O., Rossel, G., Robles, R.F., Alagon, R., Rivera, C., Quispe, C., Rojas, L., Pacheco, J.A., Solis, J., Cerna, D., Kim, J.Y., Hou, J. and Simon, R. (2010) A sweetpotato gene index established by *de novo* assembly of pyrosequencing and Sanger sequences and mining for gene-based microsatellite markers. *BMC Genomics* 11, 604.

Schneider, J. (1995) Farmer practices and sweet potato diversity in highland New Guinea. In: Schneider, J. (ed.) *Indigenous Knowledge in Conservation of Crop Genetic Resources*. Proceedings of an International Workshop held in Cisarua, Bogor, Indonesia. International Potato Center – East, Southeast Asia and the Pacific (CIP-ESEAP), Bogor, Indonesia, pp. 63–70.

Schnell, F.W. (1982) A synoptic study of the methods and categories of plant breeding. *Z. Pflanzenzüchtung* 89, 1–18.

Scott, G. (1991) Sweet potato as animal feed in developing countries: present patterns and future perspectives. Paper presented at the Food and Agriculture Organization of the United Nations (FAO) Experts Consultation on the Use of Roots, Tubers, Plantains and Bananas in Animal Feeding, 21–25 January 1991, Centro International de Agricultura Tropical (CIAT), Cali, Colombia.

Sefasi, A., Ghislain, M., Kiggundu, A., Ssemakula, G., Rukarwa, R. and Mukasa, S.B. (2013) Thidiazuron improves adventitious bud and shoot regeneration in recalcitrant sweetpotato. *African Crop Science Journal* 21(1), 85–95.

Shimada, T., Otani, M., Hamada, T. and Kim, S.-H. (2006) Increase of amylose content of sweetpotato starch by RNA interference of the starch branching enzyme II gene (IbSBEII). *Plant Biotechnology* 23, 85–90.

Solis, J. and Grüneberg, W.J. (2008) Functional genomics for the elucidation of beta-carotene and starch metabolism in sweetpotato. Final Project Report. International Potato Center (CIP), Lima. Available at: http://sweetpotatobreeder.com/SPmarkers.html (accessed 25 March 2015).

Sorensen, K.A. (2009) Sweetpotato insects: identification, biology and management. In: Loebenstein, G. and Thottappilly, G. (eds) *The Sweetpotato*. Springer Science + Business Media BV, Houten, The Netherlands, pp. 161–188.

Sseruwu, G. (2012) Breeding of sweetpotato (*Ipomoea batatas* (L.) Lam.) for storage root yield and resistance to Alternaria leaf petiole and stem blight (*Alternaria* spp.) in Uganda. PhD thesis, University of KwaZulu-Natal, Pietermaritzburg, Republic of South Africa.

Stathers, T.E., Rees, D., Kabi, S., Mbilinyi, L., Smit, N., Kiozya, H., Jeremiah, S., Nyango, A. and Jeffries, D. (2003) Sweetpotato infestation by *Cylas* spp. in East Africa. 1: Cultivar differences in field infestation and the role of plant factors. *International Journal of Pest Management* 49, 131–140.

Stevenson, P.C., Muyinza, H., Hall, D.R., Porter, E.A., Farman, D.I., Talwana, H. and Mwanga, R.O.M. (2009) Chemical basis for resistance in sweetpotato *Ipomoea batatas* to the sweetpotato weevil *Cylas puncticollis*. *Pure Applied Chemistry* 81, 141–151.

Story, R.N., Hammond, A., Murray, J., Rolston, R.H. and Labonte, D. (1996) Selection for host plant resistance in sweetpotatoes to the sweetpotato weevil. In: *Sweet Potato Research*. Louisiana Agricultural Experiment Station (LAES) Mimeo series No. 117. LAES, Louisiana State University Agricultural Center, Louisiana, pp. 71–79.

Sung, J.M. (1985a) Studies on physiological response to water stress in sweetpotato. I. The stomatal carbon assimilation of sweetpotato leaves. *Journal of the Agricultural Association of China* 129, 42–49.

Sung, J.M. (1985b) Studies on physiological response to water stress in sweetpotato. II. Osmotic adjustment in sweetpotato leaves. *Journal of the Agricultural Association of China* 129, 50–55.

Tairo, F., Musaka, S.B., Jones, R.A.C., Kullaia, A., Rubaihayo, P.R. and Valkonen, J.P.T. (2005) Unravelling the genetic diversity of the three main viruses involved in sweetpotato virus disease (SPVD), and its practical implications. *Molecular Plant Pathology* 6, 199–211.

Tarn, T.R., Tai, G.C.C., De Jong, H., Murphy, A.M. and Seabrook, J.E.A. (1992) Breeding potatoes for long-day, temperate climates. *Plant Breeding Reviews* 9, 217–332.

Thompson, A.H., Narayanin, C.D., Smith, M.F. and Slabbert, M.M. (2011) A disease survey of Fusarium wilt and Alternaria blight on sweet potato in South Africa. *Crop Protection* 30, 1409–1413.

Timberlake, C.F. and Henry, B.S. (1988) Anthocyanins as natural food colorants. *Progress in Clinical and Biological Research* 280, 107–121.

Togari, Y. (1950) A study of tuberous root formation in sweetpotato. *Bulletin of the National Agricultural Experimental Station Tokyo* 68, 1–96.

Trabalho de Inquérito Agrícola (TIA) (2012) *National Sample Survey of Agriculture in Mozambique*. Production data. Ministry of Agriculture and Rural Development, Maputo, Mozambique.

Tsuno, Y. (1971) Dry matter production of sweet potatoes and yield increasing techniques. *Fertilite* 38, 3–21.

Tsuno, Y. and Fujise, K. (1965) Studies on the dry matter production of sweetpotato. VII. The internal factors influencing photosynthetic activity of sweetpotato leaf. *Proceedings of the Crop Science Society of Japan* 28, 230–235.

Tsunoda, S. (1959) A developmental analysis of yield ability in varieties of field crops. II The assimilation system of plants as affected by the form, direction and arrangement of single leaves. *Japanese Journal of Breeding* 9, 237–244.

Tumwegamire, S. (2011) Genetic variation, diversity and genotype by environment interactions of nutritional quality traits in East African sweetpotato. PhD thesis, Makerere University, Kampala, Uganda.

Tumwegamire, S., Kapinga, R., Rubaihayo, P.R., LaBonte, D.R., Grüneberg, W.J., Burgos, G., zum Felde, T., Carpio, R., Pawelzik, E. and Mwanga, R.O.M. (2011a) Starch, sucrose, β-carotene, iron, zinc, calcium, and magnesium in East African sweetpotato [*Ipomoea batatas* (L.) Lam.] germplasm. *HortScience* 46(3), 348–357.

Tumwegamire, S., Rubaihayo, P.R., LaBonte, D.R., Diaz, F., Kapinga, R., Mwanga, R.O.M. and Grüneberg, W.J. (2011b) Genetic diversity in white- and orange-fleshed sweetpotato farmer varieties from East Africa evaluated by simple sequence repeat markers. *Crop Science* 51, 1132–1142.

Ukoskit, K., Thomson, P.G., Watson, C.E. and Lawrence, G.W. (1997) Identifying a randomly amplified polymorphic DNA (RAPD) marker linked to a gene for root-knot nematode resistance in sweetpotato. *Journal of the American Society of Horticultural Science* 122(6), 818–821.

United Nations System Standing Committee on Nutrition (UN-SCN) (2004) *Nutrition for Improved Development Outcomes.* Fifth Report on the World Nutrition Situation. UN-SCN, Geneva, Switzerland.

United States Department of Agriculture (USDA) (2015) Commodity Highlight: Sweet potatoes. *Economic Research Service Situation and Outlook Report VGS-355-SA1,* Washington DC.

Untiveros, M., Fuentes, S. and Salazar, L.F. (2006) Synergistic interaction of sweetpotato chlorotic stunt virus (*Crinivirus*) with carla-, cucumo-, ipomo-, and potyviruses infecting sweetpotato. *Plant Disease* 91, 669–676.

Uritaini, I., Saito, T., Honda, H. and Kim, W. (1975) Induction of furanoterpenoids in sweetpotato roots by the larval components of the sweetpotato weevils. *Agricultural and Biological Chemistry* 37, 1857–1862.

Utz, H.F. (1969) *Mehrstufenselektion in der Pflanzenzüchtung*. Arbeiten der Universität Hohenheim, Vol. 49. Verlag Eugen Ulmer, Stuttgart, Germany. (in German)

Utz, H.F. (1984) Calculating and maximizing the gain from selection. *Vorträge Pflanzenzüchtung* 7, 30–40.

Van Heerden, P.D.R. and Laurie, R. (2008) Effects of prolonged restriction in water supply on photosynthesis, shoot development and storage root yield in sweet potato. *Physiologia Plantarum* 134, 99–109.

Vilsoni, F. and Heinlein, M. (1982) Influence of initial inoculum levels of the reniform nematode on the growth of mung, pawpaw, pigeon pea, and sweet potato. *Fiji Agricultural Journal* 44, 67–70.

Wang, H. (1975) The breeding and cultivation of sweet potatoes. *Technical Bulletin* 26. Asian and Pacific (ASPAC) Food and Fertilizer Technology Center, Chiayi Agricultural Experiment Station, Taiwan.

Wang, M., Shi, Y., Xia, X., Li, D. and Chen, Q. (2013) Life-cycle energy efficiency and environmental impacts of bioethanol production from sweet potato. *Bioresource Technology* 133, 285–292.

Wang, Y.P., Liu, Q.C., Li, A.X., Zhai, H., Zhang, S.S. and Liu, B.L. (2003) *In vitro* selection and identification of drought-tolerant mutants in sweetpotato. *Agricultural Sciences in China* 2, 1314–1320.

Weaver, J.E. and Bruner, W.E. (1927) *Root Development of Vegetable Crops*, 1st edn. McGraw-Hill Book Company, New York.

Weber, W.E. (1979) Number and size of cross progenies from a constant total number of plants manageable in a breeding program. *Euphytica* 28, 453–456.

Wilcox, L.V. (1960) *Boron Injury to Plants*. *Agriculture Information Bulletin*, no. 211. Research Service, United States Department of Agriculture (USDA), Washington, DC.

Wilson, L.A. and Lowe, S.B. (1973) The anatomy of the root system in West Indian sweetpotato (*Ipomoea batatas* (L.) Lam.) cultivars. *Annals of Botany* 37, 633–643.

Wolfe, G.W. (1991) The origin and dispersal of the pest species of *Cylas* with a key to the pest species groups of the world. In: Jansson, R.J. and Raman, K.V. (eds) *Sweet Potato Pest Management – a Global Perspective*. Westview Press, Boulder, Colorado, pp. 13–43.

Woolfe, J.A. (1992) *Sweet Potato: an Untapped Food Resource*. Cambridge University Press, Cambridge.

Wricke, G. and Weber, W.E. (1986) *Quantitative Genetics and Selection in Plant Breeding*. de Gruyter, Berlin.

Xie, S.Q., Feng, Y.W. and Xi, L.G. (1998) On the drought resistance of local sweetpotato germplasm resources from Yunnan. *Crop Genetic Resources (China)* 1, 31–32.

Yanfu, Y., Jialan, T., Yunchu, Z. and Ruilian, Q. (1989) Breeding for early-maturing sweetpotato varieties. In: Mackay, K.T., Palomer, M.K. and Sanieo, R.T. (eds) *Sweetpotato Research and Development for Small Farmers*. SEAMEO-SEARCA, Laguna, The Philippines, pp. 67–82.

Yanggen, D. and Nagujja, S. (2005) The use of orange-fleshed sweetpotato to combat vitamin A deficiency in Uganda. A study of varietal preferences, extension strategies and post-harvest utilization. *Working Paper* No. 2005-1. International Potato Center (CIP), Lima, 80 pp.

Yen, C.T., Chu, C.V. and Sheng, C.L. (1964) Studies on the drought resistance of sweetpotato varieties. *Crop Science (China)* 3, 183–190.

Yen, D.E. (1974) The sweetpotato in Oceania. *Bishop Museum Bulletin, Honolulu* 236, 1–389.

Yen, D.E. (1976) The sweet potato. In: Simmonds, N.W. (ed.) *Evolution of Crop Plants*. Longman, New York, pp. 42–45.

Yen, D.E. (1982) Sweet potato in historical perspective. In: Villareal, R.I. and Griggs, T.D. (eds) *Sweet Potato, Proceedings of the First International Symposium*. Asian Vegetable Research and Development Center (AVRDC) Publication No. 82-172. AVRDC, Tainan, Taiwan, pp. 17–30.

Yoshida, T. (1985) Correlation between successive yield tests for agronomic characters in sweet potato. *Japanese Journal of Breeding* 35, 204–208.

Yoshida, T., Hozyo, Y. and Murata, T. (1970) Studies on the development of tuberous roots in sweet potato (*Ipomoea batatas*, Lam. Var. edulis, Mak.). The effect of deep placement of mineral nutrients on the tuber yield of sweetpotato. *Proceedings of the Crop Science Society of Japan* 39, 105–110.

You, M.K., Hur, C.G., Ahn, Y.S., Suh, M.C., Shin, J.S. and Bae, J.M. (2003) Identification of genes possibly related to storage root induction in sweetpotato. *FEBS Letters* 536, 101–105.

Zhang, D.P., Collins, W.W. and Belding, S. (1993) Improving sweet potato starch digestibility for animal feeding. *HortScience* 18, 325–326.

Zhang, D., Cervantes, J., Huamán, Z., Carey, E. and Ghislain, M. (2000) Assessing genetic diversity of sweet potato (*Ipomoea batatas* (L.) Lam.) cultivars from tropical America using AFLP. *Genetic Resoures and Crop Evolution* 47, 659–665.

Zhang, D., Rossel, J.G., Kriegner, A. and Hijmans, R. (2004) AFLP assessment of diversity in sweetpotato from Latin America and the Pacific region: its implications on the dispersal of the crop. *Genetic Resoures and Crop Evolution* 51, 115–120.

Zhang, M.S., Peng, Z.H., Xie, B., Tan, F., Zhang, Q.T., Fu, Y.F., Yang, C.X. and Yang, Y.H. (2004) Relationship between water loss rate of cut leaves and osmotic regulators under water stress and drought resistance in sweetpotato. *Scientia Agricultura Sinica* 37, 152–156.

Zhang, M.S., Liu, Z., Qi, J.L., Zhang, L.X. and Yang, Y.H. (2005) Methods of comprehensive evaluation for drought resistance in sweetpotato cultivars. *Journal of Tropical and Subtropical Botany* 13, 469–474.

Zhang, S., Zhang, S., Wang, H. and Chen, Y. (2006) Characteristics of sweet potato stem nematode in China. *Acta Phytopathologica Sinica* 36, 22–27.

Zhong, R.S. (1991) Studies on the source–sink relationship in sweetpotato. *Jiangsu Journal of Agricultural Science* 7, 44–48.

zum Felde, T., Burgos, G., Espinoza, J., Bonierbale, M. and Grüneberg, W.J. (2007) Analysis of carotenoid, iron, zinc and calcium content of potato (*Solanum phureja*) and sweet potato (*Ipomoea batatas*) using near infrared reflectance spectroscopy (NIRS). In: *Proceedings of the 13th International Conference on Near Infrared Spectroscopy (13th ICNIRS)*, 15–21 June 2007, Umeå, Sweden and Vasa, Finland. Special issue of the *Journal of Near Infrared Spectroscopy* (*JNIRS*). Available at: http://www.impublications.com/nir/page/NIR-2007 (accessed 15 September 2015).

Appendix 1: Released/Launched Sweetpotato Varieties

This appendix provides details about released/launched sweetpotato varieties over the past two decades by 15 classification variables, namely: (i) country; (ii) year of release/launch; (iii) variety type; (iv) storage root flesh colour; (v) taste type; (vi) adaptation range; (vii) CIP-code; (viii) maturity time; (ix) resistance to SPVD; (x) resistance to weevil; (xi) resistance to *Fusarium*; (xii) resistance to *Alternaria*; (xiii) resistance to nematodes; (xvi) abiotic stress resistance; and (xv) comments about special uses, resistances, names or development.

Abbreviations used for classification variables for released/launched varieties

1. Country: BD, Bangladesh; BF, Burkina Faso; BR, Brazil; BU, Burundi; CN, People's Republic of China; CU, Cuba; GH, Ghana; IN, India; JP, Japan; KE, Kenya; KR, Republic of Korea; MG, Madagascar; MW, Malawi; MZ, Mozambique; NG, Nigeria; PE, Peru; PG, Papua New Guinea; PH, Republic of the Philippines; RSA, Republic of South Africa; RW, Rwanda; TL, East Timor; TW, Taiwan; TZ, Tanzania, UG, Uganda; US-NC, USA North Carolina; US-LS, USA Louisiana; ZA, Republic of South Africa; ZM, Zambia.

2. Year of release/launch: 1992–2013.

3. Variety type: BL, breeding line; FV, farmer variety; MV, modern variety; or if not available '.' for missing value.

4. Storage root flesh colour: C, cream; DO, deep orange; DPU, deep purple; DY, deep yellow; IO, intermediate orange; LO, light orange; LPU, light purple; O, orange; OY, orange yellow; PO, pale orange; PU, purple; PY, pale yellow; W, white; Y, yellow.

5. Taste type: DS, dry and starchy; HD, high dry matter; HS, high starch; HTS, high total sugars; LD, low dry matter; LTS, low total sugars; MD, medium dry matter; MDS, moderately dry and starchy; MMS, moderately moist and sweet; MSS, medium starch and sweet; MST, moist and sweet taste; MTS, medium total sugar; SD&MS, semi-dry and medium sweet; SD&SS, slight dry and semi-sweet; SS, sweet and starchy, ST, starchy taste.

6. Adaptation: CFGS, Coastal–Forest (CF) transition and Guinea Savannah (GS) of West Africa; DST, dry subtropics; HLA, highland adaptation; HRA, high rainfall areas; HST, humid subtropical; HTL, hot tropical lowlands; SGS, short grassland savannah; SSZ, Sudano-Sahelian-Zone; STDL, subtropical dry land; TDL, temperate dry land; TGS, tall grassland savannah; TRDL&RF, tropical dry land and rice field; MUMZA, mid- and upper midland zone adaptation; WA, wide adaptation; WAD, wide adaptation to dry lands.
7. CIP-code: number or if not available '.' for missing value.
8. Maturity time: EM, early maturing in months (mths); MM, medium maturing in months (mths); LM, late maturing in months (mths).
9. Resistance to SPVD: MRVD, moderate resistance to SPVD; RVD, resistant to SPVD; SVD, susceptible to SPVD; TVD, tolerant to SPVD; or if not available '.' for missing value.
10. Resistance to weevil: MRW, moderate resistance to weevils; RW, resistant to weevils; SW, susceptible to weevils; or if not available '.' for missing value.
11. Resistance to Fusarium wilt (*Fusarium oxysporum*): RF, resistant to Fusarium wilt; SF, susceptible to Fusarium wilt; TF, tolerant to Fusarium wilt; or if not available '.' for missing value.
12. Resistance to *Alternaria*: MRAB, moderately resistant to Alternaria blight; RAB, resistant to Alternaria blight; SAB, susceptible to Alternaria blight; TAB, tolerant to *Alternaria bataticola* stem blight; or if not available '.' for missing value.
13. Resistance to nematodes: RN, resistant to nematodes; SN, susceptible to nematodes; or if not available '.' for missing value.
14. Abiotic stress resistance: DT, drought tolerant; TMD, tolerates mild dry spells; TS, tolerant to salinity; or if not available '.' for missing value.
15. Comments about special uses, names, resistances or parental material: DC&T, direct consumption and table use; DPU, dual-purpose use as a food and feed; EBA, excellent to boil as 'ampesi'; EFB, excellent for baby-foods and FDP fortification of dairy products; EFC, excellent for fried chips; EFF, excellent for French fries; EFS, excellent form and size for fresh market; EFU, excellent for *fufu*; FGT&MDMF, fairly good taste and moderate dry mouthfeel boiled roots; HF, heavy foliage; IT, industrial type; IU, industrial use (starch); LA, low adoption; OP, open pollination; RFB, moderate resistance to the sweetpotato flea beetle (*Chaetocnema confinis*); RSSR, resistant to Streptomyces soil rot (*Streptomyces ipomoeae*); WADLZ, wide adaptation to dry land zones; or if not available '.' for missing value.

Table A1. Name of released/launched sweetpotato varieties from 1992 to 2013.

Americas

Brazil: **Lapar-69** [BR 1999 MV O MST . . MM . . . . . . EFS], **Lapar-70** [BR 1999 MV W SS . . MM . . . . . . DPU&EFS], **Coquinho** [BR 2000 MV C HS . . MM . . . . . . EFS], **Princesa** [BR 2000 MV C HS . . MM . . . . RN . PU&DEFS&HF], **Brazlândia Roxa** [BR 2000 MV C HS . . MM . . . . . . EFS], **Brazlândia Rosada** [BR 2000 MV C DS . . MM . . . . . . EFS], **Brazlândia Branca** [BR 2000 MV C DS . . MM . . . . . . EFS], **Beauregard** [BR 2010 MV O MS . . . . EFS], **BRS Rubissol** [BR 2011 MV C SS . . MM . . . . . . EFS&IU], **BRS Cuia** [BR 2011 MV C . . . MM . . . . . . EFS&IU], **BRS Amélia** [BR 2011 MV IO MST . . MM . . . .], **SCS-367 Favorita** [BR 2011 MV O . . . . EFS], **SCS-368 Ituporanga**[BR 2011 MV C . . . .], **SCS-369 Águas Negras** [BR 2011 MV C . . . . EFS].

Peru: **Costanero** [PE 1992 MV LO MS TDL CIP-187016.2 . . . . . . TS.(P: DLP339 x PC_SALT87)], **Yarada** [PE 1992 MV C MS TDL CIP-187018.1 . . . . . . TS. .(P: DLP341 x PC_SALT87)], **Nacional** [PE 1992 MV W MST&HS TDL CIP-187003.1 . . . . . . TS IU (P: RCBIT-57 x PC_SALT87)], **Tacna** P[PE 1992 MV C MST TDL CIP-187019.1 . . . . . . TS .(P: CRBIN-15 x PC_SALT87)], **Caplina** [PE 1992 MV C MST&HS TDL CIP-187016.1 . . . . . TS .(P: DLP339 x PC_SALT87)], **Atacama** [PE 1992 MV C MST TDL CIP-187020.1. . . . . . TS .(P: RCBIN-17 x PC_SALT87)], **INIA-100** [PE 2001 MV DO MS TDL CIP-192033.50 . . . . . SN . .(P: NCSU 240 x PC92_5NACIONAL)], **Milagrosa** [PE 2000 FV LO HS

Continued

Table A1. Continued.

TDL&WAD . . . . DPU], **Mejorada** [PE 2005 MV LO HS TDL&WAD . . . . DPU], **Adriano** [PE 2010 MV W HS TDL CIP-105228.1 . . . . IU(P: SR02.039, CIP-102062.2 x TANZANIA, CIP-440166)]; **Alexander** [PE 2010 MV Y HS TDL CIP-105240.1 . . . . IU (P: SR02.132, CIP-102022.3 x TANZANIA, CIP-440166)], **Arne** [PE 2010 MV O MST TDL CIP-105086.1 . RSVD . . . . . .(P: SR02.178, CIP-102028.3 x INA-100, CIP-102033.5)], **Benjamin** [PE 2010 MV DO MS TDL CIP-105085.2 . SVD . . . . . EFS (P: SR02.177, CIP-102025.3 x INIA-100, CIP-102033.5)].

USA: **Beauregard** [US-LS 1986 MV O MS TDL&HTL CIP-440132 . SVD . RF . SN . RSSR&OP], **Carolina Ruby** [US-NC 1992 MV DO LD&MS HST . . . . RF . SN . RSSR&RFB& OP], **Carolina Rose** [UA-NC 1992 MV DO LD&MS HST . . . . RF . . . OP], **Covington** [UA-NC 2005 MV DO LD&MS HST . . . . RF . RN . RSSR&RFB&OP], **Hatteras** [UA-NC 2008 MV DO LD&MS HST . . . . RF . RN . RSSR&OP], **Murasaki-29** [US-LS 2008 W MV HD TDL . . . RW RF . RN . RSSR&OP], **NCPUR06-020** [US-NC 2012 MV PU DS&HD HST . . . . RF . SN . IT], **Bonita** [US-LS 2011 MV W MD TDL&HTL . . . . RF. RN . RSSR].

West Africa

Burkina Faso: **Caromex** [BF 2005 MV LO MDS SSZ CIP-440136 . MRVD . . . . . (from USA)], **199062.1** [BF 2005 MV LO MS SSZ CIP-199062.1. MRVD . . . . . (from PE)].

Ghana: **CRI-Faara** [GH 1998 MV W HD&LTS CFGS IITA-TIS-3017 MM(4mths) . . . . . . EFC&HF&DPU], **CRI-Okumkom** [GH 1998 MV W MD&MTS CFGS IITA-TIS-8266 MM(4mths) MRVD MRW . . . . EFC], **CRI-Sauti** [GH 1998 FV Y HD &LTS CFGS . MM(4mths) TVD . . . . . TMD EFC&(also named **Tanzania** or **Kenya** from MW)], **CRI-Santom-Pona** [GH 1998 MV PY HD&LTS CFGS IITA-TIS-84/0320 MM(4mths) . . . . . . EFC], **CRI-Apomuden** [GH 2005 MV O LD&HTS CFGS CIP-440254 EM(3–4mths) . . . . . . EBF&FDP&(also named **amala Sundari** from IN)], **CRI-Hi-starch** [GH 2005 MV C HD&LTS CFGS . MM(4mths) MRVD MRW . . . . EFC&EFU&IU&(probably Satsumahikari JP via CIP)], **CRI-Ogyefo** [GH 2005 FV W HD&LTS CFGS CIP-440163 MM(4mths) MRVD MRW . . . TMD EFC&EFUIU&HF&DPU&(also named **Mugande** from UG)], **CRI-Otoo** [GH 2005 FV PY HD&MTS CFGS CIP-440034 MM(4mths) MRVD MRW . . . . EFF&HF&DPU&(also named **Mogamba** from BU)], **CRI-Bohye** [GH 2012 MV PO HD&MTS CFGS CIP-199062.1 MM(4mths) MRVD MRW . . . . EFC&EFU&EFF&IU], **CRI-Dadanyuie** [GH 2012 MV W HD&MTS CFGS CIP-440170 MM(4–5mths) MRVD MRW . . . . EFC&EFFIU, **KEMB-37**/from KE], **CRI-Ligri** [GH 2012 MV PY HD&MTS CFGS CIP-400004 MM(4–5mths) MRVD MRW . . . . EFC&EFF&IU&(also named **CEMSA-74-228** from CU)], **CRI-Patron** [GH 2012 FV DY HD&MTS CFGS CIP-440034 MM(4–5mths) MRVD MRW . . . . EFC&HF&DPU&IU&(also named **Mohc** fromBU)].

Nigeria: **NIGIB-01-1** [NG 1992 MV W HD CFGS&WA IITA-TIS-87/0087 . . TW . . . . .], **NIGIB-01-2** [NG 1992 MV W HD CFGS IITA-TIS-8164 . . . . DPU&(used fried & boiled)], **NIGIB-01-3** [NG 1993 MV W HD CFGS IITA-TIS-2532.OP.1.13 . . . . used fried & boiled], **UMUSPO/1** [NG 2012 MV LO HD&SD&MS CFGS CIP-199004.2 MM(4mths) MRVD MRW . . . . OP], **UMUSPO/3** [NG 2012 MV DO MD&SD&SS GS . MM(4mths) MSVD SW . . . . (unknown clone via CIP perhaps CIP-440293)].

East Africa

Kenya: **Mugande** [KE 2001 MV W DS MUMZA . . . . . RAB . . (tracing back to Rwanda)], **SPK-004** [KE 2001 FV LO DS MUMZA CIP-441768 . MRVD . . RAB . . (also named **Kakamega**)], **KSP20** [KE 2001 MV W DS semi-arid_areas . . RVD . . RAB . . .], **Kemb-10** [KE 2001 MV Y DS WA . . MRVD . . RAB . . .], **Mwavuli** [KE 2009 FV W DS MUMZA 566682-03 . MRVD . . RAB . . .], **Bungoma** [KE 2001 FV Y DS MUMZA . LM MRVD . . RAB . . .], **Nyawo** [KE 2004 FV Y DS MUMZA . . MRVD . . RAB . . .]; **K-117** [KE 2009 FV O DS MUMZA . . MRVD . . RAB . . .], **Kabode** [KE 2013 MV O DS . CIP-100200.4 . RVD . . RAB . . (also named **NASPOT-10-O**)], **Vita** [KE 2013 MV O DS MUMZA CIP-100200.3 . RVD . . RAB . . (also named **NASPOT-9-O**)], **Kenspot-1** [KE 2013 MV Y HD HLA . LM MRVD MRW . MRAB . . DC&T], **Kenspot-2** [KE 2013 MV W MD HLA . LM MRVD MRW . MRAB . . DC&T], **Kenspot-3** [KE 2013 MV LO DS HLA . LM MRVD MRW . MRAB . . DC&T], **Kenspot-4** [KE 2013 MV O MDS HLA . LM MRVD MRW . MRAB . . DC&T], **Kenspot-5** [KE 2013 MV O MDS HLA . LM MRVD MRW . MRAB . . DC&T].

Continued

Table A1. Continued.

Tanzania: **Mavuno** [TZ 2002 FV C DS Lake_Zone . . SSPVD . . SAB . . .], **Jitihada** [TZ 2002 FV C MDS . . EM SVD . . SAB . . LA], **Simama** [TZ 2002 FV Y DS WA . . RVD . . RAB . . .], **Ukerewe** [TZ 2002 FV Y WA . . RVD . . RAB . . .], **Sinia-B** [TZ 2002 FV C DS Lake_Zone .. SVD . . . . . .], **Vumilia** [TZ 2002 . Y DS . . . RVD .. RAB . .LA], **Mataya** [TZ 2010 MV Y DS . . . RVD . . RAB . . LA], **Kiegea** [TZ 2010 MV O DS . . . SVD . . SAB LA], **Ejumula** [TZ . FV O DS TGS CIP-443750 . SVD . . MRAB, . . (landrace from Uganda)], **SPK-004** [TZ 2014 FV LO DS Lake_Zone CIP-441768 . MRVD . . MRAB . . (introduced from KE also named **Kakamega**)], **Polysta** [TZ 2014 FV C DS WA . . MRVD . . MRAB . . (also named **Polista**)], **NASPOT-1** [TZ 2014 MV Y DS Lake_Zone CIP-191133.1 . MRVD . . . . . (introduced from UG where it is SAB)].

Rwanda: **Mugande** [RW 1992 MV W DS WA . . RVD . .RAB . . .], **Kwezikumwe** [RW 1993 MV Y DS WA . . RVD . . RAB . . .], **Cacearpedo** [RW 2008 MV O MDS WA . . SVD . . RAB . . .], **SPK-004** [RW 2006 FV LO . CIP-441768 . RVD . .SAB . . (introduced from KE also named **Kakamega**)], **92-062** [RW 2004 MV O MDS WA . . SVD . . SAB . . (also named **Gihangamuhungu**)], **Ukerewe** [RW 2013 . Y DS WA . MM RVD . . RAB . . .], **2000-038** [RW 2008 . Y . . EM RVD . . RAB . . (LA in East & South RW)], **2000-040** [RW 2008 . O . . EM RVD . . RAB (LA in East & South RW)], **2000-024** [RW 2008 . Y DS . . RVD . . RAB . . (LA in East & South RW)], **RW11-17** [RW 2013 . W DS WA . MM . . . . . . (also named **Maryoha**)]; **RW11-1860** [RW . . W DS WA . . . . (also named **Giramata**)], **RW11-4923** [RW 2013 MV W DS WA . MM . . . .], **RW11-2419** [RW 2013 MV W DS WA . MM . . . . . . (also named **Izihirwe**)], **RW11-2560** [RW 2013 MV DO MDS . . MM RVD RW . . . . (also named **Terimbere**)], **RW11-2910** [RW 2013 MV O MDS . . MM RVD RW . . . . (also named **Ndamirabana**)]; **RW97-062** [RW 2013 MV DO MDS . . EM . . . . . . (also named **Gihungumuhungu**)].

Uganda: **Bwanjule** [UG 1995 FV W DS TGS CIP-440168 MM RVD MRW . RAB . DT DC&T], **New-Kawogo** [UG 1995 FV W DS TSGS CIP-441743 LM RVD MRW . SAB . DT DC&T], **Sowola** [UG 1995 MV C DS TGS CIP-441744 EM MRVD SW . RAB . . DC&T], **Tanzania** [UG 1995 FV PY DS SGS CIP-440166 MM MRVD SW . MRAB RN . DC&T], **Wagabolige** [UG 1995 FV W TGS CIP-440168 . RVD MRW SAB RN . .], **Tororo-3** [UG 1995 FV W FV TGS . . MRVD MRW . . . . DC&T], **NASPOT-1** [UG 1999 MV PY DS WA CIP-191133.1 MM MRVD SW . SAB . . DC&T], **NASPOT-2** [UG 1999 MV C MDS TGS CIP-191133.2 MM RVD SW . SAB . . DC&T], **NASPOT-3** [UG 1999 MV C DS TGS CIP-191133.3 LM RVD MRW RAB . . DC&T], **NASPOT-4** [UG 1999 MV PY DS TGS CIP-191133.4 LM RVD MRW SAB . . DC&T], **NASPOT-5** [UG 1999 MV O DS TGS CIP-191133.5 MM MRVD MRW RAB . . DC&T], **NASPOT-6** [UG 1999 MV W DS TGS CIP-191133.6 MM MRVD MRW . RAB . . DC&T], **Kakamega** [UG 2004 FV O DS TGS CIP-441768 MM MRVD SW . MRAB . . DC&T(introduced from KE also named **SPK-004**)], **Ejumula** [UG 2004 FV O DS TGS CIP-443750 . SVD SW . MRAB . . DC&T], **NASPOT-7** [UG 2007 MV O DS TGS CIP-100200.1 MM MRVD SW MRAB . . DC&T], **NASPOT-8** [UG 2007 MV O DS TGS CIP-100200.2 MM MRVD SW . MRAB . . DC&T], **NASPOT-9-O** [UG 2007 MV O DS TGS CIP-100200.3 MM RVD SW . RAB . . DC&T(also named **Vita**)], **NASPOT-10-O** [UG 2007 MV O DS TGS CIP-100200.4 MM RVD SW . RAB . . DC&T(also named **Kabode**)], **Dimbuka-Bukulula** [UG 2007 FV C MDS TSGS CIP-443752 EM SVD SW . MRAB . . DC&T], **NASPOT-11** [UG 2010 MV C DS TSGS CIP-100201 MM RVD SW . RAB . . . (also named **Tomulabula**)], **NASPOT-12-O** [UG 2014 MV O MDS TGS . EM MRVD SW . MRAB . . DPU], **NASPOT-13-O** [UG 2014 MV IO MDS TSGS . MM MRVD . . MRAB . . DC&T].

Southern Africa

Madagascar: **Rotra** [MG 1998 MV Y . MUMZA 188004 . . . . DPU], **Mahafaly** [MG 1997 MV C MD MUMZA CIP-440063 MM . . . . . . DPU&(also called TIS-2544)], **Naveto** [MG 1998 MV C MD MUMZA CIP-440131 MM . . . . . . DPU], **Mahasoa** [MG 1997 MV C MD MUMZA CIP-440034 MM . . . . . . DPU], **Riba** [MG 2000 MV O LD MUMZA CIP-420027 EM . . . . . . DPU], **Rangita** [MG 2000 MV O LD MUMZA CIP-420009 EM . . . . . . DPU], **Mavo** [MG 2002 MV PY HD MUMZA CIP-400011 MM . . . . . . DPU], **Ravo** [MG 2002 MV C HD MUMZA CIP-440004 MM . . . . . . DPU], **Mahavoky** [MG 2003 MV C HD MUMZA CIP-440163 MM . . . . . . DPU], **Mafotra** [MG 2004 MV C HD MUMZA CIP-440170 MM . . . . . . DPU], **Mendrika** [MG 2007 MV O MD MUMZA CIP-199004 EM . . . . . . DPU&EFC], **Bôra** [MG 2008 MV O MD MUMZA CIP-199062.2 EM . . . . . . DPU&EFC], **P162** [MG 2011 MV O HD . . MM . . . . . . DPU&(also called **Zambezi**)], **P163** [MG 2011 MV O HD . . MM . . . . . . DPU&(also called **Ukerewe**)], **P167** [MG 2011 FV O HD . CIP-443750 MM . . . . . . DPU&(also called **Ejumula**)].

Continued

Table A1. Continued.

Malawi: **Tainon** [MW 1999 . PO MS (suitable for HRA) . . SVD SW . . . . (from Asian Vegetable Research and Development Center (AVRDC), also named **Tainon-57**)], **Mugamba** [MW 1999 . C SS WA . . RVD TW . . . . (from CIP Nairobi, also named **Mogamba**)], **Semusa** [MW 1999 . C SS WA CIP-440034 . TVD SW . TAB . . (from CIP-Nairobi, from Cuba, also called **Cemsa-74-228**)], **Salera** [MW 2002 . W SS WA . . TVD MRW . . . . .], **Zondeni** [MW 2008 FV O MDS&HD (adapted high to mid-altitudes) CIP-443750 . TVD MRW . . . . (longer postharvest shelf life, MDMF, called **Gloria** in MZ and **Ejumula** in UG)], **Sakanantha-ka** [MW 2008 . C SS WA LU96/303 . TVD TW . . . . .], **Sungani** [MW 2011 MV C SS WA BV07/009 . TVD TW . . . . (P: LU96/374 x OP)], **Nyamoyo** [MW 2011 MV C HS WA BV07/008 . TVD TW . . . . (P: Mogamba x OP)], **Mathuthu** [MW 2011 MV IO MDS&HD (suitable for HRA) LU06/146 . TVD TW . . . . (sweet&MDMF, P: Mugamba x OP], **Kaphulira** [MW 2011 MV IO DS&HD WA LU06/428 EM TVD MRW . . . . (suitable for piece meal harvesting, sweet&DMF, P: Mugamba x OP)], **Kadyaubwerere** [MW 2011 MV O MDS&HD WA LU06/252 . TVD MRW . . . . (longer postharvest shelf life, sweet&MDMF, P: Mafutha x OP)], **Chipika** [MW 2011 MV PO SS WA LU06/527 . . . . SAB . . (suitable for the Shire Valley in medium to high temperatures, P: SPN/O x OP)], **Anaakwanire** [MW 2011 MV O MMS&HD HRA&(adapted in well rain fed areas) BV07/028 LM TVD MRW . . . . (recommended for children under five, sweet&MDMF, P: Ejumula x OP)].

Mozambique: **199062.1** [MZ 2000 MV IO MMS&HD WA CIP-199062.1 . . MRW . . . . (sweet&MDMF, P: SPV78.001.3 x OP, from Peru)], **Caromex** [MZ 2000 MV DO MD&LD WA CIP-440136 . . . . (very sweet, MDMF, P: NC-228 x NC- 234, from USA)], **CN-1448-49** [MZ 2000 . IO LD WD CIP-440181 . . SW . . . . (very sweet, MDMF, from TW)], **Japon Tremesino Selecto** [MZ 2000 MV LO MS&LD WA(in MZ) CIP-420009 . . SW . . . . (sweet&MDMF, P: JaponTresmesino x OP, from Peru)], **Kandee** [MZ 2000 MV O MS&MD WA CIP-440140 . . MRW . . . . (MDMF, P: (Yellow Yam x Nancy Hall) x Porto Rico, from USA)], **LO-323** [MZ 2000 . IO MS&LD WA CIP-440185 . . MRW . . . . (sweet&MDMF, from USA)], **Resisto** [MZ 2000 MV DO MS&MD WA CIP-440001 . . SW . . . SD (very sweet & soft mouthfeel boiled roots, P: W72 x OP, from USA)], **Tainung-64** [MZ 2000 . O MS&LD . CIP-440189 . . . . . . SD (sweet&MDMF, from TW)], **Cordner** [MZ 2006 . DO MD WA(in MZ) . . . MRW . . . . (very sweet&MDMF, introduced from Zimbabwe, country of origin: USA)], **Persistente/MGCL01** [MZ 2006 FV DO MDS&HD (adapted to central MZ) . . . RW . . . . (very sweet&MDMF)], **Amelia** [MZ 2011 MV O MDS&HD (adapted to Southern-Central MZ) CIP-106768.1 . . RW . . . . (FGT&MDMF, P: Mafutha-1 x OP)], **Bela** [MZ 2011 MV O MDS&MD WA(in MZ) CIP-106763.5 . . RW . . . . (FGT&MDMF, P: W-119 x OP)], **Cecilia** [MZ 2011 MV PO MDS&MD (adapted Southern-Central MZ) CIP-106766.1 . . RW . . . . (FGT&MDMF, P: UW119 x OP)], **Delvia** [MZ 2011 MV OY MDS&HD WA CIP-106771.1 . . RW . . . . (FGT&MDMF, P: 105369-4 x OP)], **Gloria** [MZ 2011 FV O MDS&HD (adapted Central MZ) CIP-443750 . . TW . . . . (FGT& intermediate texture of boiled roots, from UG, also called **Ejumula**)], **Erica** [MZ 2011 MV PO MDS&MD (adapted to Southern-Central MZ) CIP-106763.2 . . MRW . . . . (FGT&MDMF, P: UW119 x OP)], **Esther** [MZ 2011 MV IO MDS&HD . CIP-106770.1 . . RW . . . . (FGT&MDMF, P: MUSG-0603 x OP)], **Gaba-Gaba** [MZ 2006 . DO MS&LD WA CIP-440215 . . SW . . . . (very sweet&MDMF, from TW, also called **Tainung-65**)], **Ininda** [MZ 2011 MV O MDS&HD WA CIP-106765.1 . . RW . . . . (FGT&MDMF, P: Tacna-2 x OP)], **Irene** [MZ 2011 MV O MDS&HD WA CIP-106764.1 . . RW . . . . (FGT&DMF, P: Kakamega-7 x OP)], **Jane** [MZ 2011 MV IO MDS&HD (adapted to Central-Northern MZ) CIP-106767.1 . . RW . . . . (FGT&MDMF, P: LO323 x OP)], **Lourdes** [MZ 2011 MV IO MDS&MD WA(in MZ) CIP-106763.6 . . RW . . . . (moderately good taste & somewhat dry mouthfeel, P: UW119 x OP)], **Melinda** [MZ 2011 MV LO MMS&LD (adapted to Southern & Central MZ) CIP-106763.1 . . RW . . . . (good taste & MDMF, P: UW119 x OP)], **Namanga** [MZ 2011 MV IO MDS&MD WA CIP-106763.3 . . RW . . . . (FGT&MDMF, P: UW119 x OP)], **Sumaia** [MZ 2011 MV DO MDS&MD (adapted Southern-Central MZ) CIP-106763.4 . . RW . . . . (FGT&MDMF, P: UW119 x OP)], **Tio Joe** [MZ 2011 MV DO MS&MD WA CIP-106769.1 . . RW . . . . (MDMF, P: MUSG-0616 x OP)].

Republic of South Africa: **Ndou** [ZA 2003 . C SD&MS TDL&DST 1995-13-2 EM SVD SW TF MRAB . . .], **Monate** [ZA 2003 . C SD&MS TDL&DST 1989-17-1 EM SVD SW TF MRAB . . .], **Letlhabula** [ZA 2003 . C LD&SS TDL&DST 1985-6-3 EM SVD . SF TAB . . (high yield & quality)], **Amasi** [ZA 2004 . CO SD&MS TDL&DST 1985-6-3 . SVD . SF TAB . . (prone to sprouting)], **Phala** [ZA 2003 . C SD&MS TDL&DST 1984-2-201 . SVD . TF TAB . TMD (uniform short oblong)], **Mamphenyane** [ZA 2003 . C SD&MS TDL&DST 1984-10-340 . SVD . . TAB . . (thin vines)], **Mokone** [ZA 2003 . C SD&MS TDL&DST 1987-16-1 . SVD . TF MRAB . . (latex)], **Serolane** [ZA 2007 . YO D&MS TDL&DST 1998-12-3 . SVD . TF SAB . . (long roots)], **Khano** [ZA 2007 MV DO LD&LTS WA&TDL&DST 1999-6-1 . SVD . TF MRAB . . (soft skin, moist & not-sweet, P: Phala x OP)], **Impilo** [ZA 2008 MV LO LD TDL&DST 1998-21-1 . . . .

Continued

Table A1. Continued.

(not widely adapted, sweet&MDMF, P: Amasi x OP)], **Purple Sunset** [ZA 2009 . . SD&SS TDL&DST 2001-5-2 . SVD . TF SAB . . (attractive roots, pre-packing)], **Isondlo** [ZA 2007 . O SD&SS TDL&DST 2000-10-7 . SVD . TF SAB . . (good keeping ability, uniform)], **Bophelo** [ZA 2011 . O SD&SS TDL&DST 2001-21-1 . SVD . TF TAB . . (uniform ovate)], **W119** [ZA . . O . WA(in RSA) CIP-440004 . RVD RW MF MAB . . (MDMF, promoted but not released, from USA)], **Mvuvhelo** [ZA 2014 . C MDS TDL&DST 1999-9-4 . SVD SW TF MRAB . . (round shape)].

Zambia: **Luapula** [ZM 1993 . W DS WA&HRA . . . .], **Zambezi** [ZM 1993 MV DO MS&MD . . . SVD SW . MRAB . SD (MDMF, P: TIS2537 x OP)], **Chingovwa** [ZM 1993 . C HD (widely grown on commercial base) . . . . (used as vegetable, well grown in KE, MW & MZ)], **Lukulu** [ZM 2003 . C MD WA . . RVD . . . . . .], **Lukusashi** [ZM 2003 . LO MS&MD WA . . . . (medium sized roots)], **Lunga** [ZM 2003 . W . WA . . . MRW . . . . (high yields, prolific vine producer)], **Mulungushi** [ZM 2003 MV Y . HD WA . . . .], **Kalungwishi** [ZM 2003 . LO . . . . . MRW . . . . (medium sized roots)].

South Asia

Bangladesh: **Tripti** [BD 1985 . Y MMS STDL BARI-SP-1 MM . . . . . . (from PH)], **Kamla Sundari** [BD 1985 . O MS STDL BARI-SP-2 MM . . . . . . (from ADRDC)], **Daulapuri** [BD 1988 FV W MS&HS STDL BARI-SP-3 MM . . . . . . (local cultivar)], **BARI-SP-4** [BD 1994 MV PO MS&HS STDL . EM&MM . . . . . . (TCRC hybrid)], **BARI-SP-5** [BD 1994 MV PY MS&HS STDL . EM&MM . . . . . . (TCRC hybrid)], **BARI-SP-6** [BD 2004 . Y MS&HS STDL CIP-440258 EM&MM . . . . . . (also named **Lalkothi**, from IN)], **BARI-SP-7** [BD 2004 . Y MS&HS STDL CIP-440258 EM&MM . . . . . . (also called **Kalmegh**, from IN)], **BARI-SP-8** [BD 2008 . Y MS&HS STDL CIP-440025 EM&MM . . . . . . (also called Xushu-18, from China)], **BARI-SP-9** (BD 2008 . Y MS&HS STDL CIP-44074.2 EM&MM . . . . . . (high yielding)], **BARI-SP-10** (BD 2013 MV Y MS&HS STDL . EM&MM . . . . . . (TCRC hybrid, through OP)].

India: **Rajendra Shakarkand-47** [IN 1993 MV W SS . . MM(4mths) . MRW . . . . (tolerant to Cercospora leaf spot and frost)], **Kian** [IN 1994 MV . . . . MM(4mths) . MRW . . . .], **SreeBhadra** [IN 1996 MV W MS STDL S-1010 . . . . (high yielding, seed population from IITA CHDSS-S-1010 EM-3)], **SreeRethna** [IN 1996 MV O SS STDL X-108-2 EM(3mths) . . . . . . (progeny of S-187 x SreeVardhani)], **Gouri** [IN 1998 MV O MS STDL 85-15 MM(4mths) . SW . . . . (progeny of H-219 x H-42)], **Sankar** [IN 1998 MV PY SS STDL 85-70 MM(4mths) . . . . . . (progeny of H-219 x S-73)], **COCIP-1** [IN 1999 . Y . . . EM(3mths) . MRW . . . . (progeny of IB 90-10-20)], **Rajendra Sakarkhand** [IN 2001 MV . . . . EM(2.5mths) . MRW . . . . (tolerant to Cercospora leaf spot, low temp. and flood)], **Konkan Aswini** [IN 2000 MV . . . . EM(3.5mths) . . . . . . (high yielding)], **Narendra Shakarkand** [IN 2001 MV PY HDSS . . MM(4mths) . MRW . . . . .], **SreeArun** [IN 2002 MV W MSS STDL RS-III-3 EM(3mths) . . . .], **SreeVarun** [IN 2002 MV W MSS STDL CIP-490056 EM(3mths) . . . .], **Kalinga** [IN 2004 MV W HDSS STDL 90/704 MM(3.5mths) . . . . . . (high yielding)], **SreeKanaka** [IN 2004 MV O MS H80/168 EM(3mths) . SW . . . (progeny of S-187 x H-633)], **Goutam** [IN 2005 MV W SS STDL Pol-21-1 MM(3.5mths) . MRW . . . . .], **Sourin** [IN 2005 MV PY HDSD&SS STDL Pol-4-9 . . MRW . . . . .], **Kishan** [IN 2005 MV PY HDSS STDL Pol-13-4 MM(4mths) . MRW . . . . .], **IndiraNaveen** [IN 2006 MV C HDSS . . EM(3mths) . MRW . . . . (OP seeds of Sree Vardhini)], **IndiraNandhini** [IN 2006 MV C SS . . MM(4mths) . MRW . . . . .], **IndiraMadhur** [IN 2006 MV O SS . . MM(4mths) . MRW . . . . .], **KamalaSundari** [IN 2008 MV DO MDMST . . MM(4mths) . MRW . . . . (resistant to storage root rot, tolerant to leaf curl virus)].

West Pacific

China: **Yanshu-5** [CN 1997 MV O MST TDL . EM . . RF . . . (DC&T, excellent taste, susceptible to *Ralstonia solanacearum*)], **Suyu-303** [CN 1997 MV PY DS TDL . MM . . RF . SN . (DPU, excellent taste)], **Sushu-8** [CN 1997 MV OY MST TDL . EM . . SF . . DT (excellent taste)], **Yanshu27** [CN 1999 MV O MST TDL . . . . (DC&T, excellent taste)], **Nanshu-99** [CN 1999 MV PY DS . . . . DT (DPU, taste flavourful & sweet)], **Shangshu-19** [CN 2000 MV W DS WA . . . . RF . . . DPU], **Xushu-22** [CN 2003 MV W DS TDL . . . . IU&WADLZ], **Fushu-7-6** [CN 2003 MV OY . TDL . . . . (vegetable use, WADLZ)], **Jishu-98** [CN 2004 MV PY HS TDL . . . . . . SN . (excellent taste & yields)], Zheshu13 [CN 2004 MV OY HS TDL . . . . DT (excellent taste)], **Jishu-18** [CN 2004 MV PU MST TDL . . . . . . .DT (sensitive to waterlogging)], **Xushu-23** [CN 2004 MV PY DS TDL . EM . . . . . . (excellent taste)], **Ningzishu-1** [CN 2005 MV PU MST TDL . . . . WADLZ], **Guangshu-87** [CN 2005 MV OY DS TDL . EM . . . . . . (DC&T, EFS & taste)],

Continued

Table A1. Continued.

Xingxiang [CN 2007 MV OY DS TDL . EM . . . . . . (DC&T, EFS & taste)], **Xushu-27** [CN 2010 MV W DS TDL . . . . . . SN DT IU], **Xushu-28** [CN 2011 MV W DS TDL . EM . . . . . DT DPU&(good taste)], **Xuzishu-3** [CN 2011 MV PU HS TDL . EM . . . . RN . (anthocyanin extraction use)], **Yanshu-25** [CN 2012 MV O MST TDL . . . . FR . RN . (DC&T, EFS & taste)], **Yuzishu-7** [CN 2012 MV PU MST TDL . . . . . . RN . DC&T].

Republic of Korea: **Yulmi** [KR 1990 MV PY . . . .], **Shinyulmi** [KR 1991 MV PY . . . . DC&T], **Jungmi** [KR 1994 MV Y . . . . (DC&T & processing], **Geonmi** [KR 1995 MV PY . . . . (DC&T & processing)], **Yeonmi** [KR 1997 MV PY . . . . DC&T], **Shinhwangmi** [KR 1998 MV O . . . . (DC&T & processing)], **Jami** [KR 1998 MV PU . . . . (processing)], **Jinhonhmi** [KR 1998 MV Y . . . . IU&(DC&T & starch extraction)], **Shinchonmi** [KR 1999 MV Y . . . . DC&T], **Borami** [KR 2000 MV LPU . . . . (DC&T & processing)], **Shingeonmi** [KR 2001 MV PY . . . . IU&(DC&T & starch extraction)], **Shinjami** [KR 2001 MV PU . . . . (processing)], **Gogeonmi** [KR 2002 MV PY . . . . (DC&T & processing)], **Hayanmi** [KR 2002 MV W . . . . (edible use)], **Juhwangmi** [KR 2002 MV DO . . . . (DC&T & processing)], **Helssimi** [KR 2003 MV PY . . . . IU&(DC&T & starch extraction)], **Baiomi** [KR 2003 MV W . . . . (animal use)], **Haepymi** [KR 2004 MV PO . . . . (DC&T & processing)], **Yeonhwangmi** [KR 2005 MV DY . . . . DC&T], **Matnami** [KR 2006 MV Y . . . . (edible use)], **Morning-purple** [KR 2007 MV Y . . . . (ornamental)], **Morning-white** [KR 2007 MV Y . . . . (ornamental)], **Daeyumi** [KR 2008 MV DY . . . . IU&(bioethanol)], **Yeonjami** [KR 2008 MV LPU . . . . IU&(DC&T & starch extraction)], **Geonpungmi** [KR 2008 MV Y . . . . DC&T], **Jeonmi** [KR 2009 MV PY . . . . IU&(bioethanol & starch extraction)], **Dahomi** [KR 2012 MV PO . . . . DC&T].

East and South-east Asia

Indonesia: **Muara Takus** [ID 1995 . PO MD TRDL&RF BIS-192-Op . . . . (resistant to scab)], **Cangkuang** [ID 1998 . PY MD TRDL&RF SRIS226OPSr75 . . . . (medium resistant to *Cercospora*)], **Sewu** [ID 1998 . O LD TRDL&RF I1186-Daya-Op-Sr-8 . . . . (medium resistant to scab & *Cercospora*)], **Kidal** [ID 2001 . O MD . Inaswang-OP95-6 . . . .], **Sari** [ID 2001 O MD WA MIS104-1-Op . . . . (resistance to scab)], **Boko** [ID 2001 . PY MD TRDL MIS-146-1 . . . . (resistance to scab)], **Cilembu** [ID 2003 FV O MST TRDL&RF . . . . (honey taste, good for baking)], **Jago** [ID 2001 . W HD&HS (adaptable to poor soil fertility) CIP-B0053-9-Op . . . . . . DT (good taste, recommended for flour and starch processing, adaptable to humid tropic areas with poor soil fertility)], **Sukuh** [ID 2001 . W . HD&DS TRDL CIP-AB94001.8 . . . . . . DT (good taste, recommended for flour and starch processing, adaptable to poor soil fertility)], **Shiroyutaka** [ID 2003 . W HD TRDL&RF Kyukei-708-13-X-S684-6 . . MRW . . . . (resistant to scab, register proposal by PT Totota Bio Indonesia)], **Papua Salossa** [ID 2006 . DY MD (tropical highland of Papua – up to 1000 m above sea level (asl)) MSU99051-1 . . MRW . . . . (medium resistance to scab)], **Papua Pattipi** [ID 2006 . PY MD (tropical highland of Papua – up to 1000 masl) CIP- BB97089-12-Op . . MRW . . . . (moderate resistant to scab)], **Sawentar** [ID 2006 . C HD (tropical highland of Papua – up to 1000 masl) CIP-BB97256-9-Op . . MRW . . . . (medium resistance to scab)], **Beta-1** [ID 2009 . DO MST TRDL&RF MSU01015-07-Op . . MRW . . . . (medium resistance to scab)], **Beta-2** [ID 2009 MV O MST TRDL&RF . . . MRW . . . . (medium resistance to scab, Progeny of Kidal x CIP-BB97281-16)], **Antin-1** [ID 2013 . . MD TDLRF . . . MRW . . . DT (medium resistance to scab)].

East Timor: **Hohrae-1** (TL 2007 . PY HD TRDL&(suited to upland) CIP-B0053-9-Op . . . . (also named **Jago** in ID)], **Hohrae-2** [TL 2007. Y HD TRDL&(suited to upland)], **Hohrae-3** [TL 2007 . O MD TRDL&(suited to upland) CIP-BB97020.1-Op . . . .].

Oceania

Papua New Guinea: **B-11** [PG 1998 . W&(with purple spots) DS CFGS . . . . DPU&HF], **L-9** [PG 1992 . W . MMS . . . . TMD .], **L-16** [PG 1992 . W HD&LTS STDL . MM . . . .], **L-19** [PG 1992 . W MMS STDL . . . . TMD .], **L-46** [PG 1998 . W SD&MS STDL . . . . TMD .], **L-135** [PG 1992 . O SD&MS TDL . . . . TMD .], **L-329** [PG 1992 . W SD&MS STDL TMD . . . . TMD .], **L-676** [PG 1998 . O SD&SS TDL . . . . TMD .], **L-997** [PG 1998 . W MMS TDL . . . .], **DOY-2** [PG 1998 . W MMS CFGS . . . . DPU&HF], **KAV-79** [PG 1998 . W SD&MS HRA . . . . TMD .], **NUG-2** [PG 1998 . Y MMS TDL . MM . . . .], **NUG-5** [PG 1998 . W&(with purple spots) DS TDL . MM . . . .], **POI-6** [PG 1998 . Y MMS . . . .], **RAB-7** [PG 1998 . W SS CFGS . MM . . . .], **K-142** [PG 1998 . O MMS CFGS&HRA . MM . . . .].

Appendix 2: Predominantly Grown Sweetpotato Varieties

This appendix provides details about predominantly grown sweetpotato varieties by 14 classification variables, namely: (i) country; (ii) variety type; (iii) storage root flesh colour; (iv) taste type; (v) adaptation range; (vi) CIP-code; (vii) maturity time; (viii) resistance to SPVD; (ix) resistance to weevil; (x) resistance to *Fusarium*; (xi) resistance to *Alternaria*; (xii) resistance to nematodes; (xiii) abiotic stress resistance; and (xiv) comments about special uses, resistances, names or development.

Abbreviations used for classification variables for predominantly grown varieties

1. Country: BD, Bangladesh; BF, Burkina Faso; BR, Brazil; BU, Burundi; CN, People's Republic of China; CU, Cuba; GH, Ghana; IN, India; JP, Japan; KE, Kenya; KR, Republic of Korea; MG, Madagascar; MW, Malawi; MZ, Mozambique; NG, Nigeria; PE, Peru; PG, Papua New Guinea; PH, Republic of the Philippines; RSA, Republic of South Africa; RW, Rwanda; SB, Solomon Islands; TL, East Timor; TW, Taiwan; TZ, Tanzania; UG, Uganda; US-NC, USA North Carolina; US-LS, USA Louisiana; ZA, Republic of South Africa; ZM, Zambia.

2. Variety type: BL, breeding line; FV, farmer variety; MV, modern variety; or if not available '.' for missing value.

3. Storage root flesh colour: C, cream; DO, deep orange; DY, deep yellow; IO, intermediate orange; LO, light orange; O, orange; OY, orange yellow; PO, pale orange; PU, purple; PY, pale yellow; W, white; Y, yellow.

4. Taste type: DS, dry and starchy; HD, high dry matter; HS, high starch; LD, low dry matter; LTS, low total sugars; MD, medium dry matter; MDS, moderately dry and starchy; MMS, moderately moist and sweet; MST, moist and sweet taste; MTS, medium total sugar; SD&MS, semi-dry and medium sweet; SD&SS, slight dry and semi-sweet; SS, sweet and starchy; ST, starchy taste.

5. Adaptation: CFGS, Coastal–Forest transition and Guinea Savannah of West Africa; DST, dry subtropics; HLA, high land adaptation; HRA, high rainfall areas; HTL, hot tropical lowlands; SGS, short grassland savannah; SSZ, Sudano-Sahelian-Zone; STDL, subtropical dry land, TDL, temperate dry land; TGS, tall grassland savannah; TRDL&RF, tropical dry land and rice field; MUMZA, mid- and upper midland zone adaptation; WA, wide adaptation; WAD, wide adaptation to dry lands.

6. CIP-code: number or if not available '.' for missing value.

7. Maturity time: EM, early maturing in months (mths); MM, medium maturing in months (mths); LM, long maturing in months (mths).

8. Resistance to SPVD: MRVD, moderate resistance to SPVD; RVD, resistant to SPVD; SVD, susceptible to SPVD; TVD, tolerant to SPVD; or if not available '.' for missing value.

9. Resistance to weevil: MRW, moderate resistance to weevils; RW, resistant to weevils; or if not available '.' for missing value.

10. Resistance to Fusarium wilt (*Fusarium oxysporum*): RF, resistant to Fusarium wilt; SF, susceptible to Fusarium wilt; TF, tolerant to Fusarium wilt; or if not available '.' for missing value.

11. Resistance to *Alternaria*: MRAB, moderately resistance to *Alternaria* blight; RAB, resistant to Alternaria blight; SAB, susceptible to Alternaria blight; TAB, tolerant to *Alternaria bataticola* stem blight; or if not available '.' for missing value.

12. Resistance to nematodes: RN, resistant to nematodes; SN, susceptible to nematodes; or if not available '.' for missing value.

13. Abiotic stress resistance: RD, tolerant to drought; TMD, tolerates mild dry spells; TS, tolerant to salinity; or if not available '.' for missing value.

14. Comments about special uses, names, resistances or parental material: DC&T, direct consumption and table use; DPU, dual-purpose use as a food and feed; EBA, excellent to boil as 'ampesi'; EFB, excellent for baby-foods and FDP fortification of dairy products; EFC, excellent for fried chips; EFF, excellent for French fries; EFS, excellent form and size for fresh market; EFU, excellent for *fufu*; FGT&MDMF, fairly good taste and moderate dry mouthfeel boiled roots; HF, heavy foliage; IT, industrial type; IU, industrial use (starch); LA, low adoption; OP, open pollination; RFB, moderate resistance to the sweetpotato flea beetle (*Chaetocnema confinis*); RSSR, resistant to Streptomyces soil rot (*Streptomyces ipomoeae*); or if not available '.' for missing value.

Table A2. Currently predominantly grown farmer varieties or modern varieties – updated on 21st May 2014.

Americas

Ourinhos [BR FV C HD . . . . DPU], **Italiana** [BR FV C HD . . . . DPU], **Uruguaiana** [BR FV C HD WA . . . . EFS], **Canadense** [BR FV C HD WA . . . . EFS], **Ligeirinha** [BR FV C HD WA . . . .].

Huambachero [PE FV LO ST TDL CIP-42265 . MRVD . . . . .DPU], **Jonathan** [PE FV O MDS TDL&WDL CIP-420014 . MRVD . . . . . .], **Milagrosa** [PE MV LO ST TDL . . . . DPU], **INIA-100** [PE MV DO MST TDL CIP-192033.50 . . . . . SN . .], **Mejorada** [PE MV LO ST TDL&WAD . . . . DPU], **Benjamin** [PE MV DO MS TDL CIP-105086.1 . SVD . . . . . EFS].

Covington [USA-NC MV DO LD&MS HST . . . RF . RN . RSSR&RFB&OP], **Beauregard** [US-LS MV O MS TDL&HTL CIP-440132 . SVD . RF . SN . RSSR&OP], **Bonita** [US-LS MV W MD TDL&HTL . . . . RF. RN . RSSR].

West Africa

Safaré [BF FV W DS SSZ BF-18 . . . . . . (very uniform shape)], **Gambagre** [BF FV Y DS SSZ BF-77 . . . . (very uniform shape)], **Tiébelé** [BF FV W DS SSZ BF-13 . . . . (very uniform shape)], **Djakani** [BF FV Y DS SSZ BF-75 . . . . (very uniform shape)].

Blue-Blue [GH FV Y HD&LTS CF . . . . (fried & boiled, low perishability, also named **Mon Ami**, Tib 2/looks like Ex-Igbariam from NG)]; **Eworleworme** [GH FV W . . . .], **Kufuor** [GH FV O . . . EM(3mths) . . . . . . (Bawku, Upper East)].

Ex-Igbariam [NG FV Y HD&LTS WA . . . MRW . . . . (fried & boiled, low perishability)], **Butter Milk** [NG FV Y HD&LTS CFGS . . . MRW . . . . DPU&(fried & boiled, low perishability)].

East Africa

Bungoma [KE MV Y DS MUMZA . LM MVD . RAB . . .], **Nyatonge** [KE FV Y DS MUMZA . . MRVD . MRAB . . .], **Marooko** [KE FV C MS MUMZA . LM MRVD . MRAB . . .], **Jayalo** [KE FV W . MUMZA MM MRVD . MRAB . . .], **Bikra-Maria** [KE FV . MDS MUMZA . LM MRVD . MRAB . . .].

Mugande [RW MV W DS WA . LM RVD . RAB . . .], **Kwezikumwe** [RW MV . DS WA . EA RVD . RAB . . .], **Gihingamukungu** [RW MV O DS . EM SVD . SAB . . (also named **97-062**)], **Cacearpedo** [RW MV O MDS WA . . SVD . RAB . . .], **Gihinja** [RW FV W DS (adapted to mid-altitude in RW) . . . .].

Juhudi [TZ MV C MDS . . EM RVD .RAB . . (also named **Jitihada**)], **Polista** [TZ FV W DS Lake_Zone . LM RVD . RAB . . (also named **Polysta**)], **Ukerewe** [TZ MV C_with_O DS . . MM SVD . . RSAB . . .].

Araka [UG FV W DS SGS . EM . . MRAB . . .], **Dimbuka-Bukulula** [UG FV, W DS TSGS CIP-443752 EM SVD . MRAB . . .], **Magabali** [UG FV C DS HAA . MM . . MRAB . . .], **Tanzania** [UG FV PY DS SGS CIP-440166 EM . .MRAB . . (also named **Mwezigumu** or **Soroti** or **Mbale**)], **New-Kawogo** [UG FV W DS TSGS CIP-441743 . RVD MRW SAB . . .].

Southern Africa

Kenya [MW FV W SS WA . . . SW . . . . (poor storage shelf life, also called **Tanzania, SPN/O** in TZ, and **Chingovwa** in MZ & ZM)], **Zondeni** [MW FV O SS WA. . TSPVD MRW . . . . (longer postharvest shelf life, also named **Gloria** in MZ and **Ejumula** in UG)], **Semusa** [MW MV , S, WA. . TVD SW TAB . . (one of the highest yielding varieties, from CU, also named **Cemsa-74-228**)], **Mugamba** [MW . C SS WA . . RVD MRW . . . . (from CIP-Nairobi, also named **Mogamba**)].

Continued

Table A2. Continued.

Irene [MZ MV O MDS&HD WA CIP-106764.1 . . RW . . . . (good establishment and vigour , FGT&DMF)], **Delvia** [MZ MV OY MDS&HD WA CIP-106771.1 . . RW . . . . (FGT&MDMF)], **Sumaia** [MZ MV DO MDS&MD (adapted Southern-Central MZ) CIP-106763.4 . . RW . . . . (FGT&MDMF)], **Resisto** [MZ MV DO MS&MD WA CIP-440001 . . SW . . . SD (very sweet & soft mouthfeel boiled roots)], **Jonathan** [MZ FV O MDS WA CIP-420014 . MRVD . . . . . (good establishment, good taste)], **Namanga** [MZ MV IO MDS&MD WA CIP-106763.3 . . RW . . . . (FGT&MDMF)], **Chingova** [MZ FV C MD WA . . . RW . . . . (good establishment and vigour, also called **Kenya**, **SPN/O**, **Admarc** and **Tanzania**)].

Ndou [ZA MV C SD&MS TDL&DST&(WA in ZA) 1995-13-2 . . . .], **Monate** [ZA MV C SD&MS TDL& DST 1989-17-1 . . . .], **Impilo** [ZA MV LO SD&SS TDL&DST 1998-21-1 . . . TF TAB . . (uniform round elliptic)], **Bophelo** [ZA MV O SD&SS TDL&DST 2001-21-1 . . . . TAB . . (uniform ovate)], **Dagga** [ZA MV YO SDTDL&DST CIP-199062.1 . . . . TAB . TD (tolerant to insects, from PE].

Chingovwa [ZM FV C HD WA . . . . (commercial use, also used as vegetable, also named **Kenya** and **Tanzania**)], **Lukulu** [ZM MV C MD WA . RVD . . . .], **Mulungushi** [ZM MV Y HD WA . . . .], **Red** [ZM FV W HD WA . . . .].

South Asia

Sundori [BD FV W MS&HS STDL . MM . . . . . . (red skin, also called **Lal-Alu**)], **Mati-Alu** [BD FV W MS&HS STDL . MM . . . . . . (white skin, also called **Sada-Alu**)], **Jamalpur** [BD FV W MS&HS STDL . EM . . . . . . (white skin)].

Kanjan-Gad [IN FV C SS WA . MM(4) . . . . . . (high yielding, long tubers)], **SreeBhadra** [IN MV W MS STDL S-1010 . . . . (high yielding, seed population from IITA CHDSS-S-1010 EM-3)].

West Pacific

Xushu-18 [CN MV W DS TDL CIP-440446 . . . RF . . RD DPU&WADLZ], **Nanshu-88** [CN MV OY DS TDL CIP-440443 . . . RF . . RD DPU& WADLZ], **Beijing-553** [CN MV Y MST TDL . . . . (DC&T, EFS & taste), **Yanshu-5** [CN MV O MST TDL . EM . . RF . . . (DC&T, excellent taste & yields, susceptible to *Ralstonia solanacearum*)], **Suyu-303** [CN MV PY DS TDL . . . . RF . SN . (DPU, excellent taste)], **Sushu-8** [CN MV OY MST TDL . EM . . . SF . RD (excellent taste)], **Shangshu-19** [CN MV W DS . . . . RF . . . DPU], **Xushu-22** [CN MV W DS TDL . . . . IU&WADLZ], **Fushu-7-6** [CN MV OY . TDL . . . . (vegetable use) WADLZ], **Ningzishu-1** [CN MV PU MST TDL . . . . WADLZ], **Guangshu-87** [CH MV OY DS TDL . EM . . . . . . (DC&T, EFS&taste)], **Xingxiang** [CN MV OY DS TDL . EM . . . . . . (DC&T, small storage root, EFS taste)], **Xushu-27 [**CN MV W DS TDL . . . . . . SN RD IU], **Yanshu-25** [CN MV O MST TDL . . . . RF . SN . (DC&T, EFS & taste)], **Yuzishu-7** [CN MV PU MST TDL . . . . . . RN . (DC&T)].

East and South-east Asia

Beta-2 [ID . . . WD&(adapted to fertile soils) . . . . (good plant type, mostly planted by SP farmers in East Java since 2009, widely planted in Lombok and Barru in South Sulawesi)], **Kidal [**ID . . . . (planted by SP farmers in Kuningan West Java for the last 2 years, tuber quality as good as Beniazuma)], **Sawentar** [ID . . . .(planted by SP farmers in Kuningan West Java for the last 2 years, tuber quality as good as Kidal)], **Helaleke** [ID FV W MD&MDS . W0116 LM . . . . . . (the most highly consumed in Papua – 84%)], **Musan** [ID FV W MD&SD&MS . W0568 LM . . . . . . (for pig feed – 90%, very large size of storage roots)], **Wortel** [ID FV O MST . W0017 LM . . . . . . (for children's food)], **Papua Salossa** [ID MV . . HLA MSU99051-1 . . . . (is growing widely in areas where was 'Dilanda Kelaparan', which was drought susceptible)], **Cilembu** [ID FV O MMS&SD&MS . . . . (called honey sweetpotato, susceptible to scab, good for baking, very popular in West Java)], **Manohara** [ID FV Y MDS WA . . . . (for paste and export to Korea and Japan)], **AC-Putih** [ID FV Y MSD . . . . (for meeting the request of local company)], **Beniazuma** [ID MV Y . . . . (processing for paste and export to JP, from JP)]; **Ayamurasaki** [ID MV PU MMS . . . . (for local market, from JP)], **Ir.Melati** [ID FV W HD . . . . (high demand for local market in Malang)], **Pak-Ong** [ID FV O LD . . . . (high demand for making tomato sauce)].

Continued

Table A2. Continued.

Oceania
DOY-2 [PG FV W MMS CFGS . . . . .DPU&HF], **SILIBO** [PG FV C MS TDL&DST&CFGS . . . . DPU&HF], **L-43** [PG FV O&PY MD&LTS CFGS TLD MM . . . .], **KAISLOK** [PG FV C MMS CFGS . MM . . . . . . DPU&DPU], **K-9** [PG FV C HD&(sweet) CFGS CIP-441101 MM . . . . . . DPU&HF].
Kaulogu [SB FV W HD . . . . (high demand of local restaurant for making chips, good taste, resistant to scab)], **Bogotu** [SB FV W HD . . . .], **Vona-vona** [SB FV W . . . . (high demand of local restaurant for making chips)], **Noro** [SB FV HD . . . .], **Nambo** [SB FV W HD . . . .], **Beauregard** [SB MV O LD (widely adapted to several soil conditions) CIP-440132 . . . . (moderately resistant to scab disease, starts to be wide growing at Honiara & high demand of cuttings by farmers)].

Appendix 3: Breeding Material in the Pipeline for Release

This appendix provides details about breeding material in the pipeline for variety release by 13 classification variables, namely: (i) country; (ii) storage root flesh colour; (iii) taste type; (iv) adaptation range; (v) CIP-code; (vi) maturity time; (vii) resistance to SPVD; (viii) resistance to weevil; (ix) resistance to *Fusarium*; (x) resistance to *Alternaria*; (xi) resistance to nematodes; (xii) abiotic stress resistance; and (xiii) comments about special uses, names, resistances or parental material.

Abbreviations used for classification variables for breeding material in the pipeline for release

1. Country: BD, Bangladesh; BF, Burkina Faso; BU, Burundi; CN, People's Republic of China; CU, Cuba; GH, Ghana; ID, Indonesia; IN, India; JP, Japan; KE, Kenya; KR, Republic of Korea; MW, Malawi; MZ, Mozambique; NG, Nigeria; PE, Peru; PG, Papua New Guinea; PH, Republic of the Philippines; RSA, Republic of South Africa; RW, Rwanda; TL, East Timor; TZ, Tanzania, UG, Uganda; US-NC, USA North Carolina; US-LS, USA Louisiana; ZM, Zambia.

2. Storage root flesh colour: C, cream; DO, deep orange; DY, deep yellow; IO, intermediate orange; LO, light orange; O, orange; OY, orange yellow; PO, pale orange; PY, pale yellow; W, white; Y, yellow.

3. Taste type: DS, dry and starchy; HD, high dry matter; HS, high starch; LD, low dry matter; LTS, low total sugars; MD, medium dry matter; MDS, moderately dry and starchy; MMS, moderately moist and sweet; MST, moist and sweet taste; MTS, medium total sugar; SD&MS, semi-dry and medium sweet; SD&SS, slight dry and semi-sweet; SS, sweet and starchy, ST, starchy taste.

4. Adaptation: CFGS, Coastal–Forest transition and Guinea Savannah of West Africa; DST, dry subtropics; HLA, highland adaptation; HRA, high rainfall areas; HTL, hot tropical lowlands; MUMZA, mid- and upper midland zone adaptation; SGS, short grassland savannah; SSZ, Sudano-Sahelian-Zone; STDL, subtropical dry land; TDL, temperate dry land; TGS, tall grassland savannah; WA, wide adaptation; WAD, wide adaptation to dry lands.

5. CIP-code: number or if not available '.' for missing value.

6. Maturity time: EM, early maturing in months (mths); MM, medium maturing in months (mths).

7. Resistance to SPVD: MRVD, moderate resistance to SPVD; RVD, resistant to SPVD; SVD, susceptible to SPVD; TVD, tolerant to SPVD; or if not available '.' for missing value.

8. Resistance to weevil: MRW, moderate resistance to weevils; RW, resistant to weevils; or if not available '.' for missing value.

9. Resistance to Fusarium wilt (*Fusarium oxysporum*): RF, resistant to Fusarium wilt; SF, susceptible to Fusarium wilt; TF, tolerant to Fusarium wilt; or if not available '.' for missing value.

10. Resistance to Alternaria: MRAB, moderately resistance to Alternaria blight; RAB, resistant to Alternaria blight; SAB, susceptible to Alternaria blight; TAB, tolerant to *Alternaria bataticola* stem blight; or if not available '.' for missing value.

11. Resistance to nematodes: SN, susceptible to nematodes; or if not available '.' for missing value.
12. Abiotic stress resistance: RD, tolerant to drought; TMD, tolerates mild dry spells; TS, tolerant to salinity; or if not available '.' for missing value.
13. Comments about special uses, names, resistances or parental material: DC&T, direct consumption and table use; DPU, dual-purpose use as a food and feed; EBA, excellent to boil as 'ampesi'; EFB, excellent for baby-foods and FDP fortification of dairy products; EFC, excellent for fried chips; EFF, excellent for French fries; EFS, excellent form and size for fresh market; EFU, excellent for *fufu*; FGT&MDMF, fairly good taste and moderate dry mouthfeel boiled roots; HF, heavy foliage; IT, industrial type; IU, industrial use (starch); LA, low adoption; OP, open pollination; RFB, moderate resistance to the sweetpotato flea beetle (*Chaetocnema confinis*); RSSR, resistant to Streptomyces soil rot (*Streptomyces ipomoeae*); WADLZ, wide adaptation to dry land zones; or if not available '.' for missing value.

Table A3. Breeding material in pipe-line for variety release in 2013.

Americas

Peru: **Abigail** [PE DO MDS HTL CIP-194540.5 . . . . . RN . DPU&EFS(P: SR93.120 x OP)], **Isabel** [PE DO MDS HTL CIP-189153.18 . . . . . RN . DPU& EFS(P: YM89.158 x OP)], **Sumy** [PE DO MS HTL CIP-105523.1 . . . . DPU&EFS(P: SR02.105 x INA100)], **PZ06.120** [PE O MS HTL CIP-105058.2 . . RW . . RN . DPU(P: SR01.030 x INA100)], **CIP-VJ08.330** [PE LO MS TDL CIP-107729.9 . RVD . . . . . low_yields(P: PJ05.069 x DLP3163)], **CIP-PJ07.057** [PE DO MDS HTL . . . . . . RN . DPU&EFS EFS(P: 192131.12 x PJ05.099)].

USA: **Orleans** [US-LS O . TDL&HTL . . . . RFW . SN . RSSR], **04-175** [US-LS DO MS TDL&HTL . . . . RF . . . RSSR], **07-146** [US-LS DO MS TDL&HTL . . . . RF . RN . RSSR].

West Africa

No breeding material in the pipeline for variety release

East Africa

Rwanda: **2002-155** [RW W DS HLA . . RVD RW RAB . . DPU], **2002-166** [RW W DS HLA RW . . RVD . RAB . . DPU], **NASPOT-1** [RW C DS HLA CIP-191133.1 . RVD RWRAB . . DPU], **NASPOT-9-O** [RW O MDS HLA CIP-100200.3 . RVD RW RAB . . DPU&(also named **Vita**)], **NASPOT-10-O** [RW O MDS HLA CIP-100200.4 . RVD RW RAB . . DPU&(also called **Kabode**)].

Tanzania: **Sekondari** [TZ C DS WA . . MRVD . . . . . (in TZ recorded as RAB)].

Southern Africa

Malawi: **LU06/003** [MW C SS WA . . RVD RW . . . . .], **LU06/056** [MW C SS WA . . RVD RW . . . . .], **LU06/196** [MW C SS WA . . RVD RW . . . . .], **LU06/432** [MW C SS WA . . RVD RW . . . . .].

Mozambique: **MGCL01-17** [MZ O . . . . (good taste)], **W250-25-5** [MZ O . . . . (good taste)], **105268-10** [MZ PU HD . . . . (good taste)], **MCKSG08020-8** [MZ DO MD . . . . (good taste)], **MCKSG0825-1** [MZ O HD . . . . (good taste)], **MUSG11016-10** [MZ O HD . . . .], **MUSG11023-11** [MZ O MD . . . .], **MUSG11040-16** [MZ O MD . . . . (good taste)].

Republic of South Africa: **Mvuvhelo** [ZA C SD&SS TDL&DST 1999-9-4 . . . .(round shape)], **2004-9-2** [ZA O M&SS TDL&DST . . . . . SAB . TD (attractive roots)], **2003-23-6** [ZA O SD&SS TDL&DST . . . . (uniform, attractive skin)], **2004-9-1** [ZA O MSS TDL&DST . . . . . SAB . . (uniform oblong, attractive skin)], **2004-16-1** [ZA O MSS TDL&DST . . . . . SAB . . (uniform round elliptic, attractive skin)], **2002-8-2** [ZA O M&SS TDL&DST . . . . (long oblong, suitable for processing industry)].

Zambia: **Olympia** [ZM LO HD&MS WA . . . . DPU&(excellent form and size, P: V15 x OP)], **Kokota** [ZM . HD . . LM . . . .], **Twatasha** [ZM O HD WA . . MRVD . . . . . (pink root skin)], **Chiwokoo** [ZM DO HD&MS WA . . . . (currently being grown by farmers, P: LUS-114 x OP)], **Kanga** [ZM Y HD WA . . .].

Continued

Table A3. Continued.

South Asia
Bangladesh: **BARI-SP-12** [BD O SD&DS WA CIP-440001 MM MRVD . . . . TMD (also called **Resisto**, from USA)]. India: **CO3-4** [IN W HS&MTS STDL . . . .], **CIP-440127** [IN O MDS&MTS STDL . . . . (also called **Tsurunash** or **Tsurunashi-genji**, from JP)], **CIP-440038** [IN O MS STDL IITA-TIS-2498 . . . . (high yields)].
West Pacific
China: **Shang0829-1** [CN W DS&HD TDL . . . . RF . . . .], **Ji-08088** [CN Y DST TDL . . . . RF . RN . (edible use, EFS & taste], **Yushu-17** [SC Y DS&HD TDL . . . . RF . . . .], **E3043** [CN Y MST TDL . . . . RF . . . (edible use, EFS & taste)], **Guangzishu-8** [CN PU DS TDL . . . . RF . . . DPU(high anthocyanin, edible use)], **Fushu-24** [CN PU MST TDL . . . . RF . . . (EFS & taste)], **Yanshu-0747** [CN PU MST TDL . . . . RF . . . (edible use, EFS & taste)], **Mianzishu-9** [CN PU MST TDL . . . . RF . . . (high anthocyanin, high yields)], **Ningcaishuf-18-1** [CN . . TDL . . . . RF . . . (leaf-vegetable type, vegetable use, good taste)], **Xushu-2001** [CN PU DS TDL . . . . (high anthocyanin, edible use, EFS&taste)].
East and South-east Asia
Indonesia: **BB20413.1** [ID W HD WA CIP-W0031-Op . . . . . . DT (recommended for low to midland, adapted to poor soil fertility)]; **Wolf366.18** [ID W HD . CIP-No.105365 . . . . . . DT (adapted to poor soil fertility)], **MSU-03028-10** [ID PU MD . . . .], **RIS-03063-05** [ID PU HD . . . . (tolerant to aphids, good quality storage root, good taste)].
Oceania
Papua New Guinea: **5-ML7e** [PG Y DS&HD MUMZA&TDL . . . .], **BL8d** [PG LO DS MUMZA . . . .], **NIB0801-001** [PG W MS . . . .], **NIB0803-004** [PG W HD&DS MUMZA . . . .], **NIB0806-017** [PG W MSS MUMZA . . . .], **NIB0806-037** [PG W HD MSS MUMZA . . . .], **NIB0808-026** [PG W HD&MSS MUMZA . . . .], **NIB0812-005** [PG O MS MUMZA . . . .], **NIB0812-018** [PG O MSS MUMZA . . . .], **NIB0813-003** [PG LO MD MSS MUMZA . . . .].

2 Breeding Sweetpotato for Yield and Beta-carotene Content in Burkina Faso

K. Somé,[1,2]* T.J. Ouedraogo,[1] J. Belem,[1] K.I. Asante,[2] G. Vernon[2,3] and Y.E. Danquah[2]

[1]*Institut de l'Environnement et de Recherches Agricoles (INERA), Ouagadougou, Burkina Faso;* [2]*West Africa Centre for Crop Improvement (WACCI), University of Ghana, Accra, Ghana;* [3]*Cornell University, New York, USA*

Abstract

The potential of sweetpotato to address food security, malnutrition and poverty is acknowledged in sub-Saharan Africa. The present study was undertaken to develop varieties combining yield and quality in order to address food security and malnutrition in Burkina Faso. Eight parents (five farmers' preferred varieties with various flesh colours and three introduced orange-fleshed sweetpotato varieties) were selected to develop populations and, subsequently, to estimate heritability and genetic gain from breeding. One hundred and thirty F_1 hybrids and their eight parental clones were evaluated in three locations in an Alpha Lattice design to identify high yielding and beta-carotene rich clones with specific to wide adaptation to the local environments. Parent–offspring regression analyses and estimated genetic gain suggested that rapid progress could be attained in increasing dry matter content which was highly heritable (0.76 ± 0.03) and exhibited high genetic gain (22.60%). The same was true for beta-carotene content which was also highly heritable (0.90 ± 0.04). Low heritability (0.21 ± 0.16) associated with low genetic gain indicated that increased storage root yield improvement would be slow. Despite the substantial improvement achieved for yield and beta-carotene, further evaluations will enable better varietal recommendations to farmers.

Keywords: beta-carotene, genetic gain, genotype by environment, heritability, parent–offspring, stability, yield

2.1 Introduction

Sweetpotato improvement has been achieved successfully by alleviating some farmers' production constraints, maintaining consumers' preferences (Bashaasha *et al.*, 1995; Kapinga *et al.*, 1995; Ndolo *et al.*, 2001; Yanggen and Nagujja, 2005; Kwach *et al.*, 2010; Mukebezi *et al.*, 2010) and by variety release (Mwanga *et al.*, 2001, 2003, 2007, 2009; M.I. Andrade *et al.*, Kampala, 2010, personal communication) in eastern and southern Africa. However, in Burkina Faso, its productivity in farmers' fields is still low compared with the potential of the crop.

Breeding progress can be measured in many ways, among which are estimates of heritability and genetic gain. Parent–offspring

*koussao@hotmail.com

regression has been widely described (Kempthorne and Tandon, 1953; Foulley, 1981; Toro and Prunonosa, 1984) and used to estimate heritability in plants and animals (Lande and Price, 1989; De Araujo and Coulman, 2004; Smalley *et al.*, 2004; Singkham *et al.*, 2010). In sweetpotato breeding, Jones (1986) reported 207 studies that estimated narrow-sense heritability using parent–offspring regression. The most recent studies are heritability estimates on resistance to sweetpotato chlorotic leaf distortion (Kim *et al.*, 1996), on weevil injury and yield components (Thompson *et al.*, 1994) and on iron and zinc contents (Courtney *et al.*, 2008).

However, heritability is affected by gene frequencies within a population and by the environmental conditions under which the study is done (Falconer and Mackay, 1996). Changes in the environment or in the gene frequencies will change the relative contributions of genetic and environmental components. Therefore, generalizations should not be made concerning the heritability of a trait without considering the environment or the genetic makeup of the population (Rentz, 2002). The objectives of this study were: (i) to estimate narrow-sense heritability in newly developed populations for storage root yield, beta-carotene content, dry matter (DM) content and other important traits in the context of Burkina Faso; (ii) to estimate the genetic gain from selection of the breeding product; and (iii) to determine the phenotypic correlations between parent and offspring for these traits.

2.2 Materials and Methods

Plant materials

Five locally adapted parents (BF24, BF59, BF77, BF82 and BF92) with flesh colour varying from white to yellow were crossed with three orange-fleshed parental varieties (CIP-199062-1, Resisto and Tainung) introduced from International Potato Center (CIP)-Uganda used as males. Fifteen families were obtained with a total number of 130 F_1 individuals.

Field experiment

After the vine multiplication at the Institut de l'Environnement et de Recherches Agricoles (INERA) station of Kamboinse the F_1 progeny clones were sent for preliminary yield trial in three locations with different rainfall patterns: (i) Farakoba in the west with 900–1200mm annual rainfall; (ii) Loumbila in the centre (700–800 mm annual rainfall); and (iii) Kouare in the eastern region (700–800 mm annual rainfall). The experiment was conducted with the 130 F_1 progeny and their eight parents (a total of 138 entries) in an Alpha Lattice design of two replications and six blocks of 23 entries per block. The experiment was repeated over 2 years, from July to November 2010 and 2011 during the rainy season.

Soil was prepared in a one ridge-plot per entry. Each ridge was 1.5 m long with spacing between ridges of 1 m. Two border rows were planted to enclose the experimental area in order to avoid border effects. On each ridge-plot, the vines were planted with a spacing of 30 cm between consecutive plants. Two weeks after planting, the dead plants were replaced. Three weeks after planting the field was fertilized with NPK (14-23-14) at a rate of 350 kg/ha. The fields were weeded twice before the ground was covered by vine development. The insecticide Deltamethrin was used to protect the plants against insect pests.

Data collection

Data were collected on: (i) storage root yield; (ii) DM content; (iii) beta-carotene content; (iv) upper biomass yield; (v) number of storage roots per plant; (vi) individual storage root weight; and (vii) vine length. Three weeks after planting and at harvest, the number of surviving plants was recorded for computing adjusted yield. The vine length was measured on three plants and averaged while the total upper biomass yield per plot was recorded at harvest.

The beta-carotene content was estimated using the CIP colour chart (Burgos *et al.*, 2009). Based on the correlation that exists between the storage root colour and the beta-carotene content, the colour chart was used to record the flesh colour which was ranged from 0 to 30 while the corresponding beta-carotene content ranged from 0 to 14.37 mg/100 g of fresh storage root weight.

Data analysis

The results from the second year (2011) which was exceptionally dry were not considered as yields were very low compared with the first year. Narrow-sense heritability (h^2) was estimated from parent–offspring using the method described by Holland *et al.* (2003) according to the following formula:

$$E\left[bO\bar{P}\right] = E\left[\frac{Cov(P,O)}{Var(P)}\right] = \frac{Cov\left(\frac{1}{2}P_f + \frac{1}{2}P_m, O\right)}{Var\left[\frac{1}{2}(P_f + P_m)\right]} = \frac{\frac{1}{2}Cov(P_f,O) + \frac{1}{2}Cov(P_m,O)}{\frac{1}{4}\left[Var(P_f) + Var(P_m)\right]} = \frac{\frac{1}{2}\sigma_A^2 + \frac{1}{4}\sigma_{AA}^2}{\frac{1}{2}\sigma_P^2} = \frac{\sigma_A^2 + \frac{1}{2}\sigma_{AA}^2}{\sigma_P^2} = h_1^2 \quad (2.1)$$

Where:
$E[bO\bar{P}]$ = regression coefficient (also known as b)
$Cov(\bar{P},O)$ = covariance mid-parental offspring
P_f = female parent
P_m = male parent
O = offspring
A = additive
AA = additive by additive
h^2 = narrow-sense heritability

PROC iml (Interactive Matrix Language Procedure) in statistical software SAS version 9.1 (SAS Institute, Cary, North Carolina) and mid-parent offspring regression analyses were computed, narrow-sense heritability was estimated across locations and the mean heritability and standard error for each trait were computed. The GLM (General Linear Model) in GENSTAT 14 was also used to provide the different components of the phenotypic variance and to estimate the covariance of parent and offspring for the studied traits. PROC CORR was used to estimate Pearson's correlation coefficients among the studied trait. The genetic advance was calculated according to Falconer and Mackay (1996) from the following formula:

$$GA = ih^2V_p \quad (2.2)$$

Where:
GA = genetic advance
i = 1.76 (10% selection intensity)
V_p = phenotypic variances
h^2 = the narrow-sense heritability

2.3 Results

Mean performance

Location, family and family by location had significant effects on most of the traits studied except on beta-carotene (Table 2.1). The best performances for storage root yield, DM content, individual storage root weight and root number per plant were obtained at Kouare with values of 15.57 ± 9.08 t/ha, 30.29 ± 5.89%, 145 ± 51.12 g/plant and 3.24 ± 0.6 storage roots, respectively. In general, most of the traits except the aboveground biomass yield (19.3 ± 8.24 t/ha) had low performance at Loumbila compared with the two other locations. The storage root yield showed a close relationship to the number of storage roots produced per plant and the individual storage root weight. Except for DM and storage root number, high coefficients of variation were obtained.

Comparison of offspring average performance to their parent performance

Significant differences in yield, DM content, upper biomass yield and vine length among parents (females and males) and their offspring

Table 2.1. Average performance[a] of the F_1 populations and their parents across locations.

Location	Rtyield (t/ha)	BetaCar (mg/100 g of fresh root)	DM (%)	Biomass (kg/plot)	Rootnum (number/plant)	IRtwgt (g/plant)	VL (m)
Farakoba	13.19 ± 6.53	0.67 ± 0.06	28.53 ± 5.18	6.98 ± 1.77	2.9 ± 0.58	129.9 ± 75.3	1.96 ± 0.93
Kouare	15.57 ± 9.08	0.68 ± 0.14	30.29 ± 5.89	10.82 ± 2.87	3.24 ± 0.6	145 ± 51.12	1.25 ± 0.62
Loumbila	9.13 ± 9.42	0.79 ± 0.22	20.32 ± 4.48	19.3 ± 8.24	2.49 ± 1.08	103.6 ± 64.05	1.82 ± 0.81
Grand mean	12.63 ± 8.84	0.71 ± 0.14	26.38 ± 6.77	12.35 ± 3.04	2.88 ± 0.47	125.8 ± 39.13	1.68 ± 0.85
LSD[b]	1.40	–	0.87	6.08	0.94	78.12	0.13
Coefficient of variation (%)	65.2	192	19.7	42.7	28.3	53.9	47.3
Location	***	ns	***	***	***	*	***
Family	**	**	***	***	*	**	***
Location × family	*	ns	**	*	**	ns	*

[a]Rtyield, storage root yield; BetaCar, beta-carotene content; DM, dry matter content; Rootnum, storage root number per plant; IRtwgt, individual storage root weight; VL, vine length.
[b]LSD, least significant difference.
*, **, *** Significance of P at the 5%, 1% and 0.1% levels, respectively. ns, Not significant.

were found (Table 2.2). However, there were no differences among parents and offspring for beta-carotene content, individual root weight, and number of storage roots produced per plant. Furthermore, when comparing offspring with their parental mean, increases in performance were noticed for storage root yield, upper biomass yield, storage root number per plant, individual storage root weight and vine length of 30.12%, 105.88%, 6.86%, 17.88% and 39.34%, respectively, compared with the parental mean (MP). However, the quality parameters beta-carotene and DM content were decreased by –49.72% and –8.86%, respectively. Similarly, the comparison of offspring to the local parents (P_f for female parents), also showed increases for all the traits except for DM where an observed decrease of –11.45% was obtained. The increase in beta-carotene content in the offspring regarding the local parent mean was high (216.73%).

Family average performance

Difference in performance in the studied traits indicated a high amount of phenotypic variation among the families (Table 2.3). The storage root yield mean varied from 6.97 t/ha for the family BF82-Resisto to the highest mean of 17.14 t/ha for the family BF77-Tainung. The best yielding families were BF77-Tainumg (17.14 t/ha), BF77-CIP (17.08 t/ha), BF59-Resisto (14.71 t/ha) and BF92-CIP (14.53 t/ha). The female parent BF77 combined well with male parents to produce high yielding F_1 families. Due to the relationship between storage root yield and storage root individual weight, results from storage root individual weight followed the same order. The best families for the beta-carotene content were BF59-CIP (4.63 mg/100 g of fresh root), BF82-CIP (3.15 mg/100 g of fresh root) and BF92-CIP with 2.27 mg/100 g of fresh root. The male parent CIP199062-1 produced crossed families with high beta-carotene content. However, for the DM content the best families were BF24-CIP (33.16%), BF24-Tainung (28.25%) and BF92-Tainung (28.20%), with BF24 and Tainung-440129 being the best combiners for high DM family development. For the growth parameters (upper biomass yield and vine length), most of the local parents had combined well with Tainung-440129 to produce high biomass and long vines.

Phenotypic correlations among traits

Storage root yield (Rtyield) had significantly ($P < 0.0001$) high and positive phenotypic correlation with individual storage root weight (IRtwgt) and storage root number per plant

Table 2.2. Parents and offspring mean performance.[a]

Generation	Rtyield (t/ha)	BetaCar (mg/100 g of fresh root)	DM (%)	Biomass (kg/plot)	Rootnum (number/plant)	IRtwgt (g/plant)	VL (m)
Offspring	12.83	0.89	26.22	15.05	2.96	134.23	1.70
P_{female}	8.26	0.28	29.61	7.15	2.58	97.15	1.43
P_{male}	11.45	3.27	27.93	7.47	2.95	130.62	1.01
Mid-parent	9.86	1.77	28.77	7.31	2.77	113.87	1.22
% Increase/MP[b]	30.12	–49.72	–8.86	105.88	6.86	17.88	39.34
% Increase/P_f[c]	55.33	216.73	–11.45	110.49	14.73	38.17	18.88
All entries	***	ns	***	***	***	*	***
Offspring	***	***	***	***	***	***	***
Offspring vs ($P_f + P_m$)	**	ns	***	***	ns	ns	***

[a]Rtyield, storage root yield; BetaCar, beta-carotene content; DM, dry matter content; Rootnum, storage root number per plant; IRtwgt, individual storage root weight; VL, vine length.

[b]% Increase/MP, percentage increase in performance when comparing offspring with their parent mean (MP).

[c]% Increase/P_f, percentage increase in performance when comparing offspring to local parents (P_f for female parents).

*, **, *** Significance of P at the 5%, 1% and 0.1% levels, respectively. ns, Not significant.

Table 2.3. F_1 Family performance for yield, growth parameters and quality traits.[a]

Family	VL (m)	Biomass (kg/plot)	Rootnum (number/plant)	Rtyield (t/ha)	IRtwgt (g/plant)	DM (%)	BetaCar (mg/100 g of fresh root)
BF24-CIP	1.06 ± 0.3	17.93 ± 2.77	2.30 ± 0.4	10.65 ± 8.27	121.4 ± 21.7	33.16 ± 11.7	0.048 ± 0.04
BF24-Resisto	1.56 ± 0.8	11.93 ± 2.77	2.57 ± 0.1	11.38 ± 7.76	129.1 ± 15.2	27.90 ± 6.5	0.024 ± 0.03
BF24-Tainung	1.65 ± 0.6	14.6 ± 2.27	2.59 ± 0.3	10.73 ± 6.68	126.2 ± 13.6	28.25 ± 7.3	0.036 ± 0.14
BF59-CIP	1.55 ± 0.5	16.14 ± 4.61	3.53 ± 0.5	13.19 ± 6.79	112.2 ± 12.1	27.59 ± 9.2	4.625 ± 4.13
BF59-Resisto	1.80 ± 0.9	16.45 ± 3.55	2.65 ± 0.2	14.71 ± 9.35	165.1 ± 14.9	25.68 ± 5.0	0.042 ± 0.05
BF59-Tainung	1.66 ± 0.6	12.45 ± 3.28	3.3 ± 0.5	12.36 ± 8.68	111.7 ± 16.4	25.21 ± 5.3	0.067 ± 0.06
BF77-CIP	1.36 ± 0.5	13.53 ± 4.27	2.58 ± 1.2	17.08 ± 11.18	240.4 ± 55.9	21.62 ± 3.2	0.082 ± 0.06
BF77-Resisto	1.68 ± 1.1	14.49 ± 5.83	3.48 ± 0.3	13.28 ± 8.58	119.9 ± 19.4	27.48 ± 5.9	0.018 ± 0.02
BF77-Tainung	1.14 ± 0.5	13.73 ± 3.22	3.79 ± 0.2	17.14 ± 7.75	150.4 ± 21.3	26.29 ± 6.4	0.028 ± 0.04
BF82-CIP	1.51 ± 0.9	12.6 ± 3.44	3.42 ± 0.6	12.29 ± 8.70	109.4 ± 14.9	26.06 ± 6.6	3.156 ± 3.73
BF82-Resisto	1.87 ± 0.9	13.67 ± 3.59	2.18 ± 0.6	6.97 ± 6.76	88.3 ± 10.8	27.30 ± 6.9	0.082 ± 0.25
BF82-Tainung	2.18 ± 0.9	16.49 ± 2.69	2.84 ± 0.3	13.79 ± 9.35	147.1 ± 8.6	23.31 ± 6.6	0.717 ± 1.98
BF92-CIP	1.58 ± 0.8	15.35 ± 1.56	2.83 ± 0.4	14.53 ± 13.41	167.3 ± 32.3	26.84 ± 5.9	2.273 ± 3.44
BF92-Resisto	1.99 ± 0.8	15.47 ± 3.17	3.27 ± 0.2	12.70 ± 8.29	117.9 ± 19.9	24.73 ± 6.2	0.945 ± 2.85
BF92-Tainung	2.18 ± 1.1	20.86 ± 5.93	3.07 ± 0.24	11.06 ± 6.91	107 ± 17	28.20 ± 3.57	0.467 ± .06

[a]VL, vine length; Rootnum, storage root number per plant; Rtyield, storage root yield; IRtwgt, individual storage root weight; DM, dry matter content; BetaCar, beta-carotene content.

(Rootnum) (Table 2.4). Individual root weight and root number are components of storage root yield and therefore are structurally associated. No significant correlation was observed between storage root yield, DM content and beta-carotene content. Significant negative correlation was observed between DM content and the growth parameters: upper biomass yield and vine length.

Heritability and genetic advance

Since the parents and their offspring were grown together in three locations the heritability for each location is also computed. The results showed that the narrow-sense heritability varied from one location to another for each trait (Table 2.5). Values of zero were obtained at Kouare for biomass and individual root weight per plant. The combined estimate indicated very low heritability on a family mean basis for biomass yield (0.04), very high heritability for beta-carotene (0.90) while storage root yield, DM content, storage root number per plant, individual storage root weight and vine length had heritabilities of 0.21, 0.76, 0.41, 0.27 and 0.48, respectively. The two structurally associated traits for yield, storage root yield and storage root individual weight, had almost the same heritability of 0.21 and 0.27.

High genetic advance was observed for DM content (22%) (Table 2.5). The high value of heritability obtained for beta-carotene was coupled with a low genetic advance of 3.37%. The lowest value was obtained for vine length (0.26%) while the storage root yield, biomass, storage root number and individual storage root had moderate genetic advance of 5.85%, 7.81%, 8.03% and 6.35%, respectively.

2.4 Discussion

Knowledge of the performance of offspring compared to their parents, the estimated

Table 2.4. Phenotypic correlation between the studied traits.[a]

	BetaCar	Biomass	DM	IRtwgt	Rootnum	Rtyield
Biomass	0.229**	1				
DM	−0.127	−0.469***	1			
IRtwgt	0.089	0.083	0.165	1		
Rootnum	0.066	0.045	0.185*	0.163	1	
Rtyield	0.074	0.115	0.204*	0.78***	0.659***	1
VL	0.189*	0.291***	−0.423***	−0.137	0.139	−0.07

[a]BetaCar, beta-carotene content; Biomass, biomass yield; DM, dry matter content; IRtwgt, individual storage root weight; Rootnum, storage root number per plant; Rtyield, storage root yield; VL, vine length.
*, **, *** Significance of *P* at the 5%, 1% and 0.1% levels, respectively.

Table 2.5. Narrow-sense heritability (h^2) estimate for the different traits and per location and genetic advance from improvement.

	Heritability estimate				
	Farakoba	Kouare	Loumbila	Combined	Genetic advance (%)
Storage root yield	0.20 ± 0.25	0.58 ± 0.25	0.58 ± 0.32	0.21 ± 0.16	5.85
Dry matter	0.75 ± 0.06	0.93 ± 0.03	0.75 ± 0.14	0.76 ± 0.003	22.60
Beta-carotene	0.43 ± 0.63	0.49 ± 0.44	0.97 ± 0.02	0.90 ± 0.039	3.37
Biomass yield	0.58 ± 0.18	0	0.05 ± 0.42	0.04 ± 0.07	7.81
Root number per plant	0.62 ± 0.44	0.52 ± 0.38	0.57 ± 0.27	0.41 ± 0.21	8.03
Individual root weight	0.69 ± 0.14	0	0.55 ± 0.34	0.27 ± 0.12	6.35
Vine length	0.33 ± 0.42	0.57 ± 0.28	0.16 ± 0.07	0.48 ± 0.28	0.26

heritability and the genetic advance obtained are useful in designing a breeding programme. This was missing for a breeding programme in Burkina Faso. The F_1 population developed through this study showed extensive variability indicating good potential for conventional breeding progress.

Previous studies have shown the variation of yield, yield components and above-ground traits over environments (Jones, 1986; Collins *et al.*, 1987; Manrique and Hermann, 2000; Grüneberg *et al.*, 2005; Cervantes-Flores, 2006; Lin *et al.*, 2007, Tsegaye *et al.*, 2007; Chiona, 2009; Tumwegamire, 2011) as found in this present study. In agreement with Tumwegamire (2011), the genetic by environment interaction was not significant for beta-carotene content. However, his results were obtained on an individual genotype basis instead of family means used in this case. The absence of significance for variation in beta-carotene content suggests that evaluation for this trait could be conducted at limited environments instead of multiple environments.

Except for DM content, meaningful increase in performance was seen from parents to offspring for most of the traits studied. The female parents BF77, BF92 and the male orange-fleshed parents CIP-199062-1 and Tainung could be very important in sweetpotato improvement in Burkina Faso for these identified traits.

The observed positive correlations among storage yield, individual storage root weight and storage root number per plant can be explained by their structural and developmental relationships. In contrast to previous results (Cervantes-Flores *et al.*, 2010) no negative correlation was found between beta-carotene and DM.

In the present investigation, the very high narrow-sense heritabilities observed for beta-carotene (0.90) and DM content (0.76) suggest that breeding for increased beta-carotene and DM content is possible using the available germplasm. This high heritability for beta-carotene was associated with a low genetic advance (3.37%) revealing that non-additive gene actions on beta-carotene accumulation occurred. These results indicated that the improvement of beta-carotene could mainly be achieved through recurrent mass selection. Sardana *et al.*'s (2007) study suggested that high heritability may not necessarily lead to increased genetic gain in peas, unless sufficient genetic variability existed in the germplasm. Therefore, divergent sources of compatible parents will still be needed for the beta-carotene improvement in the Sahelian zone.

Low heritability with low genetic advance was obtained for storage root yield, individual storage root weight and biomass. According to Eid (2009) reporting heritability estimates and genetic advance on wheat, low heritability associated with low genetic advance will result in low progress for selection for these traits. Divergent values for heritability of yield have been obtained from previous studies. Jones (1986) obtained by parent–offspring regression, a relatively higher narrow-sense heritability of 0.44 and a higher value of 0.73 using the variance procedure compared with the results in this study (0.21) while Grüneberg *et al.* (2009) indicated a value of 0.5 using an accelerated breeding scheme (ABS).

DM content had higher heritability (0.76) with a relatively high genetic advance (22.60%) suggesting the presence of additive and non-additive gene action. Therefore, progress could be expected from successive mass selections and by heterosis breeding. Prediction could be made on the performance of the offspring for DM on the basis of the parental performance. These results are in agreement with those of Tsegaye *et al.* (2007) who obtained high genetic advance (22.56%) and with those of Lebot (2008) who reported an additive gene effect for DM content while Courtney *et al.* (2008) reported a value of 0.92. A high value of narrow-sense heritability of 0.65 for DM content was obtained by Jones (1986) using parent–offspring regression and a value of 0.45 using the variance-covariance procedure, whereas Grüneberg *et al.* (2009) suggested a heritability > 0.8 for DM and beta-carotene using the ABS.

Storage root number per plant and vine length had moderate narrow-sense heritabilities (0.41 and 0.48, respectively) with low genetic advance (8.03% and 0.26%, respectively). These results suggested that additive and non-additive gene action were involved indicating that low progress could be obtained through selections for these traits. Maluf *et al.* (1983) in Brazil reported a low broad-sense heritability of 0.43 (meaning low narrow-sense heritability) for vine length whereas Tsegaye *et al.* (2007) obtained in Ethiopia a value 62% of genetic advance.

Acknowledgements

The Alliance for a Green Revolution in Africa (AGRA) and the West Africa Centre for Crop Improvement (WACCI, University of Ghana) are acknowledged for the grant and scholarship for conducting this research.

References

Bashaasha, B., Mwanga, R.O.M., Ocitti p'Obwoya, C. and Ewell, P.T. (1995) *Sweetpotato in the Farming and Food Systems of Uganda: a Farm Survey Report*. International Potato Center (CIP) National Agricultural Research Organisation (NARO), Kampala, 68 pp.

Burgos, G., Carpio, R., Sanchez, C., Sosa, P., Espinoza, J. and Grüneberg, W. (2009) *Guide for Using the RHS Color Chart for Selecting for High β-Carotene Sweetpotato*. International Society for Tropical Root Crops (ISTRC), Lima.

Cervantes-Flores, J.C. (2006) Development of a genetic linkage map and QTL analysis in sweetpotato. PhD thesis, Horticultural Science, North Carolina University, Raleigh, North Carolina.

Cervantes-Flores, J.C., Sosinski, B., Pecota, K.V., Mwanga, R.O.M., Catignani, G.L., Truong, V.D., Watkins, R.H., Ulmer, M.R. and Yencho, G.C. (2010) Identification of quantitative trait loci for dry-matter, starch, and β-carotene content in sweetpotato. *Molecular Breeding* 28, 201–216.

Chiona, M. (2009) Towards enhancement of β-carotene content of high dry mass sweetpotato genotypes in Zambia. PhD thesis, African Centre for Crop Improvement (ACCI), University of KwaZulu-Natal, Pietermaritzburg, Republic of South Africa.

Collins, W.W., Wilson, L.G., Arrendell, S. and Dickey, L.F. (1987) Genotype × environment interactions in sweetpotato yield and quality factors. *Journal of the American Society for Horticultural Science* 112(3), 579–583.

Courtney, M., Mcharo, M., La Bonte, D. and Grüneberg, W. (2008) Heritability estimates for micronutrient composition of sweetpotato storage roots. *Hortscience* 43(5), 1382–1384.

De Araujo, M.R.A. and Coulman, B.E. (2004) Parent–offspring regression in meadow bromegrass (*Bromus riparius* Rehm.): evaluation of two methodologies on heritability estimates. *Canadian Journal of Plant Science* 84(1), 125–127.

Eid, M.H. (2009) Estimate of heritability and genetic advance of yield traits in wheat (*Triticum aestivum* L.) under drought conditions. *International Journal of Genetics and Molecular Biology* 1(7), 115–120.

Falconer, D.S. and Mackay, T.F.C. (1996) *Introduction to Quantitative Genetics*, 4th edn. Addison Wesley Longman, Harlow, UK.

Foulley, J.L. (1981) La regression parent–descendant dans le cas d'un nombre variable de mesures par parent. *Annales de Génétique et de Sélection Animale* 13(2), 181–188.

Grüneberg, W.J., Manrique, K., Zhang, D. and Hermann, M. (2005) Genotype × environment interactions for a diverse set of sweetpotato clones evaluated across varying ecogeographic conditions in Peru. *Crop Science* 45, 2161–2173.

Grüneberg, W.J., Diaz, F., Eyzaguirre, R., Espinoza, J., Burgos, G., Felde, T.Z. and Mwanga, R. (2009) Heritability estimates for an Accelerated Breeding Scheme (ABS) in clonally propagated crops – using sweetpotato as a model. Paper presented at the 15th Triennial International Society for Tropical Root Crops (ISTRC) Symposium, 2–6 November, Lima, Peru.

Holland, J.B., Nyquist, W.E. and Cervantes-Flores-Martinez, C.T. (2003) Estimating and interpreting heritability for plant breeding: an update. *Plant Breeding Reviews* 22, 9–112.

Jones, A. (1986) Sweetpotato heritability estimates and their use in breeding. *Horticultural Science* 21, 14–17.

Kapinga, R., Ewell, P.I., Jeremiah, S.C. and Kileo, R. (1995) *Sweetpotato in Tanzanian Farming and Food Systems: Implication for Research*. International Potato Center, Nairobi, Kenya and Ministry of Agriculture and Livestock Development, Dar es Salaam, Tanzania.

Kempthorne, O. and Tandon, O.B. (1953) The estimation of heritability by regression of offspring on parent. *Biometrics* 9(1), 90–100.

Kim, B.-K., La Bonte, D.R., Clark, C.A. and. Buteler, M.I. (1996) Heritability estimates for reaction resistance to sweetpotato chlorotic leaf distortion caused by *Fusarium lateritium*. *Hortscience* 31(3), 433–435.

Kwach, J.K., Odhiambo, G.D., Dida, M.M. and Gichuki, S.T. (2010) Participatory consumer evaluation of twelve sweetpotato varieties in Kenya. *African Journal of Biotechnology* 9(11), 1600–1609.

Lande, R. and Price, T. (1989) Genetic correlations and maternal effect coefficients obtained from offspring–parent regression. *Genetics* 122, 915–922.

Lebot, V. (2008) Sweetpotato: breeding and genetics. In: *Tropical Root and Tuber Crops: Cassava, Sweetpotato, Yams and Aroids. Crop Production Science in Horticulture Series* 17, 107–126.

Lin, K.H., Lai, Y.C., Chang, K.Y, Chen, Y.F., Hwang, S.Y. and Lo, H.F. (2007) Improving breeding efficiency for quality and yield of sweetpotato. *Botanical Studies* 48, 283–292.

Maluf, W.R., Miranda, J.E.C. and Ferreira, P.E. (1983) Broad sense heritabilities of root and vine traits in sweetpotato (*Ipomoea batatas* (L.) Lam.). *Review of Brazilian Genetics* 6(3), 443–451.

Manrique, K. and Hermann, M. (2000) Effect of G × E interaction on root yield and beta-carotene content of selected sweetpotato (*Ipomoea batatas* (L.) Lam.) varieties and breeding clones. In: *CIP Program Report 1999–2000*. International Potato Center (CIP), Lima, pp. 281–287.

Mukebezi, R., Isubikalu, P., Mwanga, R.O.M. and Orum, B. (2010) Farmers' perceptions of orange-fleshed sweetpotatoes as a food security crop in central and eastern Uganda. In: *Building Capacity for Food Security in Africa*, Second RUFORUM (Regional Universities Forum for Capacity Building in Agriculture) Biennial Meeting, 20–24 September 2010, Entebbe, Uganda. RUFORUM, Entebbe, Uganda, pp. 1637–1641.

Mwanga, R.O.M., Odongo, B. and Ocitti, O.C. (2001) Release of five sweetpotato cultivars in Uganda. *Hortscience* 36(2), 385–386.

Mwanga, R.O.M., Odongo, B., Turyamureeba, G. and Alajo, A. (2003) Release of six sweetpotato cultivars (NASPOT-1 to NASPOT-6) in Uganda. *Hortscience* 38(3), 475–476.

Mwanga, R.O.M., Odongo, B., Niringiye, C. and Alajo, A. (2007) Release of two orange-fleshed sweetpotato cultivars, 'SPK004' ('Kakamega') and 'Ejumula', in Uganda. *Hortscience* 42(7), 1728–1730.

Mwanga, R.O.M., Odongo, B., Niringiye, C., Alajo, A., Kigozi, B., Makumbi, R., Lugwana, E., Namukula, J. and Mpembe, I. (2009) 'NASPOT 7', 'NASPOT 8', 'NASPOT 9 O', 'NASPOT 10 O', and 'Dimbuka-Bukulula' Sweetpotato. *Hortscience* 44(3), 828–832.

Ndolo, P.J., Mcharo, T., Carey, E.E., Gichuki, S.T., Ndinya, C. and Maling'a, J. (2001) Participatory on-farm selection of sweetpotato varities in Western Kenya. *African Crop Science Journal* 9(1), 41–48.

Rentz, E. (2002) Estimating Additive Genetic Variation and Heritability of Phenotypic Traits. Available at: http://userwww.sfsu.edu/~efc/classes/biol710/heritability/heritability.htm (accessed 13 June 2012).

Sardana, S., Mahjan, R., Gautam, N. and Ram, B. (2007) Genetic variability in pea (*Pisum sativum* L.) germplasm for utilization. *Society for the Advancement of Breeding Research in Asia and Oceania (SABRAO) Journal of Breeding and Genetics* 39(10), 31–41.

Singkham, N., Jogloy, S., Kesmala, T., Swatsitang, P., Jaisil, P., Puppala, N. and Patanothai, A. (2010) Estimation of heritability by parent–offspring regression for high oleic acid in peanut. *Asian Journal of Plant Sciences* 9(6), 358–363.

Smalley, M.D., Daub, J.L. and Hallauer, A.R. (2004) Estimation of heritability in maize by parent–offspring regression. *Maydica* 49, 221–229.

Thompson, P.G., Schneider, J.C. and Graves, B. (1994) Genetic variance component and heritability estimates of freedom from weevil injury to sweetpotato. *Journal of American Society for Horticultural Science* 119(3), 620–623.

Toro, M.A. and Prunonosa, J.V. (1984) The use of prior information in the estimation of heritability by parent–offspring regression. *Genetics Selection and Evolution* 16(2), 177–184.

Tsegaye, E., Nigussie, D. and Devakara, S. (2007) Genetic variability for yield and other agronomic traits in sweetpotato. *Journal of Agronomy* 6(1), 94–99.

Tumwegamire, S. (2011) Genetic variation, diversity and genotype by environment interactions of nutritional quality traits in East African sweetpotato. PhD thesis, Makerere University, Kampala, Uganda.

Yanggen, D. and Nagujja, S. (2005) Analysis of producer and consumer preferences for sweetpotato varieties in Uganda: implications for research and extension efforts to promote orange-fleshed sweetpotato to combat vitamin A deficiency. *African Crop Science Conference Proceedings* 7, 807–811.

3 Development of Dual-purpose Sweetpotato Varieties through Participatory Breeding in Rwanda

D. Shumbusha,[1*] J. Ndirigwe,[2] L. Kankundiye,[2] A. Musabyemungu[2] and R.O.M. Mwanga[3]

[1]Rwanda Agriculture Board (RAB), Huye, Rwanda; [2]RAB, Kigali, Rwanda; [3]International Potato Center Sub-Saharan Africa (CIP-SSA), Kampala, Uganda

Abstract

Sweetpotato forms a major part of the diet of both rural and urban communities in Rwanda. Moreover, it is expected that the crop could become more important than it is already now, especially for farmers operating in mixed crop-livestock systems. The interest in sweetpotato as an animal feed is associated with the implementation of a policy regarding zero grazing practices as one of the ways to reduce soil erosion. This research was conducted to develop dual-purpose sweetpotato varieties through a participatory approach, using an accelerated breeding scheme. Sixty parents comprising local cultivars and introduced germplasm were used in a crossing block to generate true seeds. In total, 5380 well-established genotypes were selected from the seedling nursery and planted in an observational trial at Rubona, Karama and Ngoma, during 2011 season B (season A is the first rainy season and season B the second rainy season). Participatory evaluation and selection was conducted by 42 people, with 25 females among them and 268, 25 and 10 best clones were selected from an observational trial, preliminary and advanced trials, respectively. Selection of dual-purpose clones was based on the range of the ratio of roots to vines in terms of dry matter yields, where a clone with a range of 1.5–2.0 was classified as a dual-purpose clone. The best performing clone for root yield in 2012 season B was RW11-17, with a mean yield of 28.1 t/ha, followed by RW11-3736 with a yield of 27.5 t/ha. Clones RW11-1860 and RW11-4923 were high in root dry matter content (DMC), at 37.8% and 37.5%, respectively, whereas RW11-1860 ranked the first in vine DMC with a mean of 20.7%. In taste tests, two white-fleshed genotypes RW11-1860 and RW11-17 were ranked best by farmers, whereas clone RW11-2910 was ranked first among the orange-fleshed varieties. Six of the best performing clones, namely RW11-17, RW11-1860, RW11-2419, RW11-4923, RW11-2560 and RW11-2910, had a ratio root/vine ranging between 1.5 and 2.0, and were therefore considered as dual-purpose-use clones. These performing clones were officially released in February 2013.

Keywords: clone, dry matter, roots, vines, yield

*shudam@yahoo.fr

3.1 Introduction

Sweetpotato (*Ipomoea batatas* (L.)) ranks seventh most important food crop in the world (FAO, 2012). In sub-Saharan Africa, it is the third most important crop with nearly 90% of the total output coming from eastern and southern Africa (Ewell and Mutuura, 1994). The crop is grown for multipurpose use. Roots and vines are used both for human food and for animal feed. There has been a great deal of breeding effort focusing on developing sweetpotato varieties with high root yield for human consumption, and not on vine yield. However, the demand for new varieties producing both roots and forage has increased, especially for farmers operating in mixed crop-livestock systems. Several studies have shown that there are varieties with a potential of producing both roots and forage (Lukuyu and Agili, 2013).

The crop is one of the 12 main plant species used for human consumption throughout the world (Woolfe, 1992). Therefore, a dual-purpose sweetpotato variety should have a comparative advantage compared with varieties selected for only root or vine production alone. Research has shown that dissemination of dual-purpose sweetpotato varieties could help improve the livelihoods of smallholder farmers operating in mixed crop-livestock systems around the world. Sweetpotato is seen as a potential crop for smallholder farmers due to its high productivity and low input requirements (Claessens *et al.*, 2009).

Participatory variety evaluation and selection are the first steps in increasing the role of farmers in the crop improvement process. The advantages of involving farmers in on-station and on-farm trials in the breeding process are generally recognized. Farmers are considered to be in a better position to observe and evaluate traits that are relevant in their own farm conditions (Almekinders and Hardon, 2006). The purpose of an accelerated breeding scheme is to shorten the process of classical breeding in the framework of producing new varieties just after 3–4 years. This can be achieved by using several locations during variety testing, and involve farmers from the early breeding stages.

The main objective of the present research was to develop and release high-yielding sweetpotato varieties for human consumption and for alternative uses, with emphasis on animal feeding, through participatory breeding in Rwanda.

3.2 Materials and Methods

Parental material

Thirty farmer varieties and 30 improved clones making 60 parents were selected to be used for crossing. The farmer varieties were those considered by farmers as 'elite' or top yielding. These local varieties were collected from different districts (Muhanga, Huye and Bugesera), during October 2009. The improved clones originated from neighbouring countries and from the International Potato Center (CIP). The materials were planted in a screenhouse as well as in two new crossing blocks at Rubona, Rwanda. Crosses were made in a factorial-controlled cross design with six male parents and also in open-pollinated crosses to increase the number of possible combinations. A total of 60 parents comprising local varieties and introduced germplasm were established in a polycross at Rubona Research Station, during 2010 season B (season A is the first rainy season and season B the second rainy season). All the six selected promising clones were developed through seedling selections from open-pollinated seeds from Ejumula, 2000-203, Mugande, Ukerewe, SPK004 and Gihingumukungu from the sweetpotato programme at Rwanda Agriculture Board (RAB).

Description of trial sites

Trials were conducted in experimental fields of RAB, namely Rubona, Karama and Ngoma research stations. The three locations have a bimodal rainfall pattern. Rubona station is located at mid-altitude, at 1650 m above sea level (masl) with annual rainfall ranging between 1050 mm and 1200 mm. The annual mean temperature is 18.7°C. Karama Research

Station is located in the lowlands of Rwanda, in a semi-arid area, in a high Sweet potato virus disease (SPVD)-pressure agroecological zone, with 1400 masl, and the rainfall ranges between 700 mm and 900 mm. Ngoma Research Station is located in lowland, at 1420 masl, with rainfall ranging between 850 mm and 1000 mm (ISAR, 2000).

Development of F_1 plant material

All generated true seeds were germinated in a screenhouse at RAB – Rubona station. The seeds were therefore scarified by soaking them in concentrated (98%) sulfuric acid (H_2SO_4) for 30 min to break mechanical dormancy imposed by the seedcoat.

Field experiments

A total of 5380 well-established genotypes were selected from the seedling nursery and planted in the observational trial at Rubona, Karama and Ngoma during February 2011. A split-plot arrangement, with two replications was used, with families arranged as main plots, in a randomized complete block design (RCBD), and genotypes within each family randomized as sub-plots. Each entry in a replication was represented by a single row containing five plants. Spacing between plants in a row was 30 cm, with 80 cm between rows. Planting and management practices were similar in all locations, and weeding was carried out whenever necessary until harvest. Roots were harvested 5 months after planting, with harvest data from the three middle plants of each genotype expressed as a plot mean. One plant at either end of the row was left as a guard plant, giving a harvest area of 0.72 m^2 per plot.

In September 2011, 250 selected genotypes were advanced to preliminary trials (PT) at each of the three locations (Rubona, Karama and Ngoma) with 20 plants arranged in a two-row plot per genotype. The experimental design used was the same as those employed in the previous season. Among the 250 clones evaluated in a PT, data were recorded on genotype basis and 25 clones were selected and advanced, together with the Kwezikumwe clone as a check. Among 25 clones, the top ten performing clones were selected to be tested in national performance trials and on-farm evaluation, during September 2012 (season 2013 A).

On-farm trials for a set of ten selected clones were planted in nine sites located in five districts (Huye, Nyamagabe, Muhanga, Bugesera and Ngoma) with the purpose of: (i) introducing new performing clones for dual-purpose use in the farming systems; and (ii) evaluating their acceptability by farmers. At least four farmers were identified for each of the five districts based mainly on the willingness to host the trial, gender, experience in sweetpotato production and having sufficient land to undertake the trial.

Data collection

At harvest, data were collected on both roots and aboveground-part traits in order to select promising clones for dual-purpose use. Regarding yield-related characteristics, several root traits were considered such as marketable root weight and number and non-marketable root weight and number. Quality and nutritional traits were also recorded such as skin colour (SC) and flesh colour (FC). To characterize vines, data were collected on vine weight (VW), vine vigour (VV), SPVD and Alternaria stem blight. Percentage establishment, the first and second SPVD scores were evaluated 4, 8 and 16 weeks after planting, respectively. Diseases (SPVD and Alternaria stem blight) score were done using a scale of 1–9, where 1 stands for no visible symptoms, and 9 stands for very severe symptoms on the whole plot.

Sweetpotato storage root dry matter content (DMC) was determined using the method of Benesi *et al.* (2004). A duplicated sample of 200 g of chopped storage roots was placed in an open-topped paper bag and oven-dried at 65°C for 72 h to constant weight. After removal from the oven, samples were weighed immediately using a balance that is accurate to 0.1 g. The same procedure was used for vine DMC evaluation.

Classification of dual-purpose use varieties was done according to Leon-Velarde *et al.* (1997), based on the ratio of total dry matter yield of roots to vines (R/V), and clones were classified into five types namely:

- forage (R/V ranging from 0 to 1.0);
- low dual purpose (R/V ranging from 1.0 to 1.5);
- high dual purpose (R/V ranging from 1.5 to 2.0);
- low root production (2.0–3.0); and
- high root production (> 3.0).

Data analysis

Data were subjected to analysis of variance (ANOVA) using GenStat. A two-way ANOVA in RCBD was used. Where 'F' statistics revealed significance, means were separated using Fisher's protected least significance difference (LSD) test at $P = 0.05$.

3.3 Results and Discussion

Clones RW11-1860 and RW11-4923 produced high DMC of 37.8% and 37.3%, respectively. Six clones, namely RW11-1860, RW11-17, RW11-2419, RW11-4923, RW11-2560 and clone RW11-2910, had an R/V ratio ranging between 1.5 and 2.0, and were therefore considered as dual-purpose clones (Table 3.1). Clone RW11-17 was the best performer for root yield with a mean of 28.1 t/ha (Table 3.2). There were significant differences ($P < 0.05$) between entries (genotypes) and between locations for root yield, but the interactions between locations and entries were not significantly different.

Regarding national performance trials data, the best performing clone in root yield across locations was RW11-5091 with a mean of 25.3 t/ha, followed by clone RW11-2560 with a root yield mean of 22.2 t/ha (Table 3.3). Based on location performance, the overall location mean in Rubona, Karama and Ngoma was 21.7 t/ha, 17.5 t/ha and 15.1 t/ha, respectively.

There were significant differences ($P < 0.05$) between clones for vine weight, but no significant differences were found neither for locations nor for the interaction, location × clone. Based on the overall clone means averaged across locations, clone Kwezikumwe performed well with a vine weight mean of 15.4 kg, followed by the clone RW11-17. The clone with a lowest value for this trait was RW11-3736 with a mean of 3.8 kg (Table 3.3).

SPVD and Alternaria stem blight had no clear indication that disease incidence significantly reduced either root yield or vine yield. However, clone RW11-3736 was scored 5 during disease symptoms evaluation (Table 3.3), and therefore was more affected by SPVD compared with the other clones with SPVD score ranging from 1.3 (clones RW11-17 and RW11-2285) to 2.8 (clones RW11-2419 and RW11-3074). The SPVD mean score averaged in each location was 2.1, 2.2 and 2.8 at Rubona, Ngoma and Karama, respectively (results not shown). These results are similar to those of ISAR (2000) stating that Karama site is located in a high SPVD pressure area of Rwanda.

The clone means for marketable and non-marketable root weight (MRW and NMRW) averaged over three locations are shown in Table 3.3. The overall clone mean MRW over three locations was 6.1 kg, on a plot mean basis. Four clones out of 11 had a higher mean MRW compared with the overall mean. These clones were RW11-17, RW11-1860, RW11-2560 and RW11-5091 and their mean MRW were 7.1 kg, 6.8 kg, 8.1 kg and 9.9 kg, respectively. Referring to the results from each location, the highest location mean for this trait was 7.6 kg at Rubona, whereas the lowest clone mean of 4.7 kg was found at Ngoma. Regarding single location means, the best performing clone for MRW was RW11-2560 at Rubona with a mean of 12.0 kg.

Based on on-farm trial results, all the ten clones evaluated showed a higher mean root yield compared with the two checks. The highest on-farm fresh root yield of 20 t/ha was recorded by clone RW11-5091, followed by clone RW11-2419 (18.7 t/ha) (Table 3.4). The lowest mean on-farm root yield of 6.2 t/ha was recorded by Tanzania, a local check that farmers have been using.

The coefficients of variation (CV) for MRW (CV = 35.1%) and vine weight (CV = 35.7%) were generally high, indicating a large plot-to-plot variability (Table 3.4). This concurs

Table 3.1. Quality attributes and reaction to pests and diseases of the six sweetpotato varieties in Rwanda, February 2013.

Attribute[a]	RW11-17	RW11-1860	RW11-2419	RW11-2560	RW11-2910	RW11-4923	Local check (Kwezikumwe)
DMC (%) of roots	30.8	37.8	24.9	21.0	31.1	37.3	30.0
DMC (%) of vines	17.0	20.7	17.6	20.6	18.3	19.2	18.0
DMY of roots (t/ha)	8.3	9.5	6.3	5.4	7.9	9.4	5.4
DMY of vines (t/ha)	4.1	4.8	3.4	2.7	3.9	4.5	2.6
Ratio (R/V)	2.0	1.9	1.9	2.0	2.0	2.0	2.1
Maturity (days)	135–150	120–135	135–150	120–135	120–135	135–150	120–135
Field reaction to SPVD	MR[b]	MR	MR	MR	MR	MR	MR
Field reaction to Alternaria stem blight	MR	MR	MR	MR	MR	MR	MR
Field reaction to weevils	MR	MR	MR	MR	MR	MR	MR

[a]DMC, dry matter content; DMY, dry matter yield; R/V, ratio of root to vines; SPVD, sweetpotato virus disease.
[b]MR, moderately resistant = moderate damage.

Table 3.2. Ratio root/vine (R/V) and other related attributes[a] of ten different clones evaluated in Rwanda, 2012 B season.

No.	Clone	Vine yield (t/ha)	Root yield (t/ha)	DMCR (%)	DMCV (%)	DMYR (t/ha)	DMYV (t/ha)	Ratio (R/V)	Classification[b]
1	RW11-5091	11.1	20.7	28.8	16.1	6.0	1.8	3.3	HRP
2	RW11-3074	4.6	13.1	33.4	16.6	4.4	1.0	5.7	HRP
3	RW11-3736	15.1	27.5	23.6	16.3	6.5	2.5	2.6	LRP
4	RW11-4923	22.8	25.1	37.5	19.2	9.4	4.5	2.0	HDP
5	RW11-1860	23.3	25.3	37.8	20.7	9.5	4.8	1.9	HDP
6	RW11-2560	13.1	25.8	21.0	20.6	5.4	2.7	2.0	HDP
7	RW11-17	23.9	28.1	30.4	17.0	8.3	4.1	2.0	HDP
8	RW11-2910	21.1	25.3	31.2	18.3	7.9	3.9	2.0	HDP
9	RW11-2285	21.7	22.8	30.4	13.6	6.9	2.9	2.4	LRP
10	RW11-2419	19.2	25.3	24.9	17.6	6.3	3.4	1.9	HDP
Grand mean		17.6	23.9	29.9	17.6	7.1	3.1	2.6	

[a]DMCR, dry matter content of roots; DMCV, dry matter content of vines; DMYR, dry matter yield of roots; DMYV, dry matter yield of vines.
[b]HRP, high root production (> 3.0); LRP, low root production (2.0–3.0); HDP, high dual purpose (R/V ranging from 1.5 to 2.0).

Table 3.3. Performance[a] of promising sweetpotato clones across three locations Rubona, Karama and Ngoma, 2013 A season.

Clone	Root yield (t/ha)	MRW (kg)	MRN	NMRN	NMRW (kg)	VV	VW (kg)	SPVD	Alternaria
RW11-17	21.6	7.1	29.8	32.0	3.6	5.9	15.4	1.3	1.0
RW11-1860	20.9	6.8	29.0	26.5	3.1	5.2	12.2	2.2	1.3
RW11-2285	11.8	5.0	27.5	24.5	1.7	5.6	10.0	1.3	1.5
RW11-2419	20.3	5.6	21.5	27.7	2.1	5.8	11.7	2.8	1.3
RW11-2560	22.2	8.1	24.8	21.5	2.2	4.8	11.9	2.2	1.0
RW11-2910	20.4	5.8	17.8	27.2	3.1	4.2	12.5	2.5	2.0
RW11-3074	11.7	5.7	23.5	17.2	0.6	5.0	6.1	2.8	1.1
RW11-3736	11.8	1.6	4.0	7.0	0.4	3.0	3.8	5.0	1.7
RW11-4923	19.0	5.9	24.5	34.2	3.6	4.7	14.2	2.0	1.0
RW11-5091	25.3	9.9	24.7	13.7	0.7	4.8	13.5	2.0	1.3
Kwezikumwe	14.3	5.7	25.8	18.0	1.4	6.0	15.4	1.7	1.0
Grand mean	18.1	6.1	23.0	22.7	2.0	5.0	11.5	2.4	1.3
CV (%)[b]	16.1	19.2	32.0	32.0	32.0	16.7	32.0	35.6	36.3
$LSD_{0.05}$[c]	6.0	6.9	6.5	5.9	1.2	1.7	6.7	1.7	1.0

[a]MRW, marketable root weight; MRN, marketable root number; NMRN, non-marketable root number; NMRW, non-marketable root weight; VV, vine vigour; VW, vine weight; SPVD, sweetpotato virus disease.

[b]CV, coefficient of variation.

[c]$LSD_{0.05}$, least significant difference.

Table 3.4. Performance[a] of the ten clones and two checks in on-farm trials in Rwanda.

Clone	Root yield (t/ha)	MRW (kg)	MRN	NMRW (kg)	NMRN	SPVD	Alternaria	Vine weight (kg)
RW11-17	12.9	3.3	20.5	1.8	30.0	2.5	1.0	13.1
RW11-1860	9.8	4.7	25.5	0.5	13.0	2.0	1.0	2.6
RW11-2285	11.8	6.0	65.0	0.8	38.0	2.0	1.0	2.0
RW11-2419	18.7	7.2	40.0	0.7	19.0	2.5	1.0	5.3
RW11-2560	13.4	2.5	12.5	0.6	12.5	1.5	1.0	2.3
RW11-2910	8.31	3.9	27.5	1.2	48.5	2.0	1.0	3.1
RW11-3074	13.2	2.1	9.5	0.4	8.5	3.0	1.0	1.6
RW11-3736	11.4	4.9	49.0	0.3	24.0	2.0	1.0	3.8
RW11-4923	9.1	3.3	17.0	1.8	41.0	2.0	1.0	4.0
RW11-5091	20.0	9.6	46.0	1.0	16.0	2.0	1.0	7.0
Check (Kwezikumwe)	6.9	2.5	21.0	0.5	13.0	1.0	1.0	2.5
Local check (Tanzania)	6.2	3.7	33.0	0.5	27.0	1.0	1.0	6.7
Grand mean		4.5	30.5	0.8	24.2	2.0	1.0	4.5
CV (%)[b]	24.5	35.1	30.7	16.5	30.5	9.2	0.0	35.7
$LSD_{0.05}$[c]	6.7	4.4	31.9	1.2	16.2	1.2	0.0	7.3

[a]MRW, marketable root weight; MRN, marketable root number; NMRW, non-marketable root weight; NMRN, non-marketable root number; SPVD, sweetpotato virus disease.

[b]CV, coefficient of variation.

[c]$LSD_{0.05}$, least significant difference.

with the statement of Andrade *et al.* (2009) and the observations of Shumbusha (2011) that sweetpotato experimental trials have a large plot error for yield traits.

Based on the results from this chapter, six of the best performing clones, namely RW11- 17, RW11-1860, RW11-2419, RW11-2560, RW11-2910 and RW11-4923, were suitable for dual-purpose use, and they were recommended for release in February 2013. Among the six clones, two of them, RW11-2560 and RW11-2910 are orange-fleshed sweetpotato varieties. These new clones will be more useful than the sweetpotato varieties farmers have been using to date, especially for farmers that operate in mixed crop-livestock systems in Rwanda. These new clones could be the source of food and feed, and therefore can be suitable for Rwandese farmers where land availability is a constraint. In addition, through the use of forage as animal feed, the new clones could contribute to the reduction of soil degradation, and this is in line with the implementation of a policy regarding zero grazing practices as one of the ways to overcome environmental degradation in Rwanda. Further research would be useful, to assess cattle preferences and digestibility of these new dual-purpose clones.

Acknowledgements

This work was supported by the Alliance for a Green Revolution in Africa (AGRA), through a breeding project, Grant reference no. 2009 PASS 028. The authors are also grateful to Rwanda Agriculture Board (RAB) for having facilitated researchers during this work. Many thanks are also deserved by J. Low, J. Ininda and C. Yencho for being ever willing to support us throughout this work.

References

Almekinders, C. and Hardon, J. (2006) Bringing Farmers back into breeding. Experiences with participatory plant breeding and challenges for institutionalisation. *Agromisa Special 5*. Agromisa, Wageningen, pp. 135.

Andrade, M., Barker, I., Cole, D., Dapaah, H., Elliott, H., Fuentes, S., Grüneberg, W.J., Kapinga, R., Kroschel, J., Labarta, R., Lemaga, B., Loechl, C., Low, J., Lynam, J., Mwanga, R., Ortiz, O., Oswald, A. and Thiele, G. (2009) *Unleashing the Potential of Sweetpotato in Sub-Saharan Africa: Current Challenges and Way Forward. Working Paper* 2009-1. International Potato Center (CIP), Lima, 197 pp.

Benesi, I.R.M., Labuschagne, M.T., Dixon, A.G.O. and Mahungu, N.M. (2004) Genotype × environment interaction effects on native cassava starch quality and potential for starch use in the commercial sector. *African Crop Science Journal* 12, 205–216.

Claessens, L., Stoolvogel, J.J. and Antle, J.M. (2009) *Ex ante* assessment of dual-purpose sweetpotato in the crop–livestock systems of western Kenya: a minimum-data approach. *Agricultural Systems* 99, 13–22.

Ewell, P.T. and Mutuura, J. (1994) Sweetpotato in the food systems of eastern and southern Africa. Paper presented at *Tropical Root Crops in a Developing Economy*, the Ninth Symposium of the International Society of Tropical Root Crops, 20–26 October 1991, Accra, Ghana.

Food and Agricultural Organization of the United Nations (FAO) (2012) *FAO Statistics*. Rome.

Institut des Sciences Agronomiques du Rwanda (ISAR) (2000) *Rapport Annuel*. Butare, Rwanda.

Leon-Velarde, C., Roca, J., Arteaga, J., Quispe, L. and Parraga, A. (1997) Perspectives on sweetpotato: dual purpose varieties. In: *CIP Program Report 1995–1996*. International Potato Center (CIP), Lima, pp. 291–294.

Lukuyu, B.A and Agili, S. (2013) Exploiting Sweetpotato as an Animal Feed in East Africa. Available at: www.sweetpotatoknowledge.org (accessed 15 April 2013).

Shumbusha, D. (2011) Inheritance of root dry matter content in sweetpotato. MSc thesis, Makerere University, Kampala, Uganda.

Woolfe, J. (1992) *Sweetpotato: an Untapped Food Resource*. Cambridge University Press, Cambridge, 214 pp.

4 Development and Evaluation of New Sweetpotato Varieties through Farmer Participatory Breeding for High Altitudes in Kenya

L. Karanja,* J. Malinga, J. Ndung'u, A. Gichangi, D. Lelgut and J. Kamundia
Kenya Agricultural and Livestock Research Organization (KALRO), Njoro, Kenya

Abstract

Sweetpotato (*Ipomoea batatas* L.) plays a significant role in food and nutritional security in Kenya. However, production is constrained by: (i) Sweet potato virus diseases (SPVD) and weevils (*Cylas* spp.); (ii) shortage of clean planting materials; (iii) lack of suitable varieties for high altitude agro-ecosystems; (iv) poor postharvest handling; and (v) poor market access. In order to enhance the role of sweetpotato for food security, a breeding programme was initiated at Kenya Agricultural Research Institute (KARI)-Njoro with the objective of developing varieties with desirable genetic and agronomic attributes which include high yields, resistance to SPVD and weevils, improved food quality, and market acceptability for high altitudes. Thirty-six potential parents were identified in initial screening of 440 accessions from landraces, improved and imported accessions. Hand and polycrossing process was performed to generate a breeding nursery of 2200 seedlings. Forty-three entries were identified for advancement to preliminary yield evaluation. Out of these, ten entries were advanced to multi-location testing in five sites under the oversight of the National Performance Trials committee. The multi-location trials were conducted on a randomized complete block design with three replicates at Kabianga (LH1-1745 m above sea level (masl)), Ravine (LH3-2167 masl), KARI-Lanet (LH4-1920 masl), Lare (LH4-1900 masl) and KARI-Njoro (LH3-2166 masl). The sites were in the Central Rift Valley region. Participatory variety evaluation by researchers, extensionists, plant regulators and farmers was adopted in all the trials. Standard operating procedures and analyses as prescribed by the International Potato Center and the Kenya Plant Health Inspectorate Service were used to evaluate: (i) yields; (ii) viruses and weevils; (iii) dry matter content; (iv) beta-carotene; and (v) acceptability by consumers. The main result of the trials' evaluation was the release of five new varieties – three orange-fleshed Kenspot-4, Kenspot-5 and Kenspot-3; yellow-fleshed Kenspot-1; and white-fleshed Kenspot-2. The improved varieties have better average root yields (20 t/ha) compared with the national average root yields (12.8 t/ha); moderate field resistance to SPVD and weevils; moderate dry matter content (26–32%); and consumer acceptability. The orange-fleshed sweetpotato varieties will provide consumers and farmers with moderate provitamin A contents which will alleviate vitamin A deficiency in Kenya. Through participatory breeding,

*laurashali@yahoo.com

production capacity of sweetpotato at high altitudes has increased providing food security, farmer adoption knowledge has improved, market strategies have been determined and food quality has improved. Variety recommendation is agroecological specific.

Keywords: breeding, evaluation, farmer participatory, high altitude, *Ipomoea batatas*, orange-fleshed sweetpotato (OFSP)

4.1 Introduction

Sweetpotato (*Ipomoea batatas* L.) is grown in all tropical and subtropical areas of the world and ranks the world's seventh most important food crop (FAO, 2012). It is highly adapted to the arid and semi-arid land areas with limited rainfall (500–800 mm/year) and less soil fertility. It is grown in a wide range of environments. Sweetpotato is an important supplementary staple in southern and eastern Africa (Tumwegamire *et al.*, 2004). Sweetpotato is ranked as the crop of greatest potential impact among all the root crops in Kenya (KARI, 2012). Most of the sweetpotato in Kenya is grown in areas of 1200–1700 m above sea level (masl) and with annual precipitation of 800–1800 mm. The Lake Victoria Basin is the main sweetpotato growing area accounting for over 75% of the national production. It is followed by central Kenya and the coastal region which accounts for 20% and 5%, respectively (Munga *et al.*, 2000).

According to the Food and Agriculture Organization of the United Nations (FAO) statistics over a 10 year period Kenya grows sweetpotato on an average of 61,000 ha with an annual root production of 880,000 t (FAO, 2012). In 2012 production was 859,549 t with an average of 12.8 t/ha (Ministry of Agriculture, Kenya, 2012). Sweetpotato is grown for subsistence by about 10 million people of whom 80% are resource poor. Most farmers grow local varieties which are low yielding and rate poorly in nutritive value as a result of low beta-carotene and protein content. In the new food and nutritional policy for Kenya, sweetpotato is one of the indigenous crops envisaged to play an important role in promoting food and nutritional security in the country. The orange-fleshed sweetpotato (OFSP) in both fresh and processed forms has recently been recognized as an important food-fortification intervention in combatting vitamin A deficiency (VAD) which affects approximately 40% of reproductive women and 84% of pre-school children in Kenya. Realization of the potential of sweetpotato has been constrained by acute shortage of healthy planting material of superior varieties, viruses and weevils, and unreliable markets during surplus production (McKnight Foundation, 2005). Most local varieties are relatively low yielding, narrowly adapted and susceptible to diseases and pests.

Currently sweetpotato is transiting from subsistence to commercial production. Yields of sweetpotato production in Kenya range between 5 and 12 t/ha while the average global yield is 15 t/ha and the potential average is 30 t/ha (McKnight Foundation, 2005). Among the biological factors implicated as the causes of the low yields are virus-induced diseases which rank second to weevils causing up to 98% loss (Gibson and Aritua, 2002; Gutierrez *et al.*, 2003; Mukasa *et al.*, 2003; Fuglie, 2007). In Kenya four major viruses have been detected: (i) sweet potato feathery mottle virus (SPFMV) (family *Potyviridae*); (ii) sweet potato mild mottle virus (SPMMV) (family *Closteroviridae*); (iii) sweet potato chlorotic stunt virus (SPCSV); and (iv) sweet potato chlorotic fleck virus (family *Flexiviridae*) (Ateka *et al.*, 2004). The occurrence of sweet potato leaf curl virus, a member of the begomoviruses only reported from Kenya (Miano *et al.*, 2006) and cucumber mosaic virus (CMV) (Opiyo *et al.*, 2010) are the most recent addition to the list in Kenya. Production of the crop is also hampered by the sweetpotato weevil *Cylas* spp. with *Cylas puncticollis* and *Cylas brunneus* being of key interest in Africa.

Sweet potato virus disease (SPVD), one of the most important diseases, is induced by the synergistic interaction of two viruses: SPFMV (Sakai *et al.*, 1997) and SPCSV (Kreuze *et al.*, 2002). SPVD is especially important in East Africa (Gibson *et al.*, 1998) because it can induce synergistic interactions with other sweetpotato viruses, such as SPMMV (Tairo *et al.*, 2005; Karanja *et al.*, 2008), sweet potato virus G (IsHak *et al.*, 2003) and CMV (Cohen and Loebenstein, 1991). Sweetpotato plants infected with multiple viruses often express more severe disease symptoms and greater yield losses than those observed in single virus infections. The disease causes stunting, mottling and deformation of sweetpotato leaves. Sources of effective resistance that may be useful in breeding programmes have not been identified for SPVD.

Predominant sweetpotato varieties have been white- or yellow-fleshed varieties that contain small amounts of beta-carotene. The OFSP varieties are limited in number but are rich sources of beta-carotene. Consumers prefer varieties with higher dry matter content than is found in currently available OFSP varieties. To encourage acceptability of OFSP varieties the texture and disease resistance of the orange-fleshed varieties needed improvement. Hence a sweetpotato breeding programme was initiated at Kenya Agricultural Research Institute (KARI)-Njoro to improve resistance to viruses and weevils and develop varieties for high altitude that meet consumer preference of high dry matter and address the market needs. There have been no suitable varieties for high altitudes. Adverse and unpredictable climatic conditions in high altitude areas in Kenya have complicated agricultural production for many crops (Bourke, 2001) and hence the need for more genetically superior sweetpotato varieties. The focus on sweetpotato has been informed by the fact that the crop is readily adaptable to altitudes of up to 2400 masl (NCSPC, 2009).

The East African region has significant acreage in the highland ecosystem which is also highly populated and prone to food insecurity. To achieve the objectives for the farmer-desired attributes it was crucial to undertake collection, hybridization, selection, and farmer participatory evaluations at multi-locational trials to identify novel sweetpotato varieties. The work reported in this chapter is a contribution to this effort.

4.2 Materials and Methods

Germplasm collection, selection and maintenance

Collection of germplasm was done in 2007 from farmers' fields, Kenya Genebank, Kenya Agricultural Research Institutes (Kakamega, Embu, Katumani and Mtwapa) and the International Potato Center (CIP), Nairobi. The material was planted at KARI-Njoro field station. Characterization of 150 of the accessions using molecular tools was done to eliminate duplicates. Negative selection was done by rouging-out infected plants with routine spraying for insects and periodic rotation to maintain a virus-free field.

Crossing and evaluation

During 2007/2008 germplasm for the crossing programme was selected through farmer participatory approaches on the characterized material. Thirty-six parents preferred by farmers with orange, yellow and white flesh were selected and incorporated into the breeding programme. Two crossing blocks were established at KARI-Njoro station (2166 masl) at LH3 and Perkerra (1080 masl) at LM5 agroecological zones (AEZ). Genotype combinations with farmer preferred attributes were performed. Routine management of the breeding material was done as described in the *Sweetpotato* (Ipomoea batatas) *Breeding* manual (Wilson *et al.*, 1989).

Seeds were harvested from 56 lines consisting of both hand and polycrosses after 1 month and used to raise seedlings in the greenhouse. In 2009 a total of 2200 seedlings were selected and planted in the field from which 200 clonal materials were advanced to preliminary yield trial (PYT) and planted at two sites – Njoro and Perkerra. Single rows of 12 plants in two replications were planted at inter- and intra-row spacing of 30 cm × 100 cm. Harvesting was done after 180 days at both sites. Evaluation of PYT was done through farmer and scientists' participatory process. Forty-three superior clones were selected for advancement to National Performance Trial (NPT).

NPT evaluation

Two seasons of NPTs were done as recommended by Kenya Plant Health Inspectorate Service (KEPHIS) at five sites, namely: (i) Kabianga (LH1-1745 masl); (ii) Ravine (LH3-2167 masl); (iii) KARI-Lanet (LH4-1920 masl); (iv) Lare (LH4-1900); and (v) KARI-Njoro (LH3-2166 masl). The sites were all in the Central Rift Valley. The experimental design was the randomized complete block design with three replicates. A plot size of 4 m × 6 m with 80 plants was established. A total of ten entries (six candidates and four control varieties) were used. In all the sites harvesting was done after 180 days. Data collection on vigour, viruses and weevils was taken as recommended in the CIP breeding manual (Grüneberg *et al.*, 2010). Storage root yield, adaptability to the different AEZ, marketability and taste acceptability were determined through farmer participatory evaluation in each of the two seasons. All data were entered onto a Microsoft Excel spreadsheet and analysed with both SAS 9.3 and cloneSelector programs. On-farm trials were also conducted on 11 small-scale farms under rain-fed conditions. One on-station adaptability trial was conducted at KARI-Kakamega under another agroecozone (UM1) to evaluate the performance of new varieties.

Evaluation of SPVD disease

The varieties were evaluated for virus infection on-station and in the multi-locational trials using visual observation on a scale of 1–9 as detailed in the *CIP Manual 2010* (Grüneberg *et al.*, 2010) and using nitrocellulose membrane (NCM)-ELISA and PCR to confirm the presence of the viruses and their identity. Classifications of the relative resistance to disease and weevil damage were based on field evaluation under natural disease and weevil population pressures as described by Mwanga *et al.* (2002, 2007). Ten varieties were screened for SPVD at KARI-Njoro for 6 months in a randomized complete block design experiment replicated three times. SPVD incidence and severity were determined using a score rating of 1–9 and data were recorded. NCM-ELISA and PCR were used for detection and species identification of SPFMV, SPCSV, SPMMV and CMV according to Gutierrez *et al.* (2003) and Maruthi *et al.* (2002). Analysis was done to determine occurrence, incidence and severity.

4.3 Results

Germplasm collection, selection and maintenance

A total of 440 accessions were collected – 50 introductions and 17 mega-clones from CIP Nairobi, 80 improved clones from various NARS (National Agricultural Research System) centres, 254 accessions from Genebank, 19 landraces from farmers and 20 breeder's material from KARI-Njoro breeding programme. All had good adaptability and desirable farmer traits. Out of these 150 which were characterized enabled removal of duplicates from the germplasm.

Crossing and evaluation

By the end of 3 years, over 78,000 seeds had been harvested with 12,000 from controlled crosses and 66,000 from polycrosses, all originating from 36 parents. Seedling and clonal nurseries for approximately 2200 seedlings and 200 clones were established in the field by December 2008. In PYT evaluation, 43 superior clones were selected through farmer participatory evaluation from the clonal nursery.

NPT evaluation

Harvesting of both season one and two NPT was done through a farmer participatory process and results were analysed by a KEPHIS biometrician and verified with the cloneSelector program from CIP.

Overall means for storage root yield, quality attributes, and virus and weevils' infection are indicated for the ten varieties from all the five sites (Table 4.1). Results for

Table 4.1. Mean sweetpotato marketable root yield (t/ha), quality attributes, and pest and disease infection across the five sites – Kabianga, KARI-Lanet, KARI-Njoro, Lare and Ravine of the Central Rift Valley in Kenya over a period of two National Performance Trial (NPT) seasons.

	Mean weight of tubers (t/ha)			Quality parameters		Pest and disease	
Genotype	Marketable	Non-marketable	Total	Dry matter (%)	Beta-carotene (mg/100 g)	Virus disease (score 1–9)	Weevil damage (score 1–9)
KNSP015	28.94	9.82	38.75	21.02	5.27	4.25	0.00
Kenspot-1	23.01	8.96	31.94	29.37	0.55	4.04	0.00
Kenspot-2	20.93	2.95	23.43	26.19	0.25	4.65	0.00
Kenspot-3	18.69	6.00	24.63	32.52	1.08	3.75	0.00
Kenspot-4	17.14	6.45	23.71	30.41	3.08	4.29	0.00
Kenspot-5	14.57	6.67	21.51	27.63	4.45	3.58	0.00
Mugande	16.22	3.20	19.25	28.56	0.01	4.50	0.00
SPK013	13.46	2.75	16.38	31.12	0.01	5.58	0.00
Kemb 10	12.70	4.90	17.58	29.69	0.03	4.46	0.00
SPK004	8.32	4.06	12.05	30.67	1.41	4.29	0.00
Mean	16.03	6.49	22.50	28.18	2.09	4.24	0.00
P value	0.00	0.00	0.00	0.00	0.00	0.00	0.00
CV (%)[a]	38.63	60.87	31.75	7.14	51.19	13.86	0.00
R^{2}[b]	0.69	0.57	0.73	0.82	0.87	0.92	0.00
LSD (5%)[c]	6.15	3.93	7.10	2.00	1.06	0.58	0.00

[a]CV, coefficient of variation.
[b]R^2, index of stability of varieties.
[c]LSD, least significant difference.

the new varieties were above the national average of 12.8 t/ha. The yields of new cultivars were better than those of local control cultivars (Mugande, SPK013, Kemb 10 and SPK004).

The new varieties varied in root yield and biomass, and three of the varieties Kenspot-3, Kenspot-4 and Kenspot-5 had orange-fleshed storage roots with more pro-vitamin A. The dry matter had a mean of 28.18% with acceptability of 3.2–3.7 on a scale of 1 to 5, 1 being the worst and 5 the best.

Four viruses, namely SPFMV, SPCSV, SPMMV and CMV, were detected in the sweetpotato clones tested at NPT for the ten varieties from KARI-Njoro site. PCR was performed to verify the presence of the four viruses and the results confirmed they were present. All the varieties were infected by between two and four different viruses.

Results of the on-station trial at Kakamega showed the superiority of the new varieties (Table 4.2). The first two varieties (Kenspot-4 and Kenspot-5) were exceptionally good at the adaptive on-station trial in both yields and farmer acceptability. The large number of unmarketable roots obtained from Kenspot-5 is an indication that the variety could also be used with partial harvesting. All the five varieties had better yields than the local control varieties used (OPKAROT, ALUPEOR and OPKANYA).

Overall results of the multi-locational trial for five sites indicated Kenspot-4, Kenspot-5, Kenspot-3, Kenspot-1 and Kenspot-2 as the best varieties with the most desired traits (Table 4.3). Most of the varieties are specifically good for highland production but the results also showed that they are equally good for low altitudes. The five varieties were officially released by the Kenya Plant Variety Release Committee and Ministry of Agriculture, Livestock Development and Fisheries on 13 June 2013.

4.4 Discussion

Compared with the local controls the new improved sweetpotato varieties, Kenspot-3, Kenspot-4, Kenspot-5, Kenspot-2 and Kenspot-1 have good agronomic traits, are moderately resistant to sweetpotato viruses and are highly acceptable to farmers. In addition they have higher beta-carotene and dry matter content and are adapted to highland agro-ecozones. Since their release they have been tested outside the project area in KARI-Kakamega (UM – high virus pressure zone), Daystar University environs in Machakos, KARI-Marigat (LM5) under irrigation, Tinderet in Nandi County and Kutus in Kerugoya County. Two of the OFSP varieties, Kenspot-4 and Kenspot-5, had superior yields with being tolerant to drought, when tested among the Kenyan improved varieties. A third variety with moderate levels of beta-carotene (Kenspot-3) had comparative yields to current improved varieties. Stability tests (R^2) also showed that the new varieties were stable across the season and sites with an average R^2 index of 0.70. Results also showed that some of the varieties were site specific and their performance was better at one site than others. The cultivars have low-to-moderate levels of field resistance to SPVD. The release of these five cultivars will provide consumers and farmers with high-quality sweetpotatoes.

The most prevalent and serious sweetpotato viruses in Kenya are SPFMV and SPCSV. When both infect the plant severe SPVD is observed. This is consistent with previous reports (Ateka *et al.*, 2004; Nyaboga, 2007). The other observed viruses were SPMMV and CMV. The results emphasize the need for implementing careful indexing procedures for the new sweetpotato varieties, as they are disseminated to the growers.

4.5 Conclusion

These five varieties are expected to perform well in AEZ with low-to-moderate SPVD pressure and with well-distributed rainfall for at least 2–3 months during the growth period. Appropriate varieties, timely and adequate quality and quantity of planting materials will ensure efficient quality planting material, multiplication and dissemination systems for enhanced storage root production targeting processed and fresh-root markets. Disease and pest resistance will

Table 4.2. On-station evaluation trial at Kakamega for the five new sweetpotato varieties showing mean storage root yield, vines and virus infection.

Genotype	No. plants harvested	No. marketable roots	No. non-marketable roots	Weight of marketable roots (t/ha)	Weight of non-marketable roots (t/ha)	Vine weight (kg)	Virus disease rating (scale 1–9)	
							After 1 month	After 5 months
Kenspot-2	37.00	99.00	40.67	49.38	5.80	37.70	1	2
Kenspot-1	37.67	85.00	79.67	32.22	8.39	54.70	2	3
Kenspot-4	40.67	119.00	222.00	30.49	10.00	46.20	1	5
Kenspot-3	35.33	89.00	130.00	22.84	8.83	53.20	1	5
Kenspot-5	39.00	127.00	419.67	22.59	6.54	60.30	1	5
OPKAROT	34.00	71.00	57.33	20.99	5.06	57.80	1	1
ALUPEOR	37.00	63.00	56.33	20.74	3.09	42.20	1	6
OPKANYA	35.00	94.00	113.33	18.39	6.29	47.50	2	3
Mean	36.90	93.60	139.80	27.20	7.10	49.90		
P value	0.20	0.01	0.00	0.00	0.01	0.40		
CV (%)[a]	2.10	2.10	2.10	2.10	2.10	2.10		
LSD[b]	4.90	32.90	82.80	12.40	4.10	22.40		

[a]CV, coefficient of variation.
[b]LSD, least significant difference.

Table 4.3. The five sweetpotato varieties released and their special attributes for high altitudes in Kenya (1700–2300 masl).

Variety name/code	Year of release	Owner(s)	Maintainer and planting material source	Optimal production altitude range (masl)	Duration to maturity (months)	Tuber yield (t/ha/year)	Special attributes
Kenspot-1	2012	KARI	KARI-Njoro	1700–2300	6–7	15–25	High yielding (23 t/ha); moderate dry matter (29.4%); yellow-fleshed; and average acceptability (3.3)
Kenspot-2	2012	KARI	KARI-Njoro	1700–1900	6–7	15–46	High yielding for specific highland areas (21 t/ha) good for Kabianga area; moderate dry matter (26.2%); high acceptability (3.6); and white-fleshed
Kenspot-3	2012	KARI	KARI-Njoro	1900–2300	6–7	10–27	High dry matter (32.5%); orange-fleshed with average beta-carotene (1.08 mg/100 g); good yields (18.7 t/ha); average acceptability (3.3)
Kenspot-4	2012	KARI	KARI-Njoro	1700–2300	6–7	10–26	High dry matter (30.4); orange-fleshed with high beta-carotene (3.08 mg/100 g); average acceptability (3.2)
Kenspot-5	2012	KARI	KARI-Njoro	1700–2100	6–7	10–23	Orange-fleshed with high beta-carotene (4.7 mg/100 g); moderate dry matter (25.9%); moderately resistant to virus diseases; moderate acceptability (3.0)

reduce use of chemicals and hence protect the environment from chemical pollution. All these outputs will facilitate promotion of food security. In the absence of a seed company that deals in sweetpotato planting materials, spread of the released varieties in the counties has been mainly through farmer-to-farmer exchange or sale of planting materials and promotions by non-government organizations, schools, farmer groups and government departments. The released varieties have already reached seven counties in Kenya since their release.

4.6 Recommendations

Involvement of both farmers and agriculture extension staff through participatory variety development at all stages gave the stakeholders a sense of ownership of the varieties and hence farmer participatory activities should be promoted throughout the sweetpotato value chain. The demand for the sweetpotato planting materials increased and farmers are more willing to adopt the agronomic practices for better production. They were particularly keen on value addition and formation of marketing groups for their produce. To meet the current demand it is crucial to set up systems for seed dissemination and conservation.

Acknowledgements

We wish to acknowledge the Alliance for a Green Revolution in Africa (AGRA) for their financial support, International Potato Center (CIP), KARI-Director and KARI Centres, Genebank of Kenya and farmers for availing germplasm and Mr Charles Ngare for reviewing the manuscript.

References

Ateka, E.M., Njeru, R.W., Kibaru, A.G., Kimenju, J.W., Barg, E., Gibson, R.W. and Vetten, H.J. (2004) Identification of viruses infecting sweet potato in Kenya. *Annals of Applied Biology* 144, 371–379.

Bourke, R.M. (2001) An overview of food security in Papua New Guinea. In: Bourke, R.M., Allen, M.G. and Salisbury, J.G. (eds) *Proceedings of Papua New Guinea Food and Nutrition 2000 Conference*, ACIAR Proceedings No. 99. Papua New Guinea University of Technology, Lae, Papua New Guinea, pp. 26–30.

Cohen, J. and Loebenstein, G. (1991) Role of a whitefly-transmitted agent in infection of sweetpotato by cucumber mosaic virus. *Plant Disease* 75, 291–292.

Food and Agriculture Organization of the United Nations (FAO) (2012) FAOSTAT database. Available at: http://faostat.fao.org/ (accessed 28 April 2013).

Fuglie, K.O. (2007) Priorities for sweetpotato research in developing countries: results of a survey. *HortScience* 42, 1200–1206.

Gibson, R.W. and Aritua, A. (2002) The perspective of sweet potato chlorotic stunt virus in sweet potato production in Africa: a review. *African Crop Science Journal* 10, 281–310.

Gibson, R.W., Mpembe, I., Alicai, T., Carey, E.E., Mwanga, R.O.M., Seal, S.E. and Vetten, H.J. (1998) Symptoms, etiology and serological analysis of sweet potato virus disease in Uganda. *Plant Pathology* 47, 95–102.

Grüneberg, W.J., Eyzaguirre, R., Espinoza, J., Mwanga, R.O.M., Andrade, M., Dapaah, H., Tumwegamire, S., Agili, S., Chipungu, F.P., Attaluri, S., Kapinga, R., Nguyen, T., Kaiyung, X., Tjintokohadi, K., Carey, T. and Low, J. (2010) Procedures for the evaluation and analysis of sweetpotato trials, 2nd version. In: *CIP Manual 2010*. International Potato Center (CIP), Lima.

Gutierrez, D.L., Fuentes, S. and Salazar, L.F. (2003) Sweet potato virus disease (SPVD): distribution, incidence and effect on sweet potato yield in Peru. *Plant Disease* 87, 297–302.

IsHak, J.A., Krueze, J.F., Johansson, A., Mukasa, S.B., Tairo, F., Abo El-Abbas, F.M. and Valkonen, J.P.T. (2003) Some molecular characteristics of three viruses from SPVD-affected sweet potato plants in Egypt. *Archives of Virology* 148, 2449–2460.

Karanja, L., Malinga, J., Nyabonga, E. and Ndungu, J. (2008) Virus screening in sweet potato germplasm for farmer participatory breeding program in Central Rift, Kenya. In: *Proceedings of 11th KARI Biennial Conference and the 3rd Agricultural Forum*, 10–14 November, Nairobi, Kenya. Kenya Agricultural Research Institute (KARI), Nairobi.

Kenya Agricultural Research Institute (KARI) (2012) *Agricultural Productivity Value Chain (APVC) Report, Strategic Plan Implementation Framework 2009–2014*. KARI, Nairobi.

Kreuze, J.F., Savenkov, E.I. and Valkonen, J.P.T. (2002) Complete genome sequence and analysis of the subgenomic RNAs of sweet potato chlorotic stunt virus reveal several new features for the genus *Crinivirus*. *Journal of Virology* 76, 9260–9270.

Maruthi, M.N., Colvin, J., Seal, S., Gibson, G. and Cooper, J. (2002) Co-adaptation between cassava mosaic geminiviruses and their local vector populations. *Virus Research* 86, 71–78.

McKnight Foundation (2005) McKnight East Africa Sweetpotato Project progress report. *Annual Report*. McKnight Foundation, Minneapolis, Minnesota.

Miano, D.W., LaBonte, D.R., Clark, C.A., Valverde, R.A., Hoy, M.W., Hurtt, S. and Li, R. (2006) First report of a *Begomovirus* infecting sweetpotato in Kenya. *Plant Disease* 90, 832.

Ministry of Agriculture, Kenya (2012) An overview on the root and tuber crops. *Annual Report*. Ministry of Agriculture, Kenya, Nairobi.

Mukasa, S.B., Rubaihayo, P.R. and Valkonnen, J.P.T. (2003) Incidence of viruses and virus-like diseases of sweet potato in Uganda. *Plant Disease* 87, 329–335.

Munga, T.L., Chege, P.K., Wanjohi, B., Odhiambo, J.O., Wekesa, E.N. and Kega, U.R. (2000) On-farm sweet potato variety evaluation in the lowland coastal Kenya. *African Potato Association Conference Proceedings* 5, 131–137.

Mwanga, R.O.M., Yencho, G.C. and Moyer, J.W. (2002) Diallel analysis of sweet potatoes for resistance to sweet potato virus disease. *Euphytica* 128, 237–248.

Mwanga, R.O.M., Odongo, B., Niringiye, C., Kapinga, R., Tumwegamire, S., Abidin, P.E., Carey, E.E., Lemaga, B., Nsumba, J. and Zhang, D. (2007) Sweet potato selection releases: lessons learnt from Uganda. *African Crop Science Journal* 15, 11–23.

North Carolina Sweet Potato Commission (NCSPC) (2009) Sweet Potato Production 1976–2008. Available at: www.ers.usda.gov/media/1834605/vgs-355-sa1.pdf (accessed 15 September 2015).

Nyaboga, E.N. (2007) Characterization of transgenic sweet potato (*Ipomoea batatas* L.) lines and the response of transformants to virus inoculation. MSc thesis, University of Nairobi, Kenya.

Opiyo, S.A., Ateka, E.M., Owuor, O., Manguro, L.O.A. and Miano, D.W. (2010) Cucumber mosaic virus (CMV) infecting sweetpotato in Kenya. Sweetpotato Knowledge Portal. Available at http://www.sweetpotatoknowledge.org/ (accessed 15 April 2014).

Sakai, J., Mori, M., Morishita, T., Tanaka, M., Hanada, K., Usugi, T. and Nishiguchi, M. (1997) Complete nucleotide sequence and genome organization of sweet potato feathery mottle virus (S strain) genomic RNA: the large coding region of the P1 gene. *Archives of Virology* 142, 1553–1562.

Tairo, F., Mukasa, S.B., Jones, R.C., Kullaya, A., Rubaihayo, P.R. and Valkonen, J.P.T. (2005) Unravelling the genetic diversity of the three main viruses involved in sweet potato virus disease (SPVD), and its practical implications. *Molecular Plant Pathology* 6, 199–211.

Tumwegamire, S., Kapinga, R., Zhang, D., Crissman, C. and Agili, S. (2004) Opportunities for promoting orange-fleshed sweet potato as a mechanism for combating vitamin A deficiency in sub-Saharan Africa. *African Crop Science Journal* 12, 241–252.

Wilson, J.E., Pole, F.S., Smit, N.E.J.M. and Pita Taufatofua (1989) *Sweetpotato* (Ipomoea batatas) *Breeding*. Institute for Research, Extension and Training in Agriculture (IRETA) Publications, Apia, Western Samoa.

5 *In Vitro* Evaluation of Orange-fleshed Sweetpotato Genotypes for Drought Tolerance Using Polyethylene Glycol

S. Agili,[1*] B.N. Aggrey,[2] K. Ngamau[2] and W.P. Masinde[2]
[1]*International Potato Center Sub-Saharan Africa (CIP-SSA), Nairobi, Kenya;*
[2]*Jomo Kenyatta University of Agriculture and Technology, Nairobi, Kenya*

Abstract

In vitro techniques have been shown to be useful in identifying relatively drought-tolerant genotypes at early stages of development. In this study, drought-induced alterations in early shoot and root development of 59 sweetpotato genotypes was evaluated in the tissue culture laboratory at the Kenya Plant Health Inspectorate Services, Quarantine station, Muguga, Kenya. These genotypes were obtained from Lima, Peru and were evaluated against two known Kenyan check varieties, Marooko (drought tolerant) and K566632 (drought susceptible). Plantlets of each genotype were raised on Murashige and Skoog basal solid medium, from an original node consisting of a 0.2–0.5 cm stem segment. These were exposed to polyethylene glycol (PEG 6000) at three different concentration levels: 0, 10 and 15 g/l. The experiment was laid out in a factorial completely randomized design with three replications. Data on shoot and root growth were recorded during tissue regeneration lasting over a period of 4 weeks per regeneration cycle. Analysis of variance indicated genotypes, salt levels and salt level × genotype interaction, were highly significant ($P < 0.01$) with respect to all the traits. At 15 g/l concentration of PEG, genotypes 189135.9, 194515.5, 440024, 441724 and 440001 had roots that were longer than those of Marooko. Genotypes 194515.5, 194539.36, 441724, 441538, 189135.9, 41768, 192033.5 and 440429 recorded greater leaf expansion, greater stem length elongation, greater root and shoot growth and high dry matter production at all salt levels indicating their ability to withstand severe water stress conditions. Genotypes 189151.38, 420027, 440132, 440104, 440034, 421111, 440166 and 441755 were identified as susceptible. The technique was found to be a useful tool for screening a large number of sweetpotato genotypes within a relatively short period of time.

Keywords: drought tolerance, genotype, *in vitro* screening, polyethylene glycol salt

5.1 Introduction

Drought is one of the most common environmental stresses affecting plant growth and productivity (Boyer, 1982). Under field conditions, drought severity, timing and duration vary from year to year and a cultivar which is successful in one year, might fail in another year. The unpredictable and variable forms in which drought stress is manifest

*s.agili@cgiar.org

complicates the selection of superior plant materials as well as breeding programmes, hence the need to do *in vitro* screening. Plant cell and tissue culture has been a useful tool to study stress tolerance mechanisms under *in vitro* conditions (Baijji *et al.*, 2000). *In vitro* culture techniques minimize environmental variations due to the use of defined nutrient media, controlled conditions and the homogeneity of stress application (Sakthivelu *et al.*, 2008). In addition, the flexibility of the technique enables a large plant population and various stress treatments to be studied in a limited space and over a short period of time. One of the screening techniques based on physiological traits is the use of various osmotica to induce stress in plant tissues. Germination in mannitol and polyethylene glycol (PEG) has been suggested for drought screening (Shabir *et al.*, 2010; Geravandi *et al.*, 2011).

Several studies also support correlations between whole-plant and cell-culture responses for salt tolerance. Barlass and Skene (1981) found relative tolerance of grape (*Vitis* spp.) cultivars to salt to be the same *in vitro* and for whole plants. Likewise, Orton (1980) determined for cultivated barley (*Hordeum vulgare* L.) and a wild relative that salt tolerance at the cellular level is similar to that at the whole-plant level. Nabors *et al.* (1980) selected salt-tolerant tobacco cell lines in culture, and regenerated plants transmitted the tolerance to subsequent generations. PEG is assumed to be a non-penetrating osmotic agent that lowers the water potential of the medium and has been used to simulate drought stress in plants (Bressan *et al.*, 1981).

Selection for drought tolerance at the early seedling stage is most frequently carried out by including chemical drought-inducing molecules such as PEG 6000 in the medium. This can be used to modify the osmotic potential of nutrient solution culture and thus induce a plant water deficit in a relatively controlled manner, appropriate to experimental protocols (Khanna and Garg, 1997). The objective of the present investigation was to screen a large number of orange-fleshed sweetpotato genotypes for drought tolerance under *in vitro* conditions simulating drought stress conditions.

5.2 Materials and Methods

The *in vitro* experiment was conducted in the tissue culture laboratory of Kenya Plant Health Inspectorate Service, Quarantine station, Muguga, Kenya, located at 1° 11' 0" S, 36° 39' 0" E at an altitude of about 1950 m above sea level in 2006. The genetic materials used in this study consisted of 59 sweetpotato genotypes with contrasting beta-carotene and mineral content. These were provided by the International Potato Center (CIP) and were imported as *in vitro* plantlets from Lima, Peru. The materials were transferred into *in vitro* conditions and routinely propagated from the nodal cuttings. Each node consisted of a 0.2–0.5 cm stem segment with axillary buds with each cycle lasting 2–4 weeks. The plantlets were raised on Murashige and Skoog (MS) basal solid medium (Murashige and Skoog, 1962) containing 30 g sucrose/l and 2.8 g phytogel/l maintained at pH 5.7. These were grown under long day conditions 16 hours of light at 3000 lux and at temperatures ranging from 25°C to 28°C. The genotypes were exposed to different concentrations of PEG 6000 (Merck, Germany). MS basal medium was prepared with PEG 6000 at concentrations of 0, 10 and 15 g/l, poured into Kilner jars and autoclaved at 121°C and 15 lb/square inch for 15 min. Five cuttings per genotype, with two to three nodes each, were placed onto the media in the Kilner jars. All the planted jars were maintained under optimum culture conditions at 10 h photoperiod/day with a photon light flux density of 70 μmol/m^2/s at 28°C. The experiments were conducted in a completely randomized two- factor factorial design with three replications. The main factor was the genotypes and the sub-factor was the salt levels.

Data measurements

Harvesting was done at 65 days from the start of the study. The following data were recorded at harvest:

- **Root length (cm)** – This was determined by measuring the length of the longest root from each sample plant using a metre scale.
- **Root dry mass (g)** – Fresh root samples from plants from each jar were weighed

and heated in an oven for 48 h at 65°C until the weight became constant and these were then re-weighed to determine the dry weight.

- **Leaf area (cm²)** – The linear dimensions of length (L) and width (W) at the broadest part of the lamina of each third leaf from the bottom of the plant were measured with a ruler. The leaf area (A) was then calculated as A = L × W.
- **Shoot length (cm)** – This was determined by measuring the plants in each treatment from the surface of the medium in the jar to the tip of the tallest leaf.
- **Shoot fresh and dry mass (g)** – Fresh shoot samples separated from roots from plants from each jar were collected and weighed and then heated in an oven for 48 h at 65°C until the weight became constant. These were re-weighed to determine the dry weight.

Statistical analysis

Data were subjected to analysis of variance (ANOVA), and means were separated by a Fisher's least significance difference test (LSD) using $P = 0.05$. Water stress levels and their interactions were employed using the SAS program SAS version 8 (SAS Institute, 1999).

5.3 Results

Statistical analysis

ANOVA indicated genotypes, salt levels and salt level × genotype interaction, were highly significant ($P < 0.001$) with respect to all the traits (Table 5.1). This shows the presence of variability, different responses of genotypes to different salt intensities.

Leaf area

A significant decrease in leaf area for genotypes 420027, 440034, 440104, 194549.6 and 440643 was observed with increasing salt concentration (Table 5.2). At the higher salt concentration of 15 g/l, genotypes 189135.9, 194515.5, 441097 and 441768 recorded greater leaf expansion although this was not significantly different from that of the check. Greater mean area expansions relative to the check were noted for genotypes 189135.9, 401055, 441768 and 441097 (Table 5.2).

Root length and root dry weight

Genotypes 189135.9, 421066, 440396, 440429 and 441097 formed roots longer than those of the tolerant check although the differences were not significant. At 15 g/l concentration of PEG, genotypes 189135.9, 194515.5, 440024, 441724 and 440001 exhibited long roots (Table 5.3). Poor root growth at the same level was observed for genotypes 440031, 440286, 440025, 440132 and 420027. Genotypes 189135.9, 441538 and 441768 registered higher root weight that was significantly different from that of the check (Table 5.4). There was significant root weight reduction as stress increased. Genotypes that exhibited higher root weight at 15 g/l concentration of PEG were 189135.9, 194569.1, 440429 and 441768. These were significantly higher weights than that of the check.

Table 5.1. Summary of analysis of variance (ANOVA) showing mean square values for various variables measured during the *in vitro* drought screening of sweet potato genotypes.

Source of variation	Root length (cm)	Root dry weight (g)	Shoot length (cm)	Shoot fresh weight (g)	Shoot dry weight (g)	Leaf area (cm²)
Genotype	907.9**	20.88**	135.22**	14.19**	3.52**	27.24**
Salt level	1889.3**	10.33**	195.03**	45.53**	7.07**	90.59**
Genotype × salt level	907.9**	1.77**	18.52**	9.50**	8.02**	7.96**

**, Significant at $P < 0.001$.

Table 5.2. Effect of salt concentration on leaf area (cm^2) of 59 sweetpotato genotypes during *in vitro* screening using different concentrations of polyethylene glycol.[a]

	Leaf area (cm^2) at salt concentration					Leaf area (cm^2) at salt concentration			
Genotype	0 g/l	10 g/l	15 g/l	Mean	Genotype	0 g/l	10 g/l	15 g/l	Mean
Marooko*	5.50a	6.47a	5.67a	5.9a	440023	4.00a	4.33a	2.83a	3.7a
187016.2	2.33a	1.67a	1.50a	1.8a	440024	1.10a	8.73b	4.63c	4.8c
187017.1	3.17a	2.67a	2.33a	2.7a	440025	3.50a	5.00a	3.33a	3.9a
189123.68	4.00a	0.60b	1.67b	2.1b	440027	2.23a	5.50b	1.50a	3.1a
189135.9	7.00a	7.33a	5.20b	6.5a	440031	5.93a	5.67a	1.80b	4.5a
189140	1.90a	1.63a	1.20a	1.6a	440034	2.80a	0.87a	0.83a	1.5a
189148.21	3.83a	2.50a	3.23a	3.2a	440050	2.00a	4.20a	4.00a	3.4a
189148.65	5.10a	1.70b	2.50b	3.1b	440104	7.33a	3.87b	0.17c	3.8
189150.1	5.33a	4.33a	3.43a	4.4a	440131	5.70a	10.20b	1.47c	5.8a
189151.38	2.63a	5.67b	5.50b	4.6b	440132	4.80a	6.17a	0.00b	3.7a
192033.5	4.17a	6.33b	6.42b	5.6a	440166	2.33a	1.50a	1.67a	1.8a
194515.5	5.10a	9.50b	5.77c	6.8c	440167	1.20a	2.10a	2.37a	1.9a
194521.2	5.33a	2.10b	2.00b	3.1b	440170	2.50a	2.83a	2.93a	2.8a
194539.36	4.33a	6.00a	5.30a	5.2a	440240	4.17c	0.38b	2.17a	2.2a
194541.45	4.17a	3.00a	4.50a	3.9a	440286	5.20a	0.00b	0.00b	1.7b
194549.6	0.57a	0.60a	1.50a	0.9a	440287	6.10a	5.77a	0.00b	4.0a
194555.7	2.20a	3.77a	1.40a	2.5a	440328	8.25a	3.33b	4.20b	5.3b
194569.1	3.53a	2.50a	4.83a	3.6a	440378	2.00a	2.33a	2.50a	2.3a
194573.9	2.00a	0.82a	3.83b	2.2a	440394	1.50a	1.17a	2.50a	1.7a
400011	5.33a	4.33a	4.33a	4.7a	440396	5.00a	8.33b	4.00a	5.8a
401055	5.13a	7.17a	5.63a	6.0a	440429	3.67a	3.50a	4.50a	3.9a
420001	7.00a	1.17a	1.83b	3.3a	440643	1.17a	5.57b	0.57a	2.4a
420014	4.60a	6.13a	3.37b	4.7a	441097	5.67a	10.00b	6.50a	7.4a
420027	1.07a	1.53a	0.57a	1.1a	441538	5.67a	7.30b	5.83a	6.3a
420064	2.20a	1.83a	6.60b	3.5a	441724	3.50a	4.33a	2.00b	3.3a
421066	2.47a	2.33a	1.50a	2.1a	441725	4.13a	5.60a	6.33a	5.4a
421111	3.80a	2.83a	1.23b	2.6a	441755	0.30a	5.03b	3.43b	2.9b
422656	8.00a	2.43b	4.40c	4.9c	441768	7.03a	11.33b	3.80c	7.5a
440001	2.20a	3.00a	2.17a	2.5a	K566632**	11.23a	2.00b	1.70b	5.0b
440017	1.50a	2.33a	3.50a	2.4a					

[a]Means followed by the same letter within the rows (showing differences among different salt levels) are not significantly different ($P \leq 0.05$); *, tolerant genotype check;
**, susceptible genotype check.

Table 5.3. Effect of salt concentration on root length (cm) of 59 sweet potato genotypes during *in vitro* screening using different concentrations of polyethylene glycol.[a]

	Root length (cm) at salt concentration					Root length (cm) at salt concentration			
Genotype	0 g/l	10 g/l	15 g/l	Mean	Genotype	0 g/l	10 g/l	15 g/l	Mean
Marooko*	25.97a	22.00a	21.33a	23.1a	440023	21.00a	10.50a	13.17a	14.9a
187016.2	17.67a	14.67a	13.33a	15.2a	440024	28.77a	69.67b	40.17c	27.4a
187017.1	27.00a	26.33a	27.00a	26.8a	440025	17.83a	25.80b	3.53a	15.7a
189123.68	18.00a	16.67a	15.33a	16.7a	440027	30.33a	28.33a	16.67a	25.1a
189135.9	33.00a	33.53a	35.33a	34a	440031	5.33a	4.33a	3.83a	4.5a
189140	29.13a	28.83a	25.77a	28a	440034	29.80a	27.20a	28.07a	28.4a
189148.21	17.33a	14.80a	17.50a	16.5a	440050	13.30a	10.20a	9.83a	11.2a
189148.65	10.33a	5.67a	8.00a	8a	440104	21.97a	18.17a	22.00a	20.7a
189150.1	15.33a	7.17a	5.33a	9.28a	440131	13.63a	10.83a	9.83a	11.4a
189151.38	11.00a	17.67a	13.33a	14a	440132	17.87a	16.77a	1.87a	12.2a
192033.5	12.67a	17.67a	15.67a	15.3a	440166	18.67a	17.33a	12.00a	16.0a
194515.5	31.33a	33.33a	30.67a	31.8a	440167	30.67a	7.53b	6.50b	11.6a
194521.2	26.33a	21.67a	12.17a	20.1a	440170	20.67a	20.17a	24.00a	21.6a
194539.36	23.83a	23.67a	23.67a	23.7a	440240	21.33a	13.80a	12.63a	15.9a
194541.45	10.53a	21.50a	24.93a	18.9a	440286	11.87a	2.93a	1.43a	5.4a
194549.6	17.33a	9.00a	24.33a	16.9a	440287	27.30a	29.87a	3.50b	20.2a
194555.7	7.33a	29.00b	16.10ab	17.5ab	440328	28.33a	26.00a	17.67a	24.0a
194569.1	21.00a	4.00b	9.17b	11.4b	440378	22.00a	48.33b	12.67a	21.0a
194573.9	8.50a	4.67a	9.73a	7.6a	440394	19.67a	10.33a	5.13a	11.2a
400011	16.33a	25.00a	20.33a	20.6a	440396	36.00a	14.00b	8.73b	19.6a
401055	9.83a	6.33a	10.90a	9.0a	440429	32.33a	25.67a	21.17a	23.1a
420001	28.33a	23.67a	14.50a	22.2a	440643	16.90a	9.33a	3.80a	10.0a
420014	19.30a	16.27a	13.50a	16.3a	441097	32.50a	34.17a	25.33a	30.7a
420027	17.83a	3.60a	2.63a	8.0a	441538	31.33a	21.47a	13.67a	22.2a
420064	20.00a	23.00a	17.83a	20.2a	441724	29.33a	25.00a	29.67a	28a
421066	38.00a	16.67b	21.67b	22.1b	441725	31.33a	25.50a	19.33a	25.4a
421111	2.06a	27.27b	5.00a	11.4a	441755	8.70a	9.67a	10.83a	9.8a
422656	29.67a	25.17a	16.67a	23.8a	441768	32.67a	25.80a	22.17a	26.9a
440001	24.67a	15.00a	31.67a	23.8a	K566632**	13.67a	8.00a	4.53a	8.7a
440017	24.33a	33.67a	5.43b	21.1a					

[a]Means followed by the same letter within the rows (showing differences among different salt levels) are not significantly different ($P \leq 0.05$); *, tolerant genotype check; **, susceptible genotype check.

Table 5.4. Effect of salt concentration on root dry weight (g) of 59 sweet potato genotypes during *in vitro* screening using different concentrations of polyethylene glycol.[a]

Genotype	Root dry weight (g) at salt concentration				Genotype	Root dry weight (g) at salt concentration			
	0 g/l	10 g/l	15 g/l	Mean		0 g/l	10 g/l	15 g/l	Mean
Marooko*	1.27a	1.57a	2.20a	1.7	440023	0.87a	0.44a	0.37a	0.6a
187016.2	0.33a	0.30a	0.06a	0.2	440024	0.03a	0.87a	0.48a	0.5a
187017.1	0.33a	0.99a	0.39a	0.6	440025	1.17a	0.19a	0.03b	0.5a
189123.68	0.22a	0.38a	1.33ab	0.6	440027	0.68b	2.33a	0.04c	1.0a
189135.9	6.00b	7.87a	5.00bc	6.3	440031	0.03a	0.02a	0.01a	0.0a
189140	0.07a	0.01a	0.04a	0	440034	0.07a	0.02a	0.05a	0.04a
189148.21	0.03a	0.04a	0.02a	0	440050	0.01a	0.04a	0.42a	0.2a
189148.65	0.43a	0.02a	0.02a	0.2	440104	0.21a	0.00a	0.02a	0.1a
189150.1	0.57a	0.23a	0.13a	0.3	440131	4.33a	0.06a	0.06a	1.5a
189151.38	0.70a	2.00b	1.07ab	1.3	440132	0.04a	0.04a	0.00a	0.0a
192033.5	0.73a	1.00a	0.88a	0.9	440166	0.30a	0.63a	0.04a	0.3a
194515.5	4.77a	0.63b	3.77a	3.1	440167	0.06a	0.05a	0.04a	0.1a
194521.2	1.34a	0.64a	0.42a	0.8	440170	2.00a	1.73a	1.90a	1.9a
194539.36	1.50a	2.27a	3.17ab	2.3	440240	0.04a	0.16a	0.14a	0.1a
194541.45	1.03a	0.78a	0.65a	0.8	440286	0.34a	0.00a	0.00a	0.1a
194549.6	0.13a	0.21a	1.03a	0.5	440287	0.31a	0.40a	0.07a	0.3a
194555.7	0.49a	0.11a	0.05a	0.2	440328	2.22a	1.39a	1.73a	1.8a
194569.1	4.80b	0.01a	5.03bc	3.3	440378	1.93a	0.93b	0.83ab	1.2a
194573.9	0.39a	0.60a	0.29a	0.4	440394	2.00bc	1.50b	0.05a	1.2b
400011	0.51a	0.23a	0.60a	0.4	440396	2.13a	1.93a	1.23a	1.8a
401055	0.46a	0.81a	0.39a	0.6	440429	2.30bc	1.70b	4.37a	2.8bc
420001	1.81b	0.80a	0.62a	1.1	440643	1.10a	0.24a	0.01a	0.5a
420014	1.73a	1.73a	0.54b	1.3	441097	1.30bc	1.27b	3.53a	2.0a
420027	0.06a	0.01a	0.02a	0.0	441538	5.23bc	6.13b	3.07a	4.8bc
420064	3.15b	0.82a	0.70a	1.6	441724	0.77b	2.73a	0.90c	1.5a
421066	0.97a	0.33a	0.30a	0.5	441725	0.50a	0.04a	0.07a	0.2a
421111	0.06a	0.08a	0.17a	0.1	441755	1.93b	0.75a	0.68a	1.1a
422656	2.21a	1.21a	1.13ab	1.5	441768	5.87b	0.27a	5.37bc	3.9b
440001	1.67b	0.30a	0.20a	0.7	K566632**	0.05a	0.02a	0.04a	0.04a
440017	20.7b	1.40c	0.01a	1.4					

[a]Means followed by the same letter within the rows (showing differences among different salt levels) are not significantly different ($P \le 0.05$); *, tolerant genotype check; **, susceptible genotype check.

Shoot fresh and dry weight

High shoot fresh weight above that of the tolerant check was recorded for genotypes 189135.9, 440170, 440328 and 441538 (Table 5.5). A sharp and significant decrease in shoot fresh weight was recorded for genotypes 194541.45, 420027, K566632 and 440167 at high (15g/l) PEG concentration. In the control treatment genotypes 189135.9, 440328, 440170, 440378 and 441538 produced significantly high shoot dry matter content than the check, whereas genotypes 440429, 194539.36, 441538, 401055, 194515.5 and 189135 recorded higher shoot dry weight at 15 g/l of PEG concentration. The same genotypes recorded higher mean shoot dry weight across the salt levels that were significantly higher than that of the check. The lowest mean shoot dry weights were recorded for genotypes 420027, 440024, 440050, 440167, 440240 and 440286 (Table 5.6).

Shoot length

Increased stress at PEG concentrations of 15 g/l induced longer shoot length for genotypes 187016.2, 187017.1, 194539.36, 420064, 440378 and 441097, and shoot lengths for these genotypes were longer than that of the tolerant check. The same genotypes registered high mean shoot length values across the salt levels. Significant reduction in growth was observed for genotypes 189148.65, 194541.45 and 440286 (Table 5.7).

5.4 Discussion

High concentration of PEG severely reduced the leaf area in the potentially sensitive genotypes such as 440034, 440104, 420027, 189140 and 421111 unlike in potentially tolerant genotypes 189135.9, 194515.5, 440131, 441097 and 441768 that showed greater leaf area. Reduction in leaf area in a canopy results in drastic reduction in transpiration surface (Barta *et al.*, 2002) and subsequently results in low biomass production. This reduction may be due to inhibition of cell division as a result of water stress (Hsiao, 1973). Early detection of such genotypes with greater leaf area indicating tolerance under conditions of moisture stress can save resources in the breeding process.

High PEG concentration significantly reduced total dry matter production in susceptible genotypes 194541.45, 420014, 420027, 440167 and 440394. Genotypes 189135.9, 194515.5, 194539.36, 440027, 440429, 441538 and 401055 were observed to be relatively tolerant with high dry matter production at high PEG concentration of 15 g/l. A similar observation has been made in crops such as lucerne (Barta *et al.*, 2002). Stress affects the rate of photosynthesis thus reducing the supply of assimilates to various parts of the plant (Handa *et al.*, 1982).

In the present study, the shoot length was decreased with an increasing external water stress. Similar observations have been made in other crops (Kulkarni and Deshpande, 2007; Radhouane, 2007). The stem acts as the main reservoir of stored starch which is important during stress situations, as it assists the plant's survival, as well as at optimum yield levels. Stem parameters such as length play an important role in dry matter partitioning of plants for sustaining life during a water-stress situation (Kulkarni and Deshpande, 2006). This might explain why genotypes 187017.1 and 194539.36 exhibited high stem length elongation at high salt concentration.

The ability of the roots to continue to elongate under the situation of water stress was a remarkable character of some of the genotypes screened. Genotypes 189135.9, 194515.5, 441097, 187017.1, 440034, 441768 and 441538 were observed to have long roots and high root weights and would have the ability to survive under high moisture stress conditions. Early and rapid elongation of roots is an important indication of drought tolerance; this facilitates extraction of soil moisture from deep in the soil profile under conditions where there is limited water available (Narayan, 1991; Kim *et al.*, 2001). The rapid elongation of roots also indicates the strength and ability of a genotype to absorb moisture. These parameters are genetically governed and can be introgressed (Kulkarni and Deshpande, 2006).

The present study revealed the different responses of genotypes to various levels of PEG

Table 5.5. Effect of salt concentration on shoot fresh weight (g) of 59 sweet potato genotypes during *in vitro* screening using different concentrations of polyethylene glycol.[a]

	Shoot fresh weight (g) at salt concentration					Shoot fresh weight (g) at salt concentration			
Genotype	0 g/l	10 g/l	15 g/l	Mean	Genotype	0 g/l	10 g/l	15 g/l	Mean
Marooko*	1.63a	1.63a	2.20b	1.8a	440023	1.97b	0.87a	1.00a	1.3b
187016.2	1.10a	0.97a	0.33b	0.8a	440024	0.14b	1.36a	0.49bc	0.7b
187017.1	2.30a	1.94a	0.75b	0.7b	440025	1.07a	0.93a	0.50ab	0.8a
189123.68	1.23b	0.70a	1.43c	1.1b	440027	1.43c	4.33b	0.07a	1.9c
189135.9	6.48c	5.57b	2.83a	5b	440031	0.57a	0.42a	0.21a	0.4a
189140	0.63c	0.14b	1.90a	0.9c	440034	3.13c	2.00b	1.47a	2.2b
189148.21	0.67a	1.00a	0.97a	0.9a	440050	0.21a	0.28a	0.83b	0.4b
189148.65	0.55a	0.52a	0.23a	0.4a	440104	0.99b	0.10a	0.24a	0.4a
189150.1	1.93a	2.00a	2.13a	2.0a	440131	3.50c	0.05b	0.81a	1.5a
189151.38	1.58c	4.50b	2.30a	2.8a	440132	0.80a	0.68a	0.06b	0.5a
192033.5	1.46c	4.77b	2.19a	2.8a	440166	1.17a	1.13a	0.04b	0.9a
194515.5	4.80a	4.33a	2.73b	4a	440167	0.34a	0.26a	0.07a	0.2a
194521.2	2.95c	0.88b	1.60a	1.8a	440170	4.90a	2.83b	1.70c	3.1b
194539.36	4.07a	4.53a	5.73b	4.8a	440240	0.34c	0.16b	0.34a	0.3c
194541.45	1.64c	0.69b	0.06a	0.8b	440286	1.02b	0.06a	0.03a	0.4c
194549.6	0.73a	0.70a	1.43b	1a	440287	0.97b	1.67a	1.93a	1.5a
194555.7	1.31b	0.63a	0.17a	0.7a	440328	5.30b	1.67a	2.08a	3.0a
194569.1	1.60b	0.24a	1.67c	1.7c	440378	5.20c	2.87b	1.40a	3.2b
194573.9	1.06a	1.08a	2.50b	1.5c	440394	3.93c	2.47b	0.29a	2.2
400011	2.03b	1.27a	1.93c	1.7c	440396	3.77a	3.53a	1.03b	2.8c
401055	1.16c	2.00b	2.65a	1.9b	440429	3.77b	3.07a	3.87c	3.6c
420001	4.34c	1.03b	1.69a	2.4c	440643	2.27c	0.88b	0.13a	1.1b
420014	1.39c	0.85b	0.24a	0.8b	441097	1.90c	3.13b	4.30a	3.1b
420027	0.31a	0.21a	0.04a	0.2a	441538	5.48c	4.83b	2.24a	4.2b
420064	1.55a	1.16a	0.97ab	1.2a	441724	1.80b	6.20a	1.47bc	3.2a
421066	2.20b	1.03a	0.73a	1.3a	441725	1.29a	1.47a	1.83b	1.5b
421111	0.73a	0.81a	1.01a	0.9a	441755	0.22b	1.43a	0.61bc	0.8b
422656	4.40b	1.58a	1.65a	2.5c	441768	6.08a	5.70a	3.10b	5.0a
440001	2.50c	1.33b	0.83a	1.6b	K566632**	0.22a	0.30a	0.07a	0.2a
440017	4.43c	2.30b	0.29a	2.3b					

[a]Means followed by the same letter within the rows (showing differences among different salt levels) are not significantly different ($P \leq 0.05$); *, tolerant genotype check; **, susceptible genotype check.

Table 5.6. Effect of salt concentration on shoot dry weight (g) of 59 sweet potato genotypes during *in vitro* screening using different concentrations of polyethylene glycol.[a]

	Shoot dry weight (g) at salt concentration					Shoot dry weight (g) at salt concentration			
Genotype	0 g/l	10 g/l	15 g/l	Mean	Genotype	0 g/l	10 g/l	15 g/l	Mean
Marooko*	0.70a	0.68a	0.99a	0.8a	440023	1.04b	0.38a	0.45a	0.3
187016.2	0.31a	0.36a	0.13a	0.3a	440024	0.05a	0.54b	0.19ab	0.4
187017.1	1.05bc	1.01b	0.26a	0.8a	440025	0.48a	0.51a	0.23a	1.9
189123.68	1.23b	0.70a	1.43bc	1.1b	440027	1.43a	4.33c	0.07b	0.2
189135.9	2.79bc	2.35b	1.22a	2.1b	440031	0.25a	0.14a	0.08a	0.9
189140	0.29ab	0.05a	0.74ab	0.4a	440034	1.06a	0.91a	0.64a	0.2
189148.21	0.29a	0.44a	0.44a	0.4a	440050	0.08a	0.03a	0.35a	0.3
189148.65	0.23a	0.22a	0.10a	0.2a	440104	0.66b	0.04a	0.10a	0.6
189150.1	0.87a	0.87a	0.90a	0.9a	440131	1.43b	0.02a	0.40a	0.2
189151.38	0.67b	1.77a	0.98bc	1.1a	440132	0.35a	0.31a	0.00a	0.4
192033.5	0.86b	1.97a	0.74bc	1.2a	440166	0.44a	0.47a	0.17a	0.1
194515.5	1.99bc	1.90b	1.21a	1.7a	440167	0.14a	0.12a	0.03a	1.6
194521.2	1.26b	0.38a	0.69a	0.8a	440170	2.43b	1.27a	0.97a	0.1
194539.36	1.83bc	2.08b	2.58a	2.2b	440240	0.16a	0.04a	0.15a	0.1
194541.45	0.69a	0.30a	0.03ab	0.3a	440286	0.44a	0.00a	0.00a	0.1
194549.6	0.38a	0.32a	0.65a	0.5a	440287	0.42ab	0.69a	0.00b	0.4a
194555.7	0.44a	0.26a	0.09a	0.3a	440328	2.09b	0.69a	0.86a	1.2a
194569.1	0.68b	0.10a	0.76bc	0.5b	440378	2.13a	1.22c	0.50b	1.3c
194573.9	0.45a	0.44a	0.42a	0.4a	440394	1.65a	1.06c	0.13b	0.9b
400011	0.80a	0.69a	0.69a	0.7a	440396	1.74bc	1.67b	0.47a	1.3b
401055	0.53bc	0.55b	1.19a	0.8a	440429	1.57a	1.40a	1.74a	1.6a
420001	1.03b	0.39a	0.44a	0.6a	440643	0.50ab	0.88a	0.13b	0.5a
420014	0.58a	0.32a	0.10a	0.3a	441097	0.67b	1.47a	1.60a	1.2a
420027	0.14a	0.10a	0.05a	0.1a	441538	2.48bc	2.22b	4.35a	3.0a
420064	0.68a	0.33a	0.45a	0.5a	441724	0.77b	2.56a	0.64c	1.3b
421066	0.77a	0.47a	0.33a	0.4a	441725	0.56a	0.77a	0.77a	0.7a
421111	0.31a	0.41a	0.44a	1.0a	441755	0.18a	0.20a	0.20a	0.2a
422656	1.63b	0.67a	0.70a	0.6a	441768	2.54b	1.28a	1.28a	1.7a
440001	0.91a	0.54a	0.32ab	1.0a	K566632**	0.12a	0.03a	0.03a	0.1a
440017	1.78a	1.09c	0.13b	0.6b					

[a]Means followed by the same letter within the rows (showing differences among different salt levels) are not significantly different ($P \leq 0.05$); *, tolerant genotype check; **, susceptible genotype check.

Table 5.7. Effect of salt concentration on shoot length (cm) of 59 sweet potato genotypes during *in vitro* screening using different concentrations of polyethylene glycol.[a]

Genotype	Shoot length (cm) at salt concentration				Genotype	Shoot length (cm) at salt concentration			
	0 g/l	10 g/l	15 g/l	Mean		0 g/l	10 g/l	15 g/l	Mean
Marooko*	7.23a	5.17a	8.50a	7a	440023	5.67a	2.00b	2.67ab	3.4a
187016.2	10.33a	12.17a	14.00ab	12.2a	440024	2.80b	7.13a	6.83a	5.6a
187017.1	10.33b	13.83a	14.90a	13a	440025	2.00b	8.17a	2.77bc	4.3a
189123.68	8.00a	4.33b	6.67ab	6.3ab	440027	7.00a	10.33a	8.00a	8.4a
189135.9	15.33a	12.83a	13.17a	13.8a	440031	7.33ab	4.23a	7.67b	6.4b
189140	5.50a	5.33a	4.83a	5.2a	440034	0.23b	5.50a	4.57a	3.4a
189148.21	5.50ab	5.00a	8.63b	6.4b	440050	4.10bc	5.57b	9.00a	6.2b
189148.65	6.87b	3.00a	1.83a	3.9a	440104	14.73c	1.50b	7.57a	7.9a
189150.1	4.83a	4.33a	5.00a	4.7a	440131	7.93c	2.30b	11.60a	7.3c
189151.38	7.00a	11.00b	9.67ab	9.2a	440132	5.57bc	5.80b	1.27a	4.2b
192033.5	10.33a	14.67b	12.17ab	12.4ab	440166	9.00b	15.17a	13.17a	12.4a
194515.5	12.57a	10.00a	11.00a	11.2a	440167	5.67b	1.80a	6.33bc	4.6b
194521.2	14.00b	8.40a	8.30a	10.2a	440170	7.67b	12.33a	12.00a	10.7a
194539.36	9.33bc	10.00b	16.67a	12a	440240	6.77b	1.73a	5.40bc	4.6b
194541.45	6.67a	4.50a	1.33ab	4.2a	440286	5.83b	1.03a	1.17a	2.7
194549.6	1.83bc	2.83b	7.67a	4.1a	440287	5.50a	5.23a	3.50a	4.7a
194555.7	6.67a	4.83a	6.50a	6a	440328	17.00b	9.17a	10.97a	12.4a
194569.1	8.00b	1.50a	7.33bc	5.6b	440378	10.33a	12.00a	13.83ab	12.1a
194573.9	7.00a	7.67a	9.67a	8.1a	440394	7.00bc	6.50b	2.00a	5.2b
400011	5.67a	4.83a	5.57a	5.4a	440396	8.00a	7.33a	4.67a	6.7a
401055	8.00a	8.37a	10.17a	8.8a	440429	15.33a	9.33b	12.33ab	12.3a
420001	16.33b	8.17a	8.33a	10.9a	440643	6.90bc	8.23b	1.53a	5.4b
420014	6.70a	5.87a	5.77a	6.1a	441097	12.67a	15.67a	13.67a	14a
420027	6.70b	1.93a	2.80a	3.8a	441538	14.67a	9.50b	11.83ab	12.0a
420064	10.33bc	11.47b	18.23a	13.3b	441724	9.00a	10.67a	8.00a	9.2a
421066	11.00b	6.00a	6.33a	7.8a	441725	9.13a	4.67b	7.33ab	7.0a
421111	8.93bc	10.07b	4.77a	7.9b	441755	1.33a	1.50a	1.47a	1.3a
422656	16.33b	11.20a	9.73a	12.4a	441768	16.17a	16.00a	12.67ab	15.0a
440001	9.33a	5.33b	8.33ab	7.7a	K566632**	4.50a	3.50a	1.93a	3.3a
440017	9.00bc	10.33b	1.27a	6.9b					

[a]Means followed by the same letter within the rows (showing differences among different salt levels) are not significantly different ($P \leq 0.05$); *, tolerant genotype check; **, susceptible genotype check.

concentrations. Susceptible genotypes such as 420027, 440034, 440104, 440643, 189148.65, 194541.45, 420014 and 440131 showed significantly reduced growth parameters at the highest concentration of PEG (15 g/l). Such negative effects have been observed for susceptible genotypes in wheat (Javed, 2002) and soybean (Sakthivelu *et al.*, 2008). At the same level of stress, genotypes 189135.9, 194515.5, 194539.3 401055, 440429, 441097, 441538 and 441768 were observed to have an outstanding ability to continue root and shoot growth, indicating their ability to tolerate stress.

From this screening trial genotypes that were identified as drought tolerant included: 194515.5, 194539.36, 441724 and 441538 (dark orange); 189135.9 (orange); 441768 (light orange); 192033.5 (yellow); and 440429 (light cream). They all showed greater leaf expansion and stem length elongation, high root and shoot growth and high dry matter production at the high salt concentration level.

5.5 Conclusion and Recommendation

The *in vitro* screening method using PEG 6000 was found to be an efficient and simple method for evaluation of drought tolerance in a large number of genotypes in a very short time. It can also be used for the identification and selection of tolerant and sensitive genotypes needed for improvement. Using this method of screening, genotypes 194515.5, 194539.36, 441724, 441538, 189135.9, 441768, 192033.5 and 440429 were identified as genotypes that have outstanding ability to continue root and shoot growth under *in vitro* stress conditions at all salt levels tested indicating their ability to withstand severe water stress situations. It is recommended that field trials be done to validate this observation.

Acknowledgements

The authors express their appreciation to the 2006–2010 HarvestPlus Challenge Program, 'Orange-fleshed sweetpotato for alleviating vitamin A deficiency in the sub-Saharan Africa, South-West and South Asia' and Sweetpotato Action Security and Health in Africa (SASHA) for funding this study.

References

Bajji, M., Lutts, S. and Kinet, J.M. (2000) Physiological changes after exposure to and recovery from polyethylene glycol-induced water deficit in callus culture issued from durum wheat (*Triticum durum*) cultivars differing in drought resistance. *Journal of Plant Physiology* 156, 75–83.

Barlass, M. and Skene, K.G.M. (1981) Relative NaCl tolerances of grapevine cultivars and hybrids *in vitro*. *Zeitschrift für Pflanzenphysiologie* 102, 147–156.

Barta, A.L., Sulc, R.M., Ogle, M.J. and Hammond, R.B. (2002) Interaction between flooding or drought stress and potato leafhopper injury in alfalfa. *Plant Health Progress.* Available at: https://www.plantmanagementnetwork.org/php/elements/sum.aspx?id=2101&photo=316 (accessed 15 September 2015).

Boyer, J.S. (1982) Plant productivity and environment. *Science* 218, 443–448.

Bressan, R.A., Hasegawa, P.M. and Handa, A.K. (1981) Resistance of cultured higher plant cells to polyethylene glycol-induced water stress. *Plant Science Letters* 21, 23–30.

Geravandi, M., Farshadfar, E. and Kahrizi, D. (2011) Evaluation of some physiological traits as indicators of drought tolerance in bread wheat genotypes. *Russian Journal of Plant Physiology* 58(1), 69–75.

Handa, A.K., Bressan, R.A. and Handa, S. (1982) Characteristics of cultured tomato cells after prolonged exposure to medium containing polyethylene glycol. *Journal of Plant Physiology* 69, 514–521.

Hsiao, T.C. (1973) Plant responses to water stress. *Annual Review of Plant Physiology* 24, 519–570.

Javed, F. (2002) *In vitro* salt tolerance in wheat. II. Water relations in callus. *International Journal of Agriculture and Biology* 4(4), 465–467.

Khanna, V.K. and Garg, G.K. (1997) Somaclonal variation for plant improvement. *Acta Botany Indica* 25, 193–201.

Kim, Y.J., Shanmugasundaram, S., Yun, S.J., Park, H.K. and Park, M.S. (2001) A simple method of seedling screening for drought tolerance in soybean. *Korean Journal of Crop Science* 46, 284–288.

Kulkarni, M. and Deshpande, U. (2006) Comparative studies in stem anatomy and morphology in relation to drought resistance in tomato (*Lycopersicon esculentum*). *American Journal of Plant Pathology* 1(1), 82–88.

Kulkarni, M. and Deshpande, U. (2007) *In vitro* screening of tomato genotypes for drought resistance using polyethylene glycol. *African Journal of Biotechnology* 5(16), 1488–1493.

Murashige, T. and Skoog, F. (1962) A revised medium for rapid growth and bioassays with tobacco tissue culture. *Journal of Plant Physiology* 15, 473–497.

Nabors, M.W., Gibbs, S.E. and Bernstein, C.S. (1980) NaC1-tolerant tobacco plants from cultured cells. *Zeitschrift für Pflanzenphysiology* 97, 13–17.

Narayan, D. (1991) Root growth and productivity of wheat cultivars under different soil moisture conditions. *International Journal of Ecology and Environmental Science* 17, 19–26.

Orton, T.J. (1980) Comparisons of salt tolerance between *Hordeum vulgare* and *Hordeum jubatum* in whole plants and callus cultures. *Zeitschrift für Pflanzenphysiology* 98, 106–118.

Radhouane, L. (2007) Response of Tunisian autochthonous pearl millet (*Pemisetum glaucum* L. R.Br.) to drought stress induced by polyethylene glycol 6000. *African Journal of Biotechnology* 6(9), 1102–1105.

Sakthivelu, G., Akitha Devi, M.K., Giridhar, P., Ajasekaran, T., Ravishankar, G.A., Nedev, T. and Kosturkova, G. (2008) Drought-induced alterations in growth, osmotic potential and *in vitro* regeneration of soybean cultivars. *General Applied Plant Physiology* (special issue) 34(1–2), 103–112.

SAS Institute (1999) *SAS/STAT User's Guide*. SAS version 8. SAS Institute, Cary, North Carolina.

Shabir, H.W., Parvez, A.S., Satbir, S.G. and Singh, N.B. (2010) *In vitro* screening of rice (*Oryza sativa* L.) callus for drought tolerance. *Communication in Biometrics and Crop Science* 5(2), 108–115.

6 *Ex Ante* Evaluation of Improved Potato Varieties for Sub-Saharan Africa

U. Kleinwechter,[1*] G. Hareau,[1] M. Bonierbale,[1]
M. Gastelo[1] and D. Harahagazwe[2]
[1]*International Potato Center (CIP), Lima, Peru;* [2]*CIP, Nairobi, Kenya*

Abstract

In many parts of sub-Saharan Africa (SSA), potato plays an important role as a food security crop. Yet technological improvements to boost potato productivity have so far not been extensively utilized. Moreover, it remains unclear which potential impacts can be expected from future technological innovations in potato production in the region. To shed light on this question, a scenario of the development and diffusion of improved potato varieties for nine countries in East and Central Africa is developed and assessed. The scenario involves varieties which combine a number of improvements in pro-poor, productivity-enhancing traits and is analysed using an economic model of the world agricultural sector. Taking into account spill-over effects across markets and countries, the analysis finds positive net welfare effects at the global level from US$60 million to US$403 million. Global returns on investment are positive between 20% and 37%. Effects of the intervention on potato supply in the target countries range from 0.5% to 8.5%. Potato producers in these countries are found to benefit, but producers of other commodities and in other countries beyond the region are negatively affected. Lower market prices for potatoes and other commodities lead to welfare gains to consumers worldwide. At the level of the target countries, the improved potato varieties are found to generate returns on investment between 20% and over 70%, depending mainly on the level of adoption. The analysis shows that investing in crop improvement and variety development for SSA can be a worthwhile undertaking with high returns. It also highlights the importance of variety diffusion for the intra-regional distribution and the magnitude of the impacts and suggests putting emphasis in seed systems development to promote quick dissemination and high adoption levels.

Keywords: crop improvement, *ex ante* impact assessment, partial equilibrium model, potato, sub-Saharan Africa

6.1 Introduction

Potato (*Solanum tuberosum* L.) stands out among the world's major food crops. It is well suited for cultivation in environmental conditions where other crops may fail and lends itself to intercropping and crop rotation with other crops (FAO, 2008). Thus, potato helps to increase the availability of food and raise the aggregate efficiency of food systems. As a cash crop, potato also contributes to alleviating poverty by providing income generation

*ukleinw@posteo.net

opportunities (Scott *et al.*, 2000). It further represents an important source of energy, with a delivery of higher amounts of dry matter and protein per hectare than the major cereal crops (Burton, 1989) and a high micronutrient content.

During the food price crisis in 2007/2008, prices of potatoes were significantly less affected by the price increases in international markets (FAO, 2008), highlighting the contribution of the crop to a more stable food system. The fact that potatoes are grown in regions with high incidences of poverty, undernutrition and food insecurity underlines their particular importance (Bruinsma, 2003).

Potato is the third most important food crop in the world (Birch *et al.*, 2012). In 2009, world potato production reached 330 million t, of which 18 million t were produced in Africa (FAO, 2011). In particular sub-Saharan Africa (SSA) has witnessed a strong expansion in potato area in response to growing consumption demand from urban centres and rising relative prices (Low *et al.*, 2007).

Potato yields have kept growing constantly over the past five decades. But because of comparatively low average yield growth rates, most growth in production has come from an expansion in cultivated area (FAO, 2011). Thus, there is a potentially important role of technological innovations aimed at increasing productivity for enhancing the contribution of the crop to the food system. While it is recognized that technological improvements in potatoes have so far been an underexploited resource (Alexandratos, 1997), it remains unclear what potential impacts can be expected from future technological innovations in potato production, in particular in developing countries.

To shed light on this question, we create a case example of improved potato varieties for SSA. The exercise takes into account current and future priorities of potato research for development. A scenario of a research-driven intervention in the potato sectors of nine target countries in the region is designed. This scenario is analysed using a global economic model of the agricultural sector to obtain assessments of the impacts of the interventions on production, consumption, prices and economic welfare, while taking into consideration spill-over effects across different commodity markets and countries.

The approach not only provides a quantitative analysis of the potential impact of improved potato varieties in the region. By offering an *ex ante* impact assessment in a multi-commodity and multi-country framework, it also goes beyond former approaches to the evaluation of the economic impact of potato technologies, which have been either *ex post* (Walker and Crissman, 1996; Walker and Fuglie, 2001; Salazar *et al.*, 2009), applied only single market models (Fuglie and Thiele, 2009; Salazar *et al.*, 2009) or used more simple cost–benefit approaches (Fonseca *et al.*, 1996; Rueda *et al.*, 1996; Chilver *et al.*, 1999), thus neglecting the above-mentioned spill-over effects. Furthermore, impacts of potato technologies have been rarely considered on a (sub-)regional scale.

6.2 Methodology and Scenario

The IMPACT model

The present study uses a new version of the International Model for Policy Analysis of Agricultural Commodities and Trade (IMPACT) (Rosegrant *et al.*, 2012).[1] IMPACT has a long tradition of analyses related to projections of global food supply and demand (Pinstrup-Andersen *et al.*, 1997; Nelson *et al.*, 2010), commodity specific analyses (Scott *et al.*, 2000), regional analyses (Huang *et al.*, 1999; Pandya-Lorch and Rosegrant, 2000), or to examine specific issues related to the agricultural sector, such as biofuels (Rosegrant *et al.*, 2008) or water scarcity (Rosegrant and Cai, 2001). Evenson and Rosegrant (2003) applied the model to assess economic impacts of crop genetic improvement carried out by International Agricultural Research Centres (IARC).

IMPACT is a partial equilibrium representation of the global agricultural sector, which projects agricultural production, demand, trade flows and prices on a regional scale and provides estimates of global food security with a projection horizon from 2000 to 2050. It comprises 44 agricultural commodities, 115 geopolitical regions and

126 water basins combine to 281 food production units (FPUs). For each period t, FPU n and commodity i, agricultural production is depicted by isoelastic functions for area AC and yield YC:

$$AC_{tni} = \alpha_{tni} \times (PS_{tni})^{\varepsilon ijn} \times \prod_{j \neq i} (PS_{tni})^{\varepsilon ijn} \times (1 + gA_{tni}) \quad (6.1)$$

$$YC_{tni} = \beta_{tni} \times (PS_{tni})^{\gamma iin} \times \prod_{k} (PF_{tnk})^{\gamma ikn} \times (1 + gCY_{tni}) \quad (6.2)$$

The two functions together represent QS, the supply of each commodity for each region:

$$QS_{tni} = AC_{tni} \times YC_{tni} \quad (6.3)$$

As shown by Eqns (6.1)–(6.3), agricultural production is assumed to be a function of output prices PS, the prices PF of input k, and specific shifters gA and gCY for exogenous changes in area and yields (α and β are intercept terms of the isoelastic function; j is another index for the commodity).

On the demand side, for each region a set of functions represent different demand components, namely food, feed, biofuels, crush demand for oilseeds and other uses, which add up to total demand.

The individual regions for which supply and demand are calculated are connected to each other via trade. Net trade adds to domestic supply and stocks to equilibrate supply and demand. Global demand for each commodity is brought into equilibrium by an endogenous world market price, which determines domestic producer and consumer prices for all commodities and all regions.

A welfare module provides estimates of the welfare effects of a given scenario. The estimations consist of a traditional economic welfare analysis, comprising producer surplus, consumer surplus and net welfare effects to the society as a whole. Further, the costs of a particular intervention can be taken into account to calculate the internal rate of return (IRR) as a measure of returns on investment. For analyses of interventions from agricultural research for development (AgR4D) the approach takes into consideration the typical cost components of AgR4D projects: (i) costs incurred by IARC; (ii) costs arising to national agricultural research institutes; and (iii) costs of technology dissemination. Welfare effects are calculated at the global level taking into account all cost components, and the country level by taking into account only the costs arising within the respective country (i.e. excluding IARC costs).

To simulate the effects of production side interventions, different technologies can be distinguished at the FPU level. Each technology can occupy a specific share of the total production area dedicated to a particular crop in each FPU. Through the implementation of a logistic diffusion curve, new technologies can gradually replace traditional ones. To create a technology which has a yield advantage over an old one, the yield growth rates gCY in Eqn 6.2 are manipulated. In its standard version, the IMPACT model includes exogenous yield growth rates, which determine the value of gCY for each crop, year and FPU and which reflect the baseline development of agricultural production in what can be called a 'business-as-usual' scenario. Any factor added to these growth rates represents accelerated growth in yields.

Target areas and description of the proposed technology

The example of improved potato varieties for SSA involves a realistic case of a collaboration of an IARC, namely the International Potato Center (CIP), and National Agricultural Research Systems (NARS). Constructing this case requires: (i) the identification of target areas; (ii) the description of the characteristics of the new technology; (iii) the identification of the activities required from the beginning of the technology development to the end of the dissemination period; and (iv) the definition of a time frame. It is also necessary to include information about costs and to make assumptions on expected adoption pathways.

The target areas for the technology intervention are selected based on a targeting exercise carried out at CIP with the aim of identifying priority regions for potato research (Theisen and Thiele, 2008). Areas with high priority for potato research in Africa are concentrated in the western part (Nigeria, Cameroon, Angola), in the Great Lakes region of East and Central Africa and in Ethiopia. Based on this analysis, nine countries are selected for the intervention in the present study: Uganda, Rwanda, Burundi, Ethiopia, Tanzania, Malawi, Kenya, the Democratic Republic of the Congo (DRC) and Mozambique. Investments in potato research in these areas can be expected to generate high impacts on poverty and food security.

The proposed technology consists of improved potato varieties superior to currently available cultivars. Taking into account the characteristics of the target agroecologies and expected agronomic and socio-economic impacts, a combination of five traits is suggested: (i) higher yield potential; (ii) resistance to late blight; (iii) resistance to virus diseases; (iv) heat tolerance; and (v) suitability for processing.

The selection of target traits is based on directions for research formulated in the CIP Corporate Plan 2010 (CIP, 2010) and an assessment of research priorities made at CIP (Fuglie, 2007a; Fuglie and Thiele, 2009). In this assessment, control of late blight and viruses as well as suitability for processing have been identified as having priority for researchers and potential for poverty reduction (Fuglie, 2007a). Fuglie (2007b) identifies suitability for processing as another trait with potential for poverty reduction. This selection of target traits is also supported by Low *et al.* (2007) who call for the consideration of processing quality, late blight and virus resistance and heat tolerances in potato breeding for SSA.

Late blight (caused by *Phytophthora infestans* (Mont.) de Bary) represents the most important disease of potato worldwide (Salazar, 1996). Climate change leads to greater variation in the incidence of this disease; and increasingly new regions are affected (Haverkort and Verhagen, 2008). At the farm level, late blight resistance helps to save capital by reducing the use of fungicide, thus favouring the adoption of the technology by the poor (Thiele *et al.*, 2010).

Virus diseases represent extremely important pathogens of the potato crop (Hooker, 2001). The most important virus strains comprise potato leaf roll virus (PLRV) and potato virus Y (PVY), followed by potato virus X (PVX), potato virus M (PVM) and potato virus A (PVA). While virus diseases are seldom lethal to the plant, they lead to reductions of plant vigour and yields (Hooker, 2001). Reported effects of virus infections on yields are highly variable (Salazar, 1996), but can reach up to 90% of losses (CIP, 1996).

Beside the production of virus-free planting material, a method of control is the exploitation of genetic resistance available in the gene pool of *Solanum* spp. Genetic resistance to virus diseases increases yield stability and has the advantage of being transmitted to the progeny. Furthermore, it implies no environmental damage and has few undesirable effects on the potato plant (Salazar, 1996). Virus resistance can also enhance productivity from farm-saved seed, thereby conveying a direct benefit to poor farmers (Thiele *et al.*, 2010).

Heat tolerance helps dealing with temperature increases caused by climate change. Heat-tolerant varieties help maintain production in tropical highland regions in which temperatures might rise above critical limits. Moreover, varieties with heat-tolerance traits are suitable for a wider range of growing conditions, favouring dissemination to broader areas.

Breeding for suitability to processing takes into account a shift in consumption from fresh to processed potatoes (Walker *et al.*, 2011). Thiele *et al.* (2010) point out that due to higher prices, breeding for processing promises economic benefits to the poor.

A yield advantage of +30% of the improved varieties compared with traditional varieties is assumed. This is consistent with expert opinions about the yield and quality effects from late blight control and observed productivity effects of new potato varieties in SSA (Rueda *et al.*, 1996).

The development of the new improved potato varieties is assumed to start from CIP's LBHT (late blight, heat tolerance) breeding population and to involve one additional breeding cycle at CIP with duration of 4 years.

The advanced clones resulting from these efforts are delivered to NARS in the target countries, where further selection and improvement as well as seed multiplication during 4 years will lead to the release of a set of new varieties. Following release, dissemination of potato seeds and extension work takes place during an additional 4 years.

Based on estimates of Fuglie (2007a) of the annual cost of potato breeding at CIP of around US$3.5 million[2] and assuming that the proposed programme of technology development will cover 25% of CIP's overall breeding cost, US$0.875 million/year are required. Expenditures by NARS are assumed to be equal to CIP expenditure (Fuglie, 2007a). Together with the costs for dissemination of the new varieties, the total budget of the project accounts for US$9.1 million.

Dissemination plan and adoption pathways

The varieties will be released by NARS in the target countries and disseminated in cooperation with public sector organizations, NGOs (non-governmental organizations) and the private sector. As in the countries of SSA dissemination systems for root and tuber crops are generally weak, public subsidies for the dissemination of seed material may be necessary (Fuglie, 2007a). This need is taken into account in the assumed dissemination costs.

Assumptions on potential adoption of the new improved potato varieties are based on expert opinions and new information on adoption of modern potato varieties in Africa (Alene *et al.*, 2011). In Mozambique and Malawi, CIP varieties have only recently been released and hence past adoption of modern potato varieties of CIP origin has been very low. Further, both countries have only limited local seed production capacities. In Malawi, experts estimate the adoption of CIP-related material at below 5% to date. A similar situation can be observed in the DRC and Tanzania, where weakly developed seed programmes have provided for only low adoption so far. In Kenya, CIP-related varieties have been adopted on around 25% of the potato area, similar to Ethiopia with an estimated rate of 24%. Higher adoption rates have been achieved in Uganda and Rwanda, with 60% and almost 90%, respectively.[3]

Past adoption figures offer space for opposite expectations about future adoption potential. On the one hand, high past adoption of CIP material represent good conditions for the dissemination of new varieties. Hence, prospects for future adoption may be brighter in countries with high past adoption. On the other hand, once a set of modern varieties is in place it may be increasingly difficult to replace them with new material. This may point to lower adoption potential in countries with high levels of past adoption.

Likewise, countries with historically low adoption may provide only poor conditions for disseminating new improved varieties. However, these countries may have a high adoption potential for the future. But harnessing this potential may require more substantial investments in, for example, seed systems.

Based on these considerations, we assume a four-tier model of adoption with countries with very low (5%, Malawi, Mozambique), low (10%, DRC, Tanzania), middle (20%, Burundi, Ethiopia, Kenya, Uganda) and high (30%, Rwanda) adoption. Generally, this model can be considered as moderately conservative. In particular the potential rates for the countries with very low and low adoption rates might be higher than assumed. However, since adoption rates are a main determinant of the final impacts, three different adoption cases are analysed. The first one, termed 'high adoption', is the case just described. The second one is a 'medium adoption' case with all rates at two-thirds of the high adoption case. In the third case, 'low adoption' is assumed with all rates at one-third of the high adoption scenario. Table 6.1 presents the estimated adoption areas for the target countries after an adoption period of 10 years, calculated based on production areas in 2009.

An estimate of the cost of dissemination of agricultural technologies is provided by Fuglie (2007a). The cost of disseminating potato varieties is comparatively low, amounting to US$16/ha of the finally adopted area. It includes the value of staff time and fixed assets in extension services. Multiplication of this value with the adoption ceiling of the high adoption scenario leads to the dissemination costs of US$2.1 million. This cost is assumed for all three adoption scenarios. Although adjusting the dissemination cost to each adoption scenario would be an option, too, the highest cost scenario is kept to take into account that a given investment might not always lead to the highest desired outcomes. Thus, a rather conservative stance is taken with respect to the expected costs and benefits.

6.3 Results and Discussion: Economic and Social Value of Improved Potato Varieties for SSA

Table 6.2 shows that the introduction of the new varieties, although being a financially and spatially limited intervention, has global welfare effects. As a result from lower producer prices, the net present value (NPV) of global producer surplus over the entire simulation period declines by between US$201.78 million for the low adoption case and US$602.64 million for the high adoption case by 2050. Gains in consumer surplus are estimated to be between US$262.15 million and US$1005.35 million, a consequence of lower market prices which benefit consumers worldwide. Since losses to producers are lower than gains to consumers, global net welfare changes caused by the intervention are positive with values between US$60.37 million and US$402.71 million. This leads to calculated benefit–cost ratios of 14.08 to 93.94 and global IRR between 0.20 and 0.37.

The global welfare analysis leads to a positive assessment of the intervention. Returns on investment are in a range consistent

Table 6.1. Assumed adoption of new improved potato varieties in sub-Saharan Africa. (From FAO, 2011; own calculations.)

Country	Production area ('000 ha, 2009)	Low adoption (% of total area)	Medium adoption (% of total area)	High adoption (% of total area)	Adoption ceiling ('000 ha) (high adoption case)
Burundi	6.0	6.7	13.3	20.0	1.2
DRC[a]	20.0	3.3	6.7	10.0	2.0
Ethiopia	69.8	6.7	13.3	20.0	14.0
Kenya	153.1	6.7	13.3	20.0	30.6
Malawi	213.9	1.7	3.3	5.0	10.7
Mozambique	8.0	1.7	3.3	5.0	0.4
Rwanda	126.2	10.0	20.0	30.0	37.9
Tanzania	173.5	3.3	6.7	10.0	17.4
Uganda	101.0	6.7	13.3	20.0	20.2
Total	**871.5**				**134.3**

[a]DRC, Democratic Republic of the Congo.

Table 6.2. Global impacts of improved potato varieties in sub-Saharan Africa, 2000–2050.

Welfare and returns on investment	Low adoption	Medium adoption	High adoption
Changes[a] in producer surplus (NPV, m US$)	–201.78	–401.29	–602.64
Changes in consumer surplus (NPV, m US$)	262.15	624.28	1005.35
Net welfare change (NPV, m US$)	60.37	222.99	402.71
Cost (NPV, m US$)	4.29	4.29	4.29
Benefit–cost ratio	14.08	52.02	93.94
Net benefits (NPV, m US$)	56.09	218.70	398.43
Internal rate of return	0.20	0.31	0.37

[a]Reported changes are over baseline, represented by the IMPACT standard scenario without the enhanced technology; net present value (NPV) in 2000 constant prices.

with previous studies of the impact of improved potato varieties. Walker *et al.* (2003), for example, found that the IRR from CIP's global potato breeding programme from 1972 to 2021 ranged between 0.15 and 0.17. Fonseca *et al.* (1996) estimate an IRR of 0.26 for an improved potato variety released in Peru. Song *et al.* (1996) calculate an IRR of 1.02 for an improved variety disseminated in China. Rueda *et al.* (1996) assessed the impact of high yielding and late-blight resistant varieties in the Eastern and Central Africa Highlands and found an IRR of 0.84; using assumptions of productivity growth and adoption that were significantly higher than in this study.

The results also illustrate the role variety dissemination and adoption play for welfare effects and returns on investment. While net welfare effects and returns on investment are positive even with low levels of adoption, they vary in line with adoption rates. The magnitude of all effects is lowest in the low adoption scenario and highest with high adoption.

The different welfare effects on producers and consumers also deserve further discussion. According to the results, producers are expected to incur net welfare losses from the intervention. This is a result of gains to potato producers in the target countries and losses to producers of other commodities and in other countries, caused by lower producer prices. This underlines the fact that in a market environment increases in supply cause prices to decline (Fig. 6.1), which negatively affects all producers except those who have access to the new technology. Thus, the productivity-enhancing technology has unintended side effects on producers who cannot adopt, either because they produce other commodities, are located outside the target countries, or do not get access even though they are within the target country. The results, however, should also be considered in light of the fact that many poor households in low income countries are net food buyers (Ataman and Isik-Dikmelik, 2008; Ivanic and Martin, 2009). For those households, negative effects on the production side may be outweighed by gains on the consumption side and the intervention will most likely be poverty reducing.

Turning to the level of the individual target countries, Fig. 6.2 illustrates the effect of the improved potato varieties on potato supply. The results point to the contingency of the supply effect on the rates of adoption. While potato supply by 2050 is higher in all target countries, potato supply in the countries with middle and high adoption rates (Burundi, Ethiopia, Kenya, Rwanda and Uganda) is projected to increase by 2–3% in the low adoption case and by around 6–9% in the high adoption case. Potato production in the countries with low and very low adoption (DRC, Tanzania, Malawi and Mozambique), in contrast, increases only by between 0.5% and 1% in the low adoption case and

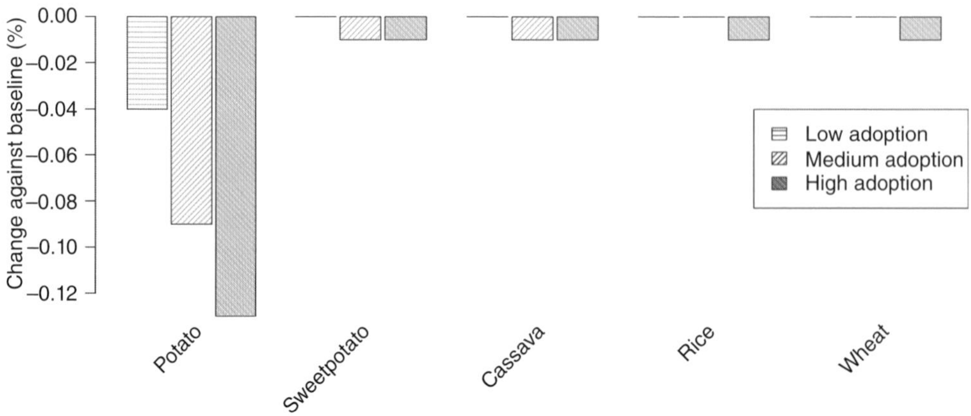

Fig. 6.1. Impacts of improved potato varieties in sub-Saharan Africa on world market prices of selected commodities by 2050 (percentage change against baseline).

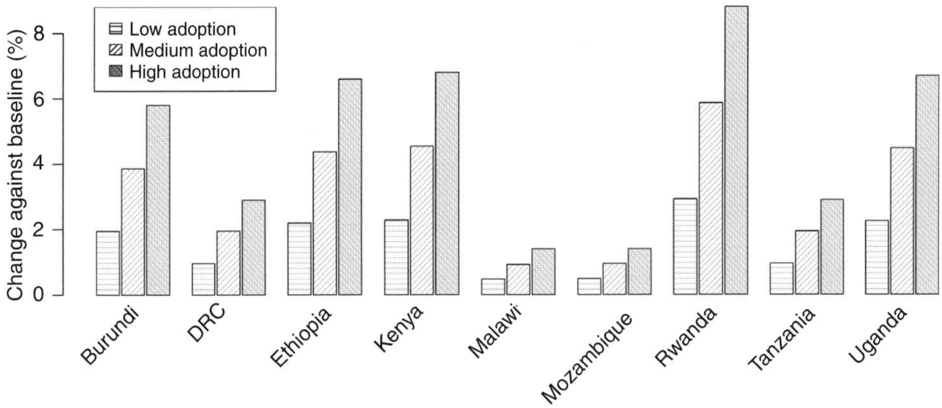

Fig. 6.2. Impacts of improved potato varieties on total potato supply in target countries by 2050 (percentage change against baseline). DRC, Democratic Republic of the Congo.

by between 1.4% and close to 3% in the high adoption case.

As already pointed out above, higher supply leads to lower market prices of potatoes and of other commodities. These price effects, however, are weak (Fig. 6.1). The decrease in world market price of potato ranges between 0.03% and around 0.1%. In the case of other commodities, prices are affected very weakly, with reductions of at most 0.01%. As shown above, however, even those weak price effects are sufficient to generate significant welfare impacts at the global scale, caused by the long time horizon and the comprehensive scope of the analysis covering all regions and markets.

Since market price effects of the intervention are limited, responses of consumers to lower prices are only marginal (Table 6.3). Per capita potato consumption increases by at most 0.11% by 2050. Differences between target countries are small, but the adoption case again plays a role. Consumption effects in non-target countries illustrate the global spill-over effects of the interventions in SSA. Due to the impact of the introduction of the new technology on world market prices, consumers in other parts of the world also benefit and increase consumption by similar margins as those in the target countries.

A case-by-case contemplation of the effects of the improved potato varieties on welfare and the returns on investment in the target countries reveals a mixed picture. The value of the net benefits generated at the country level varies greatly between countries (Fig. 6.3). Rwanda, with relatively high area of potato production and the highest levels of adoption also obtains the highest net benefits, ranging between US$20 million and US$60 million. This is followed by Kenya and Uganda, which both also have large areas and high adoption rates. In Ethiopia, which has the same adoption rates as the two former countries, the net benefits arising from the new varieties are comparable to Tanzania and Malawi. Ethiopia has only relatively small

Table 6.3. Impacts of improved potato varieties on per capita potato consumption in target and non-target countries by 2050 (percentage change against baseline).

	Low adoption (%)	Medium adoption (%)	High adoption (%)
Target countries			
Burundi	0.04	0.07	0.11
DRC[a]	0.02	0.04	0.06
Ethiopia	0.02	0.05	0.07
Kenya	0.03	0.07	0.10
Malawi	0.03	0.06	0.09
Mozambique	0.03	0.07	0.10
Rwanda	0.03	0.07	0.10
Tanzania	0.03	0.07	0.10
Uganda	0.04	0.07	0.11
Non-target countries			
China	0.04	0.07	0.11
India	0.03	0.07	0.10
Peru	0.02	0.05	0.07

[a]DRC, Democratic Republic of the Congo.

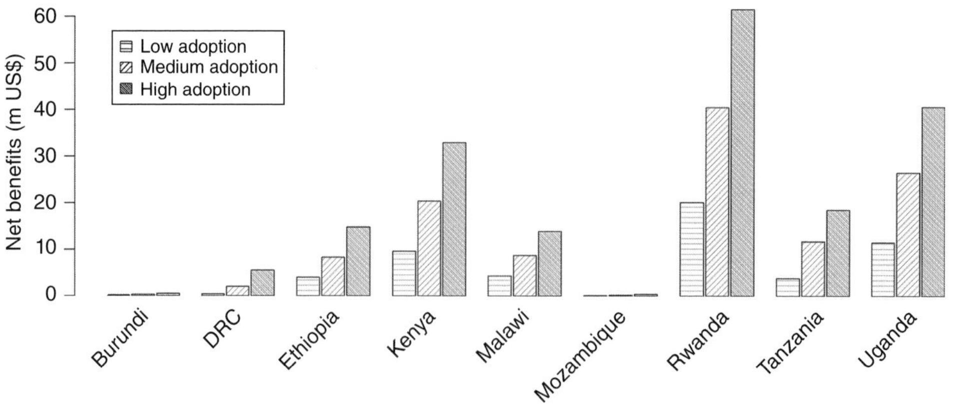

Fig. 6.3. Net benefits of improved potato varieties for sub-Saharan Africa (net present value in m US$). DRC, Democratic Republic of the Congo.

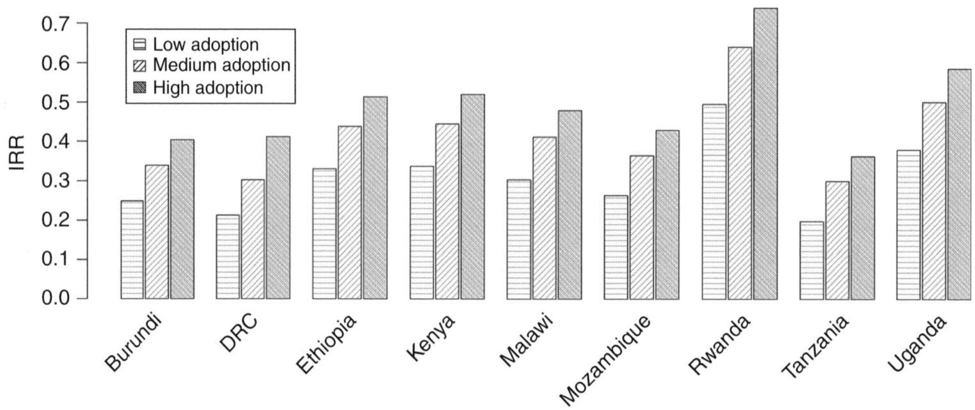

Fig. 6.4. Internal rate of return (IRR) of improved potato varieties for sub-Saharan Africa. DRC, Democratic Republic of the Congo.

areas, while Malawi can obtain higher benefits through its large areas and in spite of its low adoption rates. Burundi, Mozambique and the DRC have low net benefits due to a combination of small areas of production and low adoption rates.

While the contrasts in net benefits between the countries are easily visible, the differences between the IRR at the country level are less pronounced (Fig. 6.4). According to this measure, the best returns again can be achieved in Rwanda, with IRRs of around 0.50–0.72. For other countries, the IRR ranges between 0.20 and almost 0.60. Notably, even in those countries which rank low in terms of net benefits, positive returns can be expected.

6.4 Conclusions

The present analysis of the impact of improved potato varieties for SSA shows that the new technology has positive effects on potato supply in all target countries. Higher supply leads to lower prices and higher consumption, with spill-over effects effectuated via the market and spreading out beyond the region. The global welfare implications arising from these effects are positive: producers globally are negatively affected, but higher gains to consumers result in net welfare gains. Global rates of return are positive and comparable with findings from previous impact evaluations.

Individual target countries benefit from the improved potato varieties. Increased productivity implies higher supply and lower prices. Net welfare effects are consistently positive, but depend strongly on the extent of potato production in a given country and on the assumed adoption levels. Countries with large areas of potato production and high adoption rates tend to generate higher welfare gains. IRRs for the individual countries show a lower variation than the net benefits and generally exceed 0.2. In particular this latter result shows that improved potato varieties for SSA can generate sufficiently high returns to justify the investment.

It becomes clear that the adoption level is a principal determinant of technology impacts. Although especially the low and middle adoption cases analysed in the study take a rather conservative stance, the analysis underlines the importance of sufficiently good seed systems to ensure a quick dissemination of new varieties and to promote high rates of adoption. Especially in situations of weak national or regional seed systems, investments in this area should be seen as complementary to crop improvement.

From a methodological point of view, this study represents a showcase application of the IMPACT model for the *ex ante* assessment of agricultural technologies in developing countries. The strengths of this model comprise, among others, its global geographic coverage and its comprehensiveness in commodity coverage along with a consistent representation of supply and demand in the agricultural sector. The simultaneous depiction of supply and demand offers the advantage of being able to capture the complex market-mediated interactions across commodities and countries, which may be decisive for the impacts of particular technology interventions. Therefore, the present analysis can be seen as an important extension of previous *ex ante* assessments of potato technologies with less sophisticated models.

That said, pointing out a number of limitations of the analysis should help to stimulate future improvements. First, the assumptions on costs, adoption and the dissemination process used rely on the best information currently available. Further research, for example in the form of systematic gathering of expert opinion is required to improve these assumptions. Secondly, to arrive at better estimates about the productivity impacts of new varieties, it may be desirable to combine IMPACT with biophysical modelling tools like crop models or disease models. Finally, IMPACT relies on Food and Agriculture Organization of the United Nations (FAO) data for production consumption and trade, which is considered to be less reliable for potatoes in Africa. Any improvements in that area would be highly welcome.

Acknowledgements

The research contained in this chapter was supported by the CGIAR Research Programme on Climate Change, Agriculture and Food Security (CCAFS) and the CGIAR Research Program on Policies, Institutions, and Markets (PIM). The views expressed in this chapter cannot be taken to reflect the official opinions of CGIAR. The authors are also grateful for the support received from R. Labarta, D. Mason-d'Croz, A. Palazzo and S. Robinson and for comments of two anonymous referees. All remaining errors and omissions remain the authors' own responsibility.

Notes

[1] This section mainly draws on Rosegrant *et al.* (2012). For a more detailed documentation of the IMPACT model, the interested reader may refer to that document.

[2] This figure is the sum of the cost of research on late blight control (breeding and management), breeding for virus resistance and adaptability and breeding for processing, quality, adaptability and markets.

[3] All figures are from R. Labarta (Nairobi, Kenya, 2012, personal communication) about preliminary research results obtained in the scope of the DIIVA (Diffusion and Impacts of Improved Varieties in Africa) project (Alene *et al.*, 2011).

References

Alene, A., Yigezu, Y., Ndejunga, J., Labarta, R., Andrade, R., Diagne, A., Muthoni, R., Simtowe, F. and Walker, T. (2011) Measuring the effectiveness of agricultural R&D in sub-Saharan Africa from the perspectives of varietal output and adoption. Paper presented at the Agricultural Science and Technology Indicators (ASTI)-International Food Policy Research Institute (IFPRI)/Forum for Agricultural Research in Africa (FARA) Conference 'Agricultural R&D: Investing in Africa's Future – Analyzing Trends, Challenges and Opportunities', 5–7 December, Accra, Ghana.

Alexandratos, N. (1997) *World Agriculture – Towards 2010: an FAO Study*. Wiley, Chichester, UK.

Ataman, A.M. and Isik-Dikmelik, A. (2008) Are low food prices pro-poor? Net food buyers and sellers in low-income countries. *Policy Research Working Paper Series* No. 4642. The World Bank, Washington, DC.

Birch, P.R.J., Bryan, G., Fenton, B., Gilroy, E.M., Hein, I., Jones, J.T., Prashar, A., Taylor, M.A., Torrance, L. and Toth, I.K. (2012) Crops that feed the world 8: potato. Are the trends of increased global production sustainable? *Food Security* 4(4), 477–508.

Bruinsma, J. (2003) *World Agriculture – Towards 2015/2030: an FAO Perspective*. Food and Agriculture Organization of the United Nations (FAO), Rome.

Burton, W.G. (1989) *The Potato*. Longman, London.

Chilver, A., Walker, T., Khatana, V., Fano, H., Suherman, R. and Rizk, A. (1999) On-farm profitability of TPS utilization technologies. In: International Potato Center (CIP) (ed.) *Impact on a Changing World: Program Report, 1997–98*. CIP, Lima, pp. 213–219.

Evenson, R.E. and Rosegrant, M. (2003) The economic consequences of crop genetic improvement programmes. In: Evenson, R.E. and Gollin, D. (eds) *Crop Variety Improvement and Its Effect on Productivity: the Impact of International Agricultural Research*. CAB International, Wallingford, UK, pp. 473–498.

Fonseca, C., Labarta, R., Mendoza, A., Landeo, J. and Walker, T.S. (1996) Economic impact of high-yielding, late-blight-resistant variety Canchán-INIAA in Peru. In: Walker, T. and Crissman, C. (eds) *Case Studies of the Economic Impact of CIP-related Technology*. International Potato Center (CIP), Lima, pp. 51–63.

Food and Agriculture Organization of the United Nations (FAO) (2008) Food Outlook – Global Market Analysis. Available at: http://www.fao.org/docrep/010/ai466e/ai466e00.htm (accessed 15 June 2012).

Food and Agriculture Organization of the United Nations (FAO) (2011) FAOSTAT database. Available at: http://faostat.fao.org/ (accessed 8 November 2011).

Fuglie, K. (2007a) *Research Priority Assessment for the CIP 2005–2015 Strategic Plan: Projecting Impacts on Poverty, Employment, Health and Environment*. International Potato Center (CIP), Lima.

Fuglie, K. (2007b) Priorities for potato research in developing countries: results from a survey. *American Journal of Potato Research* 84, 353–365.

Fuglie, K. and Thiele, G. (2009) Research priority assessment at the International Potato Center (CIP). In: Raitzer, D. and Norton, G. (eds) *Prioritizing Agricultural Research for Development: Experiences and Lessons*. CAB International, Wallingford, UK, pp. 25–43.

Haverkort, A.J. and Verhagen, A. (2008) Climate change and its repercussions for the potato supply chain. *Potato Research* 51, 223–237.

Hooker, W.J. (2001) *Compendium of Potato Diseases*. APS Press, St Paul, Minnesota.

Huang, J., Rozelle, S. and Rosegrant, M. (1999) China's food economy to the 21st century. *Economic Development and Cultural Change* 47, 737–766.

International Potato Center (CIP) (1996) *Major Potato Diseases, Insects and Nematodes*. CIP, Lima.

International Potato Center (CIP) (2010) *The Strategy and Corporate Plan of the International Potato Center*. CIP, Lima.

Ivanic, M. and Martin, W. (2009) Implications of higher global food prices for poverty in low-income countries. *Agricultural Economics* 39, 405–416.

Low, J., Barker, I., Bonierbale, M., Crissman, C., Forbes, G., Lemaga, B. and Priou, S. (2007) Emerging trends and advances in potato research relevant to defining the way forward for the potato sector in sub-Saharan Africa. In: *African Potato Association Conference Proceedings*, Vol. 7. African Potato Association, Alexandria, Egypt, pp. 1–17.

Nelson, G., Rosegrant, M.W., Palazzo, A., Gray, I., Ingersoll, C., Robertson, R., Tokgoz, S., Zhu, T., Sulser, T.B., Ringler, C., Msangi, S. and You, L. (2010) *Food Security, Farming, and Climate Change to 2050: Scenarios, Results, Policy Options*. International Food Policy Research Institute (IFPRI), Washington, DC.

Pandya-Lorch, R. and Rosegrant, M.W. (2000) Prospects for food demand and supply in Central Asia. *Food Policy* 25, 637–646.

Pinstrup-Andersen, P., Pandya-Lorch, R. and Rosegrant, M.W. (1997) The World Food Situation: Recent Developments, Emerging Issues, and Long-Term Prospects. 2020 Vision Food Policy Report. International Food Policy Institute (IFPRI), Washington DC.

Rosegrant, M.W. and Cai, X. (2001) Water scarcity and food security: alternative futures for the 21st century. *Water Science and Technology* 43, 61–70.

Rosegrant, M.W., Zhu, T., Msangi, S. and Sulser, T. (2008) Global scenarios for biofuels: impacts and implications. *Review of Agricultural Economics* 30, 495–505.

Rosegrant, M.W. and the IMPACT Development Team (2012) *International Model for Policy Analysis of Agricultural Commodities and Trade (IMPACT) Model Description*. International Food Policy Institute (IFPRI), Washington, DC.

Rueda, J., Ewell, P., Walker, T., Soto, M., Bicamumpaka, M. and Berrios, D. (1996) Economic impact of high yielding, late blight resistant varieties in the Eastern and Central African Highlands. In: Walker, T. and Crissman, C. (eds) *Case Studies of the Economic Impact of CIP-related Technology*. International Potato Center (CIP), Lima, pp. 15–30.

Salazar, L. (1996) *Potato Viruses and their Control*. International Potato Center (CIP), Lima.

Salazar, L., Winters, P., Maldonado, L., Hareau, G. and Thiele, G. (2009) Assessing the impact of late blight resistant varieties on smallholders' potato production in the Peruvian Andes. *Social Sciences Working Paper* No. 2009-5. International Potato Center (CIP), Lima.

Scott, G., Rosegrant, M. and Ringler, C. (2000) Global projections for root and tuber crops to the year 2020. *Food Policy* 25, 561–597.

Song, B., Tian, W., Wang, J., Wang, C., Yan, Z., Wang, S. and Huarte, M. (1996) Economic impact of CIP-24 in China. In: Walker, T. and Crissman, C. (eds) *Case Studies of the Economic Impact of CIP-related Technology*. International Potato Center (CIP), Lima, pp. 31–49.

Theisen, K. and Thiele, G. (2008) Implementing CIP's vision: impact targeting. *Social Sciences Working Paper* No. 2008-4. International Potato Center (CIP), Lima.

Thiele, G., Theisen, K., Bonierbale, M. and Walker, T. (2010) Targeting the poor and hungry with potato science. *Potato Journal* 37, 75–86.

Walker, T. and Crissman, C. (eds) (1996) *Case Studies of the Economic Impact of CIP-related Technology*. International Potato Center (CIP), Lima.

Walker, T. and Fuglie, K. (2001) Impact assessment at the International Potato Center (CIP) in the 1990s. In: TAC Secretariat (ed.) *The Future of Impact Assessment in the CGIAR – Needs, Constraints and Options*. Proceedings of a workshop organized by the Standing Panel on Impact Assessment (SPIA). Food and Agriculture Organization of the United Nations (FAO), Rome, pp. 37–41.

Walker, T.S., Bi, Y.P., Li, J.H., Gaur, P.C. and Grande, E. (2003) Potato genetic improvement in developing countries and CIP's role in varietal change. In: Evenson, R.E. and Gollin, D. (eds) *Crop Variety Improvement and Its Effect on Productivity: the Impact of International Agricultural Research*. CAB International, Wallingford, UK, pp. 315–336.

Walker, T., Thiele, G., Suarez, V. and Crissman, C. (2011) Hindsight and foresight about potato production and consumption. *Social Sciences Working Paper* No. 2011-6. International Potato Center (CIP), Lima.

7 Durable Cisgenic Resistance to *Phytophthora infestans* in Potato, and Perspectives for Applications in Africa

G. Gheysen,[1]* B. Heremans,[1] B. Van Droogenbroeck,[2] R. Custers,[3] J.H. Vossen,[4] R.G.F. Visser,[4] E. Jacobsen,[4] R. Hutten[4] and A.J. Haverkort[4]

[1]Ghent University, Ghent, Belgium; [2]Institute for Agricultural and Fisheries Research, Merelbeke, Belgium; [3]Vlaams Instituut voor Biotechnologie (VIB), Ghent, Belgium; [4]Wageningen University and Research Center, Wageningen, The Netherlands

Abstract

Late blight caused by *Phytophthora infestans* is a major constraint in potato production. A promising strategy to combat late blight in potato is to combine different resistance genes to achieve durable resistance. Resistance genes from wild relatives can be introduced by breeding or by transformation. Single resistance genes are not durable because mutant pathogens that avoid recognition will easily be selected. Genetic engineering is a straightforward method allowing introduction of a combination of natural resistance genes into a potato cultivar without altering other agronomic characteristics. Since these genes can also be introduced by conventional breeding methods, the resulting potato plants are called cisgenic, in contrast to transgenic potatoes that have received DNA from non-crossable species. Three R genes conferring resistance to *P. infestans* (*Rpi*), *Rpi-sto1* (*Solanum stoloniferum*), *Rpi-vnt1.1* (*Solanum venturii*) and *Rpi-blb3* (*Solanum bulbocastanum*), were cloned and transformed separately or as a combination into the susceptible cultivar Désirée. The transformed clones were screened for late blight resistance using a detached leaf assay, and they were also evaluated for true-to-type performance under greenhouse conditions. To evaluate the performance of the resistance genes or combinations in realistic conditions, selected clones were tested in field trials in The Netherlands and Belgium in 2011 and 2012. This was done in comparison with the susceptible parent Désirée, and other susceptible and resistant cultivars. In both years plots were not treated with fungicides against *P. infestans*. In contrast to 2011, the summer of 2012 was humid resulting in high natural disease pressure. Nevertheless the two seasons showed similar results with clear differences between the susceptible reference clones and the genetically modified resistant clones. About 20 resistance genes against *P. infestans* have currently been mapped or cloned and more will follow. Therefore a collection can be generated of double or triple R gene-containing resistant cultivars that have the potential to make potato cultivation more sustainable. Based on the current potato cultivation area in East Africa, the effect of using cultivars with durable resistance on increasing potato yield in East Africa can be predicted.

Keywords: cisgenic, Désirée, field trial, gene stack, *Phytophthora infestans*, potato, resistance

*godelieve.gheysen@ugent.be

7.1 Introduction

Potato is the third most important food crop in the world after rice and wheat. This crop has a global production of more than 300 million t (Haverkort *et al.*, 2009). Since the 1960s, production of potato in developing countries has steadily increased and currently exceeds the output in the developed world. Although China is currently the biggest potato producer, this crop can contribute to food security in Africa because it is a locally produced crop and its price is not influenced by global markets. In sub-Saharan Africa, potato is a preferred food in many urban areas, and an important crop in the highlands of for instance Kenya and Rwanda (CIP, 2014). In a context of a growing world population and increasing demand for food, potato stands out for its productivity, with a two to four times higher food quantity and a lower water use per hectare than cereals.

A major constraint in potato crop protection is late blight caused by an oomycete, *Phytophthora infestans*. In Europe and the USA this problem is usually managed by 10–20 applications of fungicides each season. In most other countries where fungicides are too expensive and/or difficult to get, the yield loss due to late blight can easily amount to 70% (Garelik, 2002). An alternative solution for controlling late blight is use of resistant potato cultivars. Many wild *Solanum* species contain resistance (R) genes and to date over 20 such genes have been cloned (Vleeshouwers *et al.*, 2011). In the 20th century, introgression breeding has been used to introduce resistance, initially from *Solanum demissum*. Unfortunately the *S. demissum*-derived R genes failed to provide durable resistance due to the emergence of new *P. infestans* strains (Birch *et al.*, 2012). To exploit the resistance genes of other *Solanum* species in a more durable way, the suggestion is to stack several resistances in a high valued potato cultivar. However, to realize this using conventional breeding techniques is an extremely laborious and lengthy process.

A novel approach is the concept of cisgenesis, being a plant genetic engineering strategy only using natural genes that belong to the conventional breeder's gene pool for that crop (Park *et al.*, 2009). The advantages of this approach are that: (i) several R genes can be introduced swiftly in one transformation step; (ii) the R genes can be introduced into existing highly valued cultivars; and (iii) as a result the cultivar characteristics are retained and linkage drag is avoided. To impede selection of mutant *P. infestans* strains that are not recognized by R genes, the strategy is chosen to combine (or stack) at least three different R genes into a specific variety (Zhu *et al.*, 2012). Furthermore, preferentially several such stacks are designed that can be used in crop rotation, both in time and in place to adjust the management in relation to the evolving *P. infestans* strains (Haverkort *et al.*, 2008).

In the DuRPh (Durable resistance to *Phytophthora infestans*) project at Wageningen University and Research Center, several R genes against late blight are being tested singly and in combination by genetic engineering of potato (Haverkort *et al.*, 2009). This chapter summarizes results of a field trial with selected transformed lines from this project in Belgium during 2011–2012. Promising results of this strategy from this project could be applied for potato cultivation in Africa.

7.2 Materials and Methods

Plant material

The test material consisted of 26 genetically modified clones of the variety Désirée. Eight clones contained the *vnt-1.1* resistance gene (from *Solanum venturii*), eight clones had the *sto-1* gene (from *Solanum stoloniferum*) and 10 clones combined the *sto-1*, *vnt-1.1* and *blb-3* genes (*blb* from *Solanum bulbocastanum*) (Zhu *et al.*, 2012). The *vnt-1.1* clones are cisgenic, while the others contain a bacterial-derived marker gene to allow easy selection of the transformants (Haverkort *et al.*, 2009). For comparison several varieties were included in the field trial: the parent clone Désirée, the cultivars Fontane, Agria, Royal, Nicola, Bintje as susceptible and Fortuna, Bionica, Toluca, Sarpo Mira as resistant controls, respectively. Due to the lack of availability of planting material not all references could be included in both field trial years. Furthermore *S. bulbocastanum* was included as an additional control plant.

The field trial

Tubers of genetically modified clones and controls were planted in a randomized block design in general including four replications of four plants per clone. The Bintje plants were grown in rows between the blocks and cultivar Nicola was planted in a 3 m border surrounding the field trial. Appropriate weed management was done by herbicide treatment and *Alternaria* was controlled with two applications of a fungicide that does not affect *P. infestans*. The field was visited weekly to monitor late-blight disease incidence according to the key of Cox and Large (1960). Area under the disease progress curve (AUDPC) was calculated based on the formula of Bjarko and Line (1988). Data were analysed by performing analysis of variances (ANOVAs) with SAS software (SAS Inc.). The One-sample Kolomogorov-Smirnov test was used to analyse the normal distribution of the obtained results. The Tukey's multiple range comparison test was used to compare treatment means when data were normally distributed.

At the end of the field trial, all tubers were manually collected from the field and inspected for tuber rot.

7.3 Results

The field trial included 26 genetically modified clones of the variety Désirée containing either a single resistance gene, *vnt-1.1* (from *S. venturii*) or *sto-1* (from *S. stoloniferum*) or a stack of three resistance genes: *sto-1*, *vnt-1.1* and *blb-3* (from *S. bulbocastanum*). The different clones correspond to different transformation events meaning that they differ in the chromosomal location where the resistance gene(s) were inserted. Late blight was not controlled by fungicide application in the trial.

Field trial results in 2011

Spring and summer of 2011 were exceptionally dry, resulting in low late-blight disease pressure. To be able to evaluate resistance to *P. infestans*, spores from a field isolate were multiplied and used as inoculum on the Bintje plants on 15 and 20 July. The susceptible control clones Désirée, Agria and Fontane were subsequently heavily affected by late blight. The resistant controls Fortuna, Bionica, Sarpo Mira and *S. bulbocastanum* were not infected. Also the genetically engineered Désirée clones were protected against the late blight disease (Fig. 7.1). At the end of the season, Bionica and two genetically engineered clones with a single resistance gene (*sto-1*) showed minor symptoms.

Field trial results in 2012

In early July 2012 late blight spontaneously infested the field trial because spring and early summer were very humid. The disease was first observed on Bintje plants then on Nicola shortly afterwards. At the end of the season also Bionica and Toluca presented visible damage and some of the genetically engineered resistant clones showed minor damage.

7.4 Discussion

Late blight disease is a major constraint to potato production due to the lack of resistant or tolerant potato varieties. About 92% of potato grown worldwide has yields of averaging 15 t/ha which are 25 t/ha below attainable yields (Haverkort *et al.*, 2009). Assuming conservatively that only a quarter of this yield gap is due to late blight the global loss is about €12 billion per annum. This loss justifies the use of genetically engineered potato clones containing (multiple) resistance genes against *P. infestans* that are – as shown in this field trial among others – immune to late blight and do not require fungicides.

Although tubers may not be infected by late blight, foliar damage reduces

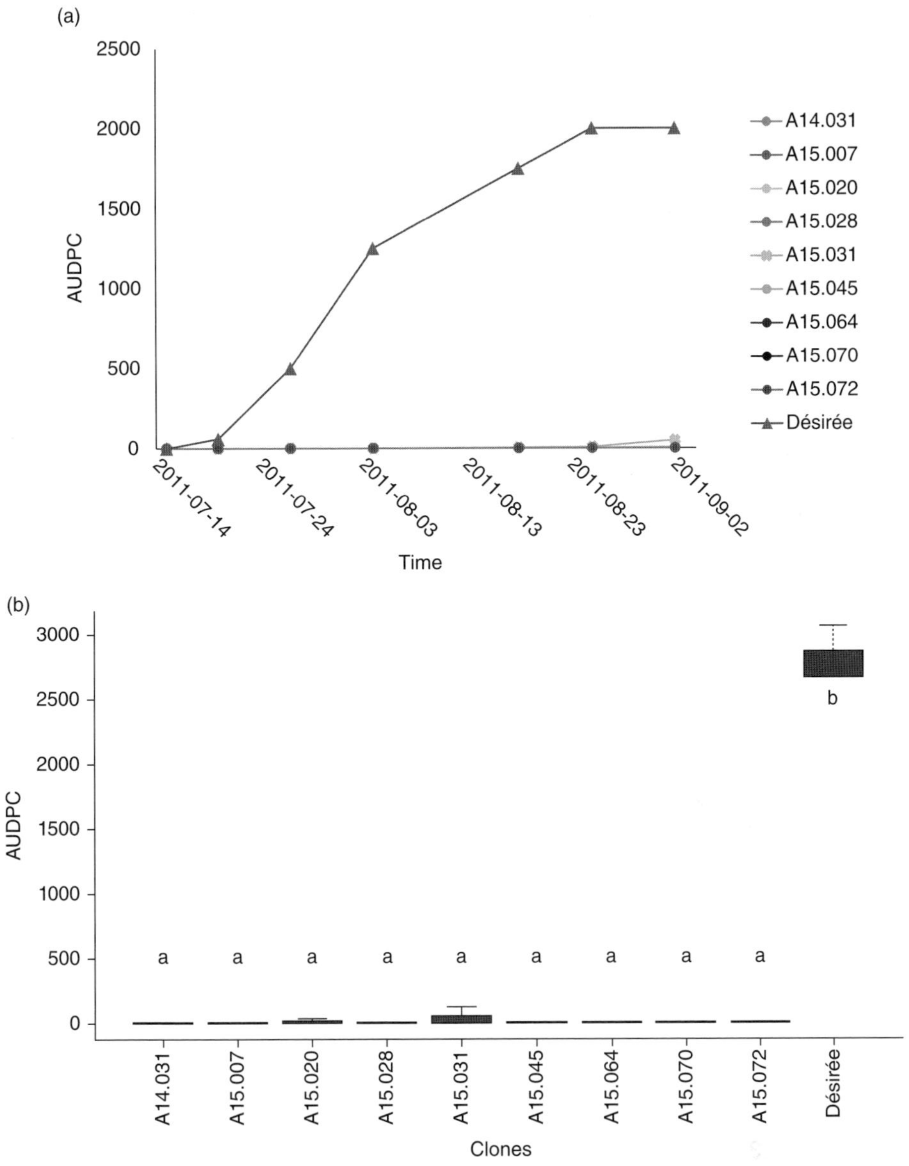

Fig. 7.1. Infection by *Phytophthora infestans* of genetically engineered clones with the *vnt-1.1* gene in comparison with the parent clone Désirée. (a) The disease severity is shown in function of time (expressed as year-month-day) during the 2011 season. AUDPC, area under the disease progress curve. Results for the parent clone Désirée are represented by filled triangles (▲) while the clones with the *vnt-1.1* resistance are shown by the symbols indicated in the key. (b) Boxplot indicating median, interquartile and entire range of AUDPC values from *P. infestans* infection of genetically engineered clones with the *vnt-1.1* gene in comparison with the parent clone Désirée. Means of AUDPC values of genotypes with a different letter differ significantly according to Tukey ($P < 0.05$).

photosynthetic area and thus ability to intercept solar radiation to form food. Haverkort and Bicamumpaka (1986) showed the yield decrease to be proportional to the reduction in intercepted radiation attributed to late blight. They also report that yield decreases up to 70% depending on varietal resistance and environmental conditions (rainfall).

Namanda *et al.* (2004) ascertained beneficial aspects of fungicides for potato production in Uganda. For the susceptible variety Victory an eightfold yield increase was realized with frequent sprays compared with the doubled yield using fewer treatments for resistant variety Rutuku. In the 1980s, commercial potato production in Rwanda was conducted without the use of fungicides yielding about 6 t/ha per season. Recently, due to more extensive use of fungicides (and fertilizers) current yields have doubled to around 12 t/ha.

Gildemacher *et al.* (2009) carried out a systems diagnosis for two countries in East Africa and estimated that late blight control, fertilizer application and improved seed are the three main factors that may substantially improve crop yields. The cultivation of late-blight resistant varieties would therefore help to improve and secure potato yields in African countries without the use of fungicides. For these resistant varieties to be accepted by the farmers, and the consumers, there is a need to consider that they are genetically engineered varieties and may be rejected. However, some surveys have shown that cisgenic plants are more positively received by the general public than genetically modified organisms (GMOs) in general (Gaskell *et al.*, 2010). Furthermore, the European Food Safety Authority (EFSA) Scientific Panel on GMOs compared the hazards of different biotechnology and breeding techniques and concluded that similar hazards can be associated with cisgenic and conventionally bred plants (EFSA Panel on GMO, 2012). Europe and the USA are considering the exemption of cisgenic plants from the scrutiny of transgenic plants. This exemption is not impossible, as illustrated by the example of protoplast fusion: an organism produced by this technique is considered as a GMO unless the fusion was between plant species that are crossable (Jacobsen and Schouten, 2007).

7.5 Conclusion and Recommendations

About 20 resistance genes against *P. infestans* have currently been mapped or cloned. Therefore a collection of resistant potato cultivars with two or three broad-spectrum R genes that have the potential to make potato cultivation more sustainable, due to durable resistance resulting in a much lower fungicide use, is possible. Cultivation of late-blight resistant varieties would improve and secure potato yields in African countries with reduced fungicide application. Resistance management, including crop monitoring and reduced use of fungicides, is recommended to avoid development of resistant late blight strains. Cisgenic potato plants are potentially more acceptable to the general public and if a lighter or no regulatory procedure was imposed, commercialization could become feasible also for small companies or farmer cooperatives.

Acknowledgements

The authors would like to thank Suxian Zhu and Maarten Nijenhuis from the laboratory of Plant Breeding in Wageningen for making the binary vector constructs. Geert Haesaert and Marc Deloose are acknowledged for critical reading of the manuscript. Part of the research was financed by the Dutch Government in the DuRPh project. Furthermore the authors are also grateful for the support of the Flemish Government during the execution of the field trial in 2011 and 2012.

References

Birch, P.R.J., Bryan, G., Fenton, B., Gilroy, E.M., Hein, I., Jones, J.T., Prashar, A., Taylor, M.A., Torrance, L. and Toth, I.K. (2012) Crops that feed the world 8: potato. Are the trends of increased global production sustainable? *Food Security* 4(4), 477–508.

Bjarko, M.E. and Line, R.F. (1988) Heritability and number of genes controlling leaf rust resistance in four cultivars of wheat. *Phytopathology* 78, 457–461.

Cox, A.E. and Large, E.C. (1960) Potato blight epidemics throughout the world. *Agriculture Handbook* No. 174. Agricultural Research Service, United States Department of Agriculture, Washington, DC, 230 pp.
European Food Safety Authority (EFSA) Panel on Genetically Modified Organisms (GMO) (2012) Scientific opinion addressing the safety assessment of plants developed through cisgenesis and intragenesis. *EFSA Journal* 10(2), 2561. Available at: http://www.efsa.europa.eu/en/efsajournal/pub/2561.htm (accessed 18 January 2015).
Garelik, G. (2002) Agriculture – taking the bite out of potato blight. *Science* 298, 1702–1704.
Gaskell, G., Stares, S., Allansdottir, A., Allum, N., Castro, P., Esmer, Y., Fischler, C., Jackson, J., Kronberger, N., Hampel, J., Mejlgaard, N., Quintanilha, A., Rammer, A., Revuelta, G., Stoneman, P., Torgersen, H. and Wagner, W. (2010) *Europeans and Biotechnology in 2010: Winds of Change?* Report to the European Commission's Directorate-General for Research. Publications Office of the European Union, Luxembourg.
Gildemacher, P., Kaguongo, W., Ortiz, O., Tesfaye, A., Woldegiorgis, G., Wagoire, W.W., Kakuhenzire, R., Kinyae, P.M., Nyongesa, M., Struik, P.C. and Leeuwis, C. (2009) Improving potato production in Kenya, Uganda and Ethiopia: a system diagnosis. *Potato Research* 52, 173–205.
Haverkort, A.J. and Bicamumpaka, M. (1986) Correlation between intercepted radiation and yield of potato crops infested by *Phytophthora infestans* in central Africa. *Netherlands Journal of Plant Pathology* 92, 239–247.
Haverkort, A.J., Boonekamp, P.M., Hutten, R., Jacobsen, E., Lotz, L.A.P., Kessel, G.J.T., Visser, R.G.F. and van der Vossen, E.A.G. (2008) Societal costs of late blight in potato and prospects of durable resistance through cisgenic modification. *Potato Research* 51, 47–57.
Haverkort, A.J., Struik, P.C., Visser, R.G.F. and Jacobsen, E. (2009) Applied biotechnology to combat late blight in potato caused by *Phytophthora infestans*. *Potato Research* 52, 249–264.
International Potato Center (CIP) (2014) Available at: http://cipotato.org/research/potato-in-east-africa/ (accessed 18 January 2015).
Jacobsen, E. and Schouten, H.J. (2007) Cisgenesis strongly improves introgression breeding and induced translocation breeding of plants. *Trends in Biotechnology* 25, 219–223.
Namanda, S., Olanya, O.M., Adipala, E., Hakiza, J.J., El-Bedewy, R., Baghsari, A.S. and Ewell, P. (2004) Fungicide application and host-resistance for potato late blight management: benefits assessment from on-farm studies in SW Uganda. *Crop Protection* 23, 1075–1083.
Park, T.-H., Vleeshouwers, V.G.A.A., Jacobsen, E., van der Vossen, E. and Visser, R.G.F. (2009) Molecular breeding for resistance to *Phytophthora infestans* (Mont.) de Bary in potato (*Solanum tuberosum* L.): a perspective of cisgenesis. *Plant Breeding* 128, 109–117.
Vleeshouwers, V.G.A.A., Raffaele, S., Vossen, J.H., Champouret, N., Oliva, R., Segretin, M.E., Rietman, H., Cano, L.M., Lokossou, A., Kessel, G., Pel, M.A. and Kamoun, S. (2011) Understanding and exploiting late blight resistance in the age of effectors. *Annual Review of Phytopathology* 49, 507–531.
Zhu, S., Li, Y., Vossen, J.H., Visser, R.G.F. and Jacobsen, E. (2012) Functional stacking of three resistance genes against *Phytophthora infestans* in potato. *Transgenic Research* 21, 89–99.

8 Exhibition Trial and Farmer Participatory Selection of New Late-blight Resistant B3C1 Potato Genotypes for Adaptation to Nigerian Conditions

C.O. Amadi,[1]* A.J. Lang,[2] E.A. Dung,[2] D.M. Lenka,[2] T.Y. Dalyop[2] and J.A. Landeo[3]

[1]National Root Crops Research Institute, Umudike, Nigeria; [2]National Root Crops Research Institute, Kuru, Nigeria; [3]International Potato Center (CIP), Lima, Peru

Abstract

Nineteen late-blight resistant B3C1 potato genotypes received from the International Potato Center (CIP) and two local checks were planted in an exhibition trial conducted in three potato-growing locations on the Jos Plateau of Nigeria in the 2007 rainy season. The field trials were situated in Bokkos, Kerang and Kuru as part of the accelerated variety selection scheme being promoted by CIP to speed up the release of new varieties and increase adoption rate. The objectives were to identify and select together with the farmers, high yielding and late-blight resistant genotypes and by so doing, popularize these genotypes in advance of their release. Eight B3C1 potato genotypes ($P < 0.05$) performed better than the local check at Kuru, while three gave higher yields than the local check at Kerang. In Bokkos, none of the B3C1 genotypes yielded greater than the local checks. Clones 392617.54, 393073.179 and 396026.103 gave tuber yields that were ($P < 0.05$) greater than the local checks in Kuru and Kerang with yields of 23.63, 25.24 and 19.79t/ha, respectively. Based on overall performance, ten of the B3C1 genotypes were selected for further evaluation. Farmers preferred genotypes with large tuber size, fewer tubers and high yield. Few of them (3% and 6%) considered tuber colour and tuber shape, respectively, as important characteristics when selecting a potato variety.

Keywords: clones, genotypes, potato, selection, tuber size, variety

8.1 Introduction

Potato is one of the most valuable staple foods due to its high nutritional value, as well as its high caloric count and yield per hectare (Van der Zaag and Horton, 1983). Although it is ranked fourth after cassava, yam and sweetpotato among the major root and tuber crops in Nigeria, it is by far the most efficient tuber crop in the country in terms of tuber yield and days to maturity. Under conditions in Nigeria, potato matures in about 75–90 days compared with 9 and 12 months for yam and cassava, respectively (Okonkwo *et al.*, 2009).

*okeyamadi2003@yahoo.com

Potatoes are an important source of food, employment and income on the Jos Plateau of Nigeria (Okonkwo *et al.*, 2009). Late blight has been known to induce significant yield losses and constitute a major threat to potato production, and while chemical control is expensive and lasts only a short time with its attendant health hazards, biological control of late blight is still in its infancy (Ke-qiang and Forrer, 2001).

Existing variety selection methods have some limitations as farmers' preferences are seldom considered but traits used as indices for selection are those deemed by the breeders as important. Participatory variety selection involves the farmers and breeders in the selection of preferred varieties (Joshi and Witcombe, 1996).

The National Root Crops Research Institute, Umudike collaborates with the International Potato Center (CIP) to evaluate potato genotypes developed by the latter to determine their adaptability to conditions in Nigeria. In the accelerated variety selection scheme, resistant breeding materials in the form of advanced clones are introduced *in vitro* from CIP headquarters to the sub-Saharan Africa (SSA) regional office in Nairobi through the Plant Quarantine Station of the Kenyan Plant Health Inspectorate Service (KEPHIS) where disease-free minitubers are produced for distribution to target countries in the region. Recipient countries also receive a multi-stakeholder accelerated participatory variety selection scheme to release varieties in 4–5 years and have at least 30 t of clean basic seed to initiate the diffusion process closely linked with their seed production schemes (CIP, 2007). Twenty-five late-blight resistant B3C1 potato genotypes (Landeo and Gastelo, 2004) were received from CIP in November, 2005. Nineteen selected genotypes (Amadi *et al.*, 2006) were included in an exhibition trial during which farmers were invited to participate in the selection at harvest.

The objectives were to identify and select, together with the farmers, high yielding and late-blight resistant genotypes and to popularize these genotypes prior to their release. It was also intended to determine farmers' preferences for a variety.

8.2 Materials and Methods

The trial was sited in Kuru, Kerang and Bokkos villages in Jos South, Mangu and Bokkos local government areas (LGAs) of Plateau State, respectively. These are among the major potato-growing LGAs in the state. The treatments consisted of 19 B3C1 genotypes laid out in a randomized complete block design replicated three times at each location. Two locally grown varieties, Nicola and BR 63-18, were included as controls. The gross and net plot sizes were 9 m^2 and 3 m^2, respectively. Recommended agronomic management practices (Amadi *et al.*, 2009) were applied in the maintenance of the trial except that fungicides were not applied. Data on foliar maturity, number of stems, early blight, late blight, number and weight of marketable, unmarketable and rotted tubers, percentage of rotted tubers, number of tubers per plant and tuber yield (in metric tons per hectare) were recorded. Late blight scores at Kuru were recorded at 8, 10 and 12 weeks after planting. However, at Bokkos and Kerang scores were recorded only at 12 weeks after planting. At harvest, 20 farmers from each of the three locations participated in the selection in which each was asked to select the best genotype based on the following criteria: (i) tuber yield; (ii) tuber number (few < 10, many 10–15, numerous > 15); (iii) tuber size (small < 35 mm, medium 35–55 mm, large > 55 mm); (iv) tuber shape; and (v) tuber skin colour. The genotype each respondent selected and criteria for selection were recorded. GENSTAT DISCOVERY edition 3 was used for statistical analysis (analysis of variance, ANOVA). The means were separated using the standard error of the mean difference at 5% level of probability.

8.3 Results and Discussion

Tuber yield and late blight incidence of some promising potato genotypes from the Kuru location are presented in Table 8.1. The result shows that at Kuru, number of tubers per plant varied with potato genotypes. Genotypes 392617.54 and 393227.66 produced a higher number of tubers compared with genotype 392657.8. Many of the B3C1 genotypes (Table 8.1) had lower late blight scores compared with the controls (Nicola and BR 63-18) indicating that these genotypes resisted the spread of late blight on their foliage. On tuber yield in Kuru (Table 8.1), genotype 392617.54 performed better than all the remaining genotypes. Genotypes 393073.179, 391011.17, 393397.164, 396026.103, 392657.8 and 393077.159 were similar in tuber yield production but were better than the controls. Variations among potato genotypes in tuber number and yield have been reported by different authors (Birhman and Kang, 1993; Amadi *et al.*, 2008).

Results from the Kerang location are presented in Table 8.2. Results were similar to that of Kuru. Genotypes 393077.54, 393280. 82, 393371.58, 396043.226, 393397.164, 396026. 103, 393382.44, 392617.54 and 393073.179 recorded lower values of late blight incidence compared with the controls. There were significant differences among the genotypes in tuber yield. Genotypes 393073.179 and 392617.54 produced tuber yields of above 20 t/ha which were higher than the yield of the best control (Table 8.2).

Table 8.1. Tuber yield and disease reaction of some B3C1 genotypes selected at Kuru.

Genotype	Early blight (%)	Late blight (%)	Rotten tubers (%)	Number of tubers per plant	Tuber yield (t/ha)
392617.54	60	40.0	4.4	8.4	23.6
393073.179	60	30.0	15.7	5.9	18.5
391011.17	70	30.0	8.8	6.8	17.8
393397.164	70	20.0	33.6	6.4	17.0
396026.103	30	40.0	14.0	6.6	14.8
393227.66	50	60.0	42.8	7.8	14.3
392657.8	50	30.0	23.0	4.9	14.2
393077.159	40	20.0	9.3	6.0	14.2
Nicola	90	100.0	1.4	5.5	8.7
BR 63-18	90	100.0	31.2	6.4	4.4
SED[a]	*	*	12.23	1.2	2.4

[a]SED, standard error of the difference of means; *, standard error of the mean (SE) for early blight = 2.28%, SE for late blight = 1.62%.

Table 8.2. Tuber yield and disease reaction of some B3C1 genotypes selected in Kerang.

Genotype	Early blight (%)	Late blight (%)	Rotten tubers (%)	Number of tubers per plant	Tuber yield (t/ha)
393073.179	50	50	0.0	6.8	25.2
392617.54	50	50	0.7	6.3	20.7
396026.103	30	40	3.7	11.0	19.8
393280.82	30	30	1.0	9.0	17.4
393397.164	50	40	4.7	6.0	15.9
393371.58	70	30	0.0	6.6	15.4
396043.226	30	40	1.3	7.5	15.2
393382.44	40	50	1.7	6.7	14.6
393077.54	80	30	0.7	4.2	14.5
393339.242	70	80	3.0	7.4	14.1
BR 63-18	90	90	0.0	7.1	12.1
Nicola	90	90	10.0	6.6	8.0
SED[a]	*	*	2.4	1.4	3.6

[a]SED, standard error of the difference of means; *, standard error of the mean (SE) for early blight = 3.32%, SE for late blight = 3.60%.

Tuber yield attributes and disease reaction of the genotypes at Bokkos are presented in Table 8.3. Both early and late blight severity was relatively high on most genotypes at this location. Tuber yield ranged from 16.1 t/ha in genotype 392657.8 to 7 t/ha in the control. However, these differences were statistically not significant (Table 8.3).

The effect of genotype and location on tuber yield and late blight are presented in Tables 8.4 and 8.5, respectively. In all the three locations and particularly at Kuru and Kerang, CIP B3C1 potato genotypes recorded low levels of late blight and higher tuber yields per hectare than the controls. Reduced foliar damage due to resistance of these genotypes to late blight perhaps allowed them to generate more photosynthates and accumulate the same in tubers over a longer period compared with the controls. This probably accounted for their higher yields relative to the control.

Table 8.3. Tuber yield and disease reaction of some B3C1 genotypes selected at Bokkos.

Genotype	Early blight (%)	Late blight (%)	Rotten tubers (%)	Number of tubers per plant	Tuber yield (t/ha)
392657.8	50	30	22.4	5.4	16.1
393073.179	50	30	12.0	6.4	15.5
396026.103	90	50	32.7	7.6	15.4
393220.54	90	70	13.0	63.1	12.6
396043.226	90	70	12.7	6.4	10.4
393077.54	70	50	3.3	3.7	9.8
392617.54	90	90	20	3.4	9.3
393382.44	90	70	16.7	5.2	9.2
393371.58	90	90	20.3	4.6	8.3
391011.17	90	90	3.3	3.9	7.9
Nicola	90	90	6.0	4.5	7.8
BR 63-18	90	90	7.7	4.3	7.0
SED[a]	*	*	9.5	1.5	4.5

[a]SED, standard error of the difference of means; *, standard error of the mean (SE) for early blight = 0.25%, SE for late blight = 0.32%.

Table 8.4. Effect of genotype and location on tuber yield (t/ha) of some B3C1 genotypes.

Genotype	Location			Mean
	Kuru	Kerang	Bokkos	
393073.179	18.5	25.3	15.5	19.7
392617.54	23.6	20.7	9.3	17.9
396026.103	14.8	19.8	15.4	16.7
392657.8	14.2	16.1	16.1	15.5
393227.66	14.3	15.5	15.5	15.1
393220.54	13.8	13.1	12.8	13.2
393397.164	16.9	15.9	5.4	12.8
393011.17	17.8	12.5	7.9	12.7
396043.226	10.7	15.2	10.4	12.1
393371.58	12.6	15.4	8.3	12.1
393077.54	11.2	14.5	9.8	11.8
Nicola	8.7	12.1	7.0	9.3
BR 63-18	4.9	7.9	7.8	6.9
Mean[a]	12.3	14.2	9.1	

[a]SED, standard error of the difference of means: Genotype = 2.0, Location = 0.8, Genotype × Location = 3.5.

Table 8.5. Effect of genotype and location on late blight severity (%) on some B3C1 potato.

Genotype	Location		
	Kuru	Kerang	Bokkos
392617.54	40.00	100.00	90.00
392657.8	30.00	30.00	30.00
393011.17	30.00	80.00	90.00
393073.179	30.00	70.00	30.00
393077.54	30.00	30.00	50.00
393220.54	20.00	70.00	70.00
393227.66	60.00	60.00	60.00
393228.67	50.00	30.00	50.00
393339.242	60.00	80.00	90.00
393371.58	20.00	30.00	90.00
393397.164	20.00	40.00	50.00
396026.103	40.00	40.00	50.00
396043.226	40.00	40.00	70.00
BR 63-18	100.0	100.00	90.00
Nicola	100.0	100.00	90.00
Mean	43.33	59.52	64.29

Farmers selected ten genotypes. These are listed in Table 8.6 according to the number of farmers that selected them as their preferred genotype; 35% and 30% of farmers selected genotype 392617.54 and 391011.17 as their most preferred genotype, respectively.

The importance farmers attached to each selection criterion is shown in Fig. 8.1. Most farmers considered tuber yield, tuber size and tuber number as the most important characteristics for selection of a potato variety. Farmers preferred genotypes with a high total yield that produced a few (five to ten), large-sized (diameter > 55 mm) tubers.

Table 8.6. B3C1 potato genotypes selected by farmers.

Genotype	Frequency[a]	Percentage[b]
392617.54	21	35
391011.17	18	30
396026.103	6	10
393220.54	6	10
393073.179	3	5
393397.164	2	3.3
392657.8	1	1.7
393227.66	1	1.7
396026.103	1	1.7
393077.54	1	1.7
Total	60	100

[a]Frequency, number of times a genotype was selected by participating farmers.
[b]Percentage =

$$\frac{\text{Number of times a genotype was selected by participating farmers}}{\text{Total number of participating farmers}} \times 100$$

8.4 Conclusion

Our investigation indicated that in all the three locations and particularly at Kuru and Kerang, CIP B3C1 potato genotypes recorded low levels of late blight and higher tuber yields per hectare than the controls. Based on overall performance and preference by farmers, ten of the B3C1 genotypes (392617.54, 391011.17, 396026.103, 393220.54, 393073.179, 393397.164, 392657.8, 393227.66, 396026.103 and 393077.54) were selected for further evaluation. Farmers preferred genotypes with large tuber size, fewer tubers and high yield.

Acknowledgements

The authors are grateful to the Executive Director of the National Root Crops Research Institute, Umudike, Dr J.C. Okonkwo and management for the permission to publish this work. We are also grateful to the International Potato Center (CIP), Lima, Peru for providing the B3C1 genotypes and

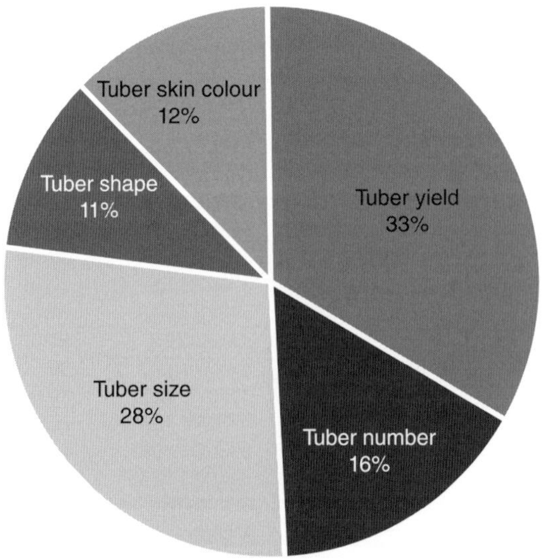

Fig. 8.1. Importance farmers attached to selection criteria.

to all the field staff of the Potato Programme, Kuru and our dear farmers who honoured our invitation to participate in the selection. The authors are also grateful to Deutsche Gesellschaft für Internationale Zusammenarbeit (GIZ) – Private Sector Development in Agriculture (PSDA) for the opportunity to present this chapter as a paper at the Ninth Triennial Conference of the African Potato Association in Naivasha, Kenya.

References

Amadi, C.O., Lang, A.J., Amadi, G., Lenka, D.M., Dung, E.A. and Landeo, J.A. (2006) Evaluation of new B3C1 late blight resistant genotypes for adaptation to Nigerian conditions. In: Asumugha, G.N., Olojede, A.O., Ikeorgu, A.O., Ano, A.O. and Herbert, U. (eds) *Proceedings of the 40th Conference of the Agricultural Society of Nigeria*. Umudike, Abia State, Nigeria, pp. 16–20.

Amadi, C.O., Ene-obong, E.E., Okonkwo, J.C. and Lenka, D.M. (2008) Preliminary studies on yield and some yield attributes of potato grown under high and low ambient temperature. *The Nigeria Agricultural Journal* 29, 65–75.

Amadi, C.O., Okonkwo, J.C., Lenka, M.D., Asumugha, G.N. and Nwosu, K.I. (2009) *A Guide to Potato Production in Nigeria*. Extension Guide No. 25. Extension Service Programme, National Root Crop Research Institute, Umudike, Abia State, Nigeria.

Birhman, R.K. and Kang, G.S. (1993) Analysis of variation and inter-relationships in potato germplasm. *Euphytica* 68, 17–26.

International Potato Center (CIP) (2007) CIP Research in Sub-Saharan Africa. Available at: http://cipotato.org/regions/sub-saharan-africa/ (accessed 15 September 2015).

Joshi, A. and Witcombe, J.R. (1996) Farmer participatory crop improvement II. Participatory varietal selection, a case study in India. *Experimental Agriculture* 32, 461–477.

Ke-qiang, C.A.O. and Forrer, H.R. (2001) Current status and prosperity on biological control of potato late blight (*Phytophthora infestans*). *Journal of Agricultural University of Hebei*. Available at: research.cip.cgiar.org. (accessed 23 June 2013).

Landeo, J.A. and Gastelo, M. (2004) *Data Collection Protocol for Potato Variety Selection in Advanced Trials in LB Resistant Clones*. Specialized technology document. International Potato Center, Lima.

Okonkwo, J.C., Amadi, C.O. and Nwosu, K.I. (2009) *Potato Production, Storage, Processing and Utilization*. National Root Crops Research Institute, Umudike, Abia State, Nigeria.

Van der Zaag, D.E. and Horton, D. (1983) Potato production and utilization in world perspectives with special reference to the tropics and sub-tropics. *Potato Research* 26, 323–362.

9 Integrative Breeding Strategy for Making Climate-smart Potato Varieties for Sub-Saharan Africa

A. Asfaw,[1]* M. Bonierbale[2] and M.A. Khan[2]
[1]*International Potato Center (CIP), Nairobi, Kenya;* [2]*CIP, Lima, Peru*

Abstract
Breeding potato (*Solanum tuberosum* L.) is becoming increasingly complicated because of the growing number of requirements for new varieties, particularly the added concern of adapting potato to climate variability, especially in regions of sub-Saharan Africa. Combining the right genes to overcome constraints of climate variability in a potato crop, together with an enhanced level of other desirable traits such as consumer and commercial preferences, yield and resistance to biotic stresses requires an integrated breeding strategy that makes use of the knowledge of scientists as well as farmers. This chapter discusses the design of a breeding strategy that incorporates adaptation traits with the commercial and home-use characteristics preferred by potato farmers for varieties to be grown in diverse environments.

Keywords: breeding design, climate variability, drought tolerance, farmer preference

9.1 Introduction

Potato is a strategic food security crop for Africa's present and future, especially in densely populated highland regions. It yields higher amounts of nutritious food per unit of land and water in less time and often under more adverse conditions than other food crops (Scott *et al.*, 2000). Total potato production in Africa is increasing (FAO, 2013). This increase has mainly resulted from expansion of the cropping area, including into warmer/drier zones less favourable to potato production, and does not reflect increasing yields on a stable total area (Low *et al.*, 2007; Gildemacher, 2012). The potato production expansion to new environments could be negatively affected by global warming linked to climate change (Thiele *et al.*, 2010). Moreover, even in traditional potato-farming ecologies of Africa, rainfall is becoming more erratic, with longer and hotter dry spells and more intense rainstorms (Haverkort *et al.*, 2013). Several places in Africa where drought is already a problem, such as the East and South, will suffer from warmer and successively drier weather linked with climate change over the next few decades (Jones and Thornton, 2003; Williams *et al.*, 2007), which could adversely impact potato production. Climate change will cause increased temperatures and higher evapotranspiration that intensify the problems for small-scale farmers who grow potato in Africa. This highlights the need to

*a.amele@cgiar.org

improve adaptation to climate variability in potato breeding efforts.

The challenge of breeding potato varieties for smallholder farmers in Africa means finding ways to cope with the variation of farmers' environments and preferences and market dynamics. Farmers face variation in climate, soils and the market in their specific conditions. On the other hand, breeders face variation in target area, target farmers, consumer preference and complexity in physiological mechanisms and inheritance of the climate change adaptation traits to breed for. Drought and/or heat tolerance are the major traits for farmers to battle the adverse impact of climate variability in potato production. For these traits in potato to contribute significantly to the well-being of farmers in Africa, they must be expressed in terms of increased yield under field conditions combined with traits that correspond to the preferences and needs of farmers and markets. This is not an easy task for a breeding programme to accomplish, but a necessary one.

To effectively use and integrate the various options for breeding climate-smart potatoes for sub-Saharan Africa (SSA), this chapter explores aspects of plant physiology and genomics-based breeding and farmer involvement (Fig. 9.1). It further discusses the downstream adoption challenges in designing a breeding strategy for climate variability and the relevance of molecular and farmer-level analysis of baseline variability available in each country for a breeding programme to approach climate variability. It also addresses genetics and plant physiology methods to identify traits characterizing drought and heat tolerance and to identify genomic regions associated with these traits in order to deliver DNA markers for marker-assisted breeding programmes. Finally, these will be used to discuss options for more effective climate-smart potato breeding for smallholder farmers in SSA.

9.2 Understanding Downstream Adoption Challenges for Breeding Climate-smart Potato Genotypes

Better understanding of target environments, either biophysical or socio-economical, is crucial for the success of any breeding programme. The breeding programme should be informed of the dynamics of adoption challenges for climate-smart potatoes and what drives the dynamics of potato farming in the target environment. This requires analysis of potato farming in context: farmers' environments, practices and market preferences for a variety construct from the breeders' point of view. This kind of analysis will give an understanding of key processes in farmers' variety and seed management and changes that are related to climate variability in variety use, perception and adaptation strategies in the potato crop.

Potato production and variety use in SSA are variable because of agroecological and socio-cultural variation. The cropping practices in growing potato might encompass sole, inter- and relay-cropping in different seasons, each with its specific features in the different regions. This makes the context of potato farming in the region dynamic, as characterized by varying conditions, varying practices and the added complexity of a changing climate. Within these diverse systems, drought and heat stresses are major environmental determinants potato farmers face now and in the future in SSA.

At a micro-regional level, drought and/or heat stress could be vegetative stress (early season), intermittent (mid-season) or terminal (end of season) in expression, but will always have large effects on potato final tuber yield during emergence, tuber initiation and growth (cf. Levy and Veilleux, 2007; Monneveux *et al.*, 2013). The effects are never uniformly distributed and vary with farmers' field conditions and growth stages of the crop. For instance, some farmers with sloping and/or thin soils would necessarily suffer more from drought on potato than those in flat or valley soils. In addition to this complexity, farmers' perceptions of environmental constraints and preferences for specific potato types may vary from region to region.

Multiple cropping practices and stress factors faced by resource-poor farmers in SSA increase the number of criteria that farmers consider in making choices about which potato genotypes to grow. These criteria are not always static or consistent across

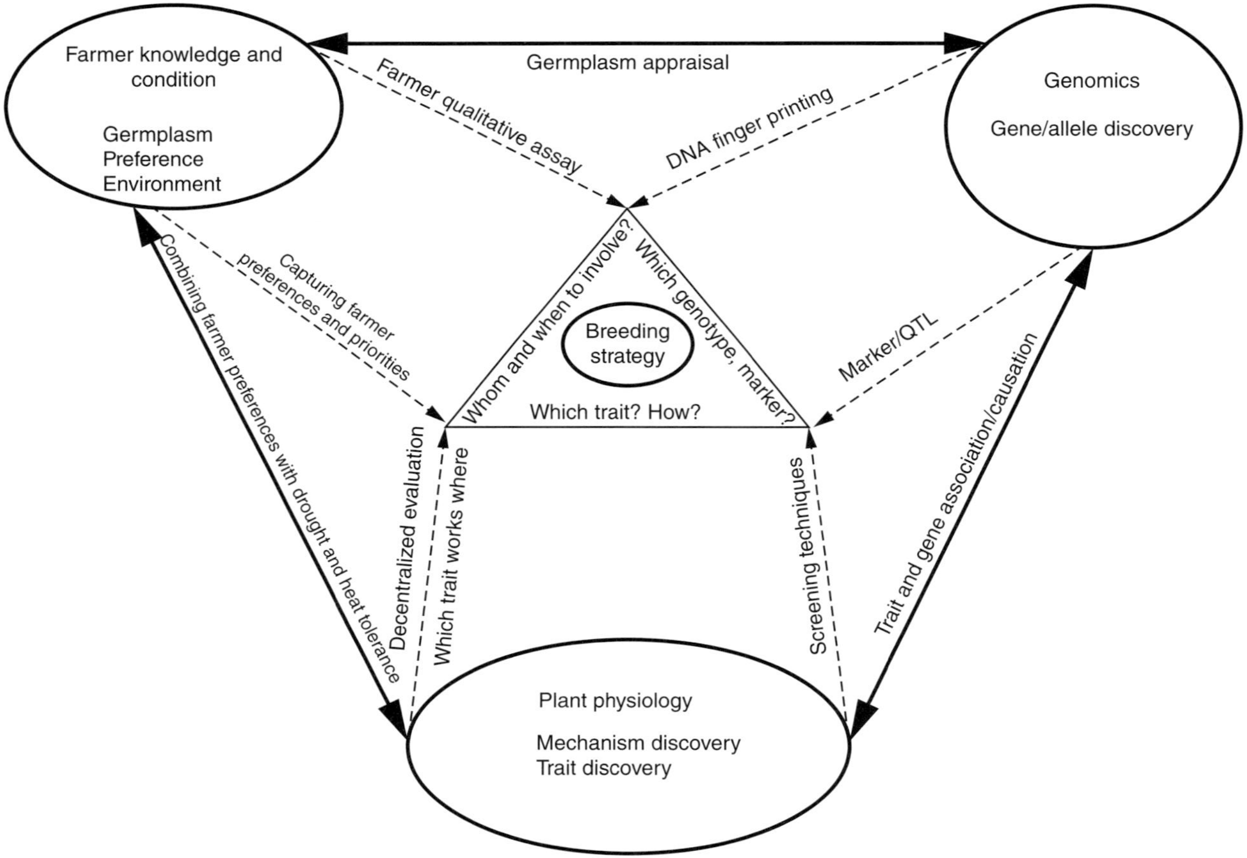

Fig. 9.1. A flow chart depicting the strategy for integrating the various options for developing climate-smart potatoes. QTL, quantitative trait loci. (Adapted from Asfaw, 2011.)

gender, location, individual or other factors, but rather co-evolve with exposure as has been seen in other crops (Asfaw *et al.*, 2012). This suggests that a breeding strategy focusing solely on traditionally grown tuber colours, shapes and sizes may therefore restrict farmers' access to novel attractive and adapted germplasm.

In general, the combination of factors such as climate change, market dynamics and shifts in farmer preference through exposure to new germplasm types represent a constantly moving target for a potato breeding programme to attempt to hit simultaneously. Farmers consistently seek for marketability, drought/heat tolerance, disease resistance, yield and culinary traits which are rarely combined in a single, agronomically superior, current farmer or improved genotype. The challenges of combining the large number of traits, the variation in which they occur, and their constant modification in response to the preferences and needs of the farmers therefore depend largely on the capacity of the breeding programme. Hence, more drought- and heat-tolerant potato varieties for a range of farmer conditions, markets and preferences can only be developed on an integrated understanding of farmers' production conditions and existing seed-system practices. On the basis of this understanding a breeder can contribute to an overall package of mechanisms that harnesses and equips farmers to adapt to their dynamic context.

9.3 Germplasm Appraisal for Breeding Climate-smart Potato

Understanding the level and structure of diversity in available germplasm resources is imperative for harnessing variation for drought or heat tolerance. Farmers' qualitative assessment, morphological phenotyping and molecular genotyping are the range of available tools for a breeding programme to uncover the level and structure of diversity in their local or potentially available germplasm base. Through an inventory of varieties grown by farmers using participatory qualitative methods, the level of visible diversity including which variety is grown by whom, where and why, and their respective desirable and undesirable characteristics could be sorted for possible targeting in the breeding programme. Farmers' knowledge of local varieties and conditions is important, but this by itself is not sufficient, as it is constrained by the invisibility of genetic variation and their limited exposure. Farmers usually have knowledge of the varietal characteristics of their varieties, but at the same time their knowledge on sources of genetic variation for potato is very limited. Moreover, it is difficult with this method to assess diversity among farmers' varieties, since identical varieties may have different names given by farmers or different varieties may have the same name.

Morphological phenotyping is another tool for a breeder to uncover the diversity of existing potato germplasm in each country. However, underestimation of genetic relationships with morphological characterization is expected as phenotype is a modest predictor of genotypic variation. Suitable molecular markers have proved to be a useful tool for detecting baseline variability in a breeding programme, for example microsatellite/SSR (simple sequence repeat) marker types were successful in detecting variability in potato (Ghislain *et al.*, 2004, 2009; Lung'aho *et al.*, 2011). These kinds of analyses are an important ingredient for a more rational use of germplasm in the region. They allow for the design of a strategic crossing plan to mine transgressive segregants based on adapted and preferred germplasm at country or region level and to harness the power of heterosis.

9.4 Exploring Mechanisms and Alleles for Breeding Climate-smart Potato Varieties

Adaptation to climate variability is not a single trait, but rather the overall manifestation of the sum of different mechanisms operating in the plant. The different mechanisms are in turn expressed by component traits which are most often governed by many genes. So it is a real challenge for plant breeders to pyramid these polygenic traits and integrate them with higher tuber yield and end-user preference. The challenge rests on the proper mechanistic understanding of each component trait and on having quick-and-easy, non-destructive, low-cost and accurate screening techniques for traits responsible for adaptation to climate variability. The questions are:

- What are these desirable traits?
- How is a trait chosen from a set of component traits?
- How, when and where is it measured?

In reality, many traits that are related to adaptation to climate variability exist in potatoes. These traits might range across an array of plant strategies to balance the need for growth and reproduction on the one hand and tolerance on the other. The traits need to be measured in managed stress environments, either in the greenhouse or in the field with standardized phenotyping protocols in a multi-replication experiment over seasons and across locations.

The phenotypic assessment from field and/or greenhouse experiments needs to be correlated with a large number of molecular markers spread across the genome using linkage mapping or association mapping. These genetic analyses would help to dissect the mechanism of inheritance of the traits related to adaptation to climate variability at the molecular level. Through such experiments, it is possible to identify a number of quantitative trait loci (QTL) or regions of the genome, the mapping position of each identified QTL, and the magnitude of the effect for traits related to adaptation to climate variability. McCord *et al.* (2011) in tetraploid potatoes and Anithakumari *et al.* (2012) in diploid potatoes identified QTL regions governing internal heat necrosis and drought tolerance, respectively, using mapping populations. These kinds of studies pave the way for marker assisted breeding for heat and drought tolerance by identifying markers tracking responsible genes in different potato populations.

Genomic selection is another genetic method coupled with high-throughput genotyping for improving populations for target traits. The functional relationship between the regions of the genome that control variation in different traits and their effect on final yield under climate variability (drought and heat stress, and non-stress conditions) should be the main focus for present and future research in potatoes. This needs careful genetic analysis that validates these traits for marker assisted breeding.

9.5 Strategies for Climate-smart Potato Breeding

Farmers harvest a lower yield in drier and hotter years compared with well-watered and cooler years provided diseases are controlled effectively. In defining a breeding strategy, the solution to overcome this reduction in yield and the kinds of varieties needed by farmers have to be clearly addressed. Understanding target environments and users, germplasm structure and traits of interest must be the basis for formulating a successful breeding programme. This understanding lies primarily on assessing the level of environmental variation and farmers' preferences, population genetic structure of the available germplasm, and the genetic architecture of adaptation traits and their response to climate variability.

The understanding of different aspects, with regards to both production and productivity, should be integrated at different stages of the cycle of a breeding programme: from setting up breeding objectives to suitable variety development and dissemination. It should start with a firm understanding of the complexities of targeting. This has to be undertaken with participation of key stakeholders, both farmers of different age and gender, and other relevant partners in the crop's production, marketing and consumption systems. The assessment of various elements and context that characterize the farmers' perspective will generate information to answer the strategic questions and facilitate defining the target traits for the new varieties. For instance, how essential is decentralization of evaluation and selection to address the variation in farmers' preferences and conditions?

A breeding strategy to confront this complexity in production ecology and variation in farmers' preferences, therefore, should generate and introduce diversity for farmers to choose from and guide varietal testing and evaluation defined by farmers' preferences, along with knowledge of climate-and-soil-based environment targeting. It is essential to expose farmers with specific criteria about desired quality to material that might be less suitable for their fixed consumption and market preferences, but that might be higher yielding in their stressful environments, in order to create awareness of the trade-offs present and the consequences of the choices farmers either explicitly or implicitly make. This must be done within the context of the stage of crop production at which the effect of drought and heat is the greatest, as well as the degree of stress that adversely impacts yield. In general, internalizing the context of production in planning the breeding programme would give a higher probability of success.

Along with understanding the complexities of targeting, a breeding strategy must also define the expectations and goals within each target. For instance, if yield is 5 t/ha under drought- and heat-stress conditions, the breeder should not target a 'yield potential' of 30 or 40 t/ha usually recorded in optimum conditions at research farms, and think instead about how to get 10 t/ha under the prevailing situation as a design question. A breeding strategy for achieving 'target yield' in target environments, and the trait(s) contributing to that yield should be the main focus. There may be some trade-offs in defining 'target yield' in relation to 'yield potential' but this should not be a major concern in a breeding programme. For instance, root traits are considered important for improving drought tolerance, but over investment in roots might draw biomass away from tubers (sink) (Beebe *et al.*, 2013). In some cases, extra investment in roots might be needed to get up to the goal of 10 t/ha. However, roots take many forms, and targeting specific root phenotypes is a difficult challenge. Roots and the balance with shoot growth and tuber development are topics that still require more work across disciplines (breeders, physiologists, soil scientists). The relative contribution and importance of other relevant traits, such as stomatal control, osmotic adjustment and transcription regulation, should also be explored. It may be necessary to breed for the right root system that uses the same biomass efficiently, with an appropriate root architecture that increases extraction of soil moisture from a greater depth, while maintaining or improving carbon assimilate partitioning to tubers in order to attain the 'target yield' in the target environment.

Fitting the right genotype into the right environment necessarily needs germplasm appraisal. Appropriate germplasm for the target environment could be selected using the criteria of various stakeholders, but which trait is responsible for drought or heat tolerance and how it is inherited requires detailed genetic analysis. A breeding strategy must therefore characterize the drought and heat tolerance of the available potato germplasm and identify the most important trait(s) related to drought and heat tolerance in the germplasm base.

Genomics is a powerful tool to dissect complex traits such as drought and heat tolerance, and to identify genomic regions significantly associated with many of a plant's responses to these stresses in target environments. Due to availability of high-throughput genotyping methods, it is possible to develop DNA markers spread across the whole genome time- and cost-effectively, however, evaluation of multiple traits in multi-environments for G × E (genotype by environment interaction) studies is time-consuming and laborious. To alleviate this problem, pilot studies can be performed to evaluate a few clones for a large number of traits related to drought- and heat-stress tolerance in multiple target environments, and multivariate analysis can be performed to identify the most important traits. After this filtering step, the larger set of germplasm can be evaluated for traits identified as most important. Then, genetic maps of bi-parental crosses and polymorphism data from the wider germplasm can be used together to identify DNA markers linked to QTL to localize chromosomal regions or candidate genes, and for association mapping.

Ultimately, these genetic studies based on genetic markers and precise phenotyping can be used to improve potatoes for drought and heat tolerance by implementing marker assisted selection or a genomic selection strategy.

A breeding strategy therefore needs to define the genetic structure of a variety to breed for. This includes either improving intra-genotypic diversity (increasing the frequency of genes controlling adaptation to drought and heat stresses in a variety) or increasing intra-varietal diversity (increasing choice for growers). To determine the genetic structure of a variety or varieties in a breeding programme, smart crossing and suitable selection methods are required. Smart crossing methods to combine physiological traits with farmers' preferences and higher tuber yield along with resistance to the predominant diseases are a core element of an integrative breeding strategy. It should systematically and empirically combine trait progenitors to see what trait combinations boost tuber yield in relevant conditions. Such breeding strategies should identify tolerant genotypes and develop a crossing plan in a suitable mating design for introgression into adapted cultivars. In the case of heat, a series of accessions of diploid species like *Solanum chacoense*, *Solanum bertheultii* and *Solanum microdontum* have been reported to possess heat tolerance (Veilleux *et al.*, 1997). Genotypes possessing drought tolerance have also been identified in Andean potatoes adapted to short day conditions (cf. Monneveux *et al.*, 2013).

The sources of tolerance could therefore originate from elite cultivars in tetraploid potatoes or germplasm in wild/cultivated diploid species. When the source of tolerance is in elite cultivars of tetraploid potatoes, an elite × elite crossing plan could be employed. Such a crossing plan should first hybridize superior cultivars (clones) in a suitable mating design and then a progeny test should be conducted to identify those with a high GCA (general combining ability; i.e. good breeding values) and then these should be used as progenitors to cross with several female clones.

Due to tetrasomic inheritance, high heterozygosity, asexual propagation and medium to low heritability for traits related to drought and heat tolerance, recurrent selection with progeny testing to identify superior progenitors is most effective and more practical in breeding tetraploid potato. By crossing locally adapted clones with heat-tolerant elite clones in a recurrent selection scheme, Benites and Pinto (2011) demonstrated genetic gain for heat tolerance.When the identified tolerance source is in diploid germplasm, introgression into tetraploid varieties through sexual polyploidization is required. For details on ploidy breeding refer to Ortize (1998). Such breeding practices generate interspecific hybrids from known heat- and drought-tolerant sources. The resulting diploid hybrids need to be screened for 2n pollen production for use as the pollen parent and crossed back with *Solanum tuberosum*. The breeding programme therefore should first generate a series of clones with unique trait combinations at a tetraploid/diploid level and see which combinations work in the field, under target conditions.

After knowing what works where, and what works across environments in improving drought or heat tolerance to attain the 'target yield', then the strategy should be to focus on genomics to localize chromosomal regions or candidate genes involved in drought or heat tolerance. This will help potato breeders to develop climate-smart potato varieties that combine drought and/or heat tolerance with other desired traits using marker-assisted selection.

Lastly, in the cyclic process of breeding, an integrative design should mainstream decentralized evaluation and selection together with farmers' and other stakeholders' involvement in the process. Potato farmers may consider many traits in evaluating their potato variety to cope with the dynamism in their production system. Many of the traits considered by farmers, such as taste, texture and attractiveness to eat, are not typically considered by breeders in their on-station data collection sheets. Moreover, findings in other crops indicate important gender difference for variety preference traits, where women clearly attached more importance to the culinary qualities than men (Asfaw *et al.*, 2012). This means that breeders and farmers (men/women) evaluate or look at traits in different ways, which might lead to valuable germplasm being discarded and less preferred varieties being developed by breeders. Therefore, breeders should collaborate with farmers and other stakeholders in developing climate-smart potato varieties, particularly in two key stages of the breeding cycle: (i) defining target traits at the time of selection of parents for the cross; and (ii) at the stage of selection of improved varieties. Such collaboration should also consider gender-differentiated preferences of producers and consumers of all ages.

Farmer knowledge combined with the scientists' point of view would give an opportunity for interaction and complementary selection criteria often overlooked by formal drought or heat breeding work. In such scientist–farmer partnerships, breeders can work closely with social scientists to assess what makes farmers attracted to certain types of variety constructs and design a strategy that catalyses rapid variety dissemination. Farmers or other relevant clients should be involved at all stages of the breeding process, scientists should help them by providing suitable variation, and selection of suitable varieties should be done in farmers' fields.

9.6 Conclusion

Breeding new climate-smart potato varieties that combine a range of mechanisms for drought and heat tolerance with farmers' and market preferred types should integrate participation, plant physiology, crop modelling and genomics tools. Involvement of farmers at all stages of the potato breeding process, together with the use of technological advancements in genomics and plant physiology is imperative. This will lead to better-targeted potato varieties suited in particular to the needs of resource-poor farmers in the harsh environments of SSA.

References

Anithakumari, A.M., Nataraja, K.N., Visser, R.G.F. and van der Linden, C.G. (2012) Genetic dissection of drought tolerance and recovery potential by quantitative trait locus mapping of a diploid potato population. *Molecular Breeding* 30, 1413–1429.

Asfaw, A. (2011) Breeding for drought tolerance by integrative design: the case of common bean (*Phaseolus vulgaris* L.) in Ethiopia. PhD thesis, Wageningen University, The Netherlands.

Asfaw, A., Almekinders, C.M.J., Blair, M.W. and Struik, P.C. (2012) Participatory approach in common bean breeding for drought tolerance for southern Ethiopia. *Plant Breeding* 131, 125–134.

Beebe, S.E., Rao, I.M., Blair, M.W. and Acosta-Gallegos, J.A. (2013) Phenotyping common beans for adaptation to drought. *Frontiers in Physiology* 4, 35.

Benites, F.R.G. and Pinto, C.A.B.P. (2011) Genetic gain for heat tolerance in potato in three cycles of recurrent selection. *Crop Breeding and Applied Biotechnology* 11, 133–140.

Food and Agriculture Organization of the United Nations (FAO) (2013) Statistical Database FAOSTAT. Available at: http://faostat3.fao.org/home/index.html (accessed 15 April 2013).

Ghislain, M., Spooner, D.M., Rodriguez, F., Villamon, F., Nunez, J., Vasquez, C., Waugh, R. and Bonierbale, M. (2004) Selection of highly informative and user-friendly microsatellites (SSRs) for genotyping of cultivated potato. *Theoretical and Applied Genetics* 108, 881–890.

Ghislain, M., Nunez, J., Herrera, M.R., Pignataro, J., Guzman, F., Bonierbale, M. and Spooner, D.M. (2009) Robust and highly informative microsatellite-based genetic identity kit for potato. *Molecular Breeding* 23, 377–388.

Gildemacher, P.R. (2012) Innovation in seed potato systems in eastern Africa. PhD thesis, Wageningen University, The Netherlands.

Haverkort, A.J. and Verhagen, A. (2008) Climate change and its repercussions for the potato supply chain. *Potato Research* 51, 223–237.

Haverkort, A.J., Frankle, A.C., Engelbrecht, F.A. and Steyn, J.M. (2013) Climate change and potato production in contrasting South African agro-ecosystem 1. Effects on land and water use efficiencies. *Potato Research* 56, 31–50.

Jones, P.G. and Thornton, P.K. (2003) The potential impacts of climate change on maize production in Africa and Latin America in 2055. *Global Environmental Change* 13, 51–59.

Levy, D. and Veilleux, R.E. (2007) Adaptation of potato to high temperature and salinity – a review. *American Journal of Potato Research* 84, 487–506.

Low, J., Barker, I., Bonierbale, M., Crissman, C., Forbes, G., Lemaga, B. and Priou, S. (2007) Emerging trends and advances in potato research relevant to defining the way forward for the potato sector in sub-Saharan Africa. In: *African Potato Association Conference Proceedings*, Vol. 7. African Potato Association, Alexandria, Egypt, pp. 1–17.

Lung'aho, C., Chemining'wa, G.N., Fu, Y.B., Shibairo, S.I., Hutchinson, M. and Paniagua, H.G. (2011) Genetic diversity of Kenyan potato germplasm revealed by simple sequence repeat markers. *American Journal of Potato Research* 88, 424–434.

McCord, P.H., Sosinski, B.R., Haynes, K.G., Clough, M.E. and Yencho, G.C. (2011) QTL mapping of internal heat necrosis in tetraploid potato. *Theoretical and Applied Genetics* 122, 129–142.

Monneveux, P., Ramirez, D.A. and Pino, M.T. (2013) Drought tolerance in potato (*S. tuberosum* L.). Can we learn from drought tolerance research in cereals? *Plant Science* 205/206, 76–86.

Ortize, R. (1998) Potato breeding via ploidy manipulation. *Plant Breeding Reviews* 16, 15–86.

Scott, G.J., Rosegrant, M.W. and Ringler, C. (2000) Global projections for root and tuber crops to the year 2020. *Food Policy* 25, 561–597.

Thiele, G., Theisen, K., Bonierbale, M. and Walker, T. (2010) Targeting the poor and hungry with potato science. *Potato Journal* 37, 75–86.

Veilleux, R., Paz, M.M. and Levy, D. (1997) Potato germplasm development for warm climates: genetic enhancement of tolerance to heat stress. *Euphytica* 98, 83–92.

Williams, J.W., Jackson, S.T. and Kutzbach, J.E. (2007) Projected distributions of novel and disappearing climates by 2100 AD. *Proceedings of National Academy of Science USA* 104, 5738–5742.

10 New Elite Potato Clones with Heat Tolerance, Late Blight and Virus Resistance to Address Climate Change

M. Gastelo,* L. Diaz, J.A. Landeo and M. Bonierbale
International Potato Center (CIP), Lima, Peru

Abstract
Potato production in developing countries is expanding to warmer environments as farmers search for income opportunities and food security. Meanwhile, climate change is already affecting weather patterns in traditional potato-growing areas, where unpredictable rains and pressure from pests and disease are increasing farmers' risk. Since 2004, the International Potato Center (CIP) has sought to develop new, more heat-tolerant generations of its tropical highland-adapted late-blight resistant population. Late-blight resistant parents were crossed with early maturing and virus resistant progenitors, and selection practised under warm temperatures, water deficit and mid-latitude conditions. During the 2005–2006 summer season (January–March) 20,000 genotypes were exposed to heat in a screenhouse at CIP's experimental station in San Ramon, a warm rain forest environment at 800 m above sea level (masl) and latitude 11° 08′ S. Selected clones were assessed in the field in the same location, where average night and day temperatures during tuberization were 21°C and 27°C, respectively; the resulting heat tolerant clones were exposed to high, endemic late blight pressure in Oxapampa (mid-elevation humid tropics) in replicated trials conducted over 4 years. Selected heat tolerant, late-blight resistant clones were evaluated for yield components in the spring–summer season in La Molina (12° 05′ S) and drought sensitivity in Majes (16° 28′ S) along the arid coast of Peru. Sixty-one advanced clones were assessed again for yield in San Ramon, La Molina and Majes and yield stability analysed. All trials were conducted in a simple lattice design, using Désirée as a heat tolerant control. Harvesting was performed at 90 days. The 61 clones were screened for resistance to potato virus X (PVX) and potato virus Y (PVY), by mechanical inoculation and grafting under greenhouse conditions. Analysis of variance for marketable tuber yield showed significant differences among clones. In San Ramon, yields were in the range of 16.01–28.43 (t/ha), significantly exceeding the control. Of the 40 mid-maturing elite clones selected from the new 'LBHT' (late blight resistant heat tolerant) population, with resistance to late blight, tolerance to heat, 11 carry extreme resistance to PVY, 25 to PVX and seven show tolerance to drought. Eleven mid-maturing clones with heat tolerance, resistance to late blight and PVY that are suitable for mid-elevation zones and climate change, and all 40 clones, are available for variety development and further use in breeding.

Keywords: earliness, heat tolerance, late blight resistance, mid-elevation adaptation, potato clones, virus resistance

*m.gastelo@cgiar.org

10.1 Introduction

The International Potato Center (CIP) has developed a new population of mid-maturing, late-blight resistant clones with improved adaptation to warm environments, resistance to viruses and good quality and agronomic traits to meet increasing market demand for fresh and processed potatoes. The new population is denominated late blight resistant and heat tolerant (LBHT) (Gastelo *et al.*, 2012) and have been produced to address climate change that is already affecting weather patterns in traditional potato-growing areas (Hijmans, 2003; Levy and Veilleux, 2007). While late blight is not a disease of warm environments (Landeo *et al.*, 1995), production of potato seed will have to take place in traditional cool environments where pressure of late blight is often high. Furthermore, adaptation to warmer conditions and dryer seasons implies the need for resistance to major potato viruses, such as potato virus X (PVX) and potato virus Y (PVY), as well as drought tolerance. We report the selection of a new set of elite clones combining resistance to late blight and viruses, heat and drought tolerance with early maturity and stable tuber yields. The products are available from CIP for variety development and use in breeding programmes (CIP, 2014).

10.2 Materials and Methods

Evaluation and selection

The LBHT population was developed using phenotypic selection with a four season cycle. In the first season, in 2005–2006, 20,000 genotypes from different cross combinations among late-blight resistant clones from CIP's B3C1 population, selected *Solanum tuberosum* varieties, and subtropically adapted CIP-bred clones with resistance to PVY and/or PVX were initially screened for heat tolerance. Selection was based on the identification of seedlings that formed tubers under greenhouse conditions in the summer season with daytime temperatures above 30°C, from January to March in CIP's experimental station in San Ramon, a warm rainforest environment at 800 m above sea level (masl) at a latitude of 11° 08′ S. Selected clones (2908) were tested in the same place under field conditions in several seasons, from July to October when average night and day temperatures during tuberization were 21°C and 27°C, respectively (Fig. 10.1). Evaluation of late blight resistance was carried out on the selected clones for four seasons under high endemic pressure at the mid-elevation humid tropic location of Oxapampa (1850 masl and 10° 37′ S). Resistance to late blight

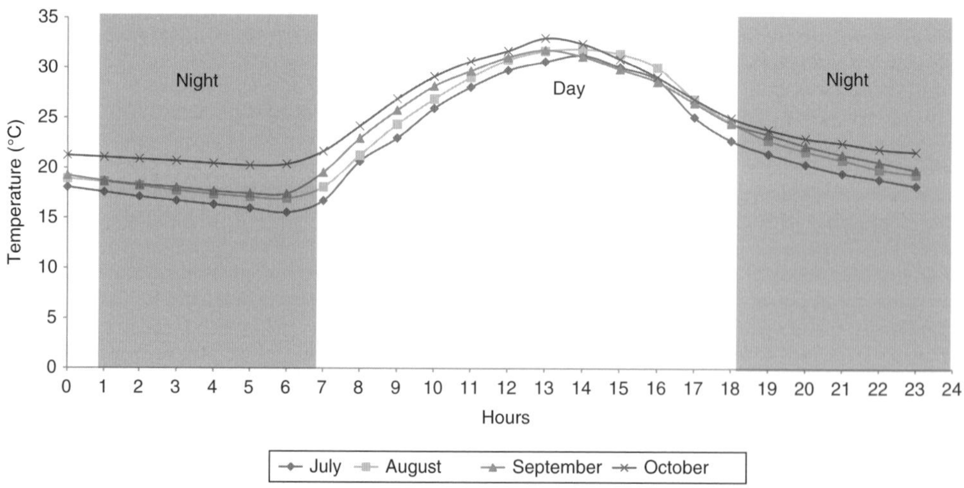

Fig. 10.1. Average day and night temperature in San Ramon in 2012.

was measured as percentage foliage affected for 6 weeks successively. Area under the disease progress curve (AUDPC) and susceptibility, using the susceptibility scale of Yuen and Forbes (2009), were calculated. Plant vigour, foliage maturity at harvest, tuber bulking, tuber appearance, marketable tuber number, marketable and total tuber yield, specific gravity, cooking and processing quality (French fries and chips) were evaluated. Additionally 90 clones were tested for heat tolerance under high temperature field conditions in the spring–summer season in La Molina (12° 05′ S) and Majes (16° 28′ S) along the arid coast of Peru (Table 10.1). Harvesting of all trials was done at 90 days after planting.

Stability analysis

In 2012–2013, 61 advanced clones resulting from selection at San Ramon, La Molina and Oxapampa (Table 10.1) were assessed for tuber yield in San Ramon, La Molina and Majes under high temperatures. All trials were conducted in a simple lattice design, with 20 plants per plot, using two varieties as controls: (i) Désirée as heat tolerant; and (ii) Amarilis as heat sensitive. The traits measured were plant vigour, foliage maturity at harvest, tuber appearance, marketable tuber number and marketable and total tuber yield. Analysis of variance for tuber yield and the additive main effects and multiplicative interaction (AMMI) model for stability analysis were performed using the statistical software R (R Core Team, 2012). The results of the AMMI model analysis were interpreted on the basis of two AMMI graphs for principal components and tuber yield.

Screening for virus resistance and drought tolerance

All clones were screened for resistance to PVX and PVY viruses following mechanical inoculation and grafting in the greenhouse and indexing by ELISA during primary and secondary infection cycles using the procedures outlined by CIP (2006).

Screening for drought tolerance was done through comparative field trials with normal and restricted irrigation in a desert environment, for selection of drought tolerant clones we use the drought tolerance index (DTI) of Cabello *et al.* (2012, 2013).

Flavour and texture were determined for 61 selected clones harvested at San Ramon in 2010.

10.3 Results

Evaluation and selection

Sixty-one clones were selected after several seasons of screening for heat tolerance, resistance to late blight, maturity (90 days), suitable agronomic traits, high tuber yield and quality for table and processing. Clones that had AUDPC values equal or less than the resistant variety Kory were selected as late blight resistant. Susceptibility of all selected clones was significantly less than the susceptible variety Yungay (Fig. 10.2).

Stability analysis

Analysis of variance for tuber yield in the experiments carried out in San Ramon, La Molina and Majes in 2012–2013 showed

Table 10.1. Selection sites in Peru.

Locality	Altitude (masl)	Latitude	Longitude	Agroecology	Major stress factor
San Ramon	800	11° 08′ S	75° 20′ W	Mid-elevation tropics	Heat
La Molina	240	12° 05′ S	76° 56′ W	Lowland tropics	Heat
Oxapampa	1850	10° 37′ S	75° 23′ W	Mid-elevation tropics	Late blight
Majes	1294	16° 28′ S	72° 06′ W	Mid-elevation tropics	Heat, drought

significant differences among genotypes ($P < 0.01$) (Table 10.2). Marketable tuber yields were 16.01–29.26 (t/ha) in San Ramon, 15.28–43.87 (t/ha) in La Molina and 10.55–43.15 (t/ha) in Majes, compared with the performance of cv. Désirée with 15.09, 22.35 and 9.28 (t/ha), respectively (Table 10.3 and Fig. 10.3).

Analysis by the AMMI model showed that genotypes (G), environments (E) and G × E interaction were highly significant ($P < 0.01$; Table 10.4), and affected tuber yield with 37.11%, 29.49% and 33.34% of the treatment sum of squares, respectively. The effect of genotype, environment and G × E interaction were very similar, indicating that variation for

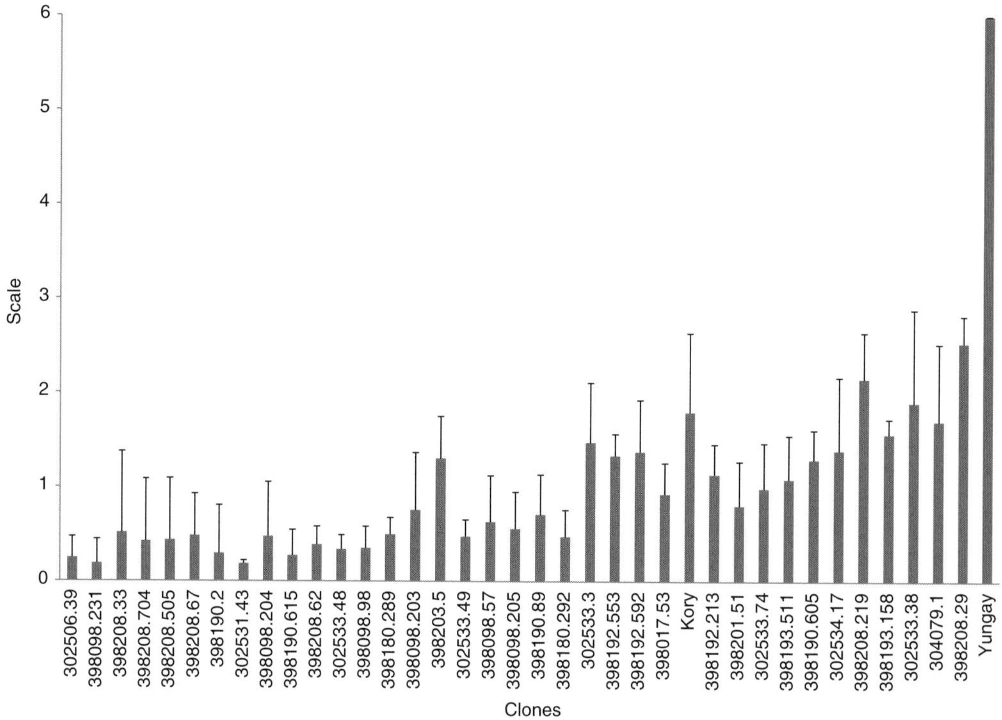

Fig. 10.2. Resistance to late blight based on the susceptibility scale of Yuen and Forbes (2009) of 34 elite clones over five seasons in Oxapampa, Peru (2009–2013).

Table 10.2. Analysis of variance for tuber yield (t/ha) in three environments during the 2012–2013 season.

		Mean square					
		Marketable tuber yield (t/ha)			Total tuber yield (t/ha)		
Source of variation	df[a]	San Ramon	La Molina	Majes	San Ramon	La Molina	Majes
Replications	1	31.245*	27.161	26.757	57.682**	10.911	57.228*
Genotypes	63	65.067**	138.491**	167.078**	68.459**	127.172**	184.397**
Block/replications	14	3.346	21.752	13.096	4.384	30.391*	13.842
Error	49	2.793	14.402	9.882	3.009	12.745	10.785
CV (%)[b]		9.8	14.5	16.8	9.5	12	16.5

[a]df, degrees of freedom.
[b]CV, coefficient of variation.
**, significant statistical differences at $P < 0.01$; *, significant statistical differences at $P < 0.05$.

Table 10.3. Tuber yield, area under the disease progress curve (AUDPC) and susceptibility index for late blight (LB) and virus resistance, heat and drought tolerance of elite clones from LBHT, 2012–2013.

		Pedigree		Marketable tuber yield (t/ha) high temperatures			Total tuber yield (t/ha) high temperatures			Total tuber yield (t/ha) under LB	AUPDC	Resistance			Tolerance[d]	
No.	Clone	Female	Male	San Ramon	La Molina	Majes	San Ramon	La Molina	Majes	Oxapampa	Oxapampa	LB[a]	PVX[b]	PVY[c]	Heat	Drought
1	398017.53	391002.6	392639.31	25.20	32.67	18.80	26.50	36.70	19.20	15.37	488	0.93		ER	T	
2	398098.203	393371.58	392639.31	20.00	35.26	11.28	21.29	41.83	11.48	24.03	199	0.75		ER	T	T
3	398098.204	393371.58	392639.31	16.07	21.24	14.02	16.81	29.98	16.65	24.93	94	0.47		ER		
4	398098.205	393371.58	392639.31	17.96	23.26	12.50	18.66	26.67	13.26	11.30	259	0.56		ER	T	
5	398098.231	393371.58	392639.31	19.35	17.26	11.57	20.37	21.37	13.28	21.85	18	0.19		ER	T	
6	398098.570	393371.58	392639.31	28.61	31.06	12.09	29.81	33.26	13.46	16.04	223	0.63	ER		T	
7	398098.98	393371.58	392639.31	14.91	15.28	12.67	16.20	16.50	13.80	12.59	166	0.35		ER	T	
8	398180.289	392657.171	392633.64	26.29	31.63	12.31	28.70	33.79	15.31	23.00	199	0.49	ER		T	T
9	398180.292	392657.171	392633.64	20.65	36.93	22.02	22.78	40.02	22.52	26.37	294	0.47	ER		T	T
10	398190.200	393077.54	392639.2	20.46	32.07	11.43	21.86	33.93	15.22	47.52	61	0.29			T	
11	398190.605	393077.54	392639.2	16.85	34.05	17.96	17.69	35.90	20.19	19.26	579	1.29			T	
12	398190.615	393077.54	392639.2	20.09	21.72	14.09	21.30	24.23	14.72	35.70	112	0.27	ER		T	
13	398190.89	393077.54	392639.2	20.00	26.35	25.39	22.69	29.49	26.99	18.44	267	0.71			T	
14	398192.213	393077.54	392633.54	19.44	32.78	15.07	21.11	34.82	16.48	27.76	512	1.13	ER		T	T
15	398192.553	393077.54	392633.54	20.28	26.02	10.55	21.48	30.66	11.30	10.70	454	1.33	ER		T	
16	398192.592	393077.54	392633.54	21.20	27.39	12.85	22.31	29.87	14.22	17.15	479	1.38	ER		T	
17	398193.158	393077.54	392633.64	29.26	26.48	16.93	30.09	31.48	17.70	14.52	617	1.56	ER		T	
18	398193.511	393077.54	392633.64	19.72	21.61	13.48	20.93	23.34	14.06	14.67	566	1.09	ER		T	
19	398201.510	393242.5	392633.64	23.43	32.19	29.69	24.54	35.05	30.65	19.41	529	0.80	ER	ER	T	
20	398203.5	393280.82	392633.64	19.91	22.55	29.79	21.76	29.10	31.69	33.60	206	1.30	ER		T	
21	398208.219	393371.58	392633.64	23.24	28.39	26.85	24.26	34.26	27.93	14.85	607	2.15	ER	ER	T	T
22	398208.29	393371.58	392633.64	25.46	41.15	31.40	26.85	44.57	31.69	11.48	905	2.53	ER	ER	T	
23	398208.33	393371.58	392633.64	20.00	36.32	23.50	20.83	38.29	24.20	30.56	18	0.51	ER		T	T
24	398208.505	393371.58	392633.64	20.28	33.26	23.41	21.29	38.19	24.87	48.63	59	0.44	ER	ER	T	
25	398208.620	393371.58	392633.64	28.43	43.87	30.46	29.54	46.54	31.26	23.11	147	0.39	ER		T	
26	398208.670	393371.58	392633.64	24.26	32.67	17.43	25.37	36.58	18.46	32.52	59	0.48	ER	ER	T	

Continued

Table 10.3. Continued.

		Pedigree		Marketable tuber yield (t/ha) high temperatures			Total tuber yield (t/ha) high temperatures			Total tuber yield (t/ha) under LB	AUPDC	Resistance			Tolerance[d]	
No.	Clone	Female	Male	San Ramon	La Molina	Majes	San Ramon	La Molina	Majes	Oxapampa	Oxapampa	LB[a]	PVX[b]	PVY[c]	Heat	Drought
27	398208.704	393371.58	392633.64	22.69	31.80	15.43	23.89	35.61	15.55	33.93	51	0.42	ER		T	T
28	302506.39	Bzura	393280.57	17.77	25.22	33.48	18.61	32.00	34.68	54.97	9	0.25			T	
29	302531.43	393280.82	396272.43	16.01	37.50	22.55	16.85	39.76	23.14	32.70	78	0.19			T	
30	302533.3	393371.159	396272.43	18.61	27.11	23.39	19.63	33.20	25.54	21.00	335	1.47			T	
31	302533.38	393371.159	396272.43	23.71	31.80	17.31	24.44	33.52	18.20	17.25	657	1.90			T	
32	302533.48	393371.159	396272.43	17.78	30.67	29.81	18.61	34.76	32.43	26.82	164	0.34			T	
33	302533.49	393371.159	396272.43	22.41	19.33	26.75	23.06	23.20	27.73	43.67	207	0.47			T	
34	302533.74	393371.159	396272.43	21.94	32.81	24.39	22.69	35.98	25.89	26.19	554	0.99			T	
35	302534.17	393371.159	396272.18	17.59	23.31	43.15	18.71	24.82	46.24	15.78	579	1.39			T	
36	304079.10	393075.54	Granola	17.41	26.41	37.91	20.00	30.94	41.83	17.11	736	1.70			T	
	Désirée			15.09	22.35	9.28	15.74	24.71	10.16			S			T	
	Amarilis			8.71	18.23	10.33	11.94	25.00	11.33	13.37	1155	3.33			NT	
	Granola			9.90	10.46	2.39	11.02	13.09	2.48			S			T	
	Yungay									3.15	2079	6.00			NT	
	Kory									9.41	488	1.79			NT	

[a]LB, late blight scored using Yuen and Forbes' (2009) scale of susceptibility to LB; S, susceptible.
[b]PVX, potato virus X; ER indicates extreme resistance.
[c]PVY, potato virus Y; ER indicates extreme resistance.
[d]T, tolerant; NT, non-tolerant.

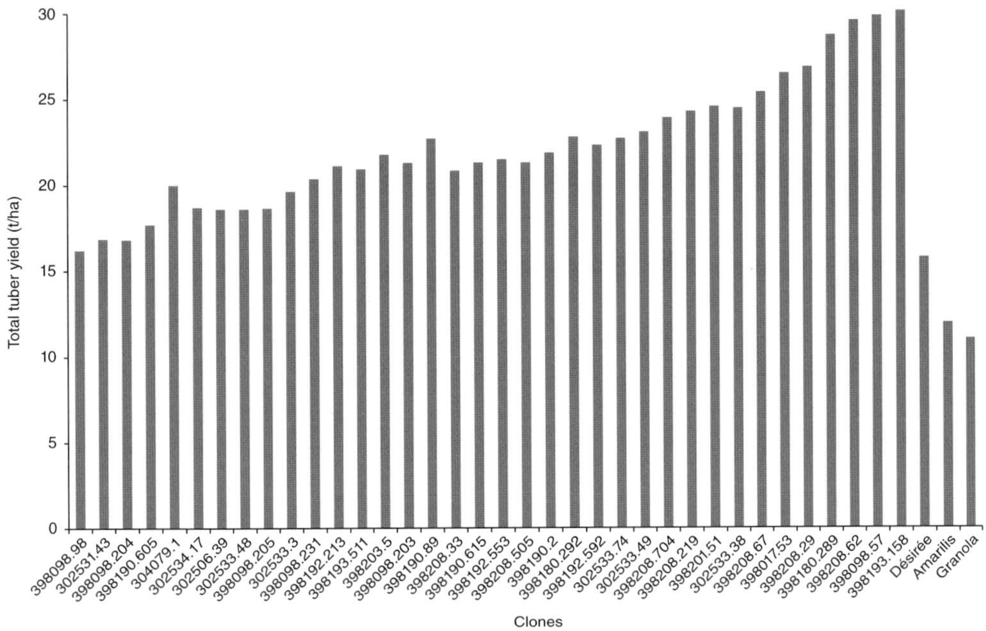

Fig. 10.3. Tuber yield of 36 elite clones from the LBHT group under conditions of San Ramon in 2012.

Table 10.4. AMMI analysis for tuber yield (t/ha) of 61 clones with resistance to late blight and heat tolerance in San Ramon, La Molina and Majes.

		Means square	
Source of variation	df[a]	Marketable tuber yield (t/ha)	Total tuber yield (t/ha)
Replications/environments	3	28.39*	41.90*
Block/replications/environments	42	12.73	16.20*
Treatments	191	153.63	177.78
Environments	2	2997.12**	5007.20**
Genotypes	63	200.92**	200.30**
Genotypes × environments	126	84.86**	89.90**
PCA1[b]	64	118.01**	134.53**
PCA2	62	51.92**	45.88**
Pooled error	147	9.03	8.80
CV (%)[c]		15.20	14.31

[a]df, degrees of freedom.
[b]PCA, principal components analysis.
[c]CV, coefficient of variation.
**, significant statistical differences at $P < 0.01$; *, significant statistical differences at $P < 0.05$.

tuber yield comes from the G × E interaction. The PC1 component was highly significant and explains 75.17% of the G × E sum of squares. In the bi-plot based on the scores of principal components PC1 and PC2 (Fig. 10.4) a group of 28 genotypes (302533.74, 302533. 38, 398208.29, 398208.505, 302531.43, 398208. 33, 398201.510, 398190.89, 398190.605, 398098. 204, 398208.620, 398180.292, 398208.670, 3987208.704, 398098.89, 398192.213, 398192. 553, 398193.511, 398190.615 and others) were stable, based on their low absolute PC1 and PC2 scores close to zero, indicating lower levels of interaction with the environment. The AMMI bi-plot for tuber yield (*x*-axis) and PC1 score (*y*-axis) (Fig. 10.5) shows the clones

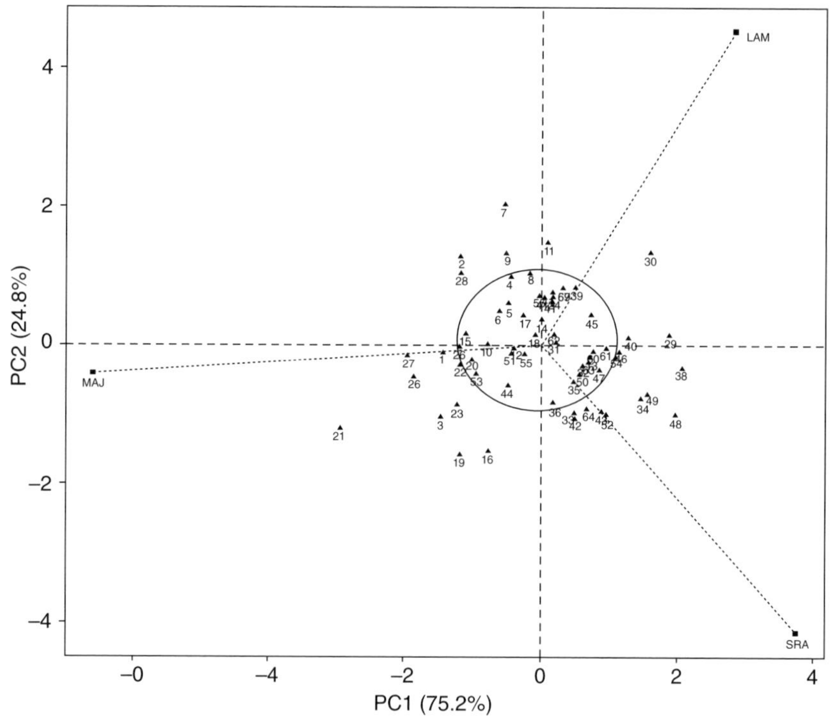

Fig. 10.4. Bi-plot from AMMI analysis showing first and second principal components (PC1 and PC2) for three locations and 61 genotypes.

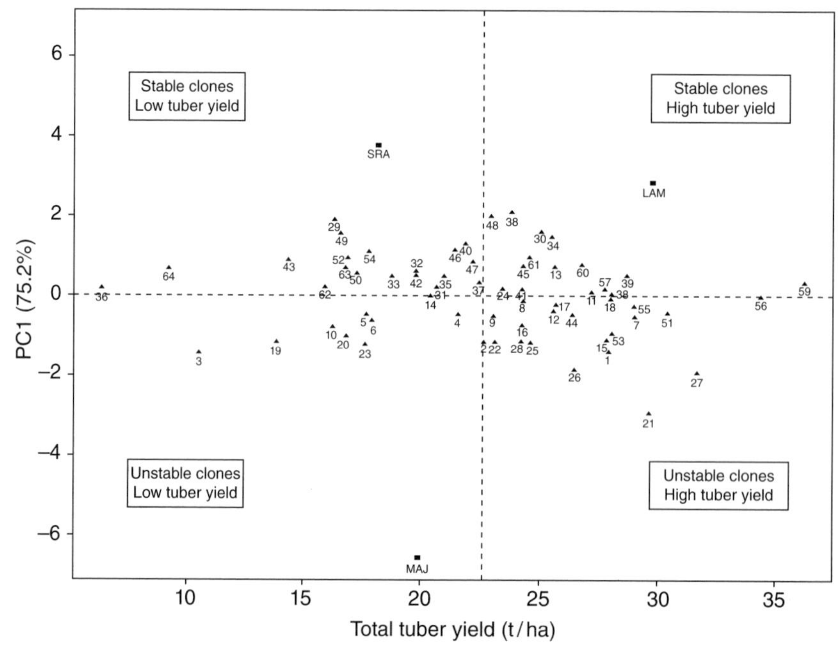

Fig. 10.5. AMMI bi-plot of tuber yield and PC1 principal component scores at three locations and 61 genotypes.

398208.620, 398208.29, 398201. 510, 398180. 292,398208. 670, 398208.704, 302533.38 and 302533.74 to have stably high tuber yield, while 398098.89, 398190.735, 302522.39, 302522.46 and 398098.119 are also stable but with low tuber yields, and others with intermediate tuber yield. A third group of clones are unstable but have high tuber yield in some environments and thus show specific adaptation, among them the clones 302534.17 and 304079.10 are well adapted to mid-elevation tropics, and 398098.289 is adapted to La Molina with lowland subtropical conditions. A fourth group of clones are less stable and have low tuber yield overall under warm conditions, although these clones can have good adaptation to highlands, where the temperatures are ideal for potato.

Virus resistance and drought tolerance

The virus resistance assays determined that 11 clones exhibit extreme resistance to PVY and 25 to PVX, five of them combining extreme resistance to PVX and PVY. In the screening for drought tolerance seven clones were identified that maintain their yields under water-deficit conditions and that showed higher values of DTI compared with the genotype used as a tolerant control.

10.4 Discussion and Conclusions

The tuber yield of 61 clones combining late blight resistance and heat tolerance is reasonably high for these conditions, given that they can be harvested at 90 days after planting in short day conditions. Forty outstanding clones have been selected for their heat tolerance and resistance to late blight. Due to their stable and high yields, 28 clones can be recommended for variety testing in mid-elevation tropical locations of Asia or Africa. Likewise, 12 clones with good yields show specific adaptation to mid-elevation or highlands in the tropics.

The test for resistance to viruses determined that 11 clones exhibit extreme resistance to PVY and 25 to PVX, five of them combining extreme resistance to PVX and PVY. These clones are highly suited for mid-elevation tropics and climate change and are available for variety development and further use in breeding.

In the screening for drought tolerance seven clones were identified that maintain their yields under water-deficit conditions and these clones showed higher values of the DTI than the tolerant control genotype.

Twenty-nine mid-maturing heat tolerant and late-blight resistant clones without extreme resistance to PVY can be recommended for highland conditions in Africa, Asia and Latin America.

Informal taste tests indicate that flavour and texture of the new clones are acceptable even when exposed to heat.

Acknowledgements

We acknowledge Elisa Mihovilovich and Rolando Cabello for their contributions in testing for virus resistance and drought tolerance, respectively.

References

Cabello, R., De Mendiburu, F., Bonierbale, M., Monneveux, P., Roca, W. and Chujoy, E. (2012) Large-scale evaluation of potato improved varieties, genetics stocks and landraces for drought tolerance. *American Journal of Potato Research* 89(5), 400–410.

Cabello, R., Monneveux, P., De Mendiburu, F. and Bonierbale, M. (2013) Comparison of yield based drought tolerance indices in improved varieties, genetic stocks and landraces of potato (*Solanum tuberosum* L.). *Euphytica* 193(2), 147–156.

Gastelo, M., Landeo, J., Díaz, L. and Bonierbale, M. (2012) Nuevos clones elites de papa con resistencia al tizón tardío y tolerancia al calor desarrollados por el CIP para enfrentar al cambio climático. Población LBHT Resumenes XXV Congreso Latinoamericano de la papa, Uberlandia, Brazil.

Hijmans, R. (2003) The effect of climate change on global potato production. *American Journal of Potato Research* 80, 271–280.

International Potato Center (CIP) (2006) Assessing potato clones for resistance to virus disease. In: *Procedures for Standard Evaluation Trials of Advanced Potato Clones. An International Cooperators' Guide*. CIP, Lima, pp. 75–93.

International Potato Center (CIP) (2014) Catalogue of Potato Varieties and Advanced Clones. Available at: www. cipotato.org/catalogue (accessed 4 August 2014).

Landeo, J., Gastelo, M., Pinedo, H. and Flores, F. (1995) Breeding for horizontal resistance to late blight in potato free of R genes. In: *Phytophthora infestans 150 Proceedings*. European Association for Potato Research (EAPR) – Pathology Section Conference, Trinity College, Dublin, Ireland. Bole Press, Dublin, pp. 268–274.

Levy, D. and Veilleux, R. (2007) Adaptation of potato to high temperatures and salinity – a review. *American Journal of Potato Research* 84, 487–506.

R Core Team (2012) *R: A Language and Environment for Statistical Computing*. R Core Team Foundation for Statistical Computing, Vienna, Austria. Available at: http://www.R-project.org (accessed 25 February 2013).

Yuen, J. and Forbes, G. (2009) Estimating the level of susceptibility to *Phytophthora infestans* in potato genotypes. *The American Phytopathological Society*, *Phytopathology* 99(6), 2.

Part II

Getting Seed Potato Systems Moving

11 Strategies to Improve Seed Potato Quality and Supply in Sub-Saharan Africa: Experience from Interventions in Five Countries

P. Demo,[1*] B. Lemaga,[2] R. Kakuhenzire,[3] S. Schulz,[4] D. Borus,[5] I. Barker,[6] G. Woldegiorgis,[7] M.L.Parker[8] and E. Schulte-Geldermann[8]

[1]International Potato Center (CIP), Lilongwe, Malawi; [2]Ethiopian Agricultural Transformation Agency, Addis Ababa, Ethiopia; [3]CIP-Tanzania, Mbeya, Tanzania; [4]CIP, Addis Ababa, Ethiopia; [5]University of Tasmania, Hobart, Australia; [6]Syngenta Foundation for Sustainable Agriculture, Basel, Switzerland; [7]Ethiopian Institute of Agricultural Research (EIAR), Addis Ababba, Ethiopia; [8]CIP Sub-Saharan Africa (CIP-SSA), Nairobi, Kenya

Abstract

Potato yields of small-scale farmers in the region fall far short of their potential, mostly due to a potent combination of inadequate supplies of high quality seed and limited awareness of better seed and crop management practices. Consequently, potato yields in sub-Saharan Africa are dismally low at 6–10 t/ha. To increase the availability of high-grade potato seed, the International Potato Center and its national partners have developed components of an innovative seed strategy, the '3 seed potato generation revolution' (3G – a seed production model), which drastically lowers the cost of production of pre-basic or 'starter' seed coupled with extension-based interventions to train smallholders to better manage their own seed on farm. Using rapid multiplication techniques (RMT), such as aeroponics or sandponics, to produce minitubers from *in vitro* plantlets, seed can be bulked in two subsequent field generations to the same quantities that under conventional practices require four to six generations. This reduces the cost of production and prevents build-up of damaging diseases in the field.

Greater involvement by the private sector in seed potato value chains offers a means to overcome the supply bottleneck that is limiting the provision of quality seed. A more efficient and responsive seed system will improve production, distribution, use and profitability for farmers. Promising RMT and an engaged private sector can provide needed capacity to broaden adoption of quality seed and accelerate availability of new varieties with greater prospect of added value. Better integration of national agricultural research and extension systems into the value chain, as well as farmer training schemes in seed management and storage, can accelerate innovation. A regional perspective can help exploit economies of scale for sharing knowledge and technology, implement creative applications of information communication technologies, advocate for farmer-friendly, seed-related regulations and policies, improve the business-enabling environment, and expand intra-regional trade for seed of the highest categories.

*p.demo@cgiar.org

Recent and ongoing interventions reveal that wide-scale adoption of these technologies, as well as improving capacities to maximize their investment in quality seed, will enable farmers to secure seed supplies and to put the entire seed supply chain onto a more sustainable path for the future. This chapter gives an overview and comparative analysis of costs and benefits from experiences from seed interventions from Ethiopia, Kenya, Malawi, Rwanda and Uganda over the past 5 years.

Keywords: potato, rapid multiplication technologies, sub-Saharan Africa, '3G seed potato revolution'

11.1 Introduction

Farmers in sub-Saharan Arica (SSA) face significant challenges to meet their year-long food requirements and generate income from agriculture. The roughly 4 million smallholder potato farmers in East and Southern Africa (ESA) from Ethiopia in the north to Mozambique on the coastal south are discovering that potato can contribute to overcoming these challenges because it is a high-yielding cash crop with a short growing season. High altitudes, a temperate climate and generally dependable rains make for near ideal growing conditions for potato – one of the few cash crops available to small-scale farmers of the highlands throughout ESA. Highland potato farmers enjoy a comparative advantage in exploiting an ever-growing, high market demand for the crop, as potato production in ESA requires the cool temperatures typical of mid- to high altitudes – roughly 1000–3000 m above sea level (masl) – to form tubers.

Potato is the fourth most important staple crop in the world. Although potato is often neglected or overlooked for agricultural development and food security in SSA, it is here where some of the largest increases worldwide in harvested areas and production have occurred over the last 20 years. Despite the generally low national consumption rates for many African nations (approximately 5–15 kg per capita/year), the crop is very important in potato-producing regions, with annual consumption often surpassing 100 kg per person. Indeed, among the extended highland farming communities of ESA, potato has high potential to contribute to poverty reduction (Demo *et al.*, 2009) as well as food and nutrition security.

There are many compelling reasons for encouraging more wide-scale adoption of potato into the livelihoods of African smallholder farmers. Potato is multifaceted as a staple crop that addresses food security and a horticultural crop for its high value per unit area of land. Potato thrives in mid- to high altitude regions, and its short cropping cycle of 3–4 months is well suited to the double cropping seasons in the African highlands, particularly in rain-fed systems. Harvestable tubers are available 60 days after the onset of the rainy season – a significant advantage over grains, which require 6–9 months. This makes potato one of the first crops that can be harvested in a given growing season, thus an important crop for the 'hunger months' – that is, a period of several months between harvests when people lack enough food to satisfy their hunger and meet their basic caloric and nutritional needs. Potato further distinguishes itself for its water use efficiency, yielding more food per unit of water than any other major crop. With a yield potential of more than 51,000 calories/ha/day in a short growing season, its productivity in terms of energy produced is the highest of all major arable crops, almost double that of wheat and rice. Potato is a cheap but nutritionally rich staple food for the fast-growing cities of SSA, contributing protein, vitamin C, B vitamins, zinc and iron to the diet. Along with groundnut, onion and carrots, the 'nutritional productivity' of potato is especially high: for every cubic metre of water applied, 5600 calories of dietary energy are produced, compared with 3860 in maize, 2300 in wheat and only 2000 calories in rice. For the same cubic metre, potato yields 150 g of protein, double that of wheat and maize; and 540 mg of calcium, double that of wheat and four times that of rice. Another advantage of potatoes is that they are largely traded locally and nationally, thus are far less susceptible to large-scale market shocks and price speculation experienced by internationally traded staples, such as grains, during international market crises (as in the food shortages of 2007–2008). As such potatoes contribute to a more stable food system and predictable source of income.

Dismally low yields, ranging from 6 to 10 t/ha in SSA, stand in the way of fully benefitting from potato. Realistically, farmers in SSA cropping robust varieties can achieve yields of 20–30 t/ha under low input conditions, above the 2010 global average of 17.4 t/ha (FAOSTAT, 2010). Demand is increasing for potato in SSA and consequently farmers repeatedly crop potato on the same plot of land. Highland farmers have few choices for cash crops other than potato; consequently, they repeatedly plant potato on the same site, which allows the build-up of diseases that decrease yields. Potato farmers in SSA commonly practise mixed farming that involves crops (maize, legumes, vegetables and sometimes tea, wheat or barley) and livestock, but due to the fast-increasing population farm sizes are continuously decreasing with average land size of approximately 2 ha leading to a strong pressure to increase the farm incomes (Shepherd and Soule, 1998; Waithaka *et al.*, 2005; CIP, 2012). Meanwhile potatoes are considered the main cash crop for highland farmers (W. Kaguongo, 2009, unpublished data; J. Obado, 2010, unpublished data), thus, they tend to grow potatoes in very close rotations or even in some cases monocropping, basically due to lack of alternative high-value cash crops and inadequate knowledge of good agricultural practices, and hence this further worsens the income situation for the farmers (Gildemacher *et al.*, 2009a; Schulte-Geldermann, 2013). As a consequence farmers are facing detrimental effects on soil fertility and farm viability and reduced yields. Moreover, due to the reliance on the performance of a few crops, farmers are facing higher risks of complete crop failures by specific abiotic and biotic stress conditions, in particular due to the build-up of crop-specific soil-borne pests and diseases (Turkensteen, 1987; Lemaga *et al.*, 2005; Gildemacher *et al.*, 2009b). The International Potato Center (CIP) recently found in a survey conducted in Kenya, Rwanda and Uganda that about 15–25% of potato farmers don't practise crop rotation at all, whereas another 20–40% only have one crop in-between potatoes (CIP, 2009–2011, unpublished data). Muthoni and Kabira (2010) showed in trials at Kenya Agricultural Research Institute (KARI) Tigoni station that widened crop rotations and organic fertilization improved soil characteristics and yield after a relatively short time. Crop diversification in terms of time and space is one of the major agronomic measures of integrated management to reduce risks of disease spread (Turkensteen, 1987; Østergård *et al.*, 2009), to increase nutrient use efficiency in low-input systems and to minimize the dependence on external inputs such as synthetic nitrogen fertilizers by the integration of legume crops.

The major bottleneck to increasing productivity is limited access to quality seed of suitable varieties, which reduces yields, food availability and farmers' incomes. The health status of seed defines the potential yield of the potato crop. Typically, farmers use unmarketable ware potato for planting, sourced from their own fields or markets, which is generally of low quality. As a result, diseases have accumulated and spread in farmer-saved seed stocks. Farmers' lack of knowledge to select quality seed is compounded with limited access to varieties with robust traits (such as drought, heat and disease tolerance, and/or biofortified with essential micronutrients, specifically iron and zinc), lack of knowledge of good agricultural practices for potato, and minimal access to storage. Although seed certification standards exist, most national policies do not recognize more practical quality standards, such as Quality Declared Planting Material (QDPM). This further limits access to quality seed (Fajardo *et al.*, 2010). Policy advocacy for more practical quality standards is required at national levels.

Gender dynamics related to potato in the ESA highlands are particularly important. It is mostly women who are responsible for household garden systems, who provide the bulk of farm labour, and often they are the ones planting and harvesting edible tubers.

Most of the traders who buy potato from the farmers to sell in markets are women as well.

Critical to improving potato production and productivity is greater supply and access to high-quality seed potato. And although opportunities exist along the seed potato value chain, two basic challenges prevent farmers from benefitting fully from potato. First is the minimally functional seed potato systems, which limit the supply of high-quality seed potato to meet increasing demand. Poor quality seed lowers yield potential and continually keeps smallholder incomes low. Secondly, good farming practices for both seed and ware potato, including the ability to select quality seed, best production practices and postharvest handling and storage, are not well understood or routinely applied. Technical assistance that centres on effective initiatives can turn this around by offering smallholder potato farmers in ESA readily accessible technologies and tools for tapping into the potato's potential to increase yields, food security and incomes.

The objectives of the interventions were: (i) to increase the availability of quality seed potato to resource-poor farmers at affordable prices; and (ii) to improve farmers' potato crop yields, food availability, incomes and livelihoods.

11.2 Improving Seed Potato Quality: The Potato '3G Seed' Strategy

The strategies used centred on the development and utilization of strong partnerships in the implementation of the CIP's '3 seed potato generation revolution' (3G – a seed production model) for improvement of seed potato quality and availability in Ethiopia, Kenya, Malawi, Rwanda and Uganda. The 3G strategy follows the seed-production chain whereby minitubers are produced using rapid multiplication techniques (RMT) from *in vitro* plantlets. Minitubers are then multiplied in the field to produce seed potato. Seed can be bulked in two subsequent field generations (hence the three generations) to the same quantities that under conventional practices require four to six generations. This reduces the cost of production *and* prevents build-up of damaging diseases in the field.

This technology package is particularly appropriate for African countries, where land shortages, the rapid degeneration of seed through disease and the cost of quality seed are key factors to consider to encourage smallholder adoption. The strategy offers a solution to break the long-standing seed bottleneck and the prospect of a sustainable seed system. The strategy must also be seen in the context of the rising demand for potato and opportunities for private-sector seed producers and multipliers.

Experience from interventions provided evidence that the 3G strategy dramatically lowers the cost of production of pre-basic or 'starter' seed. This coupled with extension-based interventions to train smallholders in better on-farm management of their own seed, significantly reduced the build-up of damaging diseases in the field and doubled yields of smallholder farmers. The strategy integrates the following key elements: (i) an increase in minituber production capacity; (ii) an increase in specialized early field multiplication capacity; (iii) a scaling up and out of decentralized seed production; (iv) implementing quality control; (v) an extensive awareness-building campaign; and (vi) it trains farmers to maintain seed quality on farm.

Increase in minituber production capacity

The '3G' seed strategy envisages the rapid multiplication of large numbers of minitubers in one generation using RMT (Fig. 11.1) The RMT method at the heart of the 3G strategy is known as aeroponics which achieves multiplication rates of over 30–50 minitubers to one *in vitro* plantlet compared with the conventional methods that yield five minitubers per *in vitro* plantlet. This high initial multiplication rate obtained from aeroponics reduces the number of field generations to bulk similar quantities of

Fig. 11.1. High seed multiplication using the technique of aeroponics (photograph taken by Paul Demo at Njuli Malawi in 2009).

seed compared with conventional methods. CIP has adapted aeroponics technology (Chiipanthenga, 2010), which is in use in Asia and at its headquarters in Peru, to lower its cost through lowering the cost of components in the system. The technology also benefits from low water and energy usage. Other RMT technologies such as sandponics and stem cuttings are also considered where more appropriate to the local conditions influenced by reliability of power supply, human capacity and climatic conditions among others. Research organizations involved in starter-seed potato production have been encouraged to facilitate the development of commercial seed multiplication by showing an open and welcoming attitude towards public–private collaboration in the starter seed-production sector.

Increase in specialized early field multiplication capacity

The first two field generations, including multiplying minitubers, are envisaged to be done by trained private seed potato producers with a back-up production by the respective public-sector institute. Therefore, the identification of highly specialized field multipliers up to the second field generation in areas suitable for seed production (where there is low aphid pressure and it is free of bacterial wilt) and having enough land to conduct a proper rotation scheme, are of paramount importance. Ideally these are located within or close to the major potato-growing regions to avoid long-distance transport to further multipliers. The given diversity of farm entrepreneurs in the considered countries led to different approaches. For example in Kenya and Tanzania, large cereal farm operations in the highlands are already engaged in seed production, or will be as they are considering seed potato production as an additional business and good rotational crop for their cereal crops. These operations proved to be highly effective due to the availability of required resources and knowledge. In contrast, such farm enterprises barely exist in Ethiopia, Malawi, Rwanda and Uganda, thus other strategies have to be considered. The focus of the strategy for these countries will be to form or strengthen specialized seed multiplier groups and/or associations perhaps where capable lead farmers exist, and provide them with technical backstopping and adequate entrepreneurial frameworks to create a sustainable business opportunity as a seed-producer organization. Farmer associations/farmer groups are the organizational structure at community level for implementing the approach. Ideally, these should be fully functional, existing and registered associations; however, in the absence of an established association, other forms of communal organizations can also be considered (farmer groups, youth clubs, women's organizations, etc.).

Scaling-up and out of decentralized seed production

To ensure better local availability the strategy involves a network of advanced farmers for further decentralized multiplication, offering business opportunities in the rural areas while improving overall seed qualities and yields of smallholder farmers purchasing the seed. Decentralized Seed Producers (DSPs) have been identified (individuals, groups, associations) and trained for further multiplication of early generation seed. As in the 3G project in Kenya, 65 small, private DSPs who have at least 3 ha of suitable land (or who can rent land for rotation purposes) have been identified by local extension services as entrepreneurial growers. These candidate DSPs must also be willing to invest in replacing their own saved seed with high-quality seed stock and in seed storage facilities.

Implementing quality control

A crucial element for the development of a functional seed system is quality control and seed traceability. A formal seed quality control system generally involves seed certification by national regulatory agencies; however, many SSA countries either do not have such systems, or they are not implemented. However, the strategy involves a strong capacity-strengthening component in the area of formal seed certification of at least very early seed multiplication generations. Nevertheless, in the near term, it is highly unlikely that formal seed certification systems can be implemented that reach seed producers in the target regions' rural areas. Many small- and medium-scale seed producers, as compared to a few large-scale commercial ones, are producing seed potatoes for further local multiplication. Two to three inspection visits and a laboratory analysis (for some countries) are required to certify a seed crop. Owing to the cost of transport and daily allowances of inspectors, the cost to certify 50 ha and 0.5 ha would hardly differ. This disproportionately increases the cost of certification per unit of seed produced. There is thus little incentive for most DSPs to pursue and practise formal seed certification. Therefore, the strategy includes quality control using QDPM standards enabling decentralized multipliers to follow quality control measures and declare seed quality based on a recognized set of standards. QDPM seed quality control is based solely on visual inspection, aiming to result in significantly better health status of the crop and increasing affordability of quality controlled seed. While perceived as a sub-standard in developing countries, it does not rely on expensive and malfunctioning certification schemes.

Extensive awareness-building campaign

Demand for seed and awareness of the value of quality seed among smallholders will be built through a variety of means. First, demonstration plots have been established with the help of the Ministry of Agriculture (MoA) extension service which will include both new varieties and demonstration of clean seed versus farmers' practice plus the effect of good management (including both late blight control and adequate fertilizer application). Secondly, following successful trials in Kenya, certified seed was distributed in small trial 5 kg, 10 kg and 25 kg bags sold at affordable prices (approximately US$0.5/kg). In Kenya, this mechanism enabled thousands of smallholder farmers to access a small amount of quality seed of improved varieties for the first time at little cost to themselves. Thirdly, linkages will be made to existing seed distribution channels currently selling seed potato of unknown sources and qualities. Fourthly, linkages between processor and producers have been examined. Seed has been marketed, partly through exercises in awareness and demand creation as described above, and through direct marketing such as the setting up of simple seed directories which list trained seed suppliers in a given district or neighbouring districts. In Kenya, MoA District

Agricultural Officers have played a critical role in identifying seed needs in their districts and passing on this information as seed orders to project seed-producing partners.

Training farmers to maintain seed quality on farm

A major component of the strategy is to employ validated and proven extension interventions, such as positive selection (Gildemacher *et al.*, 2011; Gildemacher, 2012) and small seed nurseries to help smallholder farmers maintain the health of their own farm-produced seed in between replacing their seed with quality seed perhaps every 5 years (or sooner). The innovative extension intervention system known as 'Select the Best' provides the potato producers with knowledge on how to maintain the quality of their seed potato for a longer period through positive selection. Furthermore, farmers become aware of the impact of seed potato quality on yield, and the need for regular replenishment of their seed stock with high quality seed from specialized seed growers. The process involves training farmers to recognize healthy plants which are not showing symptoms of seed-borne diseases such as virus or bacterial wilt infection. Healthy plants are marked and later harvested to provide next season's seed. The intervention is relatively low cost and requires less contact time than a conventional farmer field school. Essential in the training curriculum is that the farmers plant a demonstration experiment in which they compare the yields through using their existing seed-selection method, using positive selection or buying seed from a specialist. This provides the farmers with options to improve their seed quality.

11.3 Impacts of the Intervention to Support High Quality Seed Production

Minituber production

The introduction of the '3G' concept led to a drastic increase in the number of minitubers produced in all five countries (Uganda, Rwanda, Malawi, Kenya and Ethiopia), from 187,000 in 2003 to 2,110,000 minitubers in 2013 (Fig. 11.2). The introduction of aeroponics techniques to produce minitubers under screenhouse conditions contributed to an increase in the seed multiplication ratio by six- to tenfold compared with the conventional multiplication method that uses soil-based planting substrates.

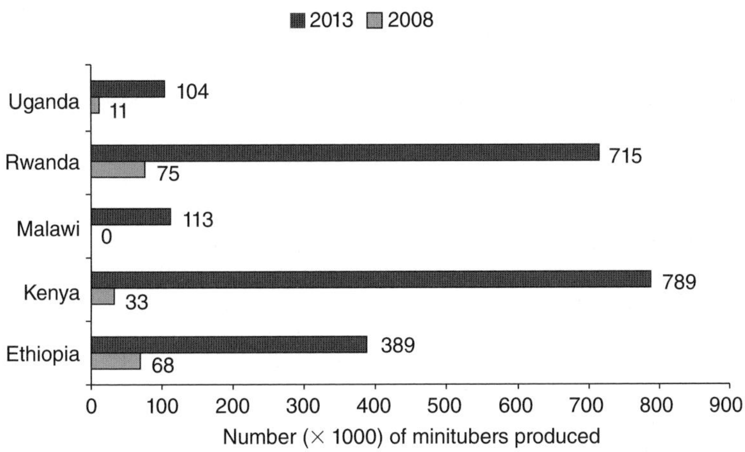

Fig. 11.2. Number of minitubers (× 1000) produced using aeroponics in five different countries in 2008 and 2013.

Seed production in Kenya

The '3G' project in Kenya significantly increased the production of certified seed potato, from less than 500 t in 2008 to more than 3000 t in 2013. It should be noted that private-sector involvement in certified seed production contributed significantly to increasing total volumes of certified seed (Fig. 11.3).

Yield and economic gains from using quality seed

To determine the extent to which seed quality contributes to the yield gap, CIP researchers conducted on-farm trials with 22 farmer groups over two seasons and one on-station trial. Besides the evaluation of yield gap caused by the seed quality, these trials served as demonstrations for farmers to create awareness of the importance of using high-grade seed. At each site, field days were held to which farmers, extension officers and other stakeholders were invited. The seed qualities used were from: (i) 3G second field generation, certified seed from Kisima Farm (corresponds to 'G3 project seed' in Fig. 11.4); (ii) basic, third field generation, certified seed from the national programme (corresponds to 'G4 basic public' in Fig. 11.4); (iii) sixth field generation, certified seed from ADC-Molo (Agricultural Development Corporation located at Molo, Kenya) (corresponds to 'G6 CF public' in Fig. 11.4); (iv) fourth to sixth generation quality-declared seed, which is seed obtained from credible farmers but produced without the normal certified seed-production regulations (corresponds to 'G5 informal multipliers' and 'G6 informal multipliers' in Fig. 11.4); (v) positive-selection (PS) seed; and (vi) typical farmers' practice (FP) seed.

The results clearly highlighted the importance of seed quality. Average yield of FP seed in on-farm evaluations was 11.1 t/ha and 9.3 t/ha in the short rains and long rains seasons, respectively. The yields were low, despite use of relatively high fertilizer inputs and strict late blight control. The low yields with FP seed clearly showed the yield gap produced by the poor quality seed. With all the varieties additional field generations led to considerable yield losses. The G3 (i.e. generation 3) seed out-yielded all the other seed quality categories significantly. On average G3 yielded 266% (213–352%), fourth generation basic seed yielded 210% (166–247%), sixth generation certified seed 201% (150–238%), fifth and sixth generation seed from informal multipliers 181% (126–229%) and 163% (107–234%), respectively, and PS seed 137% (104–203%) higher than FP seed (Fig. 11.4).

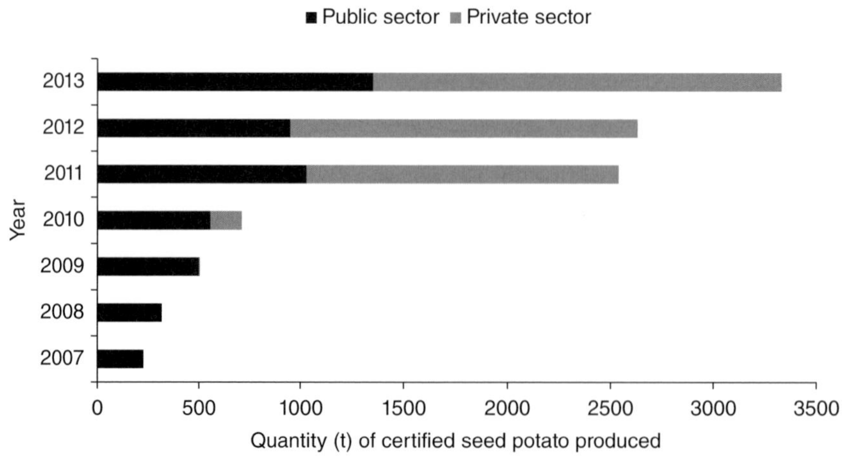

Fig. 11.3. Annual quantity (t) of certified seed potato produced in Kenya.

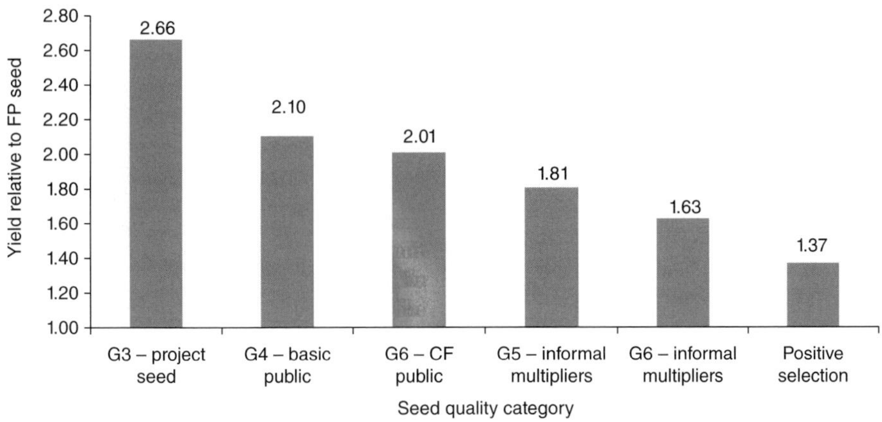

Fig. 11.4. Relative yields of seed quality categories compared with farmers' practice (FP) seed (where FP = 1), Kenya. CF, certified; G3, generation 3; G4, generation 4; G5, generation 5; G6, generation 6. The first generation is produced in the screenhouse and subsequent generations are produced in the field.

Considering the low input conditions which farmers usually practise, it is important to look at the effect of seed quality under different fertilizer input levels. Therefore to evaluate the combined effects of seed quality and plant nutrition, fertilizer (N:P:K) levels of 0 kg/ha, 45 kg/ha and 90 kg/ha were applied to G3 project seed (second field generation) and FP seed at eight sites. The results showed that tuber yields increased with increasing fertilizer levels with both seed qualities. It was remarkable that the G3 seed yielded 4.3 t/ha more without fertilizer than FP seed with 90 kg/ha N:P:K, underscoring the importance of seed quality in determining the yielding potential (Fig. 11.5).

The single most important factor for the farmers to decide using inputs is the financial return on investment. Because of high costs, farmers, especially those who are resource poor, are often reluctant to invest in quality seed and fertilizer. 3G project seed costs US$900/ha and it is unthinkable for farmers to invest so much unless they are sure of getting a higher benefit that will cover the cost and earn them good profit. Our data clearly revealed that the high quality seed (G3) fetched higher returns without the use of fertilizers than the FP seed at a fertilizer rate of 90 kg/ha N:P:K. The data further showed that the profits from use of quality seed increased with increasing levels of fertilizer application included in the trial. Farmer profits were highest when quality seed potato was used at a fertilizer rate of 90 kg/ha N:P:K. However, with use of FP seed potato the extra profit between fertilizer levels of 45 and 90 kg/ha N:P:K is only marginal and would most likely be absorbed by the costs of extra labour for fertilizer application and credit services (Table 11.1).

Food and Agriculture Organization (FAO) QDPM standards for clonal crops

Since 2012, an informal seed inspection scheme for the production of QDPM was piloted in eight woredas (districts) of Ethiopia, involving some ten farmers' seed growers' cooperatives. Simultaneously, CIP and partners discussed at national level recognition of the QDPM concept in the national seed legislation. The QDPM approach involves two inspection committees (COM 1 and COM 2) carrying out two field and one postharvest inspections. COM 1 consists of representatives of the concerned seed producer cooperatives and COM 2 of technical experts from the official, district-level R&D (research and development) system. The first field inspection (6 weeks after planting) was done by COM 1, the second field inspection (at flowering) and the postharvest inspections were done jointly by both committees. The

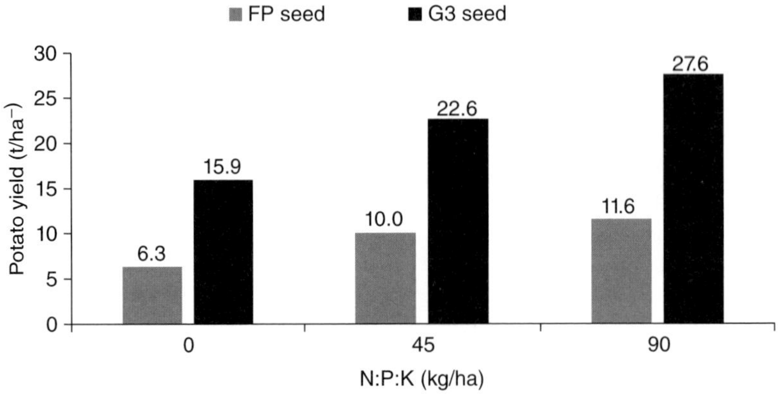

Fig. 11.5. Potato yields of farmers' practice (FP) seed compared with G3 seed at fertilization levels of 0 kg/ha, 45 kg/ha and 90 kg/ha using N:P:K (17:17:17).

Table 11.1. Combined effects of three fertilizer levels and seed quality (farmers' practice (FP) seed and G3 seed) on profit in potato production.

	Fertilizer (N:P:K) rate					
	0 kg/ha		45 kg/ha		90 kg/ha	
	FP seed	G3 seed	FP seed	G3 seed	FP seed	G3 seed
Seed cost ($/ha)	0	900	0	900	0	900
Fertilizer cost ($/ha)	0	0	175	175	350	350
Market price ($/kg)	0.15	0.15	0.15	0.15	0.15	0.15
Yield (t/ha)	6.3	15.9	10	22.6	11.6	27.6
Total income ($/ha)	945	2385	1500	3390	1740	4140
(Income - fertilizer + seed ($/ha))	945	1485	1325	2315	1390	2890

system was being piloted in eight districts in three federal states of Ethiopia, involving ten seed producer cooperatives.

In May 2013, a survey was carried out to collect feedback from stakeholders (farmers, members of the two inspection committees and woreda administrators) on their perception of the QDPM concept.

The key findings indicated that:

- The QDPM concept was generally viewed as a useful and urgently needed tool to maintain and improve seed quality.
- Peer control among cooperative members is an effective tool to maintain quality standards.
- QDPM objectives and procedures need to be clearly communicated, not only to the involved technical staff but also to the district-level administrators to create ownership and avoid misunderstandings.
- At woreda level, staff time and travel costs (per diems, transport) are serious limitations for the sustained implementation of the approach. Lasting mechanisms need to be found to address and solve these concerns.
- While some woredas volunteer to cover costs for transport and per diems alternative payment mechanisms need to be developed for the woredas that are unwilling to do so. Ideally, those activities were to become part of a woreda's regular work programme with budgets provided by the concerned line agencies.
- Most cooperatives expressed interest in covering costs in the future, once the system was going to be well established and QDPM seed commands premium

prices in the seed market. This will still require a considerable amount of awareness creation but would be a sustainable solution to funding the QDPM approach.

- Farmers using QDPM seed were able to almost double their potato yields compared with yields from their farm-saved seed.

Farm-based improvement of quality of planting material though positive-selection 'Select the Best'

Selected seed potato from healthy-looking mother plants (positive selection) was compared with the common Kenyan farmer practice of selection from the harvested bulk of potatoes (farmer selection) in 23 farmer-managed trials. Positive selection out-yielded farmer selection irrespective of the agroecology, crop management, soil fertility, variety and quality of the starter seed, with an overall average of 30% (Table 11.2). Similar farmer-managed trials were conducted in 67 sites across four districts in Malawi in 2010 and 2011. The variables measured were tuber yield, and virus and bacterial wilt incidence. Overall, the crop from PS seed out-yielded that produced from seed selected by the farmers' method by 52.2% (Table 11.3). The outperformance could be explained by the fact that crops from PS

Table 11.2. Yields from positive-selection (PS) seed compared with those obtained from planting seed from common farmer-selection (FS) seed potato using varieties Asante, Tigoni and Dutch Robyjn.

	Yield				Yield increase				
	PS (t/ha)	SD[a]	FS (t/ha)	SD[a]	(t/ha)	(%)	t-value	df[b]	*P* (two-tailed)
Asante	13.9	1.9	11.0	2.2	2.8	25	7.5	21	0.000
Tigoni	18.6	7.4	15.1	5.9	3.5	23	3.8	19	0.001
Dutch Robyjn	20.9	10.1	15.5	6.9	5.4	35	6.0	25	0.000
Farmer quality[c]	14.1	2.3	11.1	2.1	3.0	27	8.7	51	0.000
High quality[d]	30.5	6.9	23.1	4.2	7.3	32	5.2	15	0.000
All	17.9	8.0	13.9	5.8	4.0	30	8.5	67	0.000

[a]SD, standard deviation.
[b]df, degrees of freedom.
[c]Seed potatoes for these trials selected from fields planted with seed potatoes from local market or farmer fields.
[d]Seed potatoes for these trials selected from fields planted with basic seed or certified seed.

Table 11.3. Comparison of performance of potato crops grown from positive-selection (PS) seed and from seed selected using farmers' selection method with local cultivars (Rosita and Violet) during 2010 dry and 2010/2011 rainy seasons in Malawi.

		Percentage		
Districts	Number of sites	Tuber yield increase (mean ± SE)	Reduction in number of wilted plants (mean ± SE)	Reduction in number of plants with virus symptoms (mean ± SE)
Ntchisi2010/2011	9	75.8 ± 6.4	55.6 ± 9.4	33.4 ± 8.3
Dedza 2010	12	39.3 ± 5.6	73.1 ± 5.4	35.62 ± 3.6
Ntcheu 2011	11	53.9 ± 16.2	71.6 ± 6.8	36.56 ± 9.6
Ntchisi 2010	14	72.9 ± 18.0	69.8 ± 4.8	63.20 ± 6.5
Mchinji 2010	21	18.8 ± 2.6	50.6 ± 3.9	48.86 ± 3.8
Total	67	260.8	320.7	217.6
Mean ± SE		52.2 ± 10.6	64.1 ± 4.6	43.5 ± 5.6

SE, standard error.

seed had 64.1% reduction in number of wilted plants and 43.5% reduction in number of plants with virus disease symptoms (Table 11.3).

11.4 Lessons Learnt

1. Aeroponics and other RMT can be used to break the seed potato bottleneck by reducing the number of field multiplications required to bulk quality seed for supply to farmers.
2. An efficient private sector can successfully invest in quality seed potato production in SSA.
3. Clean seed is a profitable investment for farmers.
4. With proper technical advice and backstopping, farmers are able to produce clean seed in SSA.
5. Transport of bulky seed potato is a challenge, justifying the importance of a decentralized seed multiplication system.
6. Use of best seed-production practices to achieve the highest possible plant survival and highest number of tubers produced per plant is a key strategy to lower the cost of seed under an aeroponics and field production system.

11.5 Conclusions and Recommendations

1. Seed quality plays a pivotal role in improving potato yields in SSA. To overcome the supply crisis of high quality seed, serious investments are needed in the seed sector.
2. Formal regulated seed certification schemes for all planting material are unlikely to be effective or practicable in SSA smallholder potato production.
3. Certification of foundation and basic seed and implementation of community-based quality schemes (and encouragement of small private seed multipliers) is feasible and would improve access to clean planting material.
4. Seed regulations could be amended to recognize the importance of traded quality planting material derived from certified seed.

Acknowledgements

The authors of this chapter acknowledge CIP donors namely Irish Aid, USAID and the Common Fund for Commodities for financial support to CIP-led projects in which the work reported was conducted. Acknowledgements also go to the organizing committee of the African Potato Association (APA) for inviting this keynote paper and providing the opportunity for oral presentation at the Ninth Triennial APA Conference held in Naivasha, Kenya from 30 June to 4 July 2013 and sharing of the findings. All implementing partners of CIP in Kenya, Malawi, Ethiopia, Rwanda and Uganda who participated in the implementation of activities reported are acknowledged.

References

Demo, P., Pankomera, P., Connell, T. and Kumar, N. (2009) Potential of potato farming in improving the livelihoods of small scale farmers in Malawi. *African Crop Science Conference Proceedings* 9, 761–765.

Chiipanthenga, M. (2010) Performance of potato (*Solanum tuberosum* L.) genotypes in seed multiplication using tissue culture and aeroponics techniques in Malawi. MSc thesis, University of Malawi, Bunda College of Agriculture, Lilongwe, Malawi.

Fajardo, J., Lutaladio, N., Larinde, L., Rosell, C., Barker, I., Roca, W. and Chujoy, E. (2010) Quality declared planting material – protocols and standards for vegetatively propagated crops. *Food and Agriculture Organization of the United Nations (FAO) Plant Production and Protection* Paper 195. FAO, Rome, 126 pp.

FAOSTAT (2010) Available at: http://faostat.fao.org/ (accessed 22 May 2014).

Gildemacher, P.R. (2012) Innovation in seed potato systems in Eastern Africa. PhD thesis, Wageningen University, The Netherlands.

Gildemacher, P.R., Maina, P., Nyongesa, M., Kinyae, P., Gebremedhin, W., Lema, Y., Damene, B., Shiferaw, T., Kakuhenzire, R., Kashaija, I., Musoke, C., Mudiope, J., Kahiu, I. and Ortiz, O. (2009a) Participatory analysis of the potato knowledge and information system in Ethiopia, Kenya and Uganda. In: Sanginga, P.C., Waters-Bayer, A., Kaaria, S., Njuki, J. and Wettasinha, C. (eds) *Innovation Africa: Enriching Farmers' Livelihoods*. Earthscan, London, pp. 153–167.

Gildemacher, P.R., Kaguongo, W., Ortiz, O., Tesfaye, A., Woldegiorgis, G., Wagoire, W.W., Kakuhenzire, R., Kinyae, P.M., Nyongesa, M., Struik, P.C. and Leeuwis, C. (2009b) Improving potato production in Kenya, Uganda and Ethiopia: a system diagnosis. *Potato* 52, 173–205.

Gildemacher, P., Schulte-Geldermann, E., Borus, D., Demo, P., Kinyae, P., Mundia, P. and Struik, P. (2011) Seed potato quality improvement through positive selection by smallholder farmers in Kenya. *Potato Research* 54, 253–266.

International Potato Center (CIP) (2012) Roadmap for Investment in the Seed Potato Value Chain in Eastern Africa. Available at: http://cipotato.org/wp-content/uploads/2014/05/CIP_Roadmap-Final.pdf (accessed 15 September 2015).

Lemaga, B., Kakuhenzire, R., Kassa, B., Ewell, P.T. and Priou, S. (2005) Integrated control of potato bacterial wilt in Eastern Africa: the experience of African Highlands Initiative. In: Allen, C., Prior, P. and Hayward, A.C. (eds) *Bacterial Wilt Disease and the* Ralstonia solanacearum *Species Complex*. American Phytopathological Society Press, St Paul, Minnesota, pp. 145–157.

Muthoni, J. and Kabira, J.N. (2010) Effects of crop rotation on soil macronutrient content and pH in potato producing areas in Kenya: a case study of KARI Tigoni station. *Journal of Soil Science and Environmental Management* 1, 227–233.

Østergård, H., Finckh, M.R., Fontaine, L., Goldringer, I., Hoad, S.P., Kristensen, K., Lammerts van Bueren, E.T., Mascher, F., Munki, L. and Wolfe, M.S. (2009) Time for a shift in crop production: embracing complexity through diversity at all levels. *Journal of the Science of Food and Agriculture* 89, 1439–1445.

Schulte-Geldermann, E. (2013) Tackling low potato yields in eastern Africa: an overview about constraints and potential strategies. In: *Seed Potato Tuber Production and Dissemination – Experiences, Challenges and Prospects*, Proceedings of the National Workshop on Seed Potato Tuber Production and Dissemination, 12–14 March 2012, Bahir Dar, Ethiopia. Ethiopian Institute of Agricultural Research (EIAR), Amhara Regional Agricultural Research Institute (ARARI) and International Potato Center (CIP). Lima, pp. 72–80.

Shepherd, K.D. and Soule, M.J. (1998) Soil fertility management in west Kenya: dynamic simulation of productivity, profitability and sustainability at different resource endowment levels. *Agriculture Ecosystems and Environment* 71, 131–145.

Turkensteen, L.J. (1987) Survey of diseases and pests in Africa: fungal and bacterial diseases. *Acta Horticulturae* 213, 151–159.

Waithaka, M.M., Thornton, P.K., Herrero, M., Shepherd, K.D., Stoorvogel, J.J., Salasya, B., Ndiwa, N., Bekele, N., Croucher, M., Karanja, S. and Booltink, H. (2005) *System Prototyping and Impact Assessment for Sustainable Alternatives in Mixed Farming Systems in High-Potential Areas of Eastern Africa*. Final Program Report to the Ecoregional Fund to Support Methodological Initiatives. International Livestock Research Institute, Nairobi.

12 Public–Private Partnership Supporting Women-driven Seed Potato Multiplication in the Lumwana Catchment Area of North-Western Province of Zambia

A. Chalwe,[1*] S. Bwembya,[1] H. Kanema[2] and D. Subakanya[2]
[1]*Zambia Agriculture Research Institute, Solwezi, Zambia;*
[2]*Barrick Lumwana Mining Company, Solwezi, Zambia*

Abstract
The opening of mines and associated population boom is contributing to the increase in potato (*Solanum tuberosum* L.) consumption, which calls for increased production of the crop in North-Western Province of Zambia. However, the major constraint hampering expansion of potato production in the region is lack of seed for improved potato varieties. As no seed company is engaged in potato seed production in Zambia, all improved seed is imported from South Africa and Europe in the form of 'seed potatoes'. Even so, the quantity imported is not adequate to guarantee year-round production of the crop by small-scale farmers in the province. In response to the national government's strategic plan of fostering development of the agricultural sector through the establishment of public–private partnerships in Zambia, Barrick Lumwana Mining Company, Zambia Agriculture Research Institute (ZARI) and the Department of Extension of the Ministry of Agriculture and Livestock are working in partnership to support women-driven seed multiplication to promote year-round market-oriented production of potatoes in the Lumwana catchment area. In this arrangement, the mining company provides funds for the projects, while ZARI and extension agents are responsible for training and provision of technical backstopping to farmers in the project. ZARI serves as the primary multiplication site. Six women's groups were trained and attained a 1:10 multiplication ratio of seed from rain-fed upland and wetland nurseries. Gross margin analysis indicates that women in Silenda and Muyashi camps were able to obtain high revenue returns on labour either by working as a group or individually. This chapter highlights the experiences, successes and challenges encountered in the implementation of this women-driven seed multiplication and distribution project. The project provides a unique example of the social corporate responsibility in which a private company ploughs back part of the proceeds to benefit communities among which it does business.

Keywords: Barrick Lumwana, multiplication, potato, public–private partnerships, women-driven, ZARI

12.1 Introduction

Potato is the fourth most important crop in the world, but its significance varies widely from country to country (Namasiku, 1995). In Zambia potato production and consumption is increasingly becoming important as a source of income. In particular, the opening of new

*ablechalwe@yahoo.com

mines in North-Western Province and associated population increase is contributing to the increase in potato consumption in the region. Local restaurants, supermarkets, hospitals, colleges (which include Solwezi College of Education, Solwezi School of Nursing and Solwezi Trades), as well as boarding schools and mine kitchens are potential buyers of potato in Solwezi.

However, efforts to promote year-round production of potato among local farmers are hindered by lack of availability of adequate seed stocks at the required times. As such, initiatives are required to establish local seed-production centres for improved varieties which can easily be accessed by local farmers in order to increase production of the crop. Small-scale farmers in North-Western Province have been supplying potatoes to the mining communities who provide a ready market for the commodity. Despite the presence of a ready market, the quality of the available potatoes is poor. The tubers are normally small, have a short shelf life and are not good for chips.

With this background, Barrick Lumwana Mining Company and the Zambia Agriculture Research Institute (ZARI) are working together to introduce to farmers highly marketed improved varieties of potato, and provide them with the necessary skills in production, processing and marketing of the product. The project targets women since much of the production and marketing of potato in Solwezi is done by women. This chapter outlines some of the successes and challenges encountered so far in the implementation of a women-driven seed multiplication and distribution project.

12.2 Materials and Methods

Public–private partnership (PPP)

In 2010, a PPP was established with the goal of raising the living standards of small-scale potato farmers in the Lumwana catchment area through the use of improved potato seed. The PPP involved three major players, namely: (i) Barrick Lumwana copper mining company; (ii) ZARI; and (iii) the Department of Extension of the Ministry of Agriculture and Livestock. Funding for the project was provided by Barrick Lumwana, while ZARI and the Department of Extension provided training and technical backstopping on agronomic aspects of seed production to farmers. All partners contributed to continuous monitoring and provision of an advisory service.

Identifying target women's groups

Although the opening of mines in the area has opened employment opportunities for the male members of the communities, women, youths and old men still farm in the catchment area. The project aimed to empower women by offering them an alternative economic opportunity. The identification of the target group for this project was done through participatory rural appraisal (PRA) via focus group discussions and individual interviews with small-scale producers and traders within the catchment area. Two agricultural camps were identified in each of three chiefdoms and one women's group was selected from each agricultural camp (Fig. 12.1).

Training of target women's groups

Mobile training sessions were undertaken with the members of each of the identified women's groups. The women were given practical training which focused on various aspects of seed potato production (Fig. 12.2).

Sourcing seed of improved varieties and other agro-inputs

Improved varieties were sourced from South Africa through Buya Bamba Ltd in Lusaka. The project sourced a variety known as 'Red eye' which is normally found in the major supermarkets in Zambia and is highly desired for making French fries. Third generation pre-sprouted large-sized tubers were procured and distributed. The women's groups were also supported with other agro-inputs such as fertilizers, fungicides, insecticides and sprayers.

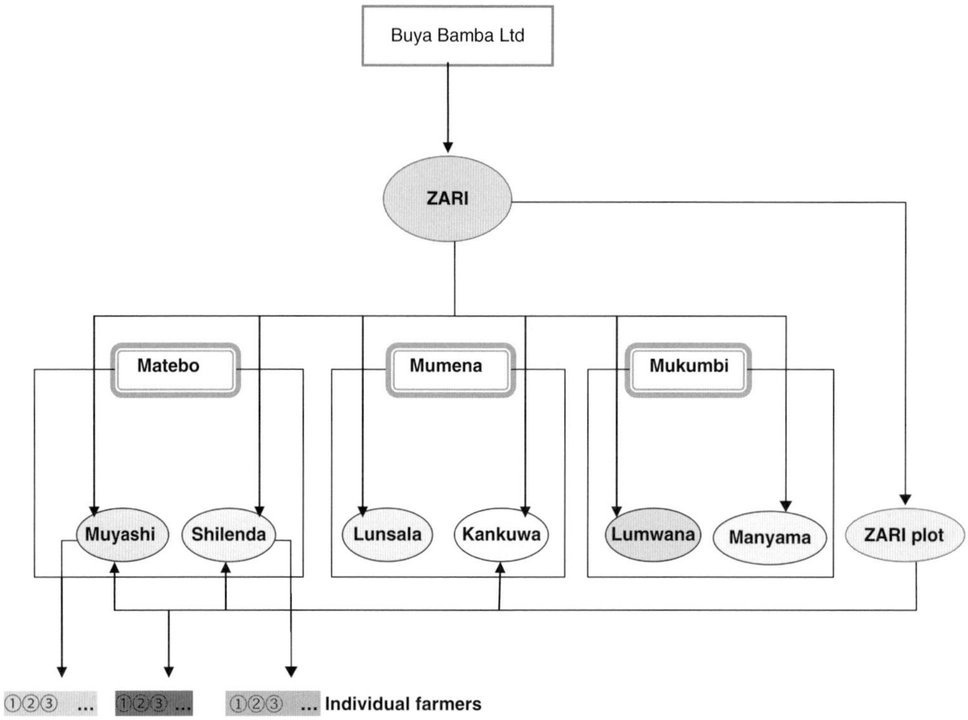

Fig. 12.1. Seed potato multiplication and distribution scheme for Lumwana potato project. Improved varieties were sourced from South Africa through Buya Bamba Ltd and were distributed to the chiefdoms (Matebo, Mumena and Mukumbi) by Zambia Agriculture Research Institute (ZARI). From each chiefdom, two agricultural camps (e.g. Muyashi and Shilenda for Matebo) were identified and from each camp a women's group was selected.

Fig. 12.2. Farmers being trained in the field.

Planting and maintenance of nurseries

Each of the six women's groups was given 75 kg of the seed to multiply during the 2011/12 rainy season. Each group planted a 22 m × 15 m plot on ridges with spacing of 80 cm between ridges and 25 cm within the ridge. Weeding was done by hand using hoes, and inorganic fertilizers were applied as recommended by Namasiku (1995).

Women's group-managed nurseries

The group members were responsible for preparing the land, planting and general maintenance of nurseries. The groups practised positive and negative selection (Gildemacher *et al.*, 2007). Routine inspections of the nurseries for diseases and pests were conducted by seed inspectors from the Seed Control and Certification Institute (SCCI) (Fig. 12.3), with a view to enable the women to produce Quality Declared Seed (QDS). After 4 months, all the seed produced by women in the group nurseries was reserved

Fig. 12.3. Plant health inspectors and farmers inspecting potato field for pests and diseases.

for distribution to individual members of the same group to plant in their own individual fields in the 2012/13 rainy season.

ZARI-managed nursery

The ZARI at Mutanda Research Station also multiplied seed in a similar way. The multiplication field was better isolated from other *Solonaceae* crops such as tomatoes and tobacco. The seed produced from ZARI nurseries was distributed to the group nurseries in order to enable the women to establish new group nurseries for the 2012/13 rainy season.

Monitoring and evaluation

Field activities were monitored by a team which included all the partners. Monitoring visits were undertaken to the multiplication fields on a monthly basis. During monitoring, discussions were held with the women's group members regarding the performances of the multiplication plots, to learn about the challenges that farmers were encountering and what improvements could be made.

Gross margin analysis

Gross margin analysis was made to determine the net returns on farmers' labour during the 4 month production period. This was done by subtracting the cost of all purchased inputs (seed, fertilizers and pesticides) from the gross income. To obtain the net monthly returns, the returns for the 4 month period were divided by four (Demo *et al.*, 2009).

Gross margin analysis was only done for two sites, namely Shilenda and Muyashi. Marketed potato yields per hectare were computed for the actual quantities of marketable-size potatoes (≥ 40 mm diameter) harvested (Demo *et al.*, 2009). From yield per hectare, the gross income per hectare was computed using the price of ZK6.00 (Zambian kwacha) per kilogram.

12.3 Results and Discussion

Effectiveness of training approach

All the farmers who had attended a Training of Trainers (TOT) course demonstrated the ability to apply the knowledge in their own individual fields correctly. Follow ups indicated that use of the TOT approach was 75% effective in transferring knowledge to the rest of the members of the women's groups. The group-managed nurseries served as field schools and information was transferred effectively through practical training of farmer to farmer. The training method proved to be cheaper and more effective than theory-based training. The farmer-field-school approach enabled women to develop the confidence and expertise through learning cycles which enabled them to make evidence-based crop management decisions with the aim of growing a healthy crop (van den Berge and Jiggins, 2006).

Effectiveness of positive selection in rain-fed potato seed multiplication

Though there were some reductions in the yield and quality of tubers with progression of seed generations (from third to fifth generation), the magnitude of the reduction was very minimal between ZARI- and group-managed nurseries at Muyashi and Shilenda. Similarly, there was only a very small reduction in the yield and quality of tubers between group-managed and individual-farmer-managed nurseries, which probably resulted from the corresponding increase in bacterial wilt and late blight in individual farmers' fields. The yields varied among individual farmers in both agricultural camps (Tables 12.1 and 12.2). The drop in tuber yield from the primary multiplication site (ZARI) to the tertiary site (farmers') indicated a significant role played by seed inspection and ZARI technical backstopping in advancing positive selection technology. Farmers were anxious to sell the potatoes, so they were harvesting progressively (piecemeal), and this contributed to the lower yields of marketable tubers in individual farmer nurseries at the end of the 4 month period. This, however, was not the case for group-managed nurseries.

Furthermore, because individual farmers were not provided with fertilizer and pesticides by the project and were expected to buy these for themselves, there is a possibility that some farmers may not have applied these inputs in the correct doses or not at all, which could also contribute to the low yields and differences observed in crop yields. Possible heterogeneity in soil fertility status and climatic conditions (data not shown) could also be another source of variation.

Regarding the extension agents, their contribution to the project was stretched and erratic since each camp officer handled many projects supported by the Ministry of Agriculture and Livestock and other non-governmental

Table 12.1. Tuber yield, gross income and returns on individual farmer's labour from the rain-fed pilot potato seed multiplication nurseries at Muyashi camp.

Farmer/group	Total tuber yield (t/ha)	Gross income[a]		Net returns on farmer's labour		Mean monthly return on farmer's labour	
		ZK	US$	ZK	US$	ZK	US$
Farmer 1	12.12	60,300.00	11,596.15	39,870.00	7,667.31	9,967.50	1,916.83
Farmer 2	10.61	48,900.00	9,403.85	28,470.00	5,475.00	7,117.50	1,368.75
Farmer 3	13.64	72,720.00	13,984.62	52,250.00	10,048.08	13,062.50	2,512.02
Farmer 4	11.21	54,900.00	10,557.69	34,470.00	6,628.85	8,617.50	1,657.21
Farmer 5	11.67	52,500.00	10,096.15	32,070.00	6,167.31	8,017.50	1,541.83
Bukwasho	15.00	72,000.00	13,846.15	51,570.00	9,917.31	12,892.50	2,479.33
ZARI	16.67	85,020.00	16,350.00	64,590.00	12,421.15	16,147.50	3,105.29
Mean	12.99	63,762.86	12,262.09	43,327.14	8,332.14	10,831.79	2083.04

[a]US$1 = ZK5.20. Purchased input costs (seed, fertilizer, pesticides) per hectare = ZK20,430.00 (US$3,928.85).

Table 12.2. Tuber yield, gross income and returns on individual farmer's and group labour from the pilot rain-fed pilot potato seed multiplication nursery at Shilenda camp.

Farmer/group	Total tuber yield (t/ha)	Gross income[a]		Net returns on farmer's labour		Mean monthly return on farmer's labour	
		ZK	US$	ZK	US$	ZK	US$
Farmer 1	10.11	48,900.00	9,403.85	28,470.00	5,475.00	7,117.50	1,368.75
Farmer 2	14.39	73,380.00	14,111.54	52,950.00	10,182.69	13,237.50	2,545.67
Farmer 3	7.56	31,260.00	6,011.54	10,830.00	2,082.69	2,707.50	520.67
Farmer 4	18.18	87,900.00	16,903.85	67,470.00	12,975.00	16,867.50	3,243.75
Farmer 5	7.56	31,680.00	6,092.31	11,250.00	2,163.46	2,812.50	540.87
Lukusa nkunde	14.5	69,900.00	13,442.31	49,470.00	9,513.46	12,367.50	2,378.37
ZARI	18.03	86,520.00	16,638.46	66,090.00	12,709.62	16,522.50	3,177.40
Mean	12.90	61,362.86	11,800.55	40,932.86	7,871.70	10,233.22	1,967.93

[a]US$1 = ZK5.20. Purchased input costs (seed, fertilizer, pesticides) per hectare = ZK20,430.00 (US$3,928.85).

organizations (NGOs). There is need to assign a full-time field-based supervisor who should be fully dedicated to the project to ensure that camp officers, who are paid allowances on a monthly basis by Barrick Lumwana Mining Company, are able to visit and provide advisory services to farmers as planned. Generally, the findings suggest that the current seed system is quite efficient but requires strengthening of the extension service to ensure affordable production of quality seed and increased yields.

Incidence of late blight and bacterial wilt

There were differences in the occurrence of the diseases at the six different sites (Fig. 12.4). Manyama and Lunsala had the highest incidence of late blight (65% and 55%, respectively), which could not be controlled by spraying with Dithane M-45. On the other hand, Muyashi and Shilenda recorded the least incidences of late blight (5% and 8%, respectively). Symptoms of bacterial wilt were also observed at all the sites, with Manyama and Luwana East recording the highest incidence of 35% and 30%, respectively, while less than 5% incidence of bacterial wilt was recorded at Shilenda, Muyashi and the ZARI station. Due to high disease incidence at Manyama and Lumwana East, the seed nurseries at these sites failed the inspection tests and were condemned. However, soil and water analyses were not performed to ascertain the source of the bacterial wilt inoculum.

Gross margin analysis shows that Bukwasho, Lukusa nkunde, Kankuwa and Lunsala women's groups had positive returns for their labour (Table 12.3), while Manyama and Lumwana East women's groups had negative returns on their labour (Bukwasho women's group is in the Muyashi camp and Lukusa nkunde women's group is in Shilenda camp). Bukwasho and Lukusa nkunde women's groups had the highest positive mean monthly return of ZK15,352.50 (US$2952.40), and ZK11,932.5 (US$2294.71), respectively.

The women at Muyashi have a longer history of growing potato in the catchment area and are therefore more experienced than their counterparts in the other women's groups. However, like other women's groups, Bukwasho women require skills on how to store seed potato tubers. Considerable losses were incurred in storage which further reduced the quantity of seed at the beginning of the next rainy season. Potato

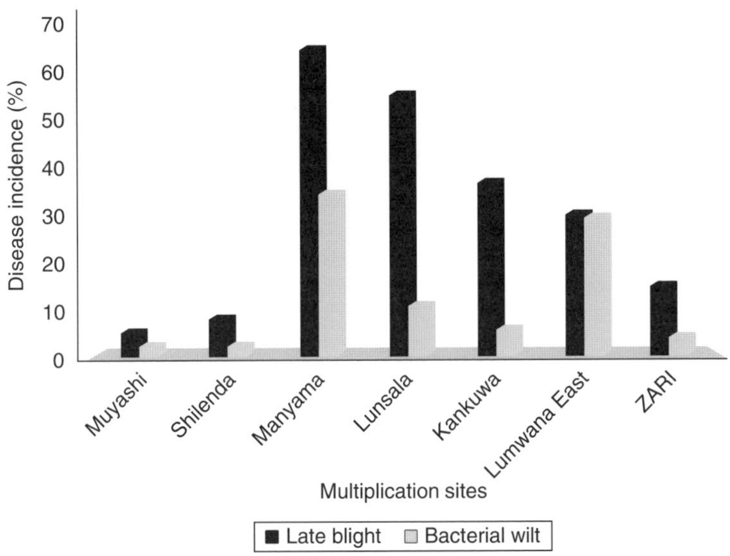

Fig. 12.4. Incidence of bacterial wilt and late blight in potato seed multiplication plots at Solwezi.

Table 12.3. Tuber yield, gross income and returns on labour of farmer groups from the pilot rain-fed potato seed multiplication nurseries managed by women's groups in Solwezi District, Zambia.

Group	Total tuber yield (t/ha)	Gross income[a]		Net returns on farmer's labour		Mean monthly return on farmer's labour	
		ZK	US$	ZK	US$	ZK	US$
Bukwasho	18.18	81,840.00	15,738.46	61,410.00	11,809.62	15,352.5	2,952.40
Lukusa nkunde	15.15	68,160.00	13,107.69	47,730.00	9,178.85	11,932.5	2,294.71
Manyama	1.52	7,320.00	1,407.69	−13,110.00	−2,521.15	−3,277.5	−630.29
Lumwana East	3.03	14,520.00	2,792.31	−5,910.00	−1,136.54	−1,477.5	−284.13
Kankuwa	6.06	27,240.00	5,238.46	6,810.00	1,309.62	1,702.5	327.40
Lunsala	9.09	46,380.00	8,919.23	25,950.00	4,990.38	6,487.5	1,247.60
ZARI	21.21	88,860.00	17,088.46	68,430.00	13,159.62	17,107.5	3,289.90
Mean	10.61	47,760.00	9,184.62	27,330.00	5,255.77	6,832.5	1,313.94

[a]US$1 = ZK5.20. Purchased input costs (seed, fertilizer, pesticides) per hectare = ZK20,430.00 (US$3,928.85).

requires good field management in terms of weeding, frequent field inspection and rouging of infected plants. The high incidence of late blight and bacterial wilt contributed to the lower yields at Manyama, Lumwana East and Kankuwa (Table 12.3) as compared with Muyashi and Shilenda (Fig. 12.4).

Ultimately, the results indicate that small-scale production of potatoes in Lumwana catchment area is a feasible and highly profitable venture which has the potential to reduce poverty among women. The returns on farmers' labour, US$24.60 (ZK127.92) and US$19.12 (ZK99.45)/day obtained from growing 0.25 ha of potatoes by Lukusa nkunde and Bukwasho women's groups, respectively, can enable women to live far above the daily poverty datum line of US$1.00/day (Demo *et al.*, 2009).

The two groups, managed to produce and distribute QDS to their members in the next rainy season. The yields varied among farmers, ranging from 8.15 t/ha to 12.12 t/ha. Some of the individual farmer-managed nurseries yielded more tubers than the group-managed nurseries, and vice versa. The ZARI-managed nursery had the highest yield of all (Table 12.3).

However, all the five farmers of Bukwasho women's group had monthly positive returns on their labour, which ranged between US$1500 and US$2500 expected from 1 ha of seed potato. The best return on farmer's labour was US$2512.02/ha which translates into a daily net return of US$83.73 (ZK435.42). Therefore, by growing 0.25 ha of seed potato, which seems to be currently manageable for women in Zambia, a farmer would expect a daily net return of US$20.93 (ZK108.85) on labour.

The returns on individual farmers' labour were highly positive for members of Lukusa nkunde women's group (Table 12.2).

The best farmer had monthly returns of US$3243.75 on labour, translating into daily returns of US$108.12 (ZK562.25)/ha, implying that by growing 0.25 ha of seed potato such a farmer in Shilenda would expect a return of US$27.03 (ZK140.56) on labour. This suggests that, either by working individually or as a group, women can still improve their livelihood by growing potatoes in the Lumwana catchment area. However, the group-managed nursery served as a field school for group members and should be promoted alongside individual farmer plots.

Challenges

In the absence of improved irrigation facilities, establishment and maintenance of off-season potato nurseries proved to be very challenging. Watering by hand using a watering can is a very burdensome and labour-tasking activity which limits expansion of production fields by small-scale potato farmers. This makes off-season farming more expensive than rain-fed production of potatoes for small-scale farmers. Worse still, the quality of water from streams or rivers can be very difficult to control and especially when it gets contaminated/infected with *Ralstonia solanacearum*.

The high market demand for potato against low production in the catchment area has prompted small-scale traders to import potatoes from Tanzania across the Tunduma/Nakonde border into the catchment area and many parts of the country. This uncontrolled movement of planting materials is likely to contribute to the spread of plant diseases, such as late blight (*Phytophthora infestans*), bacterial wilt (*R. solanacearum*) and viruses. Poor knowledge of farmers on the control of such diseases constituted a major challenge.

Lack of high yielding varieties suitable for different end uses is another limitation to increased productivity. Availability of a variety of our choice is not guaranteed at Buya Bamba Ltd at the time that small-scale farmers need it. The primary interest of this supplier is to respond to the cooperative members' demands in terms of seed, while supplies to other buyers are only made on request and are given a secondary consideration. Seed is the most expensive input for potato production. Efforts to support establishment of local seed-production schemes in Solwezi would help to reduce the cost of seed. Storage of locally produced seed constituted another serious challenge for small-scale producers and it needs urgent attention.

12.4 Recommendations

First, there is need for ZARI to develop a collection of potato germplasm comprising local and introduced accessions to support potato variety development. This is an essential step towards strengthening the national breeding programme to produce varieties with resistance to late blight, bacterial wilt and viruses and that are adaptable to local farming conditions.

Secondly, interventions through provision of a labour-saving irrigation scheme in Muyashi and Shilenda such as sinking boreholes and providing women's groups or individual potato farmers with solar-operated water pumps could make year-round farming possible, boost potato production and contribute to poverty reduction.

Finally, promising individual farmers should be identified and supported to take up seed potato production as a business in addition to introduction of improved infrastructures such as the low-cost diffused light storage (DLS) of seed potato tubers to reduce postharvest losses and ensure proper sprouting of seed potato. This technology has proved successful in Malawi and Mozambique (Paul Demo, International Potato Center, Citedze Research Station, 2013, personal communication).

Acknowledgements

The authors acknowledge financial assistance received from Barrick Lumwana Mining Company and the Ministry of Agriculture and Livestock through ZARI, without which this work could not have been possible. The Germany International Corporation and the African Potato Association (APA) are acknowledged for covering the costs towards participation in the Ninth Triennial APA Conference in Kenya.

References

Demo, P., Pankomera, P., Connell, T. and Khumar, N. (2009) Potential of potato farming in improving the livelihood of small scale farmers in Malawi. *African Crop Science Conference Proceedings* 9, 791–765.

Gildemacher, P., Demo, P., Kinyae, P., Wakahiu, M., Nyongesa, M. and Zschocke, T. (2007) *Select the Best: Positive Selection to Improve Farm Saved Seed Potatoes*. International Potato Center (CIP), Lima.

Namasiku, J. (1995) Root and tuber crops: Irish potato (*Solanum tuberosum* L.). In: Muliokela, S.W. (ed.) *Zambia Seed Technology Handbook*. Ministry of Agriculture, Food and Fisheries, Lusaka, Zambia, pp. 324–334.

van den Berge, H. and Jiggins, J. (2006) Investing in farmers – the impacts of farmer field schools in relation to integrated pest management. *World Development* 35, 663–686.

13 Risk of Uncontrolled Importation of Seed Potato from Europe to East and Central Africa: What are the Policy Options?

W. Kaguongo,[1]* I. Rwomushana,[2] I.N. Kashaija,[3] S. Ntizo[4] and J. Kabira[5]
[1]National Potato Council of Kenya (NPCK), Nairobi, Kenya; [2]Association for Strengthening Agricultural Research in Eastern and Central Africa (ASARECA), Entebbe, Uganda; [3]National Agricultural Research Organization, Entebbe, Uganda; [4]Rwanda Agriculture Board (RAB), Kigali, Rwanda; [5]Kenya Agricultural and Livestock Research Organization (KALRO), Limuru, Kenya

Abstract

Potato (*Solanum tuberosum*) is an important food and cash crop. It is among the 10 strategic staple crops for food and income security in Eastern and Central Africa (ECA). Its importance continues to rise due to increased urbanization, change in eating habits and uptake of processed potato products such as crisps and chips. This has led to a steady increase in the area under the potato crop.

Despite this, on-farm potato yields have continued to drop to about 10 t/ha compared with the potential yields of 40–60 t/ha attained by progressive farmers. This is mainly attributed to poor quality seed tubers used by farmers. On the other hand, lack of suitable processing varieties has limited expansion of potato value addition. Currently, certified or high quality seed potato accounts for less than 5% of the whole potato seed market in ECA. This scenario has encouraged the common practice among potato farmers of planting own-saved tubers from previous harvests or sourced from markets or neighbours.

Several initiatives have been undertaken by governments, stakeholders, private investors and development partners to address the challenge of scarcity of quality seed potato in the region. Introduction of rapid seed potato multiplication techniques such as aeroponics, and facilitating availability of quality declared seed and farmer-based seed quality maintenance through positive selection are among the practical solutions adopted and targeting farmers with different capacities. These initiatives have proven their potential to increase potato productivity up to threefold in a short time.

Seed potato import from European countries is also viewed as an alternative intervention. This option, however, risks clipping the development of the local seed potato sub-sector. Recent studies indicate that the local potato production risks collapse from such imports through the danger of the introduction of new diseases and pests that could be even more virulent under tropical conditions.

Diseases such as *Dickeya* spp., the A2 mating type of *Phytophthora infestans* the causal agent of late blight and bacterial ring rot are some of the dangerous diseases reported in Europe that risk being introduced into the region if importation is not regulated. European countries with low incidences of these diseases have undertaken measures such as voluntary banning while China and the USA have

*nkaguongo@npck.org

banned importation of tubers from Europe. Modern pest risk assessment allows such banning for phytosanitary reasons if there is sound scientific evidence. Therefore, ECA countries need to adopt an informed policy position that will protect small-scale farmers and local seed producers as they develop the potato sector.

Keywords: diseases, Eastern and Central Africa, imports, pests, quality seed potato

13.1 Introduction

Potato (*Solanum tuberosum*) is an important food and cash crop. It is among the ten strategic staple crops for food and income security in Eastern and Central Africa (ECA). Its importance continues to rise due to increased urbanization, change in eating habits and uptake of processed potato products such as crisps and chips. This has led to a steady increase in the area under the potato crop over the years, with smallholder farmers owning parcels of land measuring between 0.25 and 5 ha dominating the sector (over 90%) (Kinyua *et al.*, 2011).

On-farm potato yields are commonly below 10 t/ha compared with the potential yields of 40–60 t/ha attained by progressive farmers (Raemaekers, 2001; FAOSTAT, 2006). Poor quality of seed tubers used by farmers is viewed as a major yield-limiting factor while lack of suitable processing varieties has limited expansion in potato value addition. Although by the turn of century, at least 89 varieties had been released in the Association for Strengthening Agricultural Research in Eastern and Central Africa (ASARECA) sub-region with a total of 295,000 ha under improved varieties, introduction of new varieties hardly considered processing and high-temperature-tolerance attributes. This has resulted in the sub-sector attracting limited investment in potato processing and potato growing being concentrated in the highlands with an altitude range of 1500–2500 m above sea level (masl).

Currently, high quality seed potato accounts for less than 5% of the whole seed potato market in ECA. This scenario has encouraged the common practice among potato farmers of planting own-saved tubers from previous harvests or sourced from markets or neighbours (Barton *et al.*, 1997; Kaguongo *et al.*, 2008; CAPAD, 2009; UBOS, 2010). The low yields in the region are attributed to both biotic and abiotic stresses and poor management practices. Poor quality of seed tubers used by farmers is viewed as a major yield-limiting factor (Gildemacher *et al.*, 2009b). Farmers are frequently faced with several challenges that are related to lack or shortage of good quality seed potato of consumer-preferred, high yielding varieties. Of highest concern is the limited number of varieties with good processing performance in the region. This is because research institutions such as the International Potato Center (CIP) and National Agricultural Research System (NARS) in the region have given priority to breeding on traits such as disease resistance, maturity period and high yield attributes focusing on food security and yield stability but have given little attention to processing attributes (Kaguongo *et al.*, 2008).

Currently, of the varieties released in the region, only one variety, Dutch Robyjn, has suitable attributes for processing internationally acceptable chips or crisps. However, the variety is highly susceptible to diseases and pests and it is low yielding. Consequently, it is mainly grown by farmers who are assured of the market. A variety, Katchpot1, released by the National Agriculture Research Organisation (NARO) in 2006 has attributes for quality crisps but the products have not been benchmarked with other internationally marketed varieties. This bottleneck has created the need for the introduction of more processing varieties in the country. Countries such as The Netherlands, Germany, Scotland and South Africa have many varieties suitable for processing which could be accessed, evaluated through National Performance Trials (NPTs), registered and multiplied in the ECA countries. Having noticed this business opportunity, seed companies from European countries have taken advantage of these gaps and their political might to strategically position themselves and justify the need for importing seed into Kenya and the region. Lack of a well-developed local formal seed potato system is used to further justify the seed importation initiatives. However, introduction of such varieties from European countries must be done following stringent procedures and should not be left to dictates of economic laws of supply and demand, nor international trade, due to the risk of introducing dangerous diseases that could have devastating effects on the farming systems of ECA countries.

Increasing trade, globalization as well as climate change are among the most important factors contributing to the growing threat from the introduction of plant pests, which include pathogens, underlining the importance of policies and the need to keep policies up to date and relevant (Perrings *et al.*, 2005; Brasier, 2008; Waage and Mumford, 2008). The effects of plant diseases for global food supplies are potentially devastating and therefore there is need to acknowledge this threat at the political level and to devote adequate resources to their control (Strange and Scott, 2005).

13.2 Methodology

This chapter has been developed through a review of journal papers, phytosanitary procedures and legal acts of Kenya, Uganda and Rwanda, European countries' potato sub-sector, phytosanitary reports, sanitary and phytosanitary (SPS) measures and the Agreement on Technical Barriers to Trade of the World Trade Organization (WTO), and proceedings of scientists and stakeholder meetings.

13.3 Findings

Limited supply of quality seed potatoes

Quality seed potato and suitable varieties are identified as key to unlocking the potential of the potato sub-sector in the region. In spite of the many efforts to improve the seed potato system, potato farmers in eastern Africa still identify seed potato quality as their major concern within their potato production system. Improving the seed system was prioritized as an important technical intervention area to improve smallholder potato profitability (Gildemacher *et al.*, 2006).

Quality seed, which is comprised of certified seed, quality declared ('clean') seed and positively selected seed, amounts to less than 5% of seed potato in the region. This is far from adequate. Lack of farmer knowledge, limited institutional capacity, low private-sector involvement, lack of expertise and bottlenecks faced during the certification process have contributed to low production of quality seed in Kenya, and other ECA countries (Gildemacher *et al.*, 2009a; Kaguongo *et al.*, 2010, 2013).

Several initiatives have been undertaken to address the challenge of the lack of availability of certified and quality seed potato in the ECA region. In the majority of ECA countries the formal seed system is poorly developed, with limited formal institutional arrangement for multiplication and distribution of quality seed potato. Rapid seed potato multiplication techniques (aeroponics and hydroponics), quality declared seed ('clean') and positively selected seeds are among the practical solutions introduced and targeted at farmers with different capacities. These initiatives have proved their potential to increase potato productivity up to threefold in a short time and increase the number of seed multipliers (Schulte-Geldermann, 2012). However, challenges in seed inspection procedures, seed distribution, personnel and institutional capacities and financial support have limited the benefits of these interventions.

Lack of varieties with good processing attributes has caused inconsistent supply of adequate quantities of varieties to processing companies. This has discouraged would-be investors in potato processing, and hence has denied farmers additional incomes. Introduction of more processing varieties is therefore crucial in order to enhance potato processing in the region. This has greatly increased the need for introduction of new processing varieties that are high yielding and disease tolerant. Moreover, there is also high demand for the introduction of lowland and high-temperature-tolerant varieties in order to expand production of potatoes in the region and take advantage of irrigation potential, as well as counteract the effects of climate change.

Initiatives to improve the potato sub-sector

Kenya has a fairly developed potato industry with about 3.3 million people working in the sub-sector as farmers, marketing agents and processors. In the recent past, the government of Kenya, stakeholders and a number of development partners have undertaken several projects aimed at reviving the potato industry and successfully putting in place a very well-structured self-sustaining seed system. Seed development projects totalling over US$7 million have been undertaken which include: (i) the United States Agency for International Development (USAID)-funded 3G (three generation) project, for 2.5 years with a budget of over US$2.0 million for three countries (Kenya, Uganda and Rwanda); (ii) the Common Fund for Commodities (CFC) project with a budget of US$2.5 million for three countries (Kenya, Uganda and Ethiopia) of which over US$0.33 million went to seed production in Kenya; (iii) the GTZ (currently called GIZ; a German technical corporation) project on revitalization of the seed sub-sector with a budget of US$1.2 million; (iv) the government of Kenya budgetary allocation for 3 years to the seed sub-sector, a total of US$1.0 million; (v) other seed potato development projects totalling over US$1.0 million (USAID-Kenya Horticultural Competitiveness Project, CIP, ASARECA, etc.); and (vi) the private sector had invested over US$1.0 million in seed production using new production technologies by 2012.

Bottlenecks facing the seed potato industry in the ECA region

Policy issues

Development of the local seed potato industry has been curtailed by a number of bottlenecks which if addressed could lower the pressure for the need to import seed potato from Europe to the region. These bottlenecks include:

1. Lack of an elaborate plan supported by the government, development partners and stakeholders to increase and sustain seed production, distribution and use.
2. Lack of elaborate plans and processes for fast tracking introduction and release of varieties destined for special utilization and those suited to the lower elevations of 1000–1500 masl.
3. Seed potato inspection processes are inefficient often involving delays and occasionally have provided false results that discourage prospective private investors from engaging.
4. Most of the widely adopted farmer varieties have not been officially recognized and are yet to be characterized, cleaned and their quality seed made available to farmers.
5. Lack of adoption of harmonized ECA seed standards and implementation, and poor adherence to importation regulations, where they exist. Further, the regulations need to be brought up to date to reflect and address the current issues. For example in Kenya, the Plant Protection Act, Cap 324 and the Seeds and Plant Varieties Act, CAP 326 are yet to be fully implemented and where necessary they should be updated to protect and guide the growth in the potato sub-sector.

Potential risk from seed potato importation

In the recent past, seed potato import from European countries has been viewed as an alternative intervention to problems facing the potato sub-sector in the ECA region. This option, however, risks clipping the development of the local seed potato sub-sector. Studies indicate that the local potato production risks collapse from such imports through introduction of foreign diseases and pests that could be more virulent under tropical conditions. Diseases such as *Dickeya* spp., late blight mating type A2 and bacterial ring rot are some of the dangerous diseases reported in Europe that risk being introduced in the region. The following are details of some of these dangerous diseases and pests that are not found in Kenya and other countries in the ECA region.

- Potato blackleg and soft rot (caused by *Erwinia chrysanthemi or Dickeya* spp.) is a seed-borne bacterial disease that has emerged as a major threat to potato crops in Europe. A new more aggressive form, named *Dickeya solani*, is according to Toth *et al.* (2011) and Surman (2010) causing losses of between €25 and €30 million annually in the Dutch seed potato sector. A study by Tsror *et al.* (2011) indicated that out of 277 certified seed lots imported to Israel, 55 tested were *Dickeya* spp. positive, of which 49 were from The Netherlands, four from Germany, two from France and none from Scotland.
- Ring rot or bacterial ring rot is one of the most feared diseases of the potato industry, especially for seed producers. The disease is caused by the bacterium *Clavibacter michiganensis*. Information from Ohio State University states that the disease has the potential to spread quickly on a farm and can cause severe losses if left unchecked. According to Nederlandse Voedsel en Warenautoriteit (2012) bacterial ring rot was found in farms that bought seed material from the Dutch grower that supplied infected potatoes to Germany.
- Brown rot is caused by the bacterium *Ralstonia solanacearum* and is widely distributed in warm, temperate areas of the world.
- The Colorado potato beetle (*Leptinotarsa decemlineata*) was first recorded as a pest in the USA in 1859. It is widely established in much of continental Europe, having first arrived in Bordeaux in 1922, but is not present in the UK. With climate change, the risk of it extending its range and becoming established in Britain is increasing. Adult beetles as well as their larvae can strip plants of leaves and ruin an entire crop. Dome-shaped adults are 8–10 mm long, yellow, with five black stripes on each wing cover. The beetle has developed a resistance to all major insecticide classes used in the UK.
- Potato spindle tuber viroid (PSTVd) – yield losses of up to 64% have been recorded in potato crops infected by PSTVd. Interactions between other viruses may also occur.
- Other pests and diseases include potato tuber moth (*Phthorimaea operculella*), root-knot nematode (*Meloidogyne chitwoodi*) and wart disease (caused by *Synchytrium endobioticum*).

Threat and tenacity of new pests in tropical conditions

According to Elphinstone and Toth (2007), the most important means of dissemination of *Dickeya* spp. in potato is movement of latently infected seed tubers. It is observed that warmer spring and summer conditions favour the disease development (Tsror *et al.*, 2011) indicating that tropical weather conditions would make the disease more virulent in sub-Saharan Africa (SSA). This therefore means losses due to potential infections by *Dickeya* spp. are expected to be significant through seed infections if seed is exported to warm climates.

Dickeya spp. is a seed-borne pathogen and is controlled largely through seed classification in line with domestic and European Union (EU) legislation. In European countries, the seed potato classification schemes set tolerances for diseases encountered during visual inspections of growing crops and harvested tubers. By 2013 there were no specific control measures and no compulsory testing in operation for *Dickeya* spp. Unfortunaetly, *D. solani* does not have quarantine status which would bring the benefit of official testing, although that does not guarantee infections will be detected. For example, it is noteworthy that following the conventional testing method, infection levels of at least 1.5% are required to be 95% confident of detecting a disease in a 200-tuber sample from 6 t, hence infections of less than 1.5% could still remain undetected. The European countries with low incidences of such risky diseases (e.g. Scotland and the rest of the UK) have undertaken measures such as putting in place a voluntary ban, while China and the USA have banned importation of tubers from Europe. Modern pest risk analysis (PRA) allows such banning for phytosanitary reasons if there is sound scientific evidence.

In 2011 the Kenyan government through the Kenya Plant Health Inspectorate Service (KEPHIS) reached an agreement with the Dutch government on the import of seed potato. The details of the agreement between the Kenyan government and The Netherlands are not yet known by stakeholders in the industry, who are concerned that it may not stand up to scientific scrutiny of the required phytosanitary measures, including an adequate PRA meant to protect the subsector. The need to protect the potato industry and the overall economy of Kenya and other countries in the region cannot be overemphasized. In Kenya alone, over 800,000 smallholder farmers and 2.7 million marketing agents and processors derive their livelihood from the potato crop. The threat of introducing new and dangerous diseases and pests into the ECA farming system is magnified by the smallholder farming systems, which according to various studies (Barker, 2008; Huarte, 2008; Kaguongo *et al.*, 2013) involves over 95% of farmers recycling their seed compared with The Netherlands where about 99% of farmers use certified seed. Consequently, diseases would spread faster in ECA due to uncontrolled seed exchange without disease testing.

New pests and diseases will most likely lead to significant yield losses and/or high use of pesticides which will impact heavily on the farmers, leading to less production and therefore overdependence on the already burdened maize crop as a source of starch. Kenya and the region are currently faced by a serious maize disease, maize lethal necrosis disease, that is threatening maize production and the livelihood of many households (MoA, 2012).

International treaties and agreements

In January 1995 the WTO developed SPS measures that form the framework that guide governments in formulation of legislation, regulations and procedures that regulate, restrict or prevent the import and marketing of plant species or plant products. This is to ensure that trade in plants and plant products are not unnecessarily limited while ensuring it is safe. These measures aim to prevent the introduction and spread of plant pests across international boundaries or to limit the economic impact of regulated non-quarantine pests following the International Plant Protection Convention (IPPC). The IPPC is a treaty that pledges international cooperation in controlling pests of plants and plant products and in preventing their international spread, and especially their introduction into endangered areas.

The SPS agreements allow countries to set their own phytosanitary measures that are backed by sound scientific basis. The measures are applied to the extent necessary to protect human, animal or plant life or health and are formulated based on results of PRA and the assessment of the risks posed to human, animal and plant life/health or to a country. According to the Food and Agriculture Organization of the United Nations (FAO) (1996), risk assessment methods used must take the following into consideration:

- available scientific evidence;
- relevant processes and production methods;
- relevant inspection, sampling and testing procedures;
- prevalence of specific diseases or pests;
- the existence of pest- or disease-free areas;
- relevant ecological and environmental conditions and quarantine or other regulatory measures; and
- relevant economic factors, such as: (i) potential damage in terms of loss of production or sales in the event of the entry, establishment or becoming part of a disease ecology, or the spread of a pest or disease; (ii) the costs of controlling or eradicating an outbreak, and the programmes needed to manage such responses; (iii) the costs associated with

the loss of national/international markets; and (iv) the relative cost-effectiveness of alternative (SPS) measures.

The PRA process prescribes a process involving:

1. The identification and characterization of risks, for example pests and pathogens, likely to be associated with any imported product.
2. An assessment of the probability of any hazard becoming established in the importing countries' animal or plant populations, and an assessment of the impact in the face of such an eventuality.
3. The development and selection of available risk management options, such as quarantine treatment, that best suits the SPS requirements of an importing country.
4. The development of quarantine procedures that help put a risk management option in place.

A thorough PRA should therefore inform the country of the potential effects importation of seed potato into the country would have on the local seed industry, what the environmental and economic impacts would be, in case new diseases and pest are introduced, and what mitigation and remedial measures should be put in place.

Development of the local seed potato industry

The ECA countries need to adopt an informed policy position that will protect small-scale farmers and local seed producers as they develop the potato sector. In light of the risk posed by importation of seed potato from European countries to Kenya and any other ECA countries, it is prudent that a thorough PRA be transparently conducted and an informed position taken or terms of trade adopted with very clear mitigation measures following the IPPC recommendations and SPS measures. Knowledge of the potentially devastating impacts of plant diseases on food security should motivate governments to design and employ plant health legislation, under a structured policy framework, to prevent or inhibit the worst impacts.

The development and implementation of phytosanitary policies in the potato sub-sector is closely linked to reducing the risk of introduction of new devastating disease in various countries in the past and the desire for respective countries to develop their local seed potato industry. Given the fact that potato is among the ten most important food and cash crops grown by about 2.5 million farmers on about 750,000 ha in the ECA region, there is an obvious need to have a clear comprehensive policy for the crop including clear evidence based on phytosanitary procedures and guidelines. Any importation of the crop therefore should follow sound PRA, contributing to a strategy for the overall growth of the sub-sector addressing short-, medium- and long-term issues in order to ensure food security of the many smallholder farmers. Regional governments should prioritize and invest in adoptive research, breeding and support from formal institutions and facilitate development of efficient seed certification processes and disease surveillance systems in order to attract and motivate the private sector in sustainable local seed potato production. There is a need for creation and support of public–private partnership with the aim of fast-tracking adoption of innovations and technologies of rapid seed multiplication that have the capacity to increase seed production, distribution and access to farmers. Capacity-building initiatives in the area of risk assessment, implying risk identification and characterization, risk communication and risk management are crucial for providing adequate support for a flourishing seed sector.

13.4 Conclusions and Recommendations

The pathogens listed above represent some of the more important pest risks to Kenya and the ECA region in general. This chapter highlights the potential risk of seed importation from Europe to ECA. It underscores the need for policy makers in the affected countries to formulate policies and regulations that will address the seed sub-sector shortcomings without endangering the potato industry, increasing the cost of production through increased cost of crop protection, denying local investors seed business, or reducing and curtailing would-be employment creation. According to Dehnen-Schmutz *et al.* (2010), well-informed and well-designed government policies are known to work well in protecting and developing the potato industry. Before any country imports seed potatoes, a detailed and transparent PRA should be performed to evaluate the risk of importation, the potential economic and environmental impacts, and the results should be shared with local stakeholders. The PRA should involve experts and stakeholders examining published literature and analysing data and if necessary developing importation protocols.

Promotion and facilitation of the seed potato trade should aim to consider sustainability and must be sensitive to the local farming systems and be in the context of IPPC. This is important in order to avoid pushing smallholder farmers and local seed producers out of business and putting the future of the country at risk.

In Kenya, for example, it may be important to show evidence of the current lack of occurrence of the named pests (using structured surveys) which could be implemented by independent research institutes using appropriate diagnostic assays. Evidence on the risks posed to Kenyan agriculture could be achieved by looking at the response of Kenyan potato varieties after inoculation with pests under quarantine conditions (possibly in a third country). Surveys and surveillance on the distribution of existing seed-borne pest and diseases, including their respective vectors, should be combined with epidemiological and climate models to predict potential impacts of introduction of non-existing pest and disease threats. These findings could then be synthesized by KEPHIS, with assistance from CIP, to generate a PRA which could stand scrutiny by international bodies and international phytosanitary law as well as support Kenyan policy on an evidence-based approach.

Where importation of seed has already happened and stakeholders are concerned, as in the case of Kenya, the PRA results should be made available to stakeholders and other experts for positive scrutiny. Governments increasingly value the input and views of stakeholders in developing policy (MacLeod *et al.*, 2010) and can engage with stakeholders through direct correspondence and with the broader public through publishing information on official websites, where feedback is invited. This approach is important in ensuring there is good will in any initiatives and interventions by the government and development partners, hence reducing suspicion and increasing acceptance of any seed and varieties introduced into the country and the region. More importantly, it ensures that all relevant factors are considered in policy formulation.

References

Barker, I. (2008) Challenges and opportunities for improving availability of quality potato seed in Eastern and Central Africa. Paper presented at the Global Potato Conference, 9–12 December 2008, New Delhi.

Barton, D., Smith, J.J. and Kinyua, Z.M. (1997) Socio-economic inputs to biological control of bacterial wilt disease of potato in Kenya. Overseas Development Agency (ODA) Renewable Natural Resource Research Strategy (RNRRS) Crop Protection Project ZA0085. ODA, London, 23pp.

Brasier, C.M. (2008) The biosecurity threat to the UK and the global environment from international trade in plants. *Plant Pathology* 57(5), 792–808.

Confédération des Associations des Producteurs Agricoles pour le Développement (CAPAD) (2009) *Plan de Développement de la Filière Pomme de Terre*. CAPAD report. CAPAD, Burundi.

Dehnen-Schmutz, K., MacLeod, A., Reed, P. and Mills, P.R. (2010) The role of regulatory mechanisms for control of plant diseases and food security – case studies from potato production in Britain. Role of regulatory mechanisms for control of plant diseases. The University of Warwick, Warwick, UK. Available at: http://dx.doi.org/10.1007/s12571-010-0073-4 (accessed 10 May 2013).

Elphinstone, J. and Toth, I. (2007) *Erwinia chrysanthemi* (*Dickeya* spp.) The Facts. British Potato Council Publications, UK. Available at: www.potato.org.uk (accessed 14 May 2013).

FAOSTAT (2006) *FAO Statistics*. Food and Agriculture Organization of the United Nations (FAO), Rome.

Food and Agriculture Organization of the United Nations (FAO) (1996) *Guidelines for Pest Risk Analysis. International Standards for Phytosanitary Measures*. Publication No. 2. Secretariat of the International Plant Protection Convention, FAO, Rome.

Gildemacher, P.R., Maina, P., Nyongesa, M., Kinyae, P., Gebremedhin, W., Lema, Y., Damene, B., Shiferaw, T., Kakuhenzire, R., Kashaija, I., Musoke, C., Mudiope, J., Kahiu, I. and Ortiz, O. (2006) Participatory analysis of the potato knowledge and information system in Ethiopia, Kenya and Uganda. In: Sanginga, P.C., Bayer, A.W., Kaaria, S., Njuki, J. and Wettasinha, C. (eds) *Innovation Africa: Enriching Farmers' Livelihoods*. Earthscan, London, pp. 203–219.

Gildemacher, P.R., Kaguongo, W., Ortiz, O., Tesfaye, A., Woldegiorgis, G., Wagoire, W., Kakuhenzire, R., Kinyae, P., Nyogesa, M., Struik, P. and Leeuwis, C. (2009a) Improving potato production in Kenya, Uganda and Ethiopia: a system diagnosis. *Potato Research* 52, 173–205.

Gildemacher, P.R., Demo, P., Barker, I., Kaguongo, W., Woldegiorgis, G., William, W., Wakahiu, M., Leeuwis, C. and Struik, P.C. (2009b) A description of seed potato systems in Kenya, Uganda and Ethiopia. *American Journal of Potato Research* 86, 373–382.

Huarte, M. (2008) Trends in seed potato production in Latin America. Paper presented at the Global Potato Conference, 9–12 December 2008, New Delhi.

Kaguongo, W., Gildemacher, P., Demo, P., Wagoire, W., Kinyae, P., Adrade, J., Forbes, G., Thiele, G. and Fuglie, K. (2008) Farmers' practices and adoption of improved potato varieties in Kenya and Uganda. *Social Sciences Working Paper*. International Potato Center (CIP), Lima.

Kaguongo, W., Ng'ang'a, N., Muthoka, N., Muthami, F. and Maingi, G. (2010) *Seed Potato Sub-sector in Kenya: Five Year Master Plan (2009–2014)*. Study sponsored by GTZ-PSDA, United States Agency for International Development (USAID), International Potato Center (CIP) and Government of Kenya, Ministry of Agriculture. GTZ-PSDA/USAID/CIP, Nairobi.

Kaguongo, W., Maingi, G., Barker, I., Ng'ang'a, N. and Guenthner, J. (2013) The value of seed potatoes from four systems in Kenya. *American Journal of Potato Research* 91(1), 109–118.

Kinyua, Z.M., Bararyenya, A., Schulte-Geldermann, E., Obura, B.O., Kashaija, I.N., Tindimubona, S., Opio, F., Oyena, D., Rwomushana, I., Muriithi, W.M., Kinoti, J., Namugga, P., Kimoone, G., Inamahoro, M., Niko, N. and Ndayihanzamaso, P. (2011) Overcoming seed potato quality constraints to tackle food insecurity and poverty in Eastern and Central Africa in the 21st century. Paper presented at the First Association for Strengthening Agricultural Research in Eastern and Central Africa (ASARECA) General Assembly, 14–16 December 2011, Entebbe, Uganda.

MacLeod, A., Pautasso, M., Jeger, M.J. and Haines-Young, R. (2010) Evolution of the international regulation of plant pests and challenges for future plant health. *Food Security* 2, 49–70.

Ministry of Agriculture (MoA) (2012) Report on status of maize lethal necrosis disease (MLND) and general maize performance. MoA Stakeholders' Maize Tour, 2–12 July 2012. Available at: http://www.fao.org/fileadmin/user_upload/drought/docs/Maize%20Lethal%20Necrotic%20Disease%20in%20Kenya_Joint%20Assessment%20Report%20(July%202012).pdf (accessed 10 May 2013).

Nederlandse Voedsel en Warenautoriteit (2012) Netherlands Finds Potato Disease at Eight Spud Growers. By Rudy Ruitenberg, 26 April 2012. Available at: http://www.bloomberg.com/news/articles/2012-04-26/netherlands-finds-potato-disease-at-eight-spud-growers (accessed 14 May 2013).

Perrings, C., Dehnen-Schmutz, K., Touza, J. and Williamson, M. (2005) How to manage biological invasions under globalization. *Trends in Ecology and Evolution* 20, 212–215.

Raemaekers, R.H. (ed.) (2001) *Crop Production in Tropical Africa*. Directorate General for International Co-operation (DGIC), Brussels, Belgium, 1540 pp.

Schulte-Geldermann, E. (2012) Tackling low potato yields in Eastern Africa: an overview about constraints and potential strategies. In: *Seed Potato Tuber Production and Dissemination – Experiences, Challenges and Prospects*, Proceedings of the National Workshop on Seed Potato Tuber Production and Dissemination, 12–14 March 2012, Bahir Dar, Ethiopia. Ethiopian Institute for Agricultural Research (EIAR) and Amhara Regional Agricultural Research Institute (ARARI), Bahir Dar, Amhara, pp. 72–80.

Strange, R.N. and Scott, P.R. (2005) Plant disease: a threat to global food security. *Annual Review of Phytopathology* 43, 83–116.

Surman, W. (2010) Major new disease threat to potatoes. *In: Farmers Guardian 19* August 2010. Available at: https://www.fginsight.com/news/news/major-new-disease-threat-to-potatoes-6198 (accessed 15 September 2015).

Toth, K.I., van der Wolf, M.J., Saddler, G., Lojkowska, E., Hélias, V., Pirhonen, M., Tsror, L. and Elphinstone, G.J. (2011) *Dickeya* species: an emerging problem for potato production in Europe. *Plant Pathology* 60(3), 385–399.

Tsror, L., Lahkim, O., Erlich, M., Hazanovsky, B., Daniel, B., Zig, U. and Lebiush, S. (2011) Detection of *Dickeya* spp. latent infection in potato seed tubers using PCR or ELISA and correlation with disease incidence in commercial field crops under hot-climate conditions. *Plant Pathology* 61, 161–168.

Uganda Bureau of Statistics (UBOS) (2010) Statistical Abstract 10 September 2012. Available at: www.ubos.org (accessed 13 May 2013).

Waage, J.K. and Mumford, J.D. (2008) Agricultural biosecurity. *Philosophical Transactions of the Royal Society of London B* 363, 863–876.

14 Quality Seed Potato Production: Experience From the Highlands of Ethiopia

G. Woldegiorgis,[1]* G. Hailemariam,[2] B. Lemaga[3] and S. Schulz[2]
[1]*Ethiopian Institute of Agricultural Research (EIAR), Addis Ababa, Ethiopia;*
[2]*International Potato Center (CIP), Addis Ababa, Ethiopia;* [3]*Ethiopian Agricultural Transformation Agency, Addis Ababa, Ethiopia*

Abstract
Utilization of healthy planting material is a key factor to improve potato yields and to reduce the dissemination of pests and diseases. Decentralized, community-based seed production schemes have been established between 2008 and 2012 in six highland districts of Ethiopia. Pathogen-free planting material of selected varieties was multiplied on station and distributed to more than 139 seed producer cooperatives located in major seed-producing areas. More than 3390 seed potato growers were trained on clean seed potato production and postharvest management. In addition, more than 980 development agents from the Ministry of Agriculture were trained. Seed storage capacity was increased by constructing around 110 diffused light stores.

A total of 2313 t of quality seed was produced and sold to other seed and ware potato growers in different parts of the country. With the use of the improved varieties and appropriate seed production techniques, members of the seed producer cooperatives obtained yields of 32 t/ha while the national average yield is only 8 t/ha.

Experiences to date indicate that the training of farmers on the use of healthy seed tubers and other aspects of integrated disease management is crucial for sustaining the availability of clean and healthy seed tubers. Further capacity building at cooperative level is needed to strengthen business and leadership skills of farmer cooperatives to create small-scale, self-sustained seed-production enterprises. The number of farmers and cooperatives involved in seed production is expanding rapidly. To ensure that minimum seed quality standards are maintained an informal seed inspection system applying the concept of quality declared seed is currently being piloted.

Keywords: integrated disease management, quality declared seed, small-scale farmers, ware potato

14.1 Introduction

Potato (*Solanum tuberosum* L.) has a short cropping cycle and a large production per unit area in a given time. It provides more nutritious food per unit area in less time and often under more adverse conditions than other food crops due to its efficient water use. It is one of the most efficient crops in converting natural resources, labour and capital into a high quality food. Potato is a cash crop in the highlands of Ethiopia, with a great potential for raising the livelihoods of smallholders. Furthermore, potato provides

*gebregiorgis2003@yahoo.com

a cheap but nutritionally rich staple food contributing protein, vitamins, zinc and iron to the diet.

Although potato has a relatively short history of cultivation, today it is a widely grown crop in Ethiopia. It is planted in around 164,000 ha of land producing an estimated tuber yield of over 940,000 t every year (CSA, 2002). This is mainly because of the favourable climatic and edaphic conditions in many parts of the country that favour potato production. The produce can fill the gap in food supply during the 'hungry months' of September–November just before harvesting of the grain crops. In many regions of the country, it is possible to grow potato throughout the year, which offers a way to ensure a continuous supply of potato and become a reliable source of income to small-scale farmers.

Over the last century and a half, potato in Ethiopia has grown from being a garden crop grown in a few regions to a staple crop grown in many regions under different agroecological conditions. However, for nearly a century, the growth in potato production and productivity was gradual and low. The national average yield did not change for long – from about 8 t/ha to 10 t/ha (FAO, 2010), which is much lower than the world's average yield of 15 t/ha. The cultivation of this crop was characterized by traditional production and management practices that have resulted in low productivity. The major problems include: (i) shortage of good quality seed tubers; (ii) lack of adaptable and disease-resistant varieties; (iii) inappropriate agronomic practices; (iv) poor storage, transport and marketing conditions; (v) declining soil fertility; and (vi) inadequate extension services (Gebremedhin *et al.*, 2001). But the lack of high quality seed seems to explain most of the differential with the potential yields of the existing potato varieties. Increasing the availability of high quality seed at affordable prices would be a priority in order to significantly increase potato yields in the country.

14.2 Seed Potato Production

Shortage of quality seed tubers of improved cultivars is one important limiting factor to potato productivity and production. Past research experiences indicated that there is a good prospect for producing better quality seed potato on farmers' fields in collaboration with other stakeholders. Currently, few formal organizations are engaged in seed potato production as a business. However, informal seed multiplication programmes conducted by different agricultural research centres and other partner organizations were effective in disseminating improved potato varieties to farmers covering about 22% of the total seed of the improved potato used in the country (Hirpa *et al.*, 2010).

To provide disease-free planting material, a number of research activities have been conducted. Some rapid multiplication techniques (RMT, e.g. stem cuttings) were evaluated under local conditions. Tuber yield increased with increasing number of stem cuttings per hill from one to three and with closer spacing. Results revealed that the rooting abilities of stem cuttings differed with cultivar and media; fine sand was found to be the best locally available medium (Berga *et al.*, 1994). Currently, minitubers are produced under rapid multiplication for experimental purposes and for pre-basic seed in the aeroponics facilities at Bahirdar and Holetta. The conventional multiplication rate (1:3) of potato minitubers was promoted to a very high rate (1: 30) by RMT. Since the potato tuber is a living botanical organ, it loses weight and quality during storage. Ethiopia lacks proper storage facilities, and farmers are forced to sell their potato harvest at low prices during harvesting and buy seed tubers at high prices during planting. However, farmers are aware of the new seed storage technology – that is use of a diffused light store (DLS). Practical training was given to farmers in different parts of the country. Currently, 87% of the central part and 25% in the north and west are using a DLS to store their improved varieties of seed potato (Agajie Tesfaye *et al.*, 2008). Generally, better quality seed tubers are obtained with a DLS than storage in traditional dark conditions, and as a result, productivity of potato in Ethiopia increases.

14.3 Methdologies/Approaches

The activity was done in the central highlands which includes Gurage zone (Geta and Gumer districts), West Shewa zone (Tikur Inchini and Cheliya districts), South-west Shewa zone (Wonchi distict) and in East Tigray zone (Atsbi district).

Enhancing informal seed potato production and pre-scaling-up of technologies

Quality seed potato is the single most important component in improving potato productivity, considering that much of the current growth in potato production in East Africa is attributed to area expansion rather than productivity per unit area (Gebremedhin *et al.*, 2008; Gildemacher *et al.*, 2009). The main objective of the project was to improve the livelihoods of smallholder potato producers through integrated development of the seed and ware potato production and marketing chain. There were four specific objectives: (i) to increase the availability of high quality seed at an affordable price; (ii) to increase smallholder potato farmers' income by boosting potato yields through improved seed potato quality management and crop husbandry; (iii) to improve market linkages and communication between potato value chain stakeholders; and (iv) to translate the project results into national potato-sector development plans and to share the project lessons with international partners.

Stakeholder identification and consultation

The activity was implemented by the Ethiopian Institute of Agricultural Research (EIAR) in partnership with the International Potato Center (CIP) and Ministry of Agriculture (MoA).

Participatory evaluation of potato varieties

The problem of conventional research and extension has been its linear top-down approach of generating and transferring technology with little or no involvement of farmers and without proper consideration of their priorities and capacities (Kiflu and Kuma, 2002). This has brought a shift of emphasis towards participatory approaches. Improving the exchange of ideas and information among farmers, researchers and extensionists are believed to improve the technology development process for farmers, especially those in complex, diverse and marginal environments with limited resources. The Ethiopian potato programme has been conducting research to develop production technologies and to screen high yielding, widely adaptable and disease tolerant varieties for many years on the base of on-station and researcher-managed experiments which resulted in low adoption rate. This is due to lack of participation of the clients (mostly farmers) and shortage of sufficient clean seed tubers. Therefore, to encourage farmers to participate in the evaluation and selection of technologies that suit their conditions, on-farm potato variety adaptation trials were carried out using a farmer's participatory approach in the different districts of the project area. The aim was to increase farmers' awareness about the existing potato production constraints and the possible solutions at the grass-roots level, and to allow farmers to select the high yielding and disease tolerant varieties and use integrated late blight (LB) management.

Organizing farmers' groups and cooperatives

Based on results of the diagnostic survey and description of the farming systems done in the previous season, the research area was defined and farmers were selected. This was done after discussions with groups of farmers by considering the priority of community problems and objectives of the project by assessing the level of commitment of the farmer to implement the proposed plan of action. Thus, only farmers who were willing to be involved in the farmers' research group (FRG) were selected. Farmers were encouraged to allocate the plots necessary for the FRG trial and conduct seed production. Member farmers fully participated in the management of the trials: planting, fertilization, weeding and cultivation – all carried out according to research recommendations. Both farmers and researchers followed up the trials and researchers made periodic observations. In general, selection was made using the earlier set of criteria, including willingness of the farmer to be a seed potato grower, willingness to work in a group and capacity to build a DLS. Based on these criteria, 20–25 farmers were selected per site in the first season (2008). Selection and organization of the FRG continued in the following seasons using the same procedure, but with more than one FRG per site. Many farmers were willing to participate in the project after observing the varieties and the benefits of participant farmers. A total of 139 FRGs with 3390 members were organized: 65 FRGs in Gurage zone, 53 in West Shewa zone, 20 in South-west Shewa zone and one in East Tigray zone. These FRGs were organized and facilitated by the members who were most interested and motivated by the project. In general, organized farmers got additional training on potato LB management, seed and ware potato production, and storage technologies in workshops and discussions. These groups were receiving strong technical backup at a local level from the researchers and extension agents of the Bureau of Agriculture (BoA; which operates at district level) throughout the project period (Gebremedhin *et al.*, 2013).

Capacity building

Several methods were used to upgrade the capacity of the seed farmers, extension agents of the MoA and technicians from the research institutes. These methods included: (i) training of trainers; (ii) workshops for development agents; (iii) on-the-job training for researchers; (iv) demonstrations and on-site training for participant farmers; and (v) organized field days. The training included seed tuber management, postharvest management of seed, disease identification and control measures, and business skill management.

Field days and public trade and demonstration fair

The use of field days and a public trade and demonstration fair were considered key to technology dissemination in this activity. During the field days, farmers were able to show their ablity to produce clean potato seed tubers, the high yield of the improved varieties compared with the local variety, and their management skills. Moreover, farmers displayed their quality produce during the public trade and demonstration fair.

14.4 Results

Participatory evaluation of potato varieties

From the adaptation trials of different varieties across different locations it was found that varieties Gudene, Gera and Guassa followed by Jalene produced significantly higher tuber yields than the rest of the varieties included in the trails. Based on the evaluation of results the two varieties Gudene and Jalene were mostly adopted and grown widely by potato growers. From all the locations higher mean yields were obtained in the Cheliya district (Table 14.1).

In Atsbi, seven nationally released improved potato varieties (i.e. Tolcha, Digemegn, Zengena, Guassa, Gera, Jalene and Gudene) were evaluated under farmers' conditions. The three best performing varieties were Jalene, Gera and Gudene (Fig. 14.1). They were selected for their disease resistance, market quality/demand and high yields: Gudene (38.2 t/ha), Gera (38.8 t/ha) and Jalene (39.0 t/ha), respectively. This yield is very high compared with the regional average of 8.0 t/ha. Hence, participatory variety selection results have revealed the possibility of increasing farmers' income more than fourfold with the use of improved varieties compared with the local varieties

Table 14.1. Average yield (t/ha)[a] of potato variety adaptation trial in different districts. (From unpublished Common Fund for Commodities (CFC) project progress report.)

Variety	Yields (t/ha) in district: Wonchi	Cheliya	Tikur Inchini	Overall mean (t/ha)
Jalene	29.83a	35.77a	22.23bc	29.28a
Tolcha	14.37c	31.30a	16.27c	20.64b
Zengena	19.57bc	37.67a	16.70c	24.64ab
Gera	27.90ab	33.93a	28.93ab	30.26a
Guassa	28.17a	35.37a	26.83ab	30.12a
Digemegn	19.47c	35.97a	26.67ab	27.37ab
Gudene	32.47a	27.07a	32.73a	30.76a
CV (%)[b]	19.26	24.77	20.04	25.87

[a]Means followed by the same letter (a, b, c) within the rows are not significantly different ($P \leq 0.05$).
[b]CV (%), coefficient of variation.

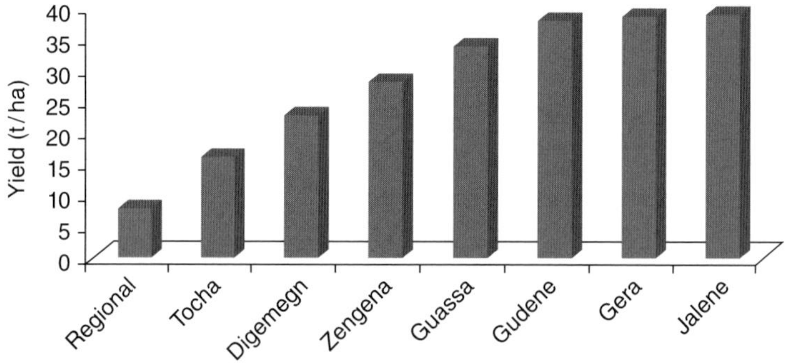

Fig. 14.1. Average yield (t/ha) of potato participatory variety selection, under farmers' conditions in Atsbi. (From Gebrehiwot *et al.*, 2013.)

that are susceptible to diseases and pests. Based on the results of the adaptation variety trials the varieties Gudene and Jalene were considered for seed production by the organized cooperatives.

Organizing farmers' groups and cooperatives

As stated in section 14.3 a total of 139 FRGs with 3390 members were organized: 65 FRGs in Gurage zone, 53 in West Shewa zone, 20 in South-west Shewa zone and one in East Tigray zone. These FRGs were organized and facilitated by the members who were most interested and motivated by the project.

In Atsbi the FRG that was formed to undertake the potato adaptation trial, which had 13 members, was changed into a cooperative. The seed producers' cooperative, Shewit Irrigation Cooperative, was established in 2007 to facilitate the smooth functioning and marketing of improved seed potato tuber in the region. The establishment of this cooperative has: (i) helped farmers reduce transaction costs such as for collection, transportation and storage of seed potato tubers; (ii) improved the bargaining power of seed producers; and (iii) facilitated the promotion and sale of seed potato. Membership of the cooperative increased from 13 to 34. Out of these, three were female-headed households. The initial capital of the cooperative was only 3750 Birr, but the capital has now grown to 1.5 million Birr, without considering individual assets (Table 14.2) (Gebrehiwot *et al.*, 2013).

According to the cooperative members, organizing them into groups has: (i) enhanced their strength; (ii) eased capacity building and support from non-governmental organizations (NGOs) and government organizations; (iii) reduced the gap in technical knowledge among members due to practical self-support in their field; (iv) improved market access for produce and inputs by reducing transaction costs; and (v) improved the food security and livelihoods of members.

Production of pre-basic and basic seed

The most feasible way in which the growing demand for potatoes can be satisfied is through increased productivity (Gildermacher *et al.*, 2009). There are known agronomic technologies that can improve potato productivity and among them quality seed is a critical component (Endale *et al.*, 2008; Gildemacher *et al.*, 2009). However, this basic component in modern potato production is in short supply (Wagoire *et al.*, 2005; Endale *et al.*, 2008) and expensive, accounting for 40–50% of potato production costs (Wagoire *et al.*, 2005). Quality seed potato is the single most important component in improving potato productivity, considering that most of the current growth in potato production in East Africa is attributed to area expansion rather than productivity per unit area (Gebremedhin *et al.*, 2008; Gildemacher *et al.*, 2009).

In most areas of Ethiopia, among local farmers, it is a common practice to save the small and inferior tubers as seed that they cannot normally sell for consumption with inherent seed degeneration because of seed-borne diseases such as viruses and bacterial wilt. It is believed that this practice has contributed to the build-up of a high level of viral diseases in the locally grown potato cultivars in Ethiopia. However, now in addition to seed tuber production going on in the research stations, informal seed production in the farmers' fields has become a major practice which helps farmers get relatively clean and healthy seed tubers in sufficient quantities for the next cropping season and the farmers have also benefited from selling their seed tubers to other users.

Quality seed of an improved potato variety is a key to increase the productivity of a potato crop. The genetic potential and other traits of a potato variety are determined or manifested by the use of healthy seed (Endale *et al.*, 2008). Unavailability of healthy seed tubers in the required quantity and quality is probably the most important factor contributing to the low yield in Ethiopia (Berga and Gebremedhin, 1994). The major focus of this activity was to increase seed and ware potato production and marketing to improve the livelihood of smallholder potato farmers who are mostly short of cash. A modest increase in cash income through

Table 14.2. Capital of Shewit Irrigation Cooperative as of November 2012. (From Gebrehiwot *et al.*, 2013.)

Description of capital[a]	Amount (Birr)
DLS (10 m × 6 m) at 70% depreciation	105,000
DLS new (12 m × 6 m)	250,000
Office and multipurpose hall	350,000
Capital goods (2000 eucalyptus trees @ 35 Birr per piece)	70,000
Saving	100,000
Saving in the form of seed	200,000
Credit to members for inputs	300,000
Collections from various sources	180,000
Total capital	1,555,000

[a]DLS, diffused light store.

improved potato farming can have a major impact on the quality of life of the whole household and the entire community in relation to health care, nutrition and education. In Ethiopia, no formal seed system exists to produce pre-basic and basic seed under a certification scheme. Research institutes produce similar quality materials to fill the gaps in seed demand, and produce and distribute the basic seed of potato. Informally, the basic seed is distributed through the farmer-based seed system where the recipient farmers multiply it at least once before it is distributed to ware potato growers. According to Berga *et al.* (2009), the system has many different advantages: (i) it increases the quantity of seed that is available to ware potato farmers with little waste of high-quality basic seed; (ii) it brings other players into the seed potato production process and reduces public monopoly; (iii) it eases the burden on public institutes in seed potato production and distribution; (iv) it is a precursor to the private-sector-led seed potato production scheme; (v) it encourages public–private partnerships in seed production; and (vi) it creates new enterprises among farmers who adopt and invest in seed potato production.

The informal seed system is the major alternative option to the formal seed systems in Ethiopia. The system largely involves on-farm seed selection and multiplication by the farmer and seed exchanges among farmers. In addition, the use of planting material as seed saved from previous harvests is not regulated and controlled. In Ethiopia, farmer-based seed production (FBSP) has been practised for seed potato. In this approach, farmers are provided with relatively clean starter planting material of the improved varieties to multiply and sell at an affordable price. The farmers are also trained by research institutes to help them build skills on basic seed potato production, including negative selection (roguing out infected plants) and positive selection (marking clean plants with a stake). In the central highlands, farmers organized in FBSP serve as sources of seed potato tuber for most of the potato-growing areas in the country. According to Berga *et al.* (2009), seed potato production by FBSP has been growing since 2004 and peaked in 2008.

Since the start of the project in 2008 and up to 2012, a total of 93.3 t of clean seed has been distributed from the Holetta Agricultural Research Centre (HARC) to the seed-producing groups established in six woredas (districts) where the project has been implemented. In addition, 49.8 t of seed has been used as revolving seed collected from the farmers to be used to scale up technology for the newly formed FRGs within the woreda. From this starter seed, 2313 t of relatively healthy seed – mainly from the varieties Gudene and Jalene – were produced to be used as seed for the producing farmers themselves (Table 14.3; note results for only five of the six districts are shown). The seed was distributed for sale in different parts of the country for ware potato production. It allows farmers to get a reasonable income to improve their livelihoods while contributing to the availability of good quality seed. Consequently, the project has contributed to solving the problem of seed in the country in general and in the project areas in particular.

Minituber production

Production of the healthy planting materials/stocks is done at HARC using different RMT for the selected varieties. Multiplication was carried out in insect-proof screenhouses. Subsequent propagation is carried out using rooted cuttings in the screenhouse to obtain enough planting material before they are planted in the open field. A total of 227,333 minitubers of several clones and released varieties were multiplied in screenhouses, which include Awash (1710 minitubers), Gudene (39,518 minitubers), Jalene (13,179 minitubers) and Belete (145,870 minitubers) (Table 14.4). Source materials for this minituber production were either *in vitro* plantlets or smaller minitubers produced from the previous season in the aeroponics units. These materials have been introduced into the centre's seed production programme and have played a major role in regenerating the stock materials.

Table 14.3. Area planted and seed tubers produced by seed farmers, 2008–2012.

	2008		2009		2010		2011		2012		Total	
Distrct	Area (ha)	Yield (t)	Area (ha)	Yield (t)	Area (ha)	Yield (t)	Area (ha)	Yield (t)	Area (ha)	Yield (t)	Area (ha)	Yield (t)
Cheliya	3.5	134.36	4.05	98.14	1.97	45.2	5.5	229.88	5.18	221.57	20.2	729.1
Tikur Inchini	2.6	40.07	2.75	36.81			4	168.74	4.5	136.9	13.85	382.53
Wonchi	5.5	182.11					4	194.96	1.78	50.58	11.28	427.63
Gumer	0.5	25.88	2.16	91.99	3.25	125.6	7.25	161.99	1	10.21	14.16	415.67
Geta	0.25	8.5	0.51	21.61	2.8	101.55	6	120.78	3.11	105.79	12.67	358.23
Total	12.4	390.9	9.47	248.55	8.01	272.35	26.8	876.35	15.6	525	72.15	2313.15
Average seed tubers (t/ha)											32.06	

Table 14.4. Numbers of minitubers produced in screenhouses planted with *in vitro* plantlets and small-sized minitubers produced from the previous season in the aeroponics unit. (From unpublished CFC project annual report.)

	Year					
Variety/clones	2008	2009	2010	2011	2012	Total
Different clones	15,041	–	1,692	6,043	–	22,776
Jalene	3,760	–	7,440	1,979	–	13,179
Guassa	1,200	–	1,080	–		2,280
Gudene	1,800	–	1,352	9,231	27,135	39,518
Belete	–	–	–	42,871	102,999	145,870
Awash	228	228	–	1,254	–	1,710
Gorebela	1,264	–	–	–	–	1,264
Zengena	516	–	–	–	–	516
Tolcha	220	–	–	–	–	220
Total	24,029	228	11,564	61,378	130,134	227,333

Recently HARC has also been multipling minitubers of the recently released potato varieties using aeroponics (Table 14.5).

Table 14.5. Number of minitubers produced under aeroponics at Holetta Agricultural Research Centre (HARC) in 2011/12. (From unpublished CFC project annual report.)

	Year of production		
Variety	2011	2012	Total
Belete	38,390	67,078	105,468
Gudene	864	1,278	2,142
Awash	4,659	–	4,659
Jalene	233	2,362	2,595
Total	44,146	70,718	114,864

DLS for seed quality enhancement

Potato is a source for high income and a better diet for small-scale rural farmers. Increasing consumption of potato demonstrates its importance for food security; however, seed storage is not easy, especially because temperature fluctuations and excessive light exposure cause its deterioration, jeopardizing availability of quality seed materials. It is crucial to guarantee a good seed storage method for communities that heavily depend on potato production. An efficient practice to overcome the storage problems of potato seed material is the use of a DLS, and any existing on-farm storage can be adapted for this use. If a farmer is able to store his or her own seed potatoes in good conditions, it enhances the probability of a good harvest the following season. The DLS method is based on the use of indirect natural light and good ventilation or air flow instead of low temperature to control excessive sprout growth and associated storage loss. DLS is a low-cost method of storing seed potatoes which has been found to extend their storage life and improve their productivity.

In the 'wealth creation' project, construction of DLSs by the seed growers in the project area was one of the activities that was successfully implemented. Since the start of the project a total of 110 DLSs have been constructed by seed multipliers with technical support from the MoA and EIAR. This is more than 1222% of the target set at the outset (which was nine). Before construction of a DLS, a 1–2 day training session was given to the farmers and development agents regarding the use, construction and maintenance of the store.

Six big DLSs with a capacity of 25–30 t, one in each of the project implementation districts, were constructed for communal use with the help of the project funding. During the construction, the project provided the corrugated sheets of iron for roofing and the nails, EIAR allocated skilled personnel for construction and the farmers provided the timber and labour so that they would feel it is their own property to use and maintain for a longer period of time.

Farmers found use of the DLS was a low-cost method of storing seed potatoes which has been found to extend their storage life and improve their productivity, providing a new opportunity for poor farmers to conserve quality seeds. It was observed that farmers who received improved potato varieties and healthy seed from the seed growers constructed their own DLS.

Seed quality assurance and seed inspection

Good quality planting material is one of the foremost prerequisites of a healthy and successful crop production venture because infected planting materials may disseminate serious pests including disease-causing pathogens. There are many seed-borne diseases of potato recorded in Ethiopia but the most important are viruses, bacterial wilt and LB. In Ethiopia for instance, bacterial wilt was spread to many of the regions in just a few years through infected seed tubers from Shashemene, a known hotspot area for bacterial wilt (Dereje and Gebremedhin, 2012). To avoid this kind of problem phytosanitary measures should be incorporated into the seed production and distribution system to ensure high quality disease-free planting materials and such a system should be properly designed to suit the circumstances of potato producers that exist in the country (Adane *et al.*, 2010).

In the ‘wealth creation’ project healthy planting material was distributed to the project participants organized into farmers’ seed producers groups or cooperatives. The seed was inspected when grown at HARC, and indexing for both virus and bacterial wilt was done occasionally. In collaboration with extension experts, the research team has also been frequently inspecting the farmers’ seed tubers. Moreover, in a few districts a survey was conducted by a virologist. It was observed that of the seed multipliers only a few of the farmers’ samples were positive for potato leaf roll virus (PLRV) and some of the samples were also positive for potato virus Y (PVY).

In collaboration with the CIP/United States Agency for International Development (USAID) project, Quality Declared Planting Materials (QDPM) of potato was introduced in Ethiopia. The QDPM scheme is intended to guide the production of clean, disease-free planting materials of vegetatively reproduced crops (FAO, 2010). Its goal is mainly to raise the physiological and phytosanitary quality of planting materials available to smallholders, and hence increase agricultural production and productivity. It is meant to be implemented primarily by seed producers at community level or field extension workers. Practical QDPM protocols and standards are used to allow easy monitoring and verification of the production and distribution process. In the absence of formal seed quality control systems the QDPM helps to produce relatively good quality seed to target users. It links the activities of research and extension to seed multiplication activities of smallholders. A field guide for inspection of QDPM was developed which was adapted from the FAO publication, *Quality Declared Planting Materials: Protocols and Standards for Vegetatively Propagated Plants* (FAO, 2010).

During consecutive meetings and a workshop involving the different stakeholders it was agreed that the quality assurance for seed inspection would be managed by a special woreda-level inspection committee (WERCOM) (where a woreda is equivalent to a district) that would be assisted by a committee of each farmer group/cooperative (COOPCOM). The system relies on the continuous supply of source seed from trustworthy sources (e.g. research centres, with technical training given by researchers and technical backstopping handled by the extension services) and finance available from various sources such as microfinance, banks and NGOs.

The woreda committee (WERCOM) consists of five members: (i) a seed agronomist from the district BoA (as the chairperson); (ii) a plant protection/agronomy expert from the district BoA (as the secretary); (iii) a horticulturist from research working in the district (optional) (as a member); (iv) the development agent closely working with the producer (as a member);

and (v) one elected representative of the cooperative/PLC (as a member). Cooperative/PLC representatives (in one woreda) elect one member to represent them, where the representation operates in rotation and he/she serves for a year. Decisions made by the WERCOM are taken by majority vote, except when research is not represented when the chairperson has the casting vote. The cooperative committee (COOPCOM) inspects all seed potato fields of members of the cooperative. The COOPCOM consists of three members elected by the members of the cooperative. It is suggested that at least one woman is included in the inspection team. Whenever necessary, decisions of the COOPCOM are taken by majority vote.

The two inspection committees carry out two fields and one postharvest inspection. The first field inspection (6 weeks after planting) is done by COOPCOM, the second field inspection (at flowering) and the postharvest inspection is done jointly by both committees.

Dissemination of technologies for increased productivity

Seed potato dissemination and promotion approaches in this project include: field days, open days, trade fairs and demonstrations. Between 2008 and 2012, 17 'field days' were held in project intervention areas to develop links between seed producers and ware potato growers in the project areas. Seven of the field days were organized by the farmers themselves (the others being organized by researchers in collaboration with other stakeholders). The field days were done during the vegetative growth stage and at the harvesting stage of the crop to create awareness of the different stages, including tolerance of the released varieties to LB and potential to yield using the improved agronomic packages. Different stakeholders, including the politicians, MoA research staff, researchers, district administrative staff and neighbouring farmers participated. Also, the project facilitated exchange visits involving these stakeholders and representatives of farmers' groups and extension staff from the different project districts.

To facilitate the promotion of quality seed potatoes the project helped the trained farmer group (i.e. the FRG) to buy or receive seed potato tubers. Some farmers were able to compare demonstration plots with seed from positive selection (PS) and farmer selection (FS) which helped the extension experts in collaboration with the project to convince and/or show the advantage of healthy seed over degenerated seed. The project also facilitated the attendance of more than eight trained extension experts of the MoA in field/open days. Representatives of trained seed potato multipliers were also facilitated to attend the field days held beyond their districts to share experiences of other farmer groups.

Ten demonstrations (two each in Cheliya, Geta, Gumer and Wonchi and one each in Tikur Inchini and Degem – Degem was not part of the initial project) on potato utilization were held and attended by 1050 people (894 females/156 males), composed of participating and non-participating farmers, agricultural development agents and health extension officers. The farmers who attended the demonstration on utilization were impressed by the technologies shown and promised to put into practice what they learned in their own households as well as to teach others.

14.5 Discussion and Recommendations

Farmers reported that the intervention enabled them to produce seed potatoes during the month of June that resulted in the potato crop being harvested in October, a month when food insecurity is usually experienced in the project areas. This bridging of the food insecurity gap therefore becomes the main achievement of the project as far as farmers are concerned.

Farmers witnessed this by saying that:

> We have been able to increase our food stock in the month of October because of potato. The varieties introduced have

enabled us to produce the crop during the high rainfall periods. We are better off in terms of food than we were before the introduction of the new potato varieties.

In Atsbi the capacity of the cooperative has increased to provide input and credit service provision to members, as the capital of the cooperative has increased from 3750 Birr in 2008 to 1.5 million Birr in 2012. It has been also reported that 29 of the 34 members have graduated from Safety net (i.e. farmers who are receiving food aid); the remaining members are candidate graduates. This is considered as one of the outcomes of the project.

Household case studies conducted with individual members of the cooperative in the different project locations showed there were many success stories and testimonies indicating that the potato production has improved the household food security, nutrition quality, income diversification and overall quality of living of the members.

Partnership and its significance to quality seed potato production

The project has been working in close collaboration with other potato projects that are being implemented in similar project intervention areas. The Common Fund for Commodities (CFC) project works closely with a USAID-funded project in Ethiopia. Among the inter-project partnerships, the project has made various contributions to this similar project and in return has received various types of in-kind support, including in the area of DLS, minitubers, logistics during planting and harvesting, and training. The two sister projects constructed one aeroponics unit each at HARC. They share costs of common equipment such as a generator for backup power, pumps and containers for spraying nutrient solutions, and training. The USAID-funded Ethiopia project has been a market for the CFC-funded project seed producers. Such partnership has helped the improvement and availability of healthy seed tubers in the country at both the research level and the farmers' level. The other partners in the project were all government institutions, including the woreda administration of the MoA, women and youth affairs, information affairs, and a private seed grower, Solagrow PLC. All these institutions supported the production of healthy seed potato in their respective woredas. In general it was learned that:

- With proper technical advice and backstopping, farmers are able to produce high quality seed through their own management.
- Farmers can handle generation 1 minitubers, which were produced under aeroponics.
- Farmers are willing to invest in clean seed, as exhibited by the high number of trained seed multipliers who are currently producing seed commercially.
- Use of quality seed leads to increased tuber yields at farm level (average of 32.12 t/ha).
- Farmers are willing to invest in the DLS as indicated by the number of DLSs constructed without support.
- Use of different media helps to create awareness on the use of improved varieties and healthy seed (as seen from the number of farmers demanding the technologies).
- Working in partnership is important for technology dissemination.
- Farmers are keen on improved technologies (i.e. other agricultural technologies).
- Farmers are not interested in 'Select the Best', which takes more time, and prefer improved seed.
- Organizing farmers into FRGs or cooperatives helps to reach more farmers in technology dissemination.

References

Adane, H., Meuwissen, M.P.M., Agajie, T., Lommen, W.J.M., Lansik, A.O., Tsegaye, A. and Struik, P.C. (2010) Analysis of seed potato systems in Ethiopia. *American Journal of Potato Research* 87, 537–552.

Agajie Tesfaye, Bedane, K., Yirga, C. and Woldegiorgis, G. (2008) Potato socioeconomics and technology transfer. In: Woldegiorgis, G., Gebre, E. and Lemaga, B. (eds) *Root and Tuber Crops: the Untapped Resources*. Ethiopian Institute of Agricultural Research (EIAR), Addis Ababa, pp. 131–153.

Berga, L. and Gebremedhin, W. (1994) Prospects of seed potato production in Ethiopia. In: Herath, E. and Lemma Dessalegne (eds) *Horticulture Research and Development in Ethiopia.* Proceedings of the Second National Horticulture Workshop, 1–3 December 1992, Addis Ababa. Ethiopian Institute of Agricultural Research (EIAR), Addis Ababa, pp. 101–119.

Berga, L., Woldegiorgis, G., Teriessa, J. and Bereke-Tsehai, T. (1994) Potato agronomy research. In: Herath, E. and Lemma Dessalegne (eds) *Horticulture Research and Development in Ethiopia.* Proceedings of the Second National Horticultural Workshop, 1–3 December 1992, Addis Ababa. Ethiopian Institute of Agricultural Research (EIAR), Addis Ababa, pp. 254–275.

Berga, L., Kakuhenzire, R., Gildemacher, P., Borus, D., Woldegiorgis, G., Barker, I., Low, J. and Ortiz, O. (2009) Current status and opportunities for improving the access to quality potato seed by small farmers in Eastern Africa. A paper presented at the 15th Triennial of the Symposium of the International Society of Tropical Root Crops, 2–6 November 2009, Lima, Peru.

Central Statistical Agency (CSA) (2002) *Statistical Report on Area and Production of Crops*. Part IIA. CSA, Addis Ababa.

Dereje, G. and Gebremedhin, W. (2012) Bacterial wilt of potato: an emerging threat to Ethiopian potato industry. Paper presented at the National Seed Potato Workshop held at Amhara Agricultural Research Institute (ARARI), 11–13 March 2012, Bahir Dar, Ethiopia.

Endale, G., Woldegiorgis, G. and Berga, L. (2008) Potato seed management. In: Woldegiorgis, G., Endale, G. and Berga, L. (eds) *Root and Tuber Crops: the Untapped Resources*. Ethiopian Institute of Agricultural Research (EIAR), Addis Ababa, pp. 53–77.

Food and Agriculture Organization of the United Nations (FAO) (2010) *Quality Declared Planting Materials: Protocols and Standards for Vegetatively Propagated Plants*. FAO Plant Production and Protection Paper 195. FAO, Rome.

Gebrehiwot, H., Woldegiorgis, G., Eyasu, A., Berga, L. and Steffen, S. (2013) Participatory potato seed production: experiences from Tigray. In: *Proceedings of the National Workshop on Seed Potato Tuber Production and Dissemination*, 12–14 March 2012, Bahir Dar, Ethiopia. Ethiopian Institute of Agricultural Research (EIAR), Addis Ababa, pp. 208–218.

Gebremedhin, W., Endale, G., Kiflu Bedane and Bekele Kassa (2001) *Country Profile on Potato Production and Utilization in Ethiopia*. National Potato Research Programme Report. Ethiopian Agricultural Research Organization, Holetta Agricultural Research Center, Addis Ababa.

Gebremedhin, W., Gebre, E. and Lemaga, B. (2008) Overview of trends in root and tuber crops research in Ethiopia. In: Woldegiorgis, G., Endale, G. and Berga, L. (eds) *Root and Tuber Crops: the Untapped Resources*. Ethiopian Institute of Agricultural Research (EIAR), Addis Ababa, pp. 1–5.

Gebremedhin, W., Kassaye, N., Atsede, S., Abebe, C. and Berga, L. (2013) Participatory potato seed production: experiences from west and southwest Shewa, and Gurage zones. In: *Proceedings of the National Workshop on Seed Potato Tuber Production and Dissemination*, 12–14 March 2012, Bahir Dar, Ethiopia. Ethiopian Institute of Agricultural Research (EIAR), Addis Ababa, pp. 162–182.

Gildemacher, P.R., Kaguongo, W., Ortiz, O., Tesfaye, A., Woldegiorgis, G., Wagoire, W.W., Kakuhenzire, R., Kinyae, P.M., Nyongesa, M., Struik, P.C. and Leeuwis, C. (2009) Improving potato production in Kenya, Uganda and Ethiopia: a system diagnosis. *Potato* 52, 173–205.

Hirpa, A., Meuwissen, M.P.M., Tesfaye, A., Lommen, W.J.M., Lansink, A.O., Tsegaye, A. and Struik, P.C. (2010) Analysis of seed potato systems in Ethiopia. *American Journal of Potato Research* 87, 537–552.

Kiflu, B. and Kuma, B. (2002) Farmer participatory research: an overview. In: Keneni, G., Gojam, Y., Bedane, K., Yirga, C. and Dibabe, A. (eds) *Towards Farmers Participatory Research: Attempts and Achievements in the Central Highlands of Ethiopia*. Proceedings of Client-oriented Research Evaluation Workshop, 16–18 October 2001, Holetta Agricultural Research Center, Holetta, Ethiopia. Ethiopian Institute of Agricultural Research (EIAR), Addis Ababa, pp. 2–11.

Wagoire, W.W., Kakuhenzire, R., Kashaija, I.N., Lemaga, B., Demo, P. and Kimoone, G. (2005) Seed potato production in Uganda: current status and future prospects. *African Crop Science Society Conference Proceedings* 7, 739–743.

15 A Possible Pathway for Developing Formal Seed Potato Production in Sub-Saharan Africa: A Case of Uganda National Seed Potato Producers' Association (UNSPPA)

R. Kakuhenzire,[1]* S. Tindimubona,[2] I.N. Kashaija[3] and B. Lemaga[4]

[1]*International Potato Center (CIP)-Tanzania, Mbeya, Tanzania;* [2]*Uganda National Seed Potato Producers' Association (UNSPPA), Kabale, Uganda;* [3]*National Agricultural Research Organisation, Entebbe, Uganda;* [4]*Ethiopian Agricultural Transformation Agency, Addis Ababa, Ethiopia*

Abstract

Seed accounts for 40–50% of the cost in potato production, however, it is one of the most neglected inputs among smallholder potato farmers in sub-Saharan Africa partly due to lack of awareness, poverty, subsistence agriculture and seed inadequacy. Recognizing this, the potato programme of the National Agricultural Research Organisation (NARO) in Uganda with the International Potato Center (CIP) started an initiative in 1996 to engage smallholder farmers in south-western Uganda to produce and distribute improved seed to other farmers in an organized manner. Assessment of this initiative in 2010 revealed evolution of a profitable, self-sustaining and farmer-managed Uganda National Seed Potato Producers' Association (UNSPPA) and efficient use of basic seed that was used to produce more seed than before. Data showed a gradual increase in basic seed procured by farmers from NARO for bulking. There was also gradual transfer of seed production, distribution and quality control costs from government institutions to farmers. Nevertheless, basic seed was inadequate and cheaper than seed from farmers. The supply gap for basic seed offers opportunities for expanding production by NARO or any investor. The study showed that farmers after joining UNSPPA adopted improved production technologies, had higher family income and better welfare. This experience from Uganda shows that the seed supply gap can be narrowed domestically and locally with high financial and social benefits. However, this is possible with consistent logistical and technical support and farmer capacity strengthening in agribusiness.

Keywords: clean seed, decentralised seed potato systems, quality declared seed

*r.kakuhenzire@cgiar.org

15.1 Introduction

Establishing availability of quality seed among smallholders in sub-Saharan Africa (SSA) has been a daunting task particularly for vegetatively propagated crops. Attempts by national governments and international agencies to improve seed availability in post-independence Africa put most emphasis on cereals and grain legumes rather than most vegetatively propagated food crops (Lowaars and Marrewijk, 1995; J.D. Alumira, unpublished data, 2000). Potato was most neglected in spite of its potential as a quality food resource because it was considered a non-African crop with limited agroecologies where it could be profitably grown. Seed accounts for 40–50% of the cost of potato production (Beukema and van der Zaag, 1990; Gildermacher *et al.*, 2011). Improving seed potato quality and availability would be one way of improving potato productivity and profitability among communities where this crop is a mainstay (Tindimubona *et al.*, 2000; Gildermacher *et al.*, 2011).

The global seed industry in the developed world has grown tremendously during the last 40 years from small to multinational corporations (Howard, 2009). Similar growth has not been realized in most developing countries and seed supply systems especially for roots and tubers are still traditional, rudimentary or non-existent. Where some seed is formally produced, it is inadequate, the supply is erratic and sometimes quality is compromised. In some countries, quality declared planting material may be accepted.

Attempts by some countries in SSA to adopt formal seed potato systems have not yielded sustainable results because most initiatives were funded and managed by government or donor projects with the hope of self-sustainability. After the end of such support, seed development initiatives would cease due to lack of entrepreneurial and commercial predisposition (Hakiza *et al.*, 2000; Rukuba *et al.*, 2000; Turiho-Ahabwe, 2000). These factors and many others led to poor availability of quality seed of vegetatively propagated crops in SSA with farmers resorting to recycled self-saved planting materials.

Recognizing these weaknesses in Uganda, the national potato programme with the International Potato Center (CIP) started an initiative to encourage a few farmers to produce quality seed potato in an informal but organized setting. The main objective of this was to build the capacity of smallholders to produce and distribute seed at the community level and eventually develop this into a formal seed scheme (Tindimubona *et al.*, 2000). Consequently, an informal seed potato production system, managed by farmers but supported by agricultural research and extension services, was piloted in Kabale and Kisoro districts in south-western Uganda between 1996 and 1998. Later, the participating farmers formally registered the Uganda National Seed Potato Producers' Association (UNSPPA) in 1999 as a company limited by guarantee without share capital (Tindimubona *et al.*, 2000).

Since its formation, the performance and sustainability of UNSPPA has never been evaluated. It is against this background that this study was undertaken and its objectives were to: (i) assess the return on investment in UNSPPA; (ii) document the potential of local farmers to produce and distribute quality seed sustainably; and (iii) evaluate the potential of such an initiative to reduce rural household poverty. The results of this study may form empirical evidence for investing in smallholders to develop a formal seed potato system in a developing country in SSA.

15.2 Methodology

Primary and secondary data were used to evaluate the UNSPPA development initiative. Data were collected at UNSPPA as an entity and from individuals that constitute it. Data were obtained from reports, company documents, books of accounts and other materials the association holds and from members using a semi-structured questionnaire in a personal interview, addressing variables that may influence a household's potato production decisions. At the time of the study, UNSPPA consisted of 40 members, however, only 27 that had been members for at least 4 years were interviewed.

Among them, 24 were from Kabale, two from Kisoro and one from Kanungu districts. The majority were from Kabale district because it produces more than 45% of potato consumed in Uganda.

15.3 Data Analysis

Financial data obtained at UNSPPA-level were analysed using MS Excel. Methods for evaluating streams of benefits and costs of the association between 1996 and 2010 such as the net present value (NPV) and the internal rate of return (IRR) were computed. The data collected at farmer level were analysed using SPSS complemented with STATA and MS Excel. Descriptive statistics, gross margin analysis and regression analyses were used to discover interactions between various factors that influence household decisions in seed potato production.

15.4 Results and Discussion

Household demographic characteristics

The 27 surveyed families consisted of 239 people having nine people per household with 54.8% males and 45.2% females (Table 15.1). Among family members, 13.8% were less than 9 years old while 86.2% were older (Table 15.1). This has an implication on agricultural family labour and production as older people may provide labour. The majority of the children among UNSPPA members were in primary to high school, 3.3% in colleges and 10.9% in universities (Table 15.1). Therefore, UNSPPA members are heavily investing in education. There were at least two people per UNSPPA household engaged in formal employment. If family UNSPPA members were to depend on this income, the dependency ratio would be very high with serious consequences on household welfare.

Farmer involvement in potato and seed potato production in Uganda

Data from the survey show that most UNSPPA members had grown potato for more than 5 years before joining the association and more than 60% had grown potato for more than 20 years. Analysis by gender composition of UNSPPA members showed that 77.8% were men and 22.2% were women. Among the women, 60% joined UNSPPA between 2001 and 2009. Overall, 40% of the present membership joined UNSPPA between 1996 and 2000. This indicates that most of the founders of the association are still active. UNSPPA is male-dominated, however, 48.1% of members consult their spouses on seed potato production decisions while in about 20% of farm families women take the major decisions for seed production.

Table 15.1. Household composition among members of the Uganda National Seed Potato Producers' Association (UNSPPA) in 2010.

	Household composition		
Demographic variable	Numbers	Mean ± standard deviation	Percentage of total
Household size	239	8.9 ± 3.3	100.0
Males	131	4.9 ± 2.4	54.8
Females	108	4.0 ± 20	45.2
Dependents less than 9 years old	33	1.2 ± 1.6	13.8
Dependents at 9 or more years old	206	7.6 ± 2.8	86.2
In primary to high school	108	4.0 ± 3.3	45.2
College	8	0.3 ± 0.6	3.3
University	26	1.0 ± 1.1	10.9
Employed	60	2.2 ± 1.8	25.1

Land ownership and utilization among seed potato producers in Uganda in 2010

The average land holding among UNSPPA members was 26.3 acres with a minimum of 3.3 acres and a maximum of 60 acres. These are above the land ownership criterion (≥ 2 acres) to be a member. Land use during the January–June 2010 cropping season showed a high proportion of land was fallowed, a desirable practice in seed potato production. A high percentage of members' land is used for forestry (Fig. 15.1) indicating that these farmers have more land than they need for food production. Among food crops, potato was allocated more land than any other crops while seed took more land (4.2%) than ware potato (3.6%) (Fig. 15.1).

The size of land in any given cropping season allocated to seed potato production among UNSPPA members was influenced by land size (23.8%), basic seed availability (22.5%) and capital for procurement of other agro-inputs (22.5%). Other influences included labour availability, expected harvest, future seed price and food needs and these accounted for 18.8%, 6.3%, 5.0% and 1.3% of members' seed production decisions, respectively.

Member investment in seed potato production in 2009 and 2010

Capital investment by UNSPPA members in 2009 and 2010 showed that farmers spent most money on seed, labour and agrochemicals, respectively (Table 15.2). The high percentage of costs being spent on seed compared with other inputs in potato production is similar to what is reported elsewhere (Beukema and van der Zaag, 1990). Fertilizers accounted for more than 50% of agrochemicals used in seed potato production (Table 15.3). This is imperative for enhancing yield in order to reduce production cost per unit weight of seed. The average labour cost was 31% of expenses indicating that these farmers provide local employment in their communities.

Profitability of quality seed potato production in an informal seed scheme

The production cost of seed potato by UNSPPA farmers between 2009 and 2011 was on average USh552.6/kg and the minimum sale price was USh900 (Table 15.4). Thus, seed potato producers have a large gross

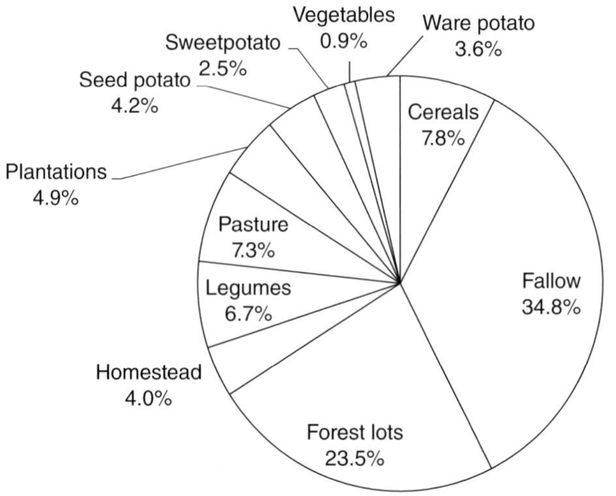

Fig. 15.1. Land utilization among members of the Uganda National Seed Potato Producers' Association (UNSPPA) in the January–June 2010 cropping season.

Table 15.2. Broad costs (USh)[a] per hectare for private seed potato production in Uganda in 2009 and 2010.

	Expenses per hectare					
	2009		2010			
Item	Cost (USh)	Percentage	Cost (USh)	Percentage	Average cost	Percentage
Seed potato	3,629,065	47.2	3,732,145	44.8	3,680,605	46.0
Farm labour	2,679,306	34.9	2,599,442	31.2	2,639,374	33.1
Agrochemicals	1,378,381	17.9	2,000,379	24	1,689,380	21.0
Total	7,686,751	100	8,331,966	100	8,009,359	100.0

[a]US$1 = USh2177.

Table 15.3. Percentage expenditure on various agrochemicals used in seed potato production in Uganda among seed producers in 2009 and 2010.

	2009		2010			
Input	Input cost (USh/ha)[a]	Percentage	Input cost (USh/ha)[a]	Percentage	Average input cost (USh/ha)[a]	Percentage
Herbicides	0	0	5,000	0.07	2,500	0.0
Foliar fertilizer	36,000	0.5	84,000	1.2	60,000	0.8
Malathion powder	147,800	1.7	198,400	2.8	173,100	2.3
Insecticides	405,500	4.7	267,450	3.8	336,475	4.3
Empty bags	673,700	7.9	694,600	9.9	684,150	8.9
Systemic fungicide	1,167,000	13.7	963,500	13.8	1,065,250	13.7
Contact fungicide	1,687,000	19.7	1,003,000	14.3	1,345,000	17.0
Granular fertilizer	4,430,000	51.8	3,790,000	54.1	4,110,000	53.0
Total	8,547,000	100	7,005,950	100	7,776,475	100

[a]US$1 = USh2177.

Table 15.4. Profitability analysis per hectare for improved seed potato production in Uganda.

Variable	2009	2010	Average
Yield (t/ha)	17.2	15.9	16.6
Total revenue (USh/ha)[a]	14,315,916	17,165,088	15,740,502
Total production cost (USh/ha)[a]	7,686,751	8,331,966	8,009,358
Gross margin (USh/ha)[a]	6,629,165	8,833,123	7,731,144
Gross margin (%)	46.3	51.5	48.9
Production cost (USh/kg)[a]	515.8	589.4	552.6
Sale price (USh/kg)[a]	832.3	1079.6	956.0
Benefit–cost ratio	0.87	1.1	0.99

[a]US$1 = USh2177.

margin that probably makes seed very expensive to ordinary farmers. The minimum price can be reduced while still keeping the gross margin by increasing productivity to reduce the production cost per kilogram of seed. This will make seed more affordable to many farmers with little effect on the gross margin and a strong effect on seed market stimulation. Sensitivity analysis revealed that the price of seed can be reduced to USh750 with little impact on profitability in view of a high benefit–cost ratio (Table 15.4).

Seed potato market for UNSPPA members

Most (92%) UNSPPA members sell their seed potato at the farm gate indicating that they do not incur marketing costs although about 8% sell their seed through the association or other dealers. The major seed buyers from UNSPPA members were farmers in the same village (92.5%) although members also sell seed to government programmes (48.1%) and NGOs (33.3%). This shows that the seed farmers have developed capacity to undertake direct marking which increases their gross margins. Selling seed to farmers within a village is more sustainable than depending on government and NGO programmes and it also achieves the mission of UNSPPA.

Utilization of revenue obtained from seed potato at household level among seed producers

The highest proportion of income from seed among UNSPPA members was used for farming expenses and children's education, respectively (Fig. 15.2). A high proportion of seed potato revenue, however, was invested in land acquisition. These are good indicators of wealth creation in a household.

Change in potato productivity and household endowment among seed potato producers in Uganda between 1999 and 2010

The highest seed multiplication rate before farmers joined UNSPPA was 1:4 bags compared with 1:7 bags by 2010 indicating improvement in potato productivity. This could be attributed to use of clean seed (Gildermacher *et al.*, 2011) and adoption of improved production technologies (Kakuhenzire *et al.*, 2001). Further improvements are revealed by changes in family endowments and investment after becoming seed producers (Table 15.5). However, the number of cows and motorcycles decreased after farmers joined UNSPPA (Table 15.5) possibly due to shifts in investment options towards specialization and profit maximization.

Changes in family welfare among seed potato producers in Uganda

In both developing and developed nations, housing is a measure of human development. The housing condition among UNSPPA before and after joining UNSPPA shows that 11.1% farmers had permanent houses in 1996 compared with 78% in 2010. Among

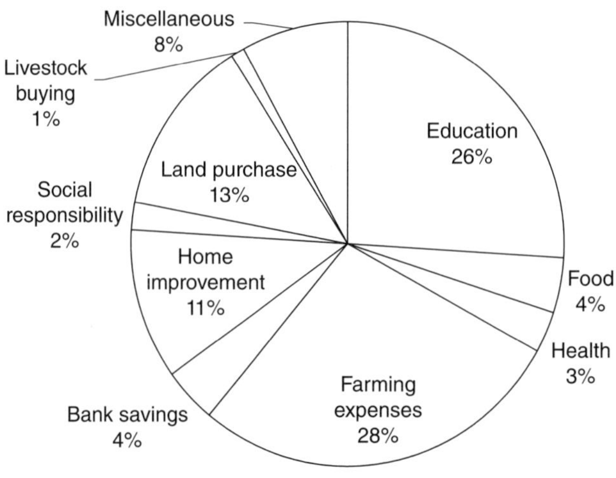

Fig. 15.2. Approximate utilization (%) of revenue from seed potato as part of family income among seed potato producers in Uganda in 2010.

Table 15.5. Changes in family endowment before and after farmers joined a seed potato producers' association in Uganda.

Variable	Number before	Number after	Change (%)
Seed potato bags produced per season	2.4	93.3	3787.5
Fertilizer bags bought per season	0.1	3.0	2900.0
Seed potato bags bought per season	1.6	19.3	1106.3
Mobile phones	0.3	2.8	833.3
Insecticide (L) used per season	0.4	2.2	450.0
Use of systemic fungicides (kg)	0.5	2.3	360.0
Number of spray pumps owned	0.6	2.5	316.7
Ware potato bags harvested per season	27.8	104.2	274.8
Number of motor cars	0.2	0.70	250.0
Sofa set	0.9	2.7	200.0
Commercial house	0.2	0.6	200.0
House in town	0.1	0.3	200.0
Number of chicken	12.5	35.3	182.4
Acres of forest	2.2	6.1	177.3
Contact fungicides (kg)	3.2	8.6	168.8
Television set	0.2	0.4	100.0
Acres of land	13.0	25.3	94.6
Radios owned	1.1	1.7	54.5
Number of bicycles	0.6	0.9	50.0
Number of living rooms	4.5	5.9	31.1
Number of goats	10.3	11.3	9.7
Number of cattle	5.5	4.5	–18.2
Motorcycle	0.1	0	–100.0

the members owning permanent houses, 44.4% of the houses currently have electrical power compared with 14% in 1996.

Farmers' opinions on government and research support to improve seed potato availability in Uganda

UNSPPA has been a key player in seed potato quality control. This will be difficult as the association grows and other actors join the business. A third party regulator without a conflict of interest will be necessary. The majority (81.4%) of the farmers wants the government to prioritize hitherto low-ranked crops like potato by increasing their budgetary support. Approximately 50% of UNSPPA members contended, and rightly so, that government should undertake seed quantity control while about 29% suggested that this function should be assigned to the National Agricultural Research Organisation (NARO) although this is outside the organization's mandate.

The potato programme has been the sole source of basic seed and has led the way in offering technical support to UNSPPA. However, farmers contend that basic seed is inadequate (51.8%), expensive (33.3%), varieties inappropriate and seed available late (26%). Between 3% and 11% of the farmers want seed on credit, NARO to improve seed grading, offer soil testing services, improve seed standards, improve seed inspection timing and increase field visits. This suggests that basic seed is insufficient, possibly over-priced, of less-preferred varieties, delivered late and is not easily accessible – all of which are business opportunities for investment.

Future outlook in quality seed potato production in Uganda

The members of UNSPPA have plans for a business-oriented outfit with a commercial focus and propose to lobby government for greater recognition as local investors in the potato sub-sector in Uganda. To achieve this, 68% of the farmers plan to invest potato revenue in land acquisition, business expansion and diversification. However, 24% of the farmers would like to build new houses or invest in better education for their children while 8% want to invest in processing, minituber production or acquire land titles. Between 1997 and 2009, the IRR and NPV of UNPPA were 336% and USh96,269,860 (US$44,221.3), respectively, indicating this organization is worth public investment.

15.5 Conclusion and Recommendation

The fresh potato market in most of SSA is generally supplied by smallholders. This is likely to remain so for a long time with farmers dependent on local seed sources. Previous attempts by some governments in SSA to produce and distribute quality seed potato did not yield sustainable outcomes. This case study from Uganda is an example of how smallholders can be engaged to develop a decentralized seed enterprise that can gradually grow into a complete seed scheme. Considering that UNSPPA members sell most of their seed to farmers in the same village, the goal of UNSPPA was achieved. The IRR and NPV of UNSPPA are healthy indicating this organization is worthy of government support. Although UNSPPA has been able to supply quality seed to other farmers, a third party regulator may be required for quality control; however, this should be implemented gradually with farmer sensitization rather than a law enforcement approach. A formal seed potato system is desirable but may not be sustainable in public institutional arrangements. In the absence of private-sector involvement, a protracted approach to develop a sustainable seed scheme may be inevitable.

References

Beukema, H.P. and van der Zaag, D.E. (1990) *Introduction to Potato Production*. Pudoc, Wageningen, 170 pp.

Gildermacher, P.R., Schulte-Geldermann, E., Borus, D., Demo, P., Kinyae, P., Mundia, P. and Struik, P.C. (2011) Seed potato quality improvement through positive selection by smallholder farmers in Kenya. *Potato Research* 54, 253–266.

Hakiza, J.J., Turyamureeba, G., Kakuhenzire, R., Odongo, B., Mwanga, R., Kanzikwera, R. and Adipala, E. (2000) Potato and sweet potato improvement in Uganda: a historical perspective. *African Potato Association Conference Proceedings* 5, 47–58.

Howard, P.H. (2009) Visualizing consolidation in the global seed industry: 1996–2008. *Sustainability* 1, 1266–1287.

Kakuhenzire, R., Hakiza, J.J., Stroud, A., Lemaga, B. and Turyahabwe, N. (2001) Small holder farmer on-farm evaluation and adaptation of improved potato cultural practices in south western Uganda. *African Crop Science Proceedings* 5, 967–973.

Lowaars, N.P. and Marrewijk, G.A. (1995) *Seed Supply Systems in Developing Countries*. Centre for Technical Cooperation and Wageningen Agricultural University, Wageningen, The Netherlands, 125pp.

Rukuba, D., Wagoire, W.W., Hakiza, J.J. and Adipala, E. (2000) Comparison of conventional and stem-cutting multiplication of potato under south western Uganda conditions. *African Potato Association Conference Proceedings* 5, 83–89.

Tindimubona, S., Kakuhenzire, R., Hakiza, J.J., Wagoire, W. and Beinamaryo, J. (2000) Informal production and dissemination of quality seed potatoes in Uganda. *African Potato Association Conference Proceedings* 5, 99–104.

Turiho-Ahabwe, G.P., Hakiza, J.J., Kakuhenzire, R.M., Bariyanga, J. and Mateeka, B. (2000) Marketing and pricing of potato seed in Uganda. From research station to farmers. *African Potato Conference Proceedings* 5, 459–462.

16 Potato Yield Variation as Affected by Virus Seed Degeneration and Growth Conditions in Tunisia

N. Khamassy,* I. Riadh and S. Boukhris-Bouhachem
National Agricultural Research Institute of Tunisia, Ariana, Tunisia

Abstract
Virus diseases mainly affect crop growth by reducing the size of the canopy, thus inhibiting the interaction of the incoming solar radiation. Additional yield reduction may be caused by effects on the radiation use efficiency or on the dry matter allocation to the tubers. Research plots were established in 2012 at the Ariana Research Station Institut National de la Recherche Agronomique de Tunisie (INRAT) to determine the effect of seed-borne potato virus Y (PVY) and poor growing conditions on the yield of cultivar Spunta (fairly good resistance to virus PVY^n) during the spring crop season (February–June). Five potato seed origins were evaluated. They have five levels of PVY infection: 0%, 2%, 4%, 8% and 50%. They were grown under poor conditions: 50% of the normal fertilization doses, 0% added organic matter and high water salinity (around 4 g/l). The results showed significant general effects of the poor growing conditions on yield reduction whatever the infection level. The differences between yield origins were statistically significant and yields were negatively affected at 8% and 50% infection levels. Thus, the PVY infection levels have a negative effect on yield losses under poor crop management. These findings confirm the previous results recorded in the 2007 season crop. Useful applications in the Tunisian national seed programme were proposed and applied.

Keywords: potato seed, seed-borne potato virus Y (PVY), Tunisian national seed programme, virus disease, yield reduction

16.1 Introduction

In Tunisia, the success of the local potato production programme is mainly impeded by the health and quality of produced seeds. Despite good management of fungal and bacterial diseases, viral disease control remains a major problem for the local seed programme. Virus symptoms are latent and become visible during subsequent crop development. In the major production areas of Tunisia, a diversity of aphids species are present all year round (Cherif and Boudhir, 1990; Khamassy *et al.*, 2007). Naturally isolated regions are few and these are usually not suitable for potato production. Due to these facts, national seed programmes cannot meet the certified seed demand for the main potato crop seasons. The situation is worse concerning production of elite class seeds. Many local elite

*khamassy.nouri@iresa.agrinet.tn

seed plots are annually rejected because virus infection rates exceed the national virus tolerance standards (Khamassy *et al.*, 2007). Recently, there have been requests to revise Tunisian Virus Standards levels since they were derived from French standards. Several authors report minimal reduction of potato yield attributed to virus diseases (van der Zaag, 1987; Fahem, 1991; Struik and Wiersema, 1999).

Compared with fungal and bacterial diseases, viruses are more important in Tunisia for phytosanitary issues (i.e. quarantine). Furthermore, Beukema *et al.* (2004) found that at a 10% and 50% potato virus Y (PVY) infection rate, yield decrease was less than 3% or 16.7%, respectively. In Tunisia, the tolerant potato cultivar Spunta exhibited no significant effect on yield decrease between 4% and 55% for PVY infection when grown under optimal growth crop management and physiological seed ageing (Khamassy *et al.*, 2007). The objectives of this study were to determine the effects of seed-borne PVY and poor growing conditions on yield of potato cultivar Spunta.

16.2 Materials and Methods

Field trials were carried out in the spring crop of 2012 at the Institut National de la Recherche Agronomique de Tunisie (INRAT) Research Station of Ariana (a northern region of Tunisia). The cultivar Spunta (grown in 80% of the potato planted area of Tunisia) known to be tolerant to PVY^n virus was used to represent different virus infection rates (Table 16.1).

These seeds were planted under poor field management: 0% organic matter, 15% of the standard fertilization and irrigated with highly saline water (water salinity around 4g/l). The experimental design was a completely randomized design with four replicates consisting of plots each planted with 100 seeds. Data collected included growth cover of the crop (60 days after planting), number of stems, number of tubers, tuber weight, total yield, commercial and non-commercial yield.

16.3 Statistical Analysis

The analysis of variance was carried out according to the general linear model (GLM) procedure developed by the Statistical Analysis Systems (SAS) Institute (SAS V6.0; SAS, Cary, North Carolina). Means and standard errors were calculated. A least significance difference (LSD) test was also used for testing significant differences between means with a confidence level of 95%.

16.4 Results and Discussion

Table 16.2 shows the mean growth and yield parameters recorded in potato affected by PVY infection rates during the spring season crop 2012. Except for cover crop, all growth and yield parameters varied significantly ($P < 0.05$) between the different origins. Stem numbers varied between 8.07/m^2 at 2% and 50% infection levels and 10.08/m^2 at 8% infection. Total yield also varied between 1.63 kg/m^2 at 50% infection level and 2.65 kg/m^2 at 2% infection. Similarly commercial yield varied from 1.33 kg/m^2 at 50% infection level to 2.47 kg/m^2 at 2% infection. Tuber weight varied from 74 g per tuber at 50% infection level to 141 g per tuber at 2% infection level.

Table 16.1. Origins of different types of seed of cultivar Spunta and their characteristics.

Codes	Seed origins	PVY secondary infection rates (%)	Average seed tuber weight (g)
01	Local super Elite (SE_L class)	0	105
02	Imported Elite (E class)	2	110
03	Imported certified (A class)	4	130
04	Local certified (A_L class)	8	145
05	Non-certified (consumption)	50	155

Table 16.2. Mean growth and yield parameters[a] in potato affected by PVY infection rates during the spring season crop 2012 at INRAT Experimental Research Station.

Origin code[b]	Virus infection rates (%)	Cover crop (%)	Stem number (m^2)	Total yield (kg/m^2)	Commercial yield (kg/m^2)	Tuberweight (g per tuber)	Commercial tubers ($number/m^2$)	Non-commercial tubers ($number/m^2$)	
								2012	2007 trials
01	0	50 a	10.00 a	2.46 a	2.34 a	103 abc	21.20 a	4.04 b	4.24
02	2	52 a	8.04 b	2.65 a	2.47 a	141 a	18.20 ab	4.08 b	4.44
03	4	48 a	8.20 b	2.42 a	2.33 a	130 ab	16.00 bc	4.00 b	4.00
04	8	49 a	10.08 a	1.82 b	1.71 b	85 bc	16.20 bc	5.40 b	4.24
05	50	34 a	8.04 a	1.63 b	1.33 b	74 c	12.08 c	10.00 a	4.05
Mean trial	12.80	46.60	8.87	2.20	2.03	106	16.73	5.50	4.19

[a]Values with the same letters are not significantly different.
[b]See Table 16.1 for details of origin.

Tuber number was also affected significantly. In fact, commercial tuber number varied from 12.08/m^2 at 50% infection level to 21.20/m^2 at 0% infection level whereas, non-commercial tuber number varied between 4/m^2 at 4% infection level and 10/m^2 at 50% infection.

In this study, deficient field management negatively affected crop growth. Ground cover and stem numbers were 50% of a normal crop growth and yield decrease was qualitatively 50% of the potential yield and qualitatively 50% of the potential tuber weight irrespective of virus infection rate. No significant yield decrease was observed between 0% and 4% PVY infection level (Table 16.2). However, at higher infection levels (8% and 50%), yield decreased significantly compared with the previous infection levels. These results confirm that PVY incidence has a minor impact on potato yield losses when a tolerant cultivar and good agricultural practices are used (Khamassy *et al.*, 2007). Therefore raising viral disease tolerance standards is possible when dealing with PVY (the most damaging virus) and Spunta cultivar.

In Tunisia there is a new seed class called 'standard seed' or 'B seed' classes with a virus infection tolerance of 15%. This has encouraged private-sector companies to start producing local (A) and (B) seed classes by the multiplication of either imported Elite (E) or certified (A) class seed from Europe since 2008. At the same time a national potato seed project is producing a local Elite seed class starting from '*in vitro*' and minituberization facilities.

Struik and Wiersema (1999) stated that growing conditions determine the degeneration rate of seed quality, and thus how many times healthy seed can be multiplied. Reliable data on the degeneration rate in the different growing regions is required to determine the most effective seed programme. When seed degeneration rate is high, local production of basic seed is not economic such as in the Tunisian case; and when the degeneration rate is low a complete programme with both production and multiplication of basic seed can be implemented, for example The Netherlands case (Struik and Wiersema, 1999). The maximum limit of virus infection rates is 10% for the certified seed (Struik and Wiersema, 1999).

According to van der Zaag (1987), a yield reduction of 5% is realized when the viruses PVY and potato leaf roll virus (PLRV) infection rate is < 20% with optimal growth conditions. Khamassy *et al.* (2007) showed that with use of cultivar Spunta and good crop management, virus infection rates as high as 50% had no significant effect on yield reduction. This study demonstrated that crop management is more a determinant than PVY infection rates on yield quantity and quality. When crop management is poor, the effect of viruses on yield reduction is amplified.

16.5 Conclusion

In countries where the degeneration rate is relatively high it still possible to produce seed with viruses tolerance around 15% (for certified seed) using a tolerant and adapted cultivar (such as Spunta in the Tunisian case). Crop management (soil preparation, fertilization, irrigation and crop protection) is a major yield factor. In Tunisia, multiplying seed potato in isolated locations is not advantageous. Besides, as the cost of basic seed is decreasing (due to hydroponics and aeroponics propagation technologies), shortening the multiplication time would be affordable and low cost certified seed could be a new opportunity for potato production promotion in the developing countries.

References

Beukema, H.P., Turkensteen, L.J. and Peeten, J.M.G. (2004) Potato Explorer: Production, Seed, Varieties, Diseases, Storage, Markets. NIVAAP, The Netherlands. CD-Rom version 2.1. January 2004.

Cherif, C. and Boudhir, H. (1990) Etude épidémiologique des virus Y et enroulement de la pomme de terre dans le Nord-est de la Tunisie. *Annales de l'Institut National de la Recherche Agronomique de Tunisie* 63, 19–25.

Fahem, M. (1991) Conséquence de la manipulation des plants et de la variation des conditions de la croissance sur le comportement d'une culture de pomme de terre en zone méditerranéenne. PhD thesis, Wageningen University, Wageningen, The Netherlands, 116 pp.
Khamassy, N., Boukhris-Bouhachem, S. and Arfaoui, I. (2007) Potato yield variation as affected by PVY in Tunisia. *African Potato Association Conference Proceedings* 7, 247–251.
Struik, P.C. and Wiersema, S.C. (1999) *Seed Potato Technology*. Pers, Wageningen, The Netherlands, 383 pp.
van der Zaag, D.E. (1987) Growing seed potatoes. In: De Bokx, J.A., Oosterveld, P., Quak, F. and van der Waat, J.P.H. (eds) *Viruses of Potatoes and Seed Potato Production*. Pudoc, Wageningen, The Netherlands, pp. 176–203.

17 Seed Potato Certification in Kenya: Prospects, Achievements and Constraints

E. Kimani,* G. Ngundo and I. Macharia
Kenya Plant Health Inspectorate Service (KEPHIS), Nairobi, Kenya

Abstract

Potato (*Solanum tuberosum* L.) is a crop of major economic importance worldwide and is considered the second most important food crop after maize in Kenya. National potato production ranges from 4.4 to 15 t/ha with an average of about 7 t/ha although yields of 40 t/ha are attainable under research conditions. The low yields are attributed to production constraints such as low soil fertility, diseases and the unavailability and high cost of inputs, mainly certified seed tubers. Improving the availability of certified disease-free seed potato of high varietal purity is therefore of paramount importance in ensuring optimum ware potato production. Kenya Plant Health Inspectorate Service (KEPHIS) along with other stakeholders have supported the implementation of various interventions aimed at alleviating constraints in the seed supply system. These include: (i) seed certification; (ii) quality assurance of seed emanating from new technologies such as aeroponics; (iii) phytosanitary certification of approved seed imports from Europe; and (iv) adoption of modern techniques such as PCR for seed health testing. Together with the recently released varieties bred by Kenya Agricultural Research Institute (KARI) and International Potato Center (CIP), farmers are set to benefit from superior seed potato varieties imported from The Netherlands.

Keywords: phytosanitary regulation, potato, quarantine, seed certification

17.1 Introduction

Potato (*Solanum tuberosum* L.) is a crop of major economic importance worldwide (FAO, 2008; Faulkner, 2013). On a global scale, potato is the fourth most cultivated food crop after wheat, rice and maize (FAO, 2008). The crop is also of paramount importance among tuber crops, ranking first in production, with an annual production of approximately 300 million t grown on about 19 million ha (FAO, 2008). Around the world, there are wide differences in potato production (FAO, 2008). Asia and Europe are the world's major potato producing regions, accounting for more than 80% of world production while Africa is the least, accounting for about 5%. Countries such as The Netherlands have a relatively high annual productivity of 46 t/ha while Africa trails at about 10 t/ha (FAOSTAT, 2011; Faulkner, 2013). The low yields are attributed to production constraints such as low soil fertility, pests and diseases and the high cost of inputs.

*ekimani@kephis.org

In Kenya, potato is the second most important food crop after maize (MoA, 2008). Potatoes are mainly cultivated in the high altitude areas (1500–3000 m above sea level) where Kenya's main staple food has no comparative advantage. These areas include: (i) the slopes around Mt Kenya, such as Meru, Embu and Kirinyaga; (ii) parts of Laikipia; and (iii) on both sides of the Nyandarua (Aberdare) range that covers parts of Nyeri, Muranga, Kiambu and Nyandarua areas. They are also grown in the highlands on Mau Escarpment (Mau Narok and Molo), Tinderet, Nandi Escarpment and Cherangani hills. Small acreages are also cultivated in Kericho and Kisii areas and isolated patches near the coast in the Taita Hills (Kirumba *et al.*, 2004).

Potatoes in Kenya are grown by about 850,000 farmers, cultivating 123,390 ha per season with an annual production of about 2,365,260 t in two growing seasons (FAOSTAT, 2011; USAID, 2012). National potato production ranges from 4.4 t/ha to 15 t/ha with an average of about 7 t/ha although yields of 40 t/ha have been achieved in Kenya under research conditions (Lung'aho *et al.*, 1997; MoA, 2008). The low yields are attributed to production constraints such as low soil fertility, diseases such as bacterial wilt and late blight as well as unavailability and high cost of inputs, mainly certified seed tubers (Ministry of Agriculture and Deustsche Gessellschaft für Technische Zusammenarbeit, 2009).

In potato production the quality of seed potatoes planted is an important determinant of the final yield and quality (Struik and Wiersema, 1999). Potato farmers can complete two seasons under rainfed conditions and three planting seasons under irrigation (each season being 3–4 months long) unlike maize, which takes up to 10 months in these areas to mature. Potato has thus become a steadier source of income as it is planted both as a cash crop and as a staple food by farmers in the country (Wambugu and Muriithi, 1982).

Due to the unavailability of certified, disease-free seed of high varietal quality, farmers mostly depend on seed from informal sources which include farm-saved (self-supply), local markets and neighbours from previous harvest of ware crops (Kaguongo *et al.*, 2008; Muthoni *et al.*, 2010). Unfortunately this practice which focuses on selling (or consuming) large tubers while retaining smaller tubers for seed purposes leads to use of poor quality seeds and often accelerates the spread of seed-borne diseases (Gildemacher *et al.*, 2009; Muthoni *et al.*, 2010, 2011).

Other factors that contribute to the poor impact of certified seed include the effect of rapid degeneration of farmer-saved seed over generations as well as the relatively high costs of certified seed. When farm-saved seed potatoes are used for several cropping cycles, without renewing the seed lot from a reliable source (such as certified seed), seed-borne diseases can cause severe yield and quality losses; the process referred to as degeneration (Gildemacher *et al.*, 2007). Seed degeneration has been shown to result from increased number of infected seed tubers after every field season, infection with multiple viruses and increased concentration of virus particles in the seed tubers (Salazar, 1996). The rate of degeneration depends on several factors such as the abundance of the vector of viruses, the number of diseased plants present and the variety grown (Salazar, 1996). This has seriously limited the impact of certified seeds and it is estimated that less than 5% of potato farmers use or have access to certified seed, which is reflected by low yields (Struik and Wiersema, 1999; Enrique and El-Badewy, 2001; Fuglie, 2007).

Yield loss can be avoided through regular replenishment of seed stocks by high quality seed potatoes multiplied by specialist growers from disease-free starter material. The specialized production skills, distribution system and quality control system required, combined with the low multiplication rate, the bulkiness and the poor shelf life of seed potatoes, all make high quality seed potatoes expensive. Seed potatoes have been shown to be a major component in potato production costs (Gildemacher *et al.*, 2011).

In a study of seed systems in three Eastern African countries, Gildemacher *et al.* (2009) revealed that farmers periodically renewed their seed stock; 41% of farmers in Kenya practice renewal while in Uganda and Ethiopia 26% and 44% of farmers do so, respectively.

On average farmers that renew their seed did so after six seasons in Kenya, seven seasons in Uganda and three seasons in Ethiopia. The most important source of seed potato in Kenya were neighbours, while in Uganda and Ethiopia the village market was the dominant source. Specialized seed growers were a source of seed in only 2% of the cases in Kenya, compared with 10% in Uganda and 16% in Ethiopia (Gildemacher *et al.*, 2009). The availability of certified disease-free seed potato of high varietal quality is therefore of paramount importance in ensuring optimum ware potato production.

Various technologies have been formulated with the aim of bridging the gap created by insufficient seed supply. For example, in a study by Gildemacher *et al.* (2011), it was observed that farmer-managed trials with positive selection, substantially increase smallholder potato productivity. This, however, may not be the best possible solution for potato farmers. The regular replacement of a farmers' seed stock with high quality seed potatoes from a formal seed multiplication system may well be more economical and cost-effective. During the same study, a number of demonstration trials in which an additional plot was planted with certified seed indicated that this can outperform positive selection under most circumstances (Gildemacher *et al.*, 2011). Investing in certified seed potato production outweighs positive selection because there is yield security, a sure market and a high return, hence use of high quality seed regularly is more economic than practising positive selection. This, however, hinges on the availability and access to these certified seeds that must be of the right variety, available in adequate quantities and at the right time (Gildemacher *et al.*, 2011).

17.2 Kenya Plant Health Inspectorate Service (KEPHIS) Activities to Promote Availability of High Quality Seed Potato

Production of seed potato in Kenya is governed under the Seeds and Plant Varieties Act (Chapter 326) and accompanying regulations. Seed potatoes should undergo certification to ensure they are free from diseases and true to type. New potato varieties must first undergo tests for distinctness, uniformity and stability (DUS) and National Performance Trials (NPTs) to determine their value for cultivation and use according to the Union for Protection of Plant Varieties (UPOV) guidelines. Varieties that qualify are officially released by the Cabinet Secretary Ministry of Agriculture, Livestock and Fisheries before official seed production and distribution begins. Over 30 potato varieties have been released in Kenya; in 2010, three high yielding potato varieties were released (i.e. Sherekea, Purple Gold and Kenya Mpya). These varieties have an average to high yield potential from 20 to 50 t/ha and mature between 3 and 4 months (Ooko and Kabira, 2011; Onditi *et al.*, 2012).

Seed potato production involves aspects such as isolation distances, standards for seed-borne diseases as well as seed tuber sizes in accordance with the International Seed Testing Association (ISTA) and Organization for Economic Cooperation and Development (OECD) standards. According to the Seeds and Plant Varieties Act, seed potato production and distribution should be done by registered seed companies; only a few are interested in pursuing the seed potato business. Certified seed potatoes have for a long time been produced primarily by the Kenya Agricultural Research Institute (KARI) and the Agricultural Development Corporation (ADC), whose capacity is limited (Kaguongo *et al.*, 2008; Riungu, 2011). However, seed potato production has been boosted by the entry of more than six seed companies who mainly obtain their foundation seed from KARI and subsequently multiply and sell it to downstream ware and seed potato growers (KARI, 2007; Muthoni *et al.*, 2010). However, KARI and the other players have over the years only been able to produce about 1% of the national certified seed requirements (Ayieko and Tschirley, 2006; Mbiyu *et al.*, 2012).

With an aim of alleviating the prevailing shortage of high quality seed of appropriate varieties, and considering the constraints affecting the industry, it became prudent to explore alternative sources of seed potato, namely seed imports. Previously, imports of seed potato from countries such as The Netherlands were permitted either in limited quantities as foundation stocks (i.e. breeders and pre-basic seed) or as tissue culture plantlets. Imports of seed potatoes were considered from The Netherlands which has a long history of high quality seed production and controls 70% of the world potato seed trade. To this end the Kenya and The Netherlands governments agreed to conduct a pest risk analysis (PRA) relating to importing seed potato from The Netherlands into Kenya. A PRA was done, consistent with International Standards for Phytosanitary Measures (ISPMs) No. 11 and No. 21, requiring the PRA to be completed before any such imports are authorized. The risk analysis was concluded in January 2012 by signing a bilateral agreement between the Ministry of Agriculture of Kenya and the Ministry of Economic Affairs, Agriculture and Innovation of The Netherlands. The agreement spelt out the stringent quality and phytosanitary conditions to be met by the exporting country (The Netherlands). It also set out a process and criteria for capacity building between the two governments to ensure that even domestic production of seed and ware potato is upgraded and improved. At the end of this process procedures for potato production, certification and quality will be improved to ensure that Kenyan potatoes can compete internationally, thus broadening the economic gains. As a result of the bilateral agreement, importation of seed potatoes has been facilitated from The Netherlands and three local seed companies have so far imported a total of 29 varieties.

Imported seed was screened for diseases of quarantine importance to Kenya and found to be free. The varieties were then entered into NPTs and grown on five sites representing all potato-growing agro-ecological zones in Kenya. The varieties have successfully completed one season of NPTs and their suitability will be determined after data from the second season are analysed. Varieties found suitable for cultivation will then be officially released, after which seed bulking and sale will commence (Centre for Development Innovation, 2013; Potato platform, 2013).

The International Potato Center (CIP) in close collaboration with KEPHIS and other stakeholders have worked on the promotion of appropriate technology in seed potato production. CIP has been able to produce and distribute quality disease-free minitubers to more than 15 African countries from their tissue culture and minituber production facility housed at the KEPHIS Plant Quarantine and Biosecurity Station (PQBS) in Muguga, Kenya (Landeo, *et al.*, 2009). CIP has also worked in partnership with KARI and privately owned companies such as the Genetic Technologies International Limited (GTIL) and Kisima Farm, to develop and promote aeroponics technology. This goes towards promoting exchange of superior germplasm, trade and improving the seed supply as a whole (Mbiyu *et al.*, 2012).

The aeroponics technology is a rapid multiplication technology able to produce large numbers of minitubers in one generation; it allows for bulking of a large number of seed potatoes. This eliminates the need for the first field generations in seed production (1, 2 and 3) thereby reducing costs and saving time. Mbiyu *et al.* (2012) reported that the aeroponics system offers several advantages over other seed production methods including the relative ease of production of disease-free material in a relatively shorter time and the improved tuber yield of seed material compared with conventional multiplication methods. The technology has been introduced in Kenya through the collaboration of various research organizations such as CIP and KARI, private companies such as Kisima Farm and GTIL as well as non-governmental organizations (NGOs) such as Farm Concern International and Wheat Foundation International. These activities have been conducted in close collaboration with KEPHIS which offers plant health testing and subsequent seed certification ensuring that the final product is of high quality.

KEPHIS has facilitated the dissemination of disease-free planting material through disease testing, which has been enhanced by the use of real-time and conventional PCR for bacterial wilt testing of seed lots (Weller *et al.*, 2000). PCR testing has been adopted as a supplement for ELISA on nitrocellulose membrane (NCM) and has the advantage of enhanced sensitivity and specificity compared with ELISA. This test is now routinely used as part of the final seed certification step to avoid disseminating seed lots latently infected with bacterial wilt. This is in addition to the field inspection standards observed for virus infection (Government of Kenya, 1991).

17.3 Conclusion

Yield losses during potato production can be avoided through regular replenishment of seed stocks by high quality seed potatoes multiplied by specialist growers from disease-free starter material. However, the specialized production skills, distribution systems, quality control system required, low multiplication rate and the bulkiness and poor shelf life of seed potatoes make high quality seed potatoes expensive and inaccessible to many farmers. KEPHIS has improved availability to farmers by facilitating the production and dissemination of disease-free planting material through seed and phytosanitary certification.

References

Ayieko, M. and Tschirley, D. (2006) Enhancing access and utilization of improved seed for food security in Kenya. *Working Paper* No. 27/2006. Tegemeo Institute of Agricultural Policy and Development, Egerton University, Nakuru, Kenya.

Centre for Development Innovation (2013) Seed Potato Development in Kenya – Phase II of the Public–Private Support Project: Introductory Note. Wageningen UR January 2013. Available at: http://www.wageningenur.nl/en/show/Seed-Potato-Development-in-Kenya-Phase-II-of-the-publicprivate-support-project.htm (accessed 8 July 2013).

Enrique, C. and El-Badewy, R. (2001) TPS: a technology for small-scale farmers in SSA. *Subproject Annual Progress Report*. International Potato Center (CIP), Nairobi.

FAOSTAT (2011) Annual Potato Production. Available at: http://faostat.fao.org/site/567/default.aspx#ancor (accessed 8 July 2013).

Faulkner, G. (2013) Essential Trends in World Potato Markets. Europatat Congress. CARLSBAD (CZ) 30 May–1 June 2013. Available at: http://www.europatatcongress.eu/docs/Taormina/Europatat_Congress_-_Guy_Faulkner.pdf (accessed 15 September 2015).

Food and Agriculture Organization of the United Nations (FAO) (2008) International Year of the Potato. Available at: http://www.potato2008.org (accessed 16 December 2012).

Fuglie, K.O. (2007) Priorities for potato research in developing countries: results of a survey. *American Journal of Potato Research* 84(5), 353–365.

Gildemacher, P., Demo, P., Kinyae, P., Nyongesa, M. and Mundia, P. (2007) Selecting the best plants to improve seed potato. *LEISA Magazine* 23(2), 10–11.

Gildemacher, P., Kaguongo, W., Ortiz, O., Tesfaye, A., Woldegiorgis, G., Wagoire, W., Kakuhenzire, R., Kinyae, P., Nyongesa, M., Struik, P.C. and Leeuwis, C. (2009) Improving potato production in Kenya, Uganda and Ethiopia: a system diagnosis. *Potato Research* 52(2), 173–205.

Gildemacher, P., Schulte-Geldermann, E., Borus, D., Demo, P., Kinyae, P., Mundia, P. and Struik, P.C. (2011) Seed potato quality improvement through positive selection by smallholder farmers in Kenya. *Potato Research* 54, 253–266.

Government of Kenya (1991) The Seeds and Plant Varieties Act (Chapter 326). The Laws of Kenya. Available at: http://www.kenyalaw.org/klr/fileadmin/pdfdownloads/Acts/SeedsandPlantVarietiesActCap326.pdf (accessed 20 June 2013).

Kaguongo, W.P., Gildemache, P., Demo, P., Wagoire, W., Kinyae, P., Andrade, J., Forbes, G., Fuglie, K. and Thiele, G. (2008) Farmer practices and adoption of improved potato varieties in Kenya and Uganda. *Social Sciences Working Paper* 2008-5. Potato International Center (CIP), Lima.

Kenya Agricultural Research Institute (KARI) (2007) *Annual Report 2006.* KARI, Nairobi.
Kirumba, W., Kinyae, P. and Muchara, M. (2004) Irish Potato Market: Survey Promotion of Private Sector Development in Agriculture. Ministry of Agriculture/Deustsche Gessellschaft für Technische Zusammenarbeit (GTZ), Nairobi.
Landeo, J.A., Barker I. and Otazu, V. (2009) Novel Approaches to Promote and Diffuse New Potato Varieties in Kenya. Available at: http://www.istrc.org/images/Documents/Symposiums/Fifthteenth/s6_landeo.pdf (accessed 15 September 2015).
Lung'aho, C., M'makwa, C. and Kidanemariam, H.M. (1997) Effect of source of mother plant, variety and growing conditions on the production of stem cuttings and subsequent yield of mini-tubers in the Kenyan potato programme. In: *Proceedings of the 4th Triennial Congress of the African Potato Association*, 23–28 February, Pretoria, South Africa. African Potato Association, Kampala.
Mbiyu, M.W., Muthoni, J., Kabira, J., Elmar, G., Muchira, C., Pwaipwai, P., Ngaruiya, J., Otieno, S. and Onditi, J. (2012) Use of aeroponics technique for potato (*Solanum tuberosum*) minitubers production in Kenya. *Journal of Horticulture and Forestry* 4(11), 172–177.
Ministry of Agriculture (MoA) (2008) *National Policy on the Potato Industry: Policy Reforms to Revitalize the Potato Industry.* MoA, Nairobi.
Ministry of Agriculture and Deustsche Gessellschaft für Technische Zusammenarbeit (GTZ) (2009) National potato taskforce final report. Ministry of Agriculture/GTZ, Nairobi.
Muthoni, J., Mbiyu, M.W. and Nyamongo, D.O. (2010) A review of seed potato systems. *Journal of Agricultural and Food Information* 11, 157–167.
Muthoni, J., Mbiyu, M.W. and Kabira, J.N. (2011) Up-scaling production of certified seed potato tubers in Kenya: potential of aeroponics technology. *Journal of Horticulture and Forestry* 3(8), 238–243.
Onditi, J.O., Nderitu, S.W.K., Landeo, J.A., Abong, G.O., Sikinyi, E.O. and Kabira, J.N. (2012) Release of three improved varieties for the expanded potato market in Kenya. *Agriculture and Biology Journal of North America* 3(5), 192–197.
Ooko, G.A. and Kabira, J.N. (2011) Suitability of three newly released Kenyan potato varieties for processing into crisps and French fries. *African Journal of Food, Agriculture, Nutrition and Development* 11, 6.
Potato platform (2013) Kenya/The Netherlands: United in Increasing Food Security. Available at: www.potatoplatformkenya.com (accessed on 7 July 2013).
Riungu, C. (2011) The small holder cash crop of the future. *Horticultural News, East African Fresh Produce Journal* 19, 15–38.
Salazar, L.F. (1996) *Potato Viruses and their Control.* International Potato Center (CIP), Lima.
Struik, P.C. and Wiersema, S.G. (1999) *Seed Potato Technology.* Wageningen University Press, Wageningen, The Netherlands.
United States Agency for International Development (USAID) (2012) Multi-Stakeholder Evaluation of Agriculture and Livestock Value Chain Activities in Kenya: USAID Compendium Report FAOSTAT. Available at: http://faostat.fao.org (accessed on 7 July 2013).
Wambugu, F.M. and Muriithi, L. (1982) *In vitro* processing of germplasm. In: Nganga, S. (ed.) *Potato Development and Transfer of Technology in Tropical Africa.* International Potato Center (CIP), Nairobi, pp. 117–118.
Weller, S.A., Elphinstone, J.G., Smith, N.C., Boonham, N. and. Stead, D.E. (2000) Detection of *Ralstonia solanacearum* strains with a quantitative, multiplex, real-time, fluorogenic PCR (TaqMan) assay. *Applied and Environmental Microbiology* 66(7), 2853–2858.

18 Adaptation and Improvement of the Seed-plot Technique in Smallholder Potato Production

Z.M. Kinyua,[1]* E. Schulte-Geldermann,[2] P. Namugga,[3] B. Ochieng-Obura,[2] S. Tindimubona,[4] A. Bararyenya,[5] I.N. Kashaija,[6] I. Rwomushana[7] and F. Opio[7]

[1]Kenya Agricultural and Livestock Research Organization (KALRO)-Kabete, Nairobi, Kenya; [2]International Potato Center Sub-Saharan Africa (CIP-SSA), Nairobi, Kenya; [3]Kachwekano Agricultural Research and Development Research Institute, Kabale, Uganda; [4]Uganda National Seed Potato Producers' Association (UNSPPA), Kabale, Uganda; [5]Institut des Sciences Agronomiques du Burundi, Bujumbura, Burundi; [6]National Agricultural Research Organization, Entebbe, Uganda; [7]Association for Strengthening Agricultural Research in Eastern and Central Africa (ASARECA), Entebbe, Uganda

Abstract

Smallholder farmers commonly use tubers from previous harvests (such as tubers that often habour tuber-borne diseases) which lead to significant yield reductions. A seed-plot technique has been developed on a pilot scale to increase on-farm availability of quality seed potato. The objectives of this study were to validate this technology under diverse management circumstances and to determine the effect of fertilizer application on seed-plot productivity as a way of increasing the gains from the technology. Validation trials were carried out in ten sites in Kenya and five sites in Uganda, where farmer groups established and managed plots by planting cultivar Asante (Victoria) at a spacing of 30 cm × 30 cm. Productivity of seed plots was statistically similar across all the sites in the two countries, indicating that the technique was easily adaptable in many areas. Fertilizer trials were carried out in Meru Central District, Kenya, with cultivar Asante and two types of fertilizer, diammonium phosphate (DAP) (18: 46: 0) and Mavuno Planting (10: 26: 10). The fertilizers were applied at low and high rates based on nitrogen supply (45 and 90 kg N/ha, respectively) in seed plots measuring 2.4 m × 1.8 m on which tubers were planted at a spacing of 30 cm × 30 cm. Plots with no fertilizer application produced a mean of 102.50 seed-sized tubers per plot, which was significantly lower than the 163.25, 162.00, 155.75 and 133.50 tubers obtained from plots that received low DAP, high DAP, low Mavuno Planting and high Mavuno Planting applications, respectively. Plots that received low DAP and low Mavuno Planting applications produced relatively higher proportions of seed-sized tubers (74.12% and 72.02%) than the DAP-high and Mavuno Planting-high plots (68.50% and 63.72%), respectively. On the basis of these findings, the seed-plot technology, with improved productivity through fertilizer application, should be widely

*kinyuazm@gmail.com

disseminated for increased availability of quality seed potato, especially in smallholder farming systems where frequent replenishment of seed stocks is hampered by financial inadequacies.

Keywords: improvement, on-farm availability, potato nursery, productivity, *Solanum*

18.1 Introduction

Potato (*Solanum tuberosum* L.) is an important food security and cash crop for both large- and small-scale farmers in East and Central Africa. The ultimate yield and quality of ware potato is largely determined by the quality of seed potato that is planted. Certified seed potato systems and novel technologies such as aeroponics exist in the region but can provide seed potato to only a small proportion of farmers (Lemaga *et al.*, 2009). Besides, the main challenge has been the high cost and related ineffectiveness of delivery of seed potato from such systems. Consequently, a majority of farmers commonly plant potato tubers recycled from their previous harvests or sourced from market places or neighbours (Gildemacher *et al.*, 2009; Gildemacher, 2012). Such tubers are often of poor quality owing to continued build up of tuber-borne diseases such as bacterial wilt caused by *Ralstonia solanacearum* and those caused by viruses (Wakahiu *et al.*, 2007; Gildemacher *et al.*, 2009). This contributes to the low yield obtained by smallholder farmers, which is below 10 t/ha compared with 40–60 t/ha achievable under enhanced production circumstances (Lung'aho *et al.*, 1997).

In order to improve on-farm availability of quality seed potato, a seed-plot technique has been developed as an innovative way of maximizing quality seed production among small-scale farmers (Kinyua *et al.*, 2001). The principle of the seed-plot technique is to maximize tuber production per unit area of limited, disease-free land through high-density planting (synonymous with a 'nursery'). Harvested potato tubers are subsequently used as seed in ware potato production fields while reserving part of the seed-plot harvest to establish another 'nursery'.

Since the technology was developed on a pilot scale, its robustness under dynamic potato production systems and the potential for its improvement has not been investigated. Productivity in potato production systems is already known to be greatly influenced by plant density and nutrient levels (Rahemi *et al.*, 2005; Zelalem *et al.*, 2009). These two factors are closely related under intensive production systems such as those for which the seed-plot technique has been developed. In order to maximize the potential benefits of the seed-plot technique, it would be useful to validate the technology under farmer crop management conditions and determine the necessary improvements on critical factors such as soil fertility. Our study, therefore, validated the seed-plot technique and determined the effect of fertilizer application on seed plot productivity.

18.2 Materials and Methods

The seed-plot technique was validated by involving diverse potato farmer groups in Kenya and Uganda. Well-sprouted tubers of cultivar Asante were planted at a spacing of 30 × 30 cm in plots measuring 1.8 m wide and varying lengths depending on availability of land and farmer group seed needs. Diammonium phosphate (DAP) (18: 46: 0) was applied at the rate of 61.7 g/m^2 (111 kg N/ha), which was the application rate during the technique development stage. In Kenya, ten seed-plot trial sites were established during the 2010 long-rains season (April–July) and repeated during the 2010 short rains (October–December). Kianthumbi farmers' group hosted three seed plots while three seed plots were hosted by Mutethia Mbuyene, two by Kibaranyaki and two by Muchicha farmers' groups. In Uganda, five seed plots were hosted for three seasons by Rushebeya, Kayanga, Kyambura, Burema and Mpangango farmers' groups. The productivity of the seed plots under the management of the various farmer groups was assessed by determining tuber yields (number and weight per unit area).

Trials to determine the effect of fertilizer application in seed plots were established using cultivar Asante in farmer-managed fields in Abothuguchi West Division, an area dominated by smallholder potato producers in Meru Central District, Kenya. Well-sprouted potato tubers were planted in seed plots measuring 2.4 m × 1.8 m at the spacing of 30 cm × 30 cm, thus accommodating eight rows of six tubers each per plot. The trials involved two fertilizer types, DAP (18: 46: 0) and Mavuno Planting (10: 26: 10), at three rates computed on the basis of the commonly recommended rate of 200 kg DAP/acre (approximately 90 kg N/ha). This gave rise to five treatments as follows: (i) no fertilizer application (0 kg N/ha); (ii) DAP low level (120 g per plot = 45 kg N/ha); (iii) DAP high level (240 g per plot = 90 kg N/ha); (iv) Mavuno low level (215 g per plot = 45 kg N/ha); and (v) Mavuno high level (430 g per plot = 90 kg N/ha). The trial was carried out at two locations in a randomized complete block design with four replicates. Data were recorded on the number of tubers and tuber yield (weight), and disaggregated by tuber size. Analysis of variance was used to detect differences among the treatments and separation of means was performed using least significance difference according to Student-Newman-Keuls Test. The data were analysed using SPSS (SPSS Statistics for Windows, version 19.0).

18.3 Results

Validation of seed-plot technique

There were no significant differences in productivity of seed plots in all the trials sites in Kenya and Uganda. The numbers of small-, medium- and large-sized tubers was not significantly different in all sites irrespective of the location in each country (Table 18.1). A similar observation was made for the total number of tubers and the total weight of harvested tubers that were obtained from the various tuber size categories. However, some sites had higher productivity of medium-sized tubers, which are considered to be most suitable for use as seed. The top five sites for production of medium-sized (seed-size) tubers were Kayanga (Uganda), Kianthumbi/Kaimenyi (Kenya), Kyambura (Uganda), Mutethia Mbuyene/Kinya (Kenya) and Burema (Uganda) that produced a mean of 1493, 1445, 1333, 1311 and 1260 tubers per plot, respectively.

The proportion of medium-size tubers was higher at the different trial sites compared with small- and large-sized tubers (Table 18.1). Tubers categorized as large in Kenya constituted a substantially high proportion of the total tubers while the same was observed for small-sized tubers in Uganda. The productivity of the seed plots in the various trial sites ranged between 28.6 tubers/m^2 and 69.6 tubers/m^2, and the yield in most of the sites was above 40 t/ha. The multiplication ratio for the high-yielding sites yielded up to six tubers per plant. However, the sites that had low yields also had poor tuber multiplication ratios (Table 18.1).

Although there were no significant differences in the productivity of seed plots in all sites, differences across seasons were significant. Season 2010A in Uganda gave a significantly higher tuber yield particularly of the medium-sized (seed) tubers than the other two seasons, a similar scenario was observed in Kenya (Table 18.2). There were no significant differences among the other four season–country combinations. However, the productivity for seed-sized tubers increased in the order 742.8, 823.6, 983.60, 1200.1 and 1721.2 tubers per plot for season 2011A-Uganda, 2010A-Kenya, 2010B-Uganda, 2010B-Kenya and 2010A-Uganda, respectively (Table 18.2).

Effect of fertilizer application on seed plot productivity

The number of tubers and the corresponding total tuber yield in plots without fertilizer was significantly lower than the fertilized plots. Subsequently, the plots with fertilizer produced more tubers than those with no fertilizer (Table 18.3). The fertilized plots did not differ significantly across all the treatments (Table 18.3). By categorizing the resultant

Table 18.1. Seed plot potato yields[a] in various sites in Kenya and Uganda.

Country	Trial site (group name or host farmer)	Small tubers No.	Small tubers % of total	Medium tubers No.	Medium tubers % of total	Large tubers No.	Large tubers % of total	Total number of tubers	Productivity (tubers/m²)	Tubers per plant	Total yield (t/ha)
Kenya	Muchicha (Munyugi P.)	96.0	5.99	872.0	54.43	634.0	39.58	1602.0	49.4	4.5	41.7
Kenya	Muchicha (Bundi H.)	112.5	7.16	963.0	61.30	495.5	31.54	1571.0	48.5	4.4	32.1
Kenya	Kianthumbi (Kimathi S.)	182.0	10.33	1241.0	70.41	218.0	12.37	1762.5	54.4	4.9	28.4
Kenya	Kianthumbi (Kaimenyi E.)	107.0	5.27	1445.5	71.14	479.5	23.60	2032.0	62.7	5.6	40.7
Kenya	Kianthumbi (Kinoti R.)	79.5	5.51	874.5	60.64	488.0	33.84	1442.0	44.5	4.0	45.1
Kenya	Mutethia Mbuyene (Wambui)	146.0	6.84	1199.5	56.16	789.0	36.94	2136.0	65.9	5.9	69.8
Kenya	Mutethia Mbuyene (Mbijiwe)	100.0	7.45	725.5	54.02	452.5	33.69	1343.0	41.5	3.7	44.6
Kenya	Mutethia Mbuyene (Kinya)	114.5	5.85	1311.0	67.02	520.5	26.61	1956.0	60.4	5.4	40.7
Kenya	Kibaranyaki (Murithi I.)	83.0	8.96	620.0	66.92	223.5	24.12	926.5	28.6	2.6	21.6
Kenya	Kibaranyaki (Agwima S.)	109.5	8.52	866.5	67.41	354.5	27.58	1285.5	39.7	3.6	30.9
Uganda	Kyambura	357.3	20.16	1333.3	75.22	82.0	4.63	1772.7	54.7	4.9	45.1
Uganda	Kayanga	596.7	26.45	1493.3	66.19	166.0	7.36	2256.0	69.6	6.3	74.1
Uganda	Rushebeya	264.0	20.35	956.7	73.74	76.7	5.91	1297.3	40.0	3.6	21.6
Uganda	Burema	607.3	29.49	1260.7	61.22	191.3	9.29	2059.3	63.6	5.7	66.4
Uganda	Mpangango	755.3	49.35	702.0	45.86	73.3	4.79	1530.7	47.2	4.3	15.1
	Significant difference (at $P = 0.05$)[b]	ns	–	ns	–	ns	–	ns	–	–	ns

[a]Tuber yields were standardized for a plot area measuring 1.8 m wide and 18 m long, giving a plant population of 360 at a spacing of 30 cm × 30 cm.

[b]ns = not significant at $P = 0.05$ according to Student-Newman-Keuls Test. Note that all the sites produced statistically similar yields for the various categories of potato tubers.

Table 18.2. Seed plot potato yields[a] in various seasons in Kenya and Uganda.

Country	Season	Numbers of tubers: Small	Medium	Large	Total	Total weight of tubers (kg)
Kenya	2010A	103.80a[b]	823.60a	626.90b	1554.30a	128.15b
Kenya	2010B	122.20a	1200.10ab	304.10a	1657.00a	99.40ab
Uganda	2010A	706.40b	1721.20b	158.80a	2586.40b	144.00b
Uganda	2010B	291.20a	983.60a	129.20a	1404.00a	92.96ab
Uganda	2011A	550.80b	742.80a	65.60a	1359.20a	58.28a

[a]Tuber yields were standardized for a plot area measuring 1.8 m wide and 18 m long, giving a plant population of 360 at a spacing of 30 cm × 30 cm.
[b]Means followed by the same letter in a column are not significantly different at $P = 0.05$ according to Student-Newman-Keuls Test.

Table 18.3. Tuber productivity[a] in seed plots with different fertilizer application regimes.

Fertilizer regime[b]	Number of tubers by size[c]: Small	Medium	Large	Total number of tubers	Total tuber yield (kg)	Average weight per tuber (g)	Tuber multiplication rate
Zero	32.25a[d]	102.50a	12.50a	147.25a	3.80a	25.82a	4.09a
DAP-Low	23.25a	163.25b	33.75ab	220.25b	10.16b	46.12b	6.12b
DAP-High	31.50a	162.00b	43.00b	236.50b	12.04b	50.93b	6.57b
Mavuno-Low	29.75a	155.75b	30.75ab	216.25b	10.35b	47.85b	6.01b
Mavuno-High	26.75a	133.50ab	49.25b	209.50b	11.14b	53.19b	5.82b

[a]The parameters are taken from plots measuring 2.4 m × 1.8 m (4.32 m^2).
[b]Zero = no fertilizer was applied at planting; DAP-Low = 120 g DAP per plot; DAP-High = 240 g DAP per plot; Mavuno-Low = 215 g Mavuno Planting per plot; and Mavuno-High = 430 g Mavuno Planting per plot.
[c]Small size = tubers of < 25 mm diameter; medium = 25–55 mm diameter tubers; and large size = tubers > 55 mm diameter.
[d]Means followed by the same letter in a column are not significantly different at $P = 0.05$ according to Student-Newman-Keuls Test.

tubers into size classes, there was no significant difference in the number of the small-sized tubers (< 25 mm diameter) between the plots where fertilizer was applied and the plots without fertilizer (control treatment). Rather, differences were observed in the number of medium-sized (25–55 mm diameter) and large-sized (> 55 mm diameter) tubers when fertilizer-applied plots were compared with the control plots. Application of fertilizer also increased the size of tubers and the tuber multiplication ratio which correlated with plots with no fertilizer producing significantly smaller and fewer tubers per plant than those in which either DAP or Mavuno Planting was applied irrespective of the rate used (Table 18.3).

Although there were no significant differences among the fertilizer regimes, the number of small-sized tubers from no-fertilizer plots tended to be higher than in fertilizer-applied plots; conversely, plots with high fertilizer application regimes yielded more large-sized tubers than the plots where no or low-level fertilizer was applied (Fig. 18.1). Plots with low DAP and low Mavuno Planting application produced relatively more seed-size (medium) tubers than the DAP-high and Mavuno Planting-high plots. Conversely, plots with high DAP and high Mavuno Planting application produced relatively more ware-size (large) tubers than the DAP-low and Mavuno Planting-low plots, respectively (Fig. 18.1). Although the two fertilizers produced a statistically similar effect on yield in the seed plots, Mavuno Planting stimulated the formation of slightly larger tubers than DAP at comparatively similar levels of nitrogen application. The proportion of medium-sized tubers from plots with DAP fertilizer was also higher than that obtained from plots with Mavuno Planting for the same nitrogen regime (Fig. 18.1).

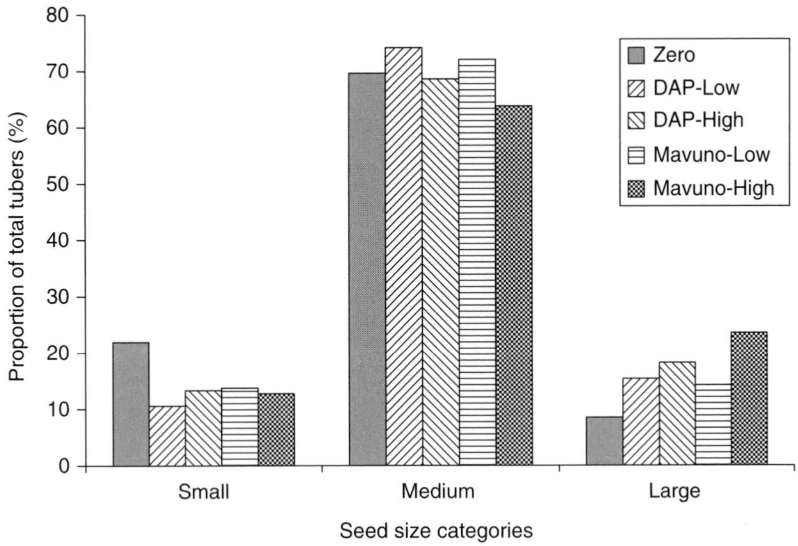

Fig. 18.1. Potato tuber sizes at different fertilizer application regimes. Zero = No fertilizer was applied at planting; DAP-Low = 120 g DAP per plot; DAP-High = 240 g DAP per plot; Mavuno-Low = 215 g Mavuno Planting per plot; and Mavuno-High = 430 g Mavuno Planting per plot. Small size = tubers of < 25 mm diameter; medium = 25–55 mm diameter tubers; and large size = tubers > 55 mm diameter.

18.4 Discussion

The seed-plot technique proved to be highly adaptable to diverse potato production systems since the yields obtained by various farmer groups in Kenya and Uganda, particularly for seed-sized tubers, were comparable. Likewise, seasons had little effect on seed potato productivity in this study, with only one exception (season 2010A in Uganda). Since the soil, weather and crop management practices in the sites where the trials were conducted were naturally different, the lack of differences in the productivity of the seed plots in those sites is a strong indication that the technology can produce the same desired effect of increasing the availability of quality seed potato in regions with climates/soils similar to those in the study areas when the basic requirements are met. Indeed, the productivity of the seed plots in the various sites was appreciably high, reaching above 60 tubers/m^2 in some areas at a spacing of 30 cm × 30 cm. This compares very well with the results obtained during the pilot development of the technology in Kenya, where cultivar Asante averaged 77.2 tubers/m^2 even though the spacing used was 20 cm × 20 cm (Kinyua *et al.*, 2001). Similarly, this study draws parallel comparisons with the findings of Rahemi *et al.* (2005) when the two planting densities were used.

The differences that arose in the proportions of tubers in the small-size and large-size categories in Kenya and Uganda can be attributed to variations in tuber classification systems in the two countries. In Kenya, small tubers (chatts) are less than 25 mm in diameter while in Uganda tubers between 30 mm and 45 mm diameter are classified as small. This implies that comparatively more tubers are classified as small in Uganda than in Kenya. Similarly, large tubers in Kenya are those with diameter above 55 mm compared with those that measure above 60 mm in Uganda. Conversely, relatively fewer tubers in Uganda than in Kenya would be categorized as large.

Application of fertilizer greatly increased productivity in seed potato plots. This is an important aspect to consider and emphasize for the purpose of increasing the productivity of seed plots. In particular, boosting soil fertility using either DAP or Mavuno Planting produced a positive effect on the total number of tubers, total yield per unit area and tuber multiplication ratio (tubers per plant). A similar trend in increased yield with application of higher levels of nitrogen and phosphorus was observed by Zelalem *et al.* (2009). These results can be used to explain the differences in the productivity of seed plots in the various trial sites in Kenya and Uganda as observed during the current study.

It was observed that seed plots with no fertilizer application produced very small and few tubers. This strongly indicates that the potential benefits of the seed-plot technique can easily increase if the technology is promoted as a package with soil fertility considerations in addition to the requirement that the seed plots be established on land that is free of soil-borne pathogens as prescribed earlier. This study, therefore, underscores the need to advocate for the boosting of soil fertility in seed plots. This is particularly critical for the success of the seed-plot technology in countries such as Uganda and Ethiopia that are already reported to rarely use soil amendments in potato production (Gildemacher *et al.*, 2009).

The stimulation of formation of larger tubers by Mavuno Planting (10: 26: 10) compared with DAP (18: 46: 0) for the same nitrogen regime (45 or 90 kg N/ha) is attributable to the extra nutrient supply from the former. Mavuno Planting is known to contain 10% potassium, 10% calcium, 4% magnesium, 4% sulfur in addition to boron, manganese, zinc, molybdenum and copper, all of which are absent in DAP. Indeed, this observation corroborates very well with the improved potato production that has been recorded upon application of potassium and calcium (Tawfik, 2001).

The seed-plot technique performed well in all the areas where it was introduced to farmer groups in Kenya and Uganda. This shows the technology, if widely adopted, can contribute to the alleviation of the chronic problem of lack of availability of seed potato in many potato-farming settings, particularly in areas where population pressure has led to small parcels of farmland. The potential of the seed-plot technique in increasing the availability of quality seed potato to farmers can be greatly enhanced through the boosting of soil nutrient levels, even at half the rate that is conventionally used in ware potato production systems. Therefore, the technique should be promoted with a deliberate soil fertility improvement component.

Acknowledgements

The planning, funding and technical backstopping for work reported here was provided by the Association for Strengthening Agricultural Research in Eastern and Central Africa (ASARECA) and supported by its regional partners. The participating farmer groups, agricultural extension officers and staff of the authors' institutions are also greatly appreciated for logistical support during the research period. Particular thanks go to Winfred Muriithi, Joseph Kinoti, Jane Ong'ele and Diana Oyena.

References

Gildemacher, P.R. (2012) Innovation in seed potato systems in Eastern Africa. PhD thesis, Wageningen University, Wageningen, The Netherlands, 184 pp.

Gildemacher, P.R., Kaguongo, W., Ortiz, O., Tesfaye, A., Woldegiorgis, G., Wagoire, W., Kakuhenzire, R., Kinyae, P., Nyogesa, M., Struik, P. and Leeuwis, C. (2009) Improving potato production in Kenya, Uganda and Ethiopia: a system diagnosis. *Potato Research* 52, 173–205.

Kinyua, Z.M., Smith, J.J., Lung'aho, C., Olanya, M. and Priou, S. (2001) On-farm success and challenges of producing bacterial wilt free tubers in seed plots in Kenya. *African Crop Science Journal* 9, 279–285.

Lemaga, B., Kakuhenzire, R., Gildemacher, P., Borus, D., Woldegiorgis, G., Barker, I., Low, J. and Ortiz, O. (2009) Current status and opportunities for improving the access to quality seed potato by smallscale farmers in Eastern Africa. In: *Tropical Roots and Tubers in a Changing Climate: A Critical Opportunity for the World*, International Society for Tropical Root Crops (ISTRC) programme and abstracts of papers, Universidad Nacional Agraria La Molina (UNALM), Lima, Peru. International Potato Center (CIP), Lima, p. 82.

Lung'aho, C., M'makwa, C. and Kidane-Mariam, H.M. (1997) Effect of source of mother plant, variety and growing conditions on the production of stem cuttings and subsequent yield of mini-tubers in the Kenyan potato programme. In: *Proceedings of the Fourth Triennial Congress of the African Potato Association*. African Potato Association, Pretoria, South Africa, pp. 275–283.

Rahemi, A., Hasanpour, A., Mansoori, B., Zakerin, A. and Taghavi, T.S. (2005) The effects of intra-row spacing and N fertilizer on the yield of two foreign potato cultivars in Iran. *International Journal of Agriculture and Biology* 7, 705–707.

Tawfik, A.A. (2001) Potassium and calcium nutrition improves potato production in drip-irrigated sandy soil. *African Crop Science Journal* 9, 147–155.

Wakahiu, M.W., Gildemacher, P.R., Kinyua, Z.M., Kabira, J.N., Kimenju, J.W. and Mutitu, E.W. (2007) Occurrence of potato bacterial wilt caused by *Ralstonia solanacearum* in Kenya and opportunities for intervention. In: *Proceedings of the Seventh Triennial African Potato Association Conference*. African Potato Association, Alexandria, Egypt, pp. 267–271.

Zelalem, A., Tekalign, T. and Nigussie, D. (2009) Response of potato (*Solanum tuberosum* L.) to different rates of nitrogen and phosphorus fertilization on vertisols at Debre Berhan in the central highlands of Ethiopia. *African Journal of Plant Science* 3, 16–24.

19 Integration of *In Vitro* Techniques in Informal Seed Production Systems of Potato in Africa

V.A. Kumar* and A. Kumar
GB Pant University of Agriculture and Technology, Pantnagar, Uttarakhand, India

Abstract

Presently, Asia and Europe are the world's major potato (*Solanum tuberosum* L.) producing regions, accounting for more than 80% of world production. Harvests in Africa and Latin America are far smaller, with productivity of around 11–12 t/ha and 16 t/ha, respectively, in comparison with North America where it is more than 40 t/ha. In Eastern Africa, potato has a high potential to raise smallholder income and improve food security. A minimal amount of seed potato is currently sourced from specialized multipliers, as farmers largely rely on farm/home-saved seed potatoes. This often makes economic sense in the absence of affordable high quality seed potatoes and limited market security. However, the common practice of farmers to use home-saved seed (informal seed) carries the risk of 'concentrating' seed-borne pests, such as bacterial wilt (*Ralstonia solanacearum*). In Africa, seed potato system interventions need to address the quality of specially multiplied and farm-saved seed potatoes simultaneously. A method known as the 'seed-plot technique' now allows smallholders with limited access to land to multiply seed tubers effectively, lessening the impact of home-saved seed practices in Africa. The seed-plot technique has been exposed to farmers in Kenya and Uganda. The 'formal' potato seed production and distribution system has been found to be insufficient to fulfil the seed requirements of growers. Another 'informal' system is also in existence and works on mutual obligation/trust among growers for promised quality/variety of potato. In both methods, integration of *in vitro* propagation technologies with traditional techniques is becoming necessary to augment availability of quality seed/planting material in sufficient quantity. The *in vitro* techniques facilitate cleaning up of virus/disease-infested material and rapid and clonal production of propagules (namely microplants (MPs), microtuber raised plants (MTPs) and microtubers (MTs)) in less time and space than are required for traditional techniques. Later, these MPs, MTPs and MTs are transplanted in the greenhouse for production of clean minitubers which possess distinct advantages and here more *in vivo* methods (e.g. use of sprouts and stem cuttings) can also be applied. The informal system of seed potato production is much more important than the formal system in most potato-growing countries. In this chapter, possibilities of integrating *in vitro* technologies with *in vivo* methods under the informal seed production chain in Africa to ensure uninterrupted supply of quality seed potato material are discussed.

Keywords: informal system, *in vitro* techniques, microtubers, minitubers, seed potato

*vandanaakumar@gmail.com

19.1 Introduction

Potato (*Solanum tuberosum* L.) is the world's single most important tuber crop, growing in about 150 countries and playing a vital role in the global food system. The tubers are highly nutritious with a high concentration of energy, high quality proteins, minerals and vitamins comparable to other high quality products of plant origin such as cereals. It is an economically important staple crop for both developed and developing countries. The top ten potato producers in the world contributing about 70% of the total production are China, Russia, India, the USA, Ukraine, Poland, Germany, Belarus, The Netherlands and France (FAOSTAT, 2008). China is the world's leading potato producer contributing 72 million t annually; followed by Russia with nearly 36 million t (Fig. 19.1).

Presently, the world potato sector is undergoing major changes. Until the early 1990s, most potatoes were grown and consumed in Europe, North America and countries of the former Soviet Union. Since then, there has been a dramatic increase in potato production and demand in Asia, Africa and Latin America, where output rose from less than 30 million t in the early 1960s to more than 165 million t in 2007. Asia and Europe are the world's major potato-producing regions, accounting for more than 80% of world production. Production in Africa and Latin America is quite small with productivity around 11–12 t/ha and 16 t/ha, respectively; in comparison, North America produces more than 40 t/ha (Table 19.1).

Current potato seed production systems in Africa

Potato production in sub-Saharan Africa has more than doubled since 1994, with 70% of that growth concentrated in Eastern Africa. Undoubtedly, potato has a high potential to raise smallholder income and improve food security in this region. Improving the quality of seed potatoes can contribute to further increase in potato productivity. Presently, minimal seed potatoes are sourced from specialized multipliers, as farmers largely rely on farm/home-saved seed potatoes in the absence of affordable high quality seed potatoes and limited market security.

Despite these gains in productivity, potato yields of small-scale farmers in the region fall short of their potential due mostly to a potent combination of inadequate supplies of high quality seed and smallholders' limited awareness of better seed management practices. Greater involvement by the private sector in seed potato value chains offers a means to unlock this yield potential

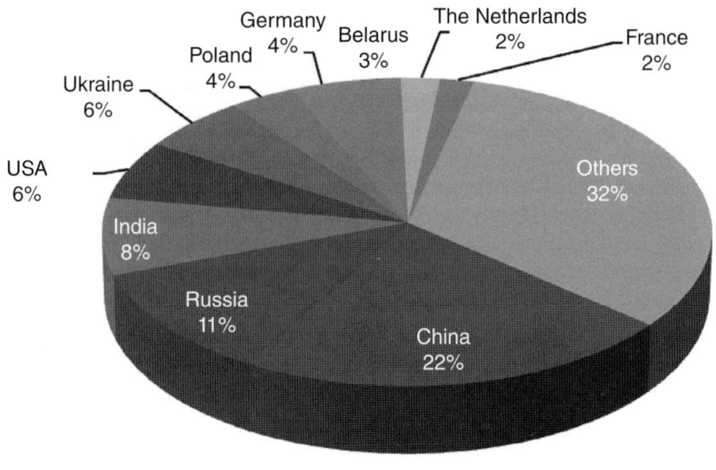

Fig. 19.1. Percentage share in global production of potatoes (From FAOSTAT, 2008).

Table 19.1. Potato production in 2007 (by region). (From FAOSTAT, 2008.)

Region	Harvested area (ha)	Quantity (t)	Yield (t/ha)
Africa	1,541,498	16,706,573	10.8
Asia/Oceania	8,732,961	137,343,664	15.7
Europe	7,473,628	130,223,960	17.4
Latin America	963,766	15,682,943	16.3
North America	615,878	25,345,305	41.2
World	19,327,731	325,302,445	16.8

by overcoming the supply bottleneck that is limiting the provision of quality seed (CIP, 2011). In Ethiopia, informal, alternative and formal seed systems coexist. In the informal system low quality seed is dominant but the formal system is too small to contribute significantly to improve the situation (Hirpa *et al.*, 2010).

19.2 Basic Requirements of a Successful Seed Potato Programme

Potato cultivation is input intensive and seed is a critical input accounting for nearly 50% of the total cost of cultivation. Therefore, development of seed stocks free from viral and other seed-borne diseases becomes vital for potato cultivation (Kumar and Kumar, 2011b).

Presently, in several northern European countries, *in vitro* technologies are combined with traditional clonal selection procedures and a sizeable proportion of basic planting material used to start clonal selection has already been replaced by *in vitro* plantlets or by minitubers. This replacement has improved the health of seed potato (Ranalli, 1997). Use of *in vitro*-raised propagules (microplants/microtubers) or minitubers developed in the greenhouse and in aphid-proof net houses/screenhouses has shown a revolutionary impact on the production of pre-basic seed where high quality seed potato tubers cannot be produced due to lack of vector-free production areas, as in southern Europe (Ranalli, 1997). The basic requirements of an integrated seed potato programme (Schmiediche, 2008) are as follows:

- An integrated seed potato programme must have *in vitro* capabilities and human resources who are competent/trained in the use of these techniques. *In vitro* techniques facilitate cleaning up of virus/disease-infected material and the cleaned-up material can be maintained *in vitro* almost indefinitely without any danger of reinfection.
- The rapid multiplication of clean material is possible through *in vitro* techniques, namely microplants (MPs), microtuber raised plants (MTPs) and microtubers (MTs).
- Material multiplied *in vitro* has to be transplanted into a greenhouse for production of clean minitubers. Use of minitubers brings certain distinct advantages in seed programme, namely:
    - o rapid propagation (one crop every 40–50 days with an average weight of 1–5 g);
    - o space requirement is minimized (1000 minitubers/m^2 of crop);
    - o suitability of minitubers for direct field planting (95% viability) with a high level of growth vigour;
    - o low production costs; and
    - o minimized requirements for tissue culture and associated costs.
- By employing appropriate technological innovations for low cost production (e.g. the use of table sugar in place of sucrose or substituting agar with tapioca during culturing, and optimizing the use of *in vitro*-produced plantlets/sprouts from stored seed tubers under field/polyhouse/net-house conditions), the cost of minitubers should be kept to a minimum.
- The greenhouse or net house for production of minitubers must be insect-proof.

Thereafter, minitubers are eventually planted in fields for larger-scale multiplication of high quality basic seed material for further multiplication.

In the context of Africa, a more efficient and responsive seed system will improve production, distribution, use and profitability for farmers. Promising rapid multiplication technologies, the three generation seed multiplication technology '3G revolution', and an engaged private sector can provide the capacity needed to broaden adoption of quality seed and accelerate availability of new varieties with more prospect of added value. Better integration of national agricultural research and extension systems into the value chain, as well as farmer training schemes in seed management and storage, can stimulate innovation. A regional perspective can: (i) help exploit economies of scale for sharing knowledge and technology; (ii) implement creative applications of information communication technologies; (iii) advocate for farmer-friendly seed-related regulations and policies; (iv) improve the business-enabling environment; and (v) expand intra-regional trade for seed of the highest quality (CIP, 2011).

Similarly, Hirpa *et al.* (2010) opined that in Ethiopia the informal seed system should prioritize improving seed quality by increasing awareness and skills of farmers, improving seed tuber quality of early generations and improving market access. The alternative and formal seed systems should prioritize improving the production capacity of quality seed by availing new varieties, designing quality control methods and improving farmers' awareness. To improve overall seed potato supply in Ethiopia, experts postulated coexistence and linkage of the three seed systems and development of self-regulation and self-certification in the informal, alternative and formal cooperative seed potato systems.

19.3 Successful Integration of *In Vitro* Technologies with *In Vivo* Methods under the Informal Seed Production Chain

Work on the Horticulture Mission for North East and Himalayan States (HMNEH) Mini Mission-I project, running at Hill Campus (Ranichauri) of GB Pant University of Agriculture and Technology for the past 6 years, has produced very encouraging results in the process of integrating *in vitro* technologies with *in vivo* methods under the informal seed production chain. The response of potato growers has been overwhelming. Since potato has been traditionally cultivated by farmers (in the hills and plains) using conventional methods, introduction of planting material developed by tissue culture (e.g. plantlets, MTs/minitubers) into their cultivation system for either seed production or ware production needed a paradigm shift. It necessitated on-farm demonstrations in their fields together with training in handling, storage and distribution of such quality material to convince them.

Our efforts towards integrating *in vitro* technologies with *in vivo* methods necessitated developing collaboration among potato growers in the hills and plains to ensure an uninterrupted supply of quality seed material of desirable variety at the right sowing time (Kumar and Kumar, 2011b). How this was achieved is summarized as follows.

In vitro technologies

Many techniques have been developed and subsequently fine tuned to speed up alternative seed production programmes using *in vitro* plantlets and MTs. *In vitro* methods followed in our laboratory for developing *in vitro* plantlets (MPs) and MTs are summarized in Fig. 19.2.

In vitro *growth of single-node cuttings*

Rapid growth of single-node cuttings has been achieved when single nodes with leaves are excised from small *in vitro* plantlets using solid or liquid culture media. The axillary bud grows quickly and within 3–4 weeks the plantlet with six or seven more nodes becomes available for further subculture.

Microtubers (MTs)

Many protocols have been developed to induce *in vitro* tubers. In our laboratory,

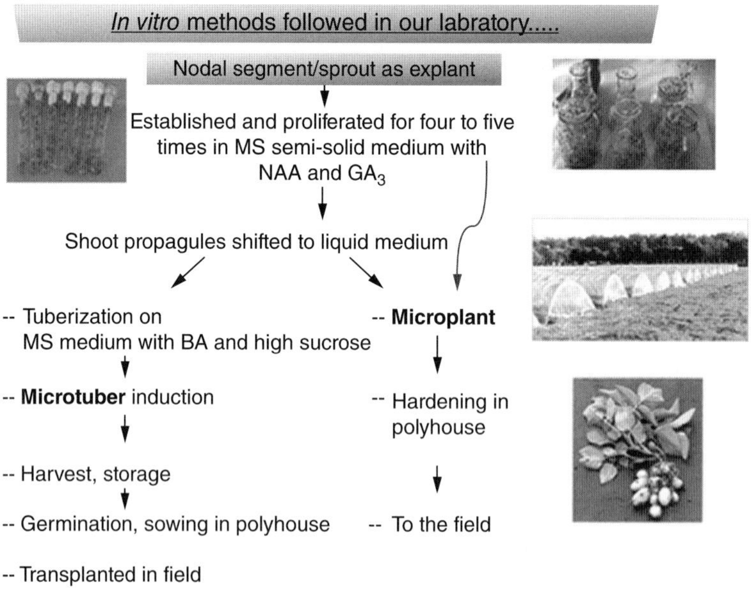

Fig. 19.2. *In vitro* methods used for the production of microplants (MPs) and microtubers (MTs). BA, benzyl adenine; GA_3, gibberellic acid; MS, Murashige and Skoog; NAA, naphthalene acetic acid.

MTs are routinely produced by taking single-node cuttings from *in vitro*-grown plantlets and growing them in a liquid tuber induction medium in darkness.

In vivo methods

Use of microplants (MPs) and microtuber raised plants (MTPs)

Year-round production of pre-nucleus potato seed stock in the form of MP and MTP *in vitro* is routinely carried out under the project referred to above (Table 19.2). Further multiplication work in the field is carried out in farmers' fields at different locations in Tehri and Haridwar districts and at the Research Station in Nainital district. Field performance of *in vitro*-raised MPs and MTPs of potato cv. Kufri Himalini as pre-nucleus seed is summarized in Table 19.2. MTs can be regenerated and stored year round, and MPs can be produced and hardened during the peak demand season both in the hills and on the plains (Kumar and Kumar, 2011a).

Multiple harvests from MPs for increasing tuber seed yield per plant

For increasing efficiency of minituber seed production, the method of multiple/repeated harvesting has been widely practised in The Netherlands as it increases growth of lateral branches on the stolons and facilitates an increase in the number of small-sized tubers per plant suitable for use as seed (Lommen and Struik, 1992). The minitubers could be produced in large quantities by repeated harvesting of tubers from *in vitro*-propagated plantlets at 4, 7 and 10 weeks after transplanting to the glasshouse/polyhouse at high plant densities. The technique of multiple/repeated harvests was successfully standardized for potato cv. Kufri Himalini; the results are summarized in Table 19.3.

Harvesting three times from the same plant was superior to harvesting two times and one time from a single plant. In addition, a suitable gradation of tuber size needed for efficiency improvement in production of seed potato has been shown (Table 19.4). More than 70% of the tubers smaller than 25 g showed good germination potential and were in accordance to the 'low input technology' proposed

Table 19.2. Performance of microtuber raised plants (MTPs) and microplants (MPs) at research (R) and farmers' (F) fields during summer and winter crop seasons as an alternative to potato tuber seed.

	Summer yield average (kg per plant (no. of plants))		Winter yield average (kg per plant (no. of plants))		
Material	Research fields	Farmers' fields	Research fields	Farmers' fields	Mean no. of tubers per plant (range)
MPs	0.085 (1667)	0.103 (190)	0.089 (508)	–	2.18 (2–4)
MTPs[a]	0.081 (2047)	0.05 (943)	0.02 (378)	0.07 (1529)	4.01 (3–9)

[a]Yield comparison for MTPs = (R) summer > (F) winter > (F) summer (average yield 0.08, 0.07 and 0.05 kg per plant, respectively).

Table 19.3. Multiple harvesting from *in vitro* developed microplants (MPs) for increasing tuber seed yield per plant.[a]

	Number of harvests		
	Three	Two	One
Mean number of tubers	116.22	91.66	42.11
Mean tuber weight (g)	1927.48	1690.46	1721.97

[a]Values shown are means of the results from ten plants. A mix of tuber sizes was obtained as in other cases of field plants.

Table 19.4. Gradation of tuber yield obtained from fields and suitable tuber size needed for efficiency improvement.[a] Data are based on harvest at research fields from microtuber raised plants (MTPs) and microplants (MPs) for 3 years.

Size	Weight range (g)	Total number	Percentage
Very big	80–140	447	5.36
Big	50–79.9	661	7.92
Medium	25–49.9	1247	14.94
Small	15–24.9	1356	16.25
Very small	7–14.9	1908	22.86
Very very small	< 7.0	2726	32.67
Total		8345	100

[a]More than 70% of the tubers smaller than 25 g showed good germination potential.

by the Central Potato Research Institute (CPRI) in which tubers of 10–20 g led to a profit of 84 paisa per rupee as compared with that of 31 paisa per rupee observed in the 'high input technology' that had used the larger seed size of 40–45 g (Kushwah and Singh, 2008).

Use of sprouts from stored tubers

Small-scale farmers can get additional yield of high quality, virus-free seed potato by just utilizing sprouts from their stored potato stock obtained by planting *in vitro* regenerated planting material. Basic studies have indicated that 4.12 tubers per plant can be harvested of which more than 53% tubers in the 10–40 g size range could be used as an additional source of seed material (Fig. 19.3a).

Additional use of double-node stem segments planted in the field

Sometimes the crop is delayed in the field and growth of plants is pretty high. Transplanting double-node segments cut from field-grown plants can produce an additional yield of high quality potato seed (Fig. 19.3b).

19.4 Success Story

An appraisal of the progress of the project funded by the HMNEH Mini Mission-I that continued for over 6 years (2006–2012) at the Hill Campus (Ranichauri) of GB Pant University of Agriculture and Technology revealed that potato growers in the hills as well as the plains of Uttarakhand showed tremendous enthusiasm for growing their own seed of early generations using MPs and MTPs. Potato growers in the hills and plains also started to collaborate among themselves, exchanging quality seed material, and strengthening the concept of the informal seed production system. During the

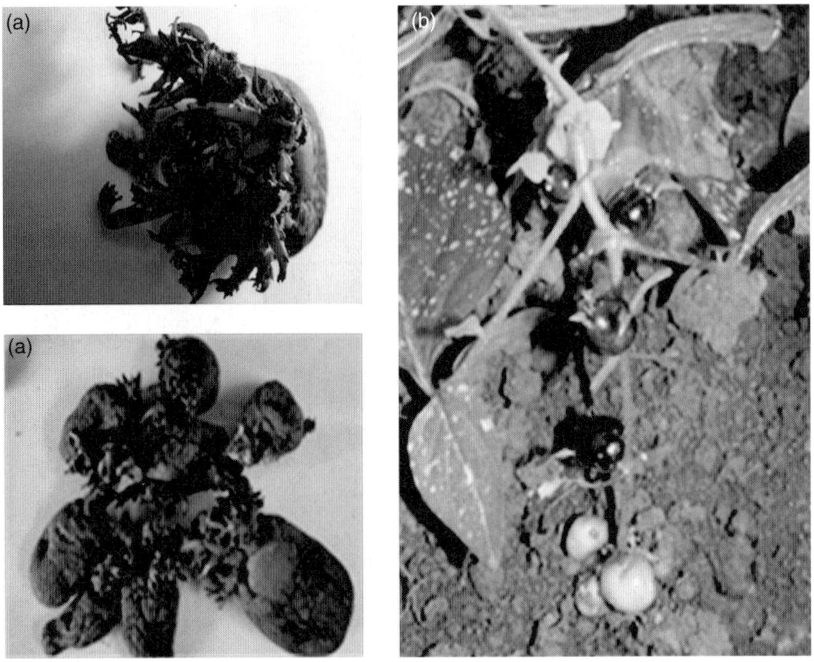

Fig. 19.3. Use of (a) sprouts and (b) stem cuttings as additional *in vivo* planting material.

entire tenure of the project, about 300 farmers (from the hills as well as the plain areas) were involved and more than 44.7 t of quality seed material was produced/exchanged among farmers (Fig. 19.4).

19.5 Model Proposed for Increasing Efficiency of Seed Production in Africa

In a study of the roadmap for investment in the seed potato value chain in Eastern Africa by CIP (2011), five mutually reinforcing core investment areas (IAs) are proposed to put the seed potato value chain interventions into practice. Three IAs make up country-level business plans:

1. Improving quality seed production and distribution.
2. Enhancing profitability of quality seed use.
3. Upgrading value chain coordination.

A further two IAs are regional and potentially cut across all five countries (Kenya, Uganda, Ethiopia, Sudan and Djibouti):

4. Promoting regional networks for sharing knowledge and best practices.
5. Growing intra-regional trade in seed.

The IAs build on a mix of value chain fundamentals and the accomplishments and lessons learned from recent projects, such as the 2 year, United States Agency for International Development (USAID)-funded 3G project. Led by the International Potato Center (CIP) and implemented in Kenya, Rwanda and Uganda, the 3G project: (i) increased access to and production of basic seed potato in both public and private sectors; (ii) successfully introduced aeroponic technology, supported its adaptation and adoption; and (iii) significantly increased production of minitubers at the national and regional scales. The project fostered private adoption of the three-generation (hence the '3G') seed multiplication strategy and improved knowledge and skills of potato production technologies and best practices,

Fig. 19.4. Potato growers in Uttarakhand grow their own seed using microplants (MPs) and microtuber raised plants (MTPs). (a) Farmers with *in vitro*-propagated plantlets; (b) planting these in the field; (c) potato crop growing in the field; (d) harvested material; (e) grading the harvested potato and storage; and (f) interactive training sessions with farmers.

leading to average yield increases of 20% for over 15,000 smallholder growers.

The following model is proposed for intervention at suitable places in the above-mentioned five IAs of the roadmap proposed by CIP (2011):

- integration of *in vitro* techniques (i.e. MP, MT and MTP production);
- increasing efficiency in *in vitro* techniques (e.g. using single-node cuttings plus delayed subculturing for further proliferation and formation of large-sized MT which would replace the minituber seed, up to 2 g);
- integrating *in vitro* and *in vivo* methods (i.e. MPs plus the aeroponic system – and harvesting small and uniform size seed from the aeroponic unit);

- increasing seed number (by using multiple harvests in the field);
- using more alternative *in vivo* methods (e.g. raising sprouts and stem segments in the polyhouse followed by field transfer);
- regulating crop timings so there are both summer and winter crops in a year;
- minimizing water usage;
- encouraging farmers to produce their own seed from as early generations as possible; and
- training farmers in handling this hi-tech material so they become part of the informal seed production system.

Further, the aeroponic system ensures uniform and repeated harvests of the desired size of potato micro/minitubers; this is gaining momentum and has the potential to virtually revolutionize potato production (Singh, 2010). In fact, a small and portable aeroponic unit placed in a polyhouse or glasshouse leads to a three- to fourfold increment in yield; ensuring eight to ten harvests per plant with minituber seed size ranging between 5 g and 10 g. Hence, harvest of minitubers is possible at an interval of every 10–15 days and a large quantity of uniform-sized seed material can be made available at a suitable sowing time.

19.6 Future Prospects

This integration of *in vitro* techniques of regenerating MPs, MTs and MTPs with *in vivo* methods holds tremendous potential as far as the supply of quality planting material in potato is concerned. In addition, MTs can be regenerated and stored all the year round and bulk supply of MP and MT can be ensured during the growing season's peak demand in the hills and plains of India.

Acknowledgements

Financial assistance provided by the Horticulture Mission for North East and Himalayan States (HMNEH) Mini Mission for running the project at Hill Campus, Ranichauri, GB Pant University of Agriculture and Technology (India) between 2006 and 2012 is duly acknowledged.

References

FAOSTAT (2008) World Potato Production and Top Potato Producers in 2007. Available at: http://faostat3.fao.org/ (accessed 10 April 2013).

Hirpa, A., Meuwissen, M.P.M., Tesfaye, A., Lommen, W.J.M., Lansink, A.O., Tsegaye, A. and Struik, P.C. (2010) Analysis of seed potato systems in Ethiopia. *American Journal of Potato Research* 87, 537–552.

International Potato Center (CIP) (2011) *Roadmap for Investment in the Seed Potato Value Chain in Eastern Africa*. CIP, Lima.

Kumar, V.A. and Kumar, A. (2011a) Field performance of *in vitro* developed micro-plants and micro tuber-raised plants of potato cv Kufri Himalini as pre-nucleus seed. In: *Proceedings of National Symposium on Technological Interventions for Sustainable Agriculture*, 3–5 May 2011, Hill Campus, Ranichauri, India. Abstract, p. 212.

Kumar, V.A. and Kumar, A. (2011b) Integrated module for potato seed production systems in India. In: *Proceedings of National Symposium on Hortibusiness – Linking Farmers with Market*, 28–31 May 2011, Dehradun, India. Abstract, pp.16–17.

Kushwah, V.S. and Singh, S.P. (2008) Low input technology for potato production. *Indian Horticulture* 58(9), 24–26.

Lommen, W.J.M. and Struik, P.C. (1992) Production of potato mini tubers by repeated harvesting: effects of crop husbandry on yield parameters. *Potato Research* 35, 419–432.

Ranalli, P. (1997) Innovative propagation methods in seed tuber multiplication programmes. *Potato Research* 40, 439–453.

Schmiediche, P. (2008) Seed program development and public–private partnerships. In: *Seed Production, Certification and Supply Systems*. Wageningen International, Wageningen, The Netherlands, pp. 1–6.

Singh, S. (2010) Aeroponics in potato seed production. Presentation made at Indian Council of Agricultural Research-funded training course 'Improved Quality Seed Potato Production through Micropropagation', 23 December 2009–3 January 2010, The Central Potato Research Institute (CPRI) Campus, Modipuram, Meerut, India.

20 Comparison Between Fluorescent Lamps and Light-emitting Diodes on *In Vitro* Growth of Potato Microplants and Subsequent *In Vivo* Performance and Minituber Production

O.H. Kwigizile,[1]* J.P. Palohuta[2] and R. Kakuhenzire[1]
[1]*Agricultural Research Institute (ARI)-Uyole, Mbeya, Tanzania;*
[2]*The Finnish Seed Potato Company Ltd, Oulu, Finland*

Abstract

Morphogenesis, growth and tuberization of potato in *in vitro* conditions can be influenced by light quality. Plant tissue culture normally uses fluorescent lamps (normal fluorescent light, NFL) to provide the red portion of the light spectrum and lots of green light, which is usually reflected from the leaves leading to low efficiency per quantum of light within the photosynthetically active radiation (PAR) range. New light sources such as light-emitting diode (LED) lamps would provide more PAR for *in vitro* plant growth in addition to being energy efficient with low heat emission. Consequently, experiments were conducted to study the effect of LEDs and NFL on *in vitro* growth and subsequent minituber production in tissue-cultured potato plantlets of three European potato varieties, Kardal, Melody and Timo. The LEDs were tested at a low (76 μmol/m^2/s) and high (121.8 μmol/m^2/s) light intensity and compared with NFL at 121.8 μmol/m^2/s as the control. For each variety, three plantlets representing one repetition in each light source were used. Results show that rate of plantlet growth, the number of nodes, internode length and average plant height after 4 weeks of *in vitro* growth were significantly ($P \leq 0.05$) affected by potato variety and light source but not their interaction. However, potato plantlets grown in LED conditions had fewer nodes than plantlets grown under NFL. Shoot biomass was significantly ($P \leq 0.05$) influenced by potato variety but not light source while root weight was significantly ($P \leq 0.05$) influenced by both potato variety and light source. Plantlets grown under low LED light intensity had a significantly lower root weight than plantlets grown under high LED light intensity. There were no significant differences in both root and shoot weight for plants grown under both NFL and LEDs at 121.8 μmol/m^2/s. Light source during micropropagation did not significantly ($P \leq 0.05$) affect minituber production in *in vivo* conditions, except potato variety Timo produced significantly ($P \leq 0.05$) more tubers per plant than Kardal and Melody. The physical benefits for use of LEDs for *in vitro* micropropagation could not be demonstrated beyond influencing the number of nodes during *in vitro* and *in vivo* propagation. Unless there are commercial benefits of LEDs in terms of energy saving in ventilation and heat emission reduction, their high cost cannot be justified against their durability or the quality of spectrum that was used in this study.

Keywords: *in vitro* light quality, micropropagation, photosynthetically active radiation (PAR), residual light effects

*wekishakwigizile@gmail.com

20.1 Introduction

Micropropagation is a tissue culture (*in vitro*) method for rapid and true-to-type multiplication of plants in artificial growth media under aseptic conditions in controlled environments. This technique is used to produce large quantities of clean planting materials of mainly vegetatively propagated crops and ornamentals which tend to have low multiplication rates in natural conditions. Besides aseptic conditions, the controlled growth environment must have a good balance of heat and light energy. In controlled plant-growth environments, the culture medium, light and temperature can be adjusted to meet specific requirements for optimum plant growth and development to enhance productivity per unit area and time. Potato, among other vegetatively propagated crops, has a low multiplication rate and seed tubers are prone to seed degeneration during repeated open-field multiplication through viral infections. Therefore, novel and cost-effective techniques need to be developed to generate large quantities of clean planting materials in a short time. The most considerable advantages offered by micropropagation include: (i) large numbers of disease-free propagules can be obtained from a single plant, tissue or organ in a short time without any change in the genotype except in rare situations; (ii) propagation can be carried out throughout the year; and (iii) the propagating material can be accommodated in a small space. Micropropagation has been successfully used in most seed-potato-producing centres to accelerate the initial stages of quality seed production. However, growth conditions have to be optimized if seed potato is to be cheaply produced.

Light is one of the environmental factors that are important for plant growth and biomass production. It controls photosynthesis, morphogenesis and photoperiodism in plant tissue culture (Chen *et al.*, 2004; Spalding and Folta, 2005; Kami *et al.*, 2010) and eventually yields (Jaimez and Rada, 2011). Well-grown plants should have internodes that are balanced with stem girth and above-root foliage. Plants cultivated in insufficient light, become etiolated, are more susceptible to pests and disease, and start reproduction later (Piszczek and Glowaska, 2008). Growth and morphogenesis of *in vitro* potato plantlets can be radically altered by different types of light sources with varying spectrum qualities (Wilson *et al.*, 1993). Tube fluorescent lamps are the most commonly used artificial light source in tissue culture for plantlet production. Fluorescent lights have the following advantages: (i) low initial cost; (ii) they provide bright, cool light; and (iii) they are more easy to maintain than fluorescent bulbs. However, tube fluorescent lamps are less durable and require regular replacement and produce more heat than is necessary in a growth room necessitating expensive ventilation. Regular lamp replacement and higher heat generation can be reduced by developing and testing more durable and energy-efficient lights that can be used in tissue-culture growth rooms. This can be partly achieved by using solid-state light-emitting diodes (LEDs). The LEDs are durable and offer a narrow-band light that can be used in a variety of horticultural and photobiological applications (Morrow, 2008; Stutte *et al.*, 2009) including tissue-culture growth rooms. Besides being more energy efficient and safe to the environment, LEDs have a controlled spectral output and offer more uniform light that ensures homogenous growth. LEDs have been successfully used for cultivation of several horticultural plant species, including lettuce (Li and Kubota, 2009; Stutte *et al.*, 2009), cucumber (Brazaityte *et al.*, 2009), spinach (Yorio *et al.*, 2001) and tomato (Brazaityte *et al.*, 2010). Red and far-red radiation in spectrum light resulted in normal and stout plants with short internodes and more balanced leaf length in tomato (Decoteau and Friend, 1991). Many studies of potato growth in tissue culture have concentrated more on testing light and photoperiodic cycles (Hayashi *et al.*, 1994) and less on light source quality. Few have considered use of LED light sources on *in vitro* growth of potato plantlets and subsequent minituber production. An experiment was therefore performed to test the efficiency of LEDs in promoting *in vitro* growth and subsequent *in vivo* minituber production of three potato cultivars.

20.2 Materials and Methods

Two experiments were conducted in the plant tissue-culture laboratory at The Finnish Seed Potato Company Ltd, one in December 2012 and the second in January–March 2013. The objective of the first experiment was to test the effect of light source and spectral strength on *in vitro* plantlet growth and development. The objective of the second experiment was to test the residual effect of different light sources during *in vitro* culture on minituber production in greenhouse conditions.

In the first experiment, 5 mm long, single nodal cuttings from 4-week-old *in vitro* plantlets of the potato cultivars Melody, Kardal and Timo were obtained using sterile micro-scissors and forceps in a laminar flow chamber. The explants were put in 20 ml test tubes (Pyrex Co.) containing 10 ml Murashige and Skoog (MS) medium (Murashige and Skoog, 1962). For each variety, 120 test tubes were prepared. To test the effect of light quality on plantlet growth, LEDs and normal fluorescent light tubes (NFLs) were used (Fig. 20.1). The NFLs were of the high performance L20 series (AP67) light spectrum that resemble fluorescent tubes in shape and size (Philips Company Ltd) and are rated at 121.8 μmol/m^2/s. Three tubes that were 150 cm long and 16 mm in diameter were used running in parallel within each growth-room shelf. LEDs were used at 76 μmol/m^2/s (low LED intensity) and at 121.8 μmol/m^2/s (high LED intensity). Each treatment combination of light quality by variety consisted of ten test tubes replicated four times. The experiment was a split plot design in randomized blocks.

For the second experiment for the *in vitro* propagation, solid modified MS medium for tissue culture was used for potato micropropagation. The modified MS basal salts were supplemented with glycine 0.2 g/l, nicotinic acid 0.05 g/l, pyridoxine HBL (B_6) 0.05 g/l,

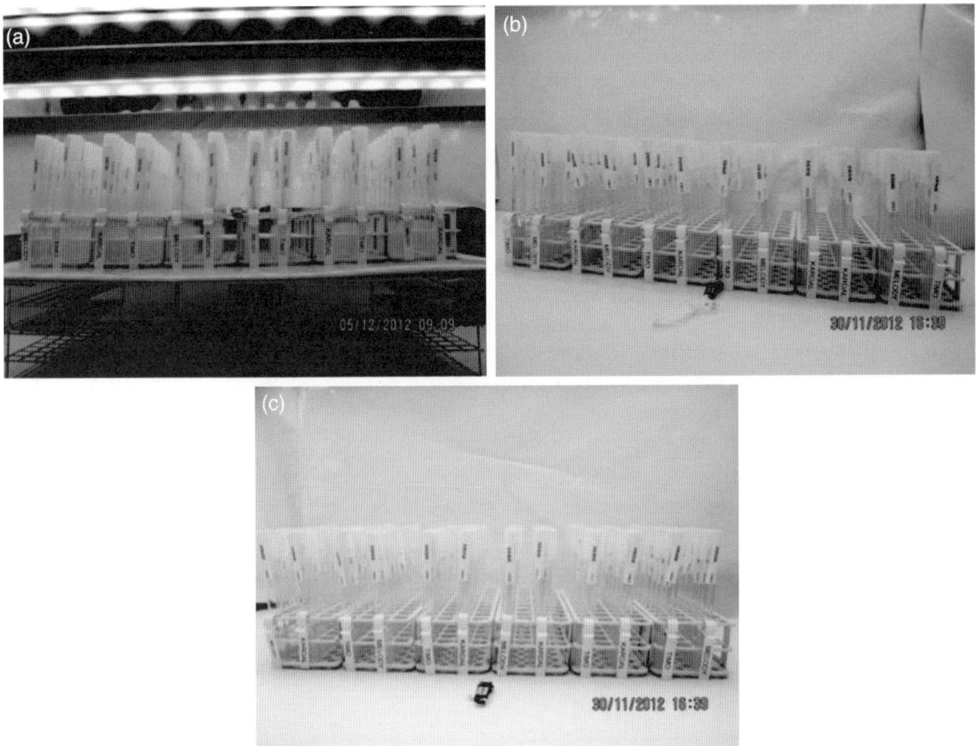

Fig. 20.1. Cultures under different light conditions: (a) high light-emitting diode (LED) intensity (121.8 μmol/m^2/s); (b) low LED intensity (76 μmol/m^2/s); and (c) normal fluorescent light (NFL) (121.8 μmol/m^2/s).

myo-inositol 0.1g/l, thiamine HBL (B_1) 0.04 g/l and sucrose 30 g/l. The pH of the medium was adjusted to 5.78 using 1M NaOH and 7.5g/l of solidifying agent (agar) was added. The culture medium (10 ml) was dispensed into 20 ml test tubes (Pyrex Co.) and autoclaved at 121°C for 25 min to sterilize it. Explants of potato cultivars Melody, Kardal and Timo were placed in these tubes (one per tube) using the same procedure as used in the first experiment. The *in vitro* cultures for each variety were placed in the growth chamber with a specific spectral treatment of 16 h of light and 8 h of darkness; the temperature was controlled at 20 ± 2°C. After 4 weeks of *in vitro* growth, eight microplants from each variety-by-light-source treatment were transplanted into pots in the greenhouse to evaluate their growth and minituber production. Commercially fertilized peat was used as the growing medium. The plants were illuminated with high pressure sodium lamps in 16 h/8 h daylight to darkness alternations. The plants were irrigated manually depending on water needs. The experiment was set in a completely randomized design.

Data collection and analysis

Data collection on *in vitro* plantlets started at 10 days after inoculation and subsequently at 7 day intervals for 4 weeks. The data collected included number of roots, number of nodes and plant height (mm). During *in vivo* cultivation data were collected on plant height, number of nodes per plant, number of tubers per plant, and minituber weight (g) by tuber size. Treatment factor effects and their interactions on measured variables were tested through analysis of variance (ANOVA) and means of significant factors compared using Fisher's protected least significant difference test at $P \leq 0.05$.

20.3 Results and Discussion

Effect of light source of *in vitro* potato plantlet growth

ANOVA revealed that fresh root weight (g) was significantly ($P \leq 0.05$) influenced by light source but highly significantly ($P < 0.001$) affected by potato variety (Table 20.1). Fresh shoot weight on the other hand was highly significantly ($P < 0.001$) affected by potato variety but not light source (Table 20.1). The interaction between potato variety and light source was not significant ($P \leq 0.05$) for both fresh root and shoot weight (Table 20.1). The number of nodes per plant was significantly ($P < 0.001$) affected by potato variety but not light source indicating that light source and quality had no effect on stem segmentation but was cultivar controlled. Potato plant height was not significantly affected by both light source and variety indicating that the plantlets grew uniformly independent of light quality and quantity.

Data further showed that both fluorescent light and LED at high intensity resulted in significantly higher fresh root weight than low LED intensity (Fig. 20.2). This means that to obtain the benefit of LEDs they should be used at the same light intensity as NFL.

Table 20.1. Effect of light source on fresh root and shoot weight (g) for *in vitro* grown potato plantlets.

Source of variation	df[a]	Fresh root weight (g)	Fresh shoot weight (g)	Number of nodes	Plantlet height (mm)
Replications	3	0.0002	0.0019	0.53	1479
Light source (L)	2	0.003*	0.0040	1.29	5976
Residual	6	0.0006	0.001	0.89	1938
Variety (V)	2	0.0251***	0.042***	24.0***	8501
L × V	4	0.0006	0.0009	1.85	4539
Residual	18	0.0005	0.0032	0.89	2688
CV (%)[b]	–	18.4	24.0	5.2	7.6

[a]df, degrees of freedom.
[b]CV, coefficient of variation.
*, $P \leq 0.05$; ***, $P < 0.001$.

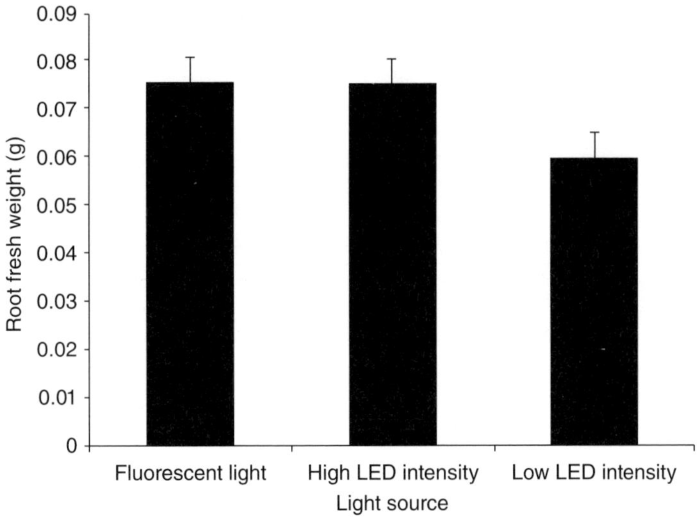

Fig. 20.2. Effect of light source on fresh root weight of tissue-cultured potato plantlets. LED, light-emitting diode. Values represent mean ± SD.

The effects of potato variety on number of nodes, fresh root weight (g) and fresh shoot weight (g) were evident. All the potato varieties used in the study were significantly different in number of nodes and fresh root weight but not fresh shoot weight, where Melody and Timo had significantly lower shoot weight than Kardal (Table 20.2). The difference among varieties is expected since this parameter may be influenced by the cultivar genetic background under similar growing conditions.

Table 20.2. Effect of potato variety on number of nodes, fresh root and shoot weight for three potato varieties grown in tissue culture under different lighting conditions.

Variety	Number of nodes	Fresh root weight (g)	Fresh shoot weight (g)
Kardal	17.08	0.07	0.21
Melody	18.08	0.10	0.14
Timo	19.08	0.05	0.16
$LSD_{0.05}$ [a]	0.55	0.01	0.03

[a]$LSD_{0.05}$, least significant difference.

Residual tissue-culture light growth conditions on *in vivo* performance on potato

In vivo growth and minituber production showed that the number of nodes was significantly ($P < 0.05$) influenced by previous light source, potato variety and interaction between the two (Table 20.3). The plant height during *in vivo* growth was independent of both experimental factors while minituber production was significantly ($P < 0.001$) influenced by potato variety but not light source (Table 20.3). Plants grown under fluorescent light were significantly taller than those grown at high LED intensity and similarly, potato plants previously grown under high LED intensity were taller than those formerly grown under low LED intensity (Fig. 20.3). Comparison of plant growth among varieties showed that Timo had significantly more nodes than Melody and Kardal, respectively (Table 20.4). This is expected since node formation may be under potato cultivar genetic control.

20.4 Conclusion and Recommendation

LEDs were found to influence formation of nodes even when used at the same intensity as NFL (Fig. 20.3). Low-intensity LEDs were not

Table 20.3. Effect of light source during *in vitro* propagation on subsequent growth and minituber production of tissue-cultured potato plantlets.

Source of variation	df[a]	Number of nodes	Plant height (mm)	Number of minitubers per plant
Light source (L)	2	4.3**	2998.0	0.9
Variety (V)	2	29.7***	5038.2	56.2***
L × V	4	1.4*	2557.5	7.0
Residual	63	0.5	977.5	5.4
CV (%)[b]	–	5.1	12.1	36.2

[a]df, degrees of freedom.
[b]CV, coefficient of variation.
*, $P < 0.05$; **, $P < 0.01$; ***, $P < 0.001$.

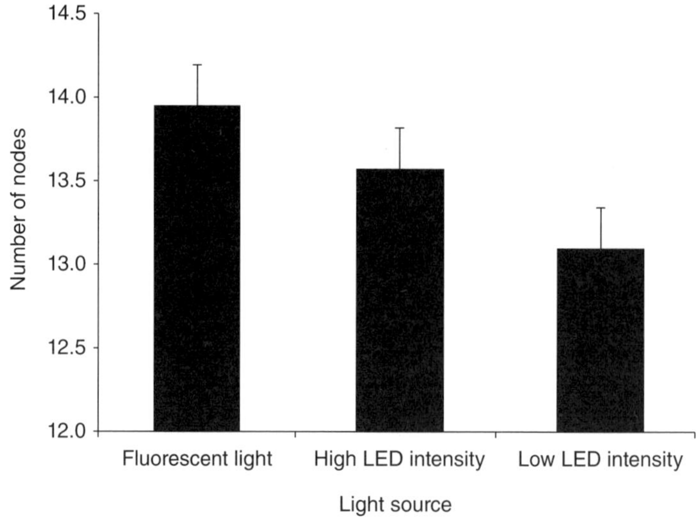

Fig. 20.3. Residual effect of light on number of nodes during *in vivo* growth of previous tissue-cultured potato plantlets. LED, light-emitting diode. Values represent mean ± SD.

Table 20.4. Effect of light source during *in vitro* growth on subsequent *in vivo* development.

	Number of nodes per potato variety			
Light source	Kardal	Melody	Timo	Mean
Fluorescent light	12.9	13.7	15.2	13.9
High LED intensity	12.2	14.3	14.2	13.6
Low LED intensity	11.8	13.4	14.1	13.1
Mean	12.3	13.8	14.5	13.5

$LSD_{0.05}$ among light source, potato variety and light source by potato variety interaction is 0.402, 0.402 and 0.696, respectively.

found to be beneficial in stimulating root and shoot growth. The quality and source of light did not affect the size of the plants and minituber production. The physical benefits for use of LEDs for *in vitro* micropropagation could not be demonstrated beyond influencing the number of nodes during *in vitro* and *in vivo* propagation. Unless there are commercial benefits of LEDs in terms of energy saving in ventilation and heat emission reduction, their high cost cannot be justified against their durability or the quality of spectrum that was used in this study.

References

Brazaityte, A., Duchovskis, P., Urbonaviciute, A., Samuoliene, G., Jankauskiene, J., Kasiuleviciute-Bonakere, A., Bliznikas, Z., Novickovas, A., Breive, K. and Zukauskas, A. (2009) The effect of light-emitting diodes lighting on the growth of cucumber transplants and after-effect on yield. *Zemdirbyste Agriculture* 96, 102–118.

Brazaityte, A., Duchovskis, P., Urbonaviciute, A., Samuoliene, G., Jankauskiene, J., Sakalauskaite, J., Sabajeviene, G., Sirtautas, R. and Novickovas, A. (2010) The effect of light-emitting diodes lighting on the growth of tomato transplants. *Zemdirbyste Agriculture* 97, 89–98.

Chen, M., Chory, J. and Fankhauser, C. (2004) Light signal transduction in higher plants. *Annual Review of Genetics* 38, 87–117.

Decoteau, D.R. and Friend, H.H. (1991) Growth and subsequent yield of tomatoes following end-of-day light treatment of transplants. *HortScience* 26, 1528–1530.

Glowaska, B. (2002) *Lycopersion esculentum* (Mill.). *Acta Scientiarum Polonorum: Hortorum Cutlus* 1(2), 93–103.

Hayashi, M., Fujiwara, K., Kozai, T., Tateno, M. and Kitaya, Y. (1994) Effects of lighting cycle on daily CO_2 exchange and dry weight increase of potato plantlets *in vitro* cultured photoautotrophically. Paper presented at the XXIV International Horticultural Congress, 21–27 August 1994, Kyoto, Japan.

Jaimez, R.E. and Rada, F. (2011) Gas exchange in sweet pepper (*Capsicum chinense* Jacq) under different light conditions. *Journal of Agricultural Science* 3, 134–142.

Kami, C., Lorrain, S., Hornitschek, P. and. Fankhauser, C. (2010) Light-regulated plant growth and development. *Current Topics in Developmental Biology* 91, 29–66.

Li, Q. and Kubota, C. (2009) Effects of supplemental light quality on growth and phytochemicals of baby leaf lettuce. *Environmental and Experimental Botany* 67, 59–64.

Morrow, R.C. (2008) LED lighting in horticulture. *HortScience* 43, 1947–1950.

Murashige, T. and Skoog, F. (1962) A revised medium for rapid growth and bioassays with tobacco tissue cultures. *Physiologia Plantarum* 115, 493–497.

Piszczek, P. and Glowaska, B. (2008) Effect of the colour of light on cucumber (*Cucumis sativus* L.) seedlings. *Vegetatable Crop Research Bulletin* 68, 71–80.

Spalding, E.P. and Folta, K.M. (2005) Illuminating topics in plant photobiology. *Plant, Cell and Environment* 28, 39–53.

Stutte, G.W., Edney, S. and Skerritt, T. (2009) Photo-regulation of bio-protectant content of red leaf lettuce with light-emitting diodes. *HortScience* 44, 79–82.

Wilson, D.A., Weigel, R.C., Wheeler, R.M. and Sager, J.C. (1993) Light spectral quality effects on the growth of potato (*Solanum tuberosum* L.) nodal cuttings *in vitro*. *In vitro Cell Development Biology* 29, 5–8.

Yorio, N.C., Goins, G.D., Kagie, H.R., Wheeler, R.M. and Sager, J.C. (2001) Improving spinach, radish, and lettuce growth under red light-emitting diodes (LEDs) with blue light supplementation. *HortScience* 36, 380–383.

21 An Analysis of Demand, Supply and Elasticities of Seed Potato in Major Producing Areas in Nigeria

D.M. Lenka,[1]* E.A. Dung[1] and G. Asumugha[2]
[1]*National Root Crops Research Institute, Kuru, Nigeria;* [2]*National Root Crops Research Institute, Umudike, Nigeria*

Abstract

The study was conducted in 2012 to assess the influence of economic variables and prices of complementary and substitute products on the demand and supply of seed potato in major potato-producing areas of Bokkos, B/ladi, Mangu, Jos south, Riyom and Jos north of Plateau State in Nigeria. Cross-sectional data on 90 seed potato producers and 90 seed potato traders were collected using a questionnaire through the use of multistage and random sampling techniques. Descriptive statistics and linear regression models were used in the analysis. The results show that farm size, income and education positively and significantly influence the demand for seed potato ($P < 0.01$). Seed price, however, negatively and significantly ($P < 0.05$) influences demand. On the supply side, results indicate that experience, gender (being a female) and household size have a positive and significant ($P < 0.05$) relationship with the value of seed potato supplied. The own-price elasticity of demand and supply for seed potato were inelastic. The elasticities of the other variables in the demand and supply functions were also less than 1 (inelastic) except that for fertilizer. The cross-price elasticity for the demand of potato and other products were < 1 and negative, while in the case of supply the cross-price elasticity was positive. From the results it is concluded that seed potato cultivation in the study area is increasing with increase in education and income of farmers. A unit change in each of the significant factors influencing demand and supply contributed to a minor shift to the demand and supply of seed potato since each of these factors is inelastic. This implies that in the long run changes in these factors will bring less than proportionate change in seed supply. These results call for policies that will provide support on pricing, regulate farm labour practices and improve access to mechanized labour, facilitate the importation of quality fertilizer and farm inputs, and provide functional education to farmers to apply modern knowledge in production and marketing.

Keywords: demand, supply, elasticity, seed potato

21.1 Introduction

Potato cultivation both as food and a cash crop in Nigeria dates back to the early 20th century with substantial production increases between 1939 and 1945 (William, 1962; Crissman *et al.*, 2007). Potato matures in about 80–90 days under Jos Plateau conditions, and the weather conditions make it possible to have four production cycles each

*lenkadaniel@yahoo.com

year. Ninety percent of the farmers cultivate the crop during the rainy season while for 10% cultivation fluctuates within the dry season (with irrigation) period of October–April (Lenka *et al.*, 2006a). It is estimated that Nigeria has a potential annual land area for potato cultivation of over 500,000 ha. Nigeria is the fourth largest producer of potato in sub-Saharan Africa in terms of output with an annual record of 840,000 t (Lutaladio and Castialdi, 2009). The average market price of ₦70,000/t (US$437.50) was recorded in 2006 implying that over ₦5 billion (US$31,250,000) circulated in the Nigerian economy from potato production (Lenka *et al.*, 2006b). Statistics from FAO (2008) estimated that 266,000 ha was cultivated under potato in 2007 with an average yield of 3.27 t/ha. FAO (2008) also reported that Nigeria has the lowest per capita potato consumption in Africa (3.27 kg). Jos Plateau accounted for over (90%) of the total output of potato production in Nigeria (Lenka *et al.*, 2010).

The low land productivity of potato is associated with numerous factors; prominent among the factors is poor quality of seed potato. This problem is attributed to high degeneration rate caused by viral infections and lack of quality seed-replacement strategies such as seed importation and the use of tissue culture facilities. Farmers therefore need high quality seed potato and are ready to pay high prices for it. The question that has remained pertinent in this seed issue is: 'What is a fair price for a quality seed potato?'. Farmers in the year 2012 paid as high as ₦360/kg (US$2.25/kg) for imported seed potato. Apart from the price of seed potato, the demand and supply of seed potato by farmers is also affected by many factors such as the availability and quality of labour, farm size, fertilizer, income, education, experience and the prices of root and tuber crops that are substitutes or complementary to seed potato. Some studies on determinants of demand and supply of root and tuber crops such as Asumugha *et al.* (2007), Okoye and Onyenweaku (2007) and Ogbonna *et al.* (2007) found evidence showing significant multidirectional influences of these factors on the supply and demand of seed.

Elasticity is defined as the responsiveness of the quantity demanded of goods or services to changes in price (Ivan, 1999; Gillespie, 2007). Price elasticities are almost always negative, only goods which do not conform to the law of demand such as Veblen and Giffen goods have a positive price elasticity of demand (Browning, 1999). In general, the demand for a good is said to be *inelastic* (or *relatively inelastic*) when the price elasticity of demand is less than 1 (in absolute value): that is, changes in price have a relatively small effect on the quantity of the good demanded. The demand for a good is said to be *elastic* (or *relatively elastic*) when its price elasticity of demand is greater than 1 (in absolute value): that is, changes in price have a relatively large effect on the quantity of a good demanded. Elasticity of demand is determined basically by three positions: whether close substitutes are available, and whether the good is a necessity or a luxury. Elasticity of supply depends on availability of inputs and time to respond to price changes. Goods are substitutes when the cross-price elasticity is positive and complements when the cross-price elasticity is negative. When the income elasticity of demand is positive the good is a normal good (i.e. the quantity demanded at any given price increases as income increases). When the income elasticity is negative the good is an inferior good (i.e. the quantity demanded at any given price decreases as income increases). On the supply side, price elasticity tends to be larger when inputs are easily available and tends to be small when inputs are difficult to obtain. The price elasticity of supply tends to be larger as producers have more time to respond to a price change. This means that the long-run price elasticity is often higher than short-run elasticity.

In this chapter we explore some of the economic and demographic factors affecting the demand and supply of seed potato in the potato-producing areas as a whole, estimate demand and supply functions, examine the demand and supply elasticities and draw policy recommendations for enhancing the demand and supply of seed potato in the major producing areas.

21.2 Methodology

Sampling procedure for data collection

The study was conducted in 2012 in the major potato-producing areas of Plateau State (i.e. Bokkos, B/ladi, Mangu, Jos south, Riyom, Jos north). Multistage and random sampling was used in selection of seed potato producers and seed potato traders in the study area. From the six major potato-producing local government areas in the state, three districts were randomly selected and from each district five farmers were randomly selected, giving a total of 90 seed potato producers. Traders were selected using the same procedure from the local markets in the areas giving a total of 90 traders.

Data collection and data analysis

Primary data were generated from the respondents through administration of a set of structured questionnaires. The information collected on seed potato producers were: (i) household characteristics and employment; (ii) household expenditure on seed potato, inputs, complements and substitutes; and (iii) prices. On the demand side data collected included: (i) the quantity and value of seed potato demanded; (ii) the price of seed potato and substitutes; (iii) disposable income; (iv) household size; and (v) experience in seed potato production. Other additional data collected included age and education of the household head, major and minor occupation, gender of the household head, farm sizes, labour, fixed inputs, and value of credit. On supply side, additional data collected included storage and value of losses during storage, membership of an association or cooperative society, value of credit, income level, and means of transportation. Secondary data were sourced from available literature and relevant research works in the area.

Conventional demand and supply functions analyses via a cross-sectional model were employed to assess the influence of explanatory variables on the value of seed potato demanded and supplied. Descriptive statistics as well as linear regression models were applied to estimate the effect of the above variables on the value of seed potato demanded and elasticity of demand and supply were derived from estimates of the linear regression models. The functional, formulae and variables applied in the analysis are listed and explained from Eqn 21.1 to Eqn 21.7 as shown below.

The demand function for seed potato is implicitly specified as:

$$y = f\,(age, edn, I, L, Fs, Exp, Psy, Pss, P, Fertkg, V_i, e) \qquad (21.1)$$

$$y = yp + Yb \qquad (21.2)$$

$$I = G_i - G_o \qquad (21.3)$$

Where:
y = quantity of seed potato demanded (t)
f = function of
yp = quantity of own seed potato (t)
Yb = quantity of seed potato purchase by growers
age = age (years)
edn = level of education of household head (years of schooling)
I = disposable income
G_i = gross income of household
G_o = income given away (₦/annum)
L = labour demanded (man days)
Fs = farm size (ha)
Exp = experience (years of farming)
Psy = price of close substitutes (yam/tuber)
Pss = price of close substitute (sweetpotato/bag)
P = price of seed potato (₦/t)
$Fertkg$ = Quantity of fertilizer demanded (in kg/ha)
V_i = other variable inputs (₦) (amount spent on pesticides and herbicides per hectare)
e = error term
The supply function is given as:

$$SS = f\,(GD, ED, I, HHS, P, PS, CR, LAB, AGE, EXP, e) \qquad (21.4)$$

Where:
SS = value of seed potato supplied (₦)
GD = gender of supplier
ED = educational level of the household head (years of schooling)

I = disposable income (₦) (as described above)
HHS = household size (number)
P = price of seed potato (₦/t)
PS = price of substitutes
CR = Value of credit
LAB = labour availability
EXP = experience (years)
AGE = age of the trader (years)
e = error term

For linear functions, price elasticity of demand can be written as applied in Parkin *et al.* (2002), Gillespie (2007) and Gwartney *et al.* (2008):

$$E_d = \frac{\delta q}{\delta p} \times \frac{p}{q} \quad (21.5)$$

$$= b \times \frac{p}{q} \quad (21.6)$$

Where:
E_d = elasticity of demand
δq = change in quantity demanded
δp = change in price
p_i = mean value of price
q_i = mean value of quantity
b_i = coefficient of the estimated variables

The formula used to calculate the coefficients cross elasticities of demand is ()

$$E_{A,B} = \frac{P_{B,1} + P_{B,2}}{Q_{A,1} + Q_{A,2}} \times \frac{\Delta Q_A}{\Delta P_B} = \frac{\delta Q_A}{\delta P_B} \times \frac{P_B}{Q_A} \quad (21.7)$$

Where:
$E_{A,B}$ = cross elasticity of demand of commodity A in relation to price of commodity B
P_B = price of commodity B at time periods 1 and 2
Q_A = mean value of quantity demanded of commodity A at time periods 1 and 2
δ = direct elasticity of one commodity
Δ = change in Q (quantity) or P (price)

21.3 Results and Discussion

The farmers cultivating seed potato in the major potato-producing areas of Nigeria were predominantly within the ages of 40–59 years old (Table 21.1). The distribution of respondents according to levels of education, farm size and farming experience showed that 51% had secondary school, 47.8% had 2.1–5 ha of farm size and 36.7% had 11–20 years of farming experience, respectively. Forty-two percent of the respondents were within the income bracket of ₦100,000–300,000 (US$625–1872) and 57.8% had income of over ₦300,000 (US$1875). The increasing trend in farm size and income patterns suggest that potato production is gradually shifting from the traditional small farm size of less than 1 ha to commercialized potato farming. The higher percentage in education and with higher incomes implies that the non-educated farmers are gradually being replaced by those with secondary school level of education and with higher incomes.

Table 21.2 shows that the most important economic variables that positively and significantly influence the demand for seed potato were education, farm size, fertilizer and income. Education provides a vital opportunity in appreciating and applying

Table 21.1. Description of farmers according to socioeconomic characteristics. (From survey data 2012.)

Variables	Frequency	Percentage
Age of producer (years)		
20–39	22	24.4
40–59	53	58.9
60–79	15	16.7
Education (years)		
0–6	31	34.4
7–12	46	51.1
> 12	15	14.5
Gender		
Female	32	35.6
Male	58	64.5
Disposable income (₦)		
100,000–300,000	38	42.2
301,000–700,000	29	32.2
> 701,000	23	25.6
Farm size (ha)		
0.1–2	17	18.9
2.1–5	43	47.8
> 5	30	33.3
Experience (years)		
1–10	36	40
11–20	33	36.7
> 20	21	23.3

knowledge and information for best production decisions and practices. Education encourages the farmers to increase their farm sizes and request a better combination of fertilizers to increase the potato yield of their farms. The coefficient for the price of seed potato is negative and significantly affects the demand for seed potato. This agrees with the a priori assumption in economics that price negatively and significantly influences the demand for seed potato. However, the quality of seed potato with farmers has degenerated over time because of lack of seed replacement and the farmers are sometimes forced to buy quality seed at exorbitant prices which contradict the law of demand. The own-price elasticity of demand (Table 21.3) for seed potato was inelastic (less than 1) implying that continuous increase in price of seed potato will bring in the long run a less than proportionate decrease in the demand for seed potato. Table 21.3 also shows that the cross-price elasticity of seed potato and other root and tuber crops, that is, cocoyam, sweetpotato and yam were negative implying that they are substitutes. Whenever the price of seed potato rises the farmers would turn their production resources to other root and tuber crops.

The key economic variables that positively and significantly influenced the supply of seed potato were household size and experience (Table 21.4). Women

Table 21.2. Demand functions for seed potato in major potato-producing areas. (From survey data 2012.)

Variables	Estimated coefficients	Standard error	*T* value
Constant	–29.29	20.59	–1.42
Age of household head (years)	6.73	4.96	1.36
Educational level (years)	21.24	10.01	2.12**
Income	0.2754	0.0610	4.52***
Labour (availability)	–34.67	27.49	–1.26
Farm size (ha)	7.6251	9.242	8.25***
Experience (years)	17.7	57.5	0.31
Cocoyam	–0.97	1.05	–0.92
Sweetpotato	–0.637	0.650	–0.98
Yam	–0.313	0.523	–0.60
Price of seed potato (₦)	–10.25	4.29	–2.39**
Fertilizer (kg)	266	59	4.51***
R^2			92
F value			98.43***

, Significant at 0.05 probability level; *, significant at 0.01 probability level.

Table 21.3. Estimated elasticities of explanatory variables on demand for seed potato. (From survey data 2012.)

Variable	Elasticity
Age (year)	–0.03
Education (year)	0.00
Income (₦)	0.37
Labour (man days)	–0.03
Farm size (ha)	0.00
Experience (year)	0.07
Cocoyam	–0.01
Sweetpotato	–0.10
Yam	–0.01
Price of seed potato (own-price elasticity)	0.08
Fertilizer (kg)	2.62

Table 21.4. Determinants of supply of seed potato in major potato-producing areas. (From survey data 2012.)

Variables	Estimated coefficients	Standard error	*T* value
Constant	96.12	75.27	1.28
Gender	1.26	3.07	0.41
Educational level	29.43	54.27	0.54
Income	0.0215	0.08	0.27
Household size	36.12	17.65	2.05**
Price of seed potato	–2.12	8.10	–0.26
Cocoyam	2.20	1.57	1.40
Sweetpotato	0.34	1.0	0.34
Yam	–0.812	0.80	–1.01
Labour cost	258	44.6	0.565
Experience	17.28	8.51	2.03**
R^2			30.4
F value			0.226

, Significant @ 0.05 probability level; *, significant @ 0.01 probability level.

in this case are the traders of seed potato in the market and they have experience in the business of seed supply. They often employ the services of their household members (children) to assist them in the trading. The elasticity of the variables in the supply function were less than 1 (inelastic) (Table 21.5) implying that in the long-run, percentage change in these variables will bring less than proportionate change in seed supply. Cocoyam and sweetpotato are substitute production activities to seed potato because the cross-price elasticity is positive, while yam production is complementary production to seed potato.

Table 21.5. Elasticities of supply of seed potato in Nigeria. (From survey data 2012.)

Variables	Elasticities
Gender	0.20
Educational level	0.01
Income	0.52
Household size	0.01
Price of seed potato	–0.02
Cocoyam	0.05
Sweetpotato	0.01
Yam	–0.02
Labour cost	0.18
Experience	0.01

21.4 Conclusion

The trend in potato production is moving towards an increase in area of land cultivated by high-income households. The demand and supply functions indicated that a unit change in each of the factors influencing demand and supply in the model contributed limited changes to the supply of seed potato and that each of these factors is inelastic relative to supply of seed potato, implying that in the long run changes in these factors will bring less than proportionate change in seed supply.

Policy implications

These results calls for policies that will provide legal protection and support on pricing, that will regulate farm labour practices and improve access to mechanized labour; facilitate the importation of quality fertilizer and farm inputs by the private sector; provide functional education to farmers to apply modern knowledge in production and marketing.

Acknowledgement

The authors acknowledge the Executive Director and Management of National Root Crops Research Institute for sponsoring this project.

References

Asumugha, G.N., Lenka, D.M., Ogbonna, M., Okoye, B.C., Dung, E., Njoku, M.E. and Nwosu, K.I. (2007) An analysis of the marketing channels and efficiency of the marketing system for yams in Nigeria. In: Nkamleu, B., Annang, D. and Bacco, N.M. (eds) *Securing Livelihoods through Yams*. International Fund for Agricultural Development (IFAD), Ibadan, Oyo State, Nigeria, pp. 71–78.

Browning, E.K. (1999) *Microeconomic Theory and Application*, 6th edn. Addison-Wesley, New York.

Crissman, C., Anderson, P., Fuglie, K., Kapinga, R., Lemaga, B., Andre, D., Graham, T., Sarath, I. and Coen, B. (2007) Trends in the potato and sweet potato sectors in sub-Saharan Africa and their contribution to the millennium development goals. In: *Proceedings of the 13th International Society for Tropical Root Crops Symposium*, ISTRC, Arusha, Tanzania, pp. 8–19.

Food and Agriculture Organization of the United Nations (FAO) (2008) New Light on a Hidden Treasure: International Year of the Potato 2008 – An end-of-year review. Available at: http://www.fao.org/potato-2008/pdf/IYPbook-en.pdf (accessed 15 September 2015).

Gillespie, A. (2007) *Foundations of Economics*. Oxford University Press, Oxford.

Gwartney, J.D., Stroup, R.L., Sobel, R.S. and David, M. (2008) *Economics: Private and Public Choice*. Cengage Learning, Boston, Massachusetts.

Ivan, P.N.G. (1999) *Managerial Economics*. Blackwell, Oxford.

Lenka, D.M., Dung, E.A., Nwauzor, E.C., Asumugha, G.N. and Panwal, E. (2006a) Analysis of potato marketing and the effects on the socio-economic status of the marketers in Plateau State. In: *Proceedings of the Farm Management Association of Nigeria (FAMAN) 20th Annual Conference*, Federal College of Forestry Research Institute, Jos Plateau. Farm Management Association of Nigeria (FAMAN), Jos, Plateau State Nigeria, pp. 210–214.

Lenka, D.M., Amadi, C.O., Okonkwo, J.C., Dung, E.A., Danbaba, A.K. and Asumugha, G.N. (2006b) *Survey of Potato Production in Nigeria*. National Root Crops Research Institute (NRCRI) Umudike Annual Report. NRCRI, Umudike, Nigeria.

Lenka, D.M., Okonkwo, J.C., Amadi, C.O., Asumugha, G.N. and Nwosu, K.I. (2010) *Potato Production, Storage and Marketing in Nigeria: Farmers' Perspectives and Practices*. National Root Crops Research Institute, Umudike, Nigeria.

Lutaladio, N.B. and Castaldi, L. (2009) Potato: the hidden treasure. *Journal of Food Consumption and Analysis, International Year of Potato (Special Issue)* 22, 491–493.

Ogbonna, M.C., Asumugha, G.N., Anyaegunam, H.N., Okoye, B.C., Onwumere, J. and Akinpelu, O.A. (2007) Resource use efficiency in cassava processing in two selected areas of Abia State. *Pakistan Journal of Social Sciences* 4, 778–781.

Okoye, B.C. and Onyeweaku, C.E. (2007) Economic efficiency of smallholder cocoyam farmers in Anambra State, Nigeria: a translog stochastic frontier cost function approach. *Agricultural Journal* 2, 535–541.

Parkin, M., Powell, M. and Matthews, K. (2002) *Economics*, 5th edn. Addison-Wesley, New York.

William, G.G. (1962) *Potato Growing in Plateau Province*. Institute for Agricultural Research, Samaru, Zaria, Nigeria.

22 An Alternative Technology for Pre-basic Seed Potato Production – Sand Hydroponics

D. Mbiri,[1]* E. Schulte-Geldermann,[1] V. Otazu,[2] R. Kakuhenzire,[3] P. Demo[4] and S. Schulz[5]

[1]International Potato Center Sub-Saharan Africa (CIP-SSA), Nairobi, Kenya; [2]International Potato Center (CIP), Lima, Peru; [3]CIP-Tanzania, Mbeya, Tanzania; [4]CIP, Lilongwe, Malawi; [5]CIP, Addis Ababa, Ethiopia

Abstract

Conventional systems of seed potato production in sub-Saharan Africa (SSA) started to have serious challenges concerning substrate sterilization particularly with the threats from *Ralstonia solanacearum* (Gildemacher, 2012). At the same time over the last two decades there has been an increase in consumption of processed potatoes in the region (Tesfaye *et al.*, 2010). One strategy for overcoming this challenge was to increase the area under production and yields per unit area. An intervention strategy for increasing yield per unit area was making disease-free planting seed potatos available to farmers. In SSA, the International Potato Center (CIP) intervened through introduction of aeroponics technology in the year 2008 under the 3G project (Landeo *et al.*, 2009). Aeroponics technology revolutionized production of pre-basic seed potato production in SSA. However, this technology requires a constant supply of electricity and specialized attendants over the growing period of the crop. In many SSA countries infrastructural challenges such as unreliable supply of electricity and expensive boiler-based sterilization methods for the growing media are a major difficulty in the adoption of this technology (Otazu, 2010). Sand hydroponics offers an alternative technology for clean pre-basic seed potato production. It uses sand as the inert material and an elevated tank to allow the flow of nutrient solution by gravity. Sand is sterilized with sodium hypochlorite which is cheaper than steam boilers (Otazu *et al.*, 2010). Sand aeroponics uses plantlets from tissue culture as well as minitubers from conventional aeroponics. Production rates in sand hydroponics are higher than the conventional system, but less than aeroponics. At Tigoni in Kenya, yields of over 500 minitubers from 1 m^2 were recorded. The minitubers produced from sand hydroponics were larger than those from conventional aeroponics. The initial installation cost and cost per season of sand hydroponics is lower than the conventional aeroponics system. In the future we believe sand hydroponics can replace the conventional system for quality pre-basic seed potato production in SSA.

Keywords: aeroponics, hydroponics, potato minitubers, sub-Saharan Africa (SSA)

*d.mbiri@cgiar.org

22.1 Introduction

Hydroponics means 'working with water', or growing plants in a water-based medium that contains no soil. Growing plants in sand or gravel without soil is not a novel concept but rather an ancient practice. However, growing seed potato in sand hydroponics is a new concept in sub-Saharan Africa (SSA). When crops are grown hydroponically, no weeding is required and there are no soil pests to control. Sand hydroponics consists of wooden boxes, plastic pots and polythene bags in a greenhouse, filled with sterilized river sand and growing plants (Naik and Karihaloo, 2007). Unlike the conventional system, sand is sterilized with ordinary sodium hypochlorite bleach. It is documented that hydroponics was practised many centuries ago in the Amazon, Babylon, Egypt, China and India where ancient men and women used dissolved manure to grow cucumber, watermelons and other vegetables in sandy riverbeds.

The 'hanging garden of Babylon' and the Aztec's floating farms were actually prototypes of hydroponic systems (Kang *et al.*, 1996; Farran and Mingo-Castel, 2006). It was the plant physiologists who started to grow plants with specific nutrients for experimental purposes and gave the name 'nutriculture'. Interest in practical application of 'nutriculture' developed in 1925 when the greenhouse industry expressed interest in its use. Greenhouse soils had to be replaced frequently to overcome problems of soil structure, fertility, pests and diseases. As a result, researchers became interested in the potential use of nutriculture to replace conventional soil culture (Wheeler *et al.*, 1990). Plants often grow faster in water-based medium because nutrients are more readily available and the root area enjoys optimum aeration. In sand hydroponics, sand allows free expansion of tubers unlike some soils which can be too compact for good tuber formation. The International Potato Center Sub-Saharan Africa (CIP-SSA) introduced sand hydroponics as a new technology for the production of clean pre-basic seed potato. The system does not require constant supply of electricity or specialized attendants over the growing period. In addition, water and nutrients flow by gravity. Sand hydroponics is an additional technology to the aeroponics which has revolutionized seed potato production in the SSA region.

22.2 Setting up a Hydroponic System

Sand hydroponics is a screenhouse-based technology and the screenhouse should have a waterproof roof to avoid entrance of rain water, pests and other pathogens of economic importance to potatoes. Sand hydroponics requires clean water free of pathogens. A dependable source of potable water is the best. Clean water that becomes turbid due to heavy rains is not sufficient and should be filtered. A chemical and microbiological analysis should provide useful information on water quality. The pH meter and an electric conductivity (EC) meter are useful tools for monitoring pH and the presence of salts in water or in the sand.

Sand is available in many places across the world. Very fine sand is not good as it becomes compacted and limits tuber development. In some areas, sand may be contaminated with heavy metals that do not go away after washing. In SSA countries, there are no recorded deposits or mining of heavy metals. Sand is sterilized by soaking it for 10 min in a solution of sodium hypochlorite at 10% concentration. Any excess bleach is removed by rinsing the sand two or three times with clean water.

If crates or wooden boxes are used to hold the sand, then the size and design will need to be considered according to the screenhouse space. Although wood is available in many places, other materials can also be used. The width and length of boxes depend on the screenhouse size as there is a need to optimize the use of screenhouse space. Enough space must be allowed for servicing the boxes with tools such as a wheelbarrow.

Pots of different sizes can be used. Black plastic bags can also be used. A plastic dish needs to be placed under each pot. The dish is useful to avoid contamination and also serves as an indicator when excess nutrient solution is delivered. The nutrient solution valve is shut when the nutrient solution is visible in the dish. Sand can be recycled after one harvest and can be used for several crops. Once harvesting is complete, all the organic materials (roots, stems and tubers) are removed and sodium hypochlorite bleach is used to re-sterilize the sand. All the procedures for the treatment with bleach are followed. The drainage system is helpful during the rinsing process. A few days must be allowed before starting to plant another crop.

Nutrient preparation

Greenhouse soluble-grade fertilizers are used. To prepare 1000 l nutrient solution in tank the following procedure can be used. Weigh and dissolve 136 g of potassium phosphate in 10 l of water in a bucket. This fertilizer does not dissolve easily and may take 1 h to dissolve completely. In another bucket containing 10 l of water, add and dissolve the following fertilizers: (i) 236 g calcium nitrate; (ii) 504 g potassium nitrate; (iii) 100 g magnesium sulfate; and (iv) 12 g Microsol B. Mix contents of both buckets and add to 1000 l of water in a nutrient reservoir. Wait at least 2 h until all the solid particles have dissolved before opening the valves to deliver the nutrients to the plants.

Sand hydroponic infrastructure

The basic components of a sand hydroponic infrastructure are: (i) a raised nutrient tank; (ii) piping; and (iii) the beds or pots. The supporting structure to lift the nutrient tank can be made of wood or metal. Metal frames are more durable, but also more expensive. The structure supporting the nutrient tank must be adequately anchored and elevated by 1.8–2.0 m in order for the nutrients to flow by gravity. The nutrient tank should be placed outside the screenhouse. This makes better use of screenhouse space and also keeps the nutrient solution in a cool place.

Piping and fittings

Distribution pipes and fittings can be of polyvinyl chloride or polyethylene material. Commonly size 16 mm feeding pipes are used. These are usually available but if not size 20 mm can also do. A filter should be installed somewhere between the nutrient tank outlet and the distribution pipe. Black canister (12.7 mm (1/2 inch) or 19 mm (3/4 inch)) filters are useful for retaining insoluble materials that may block the drippers. The filter needs to be cleaned from time to time. A general valve should also be installed as well as secondary valves that control the nutrient solution flow in each bed.

Drippers

Drippers are convenient because they allow a better control of the nutrient delivery and are relatively cheap (US$0.20 each). If drippers are not available, flat pipes can be used with incorporated drippers or holes can be made in the pipes using a drill or nails.

22.3 Lessons

In SSA per capita consumption of electricity is low compared with Western developed economies. Sand hydroponics promises an intermediate technology where the power supply will not be a challenge and community groups will be able to produce pre-basic seed potato free from *Ralstonia solanacearum*, the bacteria that causes bacterial wilt in potatoes. Bacterial wilt is a real threat to potato farming in the humid tropics (Gildemacher, 2012). The use of easily available timber and sand makes sand hydroponics a technology of choice for seed production in SSA. The cost of establishing and running a sand hydroponics seed unit is far cheaper than a conventional unit that depends on diesel/electricity boilers for substrate sterilization.

Seeds from sand hydroponics are grown on a sterile substrate and there is no effluent discharged into the environment. The bleach used for sand sterilization releases chlorine freely into the environment without any contamination or pollution. Additionally, water retention is high owing to the small size of the sand particles, allowing fewer irrigation cycles during the course of the day, unlike other systems. The initial and seasonal costs of sand hydroponics are far below those of aeroponics.

If pre-basic seed is cheaper (Maldonado *et al.*, 2008) then, the ultimate cost of the certified generations will also be cheap. This was further supported by the argument on pricing of seed in SSA by P. Demo (Demo *et al.*, 2007).

Minitubers produced by sand hydroponics are larger than those harvested from aeroponics. The larger minitubers can be cut further to increase the amount of planting material. Sand hydroponics is an 'open' ('run-to-waste') system, whereby the nutrient solution is not recycled, and this greatly reduces the likelihood of diseases, unlike aeroponics where a disease can spread very fast due to recycling of the nutrient media.

22.4 Conclusion and Recommendations

With the increasing demand for clean planting materials and quality declared seed gaining acceptance in SSA, there is a need for technologies that can meet the demand for clean pre-basic seed (Mwanga and Fuentes, 2010). Sand hydroponics is a technology based on easily and cheaply available material. Sand culture is one of the most efficient and cost-effective methods of soilless culture, and it is widely used in the dry arid regions (Ritter *et al.*, 2001) of the world such as the Middle East and therefore fits well in SSA. Sand hydroponics is new in the region and there are prospects of replicating its use in several SSA countries with seed potato changes.

References

Demo, P., Low, J. and Mwenge, J.O. (2007) Potato production in Malawi: strengths, weaknesses, opportunities and threats. In: *Proceedings of the African Potato Association Conference*, Vol 7. African Potato Association, Alexandria, Egypt, pp. 339–345.

Farran, I. and Mingo-Castel, A.M. (2006) Potato minituber production using aeroponics: effect of plant density and harvesting intervals. *American Journal of Potato Research* 83, 47–53.

Gildemacher, P. (2012) Innovations in seed systems in East Africa. PhD thesis, Wageningen, The Netherlands.

Kang, J.G., Kim, S.Y., Kim, H.J., Om, Y.O. and Kim, J.K. (1996) Growth and tuberization of potato (*Solanum tuberosum* L.) cultivars in aeroponics, deep flow technique and nutrient film technique culture systems. *Journal of Korean Horticulture Science* 37, 24–27.

Landeo, J., Barker, I. and Otazú, V. (2009) Novel approaches to promote and diffuse new potato varieties in Kenya. *International Potato Center (CIP) Working Paper*. CIP, Lima.

Maldonado, L., Thiele, G. and Otazu, V. (2008) Cost analysis of conventional quality potato seed production and production by aeroponics. In: *Alternativas al Uso del Bromuro de Metilo en la Produccion de Semilla de Calidad de Papa*. International Potato Center (CIP), Lima, pp. 45–53. (in Spanish)

Mwanga, R. and Fuentes, S. (2010) *Sweet Potato Quality Declared Planting Materials*. Food and Agriculture Organization of the United Nations (FAO), Rome, pp. 81–88.

Naik, P.S. and Karihaloo, J.L. (2007) *Micropropagation for Production of Quality Potato Seed in Asia-Pacific*. Asia-Pacific Consortium on Agricultural Biotechnology, New Delhi, p. 54.

Otazu, V. (2010) *Manual on Quality Seed Potato Production Using Aeroponics*. International Potato Center (CIP), Lima, p. 44.

Ritter, E., Angulo, B., Riga, P., Herran, C., Relloso, J. and San Jose, M. (2001) Comparison of hydroponic and aeroponics cultivation systems for the production of potato minitubers. *Potato Research* 44, 127–135.

Tesfaye, A., Lemaga, B., Mwakasendo, J.A., Nzohabonayoz, Z., Mutware, J., Wanda, K.Y., Kinyae, P.M., Ortiz, O., Crissman, C. and Thiele, G. (2010) Markets for fresh and frozen potato chips in the ASARECA region and the potential for regional trade (Ethiopia, Tanzania, Rwanda, Kenya, Burundi and Uganda). *International Potato Center (CIP) Working Paper*. CIP, Lima.

Wheeler, R.M., Mackowiak, C.L., Sager, J.C., Knott, W.M. and Hinkle, C.R. (1990) Potato growth and yield using nutrient film technique (NFT). *American Journal of Potato Research* 67, 177–187.

23 Improving Seed Health and Seed Performance by Positive Selection in Three Kenyan Potato Varieties

E. Schulte-Geldermann,[1]* P.R. Gildemacher[2] and P. Struik[3]
[1]*International Potato Center Sub-Saharan Africa (CIP-SSA), Nairobi, Kenya;* [2]*Royal Tropical Institute (KIT), Amsterdam, The Netherlands;* [3]*Wageningen University, Wageningen, The Netherlands*

Abstract

Selecting seed potatoes from healthy-looking mother plants (positive selection, PS) was compared with the common Kenyan farmer practice of selection from the harvested bulk of potatoes (farmer selection, FS) in 23 farmer-managed trials. Data was obtained from replicated researcher- and farmer-managed trials comparing PS with seed recycling, which is the common farmer practice in Kenya. Virus incidence levels were scored through double-antibody sandwich (DAS)-ELISA. The effect of PS on yield and virus incidence was investigated, and the relationship between virus incidence and yield was analysed.

PS assured lower incidences of *Potato leafroll virus* (PLRV) (39%), *Potato virus Y* (PVY) (35%) and *Potato virus X* (PVX) (35%) and out yielded FS irrespective of the agroecology, crop management, soil fertility, variety and quality of the starter seed, with an overall average of 30%. Regression analysis showed a relationship between lower virus incidence and higher yield for the varieties.

The results from this research have conclusively shown that PS is a suitable practice for maintenance of seed potato quality which can benefit all smallholder potato producers who at some stage select seed potatoes from their own fields. PS should thus be incorporated routinely in agricultural extension efforts.

Targeted controlled research to investigate additional factors (such as the effect of PS on average potato crop virus titre) would increase the understanding of the mechanisms behind PS. Current seed-potato-system management decisions are based on the assumption that degeneration as a result of tuber-borne diseases is an inevitable fact, and that regular seed renewal from a reliable disease-free source is the only way to maintain an acceptable yield potential. However, this research has shown that through PS yields of potato populations with high levels of virus infection can be improved. It would be of interest to witness potato yields over several generations of applying PS to a degenerated potato crop. This would allow one to challenge the common belief that degeneration is inevitable and irreversible in a potato population. The results obtained may confirm that as opposed to degeneration of a potato plant population over generations, also regeneration needs to be considered an option, provided ware potato farmers manage their selection process well.

Keywords: positive selection, potato, *Potato leafroll virus* (PLRV), *Potato virus X* (PVX), *Potato virus Y* (PVY), seed degeneration

*e.schulte-geldermann@cgiar.org

23.1 Introduction

The most important yield-determining factor in potato cultivation is the quality of the seed tubers used (Struik and Wiersema, 1999). The difficult availability of affordable high quality seed potatoes is the major constraint for potato production systems worldwide (Struik and Wiersema, 1999). Similarly in East Africa, the availability of affordable quality seed potato is a major obstacle for improving the profitability of potato production (Gildemacher *et al.*, 2009a; Hirpa *et al.*, 2010).

Positive seed potato selection is a simple technology to manage seed potato quality. It consists of selecting healthy-looking mother plants to obtain seed tubers for the next season's crop. Positive selection (PS) has been proved to be a promising complementary technology for smallholder farmers in Kenya, in addition to seed production and marketing by specialized seed growers (Gildemacher *et al.*, 2011). Specifically the fact that it has the potential to increase smallholder yields without monetary investment (Gildemacher *et al.*, 2012), and that it fits well into the prevailing seed sourcing strategy of smallholders, which is largely based on self- and neighbour supply (Gildemacher *et al.*, 2009b) make it a valuable complementary technology.

In this chapter data are presented from replicated researcher- and farmer-managed trials comparing PS with seed recycling, following the common farmer practice in Kenya. The chapter investigates the effect of PS on yield, virus incidence, and looks at the relationship between virus incidence and yield. The chapter discusses the proven and possible additional causes of the yield effect of PS, and provides suggestions for the direction of future research to optimize the use of the simple technology of PS in seed potato system improvement in sub-Saharan Africa (SSA).

23.2 Materials and Methods

Replicated on-farm trials

The 23 sets of PS and farmer selection (FS) seeds were used to plant replicated farmer-managed field trials at 15 different locations in the main potato-growing areas of Kenya to compare the yields from the different types of seed tubers. Each of the 23 source fields was divided into two. From half the field, tubers were sourced by practising PS, while on the other half of the field common FS was applied. PS entailed pegging healthy-looking mother plants before full flowering by farmer groups which had received a basic group-based training (Gildemacher *et al.*, 2007a). At harvest, pegged plants were harvested individually and seed-sized tubers were collected to serve as seed tubers, provided all tubers of the individual plant looked healthy. FS consisted of the selection of seed-sized tubers from the bulk of harvest potatoes from the other half of the source field, following common farmer practice.

Trials were planted either during the short rainy seasons (October 2009–January 2010) or the long rainy season (April 2010–August 2010), in a randomized complete block design layout with three or four replications of 40 plants each, at 30 cm × 75 cm distance. Fertilization was based on 90 kg N/ha supplied in the form of NPK 10:26:10 at planting in the planting hole. Late blight was controlled through a spraying regime with Ridomil and Mancozeb, adapted in response to actual disease occurrence. Further management was done according to farmer practice.

At harvest all marketable tubers (> 30 mm) were collected and weighed. For various reasons (e.g. porcupine damage, improper late-blight management or theft) not all 23 farmer-managed experiments were successful. In total 21 data sets were obtained to reliably assess yield. The pair-wise yield data were analysed by testing for a significant difference between PS and FS using a two-tailed *t*-test.

Virus infection level testing

Samples of 40 tubers were taken from both the PS and the FS seed lots for virus infection determination. Sampled tubers were tested individually for infection with *Potato virus Y* (PVY), *Potato leafroll virus* (PLRV) and *Potato virus X* (PVX) through double-antibody sandwich (DAS)-ELISA (International Potato

Center, Lima, Peru) of leaf sap obtained from plants that were grown from eyes cut after harvest and grown in aphid-free greenhouse chambers for 4–6 weeks (Casper and Meyer, 1981; Torrance, 1992).

The pair-wise virus infection data (expressed as a percentage for the individual viruses tested) and yield data were analysed by testing for a significant difference between PS and FS using two-tailed *t*-tests. Virus infection levels of PLRV, PVY and PVX were plotted against yield and trend lines were fitted using the Statistical Program for Social Scientists (SPSS) curve estimation procedure (IBM SPSS statistics 20). In addition, a combined virus infection indicator of the three viruses was calculated by taking the simple absolute sum of the number of infected sample tubers in the total sample of 40 tubers per treatment. This index can be larger than the total number of tubers in the sample and should be considered as a measure for the virus load. For this index the same plotting procedure was used.

23.3 Results

Yields

The results on yields of potatoes derived from different positive seed selection and farmer seed potato selection derived from a total of 21 farmer-managed randomized complete block trials are summarized in Table 23.1. When analysing the combined paired observations through a *t*-test, PS clearly out-yielded the FS treatment, irrespective of variety or quality of the starter material (Table 23.1).

Virus infection

The average effect of PS compared with FS on PLRV, PVY and PVX infection rate as measured through DAS-ELISA testing of plant samples is shown in Table 23.2. The results show a significant reduction in PLRV, PVY and PVX infection rate. The average measured PVY infection was 25.0% for PVY after PS, compared with 38.4% for FS. The infection rate with PLRV went down from 31.1% for FS to 19% for PS.

The relationship between yield increase and virus reduction

The yield differences observed in the individual trials were plotted against the change in virus infection (Fig. 23.1). For the varieties Tigoni and to a somewhat lesser extent Dutch

Table 23.1. Yield from positive seed selection (PS) compared with common farmer seed potato selection (FS) presented separately for the varieties Asante, Tigoni and Dutch Robyjn, and for two quality levels of the initial seed source.

	Yield[a]				Yield increase				
Name	PS (t/ha)	SD	FS (t/ha)	SD	(t/ha)	%	*P* (two-tailed)	*T* value	df[b]
Asante	13.9	1.9	11.0	2.2	2.8	25	0.000	7.5	21
Tigoni	18.6	7.4	15.1	5.9	3.5	23	0.001	3.8	19
Dutch Robyjn	20.9	10.1	15.5	6.9	5.4	35	0.000	6.0	25
Farmer quality[c]	14.1	2.3	11.1	2.1	3.0	27	0.000	8.7	51
High quality[d]	30.5	7.0	23.1	4.2	7.3	32	0.000	5.3	15
All	17.9	8.0	13.9	5.8	4.0	30	0.000	8.5	67

[a]PS, positive selection; FS, farmer selection. Each source field was divided into two. From half the field, tubers were sourced by practising PS, while on the other half of the field common FS was applied.
[b]df, degrees of freedom.
[c]Seed potatoes for these trials were selected from fields planted with seed potatoes from the local market or farmer fields.
[d]Seed potatoes for these trials were selected from fields planted with basic seed or certified seed.

Table 23.2. The effect of positive selection (PS) compared with farmer selection (FS) on the level of infection with potato leaf roll virus (PLRV), potato virus X (PVX) and potato virus Y (PVY).

	FS[a]		PS[a]		Difference (PS – FS)			
Virus	Infection (%)	SD	Infection (%)	SD	Infection (%)	Reduction (%)	*P* value	*T* value
PLRV	31.1	14.2	19.0	8.2	12.1	39	0.000	7.00
PVX	7.5	6.5	4.9	4.5	2.6	5	0.004	2.93
PVY	38.4	18.4	25.0	12.6	13.4	35	0.000	8.06

[a]Each source field was divided into two. From half the field, tubers were sourced by practising PS, while on the other half of the field common FS was applied.

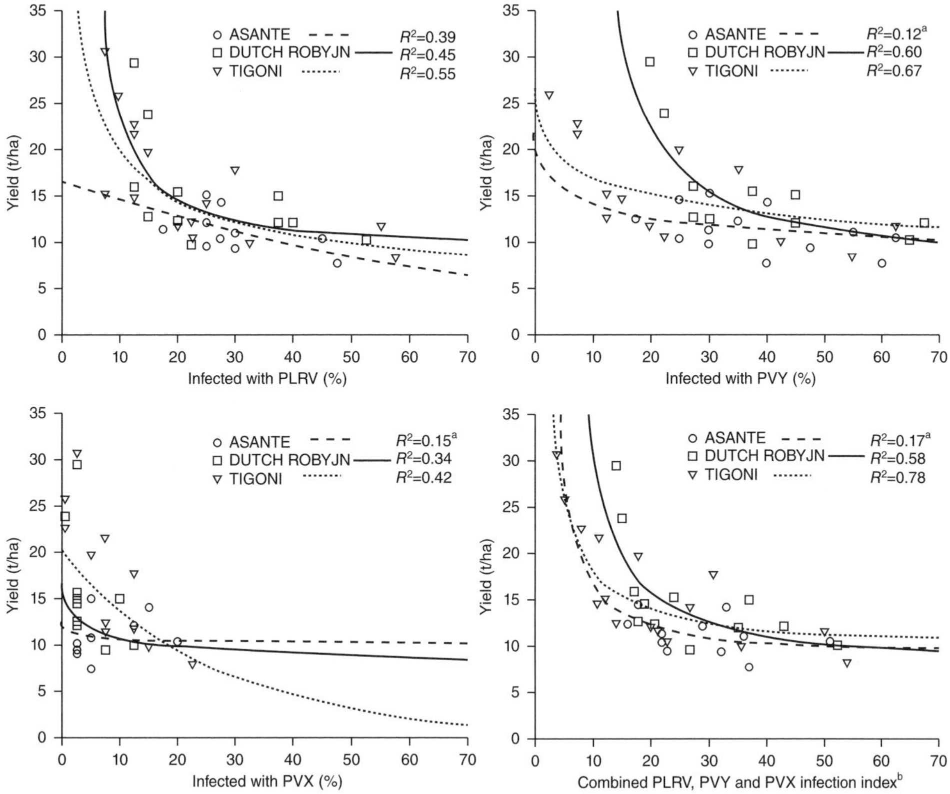

Fig. 23.1. Fresh tuber yield ('Yield') plotted against the percentage infection with PLRV, PVY or PVX and against the sum of the infection count of all three viruses for the potato varieties Asante, Dutch Robyjn and Tigoni. [a]Non-significant regression; [b]index calculated as the simple absolute sum of the number of infected tubers for each virus in the total sample of 40 tubers.

Robyjn the expected relationship of lower yields with increasing virus levels could be observed, but the shape of the regression analysis curve is not consistent with theory.

The variety Asante did not respond to a reduction in infection with PVY and PVX, and responded less than the other two varieties to a reduction in PLRV. When plotting the sum of the number of infected tubers for the three viruses in the samples against the yield, this could explain 78% and 58% of the variation for Tigoni and Dutch Robyjn, respectively, while for Asante it could only account for 17% of the variation.

23.4 Discussion

PS results in yield increase compared with FS

The results from the trials do confirm the earlier findings from farmer-managed trials (Gildemacher *et al.*, 2011, 2012) that PS is a technology that can provide smallholder producers with a significant yield advantage compared with the common practice of indiscriminate recycling of seed potatoes from their ware potato harvest. This was confirmed for three different popular Kenyan potato varieties over a wide range of agro-ecologies and (farmer) management practices. The three different varieties appeared to differ in their reaction to PS, with yield increases of 35%, 25% and 23% for Dutch Robyjn, Asante and Tigoni, respectively. Generally, PS resulted in substantial yield increases when used to source seed potatoes from farmer ware potato plots, but it also increased yields when applied on fields planted with high quality seed potatoes from formal seed sources (certified and basic seed potatoes).

PS reduces virus infection compared with common farmer practice

Both yield increases and a reduction in virus infection could be demonstrated beyond any doubt in a very wide variety of circumstances. In addition, the data show a significant relationship between yield and virus infections (Fig. 23.1), suggesting that an important mechanism behind the effect of PS is the reduction in virus infection in the plant population. The data do show the reduction in virus incidence that was measured has been a factor in the yield increases that have been observed as a result of PS, especially in the case of Tigoni and Dutch Robyjn varieties. That the results of regression analysis in the relationships are not that distinct may be the result of several factors. The different trials were implemented under different circumstances, with seed potatoes derived from different sources. This does allow for concluding that the technology is effective over a wide array of circumstances, but has also introduced variation in the trials that does to an extent obscure the effect of PS on virus levels and of virus levels on yield. Three different varieties were used, which did not respond in the same manner to PS. For Asante and Dutch Robyjn 6 data points and for Tigoni 8 data points were ultimately available to compare virus incidence levels between FS and PS. As such the comparison between yield increase and virus load reduction need to be interpreted with some caution. However, the variety Asante only showed a significant yield response to a reduction in PLRV infection, but this was less pronounced than for the varieties Tigoni and Dutch Robyjn. In addition, Asante showed no response to a reduction in PVY or PVX. This suggests that Asante harbours partial resistance or tolerance for these viruses. Differences in response to virus infection between varieties are common (Bawden *et al.*, 1948; MacKinnon and Munro, 1959; de Bokx, 1972; Radcliffe and Ragsdale, 2002). In the case of extreme resistance, however, the virus would not be detected through DAS-ELISA. The yield response of Asante to PS was similar to that of the other two varieties, while it does respond much less strongly to a reduction in PVY, PLRV and PVX incidence. This suggests that in addition to a lower incidence of PLRV, PVY and PVX, there might have been other factors that played a role in the measured effects on yield resulting from PS. The most likely explanation is that other viruses which were not tested have also been reduced in incidence as a result of PS. *Potato virus A* (PVA), *Potato virus M* (PVM) and *Potato virus S* (PVS) were found to be abundant in farmer-based seed systems in Kenya (Muthomi *et al.*, 2009), while PVA was found to be common in seed tubers sold at Kenyan rural markets (Gildemacher *et al.*, 2007b). PVA is often not considered to cause serious damage, but can cause severe symptoms in combination with PVY or PVX (Nganga and Shideler, 1982), a common combination of infections in farmer fields in Kenya (Gildemacher *et al.*,

2009b). Any endemic virus disease that would cause visible symptoms will be affected by PS, and would have contributed to the yield increase realized by PS compared with common FS.

Similarly to virus infection levels, the levels of other seed-borne diseases having an effect on general plant appearance will be affected by PS. Turkensteen (1987) identified *Erwinia* spp. bacteria (currently *Pectobacterium* spp.) and *Fusarium* spp. fungi as 'important' seed-borne pathogens in Central Africa. In addition, bacterial wilt (*Ralstonia solanacearum*) is endemic in Kenya (Wakahiu *et al.*, 2007). For the latter, however, the symptoms can hardly be mistaken and all was done to avoid bacterial wilt infection of the trials.

In addition to having an effect on the incidence of viruses on the potato plant population, it cannot be ruled out that PS results in a lower virus load of individual seed potatoes, the virus titre. Little is known about the effect of virus titre in seed tubers on the final yield. Van der Zaag (1987) reported that tubers infected late in the previous season had less severe symptoms and yield reduction than those infected earlier and mentioned that diseased tubers that had been recycled for a number of generations did worse than those having a shorter history of infection. However, no data to support this were presented. Barker and Woodford (1987) reported unusually mild PLRV symptoms in the progeny of late-infected mother plants. Interestingly enough, however, they could not show a difference in virus titre in the leaves about 7 weeks after planting the progeny tubers.

Further research to improve the understanding of the effect of PS on average potato crop virus titre, and the effect of virus titre on crop growth would be of interest. Targeted controlled research to investigate such additional factors would increase the understanding of the mechanisms behind PS. What possibly would be even more interesting is to understand the full potential of PS over several generations. The basic assumption of seed potato systems is that degeneration as a result of tuber-borne diseases is an inevitable fact, and that seed renewal from a reliable disease-free source is the pillar of maintenance of an acceptable yield potential. Through PS, however, yields of potato populations with high levels of virus infection can be improved. It would be of interest to witness potato yields over several generations of applying PS to a degenerated potato crop. This would allow one to challenge the common belief that degeneration is inevitable and irreversible in a potato population, and to assess whether as opposed to degeneration, also regeneration needs to be considered an option. Answering this question is of great importance for seed potato systems in countries where production levels are, unlike in some developed countries, far below the theoretically optimal production level. A better understanding of the rate of degeneration in relation to disease pressure, farmer management and the regeneration potential by practising PS will allow for better informed investments in seed potato programme building and the seed renewal strategy by individual potato producers. Combined economic and seed degeneration and regeneration research could contribute to this improved decision making.

Considering the highly conclusive results with regard to the effect of PS, and the fact that the technology requires only sticks or another type of marker and labour as inputs, it is very suitable for seed quality maintenance by smallholder potato farmers, who form the majority of potato farmers in SSA. In addition, positive seed potato selection can fairly easily be learned (Gildemacher *et al.*, 2012). Based on these findings it can be recommended to include PS in the training curricula and programmes of smallholder potato farmers in countries where sourcing seed potatoes from their own ware potato crop is common practice.

Acknowledgements

The research was conducted within the '3G' project, implemented by the International Potato Center in Kenya, Rwanda and Uganda, from 2008 to 2011 and was made possible by the generous support of the American people through the United States Agency for International Development (USAID).

References

Barker, H. and Woodford, J. (1987) Unusually mild symptoms of potato leafroll virus in the progeny of late-infected mother plants. *Potato Research* 30, 345–348.

Bawden, F.C., Kassanis, B. and Roberts, F.M. (1948) Studies on the importance and control of potato virus X. *Annals of Applied Biology* 35, 250–265.

Casper, R. and Meyer, S. (1981) Die anwendung des ELISA verfahrens zum nachweis pflanzenpathogener viren. *Nachrichtenblatt des Deutschen Pflanzenschutzdienstes* 33, 49–54.

de Bokx, J.A. (1972) Spread of potato virus S. *Potato Research* 15, 67–70.

Gildemacher, P., Demo, P., Kinyae, P., Wakahiu, M., Nyongesa, M. and Zschocke, T. (2007a) *Select the Best: Positive Selection to Improve Farm Saved Seed Potatoes. Trainers Manual*. International Potato Center (CIP), Nairobi.

Gildemacher, P.R., Mwangi, J., Demo, P. and Barker, I. (2007b) Prevalence of potato viruses in Kenya and consequences for seed potato system research and development. In: Khalf-Allah, A. (ed.) *Seventh Triennial African Potato Association Conference Proceedings*, Vol. 7, Alexandria, Egypt. African Potato Association, Alexandria, Egypt, pp. 84–92.

Gildemacher, P., Kaguongo, W., Ortiz, O., Tesfaye, A., Woldegiorgis, G., Wagoire, W., Kakuhenzire, R., Kinyae, P., Nyongesa, M., Struik, P. and Leeuwis, C. (2009a) Improving potato production in Kenya, Uganda and Ethiopia: a system diagnosis. *Potato Research* 52, 173–205.

Gildemacher, P., Demo, P., Barker, I., Kaguongo, W., Woldegiorgis, G., Wagoire, W., Wakahiu, M., Leeuwis, C. and Struik, P. (2009b) A description of seed potato systems in Kenya, Uganda and Ethiopia. *American Journal of Potato Research* 86, 373–382.

Gildemacher, P., Schulte-Geldermann, E., Borus, D., Demo, P., Kinyae, P., Mundia, P. and Struik, P. (2011) Seed potato quality improvement through positive selection by smallholder farmers in Kenya. *Potato Research* 54, 253–266.

Gildemacher, P., Leeuwis, C., Demo, P., Kinyae, P., Mundia, P., Nyongesa, M. and Struik, P.C. (2012) Dissecting a successful research-led innovation process: the case of positive seed potato selection in Kenya. *International Journal of Technology Management and Sustainable Development* 11(1), 67–92.

Hirpa, A., Meuwissen, M., Tesfaye, A., Lommen, W., Oude Lansink, A., Tsegaye, A. and Struik, P.C. (2010) Analysis of seed potato systems in Ethiopia. *American Journal of Potato Research* 87, 537–552.

MacKinnon, J. and Munro, J. (1959) Comparative rates of movement of potato virus X into tubers and eyes of three potato varieties. *American Journal of Potato Research* 36, 410–413.

Muthomi, J., Nyaga, J., Olubayu, F., Nderitu, J., Kabira, J., Kiretai, S., Auro, J. and Wakahiu, M. (2009) Incidence of aphid-transmitted viruses in farmer-based seed potato production in Kenya. *Asian Journal of Plant Sciences* 8, 166–171.

Nganga, S. and Shideler, F. (1982) *Potato Seed Production for Tropical Africa*. International Potato Center (CIP), Nairobi.

Radcliffe, E. and Ragsdale, D. (2002) Aphid-transmitted potato viruses: the importance of understanding vector biology. *American Journal of Potato Research* 79, 353–386.

Struik, P.C. and Wiersema, S.G. (1999) *Seed Potato Technology*. Wageningen University Press, Wageningen, The Netherlands.

Torrance, L. (1992) Developments in methodology of plant virus detection. *Netherlands Journal of Plant Pathology* 98, 21–28.

Turkensteen, L.J. (1987) Survey of diseases and pests in Africa: fungal and bacterial diseases. *Acta Horticulturae* 213, 151–159.

Van der Zaag, D.E. (1987) Growing seed potatoes. In: de Bokx, J.A. and Van der Want, J.P.H. (eds) *Viruses of Potatoes and Seed-potato Production*, 2nd edn. Pudoc, Wageningen, The Netherlands, pp. 176–203.

Wakahiu, M.W., Gildemacher, P.R., Kinyua, Z.M., Kabira, J.N., Kimenju, A.W. and Mutitu, E.W. (2007) Occurrence of potato bacterial wilt caused by *Ralstonia solanacearum* in Kenya and opportunities for intervention. In: Khalf-Allah, A. (ed.) *Seventh Triennial African Potato Association Conference Proceedings*, Vol. 7, Alexandria, Egypt. African Potato Association, Alexandria, Egypt, pp. 267–271.

24 Scaling Up the Adoption of Positive Selection and Seed-plot Techniques in Seed Potato Systems in Uganda – UNSPPA's Experience

S. Tindimubona,[1*] Z.M. Kinyua,[2] F. Opio,[3] I. Rwomushana,[3] A. Bararyenya,[4] P. Nahayo,[5] W. Kaguongo,[6] J.N. Oggema,[7] B. Biryomumaisho,[1] P. Namugga[5] and I.N. Kashaija[8]

[1]Uganda National Seed Potato Producers' Association (UNSPPA), Kabale, Uganda; [2]Kenya Agricultural and Livestock Research Organization (KALRO)-Kabete, Nairobi, Kenya; [3]Association for Strengthening Agricultural Research in Eastern and Central Africa (ASARECA), Entebbe, Uganda; [4]Institut des Sciences Agronomiques du Burundi, Bujumbura, Burundi; [5]Kachwekano Agricultural Research and Development Research Institute, Kabale, Uganda; [6]National Potato Council of Kenya (NPCK), Nairobi, Kenya; [7]Agricultural Development Corporation, Molo, Kenya; [8]National Agricultural Research Organization, Entebbe, Uganda

Abstract

Accessing quality seed potato by smallholder farmers has continued to be a major challenge to increased potato production. An intervention by the Uganda National Seed Potato Producers' Association (UNSPPA) with support from the Association for Strengthening Agricultural Research in Eastern and Central Africa (ASARECA), the International Potato Center (CIP) and the National Agricultural Research Organisation (NARO) aimed to promote the adoption of positive selection (PS) and seed-plot techniques in south-western Uganda, a major potato production zone in the country. In a participatory approach with farmers, trials were set up to compare seed potato yields produced by PS and seed obtained from farmers' own practice. Using PS, an increase of 34% or 6.8 t/ha was recorded compared with farmer's seed. This provides an economic gain with minimal or no increase in cash investment. Another approach called the seed-plot technique in which small quantities of high-quality seed are grown under intensive management on raised beds at close spacing was also demonstrated on-farm in participatory experiments involving researchers, development agents and farmers. Validations indicated that the seed-plot technique achieved two-and-a-half to three times land productivity and about 50% bacterial wilt reduction compared with conventional ridge planting. Under different programmes supported by ASARECA, CIP, Carry American Relief Everywhere (CARE) Enterprise, Environment and Equity in the Great Lakes (EEEGL), National Agricultural Advisory Services (NAADS) and NARO, these techniques have reached 5890 farmers from south-western Uganda, out of which 80% have adopted the technologies. It was revealed that such simple technologies can enhance the quality of

*sbtindi@gmail.com

seed potato used by smallholder farmers every season. It was further shown that technology adoption was achieved faster through stakeholder participatory experimentation and group learning.

Keywords: bacterial wilt, positive selection (PS), quality seed potato, seed-plot techniques, smallholder farmers, technology adoption

24.1 Introduction

Potato (*Solanum tuberosum* L.) has continued to be an important food and cash crop in East and Central Africa. It is ranked second in Kenya, sixth in Burundi and tenth in Uganda. The area under the crop has continued to increase steadily over the years, principally under smallholder production systems, which are characterized by high population density, with farmers owning parcels of land measuring between 0.25 and 5 ha. At the turn of the century, at least 89 varieties had been released in the Eastern Africa sub-region with total acreage of 295,000 ha under improved varieties.

However, a huge gap exists between on-farm potato yields, commonly below 10 t/ha, and the yields that can be attained under improved growing circumstances, in which case 40–60 t/ha is achievable (Raemaekers, 2001). National mean yields for Uganda are below 8 t/ha (UBOS, 2010). The low yields are attributed to both biotic and abiotic stresses, and poor management practices. Poor quality of seed tubers used by farmers is viewed as a major yield-limiting factor (Gildemacher *et al.*, 2007). Farmers are frequently faced with several challenges that are related to lack of good quality seed potato of consumer-preferred, high-yielding varieties. Currently, certified and disease-free seed tubers account for less than 5% of the whole seed potato market in Uganda. This scenario has encouraged the common practice among potato farmers of planting own-saved tubers from previous harvests or sourced from markets or neighbours. These tubers are often of poor health status owing to latent infections by *Ralstonia solanacearum* (causal agent of bacterial wilt), viruses and other tuber-borne pathogens (Kinyua and Smith, 2007). Past studies indicate that over 70% of farmer fields in Uganda are infected by bacterial wilt (Kaguongo *et al.*, 2010; International Fertiliser Development Centre, 2012, unpublished data).

Several initiatives have been undertaken to address the challenge of the lack of availability of certified and quality seed potato in Uganda. The formal seed potato production system is also not well developed and there is no formal institutional arrangement for multiplication and distribution of certified seed potato. Past initiatives have included adoption of rapid seed multiplication technologies and introduction and adoption of 'clean' seed which resulted in the increased number of seed multipliers. The challenges in seed inspection procedures and distribution have limited the benefits of these interventions.

In Uganda, the Uganda National Seed Potato Producers' Association (UNSPPA), a local organization which started in 1995 with the mandate of improving the availability of Quality Declared Seed (QDS) potatoes, fighting food insecurity and increasing household incomes among smallholder farmers, has continued to embrace the contribution of new innovations in the seed potato supply systems. In order to improve the availability of adequate quantities of high-quality seed potato, innovative ways of multiplying seed potato known as the seed-plot technique and positive selection (PS) have been developed for smallholder farmers. The seed-plot technique is akin to a seed potato 'nursery' and PS is conducted from the conventional fields.

With support from the Association for Strengthening Agricultural Research in Eastern and Central Africa (ASARECA), the International Potato Center (CIP), the National Agricultural Advisory Services (NAADS), Carry American Relief Everywhere (CARE) Enterprise, Environment and Equity in the Great Lakes (EEEGL) and the National Agricultural Research Organisation (NARO), UNSPPA has continued to disseminate knowledge on the seed-plot technique and PS among smallholder potato farmers and now members wish to share experiences with other stakeholders in the seed potato systems in this chapter.

24.2 Methodology

Seed-plot technique

The seed-plot technique has been demonstrated in nursery plots measuring 1.8 m wide and varying in length depending on availability of land and the seed needs of the farmers' group. Well-sprouted tubers of the variety preferred by a particular farmer are usually planted at a spacing of 30 cm × 30 cm in well-prepared plots (Kinyua *et al.*, 2010). The technique allows a farmer to produce large quantities of seed potato from a relatively small piece of land through high density planting to maximize the production per unit area. The small size of the plots also permits intensive crop management practices that lead to better quality seed potato. Watering on a small scale can be done to produce off-season seed potato. The system requires 50% less land than the conventional production system to meet on-farm seed tuber requirements. It is recommended that the 'freed' land be managed as next season's seed plot by leaving it fallow or planting a short season non-solanaceous crop (Kinyua *et al.*, 2012).

Seed plot establishment

UNSPPA acknowledges that healthy tubers for seed-plot establishment should be obtained from a reliable source, preferably a seed production station or a recognized seed potato dealer. UNSPPA witnesses that, in order to avoid the risk of infestation by bacterial wilt and other pests/pathogens, a seed plot must be established on land: (i) without a history of potato production or other solanaceous (potato family) plants such as capsicums, tomato and aubergine (brinjals or eggplant), preferably for 3 years or more; (ii) where volunteer hosts have not been allowed to establish during the non-solanaceous crop period; (iii) where no runoff water flows into such land; (iv) where potato residue/peelings have not been dumped; and (v) where no manure/compost suspected to have been made from diseased host residue has been applied or deposited. It is advisable to take a soil sample for laboratory confirmation of presence or absence of the bacterial wilt pathogen.

Validations indicate that the seed-plot technique achieved two-and-a-half to three times land productivity and about 50% bacterial wilt reduction compared with conventional ridge planting. Under different programmes supported by ASARECA, CIP, CARE EEEGL, NAADS and NARO, 5890 farmers from south-western Uganda were reached, out of which 80% have adopted the technologies and are practising the same on their individual farms.

Positive selection (PS)

Positive seed selection involves identifying healthy-looking plants in potato fields and marking them before plant senescence; these plants are then harvested first in order to obtain seed potato for the next season. 'Select the Best', or PS is the most feasible way of reducing seed degeneration as a result of virus diseases in a ware potato crop. PS will not only work against viruses and bacterial wilt, but also against other seed-borne diseases. Any serious seed-borne disease will in some way or another cause the plant to show symptoms. In other words, the plants will look different, not healthy. PS has led to an increase in seed quality production of 34% or 6.8 t/ha. This provides an economic gain with minimal, or no increase in cash investment.

24.3 Progress to Date

For a period of over 10 years, farmers have been acquiring knowledge on the seed-plot technique and PS through field demonstrations/promotion at group level. A farmer field school (FFS) approach was adopted from which individual farmers would replicate the acquired knowledge on their individual fields. UNSPPA has disseminated the mentioned technologies through FFSs reaching 5890 individual farmers, and 80% of these farmers adopted the technologies and have continued to practise them. Within the areas, and particularly with respect to participating farmers, there is increased awareness on the merits of using high-quality seed potato as well as sources of such materials. Specifically, farmers have learnt about and are practising production of their own high-quality seed potato tubers from seed plots and are carrying out positive seed selection on their farms. There are already some individual farmers who are multiplying seed potato for sale to other farmers, thereby generating income for their families and improving the availability of disease-free tubers to their localities. The seed-plot technology has also been adapted by some farmers in the project areas for bulking of planting materials of other vegetatively propagated crops such as sweetpotatoes and arrowroots.

The results will continue to be spread through farmer-to-farmer training, extension-farmer training supported by different programmes, and scaling up of the demonstrations to other potato-growing areas using future-sourced donor funding. Efforts will be made to also spread the results through printed material like pamphlets written in local languages. Because of the lack of availability of quality seed at farmer level, extension agents at the district level are now excited to help farmers to produce positively selected seed using the PS training manual (Gildemacher *et al.*, 2007).

24.4 Lessons Learnt/Experience Shared

The success of the seed-plot technique and PS technologies in maintaining and improving the quality of seed potato was highly dependent on the principle of regular training to impart or refresh the necessary knowledge and develop effective skills. Through on-farm experimentation and group-based demonstrations, farmers learnt more and faster from one another than would have been via extension agents or researchers. This approach was further strengthened through field days and agricultural exhibitions during which participating farmers expressed the gains they had received through the technologies they had learnt. UNSPPA successfully collaborated and worked closely with farmers, as is evident from the enthusiasm in activity implementation, partnership and interest shown by farmers to know how and where to buy quality seed potato to renew their seed. In terms of gender roles within participating farmer groups, males were more interested in activities that generate income (e.g. working in tea factories, forests, modern markets and commercial potato production) while females, both young and old, were more interested in social activities such as taking care of family health and education as well as small-scale subsistence farming. This was an important entry point for the seed quality improvement technologies since females were more available for practical on-farm training and demonstrations.

Acknowledgements

We (UNSPPA) are grateful to ASARECA, CIP, NAADS, CARE and NARO, and to the farmers who participated in the research and the implementation of research results for the moral, technical and financial support. Special thanks to Zachary Kinyua, Astere Bararyenya, Prossy Namugga and Wachira Kaguongo for the work done towards scaling up the technologies in the region.

References

Gildemacher, P., Demo, P., Kinyae, P., Wakahiu, M., Nyongesa, M. and Zschocke, T. (2007) *Select the Best: Positive Selection to Improve Farm Saved Seed Potatoes. Trainers Manual*. International Potato Center (CIP), Nairobi.

Kaguongo, W., Ng'ang'a, N., Muthoka, N., Muthami, F. and Maingi, G. (2010) *Seed Potato Subsector Master Plan for Kenya (2009–2014)*. Seed potato study sponsored by GTZ-PSDA, United States Agency for International Development (USAID), International Potato Center (CIP) and Government of Kenya, Ministry of Agriculture. GTZ-PSDA/USAID/CIP, Nairobi.

Kinyua, Z.M. and Smith, J.J. (2007) *Multiplication of Seed Potato in Disease-free Seed Potato in Seedplots: a Solution to Potato Bacterial Wilt and Other Pest Problems*. Revised version under Kenya Agricultural Research Institute (KARI)-Association for Strengthening Agricultural Research in Eastern and Central Africa (ASARECA) project. KARI, Nairobi.

Kinyua, Z.M., Kinoti, J., Wakahiu, M.W., Karinga, J., Nderitu, S.K., Kinyae, P.M., Kipkoech, D., Muriithi, W.M., Miruci, N.G., Otipa, M.J. and Langat, B.C. (2010) *Enhancing the Quality and Availability of Seed Potatoes to Small-scale Farmers*. Information manual (draft) on quality seed potato developed under Association for Strengthening Agricultural Research in Eastern and Central Africa (ASARECA)-funded Project ASARECA/SC/QSP/09/02. ASARECA, Entebbe, Uganda.

Kinyua, Z.M., Tindimubona, S., Namugga, P. and Bararyenya, A. (2012) *Enhancing the Production and Utilization of Quality Seed Potato: a Handbook for Farmers in East and Central Africa*. ASARECA, Entebbe, Uganda.

Raemaekers, R.H. (ed.) (2001) *Crop Production in Tropical Africa*. Directorate General for International Co-operation (DGIC), Brussels, 1540 pp.

Uganda Bureau of Statistics (UBOS) (2010) *Uganda Census of Agriculture (UCA), Agricultural Household and Holding Characteristics Report*, Volume III. UBOS, Kampala.

25 Substrate and Genotype Effects on Growth for Seed Potato (*Solanum tuberosum* L.) Production in Malawi

S.C. Zimba,* J.P. Njoloma, J.A. Nyaika, W.F. Mwase, M.F. Maliro, M.B. Kwapata and J.M. Bokosi
Lilongwe University of Agriculture and Natural Resources, Lilongwe, Malawi

Abstract

Potato farmers in Malawi and other developing countries lack healthy and quality seed potato. This is due to a limited multiplication programme to provide farmers with clean seed potato. A study to evaluate the effect of media and genotypes on growth and seed yield performance of micropropagated potato genotypes in greenhouse conditions would assist in planning for future selection of good high-yielding varieties and a rapid seed multiplication programme of pathogen-free planting materials to increase production of potato in the country. Three different types of media (vermiculite, sand and sawdust) and seven different genotypes (two local genotypes – Magalabada and Rosita; and five introduced genotypes – Up to date, Buffelspoort, Van der plank, Lady Rosetta and Bp 1 2007) were investigated. Pathogen-free potato plantlets were produced in the tissue culture laboratory and then transferred to the greenhouse for minituber production. The plantlets were planted with a 10 cm distance between rows and 10 cm between plantlets in trays filled with media. The study revealed significant ($P \leq 0.05$) variations between and among media and genotype tested on growth and yield characteristics. Genotype Rosita grown on vermiculite produced the greatest plant height (45.87 cm). Similarly plantlets grown on vermiculite performed better than those grown on the alternative media with higher minituber yield of 1740 g/m^2 for all genotypes while the yields were 850 g/m^2 and 292 g/m^2 in sand and sawdust, respectively. Among the genotypes Up to date and Lady Rosetta produced more tubers (between 12 tubers per plant and 10 tubers per plant). Thus vermiculite increases the number and size of minitubers which is a very important step for rapid multiplication of potato seed. The findings of this study provide valuable information on potato minituber production using different media and potential genotypes which would assist in scaling up seed multiplication programmes for pathogen-free planting materials.

Keywords: genotype, media, micropropagation, minituber, *Solanum tuberosum*

25.1 Introduction

Potatoes are among the ten most important food crops in the world and play nutritional, economic and industrial roles in Malawi (Chiipanthenga *et al.*, 2012). Despite potato being an important food and cash crop in Malawi, the actual yield and quality is low (MoAIFS, 2007). The present average yield ranges from 7 t/ha to 11 t/ha, but yields of

*sibozim@yahoo.co.uk

up to 40 t/ha are achievable (FAO, 2008). The major limiting factor to realizing high yields and good quality of potatoes in Malawi is lack of adequate quality seed. This is due to limited multiplication programmes to provide farmers with clean seed potato of high-yielding varieties suitable for different agroecological zones. Consequently, farmers use low quality and sometimes infected seed from the previous potato crop often from un-marketed tubers and without proper selection. This practice has led to high degeneration of seed tubers due to systemic viral and bacterial diseases resulting in poor quality and quantity of potato tubers (Demo *et al.*, 2007). As a result, national potato production is below the market demand of 88 kg per capita (FAO, 2008).

The health and quality of seed potato are important in seed technology. Moreover, the basic seed materials must be free of pathogens. Tissue culture techniques are often used at this stage to produce disease-free seeds (Altindal and Karadogan, 2010). First, clean stocks are obtained by meristem culture *in vitro* and this is followed by multiplication of the clean materials achieved through nodal cuttings. The growing plantlets are then transferred to seedbeds to produce minitubers (Farran and Mingo-Castel, 2006). There are many factors that affect growth performance of micropropagated plantlets when they are transferred to greenhouse conditions for minituber production (Farran and Mingo-Castel, 2006). These factors include substrate, plantlet age and illumination (Alsadon and Knutson, 1994).

Substrate is one of the important factors to consider for clean seed tuber (minituber) production. Although research has shown that different growth media (perlite, pumice, sand, styroform, compost, peat and sphagnum) can be used to grow *in vitro* potato plantlets in the greenhouse, the most preferred and commonly used medium is vermiculite. Vermiculite is, however, very expensive because of the energy cost for heating needed to prepare the product. Due to the high cost associated with vermiculite, other alternatives can be used such as sand, rice hulls and sawdust.

Therefore the aim of the study was to evaluate the growth performance of micropropagated potato planting materials of locally adapted and newly introduced potato genotypes in order to optimize clean seed potato production and increase its availability to farmers in Malawi.

25.2 Materials and Methods

The study was conducted in the greenhouse at Bunda College of Agriculture (now Lilongwe University of Agriculture and Natural Resources since 2013) at Lilongwe, Central Malawi. The experiment was conducted from April 2011 to November 2011 (8 months).

Plant materials

Healthy (clean) tubers of locally adapted genotypes (M.B. Kwapata, J.P. Njoloma, J.M. Bokosi, W.F. Mwase and M.M. Maliro, 2007, unpublished data) Rosita and Magalabada were collected from different growing areas in Malawi (Jenda in Mzimba, Dedza, Ntcheu, Bvumbwe and Mulanje). These genotypes were multiplied at Bunda College of Agriculture. Certified seed stock of newly introduced genotypes in plantlet form were obtained from the Agricultural Research Council (ARC-LNC) in South Africa, namely: Van der plank, Lady Rosetta, Bp 1 2007, Buffelspoort and Up to date.

Treatment and experimental design

The study had two factors. One factor was the three substrates (sand, vermiculite and sawdust) referred to as growth media in this study. The second factor was the seven genotypes: Bp 1 2007, Van der plank, Lady Rosetta, Buffelspoort, Up to date, Rosita and Magalabada. These treatments were arranged in a randomized complete block design with three replicates.

Substrate collection, sterilization and analysis

River sand was collected from the shores of Lake Malawi to obtain pure sand. Vermiculite was sourced from South Africa and sawdust was collected from carpentry shops around Lilongwe, Malawi. Aluminium jars were filled with each type of substrate, covered and autoclaved at 121°C for 1 h.

After autoclaving each substrate was then analysed for pH, total nitrogen and organic matter in the laboratory. The results of the media analysis are presented in Table 25.1.

Production of potato plantlets in greenhouse

The rooted *in vitro* plantlets were transplanted in a greenhouse in trays of 0.028 m^3 volume filled with the different substrates (sand, vermiculite and sawdust) at a volume of 0.015 m^3. The plantlets were spaced at 10 cm between plants and 10 cm between rows, corresponding to 40 plants per tray as a gross plot and 20 plants were used as a net plot. The irrigation unit was placed at the top of the tray. The inlet of this unit was connected to the water pump through a solenoid valve. The nozzles of the drip irrigation unit were fixed 10 cm equidistant apart to direct water and nutrient solution towards the root zone at required time intervals (Chiipanthenga *et al.*, 2012). NPK fertilizer was used (23:21:0 + 4S), at the rate of 1.2 g per plant every 10 days throughout the plant growth. The rate was adopted from Alsadon and Knutson (1994).

Data collection and statistical analysis

Data collection on plantlets started 3 days after transplanting and included: (i) survival percentage; (ii) plant height; (iii) plant vigour; and (iv) number of tubers produced. Survival percentage was determined by computing the number of plants surviving in different substrates at a certain period. Plant height was determined by measuring the distance from the root collar to the shoot tip. Plantlet vigour was recorded by using a plant vigour score of 1–9 developed by CIP (2008) but modified to 1–5 as follows:

- 1 = very poor (where more than half of the plants in a sample had shoots of ≤ 3 mm thick);
- 2 = poor (where more than half of the plants in a sample had shoots of ≤ 5 mm thick and a length of ≤ 10 cm);
- 3 = intermediate (where more than half of the plants in a sample had shoots of ≤ 7 mm thick and a length of ≥ 15 cm);
- 4 = vigorous (where more than half of the plants in a sample had shoots of ≤ 10 mm thick and a length of ≥ 20 cm); and
- 5 = very vigorous (where more than half of the plants in a sample had shoots of ≤ 20 mm thick and a length of ≥ 40 cm but were very firm).

Analysis of variance (ANOVA) was performed using GenStat Statistical package version 13, using the GLM (general linear model) procedure. Where significant differences were observed, treatment means were separated by the least significant difference (LSD) at $P < 0.05$.

Table 25.1. Characteristics of different substrates.

Substrate	pH	Organic carbon (%)	Nitrogen (%)
Vermiculite	7.5	0.65	0.084
Sand	7.3	0.89	0.042
Sawdust	7.6	20.05	0.070

25.3 Results and Discussion

Effect of substrates and genotypes on survival of potato plants

There were no significant differences ($P \geq 0.05$) on the interaction between media and genotypes on the survival rate of the plants (Table 25.2). Highly significant ($P \leq 0.001$) variation was observed on survival (%) among the three substrates. Survival (%) of plants planted in vermiculite differed significantly from plants observed in sand and sawdust at 91.43%, 53.57% and 21.9%, respectively. There was no significant difference

Table 25.2. Survival percentage per substrate type and genotypes 5 weeks after transplanting.

Genotypes	Survival (%)[a] in Sand	Vermiculite	Sawdust	Mean for genotype (%)
Magalabada	58.3 ± 1.67	90.0 ± 3.33	26.67 ± 0.0	58
Rosita	55.7 ± 5.00	96.7 ± 1.67	23.3 ± 8.83	58
Lady Rosetta	61.7 ± 1.67	96.7 ± 3.33	16.7 ± 4.41	58.3
Bp 1 2007	46.7 ± 4.41	90.0 ± 2.87	23.3 ± 6.67	53.3
Up to date	46.0 ± 6.67	90.0 ± 2.80	20.0 ± 5.7	52.2
Van der plank	50.0 ± 5.77	88.3 ± 4.41	23.0 ± 6.0	53.9
Buffelspoort	56.7 ± 4.41	88.3 ± 1.67	20.0 ± 2.87	55
Mean for substrate (%)[b]	53.57b	91.43a	21.90c	

[a]Values in the table are presented as mean ± standard error.
[b]Mean values followed by different letters in this row are significantly different at $P \leq 0.001$.

($P \geq 0.05$) observed among genotypes tested on survival percentage (Table 25.2).

The variation on the survival rate of plants in different substrates could be attributed to several factors which include the quality of the media among others. The results confirmed the findings of Palacios *et al.* (2009) and Dwelle and Love (2000) who reported that the survival percentage of plants under greenhouse conditions could be affected mostly by environmental factors such as growth media, temperature and water stress. Some studies have also indicated that the source of the sawdust compromises its quality and its ability to return nutrients to the plants. This observation agreed with the findings by Donnelly *et al.* (2008) who reported that the species of tree from which the sawdust is derived largely determines its quality and value for use as a growth substrate. In this study, the sawdust used was a mixture from different trees which could have an effect on plant survival rate. According to Richard *et al.* (2004) the main role of the physical characteristics of media is to have suitable air-filled porosity for efficient oxygen diffusion to enable root respiration while maintaining a favourable water content for supplying water and nutrients to the roots. This result therefore implies that careful selection and source of media should be considered for survival of plants which subsequently affects minituber yield.

Effect of substrates and genotypes on plant height

There was a significant ($P \leq 0.001$) variation observed between substrate and genotypes interaction on plant height 7 weeks after transplanting (Table 25.3). Genotypes Rosita, Magalabada, Van der plank and Up to date produced taller plants of 45.87 cm, 45.37 cm, 43.17 cm and 40.35 cm, respectively, than other genotypes that were transplanted in vermiculite. The same trend was also observed when the genotypes were transplanted in sand. Among the genotypes transplanted in sawdust Van der plank and Rosita outperformed the other genotypes producing taller plants of 17 cm and 16.6 cm, respectively.

The variation in response among genotypes could be attributed to differences in growth morphological characteristics due to genetic constitution. These findings agree with Balali *et al.* (2008) who also observed variation among genotypes in response to environmental factors. Other findings by Tadesse *et al.* (2001) indicate that potato genotypes that are hybrids from *andigena* and *tuberosa* grow faster than other genotypes in the early vegetative growing period. It was observed that Van der plank could grow fast and had vigorous vegetation as compared with other genotypes.

Among the substrates, variations could be attributed to differences in their composition. The results of the media analysis

Table 25.3. Plant height (cm) per substrate type and genotypes 7 weeks after transplanting.

Genotypes	Plant height (cm)[a] in Sand	Vermiculite	Sawdust	Mean for genotype (cm)[b]
Magalabada	26.80 ± 0.0	45.37 ± 0.0	15.45 ± 0.0	29.22ab
Rosita	27.42 ± 0.33	45.87 ± 0.38	16.60 ± 0.0	29.74a
Lady Rosetta	18.84 ± 0.08	38.60 ± 1.33	14.00 ± 0.0	23.81d
Bp 1 2007	22.92 ± 0.67	38.33 ± 0.92	15.50 ± 0.07	25.58c
Up to date	23.75 ± 0.0	40.35 ± 0.05	13.30 ± 0.0	25.80c
Van der plank	25.35 ± 0.0	43.17 ± 0.08	17.00 ± 0.0	28.51b
Buffelspoort	21.20 ± 0.7	35.58 ± 1.33	11.90 ± 0.0	22.89d
Mean for substrate (cm)[b]	23.66b	41.04a	14.83c	

[a]Values in the table are presented as mean ± standard error.
[b]Mean values followed by different letters in this column/row are significantly different at $P \leq 0.001$.

in Table 25.1 show that for sawdust the organic carbon content was about 20%. The results are supported by Öztürk and Yildirim (2010) who found that the C:N ratio of sawdust is such that it is not readily decomposed and possibly the conditions are not conducive for the growth of microbes. The high cellulose and lignin content along with insufficient nitrogen supplies create problems which can severely restrict plant growth. However, supplemental applications of nitrogen at the required rate can reduce this problem. In this study all the media used were supplemented with the same type of fertilizer delivered at the same rate. However, it can be observed that probably fertilizer used and the rate could not be enough for the sawdust medium to retain the required nutrients for the plants, causing poor growth of plants. Some studies have indicated that heat released during decomposition of sawdust can also affect plant growth and development (Öztürk and Yildirim, 2010). The results imply that to use sawdust as a possible substrate for potato minituber production would require some manipulation in terms of substrate management, such as optimizing the supplemental N or mixing with other substrates, rather than using it as the sole substrate.

Effect of substrates and genotypes on plant vigour

There were significant differences ($P \leq 0.001$) observed on interaction between substrate and genotype on plant vigour score (Table 25.4). Van der plank, Lady Rosetta and Bp 1 2007 transplanted in vermiculite had higher vigour score of 4. Significant variation was observed between the local and introduced genotypes: local genotypes Magalabada and Rosita were less vigorous in sand and sawdust growth media, having vigour score of 1, compared with most introduced genotypes. Among the genotypes tested, significant variation ($P \leq 0.001$) was observed on plant vigour. Genotype Van der plank performed better in all the growth media by having vigorous plants compared with other genotypes.

The variation in plant vigour among genotypes could be attributed to genetic differences of the genotype among other factors. Findings by Donnelly *et al.* (2008) illustrate that vigorous plants have increased survival and eventually good establishment in the greenhouse for minituber production and subsequent outdoor planting.

Effect of substrates and genotypes on number of tubers per plant

There were significant differences ($P \leq 0.001$) in the interaction between media and genotypes on the number of tubers per plant (Table 25.5). Up to date and Lady Rosetta genotypes planted in vermiculite registered the highest tuber number per plant (12 and 10, respectively). Alsadon and Knutson (1994) reported that tuber number rather than

Table 25.4. Plant vigour score per substrate type and genotype 7 weeks after transplanting.

	Plant vigour score[a] for growth in			
Genotypes	Sand	Vermiculite	Sawdust	Mean score for genotypes[b]
Magalabada	1.00 ± 0.01	2.00 ± 0.00	1.00 ± 0.00	1.33f
Rosita	1.00 ± 0.33	3.00 ± 1.60	1.07 ± 1.67	1.72e
Lady Rosetta	2.83 ± 0.17	4.83 ± 0.13	1.17 ± 1.17	2.94b
Bp 1 2007	2.00 ± 0.00	4.67 ± 0.33	1.00 ± 0.00	2.55c
Up to date	1.80 ± 0.17	3.33 ± 1.60	1.00 ± 0.00	2.04d
Van der plank	2.67 ± 0.33	4.67 ± 0.33	2.00 ± 0.00	3.11a
Buffelspoort	1.00 ± 0.00	2.33 ± 0.67	2.00 ± 0.00	1.78ef
Mean score for substrate[b]	1.76b	3.55a	1.33c	

[a]Plant vigour score used was as follows (for full details see Materials and Methods): 1 = very poor; 2 = poor; 3 = intermediate; 4 = vigorous; 5 = very vigorous. Values in the table are presented as mean ± standard error.
[b]Mean values followed by different letters in this column/row are significantly different at $P \leq 0.001$.

Table 25.5. Number of tubers (per plant) for different genotypes grown in different substrates.

	Number of tubers per plant[a] in			
Genotypes	Sand	Vermiculite	Sawdust	Mean for genotype[b]
Magalabada	5 ± 0.33	8 ± 0.00	2 ± 0.00	5.02c
Rosita	5 ± 0.78	9 ± 0.08	2 ± 0.33	5.32bc
Lady Rosetta	6 ± 0.17	10 ± 0.79	3 ± 0.44	6.13b
Bp 1 2007	4 ± 0.22	6 ± 0.29	3 ± 1.33	4.59c
Up to date	7 ± 0.29	12 ± 0.77	3 ± 0.68	6.99a
Van der plank	3 ± 0.22	4 ± 0.11	2 ± 0.15	2.76d
Buffelspoort	5 ± 0.21	7 ± 0.17	2 ± 0.47	4.60c
Mean for substrate[b]	4.94b	7.81a	2.39c	

[a]Values in the table are presented as mean ± standard.
[b]Mean values followed by different letters in this column/row are significantly different at $P \leq 0.001$.

weight is important for reporting yield under greenhouse conditions. Greenhouse conditions may permit an expression of tuber number potential but limitations in media space and possibly radiation may not permit maximum tuber size development.

Up to date and Lady Rosetta recorded a higher number of minitubers per plant indicating superiority over other genotypes. This could be attributed to the interaction between genetic constitution of the genotypes and environment. High number of minitubers per plant indicates high multiplication potential of the genotypes under greenhouse conditions. The results agree with Black (2008) that Up to date and Lady Rosetta were bred for their high yields and are the most popular varieties and make up 77 % of the potatoes grown in South Africa. These findings by Black (2008) were also reflected in this study by having these two genotypes outperforming the other genotypes. The study findings imply that Up to date and Lady Rosetta demonstrated potential traits for high yield that would assist breeders for selection and improvement.

Asghari-Zakaria *et al.* (2009) reported that nutrient uptake of plant is determined by the growth media. Vermiculite despite being expensive is still the best media indicating superiority among the media tested. This could be attributed to its ability to hold and retain nutrients for the plants. The findings by Khodadadi *et al.* (2011) suggest that generally the mixture of two or three different media is better. This therefore indicates that with good manipulation and management, other alternative media tested (e.g. sand) could do equally as well in minituber production as vermiculite.

25.4 Conclusion

The results have shown that growth performance varies among local and introduced potato genotypes under greenhouse conditions. Genotypes with high survival percentage and vigorous growth in a particular media represent a potential for establishment and growth of potato seed. Introduced genotypes grew vigorously indicating better establishment than the local genotypes. The findings of this study have shown potential genotypes that if introduced, can be used for seed multiplication. Introduced genotypes Bp 1 2007, Lady Rosetta and Up to date and Van der plank have demonstrated valuable characteristics that could assist in selection for multiplication and breeding purposes.

The results have also shown that performance of potato genotypes varies with substrate under greenhouse conditions. Among the substrates tested, vermiculite was found to be superior in supporting the plants followed by sand. Thus vermiculite has a positive effect on potato growth establishment and tuber formation. However, sand would also be a potential medium if it can be manipulated and managed by either substrate combination or nutrient supplementation. The study provided valuable information on potato growth and establishment using different substrates which will help in minituber production and scaling up programmes for making available pathogen-free seed tubers to increase potato production in Malawi.

Acknowledgements

The authors acknowledge the financial support from the Agriculture Research and Development program (ARDEP) and Bunda College of Agriculture for the MSc study period during which this chapter was written as part of the thesis work.

References

Alsadon, A.A. and Knutson, K.W. (1994) Field and greenhouse tuberization of six potato cultivars grown from *in vitro* plantlets. *Agricultural Science* 6(1), 79–86.

Altindal, D. and Karadogan, T. (2010) The effect of carbon sources on *in vitro* microtuberization of potato (*Solanum tuberosum* L.). *Turkish Journal of Field Crops* 15, 7–11.

Asghari-Zakaria, R., Maleki-Zanjani, B. and Sedghi, E. (2009) Effect of *in vitro* chitosan application on growth and minituber yield of *Solanum tuberosum* L. *Plant, Soil and Environment* 55(6), 252–256.

Balali, G.R., Hadi, M.R., Yavari, P., Bidram, H., Naderi, A.G. and Eslami, A. (2008) Effect of pot size, planting date and genotype on minituber production of Marfona potato cultivar. *African Journal of Biotechnology* 7(9), 1265–1270.

Black, V. (2008) *Hot Potato. GM Potato in South Africa – a Critical Analysis*. African Centre for Biosafety, Johannesburg, South Africa, pp. 1–365.

Chiipanthenga, M., Maliro, M., Demo, P. and Njoloma, J. (2012) Potential of aeroponics system in the production of quality potato (*Solanum tuberosum* L.) seed in developing countries. *African Journal of Biotechnology* 11(17), 3993–3999.

Demo, P., Low, W.J. and Mwenye, J.O. (2007) Potato production in Malawi: strength, weaknesses, opportunities and threats. In: Khalf-Allah, A. (ed.) *Seventh Triennial African Potato Association Conference Proceedings*, Alexandria, Egypt. African Potato Association, Alexandria, Egypt, pp. 339–345.

Donnelly, D.J., Warren, K.C. and Shirlyn, E.C. (2008) Potato microtuber production and performance: a review. *American Journal of Potato Research* 80(2), 103–115.

Dwelle, R.B. and Love, S.L. (2000) Potato Growth and Development. Available at: http://www.cals.uidaho.edu (accessed 13 May 2010).

Farran, I. and Mingo-Castel, A.M. (2006) Potato minituber production using aeroponics: effect of plant density and harvesting intervals. *American Journal of Potato Research* 83, 47–53.

Food and Agriculture Organization of the United Nations (FAO) (2008) *The International Year of Potato*. The Global Crop Diversity Trust and FAO's Plant Production and Protection Division, FAO, Rome, pp. 1–22.

International Potato Center (CIP) (2008) *Quality Seed Potato Production Using Aeroponics*. CIP, Lima. Available at: http://www.cipotato.org/publications (accessed 12 October 2011).

Khodadadi, M., Hassanapah, D., Pirovates, S.P. and Masoumi, H. (2011) Evaluation of different planting beds: effects on minituber production of potato cultivars under greenhouse conditions. *American-Eurasian Journal of Agriculture and Environmental Science* 11(3), 365–370.

Ministry of Agriculture, Irrigation and Food Security (MoAIFS) (2007) *Guide to Agriculture Production*. Government of Malawi, Lilongwe, Malawi, 51 pp.

Öztürk, G. and Yildirim, Z. (2010) A comparison of field performances of minitubers and microtubers used in seed potato production. *Turkish Journal of Field Crops* 15(2), 141–147.

Palacios, G., Abud, M., Salvador, M., Adriano, L., Dendooven, L. and Gutierrez, F.A. (2009) Influence of abuscular mycorrhizae fungi and a native diazotrophic bacteria in survival and tuberization of *ex vitro* potato plants. *Gayana Botany* 66(2), 127–133.

Richard, T.L., Veeken, A.H.M., Wilde, V. and Hamelers, H.V.M. (2004) Air-filled porosity and permeability relationships during solid-state fermentation. *Biotechnology Program* 20, 1372–1381.

Tadesse, M., Lommen, W.J.M. and Struik, P.C. (2001) Development of micropropagated potato plants over three phases. *Netherlands Journal of Agricultural Science* 49, 53–66.

26 Work of Multiple Organizations to Improve Seed Potato Health in the USA and an Example of Change to Reduce *Potato Virus Y* in Seed Potato Lots

J.L. Whitworth[1]* and P. Nolte[2]

[1]*United States Department of Agriculture Agricultural Research Service (USDA-ARS), Aberdeen, Idaho, USA;* [2]*University of Idaho, Idaho Falls, Idaho, USA*

Abstract

In the USA, seed potato improvement starts with the individual seed potato grower. The seed grower also has resources that are available from university experts and from the organizations that certify the seed potatoes. Systems that exist for the production of seed potatoes have similar but slightly different structures in individual states, although the basic principles for producing quality seed potatoes are followed in each state. Each state is signatory to a memorandum of understanding (MOU) with the national regulatory agency (United States Department of Agriculture Animal Plant Health Inspection Service, USDA-APHIS). This allows trade between states and between the USA and other countries. The MOU requires a quality manual to be developed for each certifying state. This quality manual is a procedural manual. An audit of each certifying agency is conducted to ensure compliance with the approved quality manual. The MOU is a recent document and serves to unify practices of multiple certification agencies. This allows for a set of minimum seed standards that helps to facilitate international trade. An example of how certification practices can change to improve seed health occurred in Idaho starting in 2007. At this time, an outbreak of *Potato virus Y* (PVY) necrotic strains occurred in ware fields planted with infected seed. Prior to this time, 95% of the varieties were visually inspected for PVY. The other 5% of varieties were serologically tested with ELISA because of latent symptom expression. This PVY outbreak led to a change requiring ELISA testing of all seed lots and all varieties. Over the next 4 years the number of seed lots with PVY was reduced by 10%. The change in the seed regulations was formulated by growers and university researchers and then approved by a grower advisory committee and finally a foundation seed stocks committee which consists of the certification agency and university scientists.

Keywords: ELISA testing, potato, *Potato virus Y* (PVY), quality seed potatoes, seed potato, seed potato grower, United States Department of Agriculture Animal Plant Health Inspection Service (USDA-APHIS)

*jonathan.whitworth@ars.usda.gov

26.1 Introduction

Certification of seed potatoes in the USA is done under the authority of individual states and each state has its own official certification tag that goes with every shipment of seed. At the federal level, trade between the states and between countries is regulated by the United States Department of Agriculture (USDA). The agency within the department that regulates the phytosanitary requirements is called APHIS, which stands for Animal Plant Health Inspection Service. APHIS has memorandum of understanding (MOU) with each state agency that certifies seed potatoes. Together with the state seed agencies, a programme has been developed to bring uniform minimum standards and practices to all seed produced in the USA. This programme is called the State National Harmonization Program (SNHP). The multiple agencies involved in this effort are: (i) APHIS; (ii) the National Potato Council; (iii) the United States Potato Board; (iv) the National Plant Board; and (v) the state certification agencies. Prior to this time, each agency had its own standards which were similar, but there was no minimum standard all states followed. The SNHP establishes these minimum standards and allows our government to facilitate trade based on a single set of standards.

Items covered in this agreement include quarantine pests such as potato cyst nematodes and non-quarantine-regulated pests such as viruses that cause tuber necrosis. Each state is required to produce a quality manual that includes description of seed certification procedures, pest tolerances and testing methodologies along with the methods for ensuring compliance. A brief handbook is also developed by each state as part of the quality manual that describes the standards and protocols used for disease testing. This manual and the facilities are then audited by a third party to ensure that the requirements are met. Efforts to develop the harmonization programme started in 2002 at a seed potato seminar sponsored by the National Potato Council. At that time, necrotic strains of *Potato virus Y* (PVY) had been detected in both Canada and the USA. A discussion was held as to how to survey and determine the extent of the PVY necrotic strains and how to develop an effective management plan to address viruses that cause tuber necrosis. This management plan eventually included *Potato mop top virus* and *Tobacco rattle virus* along with PVY necrotic strains known as PVY^{NTN} and $PVY^{N:O}$ (syn. $PVY^{N\text{-}Wilga}$). The survey was completed and it showed that in North America, the PVY^{O} or common strain was predominant at 61% and 67% in the USA and Canada, respectively (Gray *et al.*, 2010). Recommendations were made to implement measures to control and reduce the amount of necrotic PVY. One of the recommendations was for mandatory postharvest testing to occur for all seed potatoes. At the time, all state certification programmes required a postharvest test for seed that would be re-certified, that is seed going back into seed production. However, all seed produced for the subsequent year's ware crop was not tested in some states.

The recommendation from Gray *et al.* (2010) was that all seed produced should be postharvest tested. This is critical in areas where seed and ware potatoes are produced in close proximity as infected seed can provide a source of virus inoculum. Since the completion of the survey, work has continued with the state certification agencies to encourage serological testing of all seed lots for PVY. Research since the detection of PVY necrotic strains has shown that there is a cultivar × PVY strain interaction with visual symptoms ranging from mild to severe mosaic and various necrotic foliar symptoms (Novy *et al.*, 2010; Whitworth *et al.*, 2010, 2011). Since many postharvest inspections relied on visual symptoms, the inspections were not efficient at detection of all PVY-infected plants. The recommendation from the scientists working with the seed agencies during their annual certification meetings was to use the enzyme-linked immunosorbent assay (ELISA) method to test all seed lots for PVY during the postharvest evaluations. The purpose of the postharvest test in the USA is to provide a growout of samples from each seed lot during the off season. Plants grown from these samples are inspected and tested for viruses. This allows detection of virus that was spread during the growing season. The grower can then make a decision on whether or not to plant the seed back to produce another seed crop depending on the virus level recorded.

In 2007 in south-east Idaho, samples collected from ware fields showed different degrees of foliar mosaic symptoms and during the autumn, samples were also collected that showed tuber symptoms characteristic of potato tuber necrotic ringspot disease (PTNRD) associated with PVY. Testing of these samples indicated the presence of necrotic PVY strains such as PVY^{NTN} and $PVY^{N:O}$ (Karasev *et al.*, 2008). Subsequently, the Idaho seed certification agency held discussion with scientists, its member advisory committee and ware potato producers regarding PVY outbreaks. A decision was made to test by ELISA all postharvest samples for PVY. Results indicate since that time that this course of action has decreased the overall PVY in the system and resulted in healthier seed.

This chapter discusses the methods used for postharvest testing and an example of how ELISA-based testing all seed lots in the Idaho certification programme has resulted in a decrease in PVY and better seed health.

26.2 Materials and Methods

For the postharvest test in Idaho, a sample of 400 tubers per seed lot is collected at harvest, usually in September, by the seed grower and submitted to the certification agency (Idaho Crop Improvement Association). These samples are kept in a warming room at 26°C for approximately 2 weeks. To break dormancy and stimulate sprouting, samples are then treated with Rindite (a combination of three chemicals) (Gugerli and Gehriger, 1980). Samples are then planted in southern California in early November and when the plants reach 30–35 cm in height, leaf samples are picked from each plant and sent via express shipment to the laboratory in Idaho. At the laboratory, five leaves are combined and a 5 mm diameter cork borer is used to 'punch' through the sample and the resulting 'discs' of leaf tissue are deposited in an ELISA plate. Each plate consists of 80 samples (made up of five leaf discs combined), four healthy plant sap controls, four PVY-positive sap controls and eight buffer wells. The plate is processed according to the protocol outlined in Clark and Adams (1977). Results are read at A_{405} on a BioTek ELx808 plate reader and a test is considered to be positive if the absorbance value exceeds 0.100, as long as the background reaction is low and there are no irregularities on the plate. The percentage of PVY per sample is then calculated based on a formula which is as follows:

$$1-\left(\sqrt[5]{1-X}\right)=P \qquad (26.1)$$

Where:
P = the proportion of actual infection
X = the proportion of positive tests
This formula takes into account the probability of one to five positive leaves being in one positive ELISA test.

Prior to 2007, results for the winter test plots were based on visual inspections for 95% of the cultivars in the programme. The remaining 5% were tested by ELISA and consisted primarily of the cultivars Russet Norkotah, Shepody and Calwhite which are known to have latent PVY foliar expression. After 2007 all cultivars, regardless of mosaic symptom expression were and continue to be tested using ELISA.

26.3 Results

Since 100% ELISA-based testing started in 2007, samples infected with PVY have dropped 22 percentage points with the average over the last 3 years being 42.7% (Table 26.1). It is important to understand how the lots are figured for 'total infected' in the table. If a lot (sample) has at least one PVY-positive plant out of the whole sample, it is included in the total infected category. The line in the table that lists 'percentage of lots ineligible' relates to a re-certification tolerance of 2%. Seed that is re-certified will be planted back in the following year to produce another seed crop. Since 2007, the decline in seed lots that had over 2% PVY in them is almost half of the 2007 amount, a drop of 13 percentage points.

Table 26.1. Summary of postharvest test plots from Idaho for 2005–2012. From 2005 to 2006, testing was performed visually (mosaic symptoms), since 2007 testing has been performed by ELISA for *Potato virus Y* (PVY).

	Year							
	2005	2006	2007	2008	2009	2010	2011	2012
Total lots	604	575	551	569	625	653	722	846
Total infected (with at least one PVY-positive plant)	226	350	332	345	317	259	364	327
Percentage of samples infected	27%	61%	60%	61%	51%	40%	50%	38%
Total lots ineligible (> 2% PVY)	93	172	151	116	97	54	136	119
Percentage of lots ineligible	15%	30%	27%	20%	16%	8%	19%	14%

26.4 Discussion

The change in testing in Idaho can be partially attributed to the samples collected and tested in 2007 from fields in the Idaho Falls, Idaho, area. While the seed certification regulations do not limit where seed potatoes can be grown, this area from where the samples were taken is a ware potato growing area. Since potatoes are vegetatively propagated, seed is often the initial source of inoculums in ware potato crops. In order to remove virus problems from ware crops, efforts to reduce virus in seed are the most effective approach. In addition, other factors that helped make a change in the rules for Idaho seed was the existence of PVY strains in combination with certain cultivars that result in very mild mosaic symptoms, making removal of diseased plants from seed fields much more difficult than in the past.

To outline how multiple agencies work together to improve seed health, this example of how postharvest testing is done is a good illustration. The idea to change the postharvest protocol is either proposed by the member-growers of Idaho Crop Improvement Association or by university scientists. The proposal goes before a Potato Advisory Committee. On this committee there are three industry representatives and one university representative (a seed potato specialist). If the proposal doesn't require a rule change and is simply a protocol change, it is put into place after approval from the advisory committee. If the proposal requires a rule change, then it must be approved by the University of Idaho's Foundation Seed Stocks Committee which consists of many scientists from different academic disciplines. The state of Idaho has designated the University of Idaho as the seed certifying agency in Idaho. The university then appointed Idaho Crop Improvement Association as its duly authorized agent to administer and conduct seed certification in Idaho.

Whenever a pest situation arises that requires national involvement, such as discovery of a quarantine pest or a regulated non-quarantine pest, the agencies that work together include the state seed certification agency, the State Plant Regulatory Official (SPRO) and the State Plant Health Director (SPHD). The SPRO is employed by the state (e.g. Idaho State Department of Agriculture); the SPHD is a federal employee of the USDA-APHIS-Plant Protection and Quarantine division. If questions arise or more details are needed regarding the pest and its detection, then these three individuals communicate with each other and share information. In this manner, transparency is fostered and the ability to address the pest problem in the most efficacious way is determined. The presence of the SNHP (State National Harmonization Program) has helped to facilitate this whole process. In addition, all state seed potato certification officials participate in the Potato Association of America Seed Certification Section meeting held annually. At this meeting both US and Canadian counterparts discuss seed health and work on joint projects to produce quality seed and foster trade.

26.5 Conclusion

The ability to work across individual state certification programmes by the use of the SNHP and the involvement of industry and government officials allows production of healthier seed by reducing virus levels and having a mechanism to address pest issues that can affect whole regions and their resulting markets. For the specific aspect of meeting the challenge of reduced PVY symptoms due to cultivar × PVY strain interactions, 100% ELISA testing is essential.

References

Clark, M.F. and Adams, A.N. (1977) Characteristics of the microplate method of enzyme linked immunosorbent assay for the detection of plant viruses. *Journal of General Virology* 34, 475–483.

Gray, S., De Boer, S., Lorenzen, J., Karasev, A., Whitworth, J., Nolte, P., Singh, R., Boucher, A. and Xu, H. (2010) Potato virus Y: an evolving concern for potato crops in the United States and Canada. *Plant Disease* 94, 1384–1397.

Gugerli, P. and Gehriger, W. (1980) Enzyme-linked immunosorbent assay (ELISA) for the detection of potato leafroll virus and potato virus Y in potato tubers after artificial break of dormancy. *Potato Research* 23, 353–359.

Karasev, A.V., Meacham, T., Hu, X., Whitworth, J., Gray, S.M., Olsen, N. and Nolte, P. (2008) Identification of potato virus Y strains associated with tuber damage during a recent virus outbreak in potato in Idaho. *Plant Disease* 92, 1371.

Novy, R.G., Whitworth, J.C., Stark, J.C., Love, S.L., Corsini, D.L., Pavek, J.J., Vales, M.I., James, S.R., Hane, D.C., Shock, C.C., Charlton, B.A., Brown, C.R., Knowles, N.R., Pavek, M.J., Brandt, T.L., Gupta, S. and Olsen, N. (2010) Clearwater Russet: a dual-purpose potato cultivar with cold sweetening resistance, high protein content, and low incidence of external defects and sugar ends. *American Journal of Potato Research* 87, 458–471.

Whitworth, J.L., Novy, R.G., Stark, J.C., Pavek, J.J., Corsini, D.L., Love, S.L., Miller, J.S., Vales, M.I., Mosley, A.R., Yilma, S., James, S.R., Hane, D.C., Charlton, B.A., Brown, C.R., Knowles, N.R. and Pavek, M.J. (2010) Yukon Gem: a yellow-fleshed potato cultivar suitable for fresh-pack and processing with resistances to PVY^{O} and late blight. *American Journal of Potato Research* 87, 327–336.

Whitworth, J., Novy, R., Stark, J., Pavek, J., Corsini, D., Love, S., Olsen, N., Gupta, S., Brandt, T., Vales, M., Mosley, A., Yilma, S., James, S., Hane, D., Charlton, B., Shock, C., Knowles, N., Pavek, M., Miller, J. and Brown, C. (2011) Alpine Russet: a potato cultivar having long tuber dormancy making it suitable for processing from long-term storage. *American Journal of Potato Research* 88, 256–268.

27 *Potato Virus Y* in South Africa: Isolate Characterization and Assessment of Potato Cultivar Resistance

J.C. Visser and D.U. Bellstedt*
University of Stellenbosch, Stellenbosch, South Africa

Abstract

Potato virus Y (PVY) is one of the most serious viral threats to potato production worldwide and throughout Africa, resulting in severe yield reduction. Management of this virus is key to sustainable potato production. The virus is spread by infected tuber material and aphid vectors. In South Africa, the pathology of PVY infections has changed in recent years. Disease diagnosis based on leaf symptoms became problematic, and diagnosis by means of ELISA testing was the only way to eliminate diseased mother tuber material. The PVY^{NTN} strain has recently been found in South Africa which can cause potato tuber necrotic ringspot disease (PTNRD) and this can lead to crop failure as tubers showing necrosis are unacceptable to the consumer.

In this study we characterized the PVY strains which are currently responsible for infections in South Africa using whole genome sequencing and a novel approach to the phylogenetic analysis of these sequence data. We were able to determine that the older PVY ordinary (PVY^{O}) and PVY necrotic (PVY^{N}) strains had largely been replaced by $PVY^{N\text{-}Wilga}$ and PVY^{NTN} strains since 1995. Presently, approximately 75% of the PVY isolates in South Africa are of the $PVY^{N\text{-}Wilga}$ and PVY^{NTN} strains. To determine the resistance of potato cultivars commonly planted in South Africa against our local $PVY^{N\text{-}Wilga}$ and PVY^{NTN} isolates, we assessed the speed of viral spread of selected isolates using the quantitative real-time reverse transcriptase PCR (qRT-PCR) in mechanically infected potato plants. Tubers were assessed for necrotic symptoms. This revealed that some cultivars show higher levels of resistance to PVY infection. Future assessments of the resistance of potato cultivars planted in South Africa will be conducted using this approach.

Management of PVY in South Africa will in future be based on ongoing monitoring of PVY isolate mutation and recombination through whole genome sequencing and planting of PVY-resistant cultivars. Furthermore vector control by chemical means and planting during cooler seasons when aphid pressure is lower, and elimination of infected tubers by means of the certification scheme will together be used in a strategy to reduce the effects of PVY infections.

Keywords: crop yield reduction, potato cultivar resistance, *Potato virus Y* (PVY), PVY infection, vector control

*dub@sun.ac.za

27.1 Introduction

Potato virus Y (PVY) is one of the most important pathogens affecting potato production worldwide (Ward and Shukla, 1991; Scholthof *et al.*, 2011). PVY infection may result in 10% yield reduction up to complete crop failure. Yield reduction is dependent on several factors including strain, viral load, the time during the season at which infection occurs, potato cultivar as well as environmental temperature (Warren *et al.*, 2005). Serology, genomic sequence information and biological properties have been used to identify and classify the different strains of PVY. PVY^{O}, first described in 1931, can induce mild to severe leaf symptoms which may vary from yellowing to leaf drop (Smith, 1931). PVY^{N}, identified in the 1940s, is able to induce occasional leaf veinal necrosis and very mild tuber damage in some cases (Crosslin *et al.*, 2005). Infection of potato plants by PVY^{O} or PVY^{N} led to yield reduction but does not influence tuber quality. However, in 1984, a genetic recombinant of PVY^{N} and PVY^{O}, PVY^{NTN}, was identified (Beczner *et al.*, 1984). PVY^{NTN} can induce some leaf symptoms on sensitive cultivars but the necrotic rings on tubers induced by PVY^{NTN} are the greatest concern to seed and table growers. The formation of necrotic rings is termed potato tuber necrotic ringspot disease (PTNRD) and may vary from light and superficial rings to dark and deep rings. Another recombinant of PVY^{N} and PVY^{O}, $PVY^{N\text{-}Wilga}$, was first described in 1991 (Chrzanowska, 1991). It is known to induce mild leaf mosaic patterns but only very low PTNRD incidence. Currently $PVY^{N\text{-}Wilga}$ is divided into two sub-strains, namely $PVY^{N:O}$ (a single recombinant called A-type Wilga) and $PVY^{N\text{-}Wilga}$ (a double recombinant called B-type Wilga).

In South Africa, the occurrence of PTNRD has greatly increased, leading to significant losses in especially the seed potato market. For this reason the objective of this study was to establish the distribution of PVY strains in South Africa using gene and whole genome sequencing. Furthermore this study aims to establish a system whereby local cultivars can be graded according to the degree to which they are resistant to foliar and tuber symptoms induced by PVY^{NTN}. By establishing such a system, both commercial and subsistence farmers will be able to make informed choices with regard to cultivar selection to ensure consistently high yields in spite of high levels of viral infection.

27.2 Materials and Methods

Identification of South African isolates of PVY

During the period from 2004 to 2010, 153 leaf samples were obtained from various South African potato farms. A further 79 samples, all in the form of tubers, were obtained in 2013. Samples were screened using coat protein sequencing for strain identification (Visser and Bellstedt, 2009). To distinguish PVY^{O} strains from $PVY^{N\text{-}Wilga}$ strains, a differential multiplex reverse transcriptase PCR (RT-PCR) was employed (Lorenzen *et al.*, 2006). Twenty-seven of these isolates, selected to be representative of strains of PVY currently in South Africa, were selected for whole genome sequencing.

Plant material homogenization for RT-PCR

A crude but effective and inexpensive RNA extraction method was employed (La Notte *et al.*, 1997; Bellstedt *et al.*, 2010).

Amplification of the PVY coat protein gene and genome using RT-PCR

The coat protein gene was amplified as a single amplicon while the genome was amplified in overlapping regions by means of RT-PCR. To this end, 2 µl of homogenized leaf sample was added to 23 µl of a standard RT-PCR mixture consisting of 2.5 µl 10 × PCR-buffer, 1.25 µl 0.1 M DTT, 1.5 µl 25 mM $MgCl_2$, 0.625 µl 20 µM forward primer, 0.625 µl 20 µM reverse primer, 1 µl 5 mM dNTPs (Bioline), 0.25 µl 5 U/µl Taq™ DNA polymerase (Bioline), 0.125 µl 200 U/µl

SuperScript™ III (Invitrogen) and 15.125 µl Milli-Q® water. In order to amplify the coat protein gene, specific primers where used which bound within highly conserved regions just upstream of the 5′ end and 3′ end of the coat protein gene, respectively. Amplification of the PVY genome was achieved through the use of multiple primer combinations. Three primer sets in different combinations could be used, for the amplification of overlapping fragments of PVY^{N}, PVY^{O} and recombinant $PVY^{N\text{-}Wilga}$ and PVY^{NTN} strains. All RT-PCR amplifications were performed using the following programme: a single reverse transcription step of 48°C for 30 min, 35 cycles of DNA amplification at 94°C for 30 s, at 50°C, 55°C or 64°C (primer combination dependent) for 45 s and 72°C for 60 s followed by 72°C for 7 min in a Veriti Thermal Cycler (Applied Biosystems). Products were analysed using agarose gel electrophoresis and visualized using a UV transilluminator.

Sequencing, genome sequence analysis and alignment

Cycle sequencing was performed according to standard protocols as supplied by Applied Biosystems. The products of cycle sequencing were analysed using an ABI® 3100 Genetic Analyzer in the Central Analytical Facility, University of Stellenbosch.

Resulting electropherograms were edited using CHROMAS (v. 2.33, Technelysium, Pty, Ltd) and overlapping nucleotide sequences combined to form contiguous sequences. Isolates were identified and classified as being of a specific strain using a BLASTN search. The 27 whole genome sequences of South African PVY isolates were aligned with 29 whole genome PVY sequences as well as three outgroups downloaded from GenBank using the CLUSTAL W (v. 1.4) alignment function in BIOEDIT (v. 7.0.5.2, Tom Hall) and finer manual alignment.

Mechanical infection of potato cultivars

Disease-free first generation minitubers of the cultivars Hermes, Sifra, Mondial and Fabula were obtained from the *In Vitro* Gene Bank of the Vegetable and Ornamental Plants Research Institute, South African Agricultural Research Council, Pretoria. Hermes was included due to its extreme sensitivity to PTNRD (Browning *et al.*, 2004; Baldauf *et al.*, 2006). All plants (ten per cultivar per virus isolate) were confirmed to be PVY-free using the quantitative real-time reverse transcriptase PCR (qRT-PCR) described below. Six PVY^{NTN} isolates (GP, Z14, Z16, Z001, SS607 31, NTN-OVS) and one $PVY^{N\text{-}Wilga}$ isolate (SS607 36) were maintained in the potato cultivars from which they were isolated and used to mechanically infect the cultivars using plant sap prepared in the following manner. Infected plant material (5 g) was ground in cold PBS (10 ml) using a pre-chilled mortar and pestle. To assist in the grinding and infection process a small amount of diatomaceous earth was added to the leaf pulp. Plants were infected by gently rubbing the prepared plant sap on the terminal leaf.

Cultivar screening for PVY spread using qRT-PCR

Leaves from infected plants were harvested weekly after mechanical infection for 5 weeks. To eliminate the potential effect of wounding on subsequent viral spread, sufficient replicates of plants were included so that only one leaf per plant was harvested. The leaf material was tested using a qRT-PCR protocol designed for the specific detection of PVY (Visser, 2008). Samples were prepared as described for RT-PCR. All reactions were performed in a LightCycler 1.5 using the LightCycler® RNA Master SYBR® Green I kit (Roche Molecular Diagnostics). One microlitre of crude extract was added to 19 µl of a standard SYBR® Green RT-PCR mixture. This consisted of 7.5 µl 2.7 × LightCycler® RNA Master SYBR® Green I, 1.3 µl 50 mM $Mn(CH_3COO)_2$, 2 µl 3 µM primer mix (PVY-SGIIf – 5′-GCT TAT GGT TTG GTG CAT TG-3′, PVYSGIIr – 5′-AAT GTG CCA TGA TTT GCC TA-3′) as well as 8.2 µl PCR grade water. The cycling conditions for amplification were as follows: reverse transcription at 61°C for 20 min, RT inactivation and PCR activation at 95°C for 30 s, 55 cycles of denaturation at 95°C for 5 s, annealing at 55°C for 10 s and extension at 72°C for 10 s. Amplification was followed by a melting curve analysis consisting of denaturation at 95°C for 0 s, annealing at 65°C for 15 s and a melting step of 95°C (0.1°C/second slope) for 0 s. The rotor was cooled to 40°C upon completion of the run. An amplicon from one of the qRT-PCR runs was sequenced and aligned with the existing coat protein gene database to confirm its identity.

PTNRD assessment of infected cultivars

Tubers harvested 14 weeks after mechanical infection of the mother plants were visually assessed for PTNRD.

27.3 Results

Identification of South African isolates

RT-PCR screening of 153 potato leaf samples collected between 2004 and 2010 confirmed that 130 samples were PVY positive. Strain identification is listed in Table 27.1 showing most isolates to be the recombinant $PVY^{N\text{-}Wilga}$ or PVY^{NTN} strains.

RT-PCR screening of the 79 potato tuber samples collected in 2013 indicated that 58 samples were PVY positive. Strain identification is listed in Table 27.2 and shows that the percentage of recombinants had increased since the previous survey.

Table 27.1. Relative percentage of strains of PVY occurring in South Africa for samples tested between 2004 and 2010.

Virus strain	Number of isolates	Relative percentage
N	26	20.00
O	3	2.31
NTN	64	46.23
N-Wilga	37	28.46

Table 27.2. Relative percentage of strains of PVY occurring in South Africa for samples tested in 2013.

Virus strain	Number of isolates	Relative percentage
N	0	0.00
O	3	5.17
NTN	28	48.28
N-Wilga	26	44.83
NTN and N-Wilga	1	1.72

Sequencing, genome sequence analysis and alignment

BLAST searches revealed that six of the sequenced genomes were PVY^N. Two of these isolates (NN300 41 and NN300 60) displayed significant sequence divergence from traditional PVY^N. NN300 41 was very similar (99.5%) to SASA-61 (GenBank accession no. AJ585198), an isolate previously identified in Scotland (Barker *et al.*, 2009). NN300 60 had 83% similarity to NE-11 (DQ157180). The coat protein gene of NN300 60 had high similarity (98.8%) to Chinese isolates (XCH24 (AY841258), XCH47 (AY841267), Fengyang-8-2 (AY742729) and Mudanjiang-8-3 (AY742731)) with only 95% sequence homology to traditional PVY^N (Li *et al.*, 2006; Visser and Bellstedt, 2009). Two genomes were found to be PVY^O. Fifteen isolates were found to be PVY^{NTN} and four $PVY^{N\text{-}Wilga}$ which were identified as B-type Wilga. Advanced phylogenetic and molecular dating analysis of the whole genome sequences revealed that $PVY^{N\text{-}Wilga}$ and PVY^{NTN} were imported into South Africa (Visser *et al.*, 2012).

Cultivar screening for PVY spread using qRT-PCR

All plant material screened at the start of the trial tested negative for PVY using qRT-PCR. The time (in weeks) from mechanical infection until the plants of the different cultivars tested positive for PVY is listed in Table 27.3. The spread of PVY^{NTN} virus isolates were the most rapid in cultivar Hermes and the slowest in cultivar Fabula. The GP isolate was the slowest spreading virus isolate used in this trial, requiring 5 weeks or more to spread in all cultivars. The other viral isolates spread at intermediate rates in the different potato cultivars while the Z001 isolate spread fastest in all cultivars tested. SS607 36, the only PVY$^{N\text{-Wilga}}$ isolate included in the cultivar screening trial, spread rapidly through all of the four cultivars tested.

PTNRD assessment of infected cultivars

The occurrence of PTNRD in the tubers harvested from the various cultivars is summarized in Table 27.4. The highest occurrence of PTNRD was found in cultivar Hermes and the lowest occurrence in cultivar Fabula. Viral isolates Z001 and Z16 induced the highest occurrence of PTNRD while NTN-OVS and GP induced lower levels of PTRND. The PVY$^{N\text{-Wilga}}$ isolate, SS607 36, only induced low levels of PTNRD in all cultivars.

27.4 Discussion

Using coat protein gene sequencing and multiplex RT-PCR (Lorenzen *et al.*, 2006), we found that the predominant strain infecting South African potatoes, for the period between 2004 and 2010, was PVY^{NTN} (49.23%) with lower incidences of PVY$^{N\text{-Wilga}}$ (28.46%) and even lower incidences of PVY^{N} (20%) and PVY^{O} (2.31%). In 2013, PVY^{NTN} incidence (48.28%) had not changed much but PVY$^{N\text{-Wilga}}$ incidence (44.83%) had greatly increased. No PVY^{N} isolates and only a few PVY^{O} (5.17%) were identified. One sample was found to be infected with PVY^{NTN} and PVY$^{N\text{-Wilga}}$ strains. Although these sample sizes are small, the results do reflect a tendency observed internationally in which the occurrence of non-recombinant PVY strains has decreased while the occurrence of recombinant strains has increased dramatically.

Whole genome sequencing of 27 selected isolates confirmed that four of the six isolates previously identified by coat protein sequencing and multiplex RT-PCR as PVY^{N} were indeed PVY^{N} but that the remaining two were unique. One of these two isolates,

Table 27.3. Number of weeks up to a positive qRT-PCR test.

Potato cultivar	Virus isolate						
	Z001	Z16	SS607 31	Z14	NTN-OVS	GP	SS607 36
Hermes	2	2	2	3	> 5	5	2
Sifra	5	5	4	5	> 5	5	2
Mondial	3		5	> 5	3	5	3
Fabula	> 5	4	5	4	3	> 5	3

Table 27.4. Percentage of tubers displaying potato tuber necrotic ringspot disease (PTNRD).

Virus isolate	Potato cultivar			
	Hermes	Fabula	Mondial	Sifra
Z001	100.00	0.00	16.95	33.33
Z16	80.00	4.88	2.54	40.63
SS607 31	26.47	6.58	14.00	5.56
Z14	25.00	0.00	8.57	13.64
NTN-OVS	2.73	0.00	0.00	0.00
GP	0.00	9.68	20.00	5.66
SS607 36	3.85	0.00	0.00	12.50

NN300 41, was found to be closely related to the SASA-61 (GenBank accession no. AJ585198) strain which has not been reported from South Africa before. We previously described the NN300 60 isolate as being very similar to isolates of Chinese origin on the basis of the coat protein gene sequence (Visser and Bellstedt, 2009). Whole genome sequencing revealed that it is very similar to NE-11 (DQ157180). NE-11 was first detected in the USA in 2004, described in 2008 (Lorenzen *et al.*, 2008) and is thought to originate from China. The NE-11 variant induces leaf mosaic symptoms as well as mild PTNRD. Both SASA-61 and NE-11 are recombinants of PVY^{N} and an unknown variant of PVY of which a portion (bp 2204 – bp 2700) is shared. The presence of these unique isolates, first described in Scotland and the USA, respectively, suggests that infected plant material was imported into South Africa and emphasizes the need for strict control and sensitive screening of imports of potato material (Lorenzen *et al.*, 2008; Barker *et al.*, 2009).

Whole genome sequencing further confirmed that four isolates were $PVY^{N\text{-}Wilga}$ type B while the other 15 were typical PVY^{NTN} recombinants. Three of these isolates, GP, NTN-OVS and Z14, diverged significantly from the other PVY^{NTN} isolates. Two isolates were confirmed as being PVY^{O}.

The highly susceptible cultivar Hermes which lacks resistance to PVY^{NTN} is often used as an indicator cultivar to investigate tuber necrotic symptom development (Browning *et al.*, 2004; Baldauf *et al.*, 2006; Singh *et al.*, 2008). The high susceptibility of Hermes was confirmed as PTNRD symptoms appeared in all of the tubers of the infected plants. However, the PVY^{NTN} GP and NTN-OVS isolates failed to induce PTNRD in this cultivar suggesting that the mutations in these isolates is responsible for virulence reduction. The other three cultivars showed low or no PTNRD, indicating resistance to PVY^{NTN}. However, these trends were not uniform and it appears that specific interactions between potato cultivar and viral isolate play an important role in PTNRD expression. The isolate SS607 36, the only $PVY^{N\text{-}Wilga}$ isolate used, was only able to induce a much lower incidence of PTNRD. Other authors have also found that $PVY^{N\text{-}Wilga}$ isolates show a far lower PTNRD incidence than that induced by PVY^{NTN} infection (Chrzanowska, 1991; Boonham *et al.*, 2002a; Hu *et al.*, 2009a, 2011). PVY^{NTN}, $PVY^{N\text{-}Wilga}$ and PVY^{N} are all known to induce tuber necrotic symptoms, although with decreasing frequency, while PVY^{O} does not. Determinants of necrotic symptom induction must therefore be encoded for in the genomes of these necrotic strains but are absent in the genome of PVY^{O} strain. Since $PVY^{N\text{-}Wilga}$ contains the smallest portion of the PVY^{N} genome it stands to reason that the genetic determinant of necrosis must be encoded for in that part of the genome and which is also shared with the PVY^{N} and PVY^{NTN} isolates. This deduction was also made by other authors (Tribodet *et al.*, 2005; Hu *et al.*, 2009b; Faurez *et al.*, 2012). Recombinant isolates, PVY^{NTN} and $PVY^{N\text{-}Wilga}$ type A, have P1 and HC-Pro genes similar to that of PVY^{N}, while $PVY^{N\text{-}Wilga}$ type B isolates have recombinant (O/N) P1 genes and PVY^{N} type HC-Pro genes. Residues N339, K400 and E419, in the HC-Pro, have been identified to be responsible for the induction of leaf necrosis in tobacco (Tribodet *et al.*, 2005; Hu *et al.*, 2009b). Faurez *et al.* (2012) deduced that the amino acid residues at positions 400 and 419 are responsible for the induction PTNRD in potatoes. Analysis of the amino acid composition of HC-Pro of South African isolates of PVY revealed that critical amino acid substitutions occurred in three of the isolates (Visser, 2012). In Z16, GP and NTN-OVS there are E to K substitutions at position 419. A further E to K substitution was identified at position 410 which may also be responsible for PTNRD development in potatoes.

The PVY^{NTN} Z001 isolate spread rapidly in cultivar Hermes. However, the GP and NTN-OVS isolates spread more slowly in Hermes suggesting that the multitude of mutations in these isolates are responsible for virulence reduction. The HC-Pro has been described as being critical in systemic movement and aphid transmission (Govier and Kassanis, 1974; Peng *et al.*, 1998; Wang *et al.*, 1998). Amino acid substitutions within this protein may thus either impair or improve systemic viral movement and/or viral transmission, depending on the nature of the change. While an impaired functionality of the HC-Pro will lead to the ultimate decline of a viral line carrying such a substitution, a change leading to an improvement in aphid transmissibility and systemic movement would be an evolutionary advantage. The other three cultivars showed slower viral spread with the same virulent PVY^{NTN} isolates, indicating impaired viral movement and therefore resistance to PVY^{NTN}. It appears that specific interactions between the potato cultivar and the viral isolate play an important role in viral spread. The isolate SS607 36, which was the only PVY$^{N\text{-}Wilga}$ isolate used, spread rapidly in all cultivars.

The risk posed by PVY^{NTN} due to its ability to induce PTNRD is much greater in South Africa due to the drier and warmer conditions under which potatoes are grown than in the cooler climates found in most of Europe and the USA (Le Romancer and Nedellec, 1997; Boonham *et al.*, 2002b). For this reason, it is much more important to identify PVY-resistant potato cultivars. By selecting more resistant cultivars the spread of PVY and the occurrence of PTNRD in the local crop may be reduced. The phylogenetic analysis performed on the genomic information of South African isolates of PVY indicates that the majority of South African isolates of PVY are closely related to PVY$^{N\text{-}Wilga}$, PVY^{NTN}, SASA-61 and the PVY NE-11 strains from overseas (Visser *et al.*, 2012). As this indicates that these strains have in all likelihood been imported into this country, this study emphasizes the need for strict import regulation and the continued implementation of highly sensitive and reliable screening protocols as are currently employed by the South African Potato Certification Service.

Acknowledgements

This research was supported by funding from Potatoes South Africa and a postdoctoral fellowship from the National Research Foundation of South Africa.

References

Baldauf, P.M., Gray, S.M. and Perry, K.L. (2006) Biological and serological properties of *Potato virus Y* isolates in northeastern United States potato. *Plant Disease* 90, 559–566.

Barker, H., McGeachy, K.D., Toplak, N., Gruden, K., Žel, J. and Browning, I. (2009) Comparison of genome sequence of PVY isolates with biological properties. *American Journal of Potato Research* 86, 227–238.

Beczner, L., Horvath, J., Romhanyi, I. and Forster, H. (1984) Studies on the etiology of tuber necrotic ringspot disease in potato. *Potato Research* 27, 339–352.

Bellstedt, D.U., Pirie, M.D., Visser, J.C., De Villiers, M.J. and Gehrke, B. (2010) A rapid and inexpensive method for the direct PCR amplification of DNA from plants. *American Journal of Botany* 97, e65–e68.

Boonham, N., Walsh, K., Hims, M., Preston, S., North, J. and Barker, I. (2002a) Biological and sequence comparisons of *Potato virus Y* isolates associated with potato tuber necrotic ringspot disease. *Plant Pathology* 51, 117–126.

Boonham, N., Walsh, K., Preston, S., North, J., Smith, P. and Barker, I. (2002b) The detection of tuber necrotic isolates of *Potato Virus Y*, and the accurate discrimiation of PVY^{O}, PVY^{N} and PVY^{C} strains using RT-PCR. *Journal of Virological Methods* 102, 103–112.

Browning, I., Charlet, K., Chrzanowska, M., Dedic, P., Kerlan, C., Kryszczuk, A., Schubert, J., Varveri, C., Werkman, A. and Wolf, I. (2004) What is PVY^{NTN}? The reaction of potato cultivars to inoculation with a range of PVY isolates. Paper presented at the 12th European Association of Potato Research (EAPR), Virology Section Meeting, 13–19 June, Rennes, France. EAPR, Rennes, France, pp. 51–53. Available at: http://www.cabi.org/cabdirect/FullTextPDF/2008/20083270104.pdf (accessed 15 September 2015).

Chrzanowska, M. (1991) New isolates of the necrotic strain of *Potato virus Y* (PVY^{N}) found recently in Poland. *Potato Research* 34, 179–182.

Crosslin, J.M., Hamm, P.B., Shiel, P.J., Hane, D.C., Brown, C.R. and Berger, P.H. (2005) Serological and molecular detection of tobacco veinal necrosis isolates of *Potato virus Y* (PVY^{N}) from potatoes grown in the western United States. *American Journal of Potato Research* 82, 263–269.

Faurez, F., Baldwin, T., Tribodet, M. and Jacqout, E. (2012) Identification of new *Potato virus Y* (PVY) molecular determinants for the induction of vein necrosis in tobacco. *Molecular Plant Pathology* 13, 948–959.

Govier, D.A. and Kassanis, B. (1974) A virus-induced component of plant sap needed when aphids acquire *Potato virus Y* from purified preparations. *Virology* 61, 420–426.

Hu, X., He, C., Xiao, Y., Xiong, X. and Nie, X. (2009a) Molecular characterization and detection of recombinant isolates of *Potato virus Y* from China. *Archives of Virology* 154, 1303–1312.

Hu, X., Meacham, T., Ewing, L., Gray, S.M. and Karasev, A.V. (2009b) A novel recombinant strain of *Potato virus Y* suggests a new viral genetic determinant of vein necrosis in tobacco. *Virus Research* 143, 68–76.

Hu, X., Nie, X., He, C. and Xiong, X. (2011) Differential pathogenicity of two different recombinant PVY^{NTN} isolates in *Physalis floridana* is likely determined by the coat protein gene. *Virology Journal* 8, 207.

La Notte, P., Minafra, A. and Saldarelli, P. (1997) A spot-PCR technique for the detection of phloem-limited grapevine viruses. *Journal of Virological Methods* 66, 103–108.

Le Romancer, M. and Nedellec, M. (1997) Effect of plant genotype, virus isolate and temperature on the expression of the tuber necrotic ringspot disease (PTNRD). *Plant Pathology* 46, 104–111.

Lorenzen, J.H., Piche, L.M., Gudmestad, N.C., Meacham, T. and Shiel, P. (2006) A multiplex PCR assay to characterize *Potato virus Y* isolates and identify strain mixtures. *Plant Disease* 90, 935–940.

Lorenzen, J.H., Nolte, P., Martin, D., Pasche, J.S. and Gudmestad, N.C. (2008) NE-11 represents a new strain variant class of *Potato virus Y*. *Archives of Virology* 153, 517–525.

Peng, Y.-H., Kadoury, D., Gal-on, A., Huet, H., Wang, Y. and Raccah, B. (1998) Mutations in the HC-Pro gene of *Zucchini yellow mosiac potyvirus*: effects on aphid transmisison and the binding to purified virions. *Journal of General Virology* 79, 897–904.

Scholthof, K.-B., Adkins, S., Czosnek, H., Palukaitis, P., Jacqout, E., Hohn, T., Hohn, B., Saunders, K., Candresse, T., Ahlquist, P., Hemenway, C. and Foster, G.D. (2011) Top 10 plant viruses in molecular plant pathology. *Molecular Plant Pathology* 12, 938–954.

Singh, R.P., Valkonen, J.P.T., Gray, S.M., Boonham, N., Jones, R.A.C., Kerlan, C. and Schubert, J. (2008) Discussion paper: the naming of *Potato virus Y* strains infecting potato. *Archives of Virology* 153, 1–13.

Smith, K.M. (1931) Composite nature of certain potato viruses of the mosaic group. *Nature* 127, 702.

Tribodet, M., Glais, L., Kerlan, C. and Jacquot, E. (2005) Characterization of *Potato virus Y* (PVY) molecular determinant involved in the vein necrosis symptom induced by PVY^{N} isolates in infected *Nicotiana tabacum* cv. Xanthi. *Journal of General Virology* 86, 2101–2105.

Visser, J.C. (2008) A study of genomic variation in and the development of detection techniques for *Potato Virus Y* in South Africa. MSc thesis (Biochemistry), University of Stellenbosch, Stellenbosch, South Africa.

Visser, J.C. (2012) A study of the strain evolution and recombination of South Africa isolates of *Potato virus Y*. PhD thesis (Biochemistry), University of Stellenbosch, Stellenbosch, South Africa.

Visser, J.C. and Bellstedt, D.U. (2009) An assessment of molecular variability and recombination patterns in South African isolates of *Potato virus Y*. *Archives of Virology* 154, 1891–1900.

Visser, J.C., Bellstedt, D.U. and Pirie, M.D. (2012) The recent recombinant evolution of a major crop pathogen, *Potato virus Y*. *PLoS ONE* 7, e50631.

Wang, R.Y., Powell, G., Hardie, J. and Pirone, T.P. (1998) Role of the helper component in vector-specific transmission of potyviruses. *Journal of General Virology* 79 1519–1524.

Ward, C.W. and Shukla, D.D. (1991) Taxonomy of potyviruses: current problems and some solutions. *Intervirology* 32, 269–296.

Warren, M., Krüger, K. and Schoeman, A.S. (2005) *Potato virus Y* (PVY) and *Potato leaf roll virus* (PLRV): literature review for potatoes in South Africa. Working paper for Department of Zoology and Entomology, Faculty of Natural and Agricultural Sciences, University of Pretoria, Pretoria, South Africa.

Part III

Addressing Bottlenecks in the Sweetpotato Seed Systems

28 Can Small Still Be Beautiful? Moving Local Sweetpotato Seed Systems to Scale in Sub-Saharan Africa

M. McEwan,[1*] C. Almekinders,[2] P.E. Abidin,[3] M. Andrade,[4] E.E. Carey,[5] R.W. Gibson,[6] A. Naico,[4] S. Namanda[7] and S. Schulz[8]

[1]*International Potato Center Sub-Saharan Africa (CIP-SSA), Nairobi, Kenya;* [2]*Wageningen University, Wageningen, The Netherlands;* [3]*CIP, Lilongwe, Malawi;* [4]*CIP, Maputo, Mozambique;* [5]*CIP, Kumasi, Ghana;* [6]*Natural Resources Institute, Chatham Maritime, UK;* [7]*CIP, Kampala, Uganda;* [8]*CIP, Addis Ababa, Ethiopia*

Abstract

In sub-Saharan Africa (SSA), a range of farmer-based practices for the conservation and multiplication of sweetpotato planting material has evolved. In bimodal rainfall areas, sequential planting ensures that a ware crop is in the ground for most of the year, and vines are harvested from one crop to plant the next one. In unimodal areas with a long dry season, practices include the use of 'volunteer' planting material from sprouting roots which have been left in the ground from the previous crop. The predominant sources of planting material are from the farmer's own field or from friends or neighbours. However, these practices result in limited amounts of planting material being available at the start of the rains and contribute to the build-up of pests and diseases contributing to suboptimal root crop production. Sweetpotato breeding efforts are leading to the development of new varieties that are preferred by farmers and consumers. However, without strong linkages to seed multiplication and dissemination efforts these varieties may not quickly benefit large numbers of smallholder farmers and consumers. Increasingly there are specialized vine multipliers who have been supported by 'project' interventions. Yet, it is not clear whether and how these interventions have built on the successful elements of existing practices. Our chapter examines the literature on local seed system functioning, and the implications for crops such as sweetpotato. The chapter reviews recent efforts to multiply and disseminate sweetpotato planting material in Mozambique, Uganda, Tanzania, Malawi, Ethiopia and West Africa. New varieties and technologies have been promoted together with interventions to 'engineer' changes in the organization and coordination of the seed system. We review the country cases to gauge the extent to which successful elements of farmer-based practices for managing sweetpotato planting material have been identified and built into the process of redesigning the seed system.

A number of issues are identified for discussion. These include: (i) What are the critical points for interaction between the traditional farmer-based practices and the formal seed system?; (ii) What are the trade-offs between remaining local, and yet achieving scale?; and (iii) How can the quality of planting material be assured as we go to scale? We also assess the different drivers for the seed system, and the implications for the functions of various stakeholders and patterns of

*m.mcewan@cgiar.org

communication and coordination. The chapter concludes by highlighting gaps in our current understanding for getting sweetpotato seed systems not only moving, but working at scale.

Keywords: farmer demand, local level specialization, seed systems, sweetpotato

28.1 Introduction

After 40 years of seed sector interventions we are still faced with the challenge: Do smallholder farmers in sub-Saharan Africa (SSA) have timely access to sufficient quantities of quality seed? Despite many efforts across public, parastatal, private and civil society sectors, this question about seed provision can still be posed, particularly for vegetatively propagated crops (VPCs). Recently there has been a revival in interest in seed systems including that for sweetpotato. This has in part been driven by: (i) the recognition of the potential contribution that roots and tubers can make to diversified diets; (ii) the price rises in the cost of other staples; and (iii) the realization that increased investments in breeding and the release of improved varieties, will have limited impact on farmers and consumers unless seed multiplication and dissemination strategies can work sustainably and at scale.

As discussed in other chapters, sweetpotato (*Ipomea batatas*) and in particular the orange-fleshed varieties can play a role in strengthening food security, reducing malnutrition and combatting vitamin A deficiency. In SSA sweetpotato is grown as a staple food across a wide range of agroecological regions. Propagation is largely through the use of vine cuttings selected from the previous crop (Gaba and Singer, 2009). The crop is particularly adapted to marginal areas of low or erratic rainfall and low soil fertility; it tolerates high temperatures, is easy to propagate and maintain, and yields well even in adverse conditions. It is friendly to the environment, as chemical inputs are rarely used and it protects soil from erosion, as it closes its canopy in a short time with a well-developed root system (Woolfe, 1992). Sweetpotato requires minimal inputs, making it attractive to resource-poor households, households headed by women or people living with human immunodeficiency virus (HIV). These characteristics make sweetpotato an ideal crop in mitigating disasters and in post-conflict situations. Moreover, its short cropping season, its flexible planting and harvesting schedules, as well as its increasing commercial value, further emphasize the significance of the sweetpotato. While often considered 'a poor person's crop', the area devoted to sweetpotato is increasing throughout the region due to declining cereal yields, the cassava mosaic disease pandemic, the rapid spread of the banana wilt and lately, the threat of cassava brown streak virus.

Realizing the full potential of sweetpotato to increase the food security and nutrition of the poor requires, among others, good functioning seed systems, to effectively distribute new varieties and ensure access to high quality planting material.[1] To date, among VPCs, most attention has been paid to potato (*Solanum tuberosum*) which is a high-input cash crop in many developing countries. Seed systems of other VPCs such as sweetpotato, cassava and banana have received much less attention. These crops have high relevance for food security of the poorest in rural areas but apart from public sector support for breeding, there have been limited efforts to develop formal seed systems for multiplication and dissemination. The existing seed systems for sweetpotato have operated at a local scale, based on diverse farmer practices for selection, multiplication and exchange of planting material. However, until we have a better understanding of the dynamics of farmer demand and how to ensure a consistent supply of planting material at scale, we will be unable to address bottlenecks and take advantage of the opportunities to contribute to the transformation of food security and nutrition in SSA.

This chapter seeks to re-examine what we know about seed systems in general and the implications for sweetpotato. We will look at the specificities of a seed system for a VPC, and in particular how the characteristics of sweetpotato influence the objective and functions of seed system interventions, the types of stakeholders involved, and what this means for the end users (i.e. smallholder farmers in SSA). We take farmer demand for planting material as an entry point. The chapter first reflects on a series of demand characteristics that are of importance in crops in general, and assesses these in relation to sweetpotato seed systems. The second part reviews some of the strategies which have been used to strengthen and expand the multiplication and dissemination of sweetpotato planting material. In the final section we briefly assess some factors to consider when getting sweetpotato seed systems moving.

28.2 Reflections on the Literature: What is Relevant for Sweetpotato Systems?

A vital change in seed system thinking over the last four decades relates to the increased awareness of the importance of local seed systems. This has been partly a result of the unsuccessful efforts to get seed systems in place in developing countries using blueprints based on models for formal seed systems in developed northern countries such as the USA, The Netherlands and the UK. Such approaches (as expounded by Douglas, 1980) yielded temporary successes, but in most cases were not sustainable. Large-scale centralized government or parastatal multiplication programmes, often with the technical support of the Food and Agriculture Organization of the United Nations (FAO) and international donor finance (e.g. the FAO Seed Improvement and Development Programme; FAO, 1973) had the objective of 'meeting farmers' demand' and providing quality seed of improved varieties. However, there were various reasons why these programmes were not economically sustainable and consequently resulted in poor performance. A major argument was that in many countries the smallholder farming sector is characterized by diversity and complexity. This implied the need for formal seed programmes to deal with a large portfolio of varieties that were adapted to a range of farmers' socio-economic and agroecological conditions (Almekinders and Louwaars, 2002; Tripp, 2002). These programmes had difficulty in effectively catering for this with the result that: (i) seed was distributed for varieties which were not locally adapted; (ii) seed production was expensive; (iii) seed quality was unsatisfactory; and (iv) seed delivery was untimely.

The structural adjustment policies in the 1990s reduced national and international investment in the agriculture sector. This contributed to the closure of loss-making state or parastatal seed companies, with the expectation that the private sector would be both willing to fill the gap and lead to more competitive seed marketing and distribution practices (Cromwell *et al.*, 1992). As it became clear that a parastatal-based approach for breeding, seed multiplication and distribution was inappropriate, local seed systems came onto the radar of researchers and policy makers. At this point the focus was provision of seed for grain crops and limited consideration (if any at all) was given to VPCs, with the exception of potato.

The comparative advantages of local seed systems compared with formal seed systems have been extensively commented on, stressing in particular: (i) the local adaptation of varieties and practices; (ii) the valuable diversity which local seed systems harbour; and (iii) their ability to diffuse materials among farmers via alternative social networks and channels (Almekinders *et al.*, 1994). Since then research and development actors have been trying to find ways to build on local seed systems, link with them and integrate them into or with the formal system (Louwaars and de Boef, 2012).

From the 1990s onwards, using more integrated perspectives, seed-sector development strategies considered that the national and local private sector, specialized farmers or farmer groups supported by non-governmental organizations (NGOs), all have an important role to play in seed multiplication and distribution (e.g. Camargo *et al.*, 1989).

Most developing countries have tried this approach for one or more crops and many of these strategies included elements of participatory plant breeding or participatory varietal selection. Models that involved farmer groups or cooperatives often had the vision that these would become entrepreneurial actors. There are examples of successful entrepreneurs, in particular specialized traders (Tripp, 2002; Van Mele and Bentley, 2011), but the majority of these efforts have still not been considered as sustained successes. The participation of multinational corporations in these efforts is limited, particularly in the food-crop seed sector, with exceptions for the maize and vegetable seed sectors.

For VPCs it is only potato which has seen major seed system efforts (Devaux *et al.*, 2011; Manrique, 2013), but these have also borne few sustainable successes. Important bottlenecks which affect seed systems of VPCs are:

- bulky and perishable planting material – with implications for transport requirements, storage and for the maintenance of the planting material in the off season;
- seed contamination and degeneration which takes place through soil-borne diseases and vegetative multiplication;
- low multiplication rates, which therefore increase the amount of time needed to bulk up sufficient supply; and
- ease of propagation and common property nature of the planting material means that there is little profit incentive for commercial seed enterprises.

With renewed attention for agricultural productivity, seed systems and concern for underutilized or 'orphan' crops such as sweetpotato, what can we learn from the experiences so far? After so many failures, are the successful seed enterprises the forerunners heralding more positive developments or are they just outliers? Has the tide turned and have the bottlenecks to seed system development been overcome? Are the past failures no longer relevant? If so, what has changed: the type of initiatives or the overall context? We think that some underlying questions still merit re-examination in order to be as well prepared as we can for the new initiatives that we support, design and implement. We also need to think through how these lessons apply to sweetpotato seed systems, in particular those that priortitize food security and nutrition. This may bring in additional and different goals together with non-conventional actors, and thus alternative thinking about integrating local and formal seed systems.

28.3 Seed Demand

Ordinarily, smallholder farmers use seed from their previous harvest, seed from neighbours and relatives, or in some circumstances seed from the market or local agrochemical-input supply shops. Even for regular seed lot replacement, these sources are often sufficient or the most rational seed sources for farmers. Therefore our analysis starts out from the viewpoint that farmer demand for planting material is the most decisive driver for the development of a viable 'seed' supply system. However, 'actual' demand for seed can only be assessed when there is a sufficient supply of quality seed, which is not the case for most crops and in most countries in Africa. 'Farmers' demand' for seed is usually estimated by research and development actors on the basis of area planted with the crop and an 'expert guess' of a replacement rate. Therefore we might ask: Would a better supply of quality seed stimulate demand? Yet, from the perspective of the seed producer there is limited incentive in ensuring a stable seed supply if the seed demand is not stable. This is part of the conundrum of the interdependence between seed demand and seed supply which we need to unravel.

For sweetpotato, surveys frequently report 'shortage of planting material' as a key production constraint. For example, surveys in Lake Zone, Tanzania found that shortage of planting material was ranked third in sweetpotato production constraints (after weevil and drought), with 66% of farmers reporting this constraint (Kapinga *et al.*, 1995). Shortages of planting material are more common in areas with unimodal rainfall systems and a long dry season where there is higher risk of loss of planting material. In bimodal areas or areas where there is a longer rainy season, there is always a crop in the ground to provide planting material, but there may be higher build-up of pests and diseases thus contributing to suboptimal root production. The Programme Régional d'Amélioration de la Pomme de Terre et de la Patate Douce en Afrique Centrale et de l'Est (PRAPACE) priority setting exercise in 2003 and an International Potato Center (CIP) survey of the National Agricultural Research Institutes (NARIs) both ranked 'virus management, seed quality and supply systems' as high priority for future research and development against all other listed sweetpotato technologies (PRAPACE, 2005; Fuglie, 2006; Barker *et al.*, 2009). So, ensuring that farmers can have timely access to adequate quantities of quality planting material remains a challenge. Therefore, we consider that farmers' rationale and incentives for purchasing seed are still not fully understood.

Which then are the situations where farmers are actively seeking seed from external sources? Table 28.1 summarizes our understanding of the incentives which influence farmers' seed-purchasing behaviour, which we then discuss in more detail below.

New varieties

An obvious reason to acquire seed from formal seed sources is to get seed of new varieties. Experts usually asssume that when a succesful new variety is around, farmers are willing to pay a premium price for the seed. There is, however, limited research data to confirm this and the assumption ignores the social context and interactions among different types of farmers when sourcing seed. In any case, for farmers to be attracted to new germplasm, it needs to respond to their needs, preferences and be adapted to their agroecological conditions.

For sweetpotato, breeding work in and for SSA has only gained momentum in the last 5–10 years. While the relative importance of different varietal characteristics differs across countries, yield, root size, drought tolerance and pest and disease resistance are consistently prioritized (Table 28.2).

Table 28.1. Relative importance[a] of incentives for farmers to purchase planting material of different crop types.

Incentive	Seed-based crops: Varieties based on pure lines	Seed-based crops: Hybrid varieties	VPCs[b]
New varieties: genetic superiority of germplasm	+	+++	+
High seed health degeneration rates which affect crop yield	+	–	+++
Specialized practices required for seed production (e.g. this applies to many vegetable seeds where seed is not the harvested product; hybrid seed)	+	+++	+/–
Difficulty to store/perishability of planting material	+	+/–	++
Small volume and weight (easy to transport)	+	++	––
Low multiplication rate	++	++	++
A stable profitable market for the crop product	+/–	++	++
Demand in the consumer market for high quality (examples: (i) vegetables; and (ii) demand for grain lot purity, e.g. rice, groundnut)	++	+++	+
Poor growing season/crop failure (including disaster contexts)	+++	–	+++

[a]Relative importance is indicated as: +++ higher importance; --lesser importance.
[b]VPCs, vegetatively propagated crops.

Table 28.2. Seed demand: sweetpotato attributes preferred by farmers in Ethiopia, Ghana, Malawi, Mozambique, Nigeria, Tanzania and Uganda. (From country Sweetpotato Seed Systems Case Studies prepared for African Potato Association presentation.)

Varietal characteristics preferred by farmers[a]	Case study countries	Comments
Yield, production, pest and disease characteristics		
High yielding roots	Ethiopia, Malawi, Tanzania, Mozambique	
Large roots	Kagera	
Drought tolerance	Tanzania, Mozambique, Malawi	Particularly important in unimodal rainfall areas and areas with unpredictable climate patterns
Persistent vines	Ghana, Mozambique	
High vine yield	Ethiopia	Important in livestock systems
Resistance to diseases	Uganda, Tanzania, Mozambique	In many areas of these countries, sweetpotato virus diseases are devastating and resistance is an absolute requirement
Alternaria disease tolerant	Lake Zone, Tanzania, Uganda	This disease tends to be a problem in highland areas
Weevil resistant	Ethiopia	
Sweetpotato butterfly resistant	Ethiopia	
Root flesh colour		
White and cream	Ethiopia, Mozambique	OFSP varieties: demand increasing where varieties have been introduced
Yellow	Ghana and Nigeria	
Orange	Malawi, Burkina Faso	
Texture		
Firmness	Lake Zone, Tanzania	
Low fibre	Lake Zone, Tanzania	
Mealiness	Kagera, Tanzania	
Other organoleptic features		
High dry matter	Lake Zone, Tanzania, Ethiopia, Ghana, Nigeria, Burkina Faso, Mozambique	
Sweetness	Ethiopia, Malawi	
Moderate sweetness	Tanzania	
Leaf shape	Malawi, Mozambique	Important where leaves are used as a vegetable
Harvest and postharvest characteristics		
In-ground storability	Ethiopia, Lake Zone, Tanzania, Malawi	
Non-perishable roots	Ghana, Nigeria, Burkina Faso	
Early maturity	Ethiopia	
High market demand	Lake Zone, Tanzania	

[a]Farmer preferences are not ranked in order of importance.

Breeding efforts have taken into account these preferences and since 2005, a range of new varieties have been released in SSA, with high levels of beta-carotene, reduced virus susceptibility and improved drought tolerance. Few studies, however, with the exception of work in Tanzania (Kapinga *et al.*, 2003) for traders and in western Kenya for pregnant women and young children (SASHA, 2012), have looked at trait preferences from the perspective of other end users and products (e.g. different types of consumers, fresh root traders, processors, institutional buyers, livestock keepers). If new varieties are a key element of farmers' demand for seed, then as sweetpotato shifts from being not only a food security crop for home consumption but is also marketed and used for processing, new end users and their preferences for additional or different traits will need to be considered (e.g. shelf life, long-term storability, leaf and vine characteristics). When improved varieties are being distributed with the objective of improving food security for poorer farmers, some subsidy may be needed (e.g. through the use of targeted vouchers) to encourage risk-averse farmers to try the new varieties. This also means that when targeting particular vulnerable groups, different entry points and distribution channels may have to be considered.

Seed quality and degeneration

Decreasing yield is another important driver for farmers to purchase new seed. Decreasing yields are associated with different types of quality loss in seed, which can be related to genetic, health and physiological factors. In crops with well-developed hybrid-variety technology such as maize, sorghum and vegetables, genetic decline (or rather 'disintegration') strongly drives farmers' seed purchasing behaviour. In these crops yearly seed replacement with seed purchased from specialized sources is considered the only logical option by research and development professionals: it is as if farmers have to buy the variety 'new' every season. In grain crops such as wheat, barley, rice and grain legumes such as beans, genetic degeneration has less importance because pure line varieties prevail. However, in situations where there is market demand for consistent quality of the grain product, genetic mixtures become problematic and therefore farmers make an effort to acquire 'pure' seed lots, for example the case of sorghum in West Africa (Tripp, 2002).

Variety mixtures in the field might be considered another form of degeneration or loss of quality. For farmers, however, variety mixtures may not be problematic. On the contrary, mixtures in the field can offer advantages in variable climate conditions and disease pressure (De Haan, 2009). Various researchers have also found that farmers have remarkable knowledge and skill in varietal differentiation, on the basis of morphological characteristics of the plants and tubers. For sweetpotato, variety mixtures in the field are common, either by choice or if volunteer plants emerge from the previous crop. Variety mixtures also occur when planting is done sequentially, from whatever varieties are currently available (Badstue and Adam, 2011). Where grown for food security purposes, sweetpotato is normally harvested piecemeal and so having varieties with different maturity periods in the same field is not a disadvantage, and in terms of mitigating potential climatic, disease and pest risk is beneficial. However, if the crop is grown for commercialization and harvesting is done at one time for a specific end user (e.g. a processor), then varietal mixtures are more problematic and require sorting after harvest.

In VPCs, while genetic degeneration is not usually an important factor due to their clonally propagated character, phytosanitary and physiological considerations are of paramount importance for quality and yield decline. In SSA the most important diseases which contribute to degeneration in sweetpotato planting material are sweetpotato virus diseases, which infect either individually or in mixed infections. Sweetpotato virus diseases are spread through white fly and aphids. *Sweet potato feathery mottle virus* (SPFMV) is the most common, but is largely asymptomatic as a single infection. In mixed infections, *sweet potato chlorotic stunt virus* (SPCSV) and SPFMV combine to present as sweet potato virus disease (SPVD) with stunting, feathery vein clearing and yellowing observed (Carey *et al.*, 1999; Gaba and Singer, 2009). These diseases occur throughout SSA, albeit with differences in prevalence and strain (Gibson *et al.*, 2009).

Reduction in root yield from the complex SPVD infection is estimated at 50% or more (Loebenstein and Thottappilly, 2009). A study in China showed that the use of virus-free material (from sprouted roots) yielded 30% greater than normal planting material – with the yield reducing to the same level after five generations (Fuglie *et al.*, 1999). The evidence for reduction in root yield from the impact of single infection (e.g. SPFMV) is more ambiguous, and varies dependent on the susceptibility of the cultivar and which viruses are involved. While severity of symptoms is thought to be correlated with viral load, some viruses are symptomless or latent. The presence of a mild symptomless virus, while occurring singly may not cause significant yield decline, however, once able to combine with another virus may lead to severe visual symptoms. Landraces bred and grown by farmers in Africa together with some researcher-bred varieties (with breeding carried out in East Africa) are said to show little evidence of degeneration (Gibson *et al.*, 2009). However, local farmers' cultivars with high levels of resistance tend to be low yielding and late maturing compared with earlier maturing, high yielding, yet susceptible local cultivars or exotic introductions (Carey *et al.*, 1999).

In general, sweetpotato farmers are aware of signs that show that the planting material is poor quality, but often link the symptoms to drought rather than to pests and diseases (Gibson *et al.*, 2009). In situations where material is in scarce supply, farmers may use what they would otherwise reject. In Malawi, the existing practice was for farmers to select varieties with a strong vine so that the crop would continue growing in the early dry season (Abidin, 2013) (i.e. a varietal characteristic rather than a health characteristic).

Research and development actors and a number of projects have introduced practices to improve the health, genetic and physiological quality of sweetpotato planting material in SSA. These have included: (i) negative selection of symptomless planting material together with roguing of plants with symptoms as they appear; (ii) use of pathogen-tested tissue culture plantlets as the 'foundation' material for further multiplication, limiting the distribution of virus-susceptible varieties to low virus pressure areas; and (iii) continuing breeding work on virus resistance. The use of low-cost screen net tunnels to keep out insect vectors has proved to be very effective in maintaining virus-free planting material for at least 3 years in a virus pressure hot-spot in western Kenya. While the principal objective of root-based vine multiplication systems (e.g. the 'Triple S' or sand storage and sprouting; Namanda *et al.*, 2012) is to address the challenge of maintenance of planting material in areas with long dry seasons, this technology may also contribute to improving the health quality of planting material as the roots are only sprouted 2 months before the planting season and the subsequent planting material is not exposed to disease vectors for as long as in conventional practices.

Other diseases and pests which are soil-borne (e.g. bacterial wilts, nematodes, weevil and millipedes) are also of greater importance in VPCs than in seed for grain crops as the roots, tubers (and in the case of sweetpotato vines) are in direct contact with the soil. Planting material of VPCs is living fresh tissue; so any storage period (in the field or once harvested) provides additional opportunities for pests and diseases to proliferate and affect the quality of the planting material as compared to the seed-phase of sexually propagated crops.

Physiological factors also influence quality and degeneration. For sweetpotato the physiological vigour of the planting material is influenced by the age of the mother plant and the section of the vine which is cut and used for planting material. Generally, the older the mother plant (e.g. more than one season) the more opportunity there is for both pest and disease build up and reduced sprouting and vigour of the material selected from it. The older 'woody' part of the vine close to the ground will take longer to sprout. On the other hand plants which are too young (e.g. 2 months or less) are not physiologically mature, and the planting material will not sprout well.

Therefore, a number of questions still need to be addressed around seed quality. Currently, plant health may be more of a preoccupation of research and development practitioners, and we need a more in-depth understanding of farmers' rationale and practices for quality planting material. If there are yield benefits to using disease- and pest-free planting material, under what circumstances will farmers pay a premium for improved quality? When would the use of 'cleaned-up' planting material be economically viable (i.e. which varieties benefit from 'clean-up' and for how many seasons do they continue to produce higher yields than farmers' own selected material)? Moreover, as plant health and seed control bodies become more involved in the certification and inspection process for VPCs we should be aware that inappropriate regulation might act as a disincentive for improving seed quality.

Need for specialized seed production practices

The need for specialized conditions and practices to produce quality seed provides the basis for an ongoing demand from farmers who are not willing nor able to apply these conditons. These can include specific growing conditions, equipment, storage and packing facilities, as well as specialized practices. This creates opportunities for specialization in seed production for those who have the expertise, growing/storage conditions and are willing to invest in a seed crop. Many vegetable crops need specialized production practices in order to produce and harvest seed. Hybrid seed production similarly asks highly advanced knowledge and practices. In other situations farmer-seed producers have clear advantages to produce quality seed, for example those producing potatoes at higher altitudes where disease pressure and temperatures are more favourable or when being able to produce in the off-season (Thiele, 1999) and those that have access to cold-store facilities in warmer climates.

For sweetpotato and most other VPCs, the simplicity of vegetative multiplication means that every farmer is able to produce his or her own planting material. However, in areas with a long dry season, there is increased likelihood of loss of planting material, or farmers need to wait 2 months after the rains so that roots left deliberately or accidently sprout and produce sufficient vines to be used for planting material. In these contexts, many farmers will purchase planting material. This provides the opportunity for farmer-multipliers who have access to swamps or are able to invest in irrigation equipment and pest-and-disease management practices to maintain and multiply material for sale in time for the next rains.

Perishability and bulkiness

Unlike seed for grain crops, for most VPCs the planting material is living. For example, sweetpotato can be propagated from botanical seed, or from root slips, but most commonly in SSA it is from stem or vine cuttings from live plants. This makes the maintenance of planting material from the root harvest to the next planting season more challenging than in most seed crops. Living material (approx. 20% dry matter) is also more subject to pathogens and infection compared with dried seeds (80% dry matter). The perishability of the planting material is thus associated with the need for some degree of specialized conditions and/or practices (see previous section). For sweetpotato, specialized conservation or maintenance methods (i.e. for vines only) are predominately used in the unimodal rainfall systems, where there is an extended dry season, or where there is a significant market demand for vines.

Table 28.3 shows these vine conservation and maintenance practices in local sweetpotato seed systems. While these practices are diverse and take advantage of different localized agroecologies, it might be argued that they do not ask for highly sophisticated practices or conditions.

Living material also tends to be more bulky planting material (see Table 28.4). One to two tons of seed potatoes are needed to plant 1 ha for the ware crop. Cassava stem cutting and sweetpotato vines are less bulky than potato seed tubers or the suckers for banana, but still represent heavy and voluminous loads that farmers may need to transport from where they acquire the planting material to where they store and/or plant. For example, for sweetpotato around 34 sacks (with an estimated 1000 cuttings each) are needed per hectare, and transportation needs to be done quickly (to allow planting within 3 days of harvesting the material) in ventilated

Table 28.3. Vine conservation and maintenance practices in local sweetpotato seed systems. (Based on Abidin, 2013; Andrade and Naico, 2013; Aragaw *et al.*, 2013; Carey *et al.*, 2013; McEwan and Namanda, 2013.)

Conservation method	Case study countries	Comments (e.g. period)
Specialized vine conservation practices		
Planting of vines under enset trees. These vines have been detached from the ware crop planted in October	Ethiopia	Vines are planted during the short rains between January and March for harvesting the roots between June and August. The vines are then replanted August–October for the main rainy season and harvested again in January–March
Planting of vines under banana plantation	Kagera, Tanzania, Mozambique (banana and coconut plantations)	Central Mozambique: sweetpotato is harvested to give space for rice in the period October–January, and then sweetpotato is conserved under the banana and coconut plantations
Nursery/seed beds mulched with dry grass or maize stalks	Malawi	Watering can or treadle pump is used for irrigation
Seed beds located along river beds or wet spots	Lake Zone, Tanzania	Women are responsible for conserving planting material. Small amounts of material from the nurseries are then planted out at the beginning of the rains to generate more material during the season
Irrigated gardens	Nigeria, Ghana, Burkina Faso (northern regions)	Raised or sunken beds, or planting on the flat. Planting material is further multiplied during the latter part of the dry season using standard cuttings and spacing to obtain roots and vines
Vines planted in fertile back yards	Ethiopia	January–March
	Malawi	During winter/dry season: June–October
Use of lowlands to conserve vines	South Mozambique	Farmers use the lowlands from August to December to conserve the vines in small plots or plant around the maize, cassava or vegetable plots
Maintenance of a small area of root crop to also produce vines		
Mulching of ware crop planted in October to maintain vines until June–July the following year	Ethiopia	
Roots (accidentally or deliberately) left in soil until rains start to regenerate crop from roots	Ethiopia: January/ February–April	
	Lake Zone Tanzania	Areas deliberately left unharvested are protected from livestock
	Mozambique: roots left in the field from May to August sprout in October–January	
Piecemeal harvesting, leaving the plant in place so the vines survive	Lake Zone, Tanzania	

Continued

Table 28.3. Continued.

Conservation method	Case study countries	Comments (e.g. period)
Sequential planting for roots so vines are always available	Lake Zone, Tanzania	In bimodal rainfall areas
Existing crops and ratoons	Nigeria, Ghana, Burkina Faso (southern regions)	Areas with shorter dry season
Leaving sweetpotato plants in between cassava, maize or other crops – the shade helps prevent drying out	Lake Zone, Tanzania, Malawi	In Tanzania: bye-laws compensate for cassava plots damaged by free-ranging livestock, but not for sweetpotato plots. So women plant their sweetpotato in with the cassava to provide some protection from animals
Rotation of root crop between upland and lowland areas with vines as by-product		
Production of ware crop in wetland areas also produces vines for the following season's planting in upland areas for root production	Malawi	Most farmers are doing this in groups/ clubs related to the irrigation scheme. Clubs are promoted by the government
Transfer of crop from rain-fed upland to paddies after rice is harvested	Lake Zone, Tanzania	Considerable work is needed to prepare mounds in the paddies before planting sweetpotato (for roots and vines)
Shift between upland and low-lying areas for both root and vines	Lake Zone, Tanzania	
Vine 'storage' practices		
Leaving a bundle in a hole, termite mound or laying in a shallow trench covered lightly in soil	Lake Zone, Tanzania and Malawi	
Harvesting roots and re-planting vines in the same field before transferring to another field	Lake Zone, Malawi	

Table 28.4. Multiplication rates and bulkiness of planting material. (Based on Purseglove, 1977.)

Crop	Multiplication rate	Planting material (per ha)
Potato	6–15	1500–2000 kg
Cassava	3–4	20,000[a]
Sweetpotato	3–15	35,000[a]
Soybean	8–38	40–60 kg
Sorghum	47–60	4–12 kg
Maize	70–200	15–25 kg

[a]Number of cuttings.

sacks and in medium-size open trucks so that the material does not begin to rot. These factors make buying of planting material less attractive and highlights one of the advantages of using decentralized approaches for the multiplication and dissemination of sweetpotato planting material.

A stable and profitable market for the product

Although there have been no systematic market studies, experiences for different crops and in different situations indicate that a profitable market with stable prices for the product stimulates farmers to invest more in that crop. In Zimbabwe cash income from common beans leads farmers to invest cash in seeds and inputs whereas they do not do the same for cowpea, which has no market and is only used for home consumption (C.J.M. Almekinders, 2014, unpublished data). It seems likely that this situation would be the same for VCPs. (See also Box 28.2 Nigeria sweetpotato case study.)

There are increasing sweetpotato markets both for fresh root markets and for processing in SSA. With their development, it is likely that one-time harvesting of larger areas of sweetpotato will become more common. This may create opportunities for the specialized vine multiplier or for the use of a root-based method for vine production (see previous section).

Bad harvests and crop failure

Farmer demand for seed fluctuates considerably, not only between regions but also within a region over time. An important factor is the average yield level of the preceding growing season. After seasons with low yields or crop failures for example, farmer seed demand may be higher because more farmers were unable to save seed for the next season. In contrast, when the production conditions are favourable and the seed producers have a good seed crop, farmers also have good yields and are in a better position to save seed for the next season. This creates a situation of counter-cyclical seed demand. This forms an important challenge for the building of an economically sustainable quality seed supply.

After disasters, seed stocks in an entire region may have been lost. This then requires seed provision interventions of a different nature (i.e. which do not represent regular farmer demand for planting material). In the cases where sweetpotato plays a role in post-disaster relief, institutional buyers may tender for large amounts of planting material. It has been argued that this leads to a distortion of both supply and demand. There are examples – for example Ethiopia where cyclical drought leads to regular distributions of free planting material. Multipliers may have large institutional buyers one year and the following year there is no 'disaster' and since there is no local market for planting material their enterprise collapses (Aragaw *et al.*, 2013). Disaster relief contexts can also negatively affect farmer demand. Normally, in these situations the distribution of planting material to farmers is 'free'. Farmers have little choice about the variety, quality or timing of delivery (Sperling and Longley, 2002). This leads to a situation where farmers have little incentive to try to maintain the materials which they have received or do not consider that sweetpotato planting material has a 'value' which they should pay for. We reflect on this type of intervention in a later section.

28.4 Seed Sources and Seed Supply

Having examined some of the factors which influence farmer demand for planting material and how this plays out in the case of sweetpotato in SSA, we turn to seed supply, looking at seed sources, affordability, and how these interact with the factors which influence demand.

To date, studies indicate that different seed sources are not equally available, accessible or suitable for smallholders (Almekinders *et al.*, 1994). While we can characterize different formal and local seed sources in general terms, farmers use the seed source which suits them best when they need it. For farmers, a local informal source is sometimes superior to distant formal sources; sometimes it is the other way around. It depends on the crop, growing conditions, type of varieties, cost of the seed and the social context; and it can vary from year to year or even between seasons. The use of seeds purchased from traders at local markets may point to a bottleneck in local social relations making seed which might be available from better-off farmers in the community a less preferred option for some community members (e.g. poorer farmers) because other obligations may be bound up in the seed transaction. Most studies on farmers' acquisition or use of seed from different sources relates to seed for grain crops. It is only for potato in its centre of domestication that substantial work of this type has been done (Thiele, 1999; Almekinders *et al.*, 2009; De Haan, 2009). For other VPCs much less information is available.

Over time, sweetpotato farmers have developed their own seed practices which to a large extent reflect the influence of unimodal and bimodal rainfall patterns. Farmers source planting material from their own fields or neighbours, friends or kin (Barker *et al.*, 2009; Namanda *et al.*, 2011). Different sources may be used depending on: (i) whether cultivation is for home consumption, sale or processing; (ii) whether the farmer is seeking new varieties or requires additional or replacement material of existing varieties; (iii) seasonal conditions; (iv) the time in the season when material is being sourced; and (v) the presence of ad hoc opportunities to obtain planting material (e.g. social visits, funerals). In many societies planting material is considered a common good, and farmers can ask for or take seed freely from the fields of neighbours. Planting material is also sold; the price depending on the geographical or social proximity of the source, time in the season and presence of strong markets for roots. Where the crop is considered a woman's crop, there are strong social and kin networks for sourcing vines which may be resistant to commercialization (Badstue and Adam, 2011). In areas where the crop has become commercialized, men become more involved and there may be a greater willingness to pay because men have access to, and/or the decision-making authority over, financial resources. In parts of Ghana (Bawku), Burkina Faso (Leo, Po, Bobo) and Nigeria (Kano, Kaduna and elsewhere), where sweetpotato roots are commercialized, there is an actual commercial market for planting material with significant off-season production, integrated with dry season vegetable production (Carey *et al.*, 2013).

Seed availability and multiplication

A factor which is related to the bulky character of VPCs is the relatively low multiplication rate as compared with grain crops (Table 28.4). Multiplication rates of sweetpotato vary according to the technique used, the variety, the agroecological conditions and management practices but can range from a low multiplication rate (e.g. 1:12) using conventional multiplication to 1:50 using rapid multiplication techniques (Stathers *et al.*, 2012). This affects the quantities of planting material which are available when needed for planting the ware crop.

For sweetpotato, new multiplication methods using both vines and roots are being tried out in different countries and conditions (McEwan, 2013). Some promising methods which have been promoted include the use of rapid multiplication techniques in specially prepared seed beds. This has included testing two or three node cuttings and different plant spacing for vine multiplication (CIP, 2012). Closer spacing is more suited to contexts where high management is possible and there is access to irrigation. Adjusted conventional spacing (e.g. a 30 cm cutting with 15 cm between plants × 75 cm between ridges) has been used with farmer multipliers in Malawi, with the objective of producing both planting material and roots during the November–April hunger period (Abidin, 2013).

Sweetpotato multiplication techniques have also been tested which build on existing farmer practices to be able to maintain the planting material after the harvest until the next planting. This method is appropriate in areas with an extended dry season and has been trialled in parts of Uganda and Tanzania. Farmers can make a careful selection of small (unmarketable but undamaged) roots, which they then store in dry sand in a container during the early part of the dry season, in the home or a store. About 5–7 weeks before the expected start of the rains the roots are planted out in a protected bed, which is then watered on a regular basis. The roots sprout and can provide planting material at the start of the rains, successfully producing 40–60 cuttings per root (CIP and NRI, 2011; Namanda *et al.*, 2012). Women in particular have reported that it gives them more control over the source of their planting material, so avoiding having to expend time 'looking for vines'. The technology is also being tested in Malawi.

Seed price and affordability

Seed quality and seed price are often presented as trade-offs for farmers, and it is frequently suggested that in many situations farmers cannot afford to buy seed from formal sector sources. For sweetpotato, there is no regular supply from a formal seed source in SSA with the possible exception of the Republic of South Africa. Sweetpotato planting material is often considered a common good (i.e. freely available to whoever wants it) and is therefore 'affordable'. In some situations, while local varieties continue to be made available freely, planting material of improved varieties is bought and sold. However, for specialized multipliers there are costs involved, in particular: (i) hire of land; (ii) use of irrigation; and (iii) additional labour for bed preparation, weeding, disease management and harvesting. Therefore, a continuing challenge in many countries is how to maintain emergent vine multipliers in an environment where the commercialization of the vines is incipient, and traditional practices of vine sharing coexist. This underlines the need to appreciate that specialization may be possible only on a small scale, and to consider mixed enterprise and social models for increasing the availability of planting material.

28.5 Strategies and Approaches for Dissemination of Sweetpotato Planting Material

We now turn to review and discuss some of the different strategies which have been used to scale the dissemination of sweetpotato planting material (i.e. ensuring adequate and timely quantities of quality planting material) and the factors which have influenced the type of approach used. Three broad approaches have been used depending on the challenge being addressed, the context, and goal of the seed intervention. These are: (i) post-disaster mass multiplication and mass distribution; (ii) decentralized multiplication and distribution; and (iii) a sweetpotato seed enterprise as part of the sweetpotato value chain.

Post-disaster mass multiplication and mass distribution

Sweetpotato planting material has often been distributed with the objective of contributing to improved food security in post-disaster or post-conflict situations. The advantage of the crop is that it requires minimum inputs, matures in around 4 months and can fit into different cropping systems. The approach which has been used most in these situations is mass multiplication and mass dissemination. Under this scenario, planting material is multiplied at centralized sites, harvested and transported to central points for distribution on a predetermined day. This approach has often been used in post-disaster and post-conflict situations such as:

- after the 2001 floods in Mozambique (Andrade and Naico, 2013);
- post-conflict resettlement programmes in northern Uganda in the mid- to late 2000s (Potts, 2006); and
- in Ethiopia after recurrent cycles of drought (Aragaw *et al.*, 2013).

Some of the criticisms of this approach include: (i) the wastage and loss of material; (ii) the undetermined quality of the planting material; and (iii) that it provide farmers with limited choice (of varieties and timing for collection) and information. Distribution is normally 'free' to farmers, raising concerns as to whether farmers will value the material and make an effort to maintain it. Yet, this 'campaign' or single-shot type of dissemination may be appropriate when replacement material is needed quickly, and in bimodal rainfall areas it may be suitable for injecting new varieties into a system such that once the replacement material is in the system it will then flow from farmer to farmer and is more easily conserved or maintained than in areas with a long dry season (Stathers *et al.*, 2012).

Decentralized multiplication and distribution

The challenge of the bulky and perishable nature of sweetpotato planting material has influenced efforts to extend availability by decentralizing farmer access to planting material with general developmental objectives. This approach has been used to disseminate new varieties or cleaned up material from formal breeding programmes through the use of variations on the '1-2-3' multiplication approach (Abidin, 2013; McEwan and Namanda, 2013). This involves:

- 'primary multiplication' at research stations or on an experienced farmer's plot under researcher supervision;
- 'secondary multiplication' to bulk up the planting material, which can be under the management of experienced farmer groups or entrepreneurial individual farmers, with supervision from research or extension service providers from the public sector or NGOs; and
- 'tertiary multiplication' under farmer groups or individual farmers (decentralized vine multipliers) who normally produce both roots and vines.

The catchment area for the tertiary decentralized level depends on population density, but is usually based around a 9–12 km radius. Since the mid-2000s decentralized multiplication and distribution approaches have been used in Ethiopia, Mozambique, Kenya, Uganda, Malawi and Tanzania to target specific vulnerable groups (resource-poor farmers and farming households with pregnant women and children under 5 years old) in particular, with biofortified crops. This has required novel entry points and a different group of stakeholders to be involved to support demand creation and information education and communication activities. This has been done in collaboration with health facilities, community-based health workers, and using a partially or fully subsidized voucher system. This approach is implemented within a project time frame, through international NGOs (INGOs), NGOs and community-based organizations (CBOs), with technical support from the national research system and CIP. In Ethiopia, decentralized vine multiplication is being implemented with state-managed farmer training centres and model farmers. An example from Malawi is shown in Box 28.1 where there has been a deliberate effort to re-engineer the seed system, include non-conventional

Box 28.1. Malawi: decentralized vine multiplication (from Abidin, 2013)

In Malawi, government policy has supported agricultural diversification, and yet at the same time malnutrition remains among the highest in SSA with vitamin A deficiency rates among children less than 5 years old at nearly 60%. Sweetpotato is becoming a major food source (third after maize and cassava) and increasingly contributes to the food basket, especially in the months where maize is scarce or the price is too high for poor people. The crop is also becoming a source of cash and employment to many farmers. Dry spells are increasingly common, so there is great interest in sweetpotato. Access to clean planting material, however, remains a challenge in sweetpotato production, due to build-up of virus diseases and subsequent yield reduction; some varieties (e.g. the orange-fleshed cultivar, Kamchiputu) might have disappeared due to this. Land pressure is causing cultivation by smallholders to extend into marginal and less fertile areas thereby negatively affecting sweetpotato production and availability of clean planting material at the onset of the rainy season. Some farmers, particularly in the central and southern regions, practise intercropping and relay cropping to resolve the problem of land pressure and to secure food during the prolonged dry spells. These cultural practices also contribute to reducing the problem of shortage of sweetpotato planting material at the start of the rainy season. Maize and sweetpotato, maize and pumpkins or maize and groundnut or sweetpotato and soybeans are commonly planted together in one row/ridge. The first rains generally appear in the month of November. Short duration crops intercropped with the sweetpotato are chosen with a life cycle of 3–4months. When the first crop is harvested in February, the vines of sweetpotato are available on a relatively large scale to plant for the production of storage roots. CIP and partners through the Rooting out Hunger in Malawi with Nutritious Orange-fleshed Sweetpotato (OFSP) project have taken an integrated (seed system, demand creation and postharvest interventions) approach to the promotion of OFSP. In Malawi, the '1-2-3' seed system model comprises research, extension and trained farmers. A strong enabling policy environment promoting dietary adequacy and improvement in dietary quality among the most vulnerable groups has also supported the expansion of OFSP. Effective partnerships have contributed to the integration of OFSP interventions into the Scaling Up Nutrition (SUN) initiative.

stakeholders to achieve food security and nutrition goals and to institutionalize linkages for the flow of material through the different tiers of multiplication.

In this Malawi case, careful planning has ensured that a project intervention builds upon existing vine conservation and multiplication practices such as shifting between upland and lowland areas and using inter- and relay cropping practices to address limitations in land and climatic conditions. In Malawi an integrated agriculture-nutrition-marketing approach was used to promote: (i) increased production of sweetpotato; (ii) increased awareness of the nutritional benefits of the OFSP varieties; and (iii) a value addition component. The Malawi model has also been successful in linking into national and international policy initiatives (e.g. the SUN movement) through showing that food-based approaches are both feasible and appropriate for contributing to reduction in micronutrient deficiencies and improving nutritional well-being.

The example from Malawi also illustrates how nutritional considerations have influenced the strategy adopted. In order to stimulate demand for the vitamin-A-rich orange-fleshed varieties, this decentralized approach has been combined with a subsidized voucher system to be able to target particular vulnerable groups such as pregnant women, children under 5 years old, school children and people living with HIV. In some cases, for example Sweetpotato Action for Security and Health in Africa (SASHA) and DONATA (Dissemination of New Agricultural Technologies in Africa) project interventions in western Kenya and Rwanda, novel entry points for sweetpotato vine multiplication and dissemination have been used such as health facilities, schools, churches and mosques (SASHA, 2011, 2012; CIP, 2012). To a certain extent these development-orientated interventions have built upon existing practices of farmer-to-farmer dissemination and social networks through supporting farmer-group multiplication plots to provide vines to members, either free on a pay-back or pass-on basis; and in addition sell on a small scale to other farmers for group income generation.

Sweetpotato seed enterprise as part of the sweetpotato value chain

In many countries there are farmers who have been multiplying sweetpotato which they sell for cash or distribute by gift or barter. There are examples of where this practice has become increasingly specialized and commercialized. In parts of Ghana (Bawku), Burkina Faso (Leo, Po, Bobo) and Nigeria (Kano, Kaduna and elsewhere), where sweetpotato roots are commercialized and there is market demand for planting material, there is significant off-season production, integrated with dry season vegetable production. A short case study of commercialized informal sweetpotato vine multipliers in Nigeria is presented in Box 28.2.

In a number of countries project-based interventions have tried to support incipient commercialization in different ways. Seed is only one segment of the overall sweetpotato value chain, and as the example of local vine multipliers in Nigeria shows, where there is a large urban market demand for roots, commercialized vine multiplication is possible. In Ethiopia as shown in Box 28.3 there is the example of integrating the vine enterprise with other farm enterprises such as root production, and vines for livestock fodder. In western Kenya, a community-based organization, the Siwongo Drainage and Irrigation Group, initially supported through project funds, has transformed its status into a limited company and extended its activities from vine multiplication and processing to contracting farmers for root production. Now the Siwongo Processing Company works across the whole value chain from vine multiplication, root production and processing, with linkages to a distributor agent. Each segment provides an element of risk diversification. The company currently contracts farmers (incipient seed out-grower system) to produce planting material and or roots (Makokha, 2012). A critical factor in the growth of the company was the ability to access credit so that farmers could be paid cash for roots even when there was delayed payment for the processed flour from the distributor. In Rwanda we have seen a similar evolution and growth

Box 28.2. Nigeria: commercialized informal sweetpotato vine multipliers (from Carey *et al.*, 2013)

Until recently, sweetpotato has received little consideration in agricultural policies of most West African countries. As a result, crop production statistics have not been systematically tracked and reported and reliable information is scarce. Striking increases in production, led by a doubling of yield to 20 t/ha are reported for Burkina Faso, while yields of 2–3 t/ha are reported for Ghana and Nigeria. Even with such low production, Nigeria is reported to have the second largest production of sweetpotato in SSA, after Tanzania (FAOSTAT, 2013).

The extended rainy season in the southern regions is weakly bimodal; in these areas with a shorter dry season, planting material is selected from existing or ratoon crops. In areas where sweetpotato can be grown in three seasons (i.e. farmers have access to fadamas (low-lying irrigable plains)), vines are sourced from their own farms especially for the second and third season or from friends/neighbours. The northern regions are characterized by an increasingly short rainy season as one goes farther north, and a long, harsh dry season. Highly weathered, sandy soils and high temperatures (due to low elevation and latitude) exacerbate stresses and restrict dry season farming to irrigated areas. In dry, livestock-system-dominated areas farmers are unable to conserve their own vines due to lack of water or damage by free-grazing cattle.

Studies by Peters (2013) and others (Anyimah-Ackah, 2012; Bidzakin and Acheremu, 2012; Onumah *et al.*, 2012) indicate that sweetpotato is an increasingly important cash crop with considerable unmet demand in both urban and rural markets where it is typically consumed boiled or fried. Planting material is either produced for own use or sale in each of the countries, particularly in areas where the crop is commercially important (Peters, 2013). Vine conservation is carried out in irrigated gardens and may be in raised beds, sunken beds or on the flat; sweetpotato planting material is produced in these areas so as to be available at the onset of the rains. There are specialist vine multipliers in wetland areas or along the banks of rivers.

Examples of this are around Kano and Kaduna where there is a strong market for roots and there is also significant specialized vine production. Farmers, who are predominantly male, combine high-value horticultural crops and vine multiplication in wetland areas or along the banks of rivers. The farmers have identified a narrow window in the season when sweetpotato vines can command a high price. Their customers are local or regional farmers who come to buy vines direct from source, using private or public means of transport. The multipliers ensure that the planting material is ready on time or otherwise they risk missing the market. The quality of the planting material is unknown.

Box 28.3. Ethiopia: commercial vine multiplication as part of an integrated business model (from Woldegiorgis, 2012)

In Tigray, northern Ethiopia, sweetpotato is a newly introduced crop, and one interesting example is where a commercial agricultural enterprise (Minora Agricultural Investment Share Company) have used an integrated crop and livestock business model to support a vine multiplication enterprise. The company earns income from crops (including sweetpotato roots), horticulture, livestock and poultry; and existing irrigation facilities are used for sweetpotato vine multiplication. CIP has provided starter material for orange-fleshed varieties and in an effort to support the emergence of a sweetpotato seed system also buys vines from the company for distribution to other multipliers. Additional value from sweetpotato is derived from: (i) sale of planting material; (ii) sale of vines for livestock fodder; and (iii) utilizing sweetpotato as a rotation crop and for erosion control. The company's business model also incorporates social values through extending knowledge and practices to the community as farm employees take practices onto their own land, families and communities. Other farmers in the community have also started to grow and use vines for livestock fattening and poultry in addition to consumption of the orange-fleshed varieties by the family members, especially children.

from a CBO to a limited company and diversification into all segments of the sweetpotato value chain (Habumuremyi, 2012). Ibyiwacu Company Ltd was able to access sizeable bank loans. However, in both examples accessing formal markets (e.g. urban supermarkets with processed products) requires national Bureau of Standards Certification.

These examples of vine multiplication enterprises associated with other value chain segments show that diversification of products allows a spread of risk and vertical integration can strengthen coordination in the chain. In a mixed enterprise there can be both economic and environmental benefits. However, the enterprises need to be business driven with strong market linkages.

We have reviewed three broad sweetpotato seed system approaches from the perspective of what is working where and for who: (i) mass multiplication and mass distribution; (ii) decentralized multiplication and distribution with elements of social enterprise built in; and (iii) the seed system as part of the sweetpotato value chain. A strong profit rationale is often presented as the basis for a sustainable sweetpotato seed system, and this route has potential in areas with access to well-developed markets for roots and vines. In such areas there is a consistent demand for quality roots; root producers can be linked to the commercial supply of quality seed. Medium- to large-scale multipliers or farmer groups can have the infrastructure and organizational capacity to plan and supply seed to root producers. There may be a high demand for seed, but during a very short window. In this scenario multipliers are integrated into and operating within a market economy. However, we still need to ask the following questions:

- Are there barriers to market entry for would-be specialized multipliers?
- What happens when an intervention is targeting non-conventional seed demand and and in areas where there is limited market penetration?

28.6 Discussion

We started the chapter with a discussion around the characteristics of farmers' demand for seed, the interactions between different drivers of a seed system and how this played out for sweetpotato. As we return to discuss the question of how to get sweetpotato seed systems moving in SSA, we note the key constraints which have been identified in existing seed practices: (i) low multiplication rates; and (ii) perishability. Together these lead to limited availability of planting material. In unimodal rainfall systems with a long dry season the challenge is to maintain planting material from the root harvest period to the next planting time; in bimodal rainfall systems the challenge is disease and pest build up which can be transferred through planting material from one crop to the next. In combination these factors contribute to late planting, limited areas planted to sweetpotato, and low productivity. This in turn reduces the potential of sweetpotato to provide a source of food and income to bridge the chronic hunger period experienced by many households before cereal crops are ready for harvest.

We have seen that the complexity and diversity of existing seed practices allows for local specificity and flexibility to respond to dynamic agroclimatic conditions. Currently efforts to multiply and distribute planting material (whether based on informal farmer-based systems or project interventions) are mostly on a local and relatively small scale. Farmer demand for seed is uncertain and the quality of planting material is generally unknown – neither to farmers nor researchers.

Although in many countries sweetpotato has been regarded as a 'subsistence' food security crop there are increasing opportunities for commercialization, which will also broaden the recognition of the role which sweetpotato can play to contribute to the food security and nutritional well-being of vulnerable groups. The commercialization of the crop has a range of implications for its seed systems, for example different end users, different traits, different varieties and different types of harvesting. In turn this provides opportunities for specialist multipliers. This review of the literature, analysis of some of the experiences with VPCs and sweetpotato and reflection on some of the issues around scale leads us to three conclusions.

First, we have identified three broad approaches which are associated with the way sweetpotato is currently promoted in SSA. These types of interventions for sweetpotato seed systems are not necessarily or exclusively pursuing the objective of 'seed sector development' with the goal of improving agricultural productivity. Instead the reasons for intervention are much broader, with objectives related to improved food security, nutrition and post-disaster recovery, in particular for the most vulnerable households in rural areas. This means that we need to be more careful in assuming we understand farmers' demand for seed and we need to be more considerate in the design of interventions. Designing interventions to support the availability of sufficient quantities of sweetpotato planting material requires an understanding of the locally specific drivers (i.e. agroclimatic, market, social context and varietal preferences) and their implications for the way in which we work to support seed systems. We also need to consider how to balance economic, environmental and social sustainability.

Secondly, we have seen that a diversity of farmer practices to multiply and maintain sweetpotato planting material exists and are well adapted to local conditions. There are many examples of situations in which farmers have developed a degree of specialization and commercialization of planting material. In comparison to a VPC such as potato, the production of sweetpotato planting material requires relatively less expertise – in the sense that all sweetpotato farmers can produce their own planting material without much extra training and investment in infrastructure compared with, for example, the rapid multiplication techniques for potato. The opportunity for specialization seems to relate more to the conditions needed for the maintenance, multiplication and marketing of vines (i.e. access to different land types, access to water in the off-season, ability to invest in appropriate irrigation or to integrate into irrigated vegetable production systems, and closeness to a profitable market for sweetpotato roots). In order to protect the planting material against phytosanitary degeneration, specialized structures and practices (screenhouses, net tunnels, extra application of fungicides and pesticides) will also become increasingly relevant. These do require investment, but are not of such high capital and capacity level that they would be beyond the reach of smallholder farmers. Specialization can be local and small scale to take advantage of different agroecologies and address end-user preferences. This creates space for farmers who would otherwise not be the logical actors to engage with in seed sector development, to engage in the multiplication of sweetpotato planting materials, for example less-privileged smallholder women farmers. They may find a profitable niche from which they can reap benefits, without high levels of investment in capacity or infrastructure. The material characteristics of the root crop and its seed – its perishability, bulkiness and the common property nature of vines (i.e. they are the common property of everyone so anyone can take vines from a field) – further also underlines the economic rationale for 'staying local' and ensuring that seed production and root production areas are in close proximity.

Finally, If we allow ourselves to think along alternative lines, we see opportunities in SSA for the emergence of a seed provision system for sweetpotato that is aligned with the interventions that do not pursue seed-sector-development objectives only, but which also importantly strive to strengthen the food and nutrition security of the poorest and most vulnerable households in locally appropriate and sustainable ways. Thus we consider it unlikely that 'conventional' economically sustainable seed systems will emerge for sweetpotato. An alternative is to seek to build a large-scale pattern of small-scale enterprises, where small (and local) can still be beautiful.

For small and local to work at scale, however, we need to fix the current disconnection between breeding efforts in the 'formal' sector and 'informal' farmer-to-farmer dissemination. This will require a much better understanding of how to make linkages and partnering mechanisms across a range of stakeholders who have different objectives and are coming from different sectoral and disciplinary viewpoints.

Acknowledgements

The authors would like to thank Elias Urage of Hawassa Agricultural Research Centre, Hawassa, Ethiopia and Abiyot Aragaw of CIP, Hawassa, Ethiopia for additional contributions.

Note

[1] Throughout this chapter planting material and seed are used interchangeably.

References

Abidin, P.E. (2013) Rooting out hunger in Malawi with nutritious orange-fleshed sweetpotato. Inventory of seed systems in Malawi. Lessons learnt. *Sweetpotato Seed Systems Case Study*. International Potato Center (CIP), Blantyre, Malawi.

Almekinders, C.J.M. and Louwaars, N.P. (2002) The importance of the farmers' seed systems in a functional national seed sector. *Journal of New Seeds* 4(1–2), 15–33.

Almekinders, C.J.M., Louwaars, N.P. and De Bruijn, G. (1994) Local seed systems and their importance for an improved seed supply in developing countries. *Euphytica* 78, 207–216.

Almekinders, C.J.M., Cavatassi, R., Terceros, F., Pereira, R. and Salazar, L.F. (2009) Potato seed supply and diversity: dynamics of local markets of Cochabamba Province, Bolivia – a case study. In: Lipper, L., Anderson, C.L. and Dalton, T.J. (eds) *Seed Trade in Rural Markets. Implications for Crop Diversity and Agricultural Development*. Earthscan, London, pp. 75–94.

Andrade, M.I. and Naico, A. (2013) Sweetpotato seed systems in Mozambique. *Sweetpotato Seed Systems Case Study. Building on Traditional Seed Systems for Improved Farmer Access to Quality Planting Material*. International Potato Center (CIP), Maputo, Mozambique.

Anyimah-Ackah, E. (2012) *Sweetpotato Survey and Expert Consultation in the Central and Greater Accra Regions of Ghana*. International Potato Center (CIP), Kumasi, Ghana.

Aragaw, A., Urage, E. and Schultz, S. (2013) Sweetpotato seed systems in Ethiopia. *Sweetpotato Seed Systems Case Study. Building on Traditional Seed Systems for Improved Farmer Access to Quality Planting Material*. International Potato Center (CIP), Addis Ababa.

Badstue, L. and Adam, R. (2011) Gender and vines. Production, management and exchange of sweetpotato planting material among smallholders in the Lake Victoria region, Tanzania. *H.K. International Technical Report*. H.K. International, Nairobi.

Barker, I., Andrade, M., Labarta, R., Mwanga, R., Kapinga, R., Fuentes, S. and Low, J. (2009) Unleashing the potential of sweetpotato in sub-Saharan Africa: current challenges and way forward. *Sustainable Seed Systems. Challenge Theme Paper* 2. International Potato Center (CIP), Lima.

Bidzakin, J.K. and Acheremu, K. (2012) *Characterisation of Six Sweetpotato Production Communities in Ghana, Using Focus Group Discussions*. International Potato Center (CIP), Kumasi, Ghana.

Camargo, C.P., Bragantini, C. and Monares, A. (1989) *Seed Production Systems for Small Farmers: a Non-conventional Perspective*. International Center for Tropical Agriculture (CIAT), Cali, Colombia.

Carey, E.E., Gibson, R.W., Fuentes, S., Machmud, M., Mwanga, R.O.M., Turyamureeba, G., Zhang, L., Ma, D., Abo El-Abbas, F., El-Bedewy, R. and Salazar, L.F. (1999) The causes and control of virus diseases of sweetpotato in developing countries: is sweetpotato virus disease the main problem? In: *CIP Program Report 1997–98*. International Potato Center (CIP), Lima, pp. 241–248.

Carey, E.E., Asare, K., Some, K. and Peters, D. (2013) Sweetpotato seed systems in West Africa: the contrasting cases of Nigeria, Ghana and Burkina Faso. *Sweetpotato Seed Systems Case Study*. International Potato Center (CIP), Kumasi, Ghana.

Cromwell, E., Friis-Hansen, E. and Turner, M. (1992) The seed sector in developing countries: a framework for performance analysis. *Overseas Development Institute (ODI) Working Paper*. ODI, London.

De Haan, S. (2009) Potato diversity at height: multiple dimensions of farmer-driven *in-situ* conservation in the Andes. PhD thesis, Wageningen University, Wageningen, The Netherlands.

Devaux, A., Ordinola, M. and Horton, D. (2011) *Innovation for Development: The Papa Andina Experience*. International Potato Center (CIP), Lima.

Douglas, J.E. (1980) *Successful Seed Programs: a Planning and Management Guide*. Westview Press, Boulder, Colorado.

FAOSTAT (2013) Production Data for Sweetpotato. Available at: http://faostat3.fao.org/faostat-gateway/go/to/home/E (accessed 14 December 2013).

Food and Agriculture Organization of the United Nations (FAO) (1973) *Seed Improvement and Development Programme*. Plant Production and Protection Division, FAO, Rome.

Fuglie, K.O. (2006) *Priorities for Sweetpotato Research in Developing Countries: Results of a Survey*. International Potato Center (CIP), Lima.

Fuglie, K.O., Zhang, L., Salazar, L.F. and Walker, T. (1999) *Economic Impact of Virus-free Sweetpotato Planting Material in Shandong Province, China*. International Potato Center (CIP), Lima.

Gaba, V. and Singer, S. (2009) Propagation of sweetpotatoes, *in situ* germplasm conservation and conservation by tissue culture. In: Loebenstein, G. and Thottappilly, G. (eds) *The Sweetpotato*. Springer Science + Business Media BV, Houten, The Netherlands, pp 65–80.

Gibson, R.W., Mwanga, R.O.M., Namanda, S., Jeremiah, S.C. and Barker, I. (2009) Review of sweetpotato seed systems in East and Southern Africa. *Working Paper 2009–1*. International Potato Center (CIP), Lima.

Habumuremyi, J.M.V. (2012) Company Ibyiwacu Ltd, Rwamagana District, Rwanda. Paper presented at the Sweetpotato Support Platform for East, Central and Southern Africa: Seed Systems Consultation Meeting, 7–8 June, International Potato Center (CIP), Nairobi.

International Potato Center (CIP) (2012) Rooting out hunger in Malawi with nutritious orange-fleshed sweetpotato. *Year 3 Annual Report.* Prepared for Irish Aid. CIP, Lima.

International Potato Center (CIP) and National Resources Institute (NRI) (2011) *The Triple S System: Storage in Sand and Sprouting*. CIP and NRI, Nairobi.

Kapinga, R.E., Ewell, P.T., Jeremiah, S.C. and Kileo, R. (1995) *Sweetpotato in Tanzanian Farming and Food Systems: Implications for Research*. International Potato Center (CIP), Nairobi and Ministry of Agriculture, Tanzania.

Kapinga, R.E., Rees, D., Jeremiah, S.C. and Rwiza, E.J. (2003) Trader and consumer criteria for selection of sweetpotato varieties. In: Rees, D., Van Oirschot, Q. and Kapinga, R. (eds) *Sweetpotato Post-harvest Assessment: Experiences from East Africa*. Natural Resources Institute, Chatham, UK, pp. 23–36.

Loebenstein, G. and Thottappilly, G. (eds) (2009) *The Sweetpotato*. Springer Science + Business Media BV, Houten, The Netherlands.

Louwaars, N.P. and de Boef, W.S. (2012) Integrated seed sector development in Africa: a conceptual framework for creating coherence between practices, programs, and policies. *Journal of Crop Improvement* 26(1), 39–59.

Makokha, P.S. (2012) Siwongo Processing Company Ltd – OFSP Seed Systems. Paper presented at the Sweetpotato Support Platform for East, Central and Southern Africa: Seed Systems Consultation Meeting, 7–8 June, International Potato Center (CIP), Nairobi.

Manrique, K. (2013) Seed systems and farmer access to planting material in potato. *CGIAR Research Program – Roots, Tubers and Bananas (CRP-RTB) Working Papers.* CRP-RTB, Lima.

McEwan, M.A. (2013) Sweetpotato seed systems in sub-Saharan Africa. A literature review to contribute to the preparation of conceptual frameworks to guide practical interventions for root, tuber and banana seed systems. *CGIAR Research Program – Roots, Tubers and Bananas (CRP-RTB) Working Papers*. CRP-RTB, Lima.

McEwan, M.A. and Namanda, S. (2013) Piloting large-scale dissemination of sweetpotato planting material for food security, Marando Bora, Lake Zone, Tanzania. *Sweetpotato Seed Systems Case Study*. International Potato Center (CIP), Nairobi.

Namanda, S., Gibson, R.W. and Sindi, K. (2011) Sweetpotato seed systems in Uganda, Tanzania, and Rwanda. *Journal of Sustainable Agriculture* 35(8), 870–884.

Namanda, S., Amour, R. and Gibson, R.W. (2012) The Triple S method of producing sweet potato planting material for areas in Africa with long dry seasons. *Journal of Crop Improvement* 27(1), 67–84.

Onumah, G., Dipeolu, A. and Fetuga, G. (2012) Exploring opportunities to promote greater exploitation of the benefits of sweetpotato in representative states of Nigeria. *Report on Sweetpotato Value Chain Study.* International Potato Center (CIP), Nairobi.

Peters, D. (2013) Sweetpotato value chain and demand in Nigeria, Ghana, and Burkina Faso. *Report for the Bill & Melinda Gates Foundation (BMGF).* BMGF, Seattle, Washington, DC.

Potts, M.J. (2006) Large-scale deployment of improved sweet potatoes in sub-Saharan Africa 2003–2006. *Final Report to GTZ.* International Potato Center (CIP), Kampala, Uganda.

Programme Régional d'Amélioration de la Pomme de Terre et de la Patate Douce en Afrique Centrale et de l'Est (PRAPACE) (2005) *Strategic Research and Development Priorities for the ASARECA Potato and Sweetpotato Network (PRAPACE).* PRAPACE, Association for Strengthening Agricultural Research in Eastern and Central Africa (ASARECA), Kampala, Uganda.

Purseglove, J.W. (1977) *Tropical Crops: Dicotyledons*. Longman, London.

Sperling, L. and Longley, C. (2002) Special issue: beyond seeds and tools. *Journal of Disaster Studies, Policy and Management* 26(4), 283–287.

Stathers, T., Low, J., Mwanga, R.O.M., Carey, E.E., David, S., Gibson, R.W., Namanda, S., McEwan, M.A., Bechoff, A., Malinga, J., Katcher, H., Blakenship, J., Andrade, M., Agili, S. and Abidin, E. (2012) *Everything You Ever Wanted To Know About Sweetpotato: Reaching Agents of Change ToT Manual.* International Potato Center (CIP), Nairobi.

Sweetpotato Action for Security and Health in Africa (SASHA) (2011) *SASHA Marando Bora Project Voucher System Operation Guideline.* Catholic Relief Services, Mwanza, Tanzania.

Sweetpotato Action for Security and Health in Africa (SASHA) (2012) *Year 3 Annual Technical Report for Mama SASHA Component.* International Potato Center (CIP), Nairobi.

Thiele, G. (1999) Informal potato seed systems in the Andes: why are they important and what should we do with them? *World Development* 27(1), 83–89.

Tripp, R. (2002) *Seed Provision and Agricultural Development.* James Curry & Heinemann, Oxford.

Van Mele, P. and Bentley, J. (2011) *African Seed Enterprises.* Food and Agriculture Organization of the United Nations (FAO) and AfricaRice, Rome.

Woldegiorgis, D.M. (2012) Minora Agricultural Investment Share Company: experience from Tigray, Ethiopia. Paper presented at the Sweetpotato Support Platform for East, Central and Southern Africa: Seed Systems Consultation Meeting, 7–8 June, Nairobi, Kenya. International Potato Center (CIP), Nairobi.

Woolfe, J.A. (1992) *Sweet Potato: an Untapped Food Resource.* Cambridge University Press, Cambridge.

29 Effect of Long-term *In Vitro* Subculturing on Quality Degeneration of Sweetpotato Varieties: Morpho-anatomic Assessment and Simple Sequence Repeat (SSR) Analysis

M.C. Hundayehu,[1]* E. du Toit,[1] S.M. Laurie,[2] M. Steyn,[1] R. Greyling[2] and N. Myeza[2]
[1]*University of Pretoria, Pretoria, South Africa;* [2]*Agricultural Research Council-Roodeplaat Vegetable and Ornamental Plant Institute (ARC-VOPI), Pretoria, South Africa*

Abstract

True-to-type clonal fidelity is one of the most important prerequisites in micropropagation of crop species. The aim of this study was to assess the effect of long-term *in vitro* subculturing on the variety degeneration of three sweetpotato varieties (Monate, Mokone and Ndou). Monate was subcultured for 32 generations, whereas Mokone and Ndou were subcultured for 23 and 12 generations, respectively. The media used for subculturing contained 4.43 g/l of MS (Murashige and Skoog) salt, 30 g/l of sucrose and 2 g/l of gelrite. The plantlets were grown in a 16 h light and 8 h dark photoperiod for 30 days, and single node cuttings taken from the 30-day-old plantlets were used as the explant source to subculture the next generation. From each generation, 45 plantlets were transplanted and grown for 2 months in a glasshouse. Data on *in vitro* growth performance, as well as morphology during acclimatization were recorded. The plantlets which showed a change in morphological traits were subjected to genetic analysis by using five simple sequence repeat (SSR) primers (IB-242, IB-318, IB-255F, IB-248 and IB-255). Significantly higher growth performance, such as stem height, leaf number, internode length, as well as early root and shoot organogenesis, was observed after the 27th and 21st subcultured generations of Monate and Mokone, respectively. Furthermore, plantlets of the same variety showed differences in morphological traits such as leaf colour, abaxial leaf pigmentation, vine pigmentation, petiole pigmentation, leaf wrinkling and percentage of flowering. However, there was no correlation between the subcultured generation and the degree of morphological variability. From the five SSR loci, only IB-255F and IB-318 were able to distinguish between the three varieties. However, no allelic polymorphisms were detected between short-term and long-term propagated plantlets with the primers used in this experiment. Therefore, long-term nodal subculturing did not lead to quality degeneration of sweetpotato varieties Monate, Mokone and Ndou.

Keywords: micropropagation, morphology, organogenesis, polymorphism, subculture, true-to-type

*m.cherinet@cgiar.org

29.1 Introduction

The sweetpotato crop has a wide adaptability on marginal land and has a potential to alleviate malnutrition and enhance food security in the developing world (Diaz *et al.*, 1996). It is a highly heterozygous hexaploid crop (Ozias-Akins and Jarret, 1994) which is propagated by means of botanical seed and vine cuttings (Woolfe, 1992). To maintain genetic uniformity and higher agronomic performance, only vegetative propagation is used for commercial sweetpotato production (Lewthwaite *et al.*, 2011). However, viral diseases, sweetpotato weevil, drought and varietal degeneration are becoming a major constraint to vegetative propagation of sweetpotato. Therefore, to maintain the productivity of superior varieties, there should be a continuous reintroduction of new planting material which is free from disease, and which is physiologically vigorous and genetically true-to-type. As part of ensuring the latter, tissue culture of sweetpotato is widely used as a viable option for germplasm conservation, virus cleaning and rapid multiplication of clean foundation seed.

In a conventional sweetpotato micropropagation programme, successive nodal subculturing is used for the rapid multiplication of virus-cleaned *in vitro* stock material, as well as for long-term *in vitro* conservation of germplasm. It has generally been assumed that plants micropropagated by multiplication of organized meristems, such as stem nodes, are genetically stable (Potter and Jones, 1991; Nayak *et al.*, 2011). Similarly, *in vitro* micropropagation of sweetpotato through the use of nodal subculturing is assumed to maintain the genetic and agronomic character of the mother plant. However, some exceptions were reported where more organized tissues, such as shoot tips and stem nodes, caused more variation compared with somatic embryogenesis in banana (Israeli *et al.*, 1995) and in sweetpotato (Villordon and LaBonte, 1996). In addition, increasing the number of subcultured generations and their duration *in vitro* enhanced the rate of variation within micropropagated plants (Rodrigues *et al.*, 1998; Bairu *et al.*, 2006).

Clonal variation caused during micropropagation is one of the main sources of varietal decline or quality deterioration in vegetatively propagated plants such as sweetpotato (Villordon and LaBonte, 1995; Lin *et al.*, 2009). However, the effect of repeated nodal subculturing on the varietal degeneration of sweetpotato was not known, nor were there any techniques for detecting such variation in sweetpotato available. Therefore, in the present study the occurrence of variation after many subcultured generations was assessed. To detect the variation among the micropropagated plantlets, vegetative growth performance parameters, morphological traits, leaf epidermal patterns, stomata counts and simple sequence repeat (SSR) methods were employed.

29.2 Materials and Methods

Plant material and experimental design

Long-term subcultured *in vitro* plantlets of three different sweetpotato varieties (Monate, Mokone and Ndou) were collected from the sweetpotato germplasm collection at the Agricultural Research Council-Roodeplaat Vegetable and Ornamental Plant Institute (ARC-VOPI), Pretoria, South Africa. At the time of collection the varieties Monate, Mokone and Ndou had been subcultured for 25, 17 and four generations, respectively. These materials were further subcultured using nodal cuttings to the 32nd (Monate), 23rd (Mokone) and 12th (Ndou) generations. The composition of the growth media was MS (Murashige and Skoog) (4.43 g/l), sucrose (3 g/l) and gelrite (2 g/l). The culture tubes were placed in a light- and temperature-regulated growth chamber. The incubation temperature was 27°C, with a light/dark cycle of 16 h/8 h. The light was supplied by a white fluorescent tube which had an illumination capacity of 2000–3000 flux. On the 30th day after culturing, 15 plantlets of each variety were used as the explant source for the next generation, 15 plantlets were used for distructive data collection and 45 plantlets were hardened off for morphological studies.

Morphological assessment

After 2 months of glasshouse acclimatization, data on aerial vegetative traits were scored based on the CIP/AVRDC/IBPGR (1991) sweetpotato descriptors list. The morphological characterization considered only those traits which were related to vine characteristics and leaf characteristics. A total of 12 traits (namely, vine internode length, vine pigmentation, vine diameter, vine growth rate, plant type, petiole length, petiole pigmentation, mature leaf size, mature leaf shape, abaxial leaf vine pigmentation, foliage colour and vine tip pubescence) were evaluated in the glasshouse from January 2011 to September 2011. The score of morphological traits were then changed to binary figures based on the presence (1) and absence (0) of the specific trait. The data were then subjected to hierarchal clustering (minimum variance ward methods and the neighbour-joining method) by using SAS statistical software.

Molecular characterization with SSRs

Sweetpotato clones exhibiting different morphological traits to that of the mother plants were subjected to molecular sequencing. A total of 14 samples, two from each of the 33rd, 11th and 4th subcultured generations of Monate, two from the 26th and 4th generations of Mokone, and two from the 13th and 4th generations of Ndou were selected and reintroduced to *in vitro* conditions for shoot initiation. The aim of the reintroduction to *in vitro* conditions was to obtain new actively growing leaf samples for DNA extraction. Leaf material was collected for DNA analysis from 30-day-old *in vitro* plantlets of the different subcultured generations of Monate, Mokone and Ndou.

The DNA extractions were done by modification of the Doyle and Doyle (1987) methodology. A small leaf disc, to which a small volume of carborundum (400 grit) (SAARCHEM) was added, was ground in a 1.5 ml Eppendorf tube. Four hundred microlitres of CTAB DNA extraction buffer (2% w/v CTAB, 1.4 M NaCl, 20 mM EDTA, 100 mM TRIS (pH 8), 0.2% β-mercaptoethanol) was added to the ground leaf material. The samples were then incubated at 60°C for 20 min in a heating block, with gentle mixing at 10 min intervals. Subsequently, the samples were allowed to cool to room temperature before the addition of one volume of chloroform: isoamyalcohol (CIAA; 24:1). The samples were mixed gently for 10 min, followed by centrifugation at 10,000 **g** for 10 min at room temperature. Subsequently, 350 µl of the supernatant was transferred to a clean 2 ml Eppendorf tube and a 0.6 volume of ice-cold isopropanol was added to each tube, followed by incubation at -80°C for 10 min. The DNA was pelleted by centrifugation at 10,000 **g** for 5 min at 4°C. DNA pellets were washed with 500 µl of 70% ethanol, centrifuged at 10,000 **g** for 5 min at 4°C, the supernatant discarded and the DNA pellets air-dried before resuspension in 100 µl TE buffer (pH 8.0). The DNA concentration of each sample was determined using a Nanodrop™ Lite (Thermo Scientific) according to the manufacturer's instructions.

PCR was conducted in 0.2 ml thin-walled tubes in a PTC-200 DNA engine (Bio-Rad). The reaction mixture, in a total volume of 12.5 µl, contained 1 × reaction buffer F (Epicentre Technologies), 400 µM of each dNTP, 1.75 µl of $MgCl_2$ (3.5 mM), 0.15 µl of *Taq* polymerase (0.75 U/µl) in storage buffer B (Promega Corp.), 20 ng of template DNA, and 0.24 µM of each primer. However, primers IB-255F and IB-248 required the use of reaction buffer E (5 mM $MgCl_2$) for optimum amplification.

The PCR cycling conditions were based on Butler *et al.* (1999). The amplified DNA fragments (2 µl samples) were resolved by electrophoresis in a 6% non-denaturing polyacrylamide gel (0.4 mm thick, 38.5 cm long) in 1 × TBE buffer (89 mM Tris-borate and 2 mM EDTA, pH 8.0) for the desired time at 40 mA. The gels were run for 2 h, 1 h 20 min, 2 h 40 min, 2 h 30 min and 1 h 10 min for the SSR primers IB-242, IB-318, IB-255F, IB-248 and IB-255, respectively. These primers were adopted from Butler *et al.* (1999). Five lanes of a 10 bp ladder with marker V (Promega Corp.) were loaded on the gel. The gels were stained with silver nitrate following the procedure of Bassam *et al.* (1991). Gels were transferred to blotting paper, dried, and banding pattern images digitized with a flatbed scanner (Scan Jet Hex, Hewlett Packard). The results of SSR loci were subjected to hierarchal clustering by using SAS statistical software.

29.3 Results and Discussion

Shoot and root formation

The number of days from culturing to shoot/root formation was shortened in cultivars Monate and Mokone as a result of long-term subculturing. On day 4 after culturing, the later subcultured generation group (27th–32nd subcultured generations) of Monate showed 9–16% root formation (Fig. 29.1a) and Mokone (21st–23rd generations) showed 21–24% root formation (Fig. 29.1c). However, on day 4 after culturing there was no root formation in the earlier subcultured generations of both Monate and Mokone (Fig. 29.1a and c).

The long-term subculturing also affected the percentage of successful shoot development in Monate and Mokone. The later subcultured generations of Monate, such as generation 31 showed the highest shoot formation percentage of 90% on day 10 (Fig. 29.1b). However, the earlier subculture generation group (1st–6th generations) showed lower shoot formation (71–84%) on day 10 for both cultivars (Fig. 29.1b and d).

Similar to these results, apple plants subcultured for the more than 100 generations

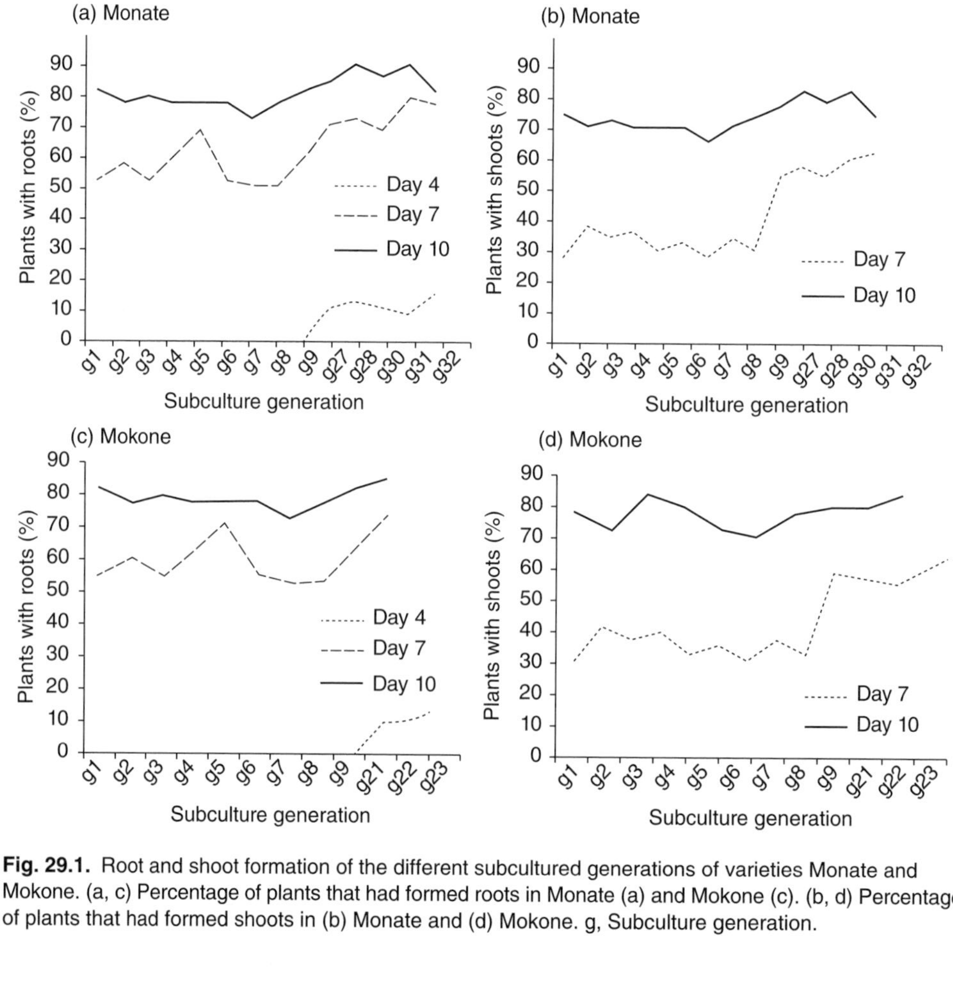

Fig. 29.1. Root and shoot formation of the different subcultured generations of varieties Monate and Mokone. (a, c) Percentage of plants that had formed roots in Monate (a) and Mokone (c). (b, d) Percentage of plants that had formed shoots in (b) Monate and (d) Mokone. g, Subculture generation.

showed higher rates of root and shoot formation (Shi *et al.*, 2007; Ying, 2011). However, the reason for change in the root and shoot formation may be either genetic or epigenetic. In studies conducted on other plants, such as *Castanea mollissima* (Hou *et al.*, 2010) and apple (Ying, 2011), long-term subculturing led to an increase in the concentration of endogenous indole-3-acetic acid (IAA). The higher concentration of endogenous IAA favoured the development of root formation. This could be the cause of change in root formation of long-term subcultured sweetpotato plants in the current experiment, but this hypothesis needs to be tested by experimentation.

Changes in leaf stomata density of subcultured generations

The effect of long-term subculturing on the density of stomata on the upper and lower leaf surfaces of sweetpotato plantlets is presented in Table 29.1. The general trend was that statistically higher stomata densities were observed on the lower (abaxial) surface of the 1st subcultured generation. No clear trend was observed for the upper (adaxial) surface. The results further showed that the ratio of the stomata density on the upper and lower surfaces of the leaf increased in the later subcultured generations of Ndou and Mokone. Leaf stomata number and epidermal structure, if they are stable in tissue culture, can be used as an additional tool for varietal identification (Savangikar and Savangikar, 2004). However, in our study the stomata number on the upper and lower part of the leaf was highly unstable over the different generations. If it was stable over different subculture generations it would have been a simpler and easier way of studying varietal conformity than other morphological and molecular techniques.

Morphological characterization of subcultured generations during acclimatization

The hierarchical classification of morphological data showed low semi-partial R^2 values among the subcultured generations of each of Monate, Mokone and Ndou (Fig. 29.2a, b and c, respectively).

The highest semi-partial R^2 value of Monate was 0.15 within 120 clones and these were grouped into two major subgroups, with 67 clones in group 1 and 53 clones in group 2 (Fig. 29.2a). The semi-partial R^2 value of the two groups ranged from 0.09 to 0.12, which showed that the dissimilarity between the groups was very small. The two groups encompass all subcultured generations, indicating statistically that there were no morphological differences among the different subcultured generations of Monate. There was no observed correlation between morphological variants and the number of

Table 29.1. The effect of long-term subculturing on the leaf stomata density (upper and lower leaf surfaces)[a] of varieties Ndou, Monate and Mokone.

Generation	Number of stomata/mm²		Generation	Number of stomata/mm²		Generation	Number of stomata/mm²	
	Upper	Lower		Upper	Lower		Upper	Lower
Ndou			Monate			Mokone		
1	8.85a	18.68a	1	12.6a	14.75a	1	6.00b	23.85a
10	7.25b	4.63b	28	8.38b	4.45b	20	5.55b	6.00b
11	9.68a	6.56b	29	8.25b	12.75a	21	8.93a	5.68b
						22	7.75a	4.75b

[a]Means followed by the same letter (a, b) in each column are not statistically significant at $\alpha < 0.05$ level when compared under Duncan's multiple range test and mean comparison.

the subculture generation. However, over all subcultured generations there was some degree of variability for a given morphological character (Table 29.2). Among all the morphological characteristics, abaxial leaf vine pigmentation and petiole pigmentation showed the highest percentage of variability (40%).

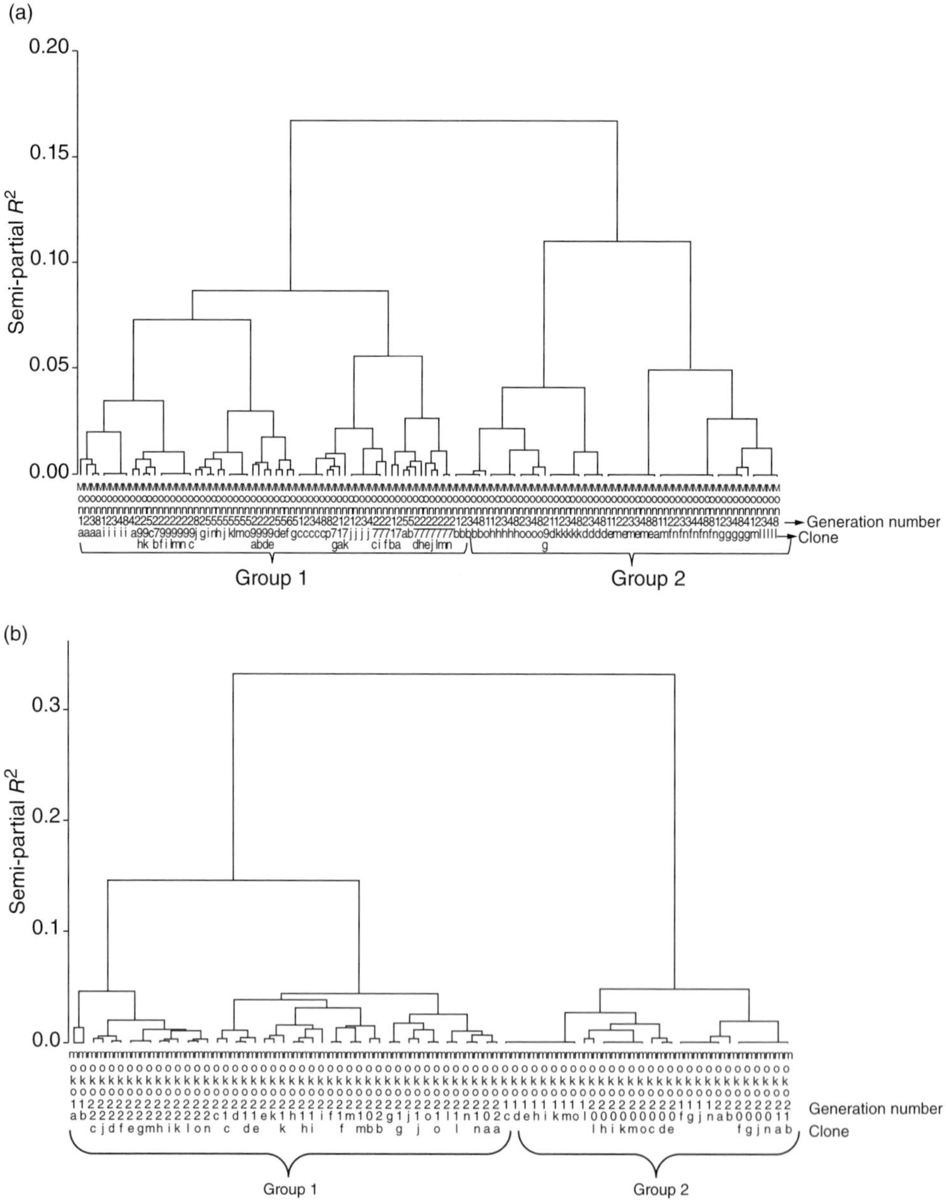

Fig. 29.2. Cluster analysis of 12 morphological characters of sweetpotato clones subcultured for many generations *in vitro*: (a) 120 clones of subcultured generations 1–9 and 27–29 of variety Monate (abbreviated to 'Mon' in figure); (b) 75 clones of the 1st, 2nd, 20th, 21st and 22nd generations of variety Mokone ('Mok' in figure); (c) 75 clones of subculture generations 8–13 of variety Ndou. The number next to the variety name indicates the generation number and the letter next to the generation number is a label given to each clone (e.g. Mon8l is a plantlet taken from the eighth subculture generation clone l).

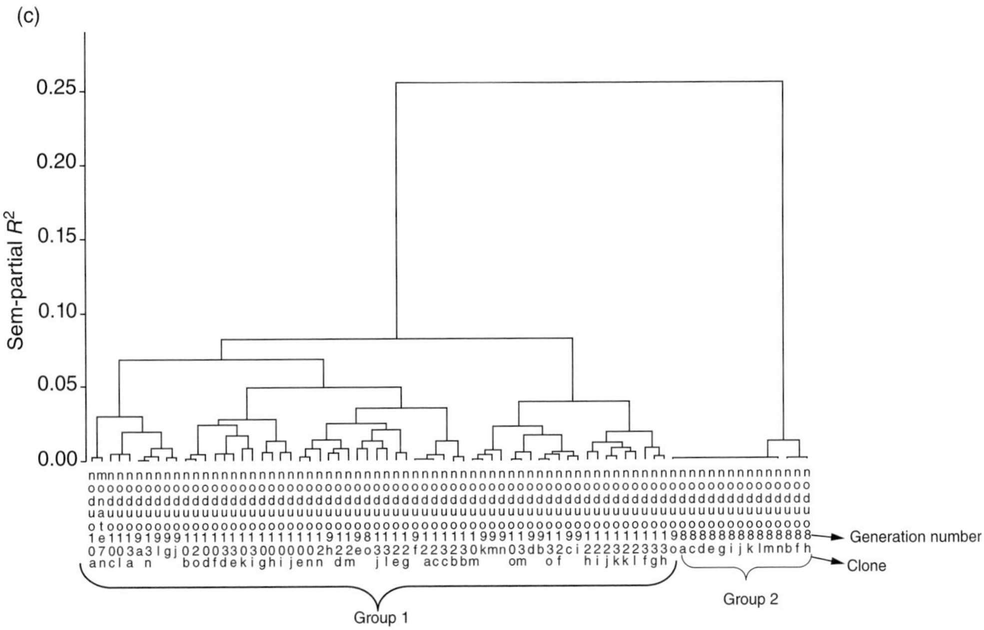

Fig. 29.2. Continued.

Table 29.2. The most dominant and variant type of six morphological characteristics of Monate during acclimatization (120 clones were evaluated for each characteristic).[a]

Morphological characteristic	Dominant type[b]	Variant type[b]
Mature leaf colour	YG (73%)	GWPA (27%)
Immature leaf colour	GWPA (73%)	YG (27%)
Abaxial leaf vine pigmentation	MRPP (60%)	PSBMR (0.13%), PSSV (27%)
Vine pigmentation	GWPN (73%)	G (27%)
Vine internode length	VS (73%)	SH (27%)
Petiole pigmentation	GPCS (60%)	GPCL (40%)

[a]Each clone was scored for the presence (1) and absence (0) of specific traits. The dominant scale was calculated by dividing the sum of frequency of presence by the total number of clones and multiplying it by 100%. If the sum of the given character is more than 50% of the total it was considered as dominant.

[b]YG, yellow green; GWPA, green with pigmented edge; MRPP, main rib partially pigmented; PSBMR, pigmented spot in the base of main rib; PSSV, pigmented spots in several vines; G, green; GWPN, green with pigmented nodes; VS, very short (< 3 cm); SH, short (3–5 cm); GPCS, green and pigmented close to the stem; GPCL, green and pigmented close to the leaf.

The highest semi-partial R^2 of Mokone was 0.3; the highest of the three varieties (Fig. 29.2b). The 75 clones evaluated in Mokone were classified into two large subgroups with 46 clones and 29 clones, respectively. The second subgroup was entirely made up by the 1st and the 2nd subcultured generations. While the first group contained clones from the 20th, 21st and 22nd subcultured generations. The semi-partial R^2 value difference between the two groups ranged from 0.05 to 1.5. Furthermore, clones of the 1st and the 2nd subcultured generations showed greater morphological similarity in comparison to the later subcultured generations.

Similarly, the cluster analysis showed that the maximum semi-partial R^2 value of Ndou was 0.25 (Fig. 29.2c). The clones from the 8th generation of Ndou exhibited different morphological characteristics and were grouped separately. The semi-partial R^2 value between the two groups of Ndou ranged from 0.02 to 0.09, which is the smallest of the three varieties.

It is an established fact that subculturing for many generations and the longer the duration between the first and last subculture generation enhances the rate of somaclonal variation in plants (Rodrigues *et al.*, 1998; Bairu *et al.*, 2006). In case somaclonal variation appeared in any of the subcultured generations, this would have caused increased variation in the subsequent subcultured generations. Opposed to this, the rate of morphological variation observed in our study appeared to be random. Rodrigues *et al.* (1998) found in banana that somaclonal variants appeared in the 5th generation at the rate of 1.3% and increased to 3.8% after the 11th generation.

Morphological characters are generally used in sweetpotato breeding programmes to detect variation among clones (Gichuru *et al.*, 2006; Manifesto *et al.*, 2010). However, for tissue-cultured plantlets the use of morphological traits to measure the level of somaclonal variation raised concern due to the effect of environmental factors on morphological characters (Mandal *et al.*, 2001) and the occurrence of unstable changes (Saifullah *et al.*, 2011). In addition, phenotypic markers are limited in number and are often developmentally regulated (Cloutier and Landry, 1994). These concerns emphasize the importance of the use of genetic markers.

Identification of genetic purity through the use of SSRs

All five SSR loci produced a reproducible banding pattern. The number of the alleles varied greatly among the SSR loci. Only two alleles were detected at locus IB-255 in all the clones and no polymorphisms were observed at this locus (Table 29.3). Thus, this locus could not be used to discriminate between the plantlets in the same generation or between the three varieties. However, three alleles were detected at locus IB-255F of which two were polymorphic. As a result, it showed the highest polymorphism percentage (67%) of all the SSR primers in this experiment. Locus IB-255F was able to discriminate among the three varieties. IB-318 also discriminated among the three varieties – five alleles were detected at this locus of which three were polymorphic (60%). Five and six alleles were detected at loci IB-242 and IB-248, respectively, with two polymorphic alleles each. The banding pattern for IB-248 for Mokone was distinct from that of Monate and Ndou (IB-248 having the same banding pattern for Monate and Ndou).

However, for all of the five primers used in this experiment, none could distinguish among plantlets of different subcultured generations in the same variety. It can be seen in Fig. 29.3 that no polymorphisms in SSR patterns were visible among the different subculture generations of the same variety. Based on the result of the gel electrophoresis patterns produced for all the five primers, the different subculture generations in the same cultivar had a monomorphic band size. According to the cluster analysis (Fig. 29.4), the highest semi-partial R^2 obtained in this analysis was 0.3 and the lowest semi-partial R^2 was 0, suggesting that varieties had genetic differences but the plantlets from the same variety, which showed differences in

Table 29.3. Alleles detected at the different loci and percentage of polymorphism.

Primer	Number of alleles detected	Number of polymorphic alleles	Percentage of polymorphism[a]
IB-255F	3	2	67%
IB-248	6	2	33%
IB-242	5	2	40%
IB-318	5	2	60%
IB-255	2	0	0%
Total	21	8	38%
Average	4.2	1.6	

[a]Percentage of polymorphism was calculated by dividing the number of polymorphic alleles per locus by the total number of alleles per locus.

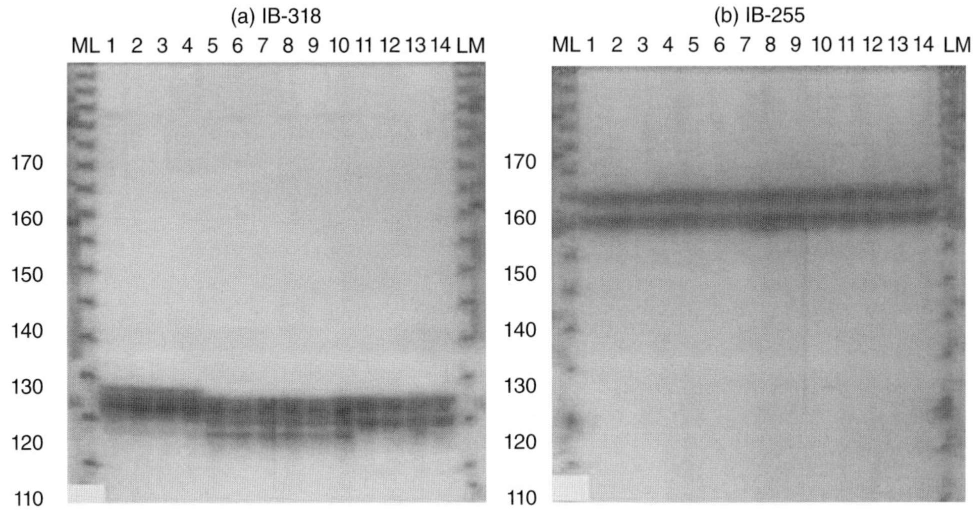

Fig. 29.3. Gel electrophoresis patterns for simple sequence repeat (SSR) primers IB-318 (a) and IB-255 (b) of the different subculture generations of Mokone, Monate and Ndou. Lane M, Marker V; lane L, 10 bp ladder; lanes 1 and 2, Mokone (4th generation); 3 and 4, Mokone (25th generation); 5 and 6, Monate (4th generation); 7 and 8, Monate (11th generation); 9 and 10, Monate (33th generation); 11 and 12, Ndou (4th generation); 13 and 14, Ndou (13th generation). The numbers on the left-hand side indicate the band length (bp) of the corresponding alleles of the marker ladder (ML).

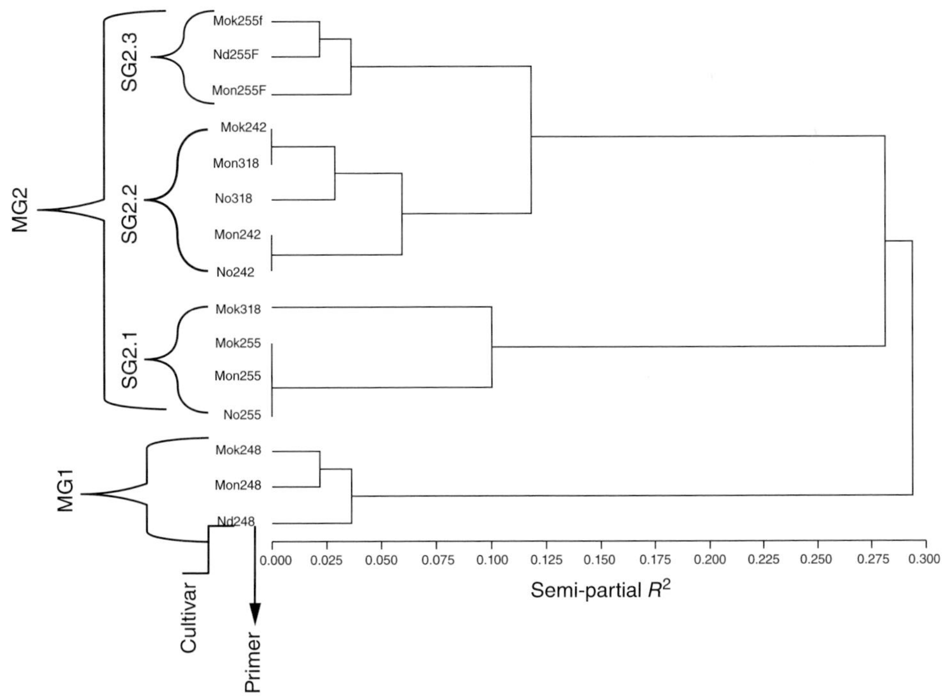

Fig. 29.4. Genetic grouping of varieties Monate, Mokone and Ndou based on the results from the five simple sequence repeat (SRR) primers. Mok, Mokone; Mon, Monate; Nd, Ndou; and the number that follows indicates the loci used. MG1, Main group one; MG2, main group two; SG2.1, 2.2 and 2.3, subgroups under main group 2.

some morphological traits, were not genetically different.

The sensitivity, reproducibility and strong discriminatory power of SSR markers (Prado *et al.*, 2010) make them particularly suitable for detecting somaclonal variation. However, their application for studying somaclonal variation in micropropagated species has been quite limited. More recently there has been some effort in using the SSR technique for detecting genetic purity of micropropagated sugarcane (Pandey *et al.*, 2012).

29.4 Conclusion

Plantlets of the same variety showed differences in morphological traits such as leaf colour, abaxial leaf pigmentation, vine pigmentation, petiole pigmentation, leaf wrinkling and percentage of flowering. However, there was no correlation between the degree of morphological variability and the number of generations of subculturing. No allelic polymorphisms were detected between early and late generations of the three varieties. The study therefore showed that long-term nodal subculture did not lead to quality degeneration of sweetpotato varieties Monate, Mokone and Ndou.

Acknowledgements

The authors wish to thank ARC-VOPI and the Forum for Agricultural Research in Africa (FARA) for their financial support towards the successful executions of this research.

References

Bairu, M.W., Fennell, C.W. and Van Staden, J. (2006) The effect of plant growth regulators on somaclonal variation in Cavendish banana (*Musa* AAA cv. 'Zelig'). *Scientia Horticulturae* 108, 347–351.

Bassam, B.J., Caetano-Anolles, G. and Gresshoff, P.M. (1991) Fast and sensitive silver staining of DNA in polyacrylamide gels. *Analytical Biochemistry* 196, 80–83.

Butler, M.I., Jarret, R.L. and La Bonte, D.R. (1999) Sequence characterization of microsatellites in diploid and polyploid *Ipomoea*. *Theoretical and Applied Genetics* 99, 123–132.

Cloutier, S. and Landry, B. (1994) Molecular markers applied to plant tissue culture. *In-vitro Cellular and Development Biology – Plant* 30, 32–39.

Diaz, J., Schmiediche, P. and Austin, D.F. (1996) Polygon of crossability between eleven species of *Ipomoea*: section *Batatas* (*Convolvulaceae*). *Euphytica* 88, 189–200.

Doyle, J.J. and Doyle, J.L. (1987) A rapid DNA isolation procedure for small quantities of fresh leaf tissue. *Phytochemical Bulletin* 19, 11–15.

Gichuru, V., Aritua, V., Lubega, G.W., Edema, R., Adipala, E. and Rubaihayo, P.R. (2006) A preliminary analysis of diversity among East African sweetpotato landraces using morphological and simple sequence repeats (SSR) markers. *Acta Horticulturae* 703, 159–164.

Hou, J., Guo, S. and Wang, G. (2010) Effects of *in-vitro* sub-culture on the physiological characteristics of adventitious root formation in microshoots of *Castanea mollissima* cv.'yanshanhong'. *Journal of Fruit Research* 21, 155–160.

International Potato Center (CIP)/The World Vegetable Center (AVRDC)/International Board for Plant Genetic Resources (IBPGR) (1991) *Descriptors for Sweetpotato*. IBPGR, Rome.

Israeli, Y., Lahav, E. and Reuveni, O. (1995) *In-vitro* culture of bananas. In: Gowen, S. (ed.) *Bananas and Plantains*. Chapman and Hall, London, pp 147–178.

Lewthwaite, S., Fletcher, P., Fletcher, J. and Triggs, C. (2011) Cultivar decline in sweetpotato (*Ipomoea batatas*). *Plant Pathology* 64, 160–167.

Lin, K., Lai, Y., Li, H., Lo, S., Chen, L. and Lo, H. (2009) Genetic variation and its relationship to root weight in the sweetpotato as revealed by RAPD analysis. *Scientia Horticulturae* 120, 2–7.

Mandal, A., Maiti, A., Chowdhury, B. and Elanchezhian, R. (2001) Isoenzyme markers in varietal identification of banana. *In-vitro Cellular and Developmental Biology – Plant* 37, 599–604.

Manifesto, M.M., Costa Ta' Rtara, S.M., Arizio, C.M., Alvarez, M.A. and Hompanera, N.R. (2010) Analysis of the morphological attributes of a sweetpotato collection. *Annals of Applied Biology* 157, 273–281.

Nayak, S., Kaur, T., Mohanty, S., Ghosh, G., Choudhury, R., Acharya, L. and Subudhi, E. (2011) *In vitro* and *ex vitro* evaluation of long-term micropropagated turmeric as analysed through cytophotometry, phytoconstituents, biochemical and molecular markers. *Plant Growth Regulation* 64, 91–98.

Ozias-Akins, P. and Jarret, R.L. (1994) Nuclear DNA content and ploidy levels in the genus *Ipomoea*. *Journal of the American Society of Horticultural Science* 119, 110–115.

Pandey, R., Singh, S., Rastogi, J., Sharma, M. and Singh, R. (2012) Early assessment of genetic fidelity in sugarcane (*Saccharum officinarum*) plantlets regenerated through direct organogenesis with RAPD and SSR markers. *African Journal of Agricultural Science* 6, 618.

Potter, R. and Jones, M.G.K. (1991) An assessment of genetic stability of potato *in vitro* by molecular and phenotypic analysis. *Plant Science* 76, 239–248.

Prado, M., Rodriguez, E., Rey, L., González, M., Santos, C. and Rey, M. (2010) Detection of somaclonal variants in somatic embryogenesis-regenerated plants of *Vitis vinifera* by flow cytometry and microsatellite markers. *Plant Cell Tissue Organ Culture* 103, 49–59.

Rodrigues, P.H.V., Tulmann, N.A., Cassieri, N.P. and Mendez, B.M.J. (1998) Influence of the number of sub-cultures on somoclonal variation in micropropagated Nanico (*Musa* spp., AAA group). *Acta Horticulturae* 490, 469–473.

Saifullah, K., Bushra, S. and Naheed, K. (2011) Establishment of genetic fidelity of *in-vitro* raised banana plantlets. *Pakistan Journal of Biotechnology* 5, 233–242.

Savangikar, V.A. and Savangikar, C. (2004) 'Quality assurance'. In: Savangikar, V.A. (ed.) *Low Cost Options for Tissue Culture Technology in Developing Countries Proceedings*. International Atomic Energy Institute, Vienna, pp. 64–69.

Shi, X., Du, G., Wang, C., Ma, B. and Ge, Y. (2007) Effects of sub-culture times on organogenesis characteristics of apple *in-vitro* shoot explants. *Acta Horticulturae* 34, 561.

Villordon, A.Q. and LaBonte, D.R. (1995) Variation in randomly amplified DNA markers and storage root yield in 'Jewel' sweetpotato clones. *Journal of the American Society of Horticultural Science* 120, 734–740.

Villordon, A.Q. and LaBonte, D.R. (1996) Genetic variation among sweetpotatoes propagated through nodal and adventitious sprouts. *Journal of the American Society of Horticultural Science* 121, 170–174.

Woolfe, J.A. (1992) *Sweetpotato an Untapped Food Resource*. Cambridge University Press, Cambridge.

Ying, L. (2011) *Organogenesis characteristics and genetic stability of apple shoot explants in vitro after being subcultured for years*. MSc thesis, Agricultural University of Hebei, China. Available at: http://www.shangxueba.com/lunwen/view/85/254928.htm (accessed 23 November 2012).

30 Shortage of Sweetpotato Planting Material Caused by Prolonged Dry Seasons in Africa: Strategies to Increase its Availability in Uganda

S. Namanda[1]* and R.W. Gibson[2]
[1]*International Potato Center (CIP), Kampala, Uganda;*
[2]*Natural Resources Institute, Chatham Maritime, UK*

Abstract

Farmers in Bukedea and Soroti districts in Uganda normally experience 3–4 months of dry weather between mid-November and March. During the dry period, vegetation of the sweetpotato crop is completely desiccated, leading to difficulties in securing vines as planting material at the onset of the rains. Failure to plant results in lack of food for vulnerable farming families, particularly before the grain harvest. The objectives of the study were: (i) to investigate the extent of lack of planting material in different agroecologies and the implications for early sweetpotato planting especially in areas that experience prolonged dry seasons; and (ii) to assess the effect of using longer cuttings and fertilizer application in high density vine beds for production of planting material.

Farmer interviews were administered in 2008 towards the end of the dry season with 44, 72, 105 and 50 farmers from Mukono, Kamuli, Bukedea and Soroti, respectively. On-farm trials on increasing production of vine areas with prolonged dry areas, compared the use of mini-cuttings (10 cm; two to three nodes) recommended by scientists, against the farmers' preferred vine length of 20 cm. Planting material was conserved and multiplied using NPK (25:5:5) fertilizer applied at 100 g/m^2 of propagation beds. On-farm trials on extending the area planted using the available vines involved shortening the vines that were planted and reducing plant densities for production of roots – this was trialled during the second 'short' rains of 2008 and 2009. The treatments were replicated three times and allocated to individual plots in a complete randomized block design.

About 58% of the farmers interviewed in Soroti reported failure to plant due to lack of planting material. Doubling the length of mini-cuttings significantly ($P < 0.001$) increased the number of normal (30 cm long) cuttings harvested from the high density vine beds. Pre-planting fertilizer application doubled the harvested number of cuttings. Further studies are needed to verify the optimum fertilizer usage.

Keywords: doubling, fertilizer, multiplication, strategies, sweetpotato, vines

*s.namanda@cgiar.org

30.1 Introduction

Importance of sweetpotato

Sweetpotato is the sixth most important world food crop after rice, wheat, potatoes, maize and cassava, and the fifth most important food crop in developing countries after rice, wheat, maize and cassava (FAOSTAT, 2010). The crop is one of the most widely grown root crops in sub-Saharan Africa (SSA); Uganda accounts for 20.5% of sweetpotato production in SSA (FAOSTAT, 2008), and sweetpotato is grown in smallholdings throughout all districts of Uganda (UBOS, 2010). In Eastern Uganda, Kamuli, Bukedea and Soroti districts produce 46.6% of sweetpotato (UBOS, 2010; MAAIF, 2011). Sweetpotato becomes a seasonal staple during the dry season when supplies of most other foodstuffs are exhausted (Hall *et al.*, 1998; CIP, 1999; Namanda *et al.*, 2011, 2013) and commercial sweetpotato production has emerged in some areas (Low *et al.*, 2009).

However, lack of sustainable seed systems (Low *et al.*, 2009), especially planting material, caused by the prolonged dry seasons (Namanda, 2012) is one of the key constraints to sweetpotato productivity in SSA. Lack of planting material is often particularly acute at the onset of the rains (Namanda *et al.*, 2003) after the long dry season has desiccated the foliage (Gibson, 2009). Methods that will ensure sufficient early availability of planting material include: (i) rapid multiplication techniques (RMT) devised by the National Agricultural Research Organisation (NARO) (Stathers *et al.*, 2005); (ii) fertilizing beds (Franklin, 1998); and (iii) using a complete fertilizer (Jonathan, 1998). Farmers prefer to plant at the onset of rains for better vine development and storage initiation (Valenzuela *et al.*, 2000), in order to harvest when there is still a poor supply to the market and prices are high (Heyd and Qaim, 2006) and to obtain early foods and better yields (Friis-hansen *et al.*, 2004; Stathers *et al.*, 2005; Peet, 2007).

There is relatively little difference in the overall yields of plant populations ranging from 25,000 to 125,000 plants/ha and plant density may be manipulated with advantage in areas where planting material is scarce (Gibson, 2009). Although, the length of the vine cuttings planted varies from location to location and in some areas from season to season, about 30 cm is recommended (Stathers *et al.*, 2005) (i.e. ideally cuttings that are 30 cm long are used to propagate the crop at the onset of the rains, but where planting material is in short supply shorter cuttings (mini-cuttings) are used (20 cm is preferred by farmers although 10 cm is recommended by scientists). Thus, in areas where planting material is scarce the following questions are raised:

- What is the optimal use of the available vines?
- How could vine production and productivity be increased and available vines manipulated to plant extended areas?

Study objectives

The objectives of the study were: (i) to investigate the extent of lack of planting material in different agroecologies and the implications for early sweetpotato planting especially in areas that experience prolonged dry seasons; and (ii) to assess the effect of using longer cuttings and fertilizer application in high density vine beds for production of planting material.

30.2 Materials and Methods

For the selected study areas the duration of dry periods and the type of rainfall experienced varies. For example:

- In Mukono dry periods are often short and broken by occasional storms.
- Kamuli experiences dry periods of around 12 weeks with occasional storms.
- Bukedea has about 17 dry weeks but with occasional unseasonal rains.
- Soroti, the most northerly district, has a long dry period of at least 18 weeks (November–mid-March), generally with no interrupting rainstorms (NaCRRI, 2010).

Questionnaire on farmers' knowledge of sweetpotato planting material

Based on field visits and knowledge of sweetpotato, a structured questionnaire was developed with mainly open-ended questions concerning farmers' planting practices. The study was administered in 2008 during the dry season (February–March), with a total of 44, 72, 105 and 50 farmers being interviewed from Mukono, Kamuli, Bukedea and Soroti, respectively.

On-farm testing the use of longer cuttings and fertilizer application to produce planting material in Soroti

The experiment compared the use of mini-cuttings (10 cm; two to three nodes) recommended by scientists, with the farmers' preferred vine length of 20 cm for planting beds for conservation and multiplication of sweetpotato planting material. In 2007, Araka, Kabode and Kakamega varieties were planted in raised beds 1.2 m wide × 2 m long and top dressed with urea fertilizer (100 g/m^2). The cuttings were planted at 20 cm between rows and 10 cm between plants and irrigated using a watering can. The treatments were replicated three times in a completely randomized block design. Beds were irrigated three times during the first week after planting, and subsequently twice weekly for 8 more weeks before the first harvesting when the main stem length was about 45 cm long. From the three middle rows (0.6 m wide × 2 m long) of each plot, 30 cm long cuttings were harvested, leaving 15 cm tall plant stumps above the ground to support tiller growth for subsequent harvests. Three harvest lots were taken at an interval of 3 weeks between each harvest lot. In the 2008 and 2009 seasons, due to the scorching effect on the plants the previous season urea treatment was substituted by the NPK (25:5:5) at the rate of 100 g/m^2. Araka (local variety) and Kabode varieties were replaced because it was difficult to obtain enough planting material. Ejumula and Kakamega varieties were used instead.

Data analysis

Data on farmers' knowledge on planting material were analysed using the Statistical Package for Social Scientists (SPSS). Analysis of variance (ANOVA) tables were generated using the GenStat program for on-farm tests.

30.3 Results

Farmers' responses on sourcing planting material and acreage normally planted at the beginning of first rains

Table 30.1 shows the percentage of farmers who failed to plant a particular area of land after the long dry season in 2007. There were none in Mukono, 62% in Kamuli, 35% in Bukedea and 100% in Soroti.

Evidently, moving from Mukono, to Kamuli, to Bukedea, to Soroti an increasing proportion bought ($P = 0.003$) planting material (Table 30.2). Overall, 14.4% of the farmers reported that they wanted to buy planting material.

Most farmers owning small plots reportedly were keen to buy planting material (Table 30.3) but failed to do so even for their small areas (Table 30.1).

The use of shorter cuttings was more popular in Bukedea and Soroti (Table 30.4), areas associated with longer dry seasons. Relatively more farmers in Mukono and Kamuli areas used cuttings longer than 30 cm than in Soroti and Bukedea ($P = 0.001$).

Producing planting material using fertilizer application in propagation beds and longer vine cuttings

Urea fertilizer was applied as a top dressing after the first harvest and 3 weeks later when the second harvest was to be done most of the plants were found to be scorched.

Urea application resulted in a less than average number of cuttings being harvested, notably due to the scorching effect on the plants after the first harvest. The variety

Kakamega produced more cuttings probably because of its spreading growth habit compared with Kabode that is semi-erect and less branching (Table 30.5).

After experiencing the scorching effect described above, urea fertilizer was substituted with a pre-planting fertilizer (NPK). Pre-planting fertilizer application (NPK 25:5:5; 100 g/m^2) doubled the harvest of cuttings (Table 30.6). NPK fertilizer was advantageous in that it did not result in the burning effect to plants under conditions of limited water supply.

Longer and shorter vine lengths were tested for production of 30 cm long cuttings during a series of three harvest lots at the interval of about 3 weeks between successive harvest lots. Planting longer cuttings for production of planting material significantly increased the number of 30 cm long cuttings harvested ($P < 0.001$) especially during the earlier harvest lots (Table 30.7).

Table 30.1. Percentage of farmers reporting a failure to plant a particular area of land at the beginning of the 2007 first rains due to lack of planting material.

	District			
Response	Mukono	Kamuli	Bukedea	Soroti
Average normal area (ha) planted by most farmers	0.25	0.13	0.38	1.55
Percentage of respondents failing to plant normal area from own source	0%	61.7%	34.9%	100%
Percentage of respondents failing to supplement with other planting material	0%	38.3%	65.1%	0%
Number of farmers asked	44	72	105	50

Table 30.2. Percentage of respondents buying planting material for particular areas for the 2007 first rains.

Area for which vines are bought	Percentage of farmers buying vines in different districts				
	Mukono	Kamuli	Bukedea	Soroti	Percentage overall
< 0.4 ha	0.0	5.6	10.2	30.0	13.0%
> 0.4 ha	0.0	0.0	1.0	6.0	1.4%
Total buying (%)	0.0	5.6	11.5	46.0	14.4%
Total not buying (%)	100	94.4	88.5	52.0	85.6%
Total number of respondents	44	72	105	50	271

Table 30.3. Percentage of respondents spending a particular amount of money on buying additional planting material in different districts.

Range total of money spent (USh)	Percentage of farmers in each district who spend this amount of money on vines			
	Kamuli	Bukedea	Soroti	Overall (%)
≤ 10,000	75.0	58.5	52.1	56.4
> 10,000–30,000	25.0	24.9	35.0	30.8
> 30,000	0.0	16.6	12.9	12.8
Total (%)	100.0	100.0	100.0	100.0

30.4 Discussion

The results clearly demonstrate that shortages of planting material at the beginning of the first rains occur as reported by Gibson (2009), in particular in the Soroti district (Table 30.1) due to pre-season longer dry periods (Namanda *et al.*, 2003). Also, normal sweetpotato planting acreage was the largest in Soroti and this is attributed to the smaller agricultural households in this district and because more oxen are raised here (UBOS, 2010). A significant ($P = 0.003$) proportion of farmers in Soroti bought vines compared with Mukono, an area with a shorter dry period, where no farmers bought vines (Table 30.2). The majority bought vines to plant up small plots (Table 30.2); either planting material was very expensive or not easily accessible to buy.

Although, generally more respondents planted 30 cm cuttings, reasonable proportions of farmers in Bukedea and Soroti reported planting vines that were less than 20 cm in length (Table 30.4). This is possibly associated with lack of vines (Stathers *et al.*, 2005; Namanda *et al.*, 2011). Farmers wanted to utilize whatever planting material available to plant extended plots (Namanda *et al.*, 2013) in order to produce an adequate food supply and access the early market (Friis-hansen *et al.*, 2004; Heyd and Qaim, 2006; Peet, 2007; Low *et al.*, 2009). Similar proportions of farmers in Mukono and Kamuli instead planted longer cuttings for the reverse reasons.

Top dressing with fertilizer (urea) generally resulted in a decrease in the number of vine cuttings produced (Table 30.5), because a post-plant fertilizer application easily burns the plants, especially if watering is inadequate, as is likely in the dry season. This chapter describes how farmers with access to water can double the amount of planting material produced by judicious use of fertilizer (NPK; Table 30.6) and, if they currently use mini-cuttings, double it again by planting 20 cm vines (Table 30.7). In most areas, vines for planting the conservation crop are in ample supply, provided by the previous rainy season's ware crop, so the use of long vines in this planting season is quite realistic. The findings confirmed those of Franklin (1998) and Jonathan (1998), that application of a preplanting fertilizer mixed into the soil at planting time roughly doubled ($P < 0.001$)

Table 30.4. Farmers using particular lengths of vine to plant their crop in Mukono, Kamuli, Bukedea and Soroti.

	Percentage of farmers using different lengths				
Length of cuttings	Mukono	Kamuli	Bukedea	Soroti	Overall
≤ 20 cm	0	1.4	28.6	18.0	14.8%
30 cm	77.2	80.6	66.7	78.0	74.1%
> 30 cm	22.8	18.0	4.7	4.0	11.1%
Number of respondents	44	72	105	50	271

Table 30.5. Effect of urea application on the total number of 30 cm cuttings harvested per 1.2 m^2 in season 2007.

	Number of cuttings harvested		
Variety	No fertilizer	Urea	Average
Araka	67	83	75
Kabode	47	45	46
Kakamega	104	69	86
Average	72	65	

Table 30.6. Average number of cuttings produced per 1.2 m² for three successive harvests with and without NPK at planting time during season 2009.

Harvest lot	Soil fertility amendment regime		Average	$LSD_{0.05}$[a]
	None	NPK		
1	11.7	35.9	23.8	10.15
2	107.2	191.7	149.5	19.05
3	78.8	179.0	128.9	38.25
Total number of cuttings	197.7	406.6	302.2	45.9

[a]$LSD_{0.05}$, least significant difference.

Table 30.7. Average number of cuttings produced per 1.2 m² for three successive harvest lots using 10 cm and 20 cm long vine cuttings and fertilizer during season 2009.

Harvest lot	Vine length (cm)		Average	$LSD_{0.05}$[a]
	10	20		
1	15.7	31.9	23.7	10.15
2	121.1	177.6	149.4	19.05
3	117.8	140.0	128.9	NS[b]
Total number of cuttings	254.6	349.5	429.4	45.9

[a]$LSD_{0.05}$, least significant difference.
[b]NS, not significant.

the number of cuttings harvested. At the current time (2011), 100 g of fertilizer applied to a bed of 1.2 m × 2 m produced about 200 more cuttings valued at about USh2000 whereas the cost of this amount of fertilizer was USh180 so it appears well worthwhile to invest in fertilizer. Also, whether applying 100 g of 25:5:5 NPK fertilizer/2.4 m² is optimal was not determined so there may be a more profitable combination. Thus, there is a need to investigate further the use of NPK fertilizer to produce planting material as well as storage roots.

It must be appreciated that there is much demand for swampy areas to produce high value vegetable crops such as cabbages and tomatoes. There is also competition for water for human and animal consumption as well as by crops, and the extra production of planting material achieved in this manner may often be necessary to make sweetpotato a competitive crop.

Using a 20 cm cutting instead of a 30 cm cutting made no significant difference to the total yield. Similarly, using two cuttings instead of three per square metre (i.e. reducing the amount of plant material used by a third) made no difference to the yield; this was also observed by the Natural Research Institute (1987). Thus, by using about half the planting material savings that cost nothing can apparently be achieved.

Planting at the beginning of the rains also allows a crop to yield more since it can grow over a longer period (Hall *et al.*, 1998; CIP, 1999; Friis-hansen *et al.*, 2004; Stathers *et al.*, 2005; Peet, 2007). Yield of sweetpotato is also closely related to the availability of sufficient water, particularly during establishment of the cuttings (Franklin, 1998), vine development and storage root initiation (Valenzuela *et al.*, 2000). Peet (2007) identified the period of within 40–60 days from planting as the critical development stage that requires adequate soil moisture.

The duration of the first rainy season in north-eastern Uganda is about 90 days and this is then followed by a dry spell of about 30–45 days (NaCRRI, 2010). If planting material is much delayed, this critical development stage may coincide with the short dry season (July in Uganda). This is an additional reason why planting sometime after the rains have started results in low yields (Yanggen and Nagujja, 2006; Namanda *et al.*, 2013).

Acknowledgements

Thanks are due to Dr Jan Low and Dr Ricardo Labarta for their valuable advice and to the farmers who collaborated with us. The activities were funded by the Reaching End Users sweetpotato project.

References

FAOSTAT (2008) Production and Area Harvested Statistics for Sweetpotato for 2007. Available at: http://faostat.fao.org/site/339/default.aspx (accessed 15 September 2015).

FAOSTAT (2010) Production and Area Harvested Statistics for Sweetpotato for 2010. Available at: http://faostat.fao.org/site/339/default.aspx (accessed 15 September 2015).

Franklin, W.M. (1998) Genetic and physiological basis for breeding and improving the sweetpotato. In: *Proceedings of the Seventh Symposium of the International Society of Tropical Root Crops (ISTRC).* ISTRC, Cotonou, Bénin, pp. 741–761.

Friis-hansen, E., Aben, C. and Kidoid, M. (2004) Smallholder agricultural technology development in Soroti District: synergy between NAADS and farmer field schools. *Uganda Journal of Agricultural Sciences* 2004(9), 250–256.

Gibson, R.W. (2009) Review of sweetpotato seed systems in Africa, especially East Africa. *Seed Systems Working Paper 2009*. International Potato Center (CIP), Lima.

Hall, A., Bockett, G. and Nahdy, S. (1998) Sweetpotato postharvest systems in Uganda: strategies, constraints and potentials. *Social Science Working Paper* 1998–7. International Potato Center (CIP), Lima.

Heyd, H. and Qaim, M. (2006) *Survey of Food Consumption and Nutritional Status in Uganda. Experiences During the First Round of Data Collection*. (Impact Assessment of Orange-fleshed Sweet Potatoes), Report submitted to the International Food Policy Research Institute (IFPRI), Harvest Plus. University of Hohenheim, Germany.

International Potato Center (CIP) (1999) *Sweetpotato Facts: Production, Utilisation, Consumption, and Feed Use*. CIP, Lima, 2pp.

Jonathan, R.S. (1998) *Guidelines for Sweetpotato Seed Stock and Transplant Production*. North Carolina Cooperative Extension Service, North Carolina State University, Raleigh, North Carolina.

Low, J., Lynam, J., Lemaga, B., Crissman, I., Barker, I., Thiele, G., Namanda, S., Wheatley, C. and Andrade, M. (2009) Sweetpotato in sub-Saharan Africa. In: Loebenstein, G. and Thottappilly, G. (eds) *The Sweetpotato*. Springer Science + Business Media BV, Houten, The Netherlands, pp. 359–390.

Ministry of Agriculture Animal Industry and Fisheries (MAAIF) (2011) *Statistical Abstract*. Agricultural Planning Department, MAAF, Kampala, Uganda, 51 pp.

Namanda, S. (2012) Current and potential systems for maintaining sweetpotato planting material in areas with prolonged dry seasons: a biological, social and economic framework. PhD thesis, Natural Resources Institute, University of Greenwich, Chatham Maritime, UK.

Namanda, S., Kapinga, R., Stathers, T., Van de Fliert, E. and Tumwegamire, S. (2003) Dissemination and commercialization of orange-fleshed sweetpotato through FFS and VITAA partnership: experiences from eastern Uganda. In: *Proceedings of the 13th Triennial Symposium of the International Society for Tropical Root Crops (ISTRC),* 10–14 November 2003, Arusha International Conference Centre, Arusha. ISTRC, Arusha, Tanzania, pp. 36–37.

Namanda, S., Gibson, R.W. and Kirimi, S. (2011) Sweetpotato seed systems in Uganda, Tanzania and Rwanda. *Journal of Sustainable Agriculture* 35, 870–884.

Namanda, S., Amour, R. and Gibson, R.W. (2013) The Triple S method of producing sweetpotato planting material for areas in Africa with long dry seasons. *Journal of Crop Improvement* 27(1), 67–84.

National Crops Resources Research Institute (NaCRRI) (2010) *Weather Data Base: Kampala, Kiige and Soroti Weather Stations*. NaCRRI, Namulonge, Uganda.

Natural Research Institute (1987) *Root Crops 308*. Available at: http://www.appropedia.org/Root_Crops_30 (accessed 15 May 2012).

Peet, M. (2007) Sustainable Practices for Vegetable Production in the South, North Carolina State University (NCSU). Available at: http://www.ibiblio.org/london/agriculture/feedback/new-links/msg00024.html (accessed 15 September 2015).

Stathers, T., Namanda, S., Mwanga, R.O.M., Khisa, G. and Kapinga, R. (2005) *Manual for Sweetpotato Integrated Production and Pest Management Farmer Field Schools in Sub-Saharan Africa*. International Potato Center (CIP), Kampala.

Uganda Bureau of Statistics (UBOS) (2010) *Uganda Census of Agriculture (UCA), Agricultural Household and Holding Characteristics Report*, Volume III. UBOS, Kampala.

Valenzuela, H., Fukuda, S. and Arakaki, A. (2000) Sweetpotato Production Guidelines for Hawaii. Available at: http://www.ctahr.hawaii.edu/oc/freepubs/pdf/RES-146.pdf (accessed 15 September 2015).

Yanggen, D. and Nagujja, S. (2006) The use of orange-fleshed sweetpotato to combat vitamin A deficiency in Uganda: a study of varietal preferences, extension strategies and postharvest utilization. *Social Sciences Working Paper* No. 2006–2. International Potato Center (CIP), Lima.

31 Evaluation and Promotion of Sustainable Farmer Seed Supply Systems for Vegetatively Propagated Crops in Central and Eastern Kenya

V.N. Gathaara,[1]* P.O. Simuyu,[2] D.W. Kilambya,[1] E.K. Muriuki,[1] E.G. Thuranira[1] and D.W. Miano[2]

[1]Kenya Agricultural and Livestock Research Organization (KALRO)-Kabete, Nairobi, Kenya; [2]University of Nairobi, Kenya

Abstract

Despite several interventions to address farmers' challenges related to seed, farmers' access to good quality seed of improved crop varieties for vegetatively propagated crops remains unsatisfactory. This chapter analyses the seed supply systems for vegetatively propagated crops in Kenya. In 2010, data were collected from randomly and purposefully selected households, farmer groups and key informants in Central (Nyeri, Mukurwe-ini and Nkubu) and Eastern (Kitui, Matuu and Embu) regions, using sweetpotato as the study crop. Semi-structured questionnaires and checklists were used during the interviews and focus group discussions. Data were analysed using Excel and the Statistical Package for Social Scientists (SPSS). Data generated from focus group discussions were used to enhance the information generated through the household and key informant interviews. The results indicated that 35.3% of the farmers' access seed from Kenya Agricultural Research Institute, 29.4% access seed through farmer-to-farmer exchange, 26.5% access seed through the farmer-to-farmer seed system with a commercial orientation and 8.8% got free seed from the Ministry of Agriculture seed supply systems. However, most (97.1%) of the farmers reported that it was important to have a formal seed system. Out of these, 60% (men and women) recommended the farmer-to-farmer seed system with the commercial orientation as the most sustainable system. This was particularly appreciated in Central Kenya where the system was already working. The system gave the farmers the opportunity to access new crop varieties, quality seed and it was available on time for planting. It allowed the farmers to take advantage of the first rains, resulting in better crop performance, improved production and food security. It also allowed the farmers, particularly those in groups, to generate income through the sale of cuttings. The government-led system, a new government initiative towards improved food security and poverty alleviation in the country had several challenges. Though the system offers farmers the opportunity to access improved/new crop varieties, the seed reached the farmers late and in a deplorable state for planting; at times after the rains had stopped and farmers having utilized all the available land. Therefore, farmer seed enterprise should be supported and promoted for sustainable agricultural production of vegetatively propagated crops for improved food security and poverty alleviation.

Keywords: commercialization, farmer seed enterprise (FSE), vegetatively propagated crops

*violet.gathaara@kalro.org

31.1 Introduction

Food security and poverty reduction are increasingly becoming a major concern given the world requirement to feed over 9 billion people. The Millennium Development Goals to which Kenya has subscribed aims at reducing the number of poor and hungry people by half, by the year 2015 (UNDP, 2003). Over the years, the country has faced increased food deficits and low agricultural productivity resulting in high levels of food-nutrition insecurity, poverty and increased reliance on food relief (Ministry of Finance and Planning, 2001). This situation has been attributed to prolonged droughts, erratic rainfall and the high cost of agricultural inputs such as seed and fertilizers and other related inputs (Omiti and Laibuni, 2012). Improved agricultural technologies/innovations and systems play an important role in addressing these challenges while also contributing to the country's economic and industrial growth through supply of raw materials and employment creation (Government of Kenya, 2007).

Seed/planting material is one of most important inputs in all crop-based agriculture. It is basic to agricultural production and a vital input to enhancing productivity (Beyene, 2010). Concomitantly, access to seed is a prerequisite to food production, security and poverty alleviation particularly in Kenya where the majority of the households (80%) depend on agriculture for a living. As seed carries the genetic potential of the variety that can be exploited by plant breeders to develop cultivars with high levels of adaptability, it becomes a determinant of the ultimate productivity of other productive inputs (Cromwell and Wiggins, 1993).

Seed systems form an integral part of production and distribution of seed and influences access by farmers to this key input in crop production. The formal seed sector which has a commercial orientation hence avails high cost seed. However, the system is mainly managed by government and/or private seed companies through contract growers. The system only supplies about 20% of the seed (Omanga and Rossiter, 2003). The informal system which encompasses the farmer seed exchange schemes is the most common system and fulfils the demand unmet by the formal seed system. Over the years, the informal seed system has served the inaccessible and resource-poor smallholder farmers (Cromwell and Wiggins, 1993). However, the system has several limitations including: (i) the risk of spreading diseases from one farm to the other; (ii) loss of genetic diversity due to increased disease incidences; and (iii) low production as a result of use of low quality seed/planting materials due to recycling (Gichuki *et al.*, 2005; KARI, 2009). Additionally, the system also lacks the capacity to supply seed of improved/new crop varieties to the farmers, limiting farmers' access to improved/new crop varieties (Sonia, 2003). Through official and non-governmental interventions, alternative systems have been tried, such as farmer seed enterprise (FSE), farmer-to-farmer seed system with a commercial orientation and supply of free seed/planting materials.

The FSE is market driven and involves decentralization of seed production with the aim of catering for area/regional-specific preferences. The system has opportunities for establishing linkages with formal institutions, production and provision of good quality seed to farmers (Sonia and Oliver, 2002). The free seed delivery system is mainly government supported and aims at providing seed/planting material to poor farmers. However, this system is not sustainable and has many challenges including late delivery of seed to farmers. Agricultural research systems have also contributed to seed delivery particularly during farmer participatory breeding and evaluation of crop varieties (Gichuki *et al.*, 2005). This study was conducted in Central and Eastern Kenya with the aim of identifying existing seed supply systems, constraints and opportunities for a sustainable seed supply system for vegetatively propagated crops.

31.2 Materials and Methods

The study was carried out in Central (Nyeri, Mukurwe-ini and Nkubu areas) and Eastern (Kitui, Matuu and Embu areas) regions of Kenya. The farmers in these areas were beneficiaries of sweetpotato and cassava seed projects either through research and/or Ministry of Agriculture (MoA) extension services in previous years (Gichuki *et al.*, 2005).

Primary and secondary data were used for the study with secondary data being generated from a literature review while primary information was sourced from individual households and interviews with key informants or focus group discussions. A total of 207 households randomly selected from all sites participated in the interviews. Key informants were purposefully selected by the MoA administration, village elders and non-governmental organization (NGO) officers working in the respective study areas. A semi-structured questionnaire and checklists were used for data collection at the household level and key informant interviews, respectively. Data generated were analysed using Excel and the Statistical Package for Social Scientists (SPSS). Data were disaggregated by gender and descriptive statistics such as frequency distribution.

31.3 Results and Discussion

Demographic information

Out of a total sample of 100 respondents (36 men and 64 women), the average age was 44 years old, with most (26.3%) having primary education, 17.8% secondary, 3.4% tertiary and 2.6% no education. It was noticeable that all the men had education unlike the women of whom 5.1% reported no education (Table 31.1).

As reported elsewhere (Hagenimana *et al.*, 1999) in 58.9% of households women were the main decision makers on variety(s) to be grown particularly of orange-fleshed sweetpotatoes and marketable varieties. However, in 28.6% of households men made the decision and in 12.5% of households the decision was made by both men and women. This has been as a result of increased understanding of the benefits of the crop both nutritionally and as a source of income. Farmers (92.1%) ranked sweetpotato very highly among the food crops in Kenya with 63.8% of households growing the crop for food, 11.1% for food diversification and/or alternative food source, and 21.7% and 4.4% for income generation and livestock feed, respectively. The orange-fleshed varieties have really gained value due to their value as a source of beta-carotene (Ukpabi and Ekeledo, 2009).

Seed production and distribution

The results indicated that farmers faced several challenges in seed production including availability of good quality seed at the onset of rains, limited access to improved varieties and lack of water for germplasm maintenance particularly in the dry areas.

Loss of germplasm by farmers was prevalent in Kitui and some parts of Yatta in Eastern Kenya due to frequent and prolonged dry spells. It was common for farmers in these areas to rely on the river bottoms for conservation of planting materials. Still, seed survival was not guaranteed meaning farmers risked having no seed to plant in the next crop season. However, in Mukurwe-ini, Nyeri and Nkubu, knowledge of the importance of good quality seed, production, multiplication and marketing among the farmers has increased over time. These challenges are not new and have been reported previously as key limitations to vegetative crop production in Kenya (Gichuki *et al.*, 2005).

Table 31.1. Respondents description by gender, age and education.

Gender category	Proportion (%)	Average age (years)	Level of education (%)			
			Primary	Secondary	Tertiary	None
Male	36	49	13.6	20.3	1.7	0
Female	64	39	39	15.3	5.1	5.1
Mean		44	26.3	17.8	3.4	2.6

Seed supply systems

The study identified four major seed supply systems, namely: (i) farmer-to-farmer seed exchange system; (ii) farmer-to-farmer seed system with a commercial orientation; (iii) research-to-farmer seed system; and (iv) government-led seed supply system comprising of free seed given to farmers through the MoA (Table 31.2).

The farmer-to-farmer seed system with a commercial orientation involved farmers buying seed (cuttings) at a cost of between KSh1 and KSh2 for 30–60 cm of sweetpotato cut by farmer groups and/or individual farmers who propagated seed for sale. This situation was mainly identified in Central (Nyeri/Mukurwe-ini) Kenya. The farmer groups/farmers were beneficiaries of previous root crop projects through various institutional and government-led projects.

The farmer-to-farmer seed exchange system was identified in all study areas. Characteristically, farmers exchange seed freely among neighbours, friends and relatives. However, it emerged that this system was changing in response to farming systems, land sizes and land use among other factors, such as climate change. Particularly, the wetlands and river beds where farmers used to conserve the planting materials have been drying up, reducing the opportunities for seed conservation, affecting the amount for free seed exchange/distribution. The farmers' level of awareness of farming as a business has also contributed to the decline of farmer-to-farmer seed exchange with commercialization of seed often being preferred.

The research-to-farmer seed supply system involves farmers buying seed direct from the Kenya Agricultural Research Institute (KARI). The farmers then multiply the seed either for their own use and/or for sale. The government-led system was reported as an on-and-off process whereby the government buys basic seed mainly from KARI, multiplies the seed through contracted growers and/or in the MoA's farms and distributes the seed particularly to poor farmers through extension in target areas. However, there are major bottlenecks to this seed delivery system including inefficiency and absence of a proper quality control mechanism.

Seeds of vegetatively propagated crops are usually perishable and therefore require a system that will minimize the duration between harvesting and planting. According to Beyene (2010), a system that would effectively support farmer-based seed production as an alternative agricultural technology transfer is particularly important in developing countries where food insecurity is critical. FSE is a better option to the informal seed supply system as it assures the farmer of access to quality seed and new/improved crop varieties for improved crop production (Sonia, 2003).

Sources of seed/planting materials

On the sources of planting materials for the long and short rains seasons, the majority of farmers (35.3%) identified KARI as the main source of seed for improved varieties with 29.4% accessing seed from other farmers, relatives and neighbours (i.e. farmer-to-farmer

Table 31.2. Seed supply systems used in relation to gender.

	Percentage of respondents[b] who were using this seed system		
Type of seed system	Husband	Wife	Both (husband and wife)
Farmer-to-farmer seed exchange system	33.3	35.7	25.0
Farmer-to-farmer seed system with commercial orientation	50.0	21.4	25.0
KARI[a]	16.7	28.6	25.0
MoA	0.0	14.3	25.0

[a]KARI, Kenya Agricultural Research Institute; MoA, Ministry of Agriculture.
[b]Respondents are categorized according to who (husband, wife or both) makes the decision as to how seed is obtained (i.e. what seed system is used).

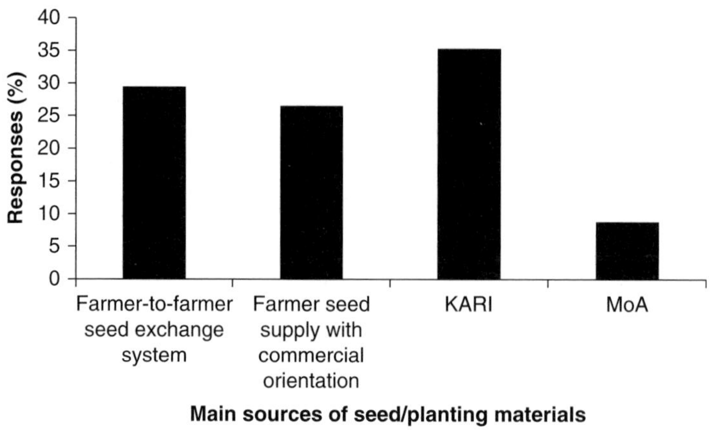

Fig. 31.1. Main sources of seed/planting materials. KARI, Kenya Agricultural Research Institute; MoA, Ministry of Agriculture.

Table 31.3. Farmers most preferred seed system by gender.

	Percentage of respondents who would prefer to use this seed system		
Seed supply system	Men	Women	Men and women
Farmer-to-farmer exchange system (free seed)	14.3	26.7	40
Cooperative system	28.6	26.7	0
Farmer-to-farmer (commercial orientation)	57.1	46.7	60

seed exchange system), 26.5% bought seed from sweetpotato farmer seed multiplication groups and/or individual farmers multiplying quality seed for business (farmer-to-farmer seed system with commercial orientation) and 8.8% got free seed from the MoA through the government-led seed supply system (Fig. 31.1).

Most (97.1%) of the farmers reported that it was important to have a formal seed system. Out of these, 60% (men and women) reported that the system which should have a commercial orientation should be farmer-led through farmer groups (Table 31.3).

Group-based seed production is complex as it involves interplay of multiple factors. For example, at entry point, choice of existing crop-based groups with the ability to multiply and sell seed to farmers could be preferable to formation of new groups and/or identifying individuals to drive the systems. In the study area, farmers considered the current price of KSh1–2 per cutting of 30 cm as affordable and reasonable. As reported elsewhere, a farmer-to-farmer seed system with a commercial orientation (farmer-to-farmer seed enterprise) ensures farmers have access to seed/planting materials of improved and desired varieties on demand and these are available in time for planting (Sonia, 2003).

31.4 Conclusion

Vegetatively propagated crops such as sweetpotato and cassava continue to play an important role in food security and poverty alleviation. Access and availability of quality seed, particularly of improved varieties, continues to be a challenge at farm level. However, efforts towards ensuring farmers have access to improved/new varieties have been made by various development agents.

KARI has been at the forefront in involving farmers through participatory research to evaluate and adopt varieties that meet their needs. Through some of these activities, farmers, mainly in groups, have been trained in seed multiplication and variety maintenance leading to seed bulking and selling to other farmers at a small fee. This has resulted in establishing a farmer-to-farmer seed supply system with a commercial orientation. The system grants farmers access to improved/new varieties and quality planting materials (cuttings) on demand, at affordable prices and at the right time for planting (Sonia, 2003). The MoA seed system is ideal especially for pro-poor farmers but has many challenges including late delivery of seed/planting materials to farmers.

The traditional farmer-to-farmer exchange seed supply has its merits. However, it mainly deals with local varieties and fails to facilitate farmers' access to improved varieties, and in addition there is the risk of spreading pests and diseases. Hence, increased demand and importance of vegetatively propagated crops for food security, livestock feed and income generation requires a well-organized, effective, affordable and sustainable seed supply system, comparable to that of high value crops as the crops continue to gain value.

31.5 Recommendations

Seed produced by farmers is still the most important seed source in the majority of developing countries (Alila and Atieno, 2006). Recognition of complementarity of the formal and informal seed system such as the FSE offers opportunities for improved seed supply particularly to poor farmers (Almekinders, 2000).

The system would ensure that farmers are provided with improved, desirable varieties and quality seed delivered to farmers on demand in the required quantities while also giving producers an opportunity for income generation (Sonia and Oliver, 2002). The fact that farmers were willing to pay for cuttings was a good indicator of the understanding of the benefits of quality seed and the value placed on seed. Farmer groups and/or individual farmers with the capacity to multiply and distribute seed can therefore be identified and trained in quality seed production, maintenance and marketing. The groups and/or farmers should be within communities to promote easy access and reduction on transport costs. This improved seed system will contribute to improved crop production, food security and incomes providing better livelihoods.

Acknowledgements

The authors acknowledge the Director at KARI for financing the study through Kenya Agricultural and Agribusiness Project (KAPAP), the Centre Director at KARI-Kabete for the facilitation and support to conduct the survey, and those who reviewed and accepted this chapter.

References

Alila, P.O. and Atieno, R. (2006) Agricultural policy in Kenya: issues and processes. Paper presented at the Future Agricultures Consortium Workshop, 20–22 March 2006, Institute of Development Studies, Nairobi, Kenya.

Almekinders, C. (2000) The importance of the informal seed sector and its relation with the legislative framework. Paper presented at GTZ-Eschborn, Management of Crop Diversity at Community Level, Eschborn, Germany.

Beyene, F. (2010) The role of NGOs in informal seed production and dissemination: the case of Eastern Ethiopia. *Journal of Agriculture and Rural Development in the Tropics and Subtropics* 2, 79–88.

Cromwell, E. and Wiggins, S. (1993) *Sowing Beyond the State: NGO and Seed Supply in Developing Countries*. Overseas Development Institute, London.

Gichuki, S.T., LaBonte, D., Burg, K., Kapiga, R. and Simon, J.C. (2005) Assessment of genetic diversity, farmer participatory breeding and sustainable conservation of East Africa sweetpotato germplasm.

Project Report. Kenya Agricultural Research Institute (KARI)-National Agricultural Research Laboratories (NARL), Nairobi.

Government of Kenya (2007) *Kenya Vision 2030. The Popular Version*. The National Economic and Social Council of Kenya, Nairobi.

Hagenimana, V., Oyunga, M.A., Low, J., Njoroge, S.M., Gichuki, S.T. and Kabira, J. (1999) *The Effects of Women Farmers' in the Adoption of Orange-fleshed Sweetpotatoes. Raising Vitamin A Intake in Kenya*. International Center for Research on Women, Washington, DC.

Kenya Agricultural Research Institute (KARI) (2009) *Annual Report*. KARI, Nairobi.

Ministry of Finance and Planning (2001) Poverty Reduction Strategy Paper for the period 2001–2004. Ministry of Finance and Planning, Nairobi.

Omanga, P. and Rossiter, P. (2003) Improving emergence seeds in Kenya: comparative analysis of direct seed vouchers and fairs. Paper presented at Effective and Sustainable Seed Relief Activities, Food and Agriculture Organization of the United Nations (FAO) conference, 26–28 May 2003. FAO, Rome.

Omiti, J. and Laibuni, N. (2012) Food security – Kenya cycle of hunger is conquerable. Here is how. *Daily Nation* (Nairobi), 14 November.

Sonia, D. (2003) Farmer seed enterprises: a sustainable approach to seed delivery. *Journal of Agriculture and Human Values* 21(4), 387–397.

Sonia, D. and Oliver, B. (2002) *Network on Bean Research in Africa*. Handbook 3. *Handbooks for Small-scale Seed Producers: a Trainer's Guide*. Occasional Publications, Series No. 37. International Centre for Tropical Agriculture (CIAT), Kampala.

Ukpabi, U.J. and Ekeledo, E.N. (2009) Feasibility of using orange-fleshed sweetpotato as an alternative to carrot in Nigerian salad preparations. *Agricultural Journal* 4(5), 216–220.

United Nations Development Programme (UNDP) (2003) Millennium Development Goals: a compact among nations to end human poverty. *Human Development Report*. UNDP, New York.

Part IV

Tackling Disease and Crop Management in Potato and Sweetpotato

32 Disease Management, Especially Viruses in Potato and Sweetpotato

J.P.T. Valkonen,[1]* J.F. Kreuze[2] and J. Ndunguru[3]
[1]University of Helsinki, Helsinki, Finland; [2]International Potato Center (CIP), Lima, Peru; [3]Mikocheni Agricultural Research Institute, Dar es Salaam, Tanzania

Abstract

Potato (*Solanum tuberosum*) and sweetpotato (*Ipomoea batatas*) are the fourth and seventh most important crop plants. Both are important subsistence crops in East Africa and complement each other since their growth requirements differ. Potato and sweetpotato are vegetatively propagated, making them prone to accumulation of viruses during cultivation, and *c.*40 and 30 viruses have been reported to infect them, respectively. In East Africa, viruses comprise the only significant disease agents in sweetpotato. The most severe yield losses are caused by dual infection of plants with *Sweet potato chlorotic stunt virus* (SPCSV, crinivirus) and other, unrelated viruses, because the RNase3 protein produced by SPCSV eliminates antiviral defence. Therefore, control of SPCSV is of utmost importance in the control of virus diseases in sweetpotato. Recent surveys have also highlighted the prevalence of begomoviruses in sweetpotato throughout Africa; significant yield losses have been reported even in symptomless single infection. In potato, *Potato leaf roll virus* (PLRV, polerovirus) and *Potato virus Y* (PVY, potyvirus) are the most widely spread and damaging viruses in potato crops. In Eastern Africa, closer knowledge on viruses affecting potato crops is relatively limited, but some recent data are available from Kenya and Tanzania. Healthy seed potatoes and sweetpotato vine cuttings available for planting are a prerequisite for successful production. Knowledge on resistance to the local viruses and virus strains is another important issue when choosing cultivars for production. Antibodies to the most common potato viruses are commercially available and their use for indexing seed potatoes is straightforward with the basic diagnostic capacity. In contrast, antibodies to sweetpotato viruses are commercially unavailable. Analysis of the small interfering RNA (siRNA) molecules resulting from degradation of viral genomes by RNA silencing, the basal antiviral defence mechanism of plants, provides a universal, novel method for detection of plant viruses and does not require advance knowledge of them. It is promising for indexing of the stock plants used for further propagation of planting material. The technology is also applied to the identification of novel viruses and analysis of virus variability throughout Africa, which will inform development of novel control methods and testing regimes to control the most harmful viruses for each region.

Keywords: potato, RNA, sweetpotato, vegetatively propagated, yield loss

*jari.valkonen@helsinki.fi

32.1 Introduction

Potato (*Solanum tuberosum* L.) and sweetpotato (*Ipomoea batatas* Lam.) are the fourth and seventh most important crop plants (FAOSTAT, 2013). Both are important subsistence crops in East Africa and complement each other since their growth requirements differ. They are vegetatively propagated, making them prone to accumulation of viruses during cultivation, and *c*.40 and 30 viruses have been reported to infect them, respectively (Valkonen, 2007; Clark *et al.*, 2012). Both crop species have their centre of evolution and geographical origin in South and Central America (Janick, 2013; Roullier *et al.*, 2013). Many viruses infecting these crops in other parts of the world have been distributed from the centre of crop origin in infected planting materials.

While international distribution of potato and sweetpotato viruses in infected plant germplasm is obvious in light of the genetic similarity of virus strains in different parts of the world (e.g. Rännäli *et al.*, 2008, 2009; Latvala-Kilby *et al.*, 2009; Santala *et al.*, 2010; Cuellar *et al.*, 2011; Cuevas *et al.*, 2012; Chiunga and Valkonen, 2013), the origin of the virus is not always obvious. One reason limiting possibilities to determine centres of virus evolution is the limited information about viruses infecting wild plants, especially perennial species that are potent virus reservoirs. This shortcoming also limits the possibilities to realize possible local sources of viruses affecting the crops. Only recently, some effort has been invested in studies on viruses infecting wild relatives of sweetpotato in East Africa (Karyeija *et al.*, 1998; Tugume *et al.*, 2008, 2010a, b).

The other main problem in controlling dispersal of viruses is related to the limited possibilities to detect all of them. The current knowledge on viruses infecting potato and sweetpotato crops in Africa is based on the conventional techniques of diagnostics relying mainly on serological and polymerase chain reaction (PCR)-based methods. Surveys of viruses in the crops growing in the field frequently report virus-like diseases not associated with any known virus detectable with the antibodies or PCR primers available (Mukasa *et al.*, 2003; Ateka *et al.*, 2004; Tairo *et al.*, 2004; Chiunga and Valkonen, 2013). These diseases of unknown aetiology suggest existence of viruses that have remained undetected despite current, sensitive diagnostic methods, whose strength and limitation lays on the target specificity. Therefore generic, highly sensitive techniques to identify viruses and their variants would be extremely useful to determine the diversity, frequency and distribution of viruses and virus strains in different agroecologies or regions. Important progress has been made recently in this respect (Kreuze *et al.*, 2009).

In the chapter we provide a brief overview on viruses reported in sweetpotato and potato crops in Africa, with emphasis on East Africa where most studies have been carried out so far.

32.2 Sweetpotato Viruses

Over 30 viruses have been found to infect sweetpotato. A recent review summarizes a wealth of the current knowledge on these viruses (Clark *et al.*, 2012) and details will not be repeated here. In Africa, the most important virus is the whitefly-transmitted *Sweet potato chlorotic stunt virus* (SPCSV; genus *Crinivirus*; *Closteroviridae*), which is able to suppress virus resistance in sweetpotato and makes the plants susceptible also to other, unrelated viruses (Karyeija *et al.*, 2000a; Cuellar *et al.*, 2009). The family *Potyviridae* members *Sweet potato feathery mottle virus* (SPFMV) (Kreuze *et al.*, 2000), *Sweet potato virus C* (SPVC) previously known as the C strain of SPFMV (Untiveros *et al.*, 2010; Tugume *et al.*, 2010a), *Sweet potato virus G* (SPVG) (Rännäli *et al.*, 2008) and *Sweet potato virus 2* (SPV2) (Tairo *et al.*, 2006) are aphid-transmissible potyviruses (genus *Potyvirus*). The vector of *Sweet potato mild mottle virus* (SPMMV, genus *Ipomovirus*) is still undetermined (Tugume *et al.*, 2010b) because transmission of SPMMV by whiteflies (Hollings *et al.*, 1976) has not been possible to demonstrate in more recent studies. All these viruses are harmful, especially when co-infecting sweetpotatoes with SPCSV. Co-infection with SPFMV and SPCSV causes heavy reduction of growth and yield, the so-called sweet potato virus disease (SPVD), which is the most devastating disease of sweetpotatoes in Africa (Gibson *et al.*, 1998; Njeru *et al.*, 2004; Mukasa *et al.*, 2006). *Sweet potato chlorotic fleck virus* (genus *Carlavirus*, family *Flexiviridae*) is rather common in East Africa (Mukasa *et al.*, 2003; Ateka *et al.*, 2004; Tairo *et al.*, 2004; Njeru *et al.*, 2008; Aritua *et al.*, 2009) and also this virus is able to synergize with SPCSV in co-infected sweetpotato plants (Untiveros *et al.*, 2007). The viruses mentioned above have an RNA genome.

Approximately half of the viruses known to infect sweetpotatoes are recently described DNA viruses (families *Geminiviridae* and *Caulimoviridae*) (Clark *et al.*, 2012). Only one caulimo-like virus was known to infect sweetpotatoes in East Africa (Aritua *et al.*, 2007), but recent studies indicate that there are at least two of them (Cuellar *et al.*, 2011). They cause more severe symptoms when co-infecting sweetpotatoes with SPCSV, which represents a rare example of synergism between DNA and RNA viruses (Cuellar *et al.*, 2011).

Begomoviruses are found in sweetpotato germplasm in South, Central and North America (Paprotka *et al.*, 2010; Barkley *et al.*, 2011; Kashif *et al.*, 2012) and have been detected in sweetpotatoes grown in East Africa (Miano *et al.*, 2006; Wasswa *et al.*, 2011). *Sweet potato pakakuy virus* (SPPV; synonyms Sweet potato badnavirus strains A and B) and the putative sweetpotato-infecting mastrevirus, sweet potato symptomless virus 1 (SPSMV) were detected in a sweetpotato cultivar grown in Peru (Kreuze *et al.*, 2009). They occur also in landrace sweetpotato varieties in Tanzania (Mbanzibwa *et al.*, 2011). Similar isolates characterized in the different continents suggest that these viruses have spread via distribution of infected, symptomless sweetpotato germplasm. It is apparent now that DNA viruses are more common in sweetpotatoes than previously thought and probably cause yield losses that have remained unnoticed.

Small RNA-based virus diagnostics

The recently developed generic, highly sensitive techniques to identify viruses and their variants are emerging as extremely useful tools. One particular approach, small RNA sequencing and assembly of viral genomes (sRSA), was first described in analysis of sweetpotato viruses (Kreuze *et al.*, 2009). sRSA is based on high throughput sequencing of small interfering RNAs (siRNAs) resulting from degradation of viral genomes by the basal antiviral defence mechanism called RNA silencing. Posterior assembly of siRNA sequences into larger contigs and similarity searches in sequence databases containing accessions of viral sequences allows detection of viruses.

The simple initial experiment led to the discovery of three new sweetpotato viruses, demonstrating the potential power of the technique (Kreuze *et al.*, 2009). Since then the technique has been applied to identify many new viruses and virus strains in sweetpotato (Untiveros *et al.*, 2010; Kashif *et al.*, 2012; De Souza *et al.*, 2013), potato (Hwang *et al.*, 2013; Kreuze *et al.*, 2013) and other plant species (Zhang *et al.*, 2011; Bi *et al.*, 2012; Fuentes *et al.*, 2012; Loconsole *et al.*, 2012a, b; Sela *et al.*, 2012; Wu *et al.*, 2012). The method is also applicable to detection of viroids in plants (Li *et al.*, 2012) and viruses of invertebrate and vertebrate animals (Wu *et al.*, 2010; Isakov *et al.*, 2011; Ma *et al.*, 2011), which shows the universal utility of the method.

Continental-scale detection of sweetpotato viruses in Africa

The International Potato Center (CIP) has started to apply the sRSA technology to identify sweetpotato-infecting viruses at a continental scale in Africa. Because the focus of sweetpotato virus research to date has been in East Africa, much less is known about the viruses and virus strains present in other parts of Africa. Results from southern Africa (Mozambique) and West Africa (Ghana and Burkina Faso) show that

SPFMV, SPVC, SPCSV, *Sweet potato leaf curl virus* (SPLCV)-related begomoviruses, SPPV-like badnaviruses and the mastrevirus SPSMV are commonly found in both regions. The frequency of begomovirus and badnavirus infections is high and highlights the need for further research into the impact that viruses may have on sweetpotato production, especially since begomoviruses can reduce yield significantly (Clark and Hoy, 2006; Ling *et al.*, 2010). Unpublished data indicate also that begomoviruses and badnaviruses infecting sweetpotato are highly variable and often occur in mixed infections.

Results have begun to shed light on a pattern of virus distribution and virus strain differentiation among different regions of Africa. For example, there are sweetpotato viruses such as SPMMV, which is common in East Africa but does not seem to occur in other sweetpotato-growing areas (Tairo *et al.*, 2005; Tugume *et al.*, 2010b). SPVG has not been found in West Africa, whereas it is reported in northern and southern Africa and many other parts of the world (Rännäli *et al.*, 2008). The East African (EA) and West African strain (WA) of SPCSV (Vetten *et al.*, 1996; Cuellar *et al.*, 2008) also show geographical separation, SPCSV-EA being found in East Africa where SPCSV-WA has not been detected (Cuellar *et al.*, 2008).

siRNA studies indicate that new alphasatellites are associated with begomovirus infections. siRNA studies have also been helpful in characterizing a new potyvirus infecting sweetpotatoes in Mozambique. This virus has been tentatively named sweet potato virus Z (SPVZ), because it corresponds to a previously published partial virus sequence from Zimbabwe (Chavi *et al.*, 1997). One sweetpotato plant was found to contain a virus belonging to the 'supergroup III' of positive-sense RNA viruses that are only distantly related to plant viruses. It is worth mentioning that because of the generic nature of sRSA, the detected virus may not necessarily infect sweetpotato, but could be infecting other organisms such as endophytic fungi associated with sweetpotato.

32.3 Potato Viruses

Potato is not a new crop in cooler areas of Africa, but potato cultivation has shown fast expansion recently. Studies on potato diseases in Africa are mainly focused on late blight and bacterial diseases, and information about incidence of viral infections is scarce. Potato viruses are transmitted to new crops in seed tubers. Vectors such as aphids transmit most potato viruses between potato plants in the field (Valkonen, 2007). Gildemacher and co-workers (2009) carried out a survey in Kenya in 2006 to detect viruses in seed potatoes sold in rural markets. They tested 1000 potato tubers for the aphid-transmissible *Potato leaf roll virus* (PLRV) (genus *Polerovirus*), *Potato virus Y* (PVY) and *Potato virus A* (PVA) (genus *Potyvirus*), and *Potato virus X* (PVX) (genus *Potexvirus*) that has no vector but is transmitted by contact (contamination). Only 27 tubers were found to be free of the four viruses. The average incidences of the four viruses were 71%, 57%, 75% and 41%, respectively, in the 11 markets surveyed in seven districts. Mixed infections with viruses were common (Gildemacher *et al.*, 2009). These results were expected because there was no seed potato certification system in place in the country and farm-own seed was used to plant potato crops.

The situation in many other African countries is expected to be similar to Kenya. In October–December 2011, viruses were surveyed in potato crops of smallholder farmers in the south-western highlands of Tanzania in the regions of Mbeya (Kawetele, Kikondo, Umalia, Uyole) and Rungwe (Mwakaleli) *c.*100 km apart (Chiunga and Valkonen, 2013). Occurrence of PLRV, PVX, PVY, PVA and the aphid-transmissible *Potato virus M* (PVM) and *Potato virus S* (PVS) (genus *Carlavirus*) was determined in several local potato varieties in 16 fields ranging from 0.2 ha to 1 ha. Potato crops were 1–3 months old when leaves were sampled. Farmers grew often mixtures of varieties. Virus-like symptoms were observed in most fields and included yellowish-green mosaic, leaf rolling and veinal necrosis. Symptoms in tubers were not studied. Leaves from ten symptomatic and three symptomless plants were sampled from each field and tested by double antibody sandwich ELISA using virus-specific antibodies, including negative and positive controls in the assay. The average incidences of PVS, PLRV, PVX and PVM were 55%, 39%, 14% and 5%, respectively, in the plants tested. PVS and PLRV were detected in all fields sampled, and PVX and PVM were found in most fields. PVA and PVY were only detected in two localities. Co-infection with PVS and PLRV was detected in 14% of the tested plants, and mixed infections involving three or four viruses were detected in 5% of the plants. Many samples were pressed on FTA cards (Ndunguru *et al.*, 2005) and analysed by reverse-transcription PCR using virus-specific primers designed to amplify the coat protein (CP) encoding region of the viruses, which re-confirmed detection of the viruses. The PCR products from three to five isolates per virus were sequenced, which revealed few if any differences among isolates of the viruses. When compared with viruses detected in potatoes in other countries and continents, the CP sequence similarities suggested that viruses might have been introduced to Tanzania through potato trade or through introducing new cultivars without adequate indexing for viruses (Chiunga and Valkonen, 2013).

As discussed concerning viral diseases of sweetpotato, it is likely that all potato viruses have not yet been detected and fully elucidated. Also here, the sRSA approach will be very useful. New viral species have been detected in potato, including new nepo-, crini-, viti-, enamo-, torrado-like- and ophio-viruses as well as an ST9-like satellite virus (CIP, 2012, unpublished). Many new viruses have been discovered in symptomless plants, and others have been found in mixed infections with known viruses and associated with symptoms. The fact that some of these new viruses are symptomless and common in the Andes may have provided a means for them to spread globally unnoticed. CIP has used sRSA also to resolve the taxonomic status of previously described viruses, for example *Potato black ring-spot virus* (PBRSV) (Souza-Richards *et al.*, 2014) and strains of *Andean potato latent virus* (APLV) resulting in the splitting of APLV into two distinct viral species, APLV and Andean potato mild mosaic virus (Kreuze *et al.*, 2013). New variants of PVY and PVX have been detected.

32.4 Control of Virus Diseases

The aforementioned results from Kenya and Tanzania suggest the need for development of virus control schemes in potato crops, including the nascent, domestic certified seed potato production. In the survey carried out by Fuglie (2007) in 47 developing countries, respondents expressed three different strategies for reducing yield losses caused by viruses in potato. One strategy was through formal seed systems providing virus-free seed.

Domestic production of healthy seed potatoes requires possibilities to maintain healthy plant stocks. Production of healthy seed potatoes from locally adapted and accepted varieties requires that clones free of viruses and other pathogens are first produced. Meristem tip culture is used as the conventional *in vitro* technique to obtain virus-free clones from infected potato plants (El-Amin *et al.*, 1994). However, some viruses invade meristems efficiently and are difficult if not impossible to eliminate mechanically by meristem tip culture. Therefore, cryotherapy is recommended for enhancement of virus elimination (Wang and Valkonen, 2009). In cryotherapy, infected cells are eliminated by the lethal effects of the ultra-low temperature of liquid nitrogen (−196°C) and/or subsequent warming; no mechanical removal is necessary. Easier excision of larger shoot tips and the higher proportion of virus-free regenerants obtained with cryotherapy are other advantages of the technique. Hence, fewer regenerated shoots need to be virus-indexed to identify those free of the pathogen (Wang and Valkonen, 2009).

However, seed systems will not work without a functional potato value chain (Rahko, 2012), which is currently not in place in most countries in sub-Saharan Africa. It will be of utmost importance to develop the local potato value chains, for which assistance of experts from the fields of social sciences and agricultural economics and politics is needed in order to develop the potato sector beyond the biological aspects.

Another approach proposed was to help farmers manage their own seed production (Fuglie, 2007). Accordingly, CIP has initiated 'Select-the-Best' projects with farmers, with the aim to help them to mark disease-free plants in the field and use the yield of those plant individuals as seed potatoes. This will be particularly important when valuable, healthy seed potatoes are available to plant the crop. The 'Select-the-Best' approach can help to delay degeneration of the crop by viruses and other pathogens when farm-own seed is used to plant the subsequent crops. The results show that positive selection can provide smallholder producers with a significant yield advantage (Gildemacher *et al.*, 2011; Schulte-Geldermann *et al.*, 2012).

A practice resembling the 'Select-the-Best' concept is adopted by subsistence sweetpotato growers who take cuttings from symptomless stems and manage to obtain a reasonable number of virus-free cuttings from a crop that is infected (Gibson *et al.*, 1997).

The most efficient, least laborious and cheapest way to control viruses is to grow virus-resistant varieties, as noted also in the survey of Fuglie (2007). There have been attempts to choose more virus-resistant or -tolerant local and introduced potato varieties for use in Kenya, and recently also in Tanzania, by local researchers in collaboration with CIP. Results in Kenya have shown that the local variety Asante is rather tolerant to PLRV, PVY and PVX, in contrast to the varieties Tigoni and Dutch Robyjn (Schulte-Geldermann *et al.*, 2012).

Recovery of certain sweetpotato landrace varieties from infection with SPVD in the field indicates a level of resistance to the disease. This type of resistance is most probably based on RNA silencing, the mechanism of antiviral defence that produces the virus-derived small RNAs utilized for diagnostics by sRSA (Cuellar *et al.*, 2009; Kreuze *et al.*, 2009). Some Namulonge Sweetpotato (NASPOT) cultivars produced in a breeding programme and released in Uganda have moderate levels of field resistance to SPVD (Mwanga *et al.*, 2002, 2009). A few sweetpotato breeding lines react to SPVD with shoot tip dieback, which is possibly due to a hypersensitive resistance response to SPFMV and/or SPCSV (Mwanga *et al.*, 2013). There are also wild species resistant to SPFMV, SPCSV or both (Karyeija *et al.*, 1998) but their utilization in breeding programmes is more challenging than the use of sweetpotato landraces. Landrace sweetpotatoes often express other traits that are preferred by farmers, which allows development of virus resistance in a genetic background that is likely to be accepted by farmers and end users. NASPOT 11 is a nutritionally improved sweetpotato cultivar with relatively good resistance to SPVD. It was released in Uganda based on a participatory breeding approach and is now known also with the name Tomulabula (Gibson *et al.*, 2011). It was derived from an open-pollinated seed of the Ugandan landrace variety New Kawogo that shows high field resistance to SPVD (Karyeija *et al.*, 2000b).

32.5 Conclusions

In East Africa, viruses comprise the only significant disease agents in sweetpotato. The most severe yield losses are caused by dual infection of plants with SPCSV and unrelated viruses, because the RNase3 protein produced by SPCSV eliminates antiviral defence (Cuellar *et al.*, 2009). Therefore, control of SPCSV is of utmost importance in the control of viral diseases in sweetpotato. SPCSV-resistant sweetpotato varieties remain as an important goal for sweetpotato breeding. Recent surveys have also highlighted the prevalence of DNA viruses in sweetpotato throughout Africa, and significant yield losses have been reported. Aetiology of these viruses, their distribution and control need to be elucidated. In potato, PLRV and PVY are the most widely spread and damaging viruses in potato crops. In East Africa, more detailed knowledge on viruses affecting potato crops is relatively limited and more surveys in different potato production areas are needed.

Healthy seed potatoes and sweetpotato vine cuttings available for planting are a prerequisite for successful production. Selection of the disease-free plants by farmers in the field for use as sources of planting material helps to limit the incidence of viruses. Production of pathogen-free clones from local sweetpotato and potato varieties can be enhanced by cryotherapy. Knowledge on resistance to the local viruses and virus strains is an important issue when choosing cultivars for production, and studying resistance requires that healthy stocks of the varieties are available.

Antibodies to the most common potato viruses are commercially available and their use for indexing seed potatoes is straightforward with the basic diagnostic capacity. In contrast, antibodies to sweetpotato viruses are commercially unavailable, and detection by serology is further complicated because many sweetpotato viruses are present in low titres and can be unequally distributed in the plant. The sRSA-based detection of viruses provides a fundamentally novel, universal method for detection of plant viruses and does not require advance knowledge of them. It is promising for indexing of the stock plants used for further propagation of planting material and can be adopted for use in African laboratories equipped to carry out molecular diagnostics. Also the technology is applied to identify novel viruses and analysis of virus variability throughout Africa, which will inform development of novel control methods and testing regimes to control the most harmful viruses for each region. In this respect, it will be important to advance surveys of viruses in wild vegetation in the proximity of potato and sweetpotato crops, in order to reveal sources of viruses that are harmful to the crops and plan schemes for elimination of the virus reservoirs.

References

Aritua, V., Bua, B., Barg, E., Vetten, H.J., Adipala, E. and Gibson, R.W. (2007) Incidence of five viruses infecting sweetpotatoes in Uganda: the first evidence of sweet potato caulimo-like virus in Africa. *Plant Pathology* 56, 324–331.

Aritua, V., Barg, E., Adipala, E., Gibson, R.W., Lesemann, D.E. and Vetten, H.J. (2009) Host range, purification, and genetic variability in *Sweet potato chlorotic fleck virus*. *Plant Disease* 93, 87–93.

Ateka, E.M., Njeru, R.W., Kibaru, A.G., Kimenju, J.W., Barg, E., Gibson, R.W. and Vetten, H.J. (2004) Identification and distribution of viruses infecting sweet potato in Kenya. *Annals of Applied Biology* 144, 371–379.

Barkley, N.A., Pinnow, D.L., Wang, M.L., Ling, K.S. and Jarret, R.L. (2011) Detection and classification of SPLCV isolates in the US sweetpotato germplasm collection via a real-time PCR assay and phylogenetic analysis. *Plant Disease* 95, 1385–1391.

Bi, Y., Tugume, A.K. and Valkonen, J.P.T. (2012) Small-RNA deep sequencing reveals *Arctium tomentosum* as a natural host of *Alstroemeria virus X* and a new putative emaravirus. *PLoS One* 7, e42758.

Chavi, F., Robertson, A.I. and Verduin, B.J.M. (1997) Survey and characterization of viruses in sweetpotato from Zimbabwe. *Plant Disease* 81, 1115–1122.

Chiunga, E. and Valkonen, J.P.T. (2013) First report on viruses infecting potato in Tanzania. *Plant Disease* 97, 1260.

Clark, C.A. and Hoy, M.W. (2006) Effects of common viruses on yield and quality of Beauregard sweetpotato in Louisiana. *Plant Disease* 90, 83–88.

Clark, C.A., Davis, J.A., Abad, J.A., Cuellar, W.J., Fuentes, S., Kreuze, J.F., Gibson, R.W., Mukasa, S.B., Tugume, A.K., Tairo, F. and Valkonen, J.P.T. (2012) Sweetpotato viruses: 15 years of progress on understanding and managing complex diseases. *Plant Disease* 96, 168–185.

Cuellar, W.J., Tairo, F., Kreuze, J.F. and Valkonen, J.P.T. (2008) Analysis of gene content in sweet potato chlorotic stunt virus RNA1 reveals the presence of the p22 RNA silencing suppressor in only a few isolates: implications for viral evolution and synergism. *Journal of General Virology* 89, 573–582.

Cuellar, W.J., Kreuze, J.F., Rajamäki, M.L., Untiveros, M., Crusado, K.R. and Valkonen, J.P.T. (2009) Elimination of antiviral defense by viral RNase III. *Proceedings of the National Academy of Sciences of the USA* 106, 10354–10358.

Cuellar, W.J., De Souza, J., Barrantes, I., Fuentes, S. and Kreuze, J.F. (2011) Distinct cavemoviruses interact synergistically with *Sweet potato chlorotic stunt virus* (genus *Crinivirus*) in cultivated sweet potato. *Journal of General Virology* 92, 1233–1243.

Cuevas, J.M., Delaunay, A., Visser, J.C., Bellstedt, D.U., Jacquot, E. and Elena, S.F. (2012) Phylogeography and molecular evolution of *Potato virus Y*. *PLoS One* 7, e37853.

De Souza, J., Fuentes, S., Savenkov, E.I., Cuellar, W. and Kreuze, J. (2013) The complete nucleotide sequence of Sweet potato C6 virus: a carlavirus lacking a cystein rich protein. *Archives of Virology* 158, 1393–1396.

El Amin, S., Valkonen, J.P.T., Bremer, K. and Pehu, E. (1994) Elimination of viruses and hypersensitivity to potato virus Y (PVY^{O}) in a local Sudanese potato stock (Zalinge). *American Potato Journal* 71, 267–272.

FAOSTAT (2013) Preliminary Production Data for 2011. Available at: http://faostat.fao.org (accessed 17 June 2013).

Fuentes, S., Heider, B., Tasso, R.C., Romero, E., Zum Felde, T. and Kreuze, J.F. (2012) Complete genome sequence of a potyvirus infecting yam beans (*Pachyrhizus* spp.) in Peru. *Archives of Virology* 157, 773–776.

Fuglie, K.O. (2007) Priorities for potato research in developing countries: results of a survey. *American Journal of Potato Research* 84, 353–365.

Gibson, R.W., Mwanga, R.O.M., Kasule, S., Mpembe, I. and Carey, E.E. (1997) Apparent absence of viruses in most symptomless field-grown sweet potato in Uganda. *Annals of Applied Biology* 130, 481–490.

Gibson, R.W., Mpembe, I., Alicai, T., Carey, E.E., Mwanga, R.O.M., Seal, S.E. and Vetten, H.J. (1998) Symptoms, aetiology and serological analysis of sweet potato virus disease in Uganda. *Plant Pathology* 47, 95–102.

Gibson, R.W., Mpembe, I. and Mwanga, R.O.M. (2011) Benefits of participatory plant breeding (PPB) as exemplified by the first-ever officially released PPB-bred sweet potato cultivar. *Journal of Agricultural Science* 149, 625–632.

Gildemacher, P., Demo, P., Barker, I., Kaguongo, W., Woldegiorgis, G., Wagoire, W., Wakahiu, M., Leeuwis, C. and Struik, P. (2009) A description of seed potato systems in Kenya, Uganda and Ethiopia. *American Journal of Potato Research* 86, 373–382.

Gildemacher, P., Schulte-Geldermann, E., Borus, D., Demo, P., Kinyae, P., Mundia, P. and Struik, P. (2011) Seed potato quality improvement through positive selection by smallholder farmers in Kenya. *Potato Research* 54, 253–266.

Hollings, M., Stone, O.M. and Bock, K.R. (1976) Purification and properties of Sweetpotato mild mottle, a whitefly born virus from sweetpotato (*Ipomoea batatas*) in East Africa. *Annals of Applied Biology* 82, 511–528.

Hwang, Y.T., Kalischuk, M., Fusaro, A.F., Waterhouse, P.M. and Kawchuk, L. (2013) Small RNA sequencing of *Potato leaf roll virus*-infected plants reveals an additional subgenomic RNA encoding a sequence-specific RNA-binding protein. *Virology* 438, 61–69.

Isakov, O., Modai, S. and Shomron, N. (2011) Pathogen detection using short-RNA deep sequencing subtraction and assembly. *Bioinformatics* 27, 2027–2030.

Janick, J. (2013) Development of New World crops by indigenous Americans. *HortScience* 48, 406–412.

Karyeija, R.F., Gibson, R.W. and Valkonen, J.P.T. (1998) Resistance to sweetpotato virus disease (SPVD) in wild East African *Ipomoea*. *Annals of Applied Biology* 133, 39–44.

Karyeija, R.F., Kreuze, J.F., Gibson, R.W. and Valkonen, J.P.T. (2000a) Synergistic interactions of a potyvirus and a phloem-limited crinivirus in sweet potato plants. *Virology* 269, 26–36.

Karyeija, R.F., Kreuze, J.F., Gibson, R.W. and Valkonen, J.P.T. (2000b) Two serotypes of Sweetpotato feathery mottle virus in Uganda and their interaction with resistant sweetpotato cultivars. *Phytopathology* 90, 1250–1255.

Kashif, M., Pietilä, S., Artola, K., Jones, R.A.C., Tugume, A.K., Mäkinen, V. and Valkonen, J.P.T. (2012) Detection of viruses in sweetpotato from Honduras and Guatemala augmented by deep-sequencing of small-RNAs. *Plant Disease* 96, 1430–1437.

Kreuze, J.F., Karyeija, R.F., Gibson, R.W. and Valkonen, J.P.T. (2000) Comparisons of coat protein gene sequences show that East African isolates of *Sweet potato feathery mottle virus* forms a genetically distinct group. *Archives of Virology* 145, 567–574.

Kreuze, J.F., Pérez, A., Untiveros, M., Quispe, D., Fuentes, S., Barker, I. and Simon, R. (2009) Complete viral genome sequence and discovery of novel viruses by deep sequencing of small RNAs: a generic method for diagnosis, discovery and sequencing of viruses. *Virology* 388, 1–7.

Kreuze, J., Koenig, R., De Souza, J., Vetten, H.J., Muller, G., Flores, B., Ziebell, H. and Cuellar, W. (2013) The complete genome sequences of a Peruvian and a Colombian isolate of *Andean potato latent virus* and partial sequences of further isolates suggest the existence of two distinct potato-infecting tymovirus species. *Virus Research* 173, 431–435.

Latvala-Kilby, S., Aura, J.M., Pupola, N., Hannukkala, A. and Valkonen, J.P.T. (2009) Detection of *Potato mop-top virus* in potato tubers and sprouts: combinations of RNA2 and RNA3 variants and incidence of symptomless infections. *Phytopathology* 99, 519–531.

Li, R., Gao, S., Hernandez, A.G., Wechter, W.P., Fei, Z. and Ling, K.S. (2012) Deep sequencing of small RNAs in tomato for virus and viroid identification and strain differentiation. *PLoS One* 7, e37127.

Ling, K.-S., Jackson, D.M., Harrison, H., Simmons, A.M. and Pesic-Van Esbroeck, Z. (2010) Field evaluation of yield effects on the USA heirloom sweetpotato cultivars infected by *Sweet potato leaf curl virus*. *Crop Protection* 29, 757–765.

Loconsole, G., Onelge, N., Potere, O., Giampetruzzi, A., Bozan, O., Satar, S., De Stradis, A., Savino, V., Yokomi, R.K. and Saponari, M. (2012a) Identification and characterization of citrus yellow vein clearing virus, a putative new member of the genus *Mandarivirus*. *Phytopathology* 102, 1168–1175.

Loconsole, G., Saldarelli, P., Doddapaneni, H., Savino, V., Martelli, G.P. and Saponari, M. (2012b) Identification of a single-stranded DNA virus associated with citrus chlorotic dwarf disease, a new member in the family *Geminiviridae*. *Virology* 432, 162–172.

Ma, M., Huang, Y., Gong, Z., Zhuang, L., Li, C., Yang, H., Tong, Y., Liu, W. and Cao, W. (2011) Discovery of DNA viruses in wild-caught mosquitoes using small RNA high throughput sequencing. *PLoS One* 6, e24758.

Mbanzibwa, D.R., Tairo, F., Gwandu, C., Kullaya, A. and Valkonen, J.P.T. (2011) First report of Sweetpotato symptomless virus 1 and Sweetpotato virus A in sweetpotatoes in Tanzania. *Plant Disease* 95, 224.

Miano, D.W., LaBonte, D.R., Clark, C.A., Valverde, R.A. and Hoy, M.W. (2006) First report of a begomovirus infecting sweetpotato in Kenya. *Plant Disease* 90, 832.

Mukasa, S.B., Rubaihayo, P.R. and Valkonen, J.P.T. (2003) Incidence of viruses and viruslike diseases of sweetpotato in Uganda. *Plant Disease* 87, 329–335.

Mukasa, S.B., Rubaihayo, P.R. and Valkonen, J.P.T. (2006) Interactions between a crinivirus, an ipomovirus and a potyvirus in coinfected sweetpotato plants. *Plant Pathology* 55, 458–467.

Mwanga, R.O.M., Yencho, G.C. and Moyer, J.W. (2002) Diallel analysis of sweetpotatoes for resistance to sweetpotato virus disease. *Euphytica* 128, 237–248.

Mwanga, R.O.M., Odongo, B., Niringiye, C., Alajo, A., Kigozi, B., Makumbi, R., Lugwana, E., Namukula, J., Mpembe, I., Kapinga, R., Lemaga, B., Nsumba, J., Tumwegamire, S. and Yencho, C.G. (2009) 'NASPOT 7', 'NASPOT 8', 'NASPOT 9 O', 'NASPOT 10 O', and 'Dimbuka-Bukulula' sweetpotato. *HortScience* 44, 828–832.

Mwanga, R.O.M., Yencho, G.C., Gibson, R.W. and Moyer, J.W. (2013) Methodology for inoculating sweetpotato virus disease: discovery of tip dieback, and plant recovery and reversion in different clones. *Plant Disease* 97, 30–36.

Ndunguru, J., Taylor, N.J., Yadav, J., Aly, H., Legg, J.P., Aveling, T., Thompson, G. and Fauquet, C.M. (2005) Application of FTA technology for sampling, recovery and molecular characterization of viral pathogens and virus-derived transgenes from plant tissues. *Virology Journal* 2, a45.

Njeru, R.W., Mburu, M.W.K., Cheramgoi, E., Gibson, R.W., Kiburi, Z.M., Obudho, E. and Yobera, D. (2004) Studies on the physiological effects of viruses on sweet potato yield in Kenya. *Annals of Applied Biology* 145, 71–76.

Njeru, R.W., Bagabe, M.C., Nkezabahizi, D., Kayiranga, D., Kajuga, J., Butare, L. and Ndirigue, J. (2008) Viruses infecting sweet potato in Rwanda: occurrence and distribution. *Annals of Applied Biology* 153, 215–221.

Paprotka, T., Boiteux, L.S., Fonseca, M.E.N., Resende, R.O., Jeske, H., Faria, J.C. and Ribeiro, S.G. (2010) Genomic diversity of sweet potato geminiviruses in a Brazilian germplasm bank. *Virus Research* 149, 224–233.

Rahko, J. (2012) Potato value chain in Tanzania. MSc thesis, University of Helsinki, Helsinki, Finland. Available at: https://helda.helsinki.fi/handle/10138/31434 (accessed 16 August 2013).

Rännäli, M., Czekaj, V., Jones, R.A.C., Fletcher, J.D., Mu, L., Davis, R., Gweyer, D.I., Coutts, B. and Valkonen, J.P.T. (2008) Molecular genetic characterization of *Sweet potato virus G* (SPVG) isolates from areas of the Pacific Ocean and southern Africa. *Plant Disease* 92, 1313–1320.

Rännäli, M., Czekaj, V., Jones, R.A.C., Fletcher, J.D., Mu, L., Davis, R. and Valkonen, J.P.T. (2009) Molecular characterization of *Sweet potato feathery mottle virus* (SPFMV) isolates from Easter Island, French Polynesia, New Zealand and southern Africa. *Plant Disease* 93, 933–939.

Roullier, C., Benoit, L., McKey, D.B. and Lebot, V. (2013) Historical collections reveal patterns of diffusion of sweet potato in Oceania obscured by modern plant movements and recombination. *Proceedings of the National Academy of Sciences of the USA* 110, 2205–2210.

Santala, J., Samuilova, O., Hannukkala, A., Latvala, S., Kortemaa, H., Beuch, U., Kvarnheden, A., Persson, P., Topp, K., Ørstad, K., Spetz, C., Nielsen, S.L., Kirk, H.G., Uth, J.G., Budziszewska, M., Wieczorek, P., Obrepalska-Steplowska, A., Pospieszny, H., Kryszczuk, A., Sztangret-Wisniewska, J., Yin, Z., Chrzanowska, M., Zimnoch-Guzowska, E., Jackeviciene, E., Taluntytė, L., Pūpola, N., Mihailova, J., Lielmane, I., Järvekülg, L., Kotkas, K., Rogozina, E., Sozonov, A., Tikhonovich, I., Horn, P., Broer, I., Kuusiene, S., Staniulis, J., Adam, G. and Valkonen, J.P.T. (2010) Detection, distribution and control of *Potato mop-top virus*, a soil-borne virus, in northern Europe. *Annals of Applied Biology* 157, 163–178.

Schulte-Geldermann, E., Gildemacher, P.R. and Struik, P.C. (2012) Improving seed health and seed performance by positive selection in three Kenyan potato varieties. *American Journal of Potato Research* 89, 429–437.

Sela, N., Luria, N. and Dombrovsky, A. (2012) Genome assembly of bell pepper endornavirus from small RNA. *Journal of Virology* 86, 7721.

Souza-Richards, R., Adams, I.P., Kreuze, J.F., De Souza, J., Cuellar, W., Dulleman, A.M., Van Der Vlugt, R.A.A., Glover, R., Hany, U., Dickinson, M. and Boonham, N. (2014) The complete genome sequence of two isolates of *Potato black ringspot virus* and their relationship to other isolates and nepoviruses. *Archives of Virology* 159, 811–815.

Tairo, F., Kullaya, A. and Valkonen, J.P.T. (2004) Incidence of viruses infecting sweet potato in Tanzania. *Plant Disease* 88, 916–920.

Tairo, F., Mukasa, S.B., Jones, R.A.C., Kullaya, A., Rubaihayo, P.R. and Valkonen, J.P.T. (2005) Unravelling the genetic diversity of the three main viruses involved in sweet potato virus disease (SPVD), and its practical implications. *Molecular Plant Pathology* 6, 199–211.

Tairo, F., Jones, R.A.C. and Valkonen, J.P.T. (2006) Potyvirus complexes in sweetpotato: occurrence in Australia, serological and molecular resolution, and analysis of the *Sweet potato virus 2* (SPV2) component. *Plant Disease* 90, 1120–1128.

Tugume, A.K., Mukasa, S.B. and Valkonen, J.P.T. (2008) Natural wild hosts of *Sweet potato feathery mottle virus* show spatial differences in virus incidence and virus-like diseases in Uganda. *Phytopathology* 98, 640–652.

Tugume, A.K., Cuellar, W.J., Mukasa, S.B. and Valkonen, J.P.T. (2010a) Molecular genetic analysis of virus isolates from wild and cultivated plants demonstrates that East Africa is a hotspot for the evolution and diversification of *Sweet potato feathery mottle virus*. *Molecular Ecology* 19, 3139–3156.

Tugume, A.K., Mukasa, S.B., Kalkkinen, N. and Valkonen, J.P.T. (2010b) Recombination and selection pressure in the ipomovirus *Sweet potato mild mottle virus* (*Potyviridae*) in wild species and cultivated sweetpotato in the centre of evolution in East Africa. *Journal of General Virology* 91, 1092–1108.

Untiveros, M., Fuentes, S. and Salazar, L.F. (2007) Synergistic interaction of *Sweet potato chlorotic stunt virus* (*Crinivirus*) with carla-, cucumo-, ipomo-, and potyviruses infecting sweetpotato. *Plant Disease* 91, 669–676.

Untiveros, M., Quispe, D. and Kreuze, J. (2010) Analysis of complete genomic sequences of isolates of the *Sweet potato feathery mottle virus* strains C and EA: molecular evidence for two distinct potyvirus species and two P1 protein domains. *Archives of Virology* 155, 2059–2063.

Valkonen, J.P.T. (2007) Potato viruses: economical losses and biotechnological potential. In: Vreugdenhil, D., Bradshaw, J., Gebhardt, C., Govers, F., MacKerron, D.K.L., Taylor, M.A. and Ross, H.A. (eds) *Potato Biology and Biotechnology*. Elsevier, Oxford, pp. 619–641.

Vetten, H.J., Hoyer, U., Maiss, E., Lesemann, D.E. and Jelkmann, W. (1996) Serological detection and discrimination of geographically diverse isolates of sweet potato sunken vein closterovirus. *Phytopathology* 86, 891A.

Wang, Q.C. and Valkonen, J.P.T. (2009) Cryotherapy of shoot tips: novel pathogen eradication method. *Trends in Plant Science* 14, 119–122.

Wasswa, P., Otto, B., Maruthi, M.N., Mukasa, S.B., Monger, W. and Gibson, R.W. (2011) First identification of a sweet potato begomovirus (sweepovirus) in Uganda: characterization, detection and distribution. *Plant Pathology* 60, 1030–1039.

Wu, Q., Luo, Y., Lu, R., Lau, N., Lai, E.C., Li, W.-X. and Ding, S.-W. (2010) Virus discovery by deep sequencing and assembly of virus-derived small silencing RNAs. *Proceedings of the National Academy of Sciences of the USA* 107, 1606–1611.

Wu, Q., Wang, Y., Cao, M., Pantaleo, V., Burgyan, J., Li, W.-X. and Ding, S.-W. (2012) Homology-independent discovery of replicating pathogenic circular RNAs by deep sequencing and a new computational algorithm. *Proceedings of the National Academy of Sciences of the USA* 109, 3938–3943.

Zhang, Y., Singh, K., Kaur, R. and Qiu, W. (2011) Association of a novel DNA virus with the grapevine vein-clearing and vine decline syndrome. *Phytopathology* 101, 1081–1090.

33 Insect Life Cycle Modelling (ILCYM) Software – A Generic Platform for Developing Insect Phenology Models, Population Analysis and Risk Mapping

H.E.Z. Tonnang,[1]* M. Sporleder,[2] H. Juarez,[2] P. Carhuapoma[2] and J. Kroschel[2]

[1]African Insect Science for Food and Health (ICIPE), Nairobi, Kenya; [2]International Potato Center (CIP), Lima, Peru

Abstract

Insect Life Cycle Modelling (ILCYM) software is an open-source computer-aided tool built on R and Java codes and linked to the uDig platform, which is a basic geographic information system (GIS). The software package consists of three modules: (i) the 'model builder'; (ii) the 'validation and simulations'; and (iii) the 'potential population distribution and risk mapping' module. ILCYM's model builder contains a library of several empirical linear and non-linear models, including the derivations of biophysical models, which have been proposed to define critical temperature effects in insects' development. Several statistical measures are incorporated in this module for estimation of parameters and comparison of models. The validation and simulations module demonstrates the application of the phenology models for estimating and simulating insect population abundance under constant and fluctuating temperatures. Outputs of the simulations are demographic parameters that include: (i) net reproduction rate; (ii) mean generation time; (iii) intrinsic rate of increase; (iv) finite rate of increase; and (v) the doubling time. Through these analyses, the biology and temperature requirements of insects are defined, and the effects of different diets or host plants in insects' demographic are assessed. The ILCYM-GIS component estimates three indices (the establishment risk index (EI), the generation index (GI) and the activity index (AI)) that guide in assessing the potential population distribution and abundance of a particular species. Several functionalities for vector (dbf to shape, raster to points, raster to polygons, extract by points) and raster analysis (merge, cut, mask, aggregate/disaggregate, re-class, describe, raster calculator) are part of the ILCYM-GIS component. Such features improve the manipulation of large datasets and help ILCYM's users in analysing and visualizing the risk assessment maps. The phenology model developed for the potato tuber moth *Phthorimaea operculella* (Zeller) (Lepidoptera: Gelechiidae) a worldwide pest of potato (*Solanum tuberosum* L.) is used to demonstrate resulting modelling outputs.

Keywords: climate change, integrated pest management (IPM), pest risk assessment, phenology model, population analysis and mapping

*htonnang@gmail.com

33.1 Introduction

In Sporleder *et al.* (2013) it was pointed out that the relationship between aspects of insects' life history and environmental variables (temperature) can be well described by process-based phenology models. Many authors (Uvarov, 1931; Brière *et al.*, 1999) have shown that the physiological response of most poikilothermic species such as insects is adapted to particular temperature ranges and temperature is often considered as a major abiotic factor influencing their abundance and to certain extent their presence in a region. In the literature three important issues have been listed for the design and analysis of temperature response experiments used to develop complete phenology models representing the life history of a species. These are: (i) analysis of development times or their inverse, development rates (Kramer *et al.*, 1991); (ii) estimation of development times at temperatures near thresholds (extremes) where excessive mortality may occur; and (iii) the relationship between individual variation and average developmental rates (Wagner *et al.*, 1984) and reproductive responses (Régnière, 1983).

Distribution modelling of biological species is an important tool for biologists to understand and predict past, current or future presence based either on ground observations or using data obtained from laboratory experiments. Applications include species under threats such as climate change, or predictions of epidemic outbreaks of organisms damaging to health or food production. The Insect Life Cycle Modelling (ILCYM) software developed by the International Potato Center (CIP), supports the development of process-based temperature-driven and age-stage structured insect phenology models and applies these models in its inbuilt geographic information system (GIS) (ILCYM-GIS) environment for insect potential population analysis and risk mapping (Sporleder *et al.*, 2013; Tonnang *et al.*, 2013). This chapter presents and discusses the main features of this open-source software and shares knowledge of insect population modelling research gained at the Agroecology/IPM (integrated pest management) sub-programme of CIP.

33.2 Materials and Methods

ILCYM's data types

Experimental data

ILCYM requires either complete life table or single-stage cohort study data collected over a wide range of constant temperatures in which the insect species develops (Sporleder *et al.*, 2013). In complete life table experiments a group of individuals of the same age is observed from the beginning of their egg stage until the death of all adults; phenological events (i.e. development to the next stage, mortality) are recorded at constant intervals. In single-stage cohort studies, individuals of the same age for each developmental stage, such as fresh eggs, freshly emerged, are exposed to different constant temperatures. The phenological events are monitored until the development to the next stage or death of all individuals in the cohort. ILCYM software only runs data in text format with the extension .txt (Fig. 33.1). Details on how different data types are used as inputs in ILCYM are described in Tonnang *et al.* (2013).

Climate data (temperature data)

The temperature data used for spatial simulations under present and future climate scenarios are obtained from WorldClim (http://www.worldclim.org/) and CCAFS (Climate Change, Agriculture and Food Security) (http://www.ccafs-climate.org) databases. They are layers (grids) with different spatial resolutions containing monthly average minimum, maximum and mean temperatures (Hijmans *et al.*, 2005; Ramirez and Jarvis, 2010; Kroschel *et al.*, 2013).

ILCYM's modelling framework

The approach used in ILCYM is to define sub-models that best describe the development and mortality in each immature life stage of an insect, with its variation between individuals in a population, as well as the senescence time and reproduction frequencies of adults according to temperature (Sporleder *et al.*, 2004, 2013).

(a) (b)

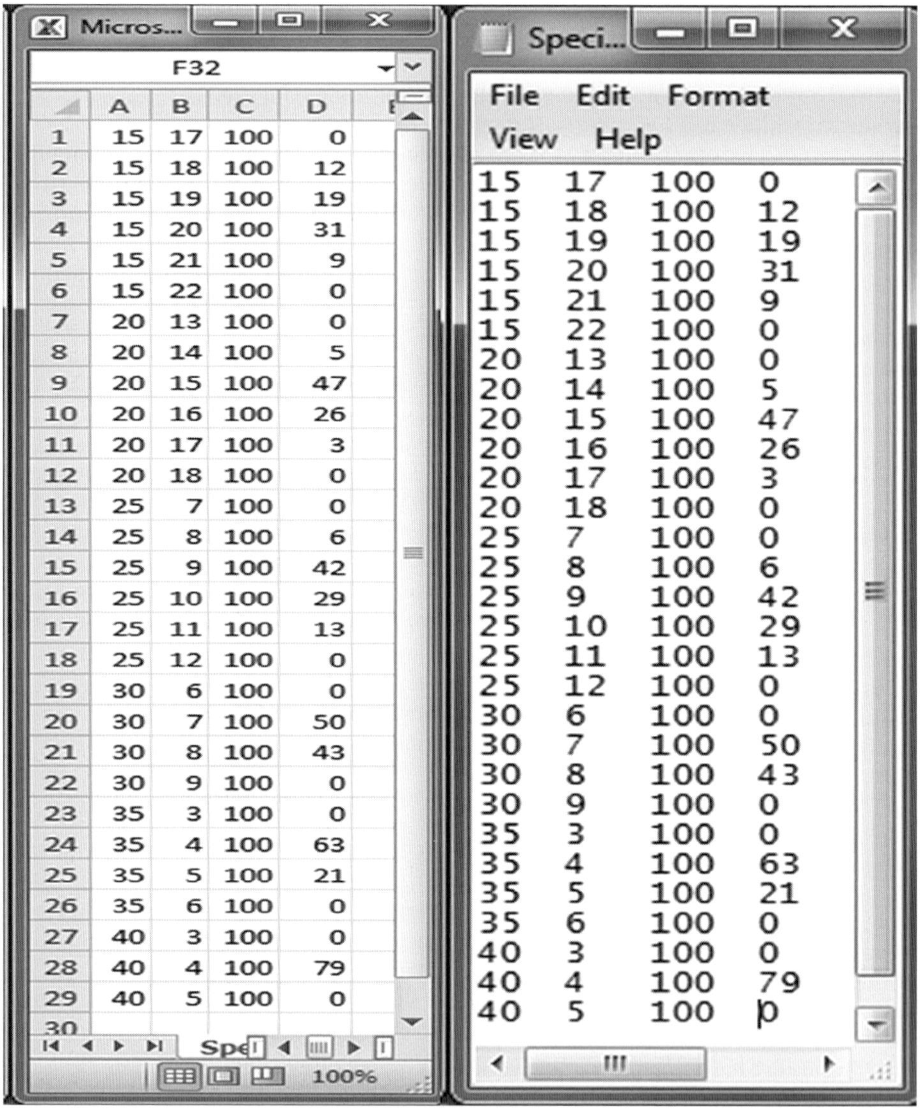

(a)

	A	B	C	D
1	15	17	100	0
2	15	18	100	12
3	15	19	100	19
4	15	20	100	31
5	15	21	100	9
6	15	22	100	0
7	20	13	100	0
8	20	14	100	5
9	20	15	100	47
10	20	16	100	26
11	20	17	100	3
12	20	18	100	0
13	25	7	100	0
14	25	8	100	6
15	25	9	100	42
16	25	10	100	29
17	25	11	100	13
18	25	12	100	0
19	30	6	100	0
20	30	7	100	50
21	30	8	100	43
22	30	9	100	0
23	35	3	100	0
24	35	4	100	63
25	35	5	100	21
26	35	6	100	0
27	40	3	100	0
28	40	4	100	79
29	40	5	100	0
30				

(b)

15	17	100	0
15	18	100	12
15	19	100	19
15	20	100	31
15	21	100	9
15	22	100	0
20	13	100	0
20	14	100	5
20	15	100	47
20	16	100	26
20	17	100	3
20	18	100	0
25	7	100	0
25	8	100	6
25	9	100	42
25	10	100	29
25	11	100	13
25	12	100	0
30	6	100	0
30	7	100	50
30	8	100	43
30	9	100	0
35	3	100	0
35	4	100	63
35	5	100	21
35	6	100	0
40	3	100	0
40	4	100	79
40	5	100	0

Fig. 33.1. Example for recording 'cohort study' data in a spreadsheet (a) and the same data saved as a txt-file (tab-delimited) (b) for use in ILCYM. The first column indicates the temperature evaluated (here a total of nine temperatures were tested); the second column indicates the evaluation intervals measured as days-after-experiment was set up; the third column indicates the number of insects used at each temperature; and the fourth column indicates the number of individuals that had developed to the next stage at each evaluation date (Tonnang *et al.*, 2013).

These sub-models are based on the experimental data obtained through temperature experiments as described in the section above. The reproduction model might include functions for different processes depending on the insect species under study (i.e. changing sex ratio in adults due to temperature, age-dependent reproduction frequencies, and temperature-dependent reproduction frequencies; Sporleder *et al.*, 2013). The overall modelling concept is factor–process based using same-shape distribution (Sharpe *et al.*, 1977) and rate summation (Worner, 1992) approaches, while considering temperature as the principal driver (factor) of the processes (Fig. 33.2).

ILCYM's modules

Model builder

ILCYM is module-based software organized in perspectives using wizard application for linking different modules and components. This tool is a complete library that includes nine functions for development time, 59 sub-models for developmental rate, 31 senescence sub-models, 45 mortality sub-models, 39 sub-models for total oviposition and six accumulative functions for relative oviposition. The model builder in ILCYM helps the user to build, manage and automate insect phenology models. It is a flexible interface for creating, visualizing, running, modifying, documenting and sharing models. Details on the merits of the model builder are described in detail in Tonnang *et al.* (2013).

Validation and simulation

The validation process in ILCYM is conducted by stochastic simulation. ILCYM stochastically simulates a user-defined number of life tables, each with a user-defined number of individuals, through rate summation and random determination for each individual's survival, and development to the next stage and sex. Simulation is also being performed in deterministic form in ILCYM (Sporleder *et al.*, 2013). During this process, cohorts are updated in daily intervals. Within each

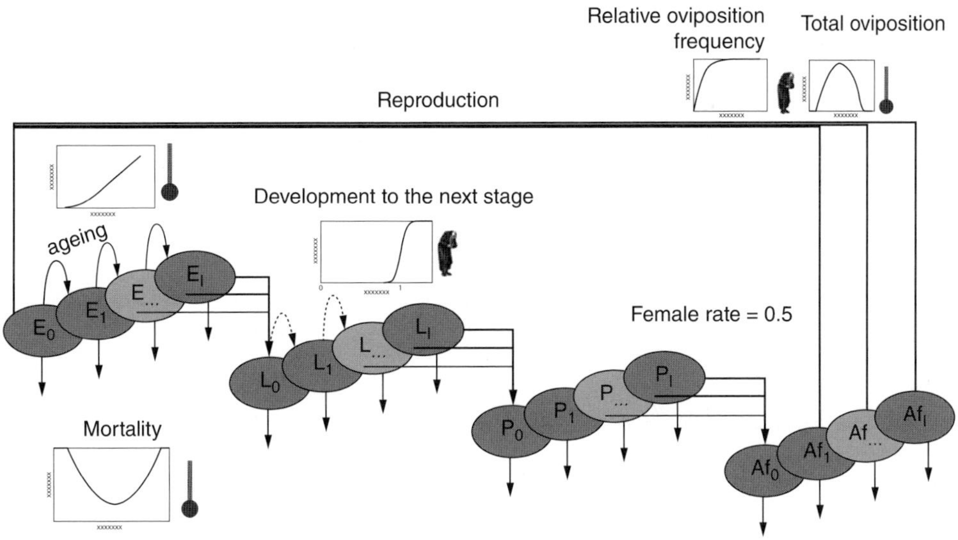

Fig. 33.2. A schematic picture of an ILCYM temperature-based phenology/population model using a rate summation and cohort up-dating approach. The population is structured in different life stages represented in a 'box car train' (i.e. eggs [E], larvae [L], pupae [P] and adults [Af = adult females]) and into groups of individuals of the same age within each life stage (i.e. cohorts) represented in a 'box car' (i.e. E_0, E_1, E_2, …, E_l). When running the model, each 'box car' is up-dated in daily intervals (i.e. the daily development rate is added to the physiological age (rate summation) and the number of individuals in the cohort is minimized by the daily mortality and the proportion of individuals that develop into the next life stage). (From Sporleder *et al.*, 2009.)

life stage the individuals are ageing. The physiological age of all individuals in one age class is calculated by rate-summation of the daily development rate using development rate sub-models. From each age class the daily mortality rate is subtracted. Age-dependent development from one stage to the next is calculated using the distribution curve of development. Reproduction is a function of temperature-dependent total reproduction per female and age-dependent relative reproduction frequency. Both simulations feature in ILCYM estimates of demographic parameters.

Potential population distribution and risk mapping

Potential population distribution and risk mapping in ILCYM are based on (daily/monthly) minimum and maximum temperatures that use a 15 min time-step length for considering the within-day temperature variability (Sporleder *et al.*, 2013). A cosine function is applied for calculating the temperature-dependent quantity in 15 min time steps. For selected study regions, temperature data for 1 year with their respective geographical coordinates are extracted from the database. The information is organized in matrices using the longitude as the columns and latitude as the rows. Thereafter, a point object is created for each geographical coordinate (longitude and latitude) in the form of a table with two columns (the first column includes the minimum temperatures and the second the maximum temperatures of the year), which is directly used for spatial simulation (Sporleder *et al.*, 2013). The program calculates stage-specific daily development and mortality rates for each immature development stage of the species, as well as evaluating adult senescence rates for females and males and total fecundity per female according to temperature. The obtained results are used to estimate three risk indices (the establishment risk index (EI), the generation index (GI) and the activity index (AI)) that guide in assessing the potential population distribution and abundance as described in Kroschel *et al.* (2013) for the potato tuber moth.

33.3 Results

Development and its distributions

Figure 33.3 shows ILCYM's window and analysis of the cumulative frequency distribution of developmental time for the potato tuber moth at different temperatures. In this case, the developmental time decreased significantly with increasing temperatures within the evaluated temperature ranges. There is no recommended standard density function for representing this variable. However, it is generally recognized that the density functions for various species are similar in shape and asymmetric about their mean times. Developmental density curves of insects are generally skewed to the right.

A linear model (dashed line) was fitted to data for median egg developmental rates (see Fig. 33.4). The slope and intercept derived from this model are used to calculate the theoretical lower developmental threshold temperature for the species at immature life stages. Based on these thresholds, the median period for each life stage is estimated. The sub-model (Fig. 33.4) is a four parameter mathematical expression that best described median developmental rates for the selected egg stage of the potato tuber moth.

Immature mortality and adult life span and fecundity

As the mortality of insects' increases at lower and higher temperatures, the majority of sub-model curves describing the effect of temperature on mortality of immature life stages have a form close to a U shape. Plotting the female life span (per day) against temperature models are similar to the functions used for describing the development rate. Generally, longevity of insect adults is expected to decrease significantly with increasing temperatures.

ILCYM provides non-linear sub-models to describe the age-specific fecundity frequency, which supports estimating the lower temperature threshold at which oviposition of a species still may occur (Fig. 33.5).

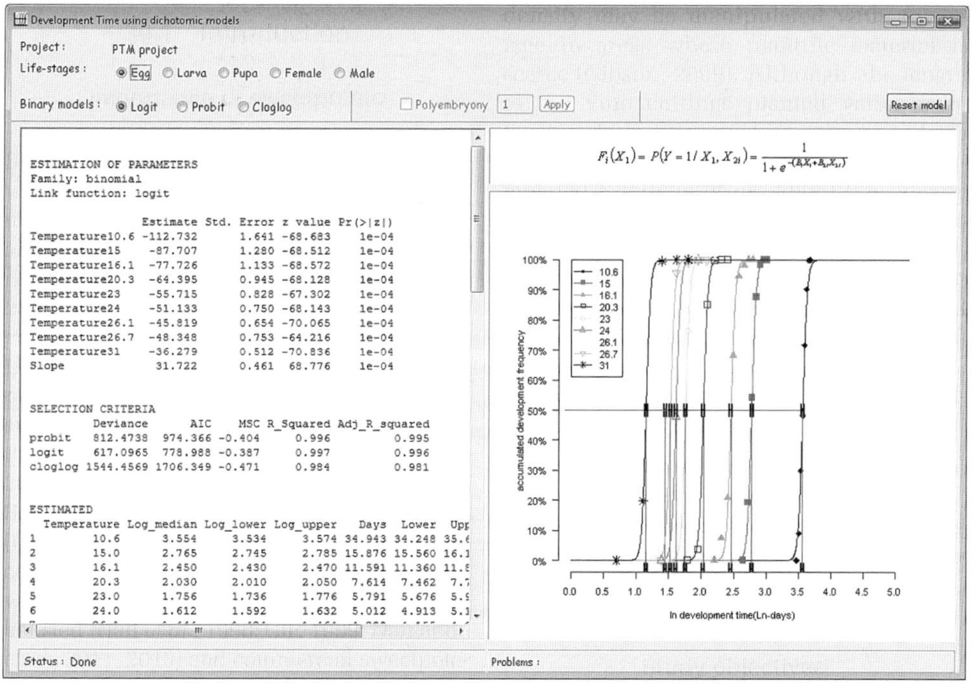

Fig. 33.3. ILCYM's software outputs showing mathematical expression and parameter values used to describe egg developmental times for the potato tuber moth (*Phthorimaea operculella*). The ANOVA (analysis of variance) table and selection criterion for comparing the best-fit distributed functions are shown. The curves representing the accumulated development frequency against development time (in days) at different temperatures are displayed.

Using the best-fit model obtained, the average fecundity per female can be estimated. The selected sub-model also can guide in determining the number of days used by the female to oviposit a certain percentage of its eggs. The highest and lowest values of fecundity are also estimated. In general, the fecundity increases with rising temperatures.

Overall phenology model

The overall phenology model developed for the species under investigation is presented in the project summary (Fig. 33.6); it shows for each life stage, the parameters and sub-models selected during project development. It is recommended that ILCYM's users open this file after completing the development of the phenology model, to check that all the accepted functions and sub-models are properly saved and ready for simulations.

Demographic parameters

Outputs of the simulations (stochastic and deterministic) are life table parameters that include: (i) intrinsic rate of increase; (ii) gross reproduction rate; (iii) finite rate of increase; (iv) net reproduction rate; (v) mean generation time; and (vi) the doubling time. Users may decide to simulate the population growth parameters at different constant temperatures (Fig. 33.7).

The net reproduction rate is expected to be significantly different at all temperatures. Similar behaviour may apply to the intrinsic rate of natural increase. The quantity may increase almost linearly with increasing temperature to reach a maximum at a particular temperature and then decrease at the upper temperature limits for development. In most cases the curve presents an asymmetrical dome-shaped pattern. Sometimes, the finite rate of increase may have a shape similar to

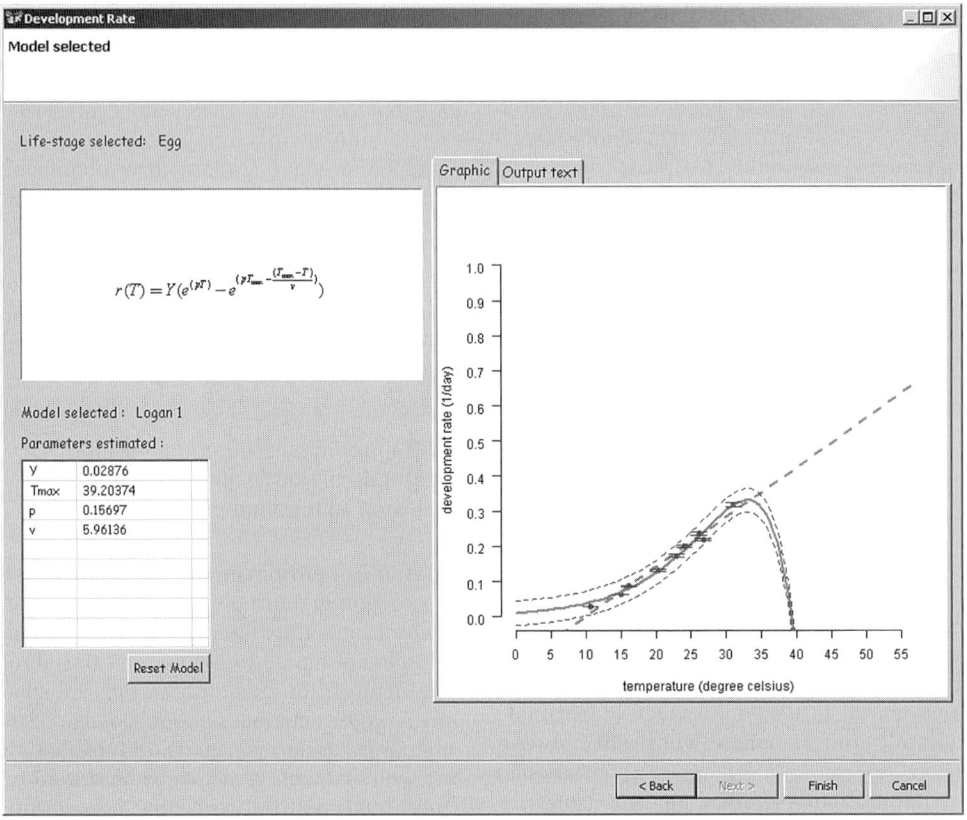

Fig. 33.4. ILCYM's software outputs showing mathematical expression and parameter values used to describe the potato tuber moth (*P. operculella*) egg developmental rate. The graph representing the development rate (1/day) against temperature (degree Celsius) is displayed.

the intrinsic rate of population increase. The values for the shortest and longest mean generation time can also be estimated at particular temperature.

Distribution maps

Applying and linking the overall phenology model with the GIS module in ILCYM produces maps indicating the potential establishment and abundance of insect species representing the three risk indices (Fig. 33.8). For the GI, the software estimates the number of generations per year the insect species could potentially complete, based on the prevailing temperatures. The process of determining the degree to which the simulated regional or global maps are accurate is conducted through comparisons of observed occurrence data and simulated outputs, as described by Kroschel *et al.* (2013). Such an approach in validating maps is original (a different data type and format is used) and has to be done for each species under investigation and according to available data sets.

33.4 Discussion

The ILCYM software has implemented a novel approach capable of inputting biological data obtained experimentally in the laboratory on the life history of insects at different temperatures and yielding complete phenology models as well as global and regional

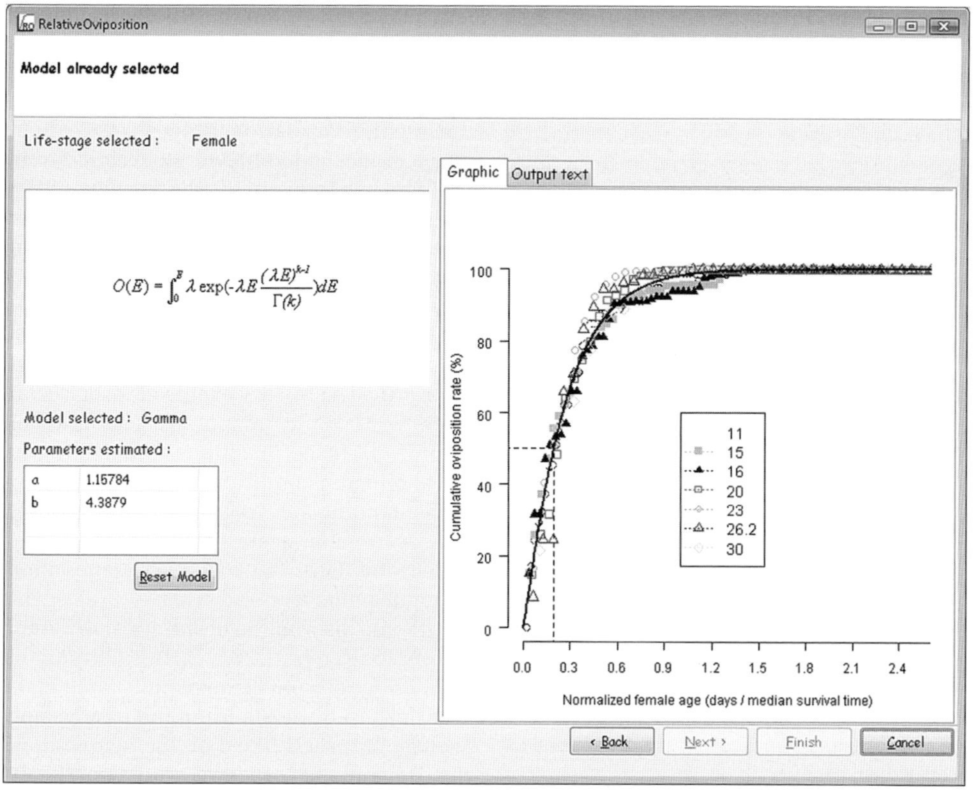

Fig. 33.5. ILCYM's software outputs showing mathematical expression and parameter values used to describe the potato tuber moth (*P. operculella*) female relative oviposition frequency. The graph representing the cumulative oviposition rate (%) against normalized female age (days/median survival time) at different temperatures (degree Celsius) is displayed.

risk maps. ILCYM's starting point for developing phenology models is fitting dichotomy or exponential functions to development time rather than to the developmental rate. Bentz *et al.* (1991) showed that fitting developmental rate curves directly to the reciprocal of laboratory development time data results in large errors. These errors are due to transformation of the errors' variance that result from direct use of the reciprocal of the laboratory data which does not occur when curves are fitted directly to development time data. Phenology models obtained through fitting to laboratory development times rather than rates offer a clear advantage due to greater accuracy at extreme temperatures (lower and upper limits).

Variation in insect development is found in both the laboratory and the field. To account for variability ILCYM applies the distributional concept developed by Sharpe *et al.* (1977), which assumes that variation in developmental rates at constant temperature is the result of variability in the concentration of rate-controlling enzymes and developmental rate variance is proportional to the mean developmental rate. With these features, ILCYM allows the development and analysis of individual-based and cohort-based phenology models. In individual-based models, development, survival and oviposition are modelled for a collection of individuals, each having its own assigned traits and going through successive life stages at its own individual pace. In contrast, in cohort-based modelling, ILCYM considers the insects in groups or cohorts.

In reality, fitting complex mathematical expressions using either maximum likelihood

Fig. 33.6. ILCYM's project summary file. It contains the overall phenology model that includes all best selected and saved mathematical expressions and their estimated parameters for representing the life history (developmental time, development rate, mortality, adult senescence and reproduction) of the species under investigation. This file is used for simulations.

estimation or least-squares estimation and treatment medians is difficult and sometimes leads to issues of over parameterization and lack of convergence (Schoolfield *et al.*, 1981). ILCYM offers some solutions to the problem: users may be requested to increase the amount of data or the inbuilt software features allow the use of censored data which may make the analysis possible. In some cases sub-model convergence with certain initial parameters may remain an issue. However, ILCYM provides an option to automatically modify the initial parameters of sub-models to ease the convergence. In addition, two fitting algorithms are incorporated in this software. Inclusion in ILCYM of an explicit likelihood framework with common error assumptions allows comparison of multiple sub-models, potentially representing different physiological hypotheses, via the Akaike International Criterion.

Validation used as a generic term is often defined as the degree to which a model output matches independent data sets (Rykiel, 1996). However, ILCYM applies validation as an assurance of the appropriateness of the model for its intended use. This is because during the validation process, the phenology model is simulated stochastically to predict development of the species under fluctuating temperatures and to compare this to experimental results obtained in similar conditions. Once the outcome is satisfactory based on ILCYM's inbuilt statistics, the obtained phenology models present a vast range of utilities. They are great instruments for scaling life history events of insects over wide spatial and temporal scales and for examining insect reactions to changing or novel climatic conditions and for providing understanding for ecological investigations at a reasonable scale of host plant–insect synchrony. They can be used for population dynamics analyses and used to project the geographic distribution of species at different ranges (Safranyik *et al.*, 2010). Outputs of the simulations are demographic parameters that help in understanding the biology and ecology of insects based on the temperature requirements of insects as well as of the effects of different diets or host plants.

According to Sporleder *et al.* (2013) climate change is inevitably causing serious challenges for the development and interpretation of pest risk maps. In this context, proper incorporation of features in ILCYM for climate change analysis was crucial. Important difficulties in the existing tools are the type of data resolution for mapping and the selection of appropriate emission scenario (Sporleder *et al.*, 2013). In ILCYM it was opined that the coarse resolution of 100 km or 200 km (or even more) of general circulation model (GCM) projections is simply not realistic for assessing the distribution of herbivore insects on a landscape level, particularly in the tropics, where orographic and climatic conditions vary significantly across relatively small distances (Sporleder *et al.*, 2013). Ramirez and Jarvis (2010) indicated that changes in topography and climate variables are not the only factors accounting for variability but also soils and socio-economic drivers, which vary over small distances, influence ecosystems and

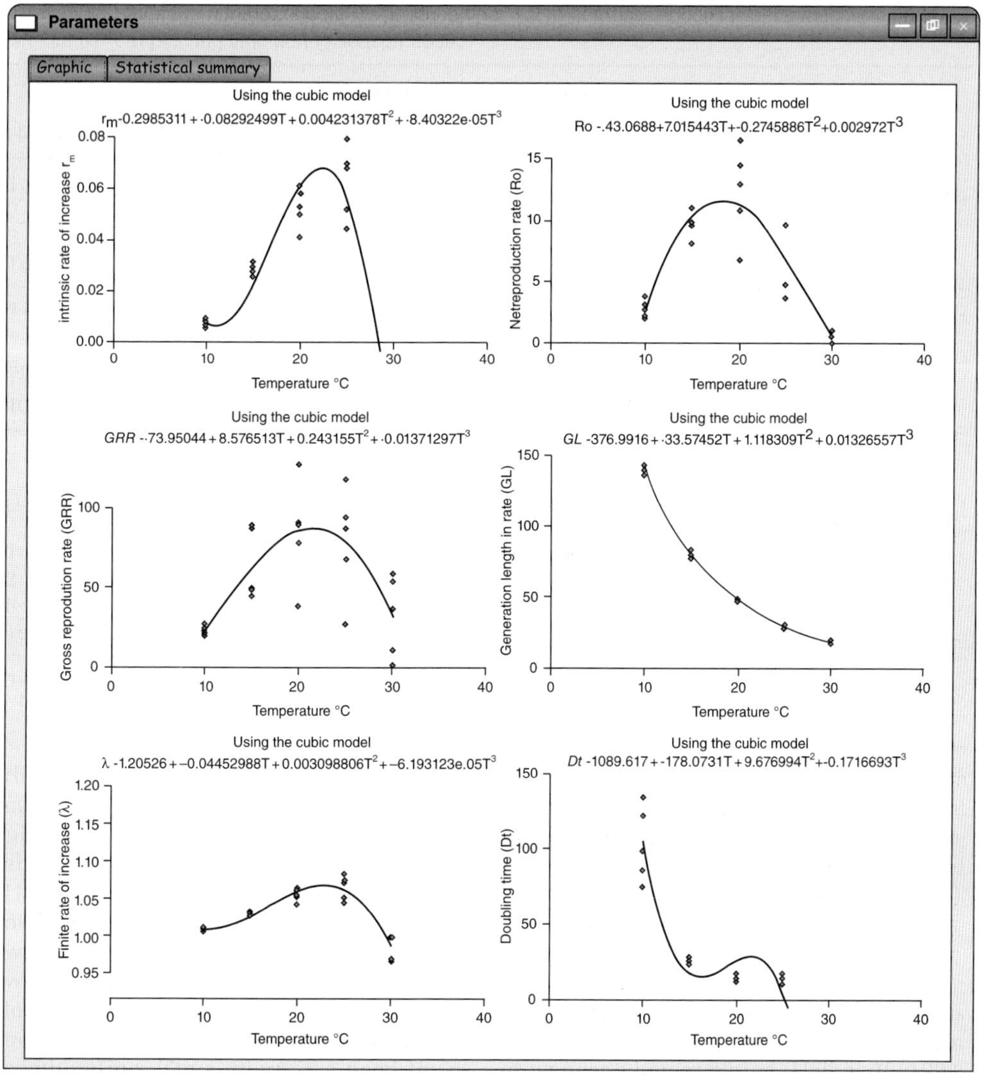

Fig. 33.7. ILCYM's outputs for stochastic simulations conducted at different constant temperatures. The values represent the demographic parameters (net reproduction rate, mean generation time, intrinsic rate of increase, finite rate of increase and the doubling time) obtained at the respective temperatures. Polynomial curves are fitted to the values of each parameter for capturing the trend. Mathematical expression representing the curves is provided.

increase uncertainties, making the forecasting and assessment models more inaccurate and complicated to calibrate. These are the reasons that stimulated ILCYM developers to use downscaled temperature datasets rather than direct outputs from GCM. The software offers options for inputting different downscale datasets of any emission scenario (A1, A2, B1 and B2) of the year 2020, 2030, 2040, …, 2100 (Sporleder *et al.*, 2013).

ILCYM requires input data (life table) of the appropriate quantity and quality, the collection of which is quite demanding and time consuming. This might be seen as a limiting factor, and deter some researchers from using the software for their studies. When using ILCYM for

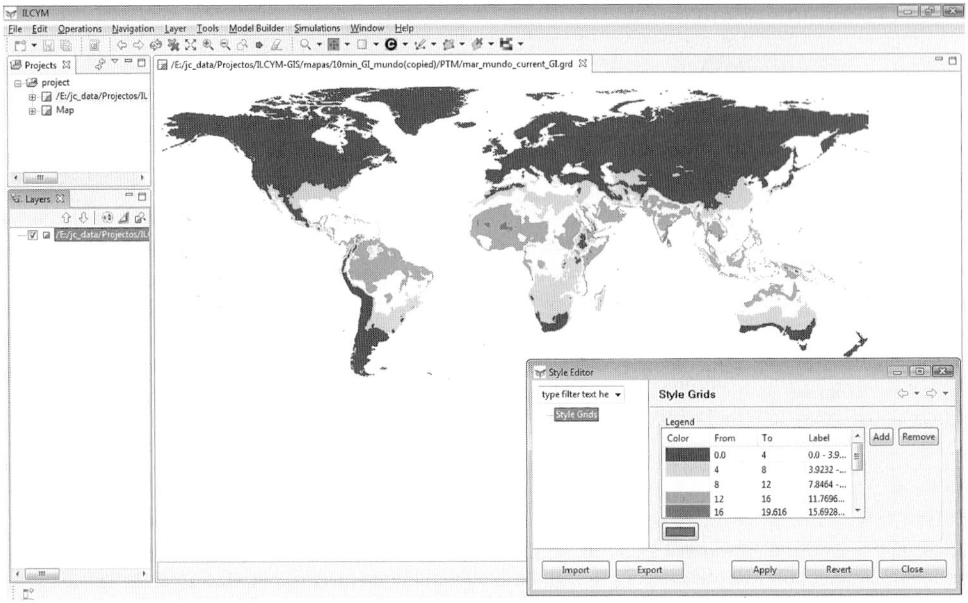

Fig. 33.8. ILCYM's 'risk mapping' interface; here showing the generation index (GI) simulated for the potato tuber moth (*P. operculella*) globally. The number of generations the pest might develop per year is visualized by different colours (reproduced here in black and white); the appearance of the map (ranges, colours) can be modified using the 'Style Editor'. In the same way, ILCYM produces maps for the species establishment and activity.

determining a species' potential environmental suitability, the program specifically guides the user to understand the temperature effects that limit the geographical distribution of the species and to identify the temperature optima that favour population growth. This approach therefore takes temperature as the only influencing variable; it is therefore recommended that effects of other abiotic factors (i.e. rainfall, humidity) as well as species-specific limiting factors and environmental stresses should be quantified and included in the concept for better accuracy.

Acknowledgement

The authors gratefully acknowledge the financial support for developing the ILCYM software that was provided by the Federal Ministry of Co-operation and Development (BMZ), Germany, for the project 'Predicting climate-change-induced vulnerability of African agricultural systems to major insect pests through advanced insect phenology modelling, and decision aid development for adaptation planning'.

References

Bentz, B.J., Logan, J.A. and Amman, G.D. (1991) Temperature-dependent development of the mountain pine beetle (Coleoptera: Scolytidae) and simulation of its phenology. *The Canadian Entomologist* 123, 1083–1094.

Brière, J.F., Pracros, P., le Roux, A.Y. and Pierre, J.S. (1999) A novel rate model of temperature-dependent development for arthropods. *Environmental Entomology* 28, 22–29.

Hijmans, R.J., Cameron, S.E., Parra, J.L., Jones, P.G. and Jarvis, A. (2005) Very high resolution interpolated climate surfaces for global land areas. *International Journal of Climatology* 25, 1965–1978.

Kramer, D.A., Stinner, R.E. and Hain, F.P. (1991) Time versus rate in parameter estimation of nonlinear temperature-dependent development models. *Environmental Entomology* 20, 484–488.

Kroschel, J., Sporleder, M., Tonnang, H.E.Z., Juarez, H., Carhuapoma, P., Gonzales, J.C. and Simon, R. (2013) Predicting climate-changed-caused changes in global temperature on potato tuber moth *Phthorimaea operculella* (Zeller) distribution and abundance using phenology modeling and GIS mapping. *Journal of Agricultural and Forest Meteorology* 170, 228–241.

Ramirez, J. and Jarvis, A. (2010) Downscaling global circulation model outputs: the delta method decision and policy analysis. *Working Paper* 1. International Center for Tropical Agriculture (CIAT), Cali-Palmira, Valle del Cauca, Colombia.

Régnière, J. (1983) An oviposition model for the spruce budworm, *Choristoneura fumiferana* (Lepidoptera: Tortricidae). *The Canadian Entomologist* 115, 1371–1382.

Rykiel, E.J., Jr (1996) Testing ecological models: the meaning of validation. *Ecological Modeling* 90, 229–244.

Safranyik, L., Carroll, A.L., Régnière, J., Langor, D.W., Riel, W.G., Shore, T.L., Peter, B., Cooke, B.J., Nealis, V.G. and Taylor, S.W. (2010) Potential for range expansion of mountain pine beetle into the boreal forest of North America. *The Canadian Entomologist* 142, 415–442.

Schoolfield, R.M., Sharpe, P.J.H. and Magnuson, C.E. (1981) Non-linear regression of biological temperature-dependent rate models based on absolute reaction-rate theory. *Journal of Theoretical Biology* 88, 719–731.

Sharpe, P.J.H., Curry, G.L., DeMichele, D.W. and Cole, C.L. (1977) Distribution model of organism development times. *Journal of Theoretical Biology* 66, 21–38.

Sporleder, M., Kroschel, J., Gutierrez Quispe, M.R. and Lagnaoui, A. (2004) A temperature based simulation model for the potato tuberworm, *Phthorimaea operculella* Zeller (Lepidoptera; Gelechiidae). *Environmental Entomology* 33, 477–486.

Sporleder, M., Chavez, D., Gonzales, J.C., Juarez, H., Simon, R. and Kroschel, J. (2009) ILCYM – Insect life cycle modeling: software for developing temperature-based insect phenology models with applications for regional and global pest risk assessments and mapping. In: *15th Triennial Symposium of the International Society for Tropical Root Crops (ISTRC)*, 2–6 November, International Potato Center (CIP), Lima, Peru. ISTRC, Lima, pp. 216–223.

Sporleder, M., Tonnang, T.Z.H., Carhuapoma, P., Gonzales, J.C., Juarez, H. and Kroschel, J. (2013) Application and use of insect phenology modeling for invasive species for regional and global risk assessments under future climate change scenarios. In: Peña, J. (ed.) *Potential Invasive Pests of Agricultural Crops*. CAB International, Wallingford, UK, pp. 412–426.

Tonnang, E.Z.H., Juarez, H., Carhuapoma, P., Gonzales, J.C., Mendoza, D., Sporleder, M., Simon, R. and Kroschel, J. (2013) ILCYM – *Insect Life Cycle Modeling.* A software package for developing temperature-based insect phenology models with applications for local, regional and global analysis of insect populations and mapping. International Potato Center (CIP), Lima, 193 pp.

Uvarov, B.P. (1931) Insects and climate. *Transactions of the Entomological Society of London* 79, 1–247.

Wagner, T.L., Wu, H.I., Sharpe, P.J.H. and Coulson, R.N. (1984) Modeling distributions of insect development time: a literature review and application of the Weibull function. *Annals of the Entomological Society of America* 77, 475–487.

Worner, S.P. (1992) Performance of phenological models under variable temperature regimes: consequences of the Kaufmann or rate summation effect. *Environmental Entomology* 21, 689–699.

34 *Phytophthora infestans* Population Changes in Kenya Pose Challenges to Existing Potato Blight Control Strategies

M. Nyongesa,[1]* C. Lung'aho,[2] L. Wasilwa,[3] M. Mbiyu,[1] J. Onditi[1] and S. Otieno[1]

[1]*Kenya Agricultural and Livestock Research Organization (KALRO), Limuru, Kenya;* [2]*International Potato Center Sub-Saharan Africa (CIP-SSA), Chimoio, Mozambique;* [3]*KALRO, Nairobi, Kenya*

Abstract

Potato blight caused by *Phytophthora infestans* remains the leading biotic constraint to production of potato (*Solanum tuberosum*), the second most important food crop in Kenya. Until recently, the local population of *P. infestans* in Kenya was a typical US-1 clonal lineage of A1 mating type associated with the worldwide dispersal of the pathogen in the 1970s. Up to this point, blight management strategies were based on key phenotypic characteristics of the US-1 clonal lineage, namely fungicide sensitivity and virulence to the potato host. For example, although large-scale resistance to phenylamide fungicides (e.g. metalaxyl) has been widely reported globally, field resistance to metalaxyl is yet to be reported in Kenya. Results from a study employing molecular markers on isolates of *P. infestans* obtained from Kenya have indicated occurrence of new genotypes in high frequencies. Whether the recently occurring genotypes will upset blight control strategies already in place is currently unknown. Based on what is known about these newly occurring genotypes, this chapter puts into perspective the implications of these changes to blight control. The chapter also relates these changes to available data on trends in *P. infestans* populations in the region. Findings indicate that efforts to mitigate the impact posed by these changes are likely to be hampered by cross-border trade in potatoes as well as shifts in weather patterns. A programme comprising regular monitoring of the composition of local *P. infestans* populations, and regular screening of local isolates for sensitivity to fungicides and their pathogenicity on existing potato germplasm is needed to ensure that strategies to combat potato blight evolve in tandem with emerging pathogen population patterns.

Keywords: integrated pest management (IPM), late blight management, metalaxyl resistance, *Phytophthora infestans*, population changes, *Solanum tuberosum*

34.1 Introduction

Since the mid-19th century blight instigated famine in Ireland that led to massive starvation and up to 1.5 million people emigrating (Bourke, 1993), blight remains a menace to potato farming globally. Efforts to control blight based on either host resistance or fungicides

*moses.nyongesa@kalro.org

continue to be undermined by the rapidly changing populations of the blight pathogen *Phytophthora infestans* (Mont.) De Bary. Notably, conventional breeding of potato for resistance to blight conferred by single major genes from wild *Solanum* species failed to get traction when virulent strains soon emerged to which the new varieties became susceptible (Black *et al.*, 1953; Black, 1970). The alternative was to implement breeding strategies aimed at achieving host resistance conditioned by multiple major genes in a single variety. Yet, these efforts too have been negated by occurrence of highly variable strains of *P. infestans*, some of which adapt to the hosts leading to breakdown of resistance and eventual disease epidemics. To date, durable host resistance to blight remains largely elusive. As such, fungicides are still a conventional and the preferred method of managing potato blight to support cultivation of susceptible but popular potato cultivars in many counties in the world, including those in the developing world (Kaguongo *et al.*, 2008). One of the disadvantages associated with this strategy is the tendency for strains within *P. infestans* populations to develop resistance against certain fungicide formulations (e.g. phenylamides) following frequent use (Gisi *et al.*, 2011).

34.2 Sources of Variation Globally

Over the last two decades, the philosophy of blight control has been broadened to encompass an understanding of pathogen population dynamics including the forces driving genetic changes. Evidence shows that until the early 1990s, *P. infestans* isolates obtained from regions outside Mexico belonged to the US-1 clonal lineage of the A1 mating type (Goodwin *et al.*, 1992). Accordingly, propagation by the organism in those regions was only by clonal means characterized by limited genetic variation. This situation has changed in many regions in the world where displacements of the 'old late blight' by new genetically more variable strains (including A2 mating types) has occurred (Spielman *et al.*, 1991; Fry *et al.*, 1993). Authors have alluded to there being a conspicuous relationship between these blight changes with movement of potato stocks along international trading channels for fresh potato (Spielman *et al.*, 1991; Fry *et al.*, 1993). This scenario illustrates changes in populations occurring as a result of arrival of foreign clonal genotypes. However, further population changes are inevitable when the newly arrived genotypes are sexually compatible with pre-existing ones giving rise to unique genotypes via sexual reproduction (Drenth *et al.*, 1994). The present situation in north-western European countries best reflects this source of variation in *P. infestans* populations (Hannukkala *et al.*, 2007; Fry *et al.*, 2009; Brurberg *et al.*, 2011). Mutations associated with over-reliance on fungicides with a single target site in *P. infestans* like metalaxyl have also been another source of variation in predominantly clonal populations (Gisi *et al.*, 2011).

34.3 Sources of Variation within the Kenyan *P. infestans* Population

With regards to sexual reproduction, much of what has underpinned the population dynamics of *P. infestans* in European countries has remained excluded in the Kenyan case. All the population studies, including the most recent one that employed simple sequence repeat (SSR) markers, have failed to show the occurrence of any A2 mating types in the country (Hohl, 1998; Vega-Sanchez *et al.*, 2000; Pule *et al.*, 2013). It follows that sexual reproduction cannot be invoked to account for the changes occurring within the local population. Instead, assessments involving the typical factors driving changes within clonal populations become plausible. These sources of variation may include transfer of novel genotypes via potato tubers, fungicide-associated mutation and host-resistance-induced mutation. Previous authors have shown that although Kenyan potato growers use fungicides as the priority intervention in combatting potato blight, a majority of them use negligible amounts of these products compared with counterparts in European countries (Nyankanga *et al.*, 2004). Arguably, the logic that the amount of fungicides used by

Kenyan farmers in a typical growing season in a single farm does not reach the threshold necessary to induce pathogen mutation, is therefore plausible. Further, this view is consistent with findings of previous studies that have seldom reported occurrence of metalaxyl-insensitive strains in Kenya (Olanya *et al.*, 2001).

There are scant data to support a conclusive determination of the actual causes of population changes within *P. infestans* in Kenya. However, it can be inferred from what is presently known that the observed changes in *P. infestans* are due in part to introduction of foreign strains and partly as a result of adaptation by native genotypes to the primary host. Introduction of foreign strains can be attributed to the free movement of potato between countries within the East African region, while adaptation of native genotypes is premised on the evidence from studies of isolates from the region showing specificity of *P. infestans* isolates to either potato or tomato probably due to adaptation through mutation (Kori, 1972; Erselius *et al.*, 1999).

34.4 *P. infestans* Population Studies in Kenya

Since Kori's (1972) study of isolates from Kenya and Uganda, work to characterize the population of *P. infestans* in Kenya, and indeed the entire East African region, has been erratic, with at most a single study for every decade. Consistent with global trends, methods employed have evolved with advances in technology. Kori (1972), for instance, limited his study to the specific virulence of isolates on potato and tomato, and made inferences suggesting host specificity within the population. Hohl (1998) worked on isolates obtained from diverse regions in Kenya and Uganda using a mating type test (with the A2 mating type tester isolate from Cornell University) and metalaxyl sensitivity analysis which revealed an exclusively A1 mating type and a generally metalaxyl-sensitive population.

Following on from this was a more detailed population analysis by Vega-Sanchez *et al.* (2000), made possible by the application of the more robust protein- and DNA-based techniques: allozyme (Forbes *et al.*, 1998; Goodwin *et al.*, 1998) and mitochondrial DNA (mtDNA) (Griffith and Shaw, 1998), respectively. More recently, Pule *et al.* (2013) undertook an analysis of isolates, from eight countries in eastern and southern African countries including Kenya, by complementing the old markers (mating type, metalaxyl sensitivity, mtDNA) with the new SSR markers which possess a superior resolving power. A significant finding of this study was the discovery of a unique strain KE-1 among the Kenyan isolates which although having an A1 mating type, varied from the typical US-1 clonal lineage and was metalaxyl insensitive.

One prominently common feature among the studies so far has been the population in the region is a clonal lineage of the A1 mating type (Kori, 1972; Vega-Sanchez *et al.*, 2000; Olanya *et al.*, 2001; Pule *et al.*, 2013). In between the studies, management strategies have been designed on the assumption that the populations remained substantively unchanged. This is despite the evidence from temperate regions suggesting that even for clonal populations, genotype compositions vary between cropping seasons due to factors such as differences in fitness between pathogen genotypes (Montarry *et al.*, 2010).

34.5 Current Strategies for Blight Control in Kenya

Insufficient knowledge about blight population dynamics is an obvious shortcoming hampering the success of the blight management efforts in Kenya and thus constraining potato production. These efforts have for long remained inseparable from the national potato variety development programme whose principal thrust has been to produce high-yielding blight-resistant varieties (Lung'aho *et al.*, 2001, 2006). The upshot has been major gains such as nearly a dozen varieties with high resistance to blight being released and adopted by growers, including three new varieties (Kenya Mpya, Purple Gold and Sherekea) released in 2010 (Onditi *et al.*, 2012).

However, more would be realized if the deployment of host resistance was supported by a rigorous and regular characterization of the national blight populations.

Within the context of low research funding from the national government, projects on breeding for host resistance receive priority perhaps due to the spin-off of additional varieties seen as returns on research investments. In contrast, projects on pathogen studies are considered basic research whose outputs are abstract and difficult to justify. The recourse has seen support coming through collaborative initiatives involving fungicide producers with local operations, universities and international institutions. Such support is mostly short term and usually insufficient to support a sustainable programme on *P. infestans* population monitoring as a key component in the overall national blight management strategy.

On the basis of what is known, access to blight-tolerant potato varieties is not sufficient to assure successful potato farming when faced with risks of severe blight epidemics under smallholder conditions (Ojiambo *et al.*, 2001; Olanya *et al.*, 2001). Agronomic practices adopted have an obvious effect on the ultimate yield achieved. Evidently, a combination of elements of integrated pest management (IPM) including good cultural practices and different spray regimes optimized to complement the various levels of host tolerance to blight are required to deliver the yield potential of a variety (Namanda *et al.*, 2004; Nyongesa *et al.*, 2005). Also required are disease support systems (DSSs) to guide the timing and optimized application of fungicides.

34.6 Way Forward for Late Blight Research in Kenya

As a direct outcome of climate change, countries in the temperate regions are warmer while those in the tropical belt are getting wetter (Forbes and Simon, 2007). This shift in climate patterns poses serious agricultural challenges, especially in smallholder potato farming. Long rainy seasons provide optimal conditions for blight epidemics especially when blight-susceptible varieties are grown (Hannukkala *et al.*, 2007). The emergence of new blight strains with insensitivity to commonly used fungicides signifies a particularly unique risk of blight epidemics that defy fungicide control in regions that rely on metalxyl-based products.

What is presently unknown is how the available varieties respond to the new blight strains, including KE-1, whose frequencies are generally believed to be on the rise. The prognosis is that although available resistant varieties will continue to perform satisfactorily under severe blight challenges in the short run, they may eventually lose their resistance. This is predicated on the assumptions that these varieties are of diverse genetic backgrounds and therefore will not uniformly collapse under attack from the newly emerged blight strains and that the strains will eventually circumvent host resistance (McDonald and Linde, 2002). Consequently, several interventions become urgent.

The first intervention is pathogen population characterization, followed by screening of dominant blight genotypes against the various potato cultivars to determine the levels of interactions.

Secondly, adoption of quicker ways of exploiting host resistance and introgression into cultivated varieties are necessary so as to keep pace with the fast-changing pathogen population. For example, compared to use of conventional breeding methods, marker assisted breeding (MAS) approaches that employ molecular markers linked to desirable traits would significantly reduce the time it takes to develop and release a variety. Of interest to blight control are recent advances in exploitation of novel sources of resistance to blight through genetic modification and cisgenics approaches towards durable resistance (Haverkort *et al.*, 2009; Zhu *et al.*, 2012). Since the establishment in 2009 of the National Biosafety Authority in Kenya, the heavily polarized views on genetic modification technology have been replaced by more informed and balanced debates. The significance of this is reduced hostility to transgenic-technology-derived tools for managing plant disease challenges.

These technologies can be harnessed for improvement of local popular yet susceptible varieties like Kerr's Pink and Désirée.

Thirdly, an equally essential aspect to successful blight management is to accurately predict onset of the disease. It is crucial that the first fungicide spray to a crop is made upon detection of disease onset as this leads to better disease control compared with delayed fungicide applications. Several sophisticated DSSs (e.g. NEGFRY and PLANT-Plus) are available to guide the timing of fungicide sprays, mainly in the developed world. In contrast, systems in Kenya upon which to base reliable disease forecasts are non-existent. Effort should be focused on developing an inexpensive DSS that depends on basic data to generate blight disease alerts and to provide variety-specific fungicide spray programmes involving newer formulations that do not lead to selection of fungicide-resistant pathogen strains.

References

Black, W. (1970) The nature of inheritance of field resistance to late blight (*Phytophthora infestans*) in potatoes. *American Potato Journal* 47, 279–288.

Black, W., Mastenbroek, C., Mills, W.R. and Peterson, L.C. (1953) A proposal for an international nomenclature of races of *Phytophthora infestans* and of genes controlling immunity in *Solanum demissum* derivatives. *Euphtytica* 2(1), 173–240.

Bourke, P.M.A. (1993) *'The Visitation of God'? The Potato and the Great Irish Famine*. Lilliput Press, Dublin.

Brurberg, M.B., Elameen, A., Le, V.H., Naerstad, R., Hermansen, A., Lehtinen, A., Hannukkala, A., Nielsen, B., Hansen, J., Andersson, B. and Yuen, J. (2011) Genetic diversity of *Phytophthora infestans* populations in the Nordic European countries reveals high genetic variability. *Fungal Biology* 115, 335–342.

Drenth, A., Tas, I.C.Q. and Govers, F. (1994) DNA fingerprinting uncovers a new sexually reproducing population of *Phytophthora infestans* in the Netherlands. *European Journal of Plant Pathology* 100, 97–107.

Erselius, L.J., Vega-Sanchez, M.E., Rodriques, A.M., Bastidas, O., Hohl, H.R., Ojiambo, P.S, Mukalazi, J., Vermeulen, T., Fry, W.E. and Forbes, G.A. (1999) Host specificity of *Phytophthora infestans* on tomato and potato in Uganda and Kenya. In: *CIP Program Report 1997–1998*. International Potato Center (CIP), Lima, pp. 49–55.

Forbes, G.A. and Simon, R. (2007) Implications of a warmer, wetter world on late blight pathogen. How CIP efforts can reduce risk for low input potato farmers. *SAT ejournals* 4(1). Available at: http://www.icrisat.org/journal/SpecialProject/sp4.pdf (accessed 16 February 2015).

Forbes, G.A., Goodwin, S.B., Drenth, A., Oyarzun, P., Ordoñez, M.E. and Fry, W.E. (1998) A global marker database for *Phytophthora infestans*. *Plant Disease* 82, 811–818.

Fry, W.E., Goodwin, S.B., Deyer, A.T., Matuszak, J.M., Drench, A., Tooley, P.W., Sujkowski, L.S., Koh, Y.J., Cohen, B.A., Spielman, L.J., Deahl, K.L., Inglis, D.A. and Sandlan, K.P. (1993) Historical and recent migrations of *Phytophthora infestans*: chronology, pathways and implications. *Plant Disease* 77, 653–661.

Fry, W.E., Grunwald, N.J., Cooke, D.E.L., McLeod, A., Forbes, G.A. and Cao, K. (2009) Population genetics and population diversity of *Phytophthora infestans*. In: Lamour, K. and Kamoun, S. (eds) *Oomycete Genetics and Genomics: Diversity, Interactions and Research Tools*. Wiley, Hoboken, New Jersey, pp. 139–164.

Gisi, U., Walder, F., Resheat-Eini, Z., Edel, D. and Sierotzki, H. (2011) Changes of genotype, sensitivity and aggressiveness in *Phytophthora infestans* isolates collected in European countries in 1997, 2006 and 2007. *Journal of Phytopathology* 159, 223–232.

Goodwin, S.B., Spielman, L.J., Matuszak, J.M., Bergeron, S.N. and Fry, W.E. (1992) Clonal diversity and genetic differentiation of *Phytophthora infestans* populations in Northern and Central Mexico. *Phytopathology* 82, 955–961.

Goodwin, S.B., Smart, C.D., Sandrock, R.W., Deahl, K.L., Punja, Z.K. and Fry, W.E. (1998) Genetic change within populations of *Phytophthora infestans* in the United States and Canada during 1994 to 1996: role of migration and recombination. *Phytopathology* 88, 939–949.

Griffith, G.W. and Shaw, D.S. (1998) Polymorphisms in *Phytophthora infestans*: four mitochondrial haplotypes are detected after PCR amplification of DNA from pure cultures or from host lesions. *Applied and Environmental Microbiology* 64, 4007–4014.

Hannukkala, A.O., Kaukoranta, T., Lehtinen, A. and Rahkonen, A. (2007) Late-blight epidemics on potato in Finland, 1933–2002: increased and earlier occurrence of epidemics associated with climate change and lack of rotation. *Plant Pathology* 56, 167–176.

Haverkort, A.J., Struik, P.C., Visser, R.G.F. and Jacobsen, E. (2009) Applied biotechnology to combat late blight in potato caused by *Phytophthora infestans*. *Potato Research* 52, 249–264.

Hohl, H.R. (1998) On A2 mating safari in East Africa. Low levels of oospore in mating and high levels of oospore in selfing. *Final Report*. International Potato Center Sub-Saharan Africa (CIP-SSA), Nairobi.

Kaguongo, W., Gildemacher, P., Demo, P., Wagoire, W., Kinyae, P., Andrade, J., Forbes, G., Fuglie, K. and Thiele, G. (2008) Farmer practices and adoption of improved potato varieties in Kenya and Uganda. *Social Sciences Working Paper 2008-5*. International Potato Center (CIP), Lima, 85 pp.

Kori, J. (1972) A survey of races of *Phytophthora infestans* in East Africa and development of late blight resistant varieties. MSc thesis, Makere University, Kampala, Uganda.

Lung'aho, C., Njenga, D.N. and Kabira, J.N. (2001) *Potato Late Blight*. National Potato Research Centre, Tigoni, Kenya.

Lung'aho, C., Nderitu, S.W.K., Kabira, J.N., El-Bedewy, R., Olanya, O.M. and Walingo, A. (2006) Yield performance and release of four late blight tolerant potato varieties in Kenya. *Journal of Agronomy* 5(1), 57–61.

McDonald, B.A. and Linde, C. (2002) Pathogen population genetics, evolutionary potential, and durable resistance. *Annual Review of Phytopathology* 40, 349–379.

Montarry, J., Andrivon, D., Glais, I., Corbiere, R., Mialdea, G. and Delmotte, F. (2010) Microsatellite markers reveal two admixed genetic groups and an ongoing displacement within the French population of the invasive plant pathogen *Phytophthora infestans*. *Molecular Ecology* 19, 1965–1977.

Namanda, S., Olanya, O.M., Adipala, E., Hakiza, J.J., El-Bedewy, R., Baghsari, A.S. and Ewell, P. (2004) Fungicide application and host-resistance for potato late blight management: benefits assessment from on-farm studies in S.W. Uganda. *Crop Protection* 23, 1075–1083.

Nyankanga, R.O., Wien, H.C., Olanya, O.M. and Ojiambo, P.S. (2004) Farmers' cultural practices and management of potato late blight in Kenya highlands: implications for development of integrated disease management. *International Journal of Pest Management* 50(2), 135–144.

Nyongesa, M.W., Gildemacher, P.R., Wakahiu, M.W., Demo, P. and Lung'aho, C. (2005) Optimizing fungicide spraying regimes for integrated management of potato late blight in Kenya. In: Tenywa, J.S., Adipala, E., Nampala, P., Tusiime, G., Okori, P. and Kyamuhangire, W. (eds) *African Crop Science Conference Proceedings* Vol. 7, 5–9 December 2005, Entebbe, Uganda. African Crop Science Society, Kampala, pp. 337–342.

Ojiambo, P.S., Namanda, S., Olanya, O.M., El-Bedewy, R., Hakiza, J.J., Adipala, E. and Forbes, G. (2001) Impact of fungicide application and late blight development on potato growth parameters and yield in the tropical highlands of Kenya and Uganda. *African Crop Science Journal* 9, 225–234.

Olanya, O.M., Adipala, E., Kakiza, J.J., Kedera, J.C., Ojiambo, P., Mukalazi, J.M., Forbes, G. and Nelson, R. (2001) Epidemiology and population dynamics of *Phytophthora infestans* in sub-Saharan Africa: progress and constraints. *African Crop Science Journal* 9, 185–193.

Onditi, J.O., Nderitu, S.W.K., Landeo, J.A., Abong', G.O., Sikinyi, E.O. and Kabira, J.N. (2012) Release of three improved varieties for the expanded potato market in Kenya. *Agriculture and Biology Journal of North America* 3(5), 192–195.

Pule, B.B., Meitz, J.C., Thompson, A.H., Linde, C.C., Fry, W.E., Langenhoven, S.D., Meyers, K.L., Kandolo, D.S., van Rij, N.C. and McLeod, A. (2013) *Phytophthora infestans* populations in central, eastern and southern African countries consist of clonal lineages. *Plant Pathology* 62(1), 154–165.

Spielman, L.J., Drenth, A., Davidse, L.C., Sujkowski, L.J., Gu, W.K., Tooley, P.W. and Fry, W.E. (1991) A second world-wide migration and population displacement of *Phytophthora infestans*? *Plant Pathology* 40, 422–430.

Vega-Sanchez, M.E., Erselius, L.J., Rodriguez, A.M., Bastidas, O., Hohl, H.R., Ojiambo, P.S., Mukalazi, J., Vermeulen, T., Fry, W.E. and Forbes, G.A. (2000) Host adaptation to potato and tomato within the US-1 clonal lineage of *Phytophthora infestans* in Uganda and Kenya. *Plant Pathology* 49, 531–539.

Zhu, S., Li, Y., Vossen, J.H., Visser, R.G.F. and Jacobsen, E. (2012) Functional stacking of three resistance genes against *Phytophthora infestans*. *Transgenic Research* 21(1), 89–99.

35 Root Knot Nematodes and Soft Rot *Enterobacteriaceae*, Two Emerging Problems of Potatoes

L.N. Moleleki*
University of Pretoria, Pretoria, South Africa

Abstract

The soft rot *Enterobacteriaceae* (SRE) and root knot nematodes (RKN) are major pathogens of potatoes and other crops globally; one of the most threatening emerging species of SRE is the newly identified *Dickeya solani* biovar 3. Using advanced molecular-based technologies for identification and evaluation of genetic diversity of many pathogens, new and threatening species of RKN and SRE are constantly being identified. In our research, PCR-based diagnostic tools have been used to conduct surveys of the different SRE and RKN infecting potatoes in South African potato fields. Results from the survey indicate that there are other SRE which had previously not been reported and are emerging problems of potatoes globally. These include *Pectobacterium wasabiae* which was identified in some potato-producing farms in South Africa. A countrywide survey of RKN using PCR-based diagnostic assays identified the presence of the three main tropical species, namely *Meloidogyne javanica* (24% of the samples tested), *Meloidogyne incognita* (23%) and *Meloidogyne arenaria* (17%). The study further established the presence of more temperate species such as *Meloidogyne chitwoodi* (3%) and *Meloidogyne hapla* (1%) in the samples tested. Of significant interest was the identification of *Meloidogyne enterolobii* (constituting 13% of the tested samples) in some potato-producing regions. A phylogenetic tree was constructed using partial sequences of the mitochondrial DNA and intergenic spacer (IGS) region to determine and compare the genetic diversity of South African RKN to those isolated from other parts of the world. In the survey, RKN samples were also profiled according to cultivar from which they were isolated. The highest number of RKN were isolated from Mondial (61%) followed by Up to date (14%). RKN samples were isolated from other cultivars including Sifra, Buffelspoort, Argos, Fianna, Valor, BP1 and Van der Plant (1–5%). RKN are known to form synergies with other pathogens in the soil, leading to increased incidences of disease. To determine whether such synergy exists between RKN and SRE, a *Pectobacterium* strain tagged with reporter fluorescent mCherry protein was developed. With the aid of this tool, the presence of RKN in the soil was shown to facilitate entry of *Pectobacterium* spp. into potato tubers, leading to latent infection.

Keywords: *Dickeya*, emerging, *Meloidogyne, Pectobacterium*, soft rot *Enterobacteriaceae* (SRE)

*lucy.moleleki@up.ac.za

35.1 Introduction

The potato industry in South Africa accounts for more than 50,000 ha of cultivated land yielding approximately 2 million t of potatoes annually (Potatoes South Africa, 2012). It is made up of four different sectors, namely seed, table, export and processing, distributed across 16 different potato-growing regions. Like other crop plants, potato production is vulnerable to many pests (Potatoes South Africa, 2012). Without doubt the success of the entire industry is reliant on production of high quality, disease-free seed potatoes. However, the threat of pathogens, particularly seed-borne pathogens, is more acute in the seed production industry. Hence, failure to detect and accurately diagnose these pathogens has dire implications for potato production in general. Therefore, the use of molecular-based diagnostic approaches for the potato industry to report new 'emerging' highly aggressive pathogens in various parts of the world is imperative to be able to screen and identify presence of these pathogens in South Africa.

The soft rot *Enterobacteriaceae* (SRE) and root knot nematodes (RKN) are two major pathogens of potatoes and other crops globally; one of the most threatening emerging species of SRE is the newly identified *Dickeya solani* biovar 3 (Toth *et al.*, 2011). With the advancing use of molecular-based technologies for identification and evaluation of genetic diversity of many pathogens, new and threatening species of RKN and SRE are constantly being identified. In this study, the use of PCR-based methods was employed to determine the diversity of SRE and RKN in the South African potato industry. Additionally, the potential interaction between members of these two groups was established.

35.2 Materials and Methods

Identification and genetic diversity of RKN species

Potato tubers showing symptoms of either RKN infection or soft rot were submitted to the host–pathogen interactions laboratory at the University of Pretoria. Second stage juveniles (J2) of the RKN were isolated from 100 g potato peel using the centrifugal floatation method (Bezooijen, 2006). DNA was extracted from J2s using the Worm Lysis Buffer (MLB) (Castagnone-Sereno *et al.*, 1995). PCR amplification was carried out using primers 194 (5′-TTAACTTGCCAGATCGGACG-3′) and 195 (5′-TCTAATGAGCCGTACGC-3′) to amplify the intergenic spacer (IGS) region of the ribosomal DNA (Blok *et al.*, 1997) and primers C2F3 (5′-GGTCAATGTTCAGAAATTTGTGG-3′) and 1108 (5′-TACCTTTGACCAATCACGCT-3′) to amplify the region located between the 3′ region of the cytochrome oxidase II (COII) of the mitochondrial DNA (mtDNA) and the 5′ IGS region of the 16S rRNA gene (Powers and Harris, 1993). All primers were sourced from Inqaba Biotechnologies. Amplifications were performed in a final volume of 25 μl containing 25 ng DNA, 200 μm dNTPs (Fermentas), 0.4 μm each forward and reverse primers, 0.5 U Taq DNA polymerase (Fermentas) and 10 × Taq DNA polymerase reaction buffer with 20 mM $MgCl_2$ (Fermentas). For IGS amplification reactions, the following temperature profiles were used: 94°C for 2 min, followed by 45 cycles of 94°C for 30 s, 50°C for 30 s and 72°C for 90 s, with a final extension at 72°C for 10 min. The COII PCR reactions were set up at 94°C for 2 min, followed by ten cycles of 94°C for 10 s, 48°C for 30 s and 68°C for 2 min. The next 25 cycles were set up at 94°C for 10 s, 48°C for 30 s with a final extension at 72°C for 10 min. All PCR amplifications were carried out using a Biometra Analytica Jena thermocycler. PCR products were purified and sequenced at the University of Pretoria (Onkendi and Moleleki, 2013a, b). Preliminary identification of isolates was done according to the BLAST algorithm. Thereafter, sequences were manually edited and aligned and thereafter phylogenetic trees were constructed.

Identification of soft rot coliform

Potato tubers samples showing soft rot symptoms were obtained and SRE isolated according to Hyman *et al.* (2001). To elute the bacteria, 1 g of rotting potato tuber tissues was briefly placed in 1 ml of water and thereafter the eluted bacteria were plated on to crystal violet pectate (CVP) medium using a sterile loop; cavity-forming colonies were purified several times. DNA was isolated from individual cavity-forming colonies, followed by PCR amplification using primers designed to anneal to two housekeeping genes, *malate dehydrogenase* (mdh2_F GCGCGTAAGCCGGGTATGGA and mdh4_R CGCGGCAGCCTGGCCCATAG) and *glyceraldehyde-3- phosphate dehydrogenase A* (gapA_F ATCTTCCTGACCGACGAAACTGC and gapA_R ACGTCATCTTCGGTGTAA CCCAG) (Moleleki *et al.*, 2012). The PCR reaction was performed in a total volume of 25 µl consisting of 10 × DreamTaq buffer (supplemented with 20 mM $MgCl_2$, 2.5 mM dNTPs, 10 µM each forward and reverse primer), 0.5 U DreamTaq Polymerase (Fermentas) and 100 ng DNA template. PCR amplification was conducted using a Biometra Thermocycler (Germany) with the following thermal regime: initial denaturing for 2 min at 95°C, 30 cycles of denaturing at 94°C for 30 s, followed by annealing at 62°C for 45 s, and elongation at 72°C for 90 s followed by a final extension step at 72°C for 7 min. The resulting amplicons were sequenced at the University of Pretoria using standard procedures (Sambrook and Russell, 2001), sequences were trimmed and aligned then alignment used to construct phylogenetic trees.

Determination of potential interaction of RKN and SRE using fluorescent mCherry

Fluorescent mCherry-tagged *Pectobacterium carotovorum brasiliense* (*Pcb*) was co-incubated with 600 second stage juveniles (per experiment) *in vitro* at 4°C, 28°C and 37°C for 24 h, 48 h and 72 h post-incubation. Nematodes were then washed. Scanning electron microscopy and confocal microscopy were used to determine attachment according to methods outlined by Mongae *et al.* (2013). The total colony forming units of attached bacteria was determined per gram of J2.

35.3 Results

A total of 78 composite potato tuber samples infected with RKN were obtained for the study; second stage juvenile RKN were isolated from infected tubers and DNA was isolated from five juveniles per composite sample. DNA sequences of the second stage juveniles were first identified based on highest homology to RKN samples deposited in GenBank. After this initial identification, consensus sequences with the same identity were aligned to 17 sequences obtained from GenBank and a phylogenetic tree was constructed.

Our study further established presence of the more temperate species such as *Meloidogyne chitwoodi* (3% of the samples tested) and *Meloidogyne hapla* (1%); of significant interest was the identification of *Meloidogyne enterolobii* (13% of the tested samples) in some potato-producing regions. This is a highly virulent species known to break resistance such as that encoded in the *Mi*-gene of tomatoes. In the survey, RKN samples were also profiled according to cultivar from which they were isolated. The highest number of RKN were isolated from Mondial (61%) followed by Up to date (14%). RKN samples were isolated from other cultivars including Sifra, Buffelspoort, Argos, Fianna, Valor, BP1 and Van der Plant (1–5%). RKN are known to form synergies with other pathogens in the soil, leading to increased incidences of disease.

In Fig. 35.1, the South African RKN isolates are highlighted with the prefix 'MeloSA'. Initial BLAST identification of each of the populations was confirmed in the constructed tree. However, surprisingly, it was observed that each of the South African populations were highly homogenous. The phylogenetic tree derived from mtDNA sequences separated the tropical *Meloidogyne* species into one clade, which was completely distinct from the temperate species. Closely related apomictic species (*Meloidogyne javanica, Meloidogyne arenaria* and *Meloidogyne incognita*) grouped together with a 100% bootstrap support. The automictic species (*M. chitwoodi* and *Meloidogyne fallax*) also grouped together in a separate clade supported by a bootstrap value of 98%. *M. hapla* formed a separate clade and was closer to the automictic species

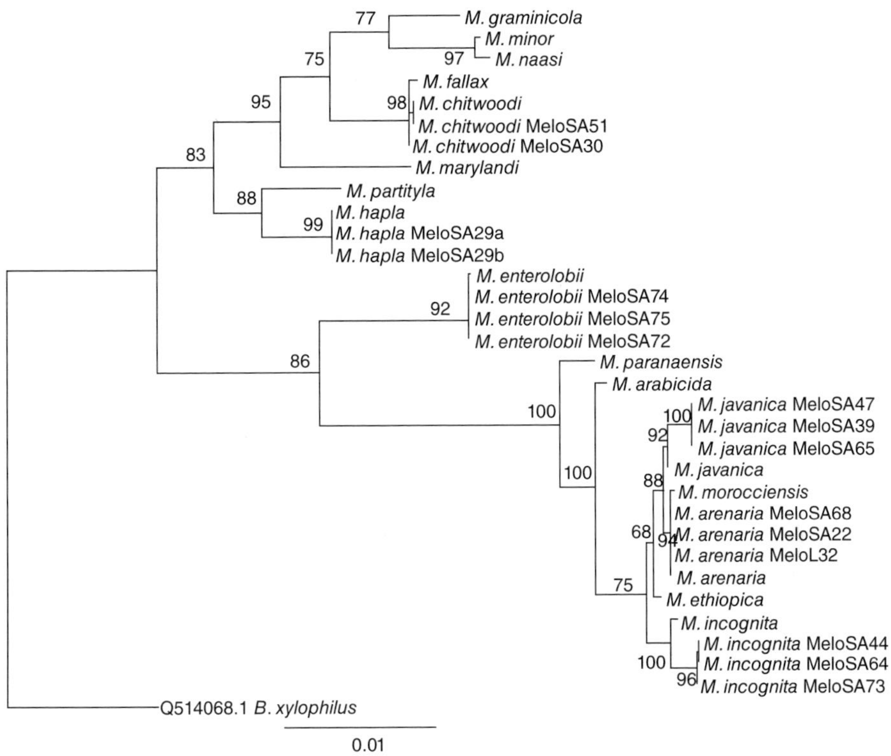

Fig. 35.1. Maximum likelihood tree constructed using alignment of partial gene sequences of the mtDNA region of South African *Meloidogyne* isolates and sequences obtained from GenBank. *Bursaphelenchus xylophilus* was used as an outgroup. Bar, 0.01 nucleotide substitutions per site.

than the apomictic species forming a clade supported by a bootstrap value of 83%. Interestingly, *M. enterolobii*, which is also a tropical species, formed an independent clade which was much closer to the tropical than the temperate species. These results were supported by similar tree constructed using the IGS region.

To screen for the diversity of soft rot pathogens in South Africa, samples of infected potato tubers were routinely obtained from potato fields during the 2011–2013 period. SRE were isolated from rotting potato tissue and plated on CVP (Hyman *et al.*, 2001). After several rounds of purification, cavity-forming colonies were characterized by amplifying partial sequences of the two house-keeping genes, *mdh* and *gapA*. The amplicons were sequenced and initial identification of all isolates was made based on BLAST. Initial BLAST analysis identified approximately 52% of the SRE coliforms as *Pectobacterium carotovorum brasiliense* (*Pcb*) and the other 48% as *Pectobacterium wasabiae*. Identification of *P. wasabiae* in South Africa was particularly significant as this constitutes the first report of this pathogen in South Africa. To further confirm these identities, four sequences of each of the newly identified South African SRE were used to construct a phylogenetic tree using sequences obtained from GenBank for comparison.

In Fig. 35.2, the *Pectobacterium* species formed a separate major group made up of five clades, namely: (i) *P. c. brasiliense* (*brasiliensis*); (ii) *P. c. carotovorum*; (iii) *Pectobacterium atrospecticum*; (iv) *P. wasabiae*; and (v) *Pectobacterium betavasculorum*. As expected, members of the genus *Dickeya* formed a separate clade supported by a high bootstrap value of 100%. Interestingly, the *Pcb* isolates did not form a monophyletic

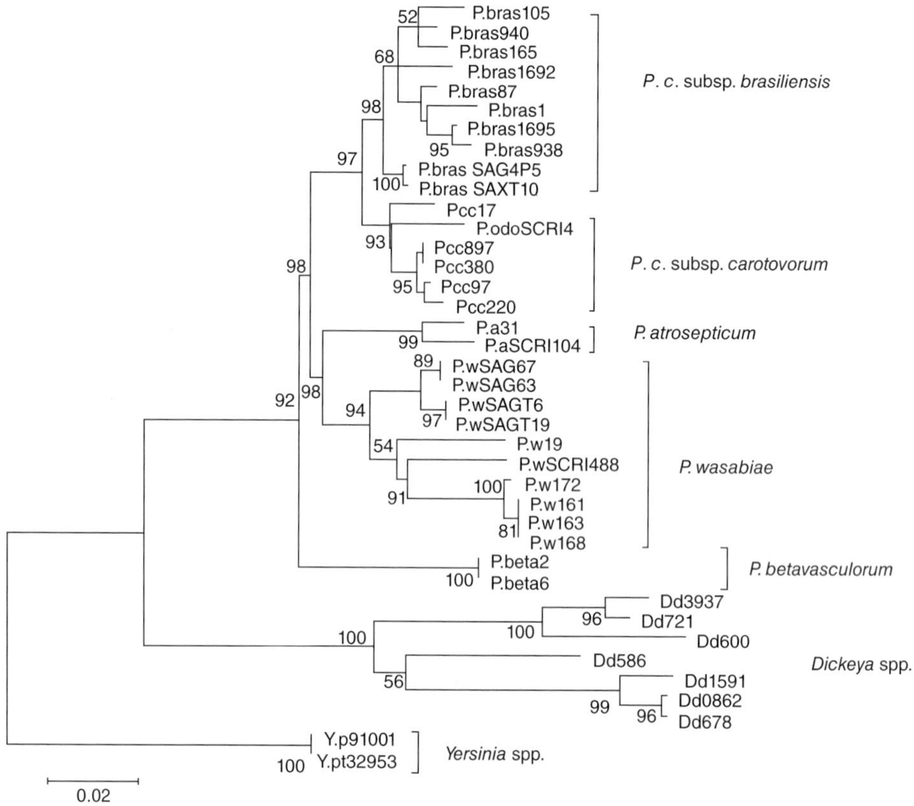

Fig. 35.2. A concatenated maximum likelihood phylogenetic analysis of *Pectobacterium* spp. strains from South Africa based on *gapA* partial gene sequences. Numbers represent bootstrap support values greater than 50% from 1000 replicates using heuristic searches. *Yersinia* spp. were used as an outgroup. Bar, 0.02 nucleotide substitutions per site.

group suggesting that there is high diversity within this group of pathogens. The South African *P. wasabiae* isolates grouped together within the *P. wasabiae* clade and were separated from the USA isolates 19 (P.w19 on Fig. 35.2) and the type strain SCRI 488.

The SRE interact with RKN juveniles *in vitro*

The interaction between SRE and RKN was determined *in vitro* by first transforming an mCherry fluorescent protein into a *Pcb* strain. The mCherry-tagged strain was co-inoculated with 100 RKN juveniles at 4°C, 28°C and 37°C for 24 h, 48 h and 72h. Thereafter, the juveniles were washed to remove superficially attached bacteria and patted dry on a paper wipe; attached RKN juveniles were viewed under scanning and confocal microscopes (Fig. 35.3).

Attachment of mCherry-tagged *Pcb* was further verified by grinding 1 g of RKN, serial dilution and plating on to nutrient agar plates supplemented with tetracycline. Optimum attachment was observed at 28°C. There did not appear to make any difference in the amount of *Pcb* attachment to RKN juvenile coats.

35.4 Discussion

Potato production is hampered by the presence of different pests; it is a particular

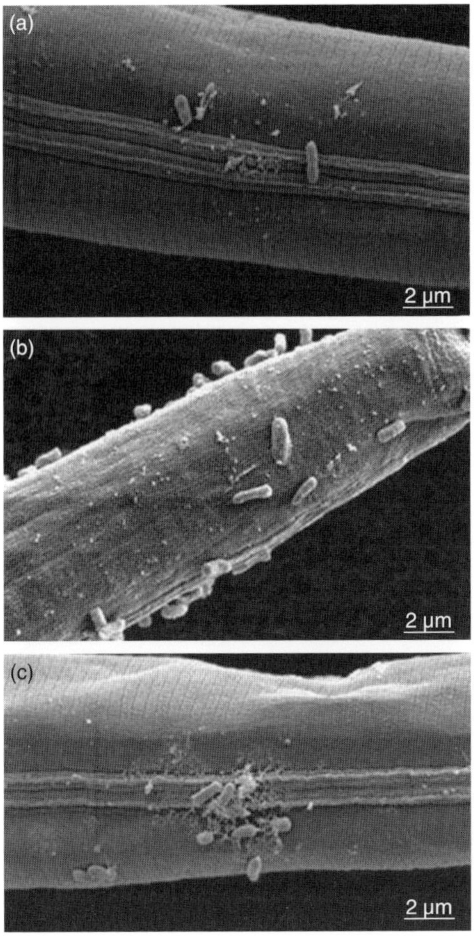

Fig. 35.3. Scanning electron microscopy micrographs showing attachment of *Pcb*_mCherry cells on the surface of second stage root knot nematodes (J2s). *Meloidogyne incognita* J2s were co-incubated with overnight cultures of *Pcb*_mCherry in minimal medium for 72 h at (a) 4°C, (b) 28°C and (c) 37°C.

problem in seed-potato production. Accurate identification of most pathogens has historically been challenging due to traditional approaches such as the use of symptoms, phenotypic and morphological features of the pathogens and similar methods. This often led to incorrect identification of infecting pathogens. Additionally, reliance on the presence of symptoms often resulted in distribution of infected seed material. The recent developments in molecular-based identification methods have increased the sensitivity of detecting pathogens on seed tubers and the accuracy with which they are identified.

In this study, PCR-based detection methods were used to determine the population structure of RKN and SRE infecting potatoes in South Africa. The use of PCR combined with sequencing was a formidable tool that allowed not only detection and identification but also determining the genetic diversity of the two groups of pathogens. It was noteworthy that, apart from the previously reported *Pcb* (Duarte *et al.*, 2004), this study reported for the first time the presence of a new species *P. wasabiae. P. wasabiae* was first identified in horseradish in Japan. However, it has now been reported in potatoes in New Zealand, the USA, Canada and Iraq. This study constituted the first report of *P. wasabiae* in South Africa (Ma *et al.*, 2007; Kim *et al.*, 2009; Baghaee-Ravari *et al.*, 2011).

In the RKN survey, the expected tropical species *M. javanica*, *M. incognita* and *M. arenaria* were identified. Furthermore, the presence of temperate species such as *M. chitwoodi* and *M. hapla* was established. Perhaps the most disturbing was the identification of *M. enterolobii* in seed-potato-producing fields. *M. enterolobii* is a highly aggressive species of RKN known to break *Mi*-resistance found in most solanaceous crops. Identification of these highly damaging RKN species coupled with increasing withdrawal of various nematicides from the market prompted the study of potential interactions between RKN and SRE. The rationale was based on the fact that as more nematicides are withdrawn, there is likely to be a significant increase in the number of RKN in the soil. Given that RKN feed on roots and potato tubers by thrusting their stylets into plant tissue causing wounds, we reasoned that these wounds can be used by secondary pathogens such as SRE that may be present in the soil. Additionally, SRE could attach to RKN and be transported from an infected host to a new host. To test this hypothesis, an *in vitro* study was conducted to determine whether SRE can attach to RKN juveniles. Attachment was observed to occur at the optimum temperature of 28°C. This attachment will need to be further investigated in the glasshouse to establish the real impact in potato production, however, this in itself raises concerns that increasing numbers of RKN could lead to increased incidences of other soil- and seed-borne diseases such as the SRE.

Acknowledgements

This work was funded by Potatoes South Africa and the Technology and Human Resources Industry Programme.

References

Baghaee-Ravari, S., Rahimian, H., Shams-Bakhsh, M., Lopez-Solanilla, E., Antunez-Lamaz, M. and Rodriguez-Penzuela, P. (2011) Characterization of *Pectobacterium* species from Iran using biochemical and molecular methods. *European Journal of Plant Pathology* 129, 413–425.

Bezooijen, J. (2006) *Methods and Techniques for Nematology*. Wageningen UR, Wageningen, The Netherlands.

Blok, V.C., Phillips, M.S. and Fargette, M. (1997) Comparison of sequences from the ribosomal DNA intergenic region of *Meloidogyne mayaguensis* and other major tropical root-knot nematodes. *Journal of Nematology* 29, 16–22.

Castagnone-Sereno, P., Esparrago, G., Abad, P., Leroy, F. and Bongiovanni, M. (1995) Satellite DNA as a target for PCR-specific detection of the plant-parasitic nematode *Meloidogyne hapla*. *Current Genetics* 28, 566–570.

Duarte, V., de Boer, S.H., Ward, T.L. and de Oliveora, A.M.R. (2004) Characterisation of atypical *Erwinia carotovora* subsp. *atroseptica* causing blackleg of potato in Brazil. *Journal of Applied Microbiology* 96, 535–545.

Hyman, L.J., Sulivan, L., Toth, I.K. and Pérombelon, M.C.M. (2001) Modified crystal violet pectate medium (CVP) based on a new polypectate source (Slendid) for the detection and isolation of soft rot *Erwinias*. *Potato Research* 44, 265–270.

Kim, H.S., Ma, B., Perna, N.T. and Charkowski, A.O. (2009) Phylogeny and virulence of naturally occurring type III secretion system-deficient *Pectobacterium* strains. *Applied and Environmental Microbiology* 75, 4539–4549.

Ma, B., Hibbing, M.E., Kim, H.S., Reedy, R.M., Yedidia, L. and Breuer, J. (2007) Host range and molecular phylogenies of the soft rot enterobacterial genera *Pectobacterium* and *Dickeya*. *Phytopathology* 97, 1150–1163.

Moleleki, L.N., Onkendi, E., Mongae, A.O. and Kubheka, G.C. (2012) Characterisation of *Pectobacterium wasabiae* causing tuber soft rot and blackleg diseases in South Africa. *European Journal of Plant Patholology* 135, 279–288.

Mongae, A., Kubheka, G. and Moleleki, L. (2013) The use of fluorescent protein tagging to study the interaction between soft rot *Enterobacteriaceae* and root knot nematodes. *Letters in Applied Microbiology* 56, 258–263.

Onkendi, E. and Moleleki, L.N. (2013a) Distribution and genetic diversity of root knot nematodes (*Meloidogyne* spp.) in potatoes in South Africa. *Plant Pathology* 62, 1184–1192.

Onkendi, E. and Moleleki, L.N. (2013b) Detection of *Meloidogyne enterolobii* in potatoes in South Africa and phylogenetic analysis based on intergenic region of the mitochondrial DNA. *European Journal of Plant Pathology* 136, 1–5.

Potatoes South Africa (2012) Available at: www.potatoes.co.za (accessed 22 September 2012).

Powers, T.O. and Harris, T.S. (1993) A polymerase chain reaction method for identification of five major *Meloidogyne* species. *Journal of Nematology* 25, 1–6.

Sambrook, J. and Russell, D. (2001) *Molecular Cloning: a Laboratory Manual*, 3rd edn. Cold Spring Harbor Press, New York.

Toth, I.K., van der Wolf, J.M., Saddler, G., Lojkowska, E., Hélias, E. and Pirhonen, M. (2011) *Dickeya* species: an emerging problem for potato production in Europe. *Plant Pathology* 60, 385–399.

36 Can Mineral Oil Protect Seed Potato Against Aphid Transmission of *Potato Virus Y*?

S. Boukhris-Bouhachem,[1]* M.H. Sellami,[1] I. Chaieb,[1] R. Souissi[1] and M. El Fahem[2]
[1]*National Agricultural Research Institute of Tunisia, Ariana, Tunisia;*
[2]*Groupement Interprofessionnel des Légumes (GIL), Tunis, Tunisia*

Abstract

Potato virus Y (PVY) is a limiting factor for the production of certified seed potato in Tunisia. PVY^{NTN} is the most important reason for downgrading seed potato lots of variety Spunta to lower classes and for rejection during certification. Transmission efficiencies of aphid species for different PVY strains are established for the most abundant ones. In total about 15 aphid species are known to be vectors of PVY with different efficiencies. There are five aphid species that colonize potatoes: *Aphis gossypii*, *Macrosiphum euphorbiae*, *Myzus persicae*, *Aphis fabae* and *Aulacorthum solani*. Eleven other aphid species that do not colonize potatoes seem to be far more important in PVY transmission: *Aploneura lentisci*, *Brachycaudus helichrysi*, *Brachycaudus cardui*, *Hyalopterus pruni*, *Hyperomyzus lactucae*, *Acyrthosiphon pisum*, *Rhopalosiphum padi*, *Rhopalosiphum maidis*, *Aphis spiraecola*, *Cavariella aegopodii* and *Lipaphis erysimi*. Therefore, to limit PVY propagation insecticide sprays are used to control aphids. However, intense use of aphicides has led to the selection of resistant aphid populations. As an option for PVY control, a few alternatives are available for seed potato production, among them mineral oil whose efficacy has been widely reported. Nevertheless, the use of mineral oils for agricultural aphid management is still very limited in Tunisia, probably due to the tendency of seed growers to believe that oils may cause: (i) phytotoxicity (while this is a problem only when oils are used with fungicides); and (ii) the loss of yield. Noting this problem, we evaluated a strategy consisting of use of oil to reduce aphid multiplication and to protect plants against aphids. This chapter compares the effect of mineral oil with and without insecticides on PVY infection. Results indicate that infectious aphids cannot transmit PVY in plants with oil in the laboratory, and also that healthy aphids cannot acquire viruses on oil-treated plants infected by PVY. Oil treatment of infected plants limited PVY propagation to 2.5% while plants not treated with oil showed a level of 9.5%. Finally, there were no significant yield losses in potato plots treated with oil (with yields of 0.28–0.33 kg per plant) compared with non-treated plots (yields of 0.26–0.30 kg per plant) under field conditions. This work illustrates that the bias farmers have against oils as an option for control of aphids is unfounded.

Keywords: aphids, insecticide, mineral oil, potato, PVY, traps, yield

*bouhachem.sonia@iresa.agrinet.tn

36.1 Introduction

Tunisian seed potato production is faced with virus contamination that cause a decline of 12% in the seed potato areas because they exceed 5% infected plants (GIL, 2003). *Potato virus Y* (PVY) is one of the most aggressive viral diseases of potato. This is because not only does it spread rapidly as it is vectored by many aphid species, but also because it leads to high potato yield losses of 30–70% of the production in the absence of control measures (Van der Zaag, 1987). Mineral oils are well known for control of insects, especially against aphids (Cornu and Gehriger, 1981; Martín-López *et al.*, 2006, Al-Mrabeh *et al.*, 2010). They reduce aphid infestation on plants and therefore the transmission of viral disease (Simons and Zitter, 1980). Vanderveken (1977) suggested several hypotheses for the effect of mineral oils on aphid transmission of plant viruses. One is that oil might modify the charge of the stylets, thus impeding attachment of virus particles and adsorption or elution by aphid mouthparts (Bradley *et al.*, 1962). The second is that the inhibitory properties of oils might result from the effective mediation of these processes by the potyvirus-coded protein 'helper component' (Berger and Pirone, 1986). However, other research suggests that the mineral oil alters the activities of the stylets as they penetrate the plant surface (Simons *et al.*, 1977).

Many reports have demonstrated that mineral oil sprayed on the seed potato crop can control PVY (Simons and Zitter, 1980; Gibson *et al.*, 1988; Merlet and Le Hingrat, 1996; Rolot, 2005; Wróbel, 2007). Nevertheless, the use of mineral oils for agricultural aphid management is still very limited in Tunisia, probably due to the tendency of the seed potato producers associating them with phytotoxicity, which is a problem only when mineral oils are used with fungicide. They also suspect use of mineral oils leads to yield loss. However, mineral oils are used in citrus nurseries against leaf miners with no evidence of adverse effects to crops. The objective of the present study was to evaluate the effect of the mineral oils as a preventive treatment against PVY infection of potato.

36.2 Materials and Methods

Two experiments were performed. The first experiment was to assess the effect of treatment with insecticide only (Treatment 1) in comparison with insecticide + oil (Treatment 2) on the level of PVY infection in the late season and also to determine the potato yield in plots treated with oil relative to plots without oil. The second experiment was to explore the possibility of acquisition and transmission of PVY by the aphid vector *Myzus persicae* (Boukhris-Bouhachem *et al.*, 2011) after the oil application on potato plants.

Experiment 1

The experiment was carried out between February and May 2008, at the Groupement Interprofessionnel des Légumes (GIL) station of Korba (Cap Bon) in the seed potato crop. The experiment was set up as two adjacent plots of 5000 m^2 each containing about 2000 plants. The potatoes were of the variety Spunta imported from France as an 'Elite' class seed potato known to be of high quality (< 2% viruses). Randomized complete block experimental designs with four replicates were adopted throughout and unsprayed guard rows of 1 m were used. In addition four plots served as an untreated check. A 20 l manual sprayer was used to administer treatments (insecticides and insecticides + mineral oil) after the crop emergence. Three sprays using the products Thiamethoxam (Synara, Syngenta Agro), Imidacloprid (Confidor, Bayer Crop Science) and Pymetrozine (Chess, Syngenta) were carried out on 7 April 2008, 21 April 2008 and 5 May 2008, respectively, to cover the crop against aphid infestation. The dates of foliar application were chosen based on aphid catch data (using a suction trap) for the Cap Bon region since 2002 that shows the peak dates (Fig. 36.1). Treatment 2 consisted of the same insecticides with the addition of mineral oil (Citrole, Total Solvent, 6% mineral oil and 94% water). Experiments evaluated the efficacy of three applications of different products on the sanitary quality of seed potatoes. The two

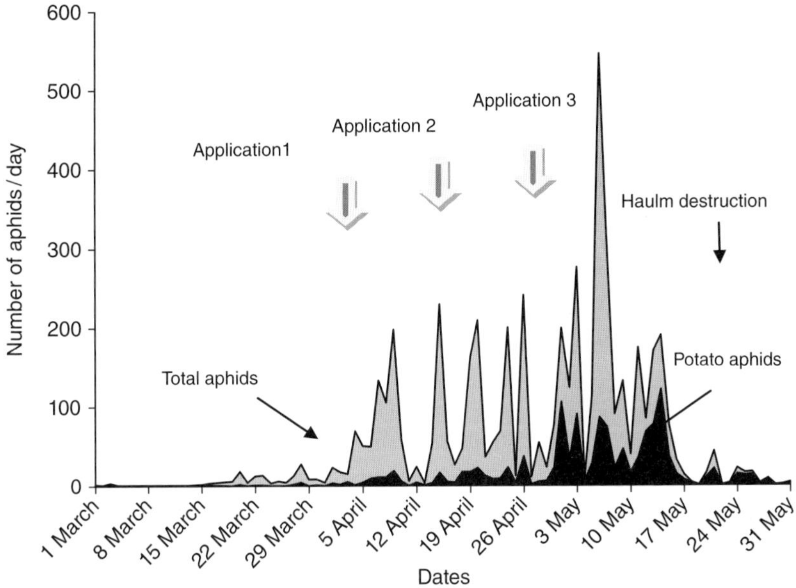

Fig. 36.1. Aphid flight data at Soliman, 2008 indicating dates of foliar application of treatments. 'Potato aphids' (black) are aphids that colonize and multiply on potato while 'Total aphids' (grey) include all the aphids captured in the traps (living on other plants as well as potato).

treatments, Treatment 1 and Treatment 2, on aphid populations were evaluated after calculation of the percentage of aphid reductions compared with the control using the Abbott formula:

$$\text{Aphid reduction} = (\text{Control} - \text{Treated})/\text{Control} \quad (36.1)$$

Field test

For each assessment, the number of aphids collected from 30 leaves on each plant per plot was recorded. Four replications were sampled at random and counting was performed with the aid of a binocular microscope in the laboratory.

Estimation of the yield and the PVY infection in late season

The effect of mineral oil on potato yield was estimated from the weight of tubers harvested from 100 plants randomly sampled from each plot. For the PVY infection, 200 tubers were sampled in each experiment field and tested using double antibody sandwich ELISA (DAS-ELISA) using a PVY polyclonal antibody (Bioreba, Basel, Switzerland).

Experiment 2

The experiment was conducted during 2012 in the plant protection laboratory at the Institut National de la Recherche Agronomique de Tunisie (INRAT) on potato plants obtained from *in vitro* culture (four replications of four small plots each with ten potato plants). A separate set of ten potato plants without oil was included as a control for each replication. Mineral oil was uniformly applied as a 6% suspension in water (Citrole) using a 5 l sprayer when the plants were at the four-leaf stage. Potatoes coated with oil were left 24 h before aphid inoculation, as cited by Gibson *et al.* (1988).

Aphid rearing

M. persicae were collected from several regions of Tunisia (Cap Bon, Béja, Kairouan) and reared on pepper (*Capsicum annuum*) in a growth chamber under controlled conditions of temperature 23°C ± 2°C and photoperiod regime of 16 h light: 8 h dark.

Potato inoculation

Aphids were starved for 2 h in plastic Petri dishes before they were allowed to acquire the virus by feeding for 2 min on tobacco (*Nicotiana tabacum* var. Xanthi) infected with the PVY^{NTN} strain. At inoculation, each potato plant received three aphids to transmit the virus. After 24 h, the aphids were killed with an insecticide (Imidacloprid) following the manufacturer's instructions. Potato leaf samples were collected 15 days after inoculation and tested for PVY infection by DAS-ELISA.

Statistical analysis

Data from the first experiment were set up according to a randomized complete block design. The dependent variable was first evaluated for normal distribution according to Shapiro-Wilk's test. Statistical analyses were performed through the GLM (general linear model) procedure of SAS/STAT. Duncan's test at 0.05 probability level was used as a mean separation test. PVY-infected plants in experiments 1 and 2 were compared with the control group (not sprayed by mineral oil) using chi-squared (χ^2) analysis.

36.3 Results and Discussion

In the first experiment, we observed that the development of aphids was almost identical in the two plots (with and without oil) (Table 36.1). Treatment with oil did not change the behaviour of aphids. This is in accordance with Bradley *et al.* (1962) who reported no changes in the feeding behaviour of aphids in the presence of mineral oil. Another study (Hein, 1972) has shown that aphids had no preference or aversion to the treated leaf material. In contrast, the mean number of aphid colonies (of three repetitions) for Treatment 1 (insecticide only) and Treatment 2 (insecticide + oil) were significantly reduced in comparison with their respective controls (untreated plots) according to Duncan's test ($P < 0.05$) (Table 36.1). Application of the insecticide alone and the mineral oil with insecticide at a concentration of 6% reduced aphid infestation by 82% and 80%, respectively (Table 36.2).

It was observed that, with three applications (one every 2 weeks) there was a significant difference between the PVY infections/200 tubers in Treatment 1 (insecticide only) and Treatment 2 (insecticide + oil) resulting in 9.5% and 2.5% inoculation, respectively (Treatment 1 = 19 tubers out of 200 infected, i.e. 9.5% PVY inoculation, while Treatment 2 = 5 tubers out of 200 infected, i.e. 2.5% PVY inoculation) (Table 36.1). These results suggest that oil application as

Table 36.1. Impact of treatments on aphid populations, 2008.

	Mean number of aphid colonies (on the dates shown)[a]						PVY infections/200 tubers harvested on 12 October[b]
	7 April (sprayed with Thiamethoxam)	14 April	21 April (sprayed with Imidacloprid)	28 April	5 May (sprayed with Pymetrozine)	12 May	
Treatment 1 (insecticide only)	9	6	25	13	31	27	19(+)
Control 1	35	43	215	102	103	50	24(+)
Treatment 2 (insecticide + oil)	21	9	12	15	34	26	5(+)
Control 2	66	52	193	89	158	79	8(+)

[a]Mean number of aphid colonies (of three repetitions) for Treatment 1 and Treatment 2 were significantly different from their controls (Control 1 and Control 2, respectively) according to Duncan's test ($P < 0.05$).
[b](+), Positive samples by DAS-ELISA. PVY infections/200 tubers for Treatment 1 and 2 were significantly different ($P < 0.05$).

a preventative treatment is a good method to control PVY.

Furthermore, it is important to note that, oil application at the rate of 6% did not have any phytotoxicity on treated plots. The results also revealed that there is no significant difference between the potato yields in the two treatments, 1 and 2 (Fig. 36.2).

Data from the second experiment showed that the oil inhibits the transmission of PVY by *M. persicae* after acquiring the virus on infected tobacco plants, compared with the control (95% infection). The presence of oil reduced the inoculation efficiency of *M. persicae* and lead to a strong inhibition of PVY transmission by aphids (Table 36.3). Also, there was reduced ability by healthy aphids to acquire viruses from oil-treated PVY-infected plants. On the contrary, infected aphids could transmit PVY to healthy potato plants untreated with oil.

It can be suggested that the healthy aphids failed to acquire viruses from the oil-treated PVY-infected plants because of retention of the viral particles in the oil coat of the treated plants, preventing the virus from being sucked up by the aphid with the plant sap (Gibson *et al.*, 1988). The significant reduction of PVY by mineral oil agrees with the findings of Simons *et al.* (1977).

On the basis of these findings, 6% oil emulsion may be recommended as preventive treatment for use with insecticide. As observed, the rates of infected plants are reduced by 80% by insecticide + oil application, greatly improving the health quality of the produce. In the laboratory, the contribution of mineral oils to reduction of PVY infection is greater even without insecticides. Oil treatments of infected plants limit PVY propagation at less than 5% (required by legislation of seed production)

Table 36.2. Effect of two treatments (Treatment 1, insecticide; Treatment 2, insecticide + oil) on aphid populations at Korba 2008.

Treatments	Reduction in aphids (%)[a]
Treatment 1	82a
Treatment 2	80a

[a]Reduction in aphids compared with the control as calculated with the Abbott formula. Means followed by common letters are not significantly different at 5% level.

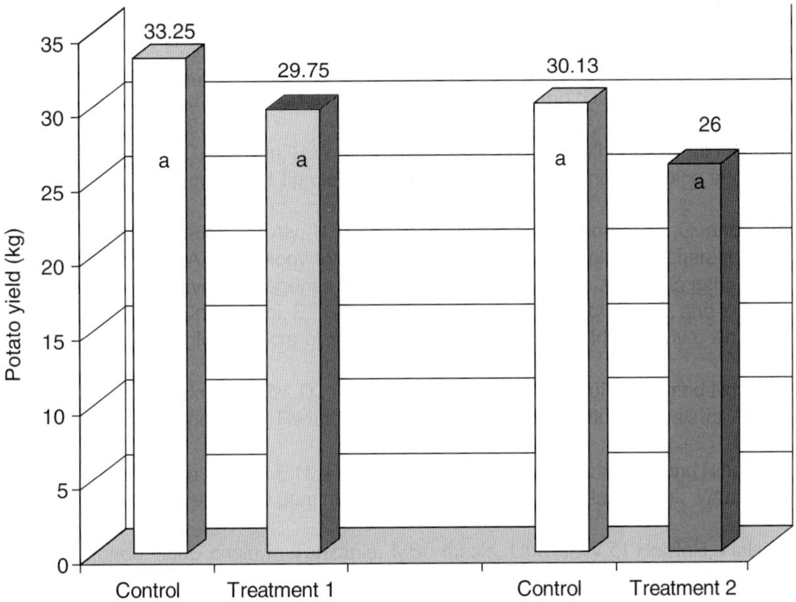

Fig. 36.2. Comparison of potato yield in the two treatments (Treatment 1, insecticide; Treatment 2, insecticide + oil) at Korba 2008. There is no significant difference between bars with the same letter.

Table 36.3. Ability of *M. persicae* to transmit PVY between potato plants in the presence/absence of mineral oil treatment (ELISA results[a] after 15 days inoculation of healthy potato plants).

	Transmission by healthy aphids from oil-treated infected potato to healthy potato[b]	Transmission by infected aphids feeding on healthy potato plants	
		With mineral oil treatment	Without oil treatment
Number of infected potato plants	0 (+)/40	0 (+)/40	38 (+)/40

[a](+), Positive samples by DAS-ELISA.
[b]Healthy aphids fed on oil-treated PVY-infected potato for 2 min and were then transferred to feed on healthy potato plants to test if the aphids could transmit PVY to the healthy plants.

while plants that were not treated with oil show a level of 9.5% infection. The efficiency of the oil may be preserved by applying a uniform coat to the leaves. Better results were obtained in the laboratory with oil because leaves could be given a good coat compared with the field application. A uniform oil application is needed to limit PVY transmission. Finally there were no significant yield losses in potato fields treated with oil compared with non-treated fields and no phytoxicity was observed. This work shows that farmers' bias against oils as a control option for aphids is unfounded.

Acknowledgements

The authors are grateful to the all the staff of GIL (Groupement Interprofessionnel des Légumes, the international organization for vegetables including seed potato production in Tunisia) for their technical support.

References

Al-Mrabeh, A., Anderson, E., Torrance, L., Evans, A. and Fenton, B. (2010) *A Literature Review of Insecticide and Mineral Oil Use in Preventing the Spread of Non-persistent Viruses in Potato Crops*. Potato Council, Agriculture and Horticulture Development Board, Stoneleigh Park, Kenilworth, UK, 65 pp.

Berger, P.H. and Pirone, T.P. (1986) The effect of helper component on the uptake and localization of potyviruses in *Myzus persicae*. *Virology* 153, 256–261.

Boukhris-Bouhachem, S., Rouzé-Jouan, J., Kerlan, C., Glais, L. and Hullé, M. (2011) Transmission of PVY^{NTN} by major aphid species dispersed in Tunisian potato crop under controlled conditions. *Plant Pathology Journal* 10(1), 22–28.

Bradley, R.H.E.W., Wade, C.V. and Wood, F.A. (1962) Aphid transmission of potato virus Y by inhibited by oil. *Virology* 18, 327–329.

Cornu, P. and Gehriger, W. (1981) The protection of potatoes against infection by virus Y by means of mineral oil treatments. *Revue Suisse d'Agriculture* 13, 97–102.

Gibson, R.W., Payne, R.W. and Katis, N. (1988) The transmission of potato virus Y by aphids of different vectoring abilities. *Annual Review of Biology* 113, 35–43.

Groupement Interprofessionnel des Légumes (GIL) (2003) *Annual Report*. Ministry of Agriculture, Tunis, Tunisia. 42 pp.

Hein, A. (1972) Unterschungen zur Wirkung von Ölen bei der virusübertragung durch blattläuse. I Wirkung von Ölen auf mechanische Virusübertragungen. *Phytopathology* 74, 126–130.

Martín-López, B., Varela, I., Marnotes, S. and Cabaleiro, C. (2006) Use of oils combined with low doses of insecticide for the control of *Myzus persicae* and PVY epidemics. *Pest Management Science* 62, 372–378.

Merlet, J. and Le Hingrat, Y. (1996) Production de plants. In: Rousselle, P., Robert, Y. and Crosnier, J.C. (eds) *La Pomme de Terre*. Institut National de la Recherche Agronomique (INRA), Paris, pp. 415–448.

Rolot, J.L. (2005) Analyse des facteurs régulant la dissémination du virus Y de la pomme de terre (PVY) en vue de stratégies de lutte raisonnées. Thèse de Doctorat, Faculté Universitaire des Sciences Agronomiques de Gembloux, Gembloux, Belgium, 252 pp.

Simons, J.N. and Zitter, T.A. (1980) Use of oils to control aphid-borne viruses. *Plant Disease* 64, 542–546.
Simons, J.N., McLean, D.L. and Kinsey, M.G. (1977) Effects of mineral oil on probing behaviour and transmission of stylet-borne viruses by *Myzus persicae*. *Journal of Economic Entomology* 70, 309–315.
Van der Zaag, D.E. (1987) Yield reduction in relation to virus infection. In: De Bokx, J.A. and Van der Want, J.P.M. (eds) *Viruses of Potatoes and Seed Potato Production*. Pudoc, Wageningen, The Netherlands, pp. 146–150.
Vanderveken, J.J. (1977) Oils and other inhibitors of non-persistent virus transmission. In: Harris, K.F. and Maramorosh, K. (eds) *Aphids as Virus Vectors*. Academic Press, New York, pp. 435–454.
Wróbel, S. (2007) Effect of mineral oil on *Myzus persicae* capability to spread of PVY and PVM to successive potato plants. *Journal of Plant Protection Research* 47, 4.

37 The Farming Systems of Potential Potato Production Areas of Chencha, Southern Ethiopia

W. Mazengia,[1]* R. Schulte,[2] Y. Tadese,[1] D. Griffin,[3] S.Schulz[4] and P.C. Struik[1]
[1]Wageningen University, Wageningen, The Netherlands; [2]Teagasc, Johnstown Castle, Ireland; [3]Teagasc, Oak Park, Ireland; [4]International Potato Center (CIP), Addis Ababa, Ethiopia

Abstract

A survey was conducted in 2012 to assess the diversity of farming systems in the potential potato production areas of Chencha in southern Ethiopia. It was part a PhD research study on evaluation and optimization of sustainability of farming systems which is part of an initiative to make the area a potato centre of excellence. Household surveys (n = 57) and farmers' group discussions were used to collect data from 21 villages. A mixed farming system was the prevailing system in the area where crops and livestock are simultaneously grown. The major crops grown in the area were potato, enset, wheat, barley and kale. Most crops were grown as food and cash crops. Barley and enset were mainly grown for household consumption. Farmers used improved varieties mainly for potato (73% of respondents) and wheat (77%). There were two cropping seasons per calendar year and crop rotation is common in the area. The dominant rotation system practised by 95% of the respondents was planting potato followed by wheat or barley. The common intercropping practices were mixed intercropping of barley with lentil and wheat with linseed. Based on local classification there were 12 soil types. The most common ones were Modo (dark loam), Gobo (red clay) and Kalta (brown clay). Fertilizers used were farmyard manure (97% of households), compost (40%), urea (78%) and diammonium phosphate (DAP) (81%). Farmers used low rates of inorganic fertilizers due to shortage of cash. Most of the households (55%) obtained a cash income from agricultural activities, mainly from crop production. The most important off-farm activity was weaving. Household food demands were met from own farm and external sources. Constraints of the farming system that need research and policy interventions include soil fertility depletion, low productive crop varieties, shortages of land, improved seed and cash.

Keywords: Chencha, Ethiopia, farming system, potato, *Solanum tuberosum*

37.1 Introduction

Potato has been cultivated in Ethiopia since 1858. Its production area has reached 59,504 ha in the country and 8978 ha in the Southern Nations Nationalities and Peoples' Region (SNNPR) in the main cropping season (CSA, 2012). Chencha is one of the areas in SNNPR where potato is one of the most important crops. Shortage of seeds of improved

*waga966@yahoo.com

varieties is one of the bottlenecks of potato production in the area. Non-governmental organizations such as Vita and World Vision have recently been involved in disseminating improved potato varieties. Vita has capacitated and organized farmers to produce and sell seed potato tubers at the local level. Recently, a research-based development project has been launched to make Chencha a centre of excellence for seed potato with the technical and financial support of Vita, Teagasc and Wageningen University. One of the aims of the project is to ensure sustainability of the farming system while producing potato in Chencha. A farming system is a system where different farm enterprises that ensure growth, stability and overall productivity are integrated (Taunk and Shrivastava, 2007). Potato is one of the components of the farming systems in Chencha and its production is linked with other prevailing farm and off-farm enterprises. Therefore, there was a need to conduct an initial survey to understand the major crop production system and the bottlenecks of production in the potential potato production areas of Chencha. This chapter presents the findings of this survey.

37.2 Materials and Methods

The study area

The study area was Chencha which is a district in the SNNPR of Ethiopia. The area of the district is 373.5 km^2 with a human population density of 388 persons/km^2 (CSA, 2011). The altitude ranges from 2000 to 3000 m above sea level (masl). The agroecology is classified as highland (> 2500 masl) and midland (2000–2500 masl). The district is divided into 45 lower administrative units which are locally known as kebeles. Potato is produced in all kebeles. However, the potential potato production kebeles are 33 out of which five were selected for the study in consultation with experts of Chencha Office of Agriculture (COoA). Four of these kebeles represent the highland whereas one kebele represents the midland agroecology. A total of 21 villages were selected from all kebeles. Each kebele was represented by two to six villages (Table 37.1).

Farmer selection

A total of 57 households were selected randomly from all selected kebeles. Care was taken to make sure that different wealth classes and genders were included. Nine to 14 households were selected to represent each kebele. Local classification of wealth classes was adopted for selection. Indicators used to classify wealth status were: (i) type and number of houses; (ii) number of cattle and horses; (iii) size of land holding; (iv) age of owned enset plants; and (v) amount of bamboo and eucalyptus trees. The proportions of selected rich, medium and poor households were 30%, 40% and 30%, respectively.

Data collection and analysis

Individual household interviews were conducted using a questionnaire to collect data on crops, soils, and sources of income, food and inputs. The interviews were conducted with the household head from October to November 2012. Group discussions, key informant interviews, secondary data and experts' estimates were also used to collect

Table 37.1. Selected kebeles and number of villages and their agroecology.

Kebeles	Number of villages	Altitude[a] (masl)	Agroecology
Losha	4	2749	Highland
Yuera	5	2600	Highland
Laka	6	2578	Highland
Tegecha	4	2250	Midland
Gendo Gembela	2	2640	Highland

[a]The altitudes were taken using a GPS (global positioning system) where the interviews were conducted and might not show the range of altitudes in a kebele.

some data such as productivity and constraints. The feeling method and visual observation were used to identify soil texture and colour, respectively. The household data were analysed using SPSS software.

37.3 Results and Discussion

Rainfall pattern

The rainfall pattern is bimodal (Fig. 37.1) resulting in two cropping seasons. The first season which is locally known as *Belg* extends from March to May whereas the second season called *Meher* is between June and October. The rainfall data from the COoA compound (unpublished) revealed that the average monthly rainfall was at a peak (179 mm) in April in the *Belg* season although the total rainfall received was higher in *Meher* (506 mm) than in *Belg* (327 mm). The average total annual rainfall varies over years ranging from 830 mm to 1679 mm in 2003 and 2006, respectively (Fig. 37.2). Farmers complained that the distribution of the rainfall has changed over time and affected the performance of crops. Rainfall amount and distribution are commonly unpredictable in Ethiopia limiting the choice of crops, varieties, planting times and production (Belay, 2003).

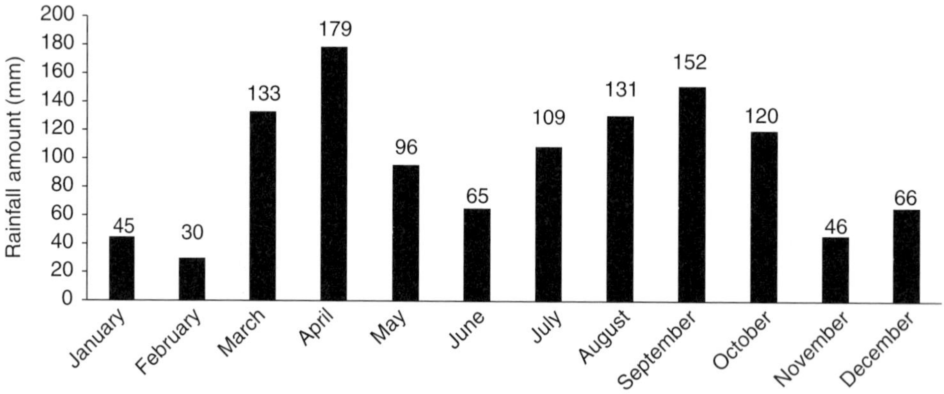

Fig. 37.1. Monthly average (8 years) rainfall at the Chencha Office of Agriculture (COoA) compound.

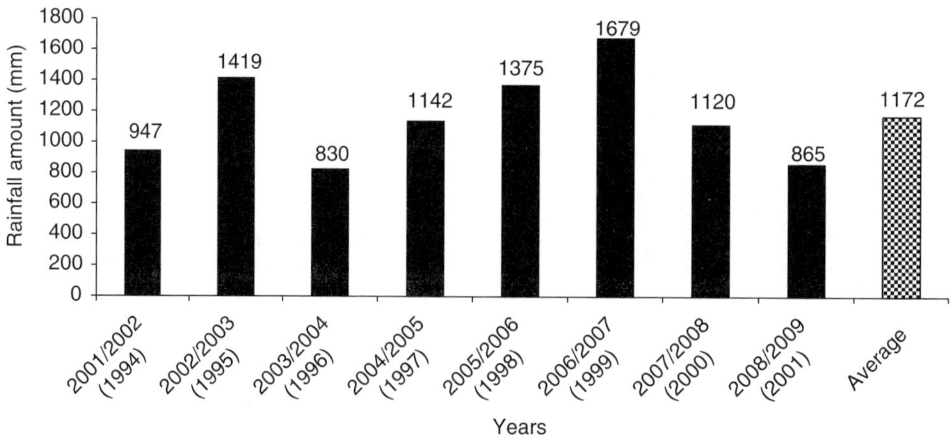

Fig. 37.2. Annual total rainfall in different years at the COoA compound.

Crop husbandry

Types of crops grown and their importance

There were 28 crop types grown in different agroecologies of the district (Table 37.2). The diversity of crop types included cereals, pulses, root and tuber crops, vegetable and fruit crops, oil and stimulant crops. The number of crops grown per household ranged from five to 14. However, most of the households grew seven to eight crops. The major crops grown by most (60–100%) of the households were potato (*Solanum tuberosum*), enset (*Enset ventricosum*), wheat (*Triticum aestivum*), barley (*Hordeum vulgare*), kale (*Brassica oleracea*), faba bean (*Vicia faba*) and apple (*Malus domestica*) (Fig. 37.3).

Farmers mainly used improved varieties for a few crops including potato (73%), wheat (77%), maize (72%) and barley (21%) (Table 37.3). Few farmers used improved varieties of faba bean and field pea. Rich farmers were more inclined to use improved varieties of relatively more crops than medium and poor farmers. For instance, among the potato growers 100%, 86% and 33% of rich, medium and poor farmers, respectively, used improved varieties of potato. The major sources of seeds were own farm and external sources for all wealth and gender classes. Female-headed households depended more often on their

Table 37.2. Types of crops grown in different agroecologies of Chencha district.

Crops grown in the highlands only	Crops grown in the highlands and midlands	Crops grown in the midlands only
Faba bean (*Vicia faba*)	Wheat (*Triticum aestivum*)	Maize (*Zea mays*)
Fenugreek (*Trigonella foenum-graecum*)	Barley (*Hordeum vulgare*)	Sorghum (*Sorghum bicolor*)
Field pea (*Pisum sativum*)	Triticale (× *Triticosecale*)	Tef (*Eragrostis tef*)
Lentil (*Lens culinaris*)	Potato (*Solanum tuberosum*)	Haricot bean (*Phaseolus vulgaris*)
Chickpea (*Cicer arietinum*)	Enset (*Enset ventricosum*)	Sweetpotato (*Ipomoea batatas*)
Arisaema schimperianum	*Plectranthus edulis*	Coffee (*Coffea arabica*)
Beetroot (*Beta vulgaris*)	Kale (*Brassica oleracea*)	Sugarcane (*Saccharum officinarum*)
Carrot (*Daucus carota*)	Cabbage (*Brassica oleracea*)	
Linseed (*Linum usitatissinica*)	Garlic (*Allium sativum*)	
Plum (*Prunus domestica*)	Onion (*Allium cepa*)	
Apple (*Malus domestica*)		

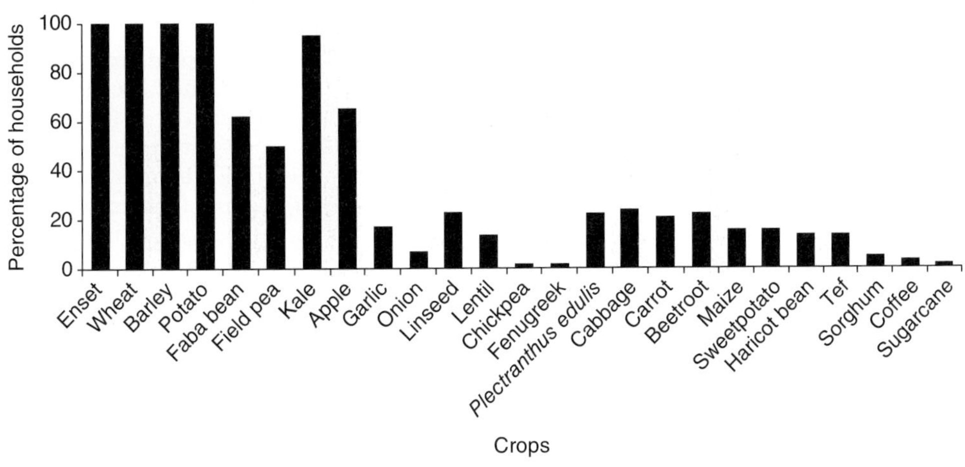

Fig. 37.3. Percentage of households cultivating different crops in the Chencha area.

Table 37.3. Percentage of households using improved varieties of crops and seed sources per wealth status and gender.

Crops and seed sources	Percentage of respondents and number of samples (*n*) by wealth status and gender										Total (%)
	Wealth status						Gender				
	Rich		Medium		Poor		Male		Female		
	%	*n*	%	*n*	%	*n*	%	*n*	%	*n*	
Crops											
Potato	100	17	86	22	33	18	–	–	–	–	73
Wheat	94	17	82	22	56	18	–	–	–	–	77
Barley	29	16	27	22	6	18	–	–	–	–	21
Faba bean	12	16	6	18	0	13	–	–	–	–	6
Field pea	6	16	0	18	0	13	–	–	–	–	2
Maize	100	1	50	4	67	3	–	–	–	–	72
Seed sources											
Own farm	26	5	17	3	28	6	14	12	33	2	24
Own farm and external	74	13	83	19	72	12	86	34	67	10	76

own farm as the seed source than male-headed households. Assessment reports indicate that only 4.6% and 8.4% of sample farmers from SNNPR (n = 852) and Ethiopia (n = 4587), respectively, use improved varieties of crops (Amsalu, 2006), whereas the majority (80% and 75%, respectively) use local cultivars.

Purposes of growing crops

The main purpose of growing crops was either as sources of food (household consumption) or income or both depending on the crop type and wealth class. The majority of rich and medium households grew enset, wheat, potato and sweetpotato as sources of food and cash whereas the majority of poor households grew these crops as sources of food. Barley and haricot bean were mainly grown for food whereas apple was produced mainly as a cash source by all wealth classes. Vegetables were grown mostly by rich and medium farmers for home consumption and the market. Generally, poor farmers grew most of their crops primarily to meet their household's food demands implying that poor farmers do not have surplus produce to sell. About 56–85% of households in SNNPR use most of their crop products for home consumption (CSA, 2010). Relative to other crops, wheat, onion and coffee were largely used for sale by 30%, 42% and 55% of households, respectively. Potato was used for food, seed and sale by about 61%, 27% and 11% of households, respectively.

Overall importance of crops

Results of ranking the overall importance of crops within the wealth classes revealed that enset was the most important crop for all wealth classes for the main reason that enset is resilient to weather fluctuation and its product is available throughout the year in all households. Seifu (1996) also stated that enset is the best adapted crop in areas where land holding is small because of its high productivity. Potato was the second most important crop for rich and medium farmers while it was the third most important crop (like wheat) for poor farmers whose second most important crop was barley. The third and fourth most important crops for rich and medium farmers were barley and wheat, respectively, while the third most important crop for poor farmers was potato or wheat. Overall, the most important crops in the area were enset (84%), barley (40%), potato (36%) and wheat (28%) irrespective of wealth classes (data not shown).

Cropping systems

Sole cropping was most common in all study areas with some practices of intercropping. The most common intercropping practices in highland areas were mixed cropping of barley with lentil and wheat with linseed. Other intercropping practices were wheat mixed with a small proportion of barley, enset with kale and apple with either kale or garlic. *Arisaema schimperianum* was not usually grown as a sole crop because it takes a long time to mature. Therefore, it was intercropped with different crops except field pea (whose yield is tremendously affected by intercropping) and triticale which highly affected *A. schimperianum*. There was a little practice of mixed cropping of faba bean with a small proportion of field pea. The main reasons for intercropping were shortage of land and to utilize the wide open spaces of permanent crops. *Plectranthus edulis* was not intercropped with any other crop as it is easily affected through competition with other crops. Maize–haricot bean intercropping was also common in the midland areas.

Double cropping and crop rotation in two seasons per year were very common practices in all areas. There were different practices of crop rotation arrangements. The most dominant double cropping or rotation system practised by about 95% of the respondents was planting potato followed by either wheat or barley. Other important practices included barley followed by either potato or wheat, faba bean, or field pea. Usually the same crop was not successively double cropped although sometimes barley, potato and sweetpotato were double cropped. Most vegetables were planted in home gardens and they usually succeeded each other. Pulse crops were commonly sown as the succeeding crop in double cropping.

Sowing dates

Some crops such as potato, barley and kale are sown in both *Belg* and *Meher* seasons (Table 37.4). Potato is grown more in the *Belg* season. *P. edulis* can be planted in the beginning of the two seasons. Kale has a wide range of sowing dates extending from February to September but the best time is the beginning of the *Meher* season. Most crops in the highland areas are sown in the *Meher* season predominantly in August whereas most crops in the midlands are sown in the *Belg* season. Enset can be planted before the onset of rainfall as it is less affected by a delay of rainfall. Vegetables could be produced either using irrigation during the off season or as rain fed. The first 1 or 2 months before planting (transplanting) are seedling-raising times particularly for cabbage. *A. schimperianum* is usually planted immediately after harvest and there is no specific planting date as the crop is usually grown as a volunteer. Generally, farmers determine specific sowing dates of most crops depending on the onset of the rainfall.

Seeding rates, harvesting dates and productivity of crops

Variations were observed with the seeding rates used by farmers across villages for the same crops (Table 37.5). The seeding rates used for local cultivars of potato were 600 kg/ha and 1600 kg/ha in Gendo Gembela and Laka areas, respectively. However, farmers who had access to improved varieties of potato with training used seeding rates similar to the national recommendation which is 1800–2000 kg/ha (MoA, 2006). The seeding rate used for wheat in Gendo Gembela area was 88 kg/ha whereas in Laka it was 125 kg/ha and 160 kg/ha for improved and local varieties, respectively. The national recommendation is 125–175 kg/ha. The seeding rates used for faba bean and field peas were low but that of barley was near to the recommended rate.

Potato is ready to be harvested within 3–4 months (Table 37.5). Most cereal crops were harvested within 4 months with exceptions of triticale and sorghum which were harvested in 9 months. Depending on the cultivar type and management, enset matured between 36 and 120 months. However, most cultivars matured between 60 and 84 months and most cultivars grown for

Table 37.4. Sowing dates of different crops throughout a year in Chencha. (From group discussion with farmers from Gindo Gembela (GG), Losha (Lo), Laka (La) and Tegecha (Te).)

Crop type	Jan	Feb	Mar	Apr	May	Jun	Jul	Aug	Sep	Oct	Nov	Dec
Enset		La	La									
Wheat							La	GG, Lo, La				
Barley												
Potato												
Faba bean							La	GG, Lo				
Field pea												
Kale												
Linseed												
Lentil												
Plectranthus edulis		GG			Lo							
Arisaema schimperianum												
Cabbage, carrot, beetroot, onion and garlic		Transplant – rain fed GG, Lo, La		La					Raise seedlings		Transplant – for irrigation	Raise seedlings
Apple							GG, Lo	La				
Maize			Te									
Sweetpotato	Te	Te	Te	Te	Te	Te	Te	Te	Te	Te	Te	Te
Haricot bean			Te									
Tef			Te					Te				
Sorghum			Te									
Coffee				Te								

Table 37.5. Seeding rate, harvesting dates and productivity of major crops across locations in Chencha.

	Seeding rate (kg/ha)		Harvesting dates (months)			Yield (kg/ha)			
						Estimated by farmers		Annual report[a]	
Crop type	Gendo Gembela	Laka	Gendo Gembela	Losha	Laka	Gendo Gembela	Laka	*Belg*	*Meher*
Potato	600–2,000	1,600–2,000	3–4	4	3	7,000, 2,000 (local)	24,000, 12,000 (local)		18,500
Wheat	88	125–160	4	4	5	1,200–2,200	1,200		2,600
Triticale	–	–	9	–		1,600	2,000		
Barley	120	140	4	4	5	1,600	1,600	3,690	(152)
Maize	–	–	–	–	5–6	–	–		
Sorghum	–	–	–	–	9	–	–		
Tef	–	–	–	–	3.5	–	–		(450)
Faba bean	40	–	4	4	5	500	–		940
Field pea	80	–	4	4	5	300	–		840
Lentil	–	–	4	4	5	–	–		
Linseed	–	–	4	4	5	–	–		
Haricot bean	–	–	–	–	3	–	–		
Enset[b]	1 × 1m	2 × 2m	36–72	60	36–84	–	–	7,000 (11,464)	
Kale	–	–	3	3	3	–	–		(7,275)
Cabbage, carrot, beetroot, onion and garlic	–	–	3	3	4–5[c]	–	–	19,000	19,000
Plectranthus edulis	–	–	7	7	9	–	–		
Arisaema schimperianum	–	–	36	≥36	≥24	–	–		
Sweetpotato	–	–	–	–	4–6	–	–	18,000	
Apple	–	–	36–60	36–60	36–60	–	–	(6,857)	
Coffee	–	–	–	–	36	–	–		

[a]Annual (2012) report of COoA (unpublished). Numbers in parentheses were taken from report (2012) of Gendo Gembela Development Office.
[b]Enset yield estimates did not consider growing years (average = 6).
[c]Onion and garlic are harvested in 5 months while other vegetables are harvested in 4 months.

their corm product matured in 36 months. The first fruit setting of apple ranges from 36 to 60 months depending on the crop management.

The productivity of potato was variable across locations (Table 37.5). In Gendo Gembela, it ranged from 2000 to 7000 kg/ha for local cultivars and improved varieties, respectively. This productivity of potato was lower than its productivity in the 2012 cropping season both at the national (7989 kg/ha) and the regional (9015 kg/ha) level (CSA, 2012). But, the productivity of potato in Laka was relatively high (i.e. 12,000 kg/ha and 24,000 kg/ha for local and improved varieties, respectively). The productivity of potato in Chencha was very high (18,500 kg/ha) as reported by COoA. The productivity of wheat ranged from 1200 to 2200 kg/ha for local and improved varieties, respectively, while the national and regional productivity was about 2000 kg/ha in 2012 (CSA, 2012). The productivity of enset ranged from 1167 to 1911 kg/ha/year. However, enset yield can reach 7000 kg/ha/year (Hiebsch, 1996).

Tillage and weeding frequency

Frequency of ploughing for the same crops was more or less similar in different areas (Table 37.6). The differences observed were related mainly to the methods of ploughing and whether the land was previously virginal or cultivated. More frequent ploughing is required for oxen ploughing and virginal land. Hand ploughing turns the soil relatively deeper and is used for crops such as enset. The most common ploughing frequency for most crops including potato was four times. The frequency of ploughing for field pea, faba bean and linseed was usually once.

The frequency of weeding varies per crop type and depending on weed growth and method of planting (Table 37.6). A potato field is weeded three times in most areas, however, it varies from one to three times for row planting and local practice, respectively, in Laka. The frequency of weeding for young perennial crops such as enset and apple are higher than old plant stands. It is not common to weed the fields of *A. schimperianum* when it is grown as a sole crop. Wheat and barley plots are weeded once in all locations. Generally, the target of weeding in all areas is mainly to use weeds as feed for livestock and not to avoid the adverse effect of weeds (Farm Africa, 1994).

Soil types and fertility management

Based on local classification, there were different types of soils in the study areas. However, the most common occurring soil types across locations were Modo (69%), Gobo (67%) and Kalta (50%). The colour of most soil types were red and brown. The soils texture was predominantly clay (data not shown).

Fertilizer types used in the study areas were diammonium phosphate (DAP) (81% of households), urea (78%), farmyard manure (97%) and compost (40%). DAP was used by 100%, 86% and 56% of rich, medium and poor farmers, respectively (data not shown). Practices of using urea by wealth class were more or less similar to the practices of DAP. More households used farmyard manure than either DAP or urea in all wealth classes. Practices of use of inorganic fertilizers were negatively correlated with practices of organic fertilizers particularly compost ($r = -0.81$ and -0.88 for DAP and urea, respectively). Farmers also use leaves of *Croton macrostaches*, *Erythrina abyssinica* and *Hagenia abyssinica* to reduce soil acidity in Chencha (Wassie and Shiferaw, 2009).

Most farmers used low amounts of chemical fertilizers due mainly to a shortage of cash. Most rich households used 100 kg DAP (39%) and 50 kg urea (44%) per year for all crops grown with inorganic fertilizers (data not shown). Most medium households used 50 kg and 25 kg DAP and urea, respectively. The majority of poor households did not apply DAP and urea fertilizers. Generally, the average amounts of fertilizer used per household were 63.5 kg and 29.1 kg DAP and urea, respectively.

Farmers used different types of fertilizers for different crops. Most rich and medium households used more inorganic fertilizer than organic for some crops such as potato and wheat. Most poor farmers applied either

Table 37.6. Ploughing and weeding practices for different crops across locations in Chencha.

	Frequency of ploughing over location[a]			Frequency of weeding over location		
Crop type	Gendo Gembela	Losha	Laka	Gendo Gembela	Losha	Laka
Potato	4	4	4	3	3	1–3
Enset	3	1	1–2	[b]	[b]	1–2
Wheat	4	4	3–4	1	1	1
Barley	4	4	3–4	1	1	1
Faba bean	1	1	2	1	1	1
Field pea	1	1	2	1	0	0
Kale	4	1	3	3	–	4–5
Linseed	4	4	1	1	–	1
Lentil	4	4	1	1	–	1
Plectranthus edulis	4	1	3	3	2	3
Arisaema schimperianum	1	1	1	0	0	0
Cabbage, carrot, beetroot, onion and garlic	4	4	3	1	–	3
Apple	4	4	3	–	–	1
Maize	–	–	3	–	–	2
Sweetpotato	–	–	3	–	–	2
Haricot bean	–	–	3	–	–	1
Tef	–	–	4	–	–	2
Sorghum	–	–	3	–	–	1
Coffee	–	–	–	–	–	1

[a]Includes seed bed preparation and seed covering.
[b]Depends on weed growth.

inorganic or organic fertilizer alone for potato but organic fertilizer alone for wheat. Most maize growers used inorganic fertilizer for the crop. At the national level, tef, wheat and maize account for the majority of inorganic fertilizer use (Spielman *et al.*, 2011). Organic fertilizer alone was more often used for some crops such as barley, enset, vegetables (cabbage, beetroot and carrot) and apple. Crops such as faba bean, field pea, sweetpotato and tef were usually grown without any fertilizer by most of the households.

Constraints of crop production

The most important challenge in all locations was enset disease (Table 37.7) probably caused by bacterial wilt which is widely spread in the highlands (Quimio and Mesfin, 1996). Enset bacterial wilt has been a problem in Ezzo Tula Kebele (Farm Africa, 1994). The second important constraint in Losha was potato late blight (*Phytophthora infestans*) which was the third and fourth important constraint in other areas. Millipedes were the other important problem for potato production in all locations. Prevalence of potato diseases and insect pests were also important problems in the country as a whole (MoA, 2011). Similarly, an assessment study conducted in 2008 by South Agricultural Research Institute (SARI) showed that crop disease, including enset and potato disease, were the second priority problems in Doko Shaye Kebele of Chencha. Yellow rust was a challenge for newly introduced wheat varieties in all locations. When it is severe, yellow rust (caused by *Puccinia striiformis*) can result in a yield loss of 100% on wheat (MoA, 2012). Disease occurrence was one of the problems of using improved varieties in many areas of the country (Amsalu, 2006). Weeds were not considered as a problem in most locations. The reason might be because of the importance attached to weeds as a livestock feed. *Guizotia scarab*, however, was

Table 37.7. List of constraints of crop production and their ranks over locations in Chencha.

Constraints	Ranks[a]		
	Gendo Gembela	Losha	Laka
Enset bacterial wilt	1	1	1
Potato diseases	4	2	3
Millipedes on potato	5	4	4
Yellow rust on wheat	–	6	7
Aphid on field pea and kale	6	–	x
Weeds	–	–	x
Decline in soil fertility	–	5	6
Unpredictable rainfall	3	–	2
Shortage of improved seeds	2	3	x
Low productivity of local cultivars	–	–	x
High cost of fertilizer	–	–	8
Land shortage	–	–	x
High cost of grain	x	–	x
Shortage of oxen	–	x	x
Shortage of labour	x	x	9
Vertebrate pests (monkeys/apes and/or porcupine)	x	x	5

[a]A constraint ranked as 1 is the most important constraint to crop production; –, the problem exists in the area but is not ranked; x, not reported as problem.

considered a problem in cereals only in one location. Porcupine and apes were problematic in Laka and Tegecha.

Shortage of improved crop variety seed and unpredictable rainfall were the second most important constraints in Gendo Gembela and Laka areas. Shortage of improved and quality seed was one of the problems of the potato value chain in SNNPR (Bezabih and Mengistu, 2011). Such seed shortages can be partly alleviated by involving farmers and producing seeds on farmers' fields (Agdew *et al.*, 2012). The high cost of chemical fertilizers was also a challenge while decline in soil fertility is aggravating over time. High cost of fertilizer was also a key problem at the national level (Amsalu, 2006). Labour shortage was a problem in Laka where about 62% of the households were female-headed households whose spouses migrated mainly to Addis Ababa for weaving, resulting in a shortage of labour for the women. Land shortage was also a problem of most households.

Sources of food and income

Most households (76%) fulfilled their food demand from their own farm products and external sources (Table 37.8). Poor households were more dependent on external food sources than rich and medium households. Crop products were the main on-farm income sources practised by most households, particularly by rich and medium and male-headed households. Potato was one of the main income sources particularly for rich households. Weaving was a good off-farm income source mainly practised by poor and male-headed households. Trading was more practised by rich and female-headed households. Generally, crop products were ranked as the most important source followed by weaving (Table 37.8). Comparison of the overall on-farm and off-farm income sources also revealed that on-farm activities were very important to rich and medium households whereas off-farm income sources were more important to poor households (data not shown). A diagnostic survey conducted in Ezzo Tula Kebele also showed the most important income sources of farmers were crops, remittances (remittance is money obtained from relatives in other places) and animals in that order (Farm Africa, 1994).

Table 37.8. Sources of food and income for households (%) within gender and wealth class.

	Gender		Wealth status			
	Male	Female	Rich	Medium	Poor	Total
Food sources						
Own farm	24	25	33	27	11	24
Own farm and external sources	76	75	67	73	89	76
Income sources						
Crop products	93	75	100	100	67	90
Livestock and their products	76	58	78	77	61	72
Weaving	50	17	17	45	67	43
Trading	17	25	44	9	6	19
Selling labour	11	0	0	0	28	9
Remittance	0	25	6	0	11	5
Hiring out oxen	2	0	0	0	6	2
Ranking primary income source						
Crop products	54	58	78	64	22	55
Livestock and their products	2	0	0	4.5	0	2
Weaving	30	8	6	23	50	26
Trading	9	8	11	9	6	9
Selling labour	4	0	0	0	11	3
Remittance	0	25	5.6	0	11	5

37.4 Conclusion and Recommendations

The farming system of Chencha is a mixed farming system where crops and livestock are simultaneously grown. The major crops grown in the area were potato, enset, wheat, barley and kale. The most common soil types were Modo, Gobo and Kalta which are more of a clay type of soil. Crop products were the main sources of income for most farmers. However, crop production was constrained by different challenges that need research and policy interventions. The following are the main constraints and suggestions to overcome these:

- Enset bacterial wilt was the key problem in the area. However, there is no reliable solution to the problem at the moment. Therefore, use of cultural practices and sanitary measures could be important measures to control enset bacterial wilt (Fikire *et al.*, 2012). A relatively tolerant variety called Mazea can be used in areas where the disease is severe.
- Specific potato pest types should be identified to plan proper control measures. Depending on the disease type, tolerant varieties with proper agronomic practices (e.g. crop rotation) and chemical treatments could be used. Practices which are successful to control insect pests elsewhere in the country could be adapted and used in Chencha. Vertebrate pests such as porcupine can be effectively controlled by using different local trapping methods through community participation (Leulseged *et al.*, 2012).
- Crop varieties should be evaluated at local level before being widely disseminated, to make sure that they are adapted to the climate, tolerant of the major diseases and accepted by the society. Tolerant varieties and chemicals could be used to control the yellow rust problem on wheat.
- Weeds could be managed through proper ploughing, cultivation and rotation. However, most farmers considered weeds as one of the main sources of livestock feed. Enhancing other sources of livestock feed may help to control weeds in time.
- Proper practices should be in place to restore soil fertility. Optimum and economical fertilizer rates should be determined and used with efficient

management methods. The current integrated soil and water conservation work of the government should be strengthened to cover all sites. The physical and chemical characteristics of locally classified soil types should be known.

- Practices that improve water use efficiency at farm level could be options to overcome the challenge of intermittent dry spell(s).
- Shortage of seeds of improved crop varieties is a problem that occurs throughout the country. One way of solving the challenge might be to strengthen on-farm production of seeds by farmers of at least the self-pollinating crops.
- Introduction of better on-farm and off-farm income-generating activities would solve problems related to the high cost of fertilizers and shortages of oxen, labour and land. The practice of using available horses and the method of single ox ploughing would also help to partly alleviate the oxen shortage.
- Farmers grow many types of crops using different management practices on relatively small areas of land. However, there is no information on the efficiency of these agricultural components in terms of production, economics and labour use. Therefore, for sustainability of production there is a need to know the efficiency of local management practices that are environmentally friendly.

Acknowledgements

The authors acknowledge the financial support of Teagasc, Vita and Wageningen University and the chapter reviewers. The contributions of staff members of COoA and Mr Seifu Damte from Chencha Vita Office are also highly appreciated.

References

Agdew, B., Waga, M., Fikire, H. and Shimekit, M. (2012) On-farm pre-basic and basic seed production by Ediget Farmers Seed Multiplication Cooperatives Union: practices and lessons from basic seed multiplication farmers of Silti and Soddo Woredas, SNNPR. In: Dauro, D. (ed.) *Proceedings of On-farm Pre-basic and Basic Seed Production in Siltie and Guraghe Zones of SNNPR*, 15 June 2012. Self Help Africa and Hawassa Research Centre/South Agricultural Research Institute (SARI), Hawassa, Ethiopia, pp. 11–38.

Amsalu, A. (2006) Assessment of crop production and natural resources management extension packages in Ethiopia (1994/95–2003/04). *Ethiopian Economic Association/Ethiopian Economic Policy Research Institute, Issue Paper Series* No. 01/2006, December 2006. Ethiopian Economic Association/Ethiopian Economic Policy Research Institute, Addis Ababa.

Belay, S. (2003) Agroclimatic analysis in relation to crop production. In: Engida, M. (ed.) *Proceedings of the National Sensitization Workshop on Agrometeorology and GIS*, 17–18 December 2001. Ethiopian Agricultural Research Organization (EARO), Addis Ababa, pp. 69–85.

Bezabih, E. and Mengistu, N. (2011) *Potato Chain Analysis and Development in Ethiopia: Case of Tigray and Southern Nations Nationalities and Peoples Regions*. International Potato Center (CIP)-Ethiopia and United States Agency for International Development (USAID), Addis Ababa.

Central Statistics Authority of the Federal Democratic Republic of Ethiopia (CSA) (2010) *Agricultural Sample Survey 2009/2010.* Volume VII. *Report on Crop and Livestock Products Utilization (Private Peasant Holding Meher Season). Statistical Bulletin* No. 468. CSA, Addis Ababa.

Central Statistics Authority of the Federal Democratic Republic of Ethiopia (CSA) (2011) *Statistical Abstract 2011*. CSA, Addis Ababa.

Central Statistics Authority of the Federal Democratic Republic of Ethiopia (CSA) (2012) *Agricultural Sample Survey 2011/2012.* Volume I. *Report on Area and Production of Major Crops (Private Sector Meher Season)*. CSA, Addis Ababa.

Farm Africa (1994) *Report of Diagnostic Survey Using Rapid Rural Appraisal (RRA) Techniques of Ezzo Tula Peasant Association in Chencha woreda of Semen Omo Zone*. Report based on the works of participants in RRA training held on 4–16 October 1994 at Arbaminch town organized by Farm Africa. Farm Africa, Awassa, Ethiopia.

Fikire, H., Tariku, H. and Endale, H. (2012) Research achievements, experiences and future directions on bacterial wilt of enset. In: Mohammed, Y. and Tarik, H. (eds) *Enset Research and Development in Ethiopia: Proceedings of Enset National Workshop,* 19–20 August 2010, Wolkitie, Ethiopia. Ethiopian Institute of Agricultural Research (EIAR), Addis Ababa, pp. 64–73.

Hiebsch, C. (1996) Yield of *Ensete ventricosum*: a concept. In: Tsedeke, A., Heibesch, C., Brandt, S.A. and Seifu, G. (eds) *Enset-based Sustainable Agriculture in Ethiopia. Proceedings of the International Workshop on Enset*, 13–20 December 1993, Addis Ababa, Ethiopia. Ethiopian Institute of Agricultural Research (EIAR), Addis Ababa, pp. 15–35.

Leulseged, B., Waga, M. and Mikias, Y. (2012) Mole rat and porcupine as production threats and their options. In: Mohammed, Y. and Tarik, H. (eds) *Enset Research and Development in Ethiopia: Proceedings of Enset National Workshop,* 19–20 August 2010, Wolkitie, Ethiopia. Ethiopian Institute of Agricultural Research (EIAR), Addis Ababa, pp. 110–129.

Ministry of Agriculture (MoA) (2006) *Crop Variety Register*. Issue No. 9. MoA, Addis Ababa.

Ministry of Agriculture (MoA) (2011) *Crop Variety Register*. Issue No. 14. MoA, Addis Ababa.

Ministry of Agriculture (MoA) (2012) *Wheat Production Manual*. MoA and Agricultural Transformation Agency, Addis Ababa.

Quimio, A.J. and Mesfin, T. (1996) Diseases of enset. In: Tsedeke, A., Heibesch, C., Brandt, S.A. and Seifu, G. (eds) *Enset-based Sustainable Agriculture in Ethiopia. Proceedings of the International Workshop on Enset,* 13–20 December 1993, Addis Ababa, Ethiopia. Ethiopian Institute of Agricultural Research (EIAR), Addis Ababa, pp. 188–203.

Seifu, G. (1996) Enset research in Ethiopia: 1976–1984. In: Tsedeke, A., Heibesch, C., Brandt, S.A. and Seifu, G. (eds) *Enset-based Sustainable Agriculture in Ethiopia. Proceedings of the International Workshop on Enset,* 13–20 December 1993, Addis Ababa, Ethiopia. Ethiopian Institute of Agricultural Research (EIAR), Addis Ababa, pp. 204–220.

Spielman, D.J., Dawit, K. and Dawit, A. (2011) Seed, fertilizer and agricultural extension in Ethiopia. *International Food Policy Research Institute, Ethiopia Support Programme II, Working Paper* 020. International Food Policy Research Institute (IFPRI), Addis Ababa.

Taunk, S.K. and Shrivastava, G.K. (2007) Economic analysis of complex interaction in cropping system and allied enterprise of farming systems to augment more income. In: Donatelli, M., Hatfield, J. and Rizzoli, A. (eds) *Farming Systems Design 2007.* Book 1: *Farm-regional Scale Design and Improvement*. International Symposium on Integrated Analysis on Farm Production Systems, 10–12 September 2007, Catania, Italy. The University of Catania, Catania, Italy, pp. 156–157.

Wassie, H. and Shiferaw, B. (2009) Mitigation of soil acidity and fertility decline challenges for sustainable livelihood improvement: evidence from southern region of Ethiopia. In: Atlaw, A., Saba, Y., Alemu, M. and Minale, K. (eds) *Proceedings of the National Conference on Sustainable Land Management and Poverty Alleviation,* December 2009, Addis Ababa, Ethiopia. Co-organized by Environmental Economics Policy Forum for Ethiopia at the Ethiopian Development Research Institute, Sustainable Land Use Forum, Oromia Agricultural Research Institute and College of Agriculture at Hawassa University, Addis Ababa, pp. 131–143.

38 Aphids Infesting Potato in Kenya

H.K. Were,[1]* F.M. Olubayo,[2] J. Kabira,[3] J. Aura[2] and L. Torrance[4]
[1]Masinde Muliro University of Science and Technology, Kakamega, Kenya; [2]University of Nairobi, Nairobi, Kenya; [3]Kenya Agricultural and Livestock Research Organization (KALRO), Limuru, Kenya; [4]The James Hutton Institute, Dundee, UK

Abstract

Aphid-transmitted viruses probably cause greater economic loss in potato production than all other insect-related damage. Some 40 virus species are known to infect potato, and of these 13 are aphid transmitted. Monitoring of aphid populations in potato fields is therefore essential to determine areas with low aphid occurrence suitable for seed potato production and the right time for haulm destruction. While plenty of information on such areas is available in traditional seed-potato-producing countries of America and Europe, there is none available for Kenya. The current study determined the best locations for seed potato multiplication by monitoring aphids (from May 2009 to March 2010) in major potato-producing areas of Kenya. Two methods, yellow water traps (YWTs) and leaf-aphid counts, were utilized. Ten aphid species, *Aphis gossypii* (Glover), *Aphis fabae* (Scopoli), *Aulacorthum solani* (Kaltenbach), *Acyrthosiphon pisum* (Koch), *Brevicoryne brassicae* L., *Cavariella aegopodii* (Scopoli), *Macrosiphum euphorbiae* (Thomas), *Myzus persicae* (Sulzer), *Myzus ascalonicus* (Doncaster) and *Rhopalosiphum maidis* (Fitch) were caught in YWTs while four aphid species, *A. fabae*, *A. gossypii*, *M. persicae* and *M. euphorbiae* were also found colonizing potato leaves. The populations of the aphid species varied significantly, *R. maidis* had the greatest numbers, followed by *B. brassicae*, *A. gossypii*, *M. euphorbiae*, *M. persicae* and *A. fabae* in this order while *A. solani* was the least abundant. The populations of *A. fabae*, *A. solani*, *C. aegopodii* and *M. ascalonicus* did not vary between seasons and sites but *A. gossypii*, *A. pisum*, *B. brassicae*, *M. euphorbiae*, *M. persicae* and *R. maidis* populations varied significantly between the five sites, and between the three seasons. Although aphid numbers recorded at all sites were above the threshold for virus transmission, not all the species caught were efficient virus vectors. In conclusion, Njabini consistently had low aphid numbers and may be the site most suitable for seed potato multiplication.

Keywords: aphid, Kenya, leaves, monitoring, potato, viruses, yellow water traps (YWTs)

38.1 Introduction

Potato (*Solanum tuberosum* L.) is Kenya's second most important food crop after maize. It is both a staple food and a cash crop for many rural and urban families. The total annual production is approximately 1.4 million t (Anon., 2009) from a total area of about 159,994 ha giving an average yield of about 9.5 t/ha (FAOSTAT, 2013), which is well below the

*werehkde@yahoo.com

average of 40 t/ha obtained in well-developed production systems.

Insect pests cause tuber yield or quality reductions of 30–70% if not controlled (Raman and Radcliffe, 1992). Insects with piercing-sucking mouthparts damage potatoes by direct physical injury, extracting phloem, injecting toxic saliva and transmitting pathogens or facilitating their establishment in the host. Although potato has considerable ability to compensate for early season loss of foliage by defoliators, there is little compensation for adverse effects on plant health (e.g. disruption of nutrient transport, reduced photosynthetic efficiency or infection with plant pathogens). Direct physical injury by sap-feeding insects often results in premature leaf senescence. However, aphid-transmitted viruses are thought to cause greater economic losses in potato (through yield reduction) than direct damage by aphids. Some 40 virus species are known to infect potato, and of these 13 are aphid-transmitted (Salazar, 1996).

In many potato production systems, the most commonly occurring aphid-transmitted viruses are *Potato leaf roll virus* (PLRV) and *Potato virus Y* (PVY). Planting seed tubers with a high incidence of secondary (tuber-borne) PLRV or PVY infection can result in yield losses exceeding 50–80% (Beemster and de Bokx, 1987; Basky, 2002). Primary (current season) infection with either PLRV or most PVY strains tends to have little effect on tuber yield. However, primary infection with PVY can affect tuber quality in certain cultivars by causing a phloem defect termed 'net necrosis' that develops in storage while recombinant strains of PVY can induce necrotic lines in tubers of sensitive cultivars. Other aphid-transmitted viruses that occur in Kenya include the potyvirus, *Potato virus A* (PVA) and the carlaviruses, *Potato virus M* (PVM) and *Potato virus S* (PVS) (Were *et al.*, 2014).

Non-persistent viruses such as potyviruses and carlaviruses are transmitted by a large number of aphid species in a non-persistent manner where a virus is acquired in a few minutes while the aphid probes (punctures) the leaves with its stylet and infectivity is lost readily after probing on a healthy plant. On the contrary, persistent viruses such as PLRV are transmitted in a specific manner and require longer acquisition periods; they are transmitted by a few aphid species that feed and colonize the crop (Ng and Perry, 2004).

Currently, aphid control strategies for large-scale seed potato multipliers in Kenya regularly rely on pesticide applications. However, for small-scale farmer-based seed producers (unregistered growers and suppliers of seed mainly in their immediate localities) no aphid control is practised because the capacity of farmers to recognize and manage aphids and virus diseases is very low (Kibaru, 2003; Kabira *et al.*, 2006) and yet these farmers produce 99% of seed potato tubers (KARI, 2007). Although there are many factors involved in seed potato degeneration, aphid vectored viruses are a major cause. While much information is available on aphid occurrence in traditional seed-potato-producing countries in Europe and America, there is very little such information available for Kenya. This study investigated aphid incidence in five major potato-producing areas in Kenya with the aim of determining suitable location(s) with low aphid incidence for seed potato multiplication.

38.2 Materials and Methods

Monitoring of aphid populations was carried out from May 2009 to March 2010 at the following specific locations:

- Meru (elevation: 2300 m above sea level (masl); latitude: 0.05080° N; longitude: 37.225730° E);
- Njabini (elevation: 2545 masl; latitude: 0.74235° S; longitude: 36.64651° E);
- Mt Elgon (elevation: 2630 masl; latitude: 0.47362° S; longitude: 35.69903° E);
- Molo (elevation: 2630 masl; latitude: 0.47276° S; longitude: 35.70024° E); and
- Nairobi (elevation: 1.794 masl; latitude: 1.15501° S; longitude: 36.46278° E).

Sampling was done weekly on two seed potato farms (one field in each farm) at each site using two methods. In the first method, wingless (*apterae*) aphids colonizing potato leaves were sampled from the third week after emergence of the crop until the haulms started yellowing. Three leaves per plant (from top, middle and bottom of the plant) were picked randomly from ten potato plants in each field and placed in labelled paper bags in a cool box and taken to the laboratory for identification and counting as described by Raman (1984).

In the second method, four circular yellow water traps (YWTs) (each measuring 40 cm in diameter × 10 cm in depth) were placed on a platform 60 cm above the ground and at a spacing ranging from 8 m to 15 m depending on the size of the field. The traps were filled with water up to a height of 5 cm. To remove surface tension and to prevent aphids from escaping, a few drops of detergent were added to the water as described by Moëricke (1951). Winged (*alate*) aphids caught in YWTs were collected on a weekly basis continuously from May 2009 until the end of March 2010. Trap catches were placed in a bottle containing 70% ethanol and taken to the lab for identification and counting. The aphids were identified to species level by examination under stereo-telescopic dissecting microscope (Zoom NTB-3A) based on aphid colour, lateral abdominal spiracles, antennal tubercles, shape of siphunculi and dorsal abdominal pigmentation with the help of appropriate keys (Macgillivray, 1979; Blackman and Eastop, 1984; Teulon, 1999). Data for the Scottish aphid suction traps was kindly provided by Science and Advice for Scottish Agriculture (SASA), Edinburgh.

Statistical analysis

Data were analysed by the computer software Statistical Analysis System (SAS) version 9.1. Summary statistics including means and ranges of each aphid species were generated using proc means. Two-way analyses of variance for the effects of location and season on log-transformed aphid populations were conducted using proc glm; two-way interactions between the factors were adopted whenever they were significant, followed by mean separation using the Lsmeans procedure. Proc corr was used to analyse for correlations between aphid incidence and time in weeks or altitude; as well as the correlation between the numbers of different aphid species that were identified. Relative numbers of the identified aphids were analysed using proc genmod with pairwise comparison using the Lsmeans procedure.

38.3 Results

Ten aphid species, the cotton aphid *Aphis gossypii* Glover, the black bean aphid *Aphis fabae* (Scopoli), foxglove aphid *Aulacorthum solani* (Kaltenbach), the pea aphid *Acyrthosiphon pisum* (Koch), the cabbage aphid *Brevicoryne brassicae* L., the carrot aphid *Cavariella aegopodii* (Scopoli), the potato aphid *Macrosiphum euphorbiae* (Thomas), the green peach aphid *Myzus persicae* (Sulzer), the shallot aphid *Myzus ascalonicus* (Doncaster), and the corn leaf aphid *Rhopalosiphum maidis* (Fitch), were caught in YWTs. The total numbers of aphids caught at each site differed significantly ($P = 0.05$) (Table 38.1; Fig. 38.1a). Total aphid population in YWTs was largest in Mt Elgon followed by Meru, Molo and smallest in Njabini. In all sampled areas, the population of aphids increased with growth of the crop then declined as the crop aged.

The relative numbers of the different aphid species caught in YWTs also varied significantly ($P = 0.05$) among the sampled sites. In general, the most abundant aphid species caught in all areas was *R. maidis*, followed by *B. brassicae*. *M. euphorbiae* was the least abundant aphid species in all sites, except Mt Elgon where *A. fabae* occurred least. The numbers of *A. fabae*, *A. solani*, *C. aegopodii* and *M. ascalonicus* did not vary between seasons and locations. In contrast, the numbers of *A. gossypii*, *A. pisum*, *B. brassicae*, *M. euphorbiae*, *M. persicae* and *R. maidis* varied significantly ($P = 0.05$) among the five areas, and between the different seasons (Table 38.1).

Fewer aphids were caught in YWTs in the main growing season in three locations with most aphids found in the second growing season (October–January) in Mt Elgon, Molo and Nairobi. In contrast, in Meru and Njabini, fewer migrant aphids were caught

Table 38.1. Statistical analysis of variation in aphid numbers between seasons and sites.

Aphid species			A.g.[a]	A.f.[b]	A.p.[c]	A.s.[d]	B.b.[e]	C.a.[f]	M.e.[g]	M.a.[h]	M.p.[i]	R.m.[j]	Total
							F values						
Factors													
Site			7.7***	1.3	1.9	0.3	11.1***	0.4	5.4***	1.6	8.1***	17.0***	16.6***
Season			24.3***	1.5	9.6***	0.4	41.1***	0.4	0.8	0.5	15.1***	46.7***	26.2***
Site × Season			2.0*	1.7	2.1*	0.5	4.9***	1.7	3.0*	1.4	5.4***	11.3***	7.6***
							Means						
Overall			10	1	0	0.02	14	1	4	0	4	32	66
Season		Off	3 c	0	0.00 b	0	35 a	0	3	0	7 a	78 a	127 a
		First	18 a	1	0.04 b	0.01	13 b	0	5	0	4 a	44 a	86 b
		Second	6 b	1	0.40 a	0	7 c	1	4	0	2 b	7 b	29 c
Site		Meru	12 a	3	1 a	0	10 b	3	5 b	0	5 a	52 a	98 a
		Molo	6 a	0	0 a	0	13 b	0	12 a	0	6 a	32 b	69 b
		Nairobi	19 a	1	0 a	0.01	5 b	0	2 b	0	3 a	17 bc	47 c
		Njabini	3 b	1	0 a	0	8 b	0	1 b	0	1 b	6 c	20 d
		Elgon	8 a	1	0 a	0	37 a	0	3 b	0	6 a	59 a	115 a
Site × Season	Meru	Off	3 c	1	0 b	0	29 a	0	3 bcd	0	5 cd	234 a	275 a
		First	18 a	0	0 b	0	5 def	0	1 d	0	2 def	23 d	48 g
		Second	12 a	5	1 a	0	6 de	5	8 b	0	7 bc	19 e	64 e
	Molo	Off	5 bc	0	0 b	0	17 bc	0	4 bcd	0	6 cd	18 ef	51 f
		First	10 ab	1	0 b	0	26 b	0	24 a	0	12 ab	92 b	165 c
		Second	4 c	0	0 b	0	4 ef	0	8 b	0	3 de	3 g	22 i
	Nairobi	Off	4 c	0	0 b	0	16 cd	0	5 bc	0	11 ab	13 ef	50 g
		First	36 a	1	0 b	0.02	4 ef	0	2 cd	0	3 de	29 c	76 d
		Second	5 bc	0	1 a	0	2 f	0	1 d	0	1 def	4 fg	14 k
	Njabini	Off	0 c	0	0 b	0	36 b	0	1 d	0	1 def	5 fg	43 h
		First	4 c	0	0 b	0	2 f	0	2 d	0	1 def	6 fg	15 j
		Second	3 c	1	0 b	0	6 de	0	1 d	0	0 f	8 ef	18 i
	Elgon	Off	3 c	0	0 b	0	75 a	0	3 bcd	0	14 a	118 b	215 a
		First	17 a	1	0 b	0	50 b	0	6 b	0	9 abc	122 ab	204 b
		Second	5 bc	1	0 b	0	17 cd	0	1 d	0	1 ef	3 g	29 hi
Range (for all minimum = 0)			510	148	19	1	610	196	205	4	92	846	846
Relative populations (%); *n* = 32,169[§]			14.8 c	1.5 f	0.3 h	0.003 j	21.0 b	0.9 g	6.6 d	0.1 i	5.9 e	48.9 a	100
Correlation with time in weeks (*R* values)			−0.2***	+0.04	+0.2***	−0.06	−0.07	+0.05	+0.01	−0.04	−0.06	−0.2***	−0.07***[α]
Correlation with altitude (*R* values)			−0.1*	−0.08	−0.1*	−0.03	+0.2***	−0.08	−0.04	−0.05	−0.03	−0.02	−0.0004[†]

[a]*Aphis gossypii*, [b]*Aphis fabae*, [c]*Acyrthosiphon pisum*, [d]*Aulacorthum solani*, [e]*Brevicoryne brassicae*, [f]*Cavariella aegopodii*, [g]*Macrosiphum euphorbiae*, [h]*Myzus ascalonicus*, [i]*Myzus persicae*, [j]*Rhopalosiphum maidis*.

***$P \leq 0.001$, **$P \leq 0.01$, *$P \leq 0.05$ (indicating a significant effect); means (column wise) and percentages (rows) with the same letter are not significantly different ($P > 0.05$).

[§]Degree of freedom = 11, $\chi^2 = 77415$, $P < 0.\ 0001$; [α]Overall correlation between time in weeks[α] or altitude[†] with number of aphids regardless of the species.

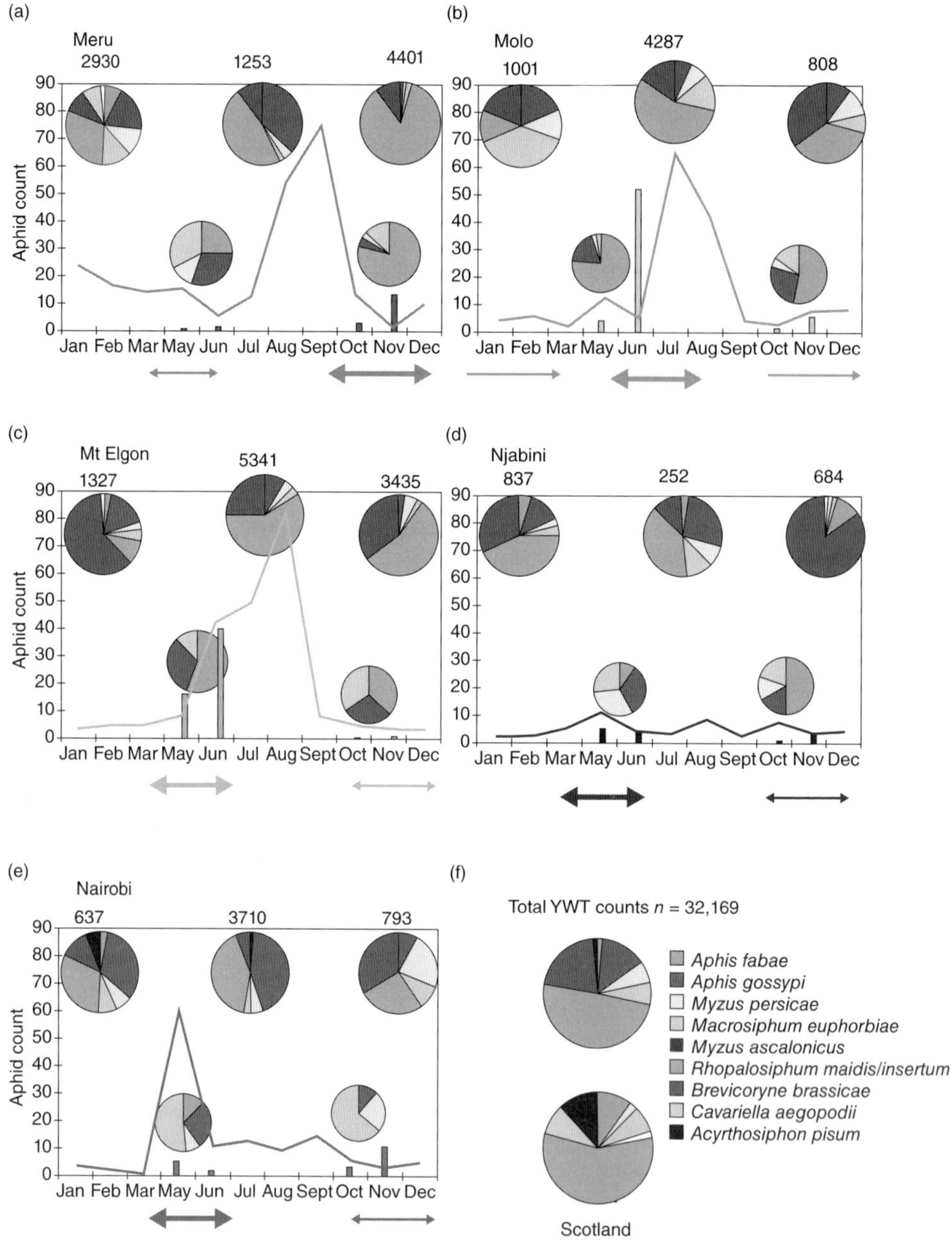

Fig. 38.1. Total aphid numbers and species recorded from each of the five sites: (a) Meru, (b) Molo, (c) Mt Elgon, (d) Njabini and (e) Nairobi. Within each panel is displayed: average total numbers of aphids caught per yellow water trap (YWT) from May 2009 until March 2010, shown as a line; average total numbers of aphids per leaf on the potato crop in May–June and October–November, shown as bars; the species composition caught in YWTs during the main and second growing seasons and off seasons (presented in the three large pie charts from left to right, respectively); the total number of aphids counted is shown above each pie chart; and the species composition on the leaves of the potato plants, shown in the smaller pie charts above the bars in the growing seasons. The thick double-headed arrows denote the main growing season and the thin ones, the second season in each location. On the bottom right are two pie charts showing the species composition from Kenya and the equivalent aphid species from Scottish suction traps.

in the second growing season (March–July and October–December, respectively) than the main season. Aphid monitoring during the off seasons (where no potatoes were being grown) showed that total aphid numbers were lowest in Nairobi, Njabini and Molo (total numbers of 684–808) compared with Meru and Mt Elgon (total numbers of 3435–4401) (Fig. 38.1c).

Except for *B. brassicae*, there was a negative correlation between aphids and altitude an indication that the higher the altitude the less the aphids that can survive.

On potato leaves, four aphid species, *A. fabae*, *A. gossypii*, *M. persicae* and *M. euphorbiae* were found in all areas sampled (Fig. 38.1b and d). During the entire study period the mean total number of aphids per three leaves was lower in Njabini (3.4) followed by Meru (4.9) and Nairobi (5.4) while Mt Elgon and Molo had the highest mean total numbers of 14.5 and 15.7, respectively. In the main growing season, mean total number of aphids per three leaves increased to a maximum in June in Molo and Mt Elgon (but in Meru the numbers were high in November) whereas Njabini and Nairobi had relatively low numbers in May and June. In contrast during the second growing season, fewer aphids colonized leaves in Njabini, Mt Elgon, Meru and Molo but aphid numbers were greater in Nairobi (Fig. 38.1e).

A. fabae was the most numerous aphid species in Molo (73%), Meru (72%) and Mt Elgon (60%) but in Njabini and Nairobi, *A. gossypii* (30.2%) and *M. euphorbiae* (60%), respectively, were most numerous (Fig. 38.1d). Although *M. persicae* (the most efficient vector of PLRV) was the least abundant species found on leaves in most areas, it was caught in larger numbers in the YWTs (mostly) in Molo and Mt Elgon followed by Meru but least numbers were caught in Njabini.

38.4 Discussion

Aphids are of major importance in agriculture, as many species damage crops by feeding on the plant sap, and by transmitting plant virus diseases. In this work, the first comprehensive study of aphids associated with potato in Kenya, ten aphid species were caught in YWTs placed in potato fields, namely *A. gossypii*, *A. fabae*, *A. solani*, *A. pisum*, *B. brassicae*, *C. aegopodii*, *M. euphorbiae*, *M. persicae*, *M. ascalonicus* and *Rhopalosiphum maidis*. Four of these species: *A. gossypii*, *A. fabae*, *M. euporbiae* and *M. persicae* were also found reproducing on potato leaves. The last wide-scale aphid study was done by Eastop (1952) who recorded the incidence of general polyphagous aphids in Kenya. Other workers (Muthomi *et al.*, 2009; Olubayo *et al.*, 2010) who monitored aphid incidence in particular regions found only four species of aphids, *M. euphorbiae*, *M. persicae*, *A. gossypii* and *A. fabae*, both in YWTs and on the (potato) leaves. The reason for this discrepancy in the YWT data may be that in the previous studies incidence was monitored over a smaller area and for a shorter duration and the incidence of migrant aphids is known to be affected by the types of crops grown in neighbouring fields.

Those species colonizing potato are well-known polyphagous aphids that could switch between crops to maintain their populations. It is recommended that crops are sprayed with insecticide to protect against damage by aphids when a threshold level of ten wingless aphids per 100 leaves (Raman, 1984; Capinera, 2001; Thomas, 2002) or five winged aphids per YWT/week (Raman, 1984; Hanafi *et al.*, 1989; Hanafi, 1992) is reached. Although aphid numbers recorded at all sites were above this threshold, not all the species caught were efficient virus vectors and the species of aphid as well as total numbers should be taken into consideration. For example, the most abundant aphid species *R. maidis*, *B. brassicae* and *A. gossypii* are vectors of non-persistent viruses whereas *M. persicae* which was found least often is an efficient vector of PLRV. Njabini consistently had the lowest mean total of aphid numbers in both growing seasons which is an indication that it may be best suited of all the locations for basic seed multiplication in Kenya. However, a comparison of sites by season shows that some had relatively low aphid numbers in the second season.

In particular, Molo, Meru and Nairobi might be suitable for seed production in the second growing season if aphids were controlled.

The most abundant aphid species caught in YWTs in Njabini, Molo and Mt Elgon was *R. maidis*, followed by *B. brassicae*; these aphids are known to transmit non-persistent viruses. The presence of *R. maidis* is a likely indication of the presence of cereal crops nearby, whereas the presence of *B. brassica* in all areas and seasons suggests that vegetable brassicas are common in Kenya all year round. Although *R. maidis* and *B. brassicae* were not found colonizing potato, as would be expected from their host range, their transmission efficiency of the tuber necrotic strain of PVY (PVY^{NTN}) was found to be 68% and 2%, respectively (Boukhris-Bouhachem *et al.*, 2011) and this together with their presence in large numbers can lead to the rapid spread of non-persistent viruses by their winged forms in both ware and seed potato crops.

In most of the areas where *A. fabae* was found on potato, beans are also grown and this proximity could lead to better adaptation of this species to potato as the ecology will select for any individuals that can switch between these hosts. *A. gossypii* does not colonize plants in cooler temperate regions and yet it is polyphagous and colonizes potato in warmer areas. The same applies to *R. maidis* and *B. brassicae*, which were found in areas that traditionally grow cereals and brassicas close to potatoes, yet these aphids do not adapt to colonize potato as do *A. fabae* and *A. gossypii*. Bell *et al.* (2012) observed that host alternation in aphids is a significant trait that explains gross annual patterns of site occupancy, site continuity and abundance and that host-alternating species have to produce a flood of migrants to effect the move from the primary to the secondary host and vice versa.

In Kenya there will be very little capacity for aphid species to produce a sexual generation as, even in the presence of a suitable host plant, the daylight changes that induce the sexual cycle are likely to be absent. Any specialized phenotypes have either arisen away from the regions and were introduced, or they have been generated by mutation and selection on the very large asexual population (The Aphid Room, 2012). This seems to have happened with the cotton aphid and the bean aphid that now colonize potato and are replacing traditional potato aphid species in many areas.

A comparison of aphid species dominant in Kenya with the same or equivalent species in Scotland (a major producer of seed potatoes) (Fig. 38.1) shows that in both countries, the numerically dominant aphid species come from monocotyledons (presumably maize in Kenya and in Scotland, grasses). The next most common species are quite different for the two countries, in Kenya *B. brassicae* is very common and widespread (21%), whereas in Scotland it is far less common (1%) than *A. fabae* (11%), *A. pisum* (11%) and *C. aegopodii* (9%). The current reference values (Boukhris-Bouhachem *et al.*, 2011) suggest *B. brassicae* (2%) is less of a threat for PVY spread than *R. maidis* (68%) and in Njabini, *B. brassicae* occurs almost as frequently as *R maidis*, implying less risk of transmitting non-persistent viruses in this area.

The spread of viruses transmitted in a non-persistent manner in the field is known to be largely carried out by non-colonizing aphid species landing in large numbers on the crop. In this study, YWTs were used to capture and provide an estimate of aphids landing on the crop. Aphids that transmitted non-persistent viruses such as *R. maidis* were more abundant than *M. persicae*, an efficient transmitter of PLRV. In a companion study on virus incidence in potato crops in Kenya (Were *et al.*, 2014), we found that the non-persistently aphid-transmitted PVY was detected in more plant samples in all areas than the persistently transmitted PLRV.

This year-long study has provided detailed information on virus–vector aphid population dynamics that suggests areas in the East of Kenya, particularly Njabini, have low aphid incidence and may be well suited to multiplication of basic seed stocks for seed potatoes. Further work is required to confirm these findings by monitoring virus incidence in basic seed grown over successive field generations to estimate how quickly stocks degenerate.

Acknowledgements

Financial support was provided by the Monsanto Fund and the work of L. Torrance by the Scottish Government's Research and Analytical Services (RESAS) Division. We thank Dr B. Fenton for his help in data analysis, and J. Pickup and F. Highet of SASA, Edinburgh, UK for providing the Scottish suction trap data.

References

Anon. (2009) Mainstreaming the potato crop from orphan crop status. Study towards a seed master plan. Presented by the Ministry of Agriculture on 8 May 2009 at Silver Springs Hotel, Nairobi, Kenya.

Basky, Z. (2002) The relation between aphid dynamics and two prominent potato viruses (PVY and PLRV) in seed potatoes in Hungary. *Crop Protection* 21, 823–827.

Beemster, A.B.R. and de Bokx, J.A. (1987) Viruses of potatoes and seed-potato production. In: de Bokx, J.A. and van der Want, J.P.H. (eds) *Viruses of Potatoes and Seed-Potato Production*. Pudoc, Wageningen, The Netherlands, p. 84.

Bell, J.R., Taylor, M.S., Shortall, S., Welham, C.J. and Harrington, R. (2012) The trait and host plant ecology of aphids and their distribution and abundance in the United Kingdom. *Global Ecology Biogeography* 21, 405–415.

Blackman, R.L. and Eastop, V.F. (1984) *Aphids on the World's Crops, an Identification Guide*. Wiley, New York, 466 pp.

Boukhris-Bouhachem, S., Rouze-Jouan, J., Souissi, R., Glais, L. and Hulle, M. (2011) Transmission efficiency of the strain PVYMTN by commonly captured aphids in Tunisian potato fields. *Plant Pathology Journal* 10, 22–28.

Capinera, J. (2001) *Hand Book of Vegetable Pests*. Academic Press, New York.

Eastop, V.F. (1952) Notes on East African aphids: 11 polyphagous species. *East African Agricultural and Forestry Journal* 18, 184–186.

FAOSTAT (2013) Available at: http://faostat.fao.org/site/339/default.aspx (accessed 6 April 2013).

Hanafi, A. (1992) Spread and control of potato leaf roll virus in potato: implications for management of seed potatoes in the Souss valley, Morocco. PhD thesis, University of Minnesota, Minnesota, 63 pp.

Hanafi, A., Radicliffe, E.B. and Ragsdale, D.W. (1989) Spread and control of potato leaf roll virus in Minnesota. *Journal of Economic Entomology* 82, 1201–1206.

Kabira, J.N., Wakahihu, M., Wagoire, W., Gildemacher, P. and Lemaga, B. (2006) *Guidelines for Production of Healthy Seed Potatoes in East and Central Africa*. Kenya Agricultural Research Institute (KARI), Nairobi.

Kenya Agricultural Research Institute (KARI) (2007) *Annual Report 2006*. National Potato Research Center, Tigoni, Kenya.

Kibaru, A. (2003) Management of aphids and aphid transmitted viruses in seed potato production in Kenya. MSc thesis, University of Nairobi, Kenya.

Macgillivray, M.E. (1979) Aphids infesting potatoes in Canada: life cycle and field key. *Information Services, Agriculture Canada, Publication 1678*. Agriculture Canada, Ottawa, 14 pp.

Moëricke, V. (1951) Eine farbfalle zur kontrolle des fluges von blattläusen, insibesondere der pürishblattlaus, *Myzodes persicae* (Sulzer). *Nachrischtenblatt für den Deutschen Phlanzenschutzdienst* 3, 23–24.

Muthomi, J.W., Nyaga, J.N., Olubayo, F.N., Nderitu, J.H., Kabira, J.N., Kiretai, S.M., Aura, J.A. and Wakahiu, M. (2009) Incidence of aphid transmitted viruses in farmer based seed potato production in Kenya. *Asian Journal of Plant Sciences* 8, 166–171.

Ng, J.C.K. and Perry, K.L. (2004) Transmission of plant viruses by aphid vectors. *Molecular Plant Pathology* 5, 505–511.

Olubayo, F., Kibaru, A., Nderitu, J.H., Njeru, R. and Kasina, M. (2010) Management of aphids and their vectored diseases on seed potatoes in Kenya using synthetic insecticides, mineral oil and plant extract. *Journal of Innovative Development Strategy* 4, 1–5.

Raman, K.V. (1984) Monitoring aphid populations. *CIP Slide Training Series* IV-2. International Potato Center (CIP), Lima, 12 pp.

Raman, K.V. and Radcliffe, B.E. (1992) Pest aspects of potato production, part 2. Insect pests. In: Harris, P.M. (ed.) *The Potato Crop. The Scientific Basis for Improvement*, 2nd edn. Chapman and Hall, London, pp. 476–506.

Salazar, L.F. (1996) *Potato Viruses and their Control*. International Potato Center (CIP), Lima.

Teulon, D.A.G. (1999) Illustrated multiple entry key for winged aphids in New Zealand. *Crop and Food Research Report* No. 614. Crop and Food Research, Lincoln, New Zealand.

The Aphid Room (2012) Aphid Invasion? Available at: http://theaphidroom.wordpress.com/2012/02/02/aphid-invasion/ (accessed 2 February 2013).

Thomas, C. (2002) Bug vs bug crop scouting. In: *Integrated Pest Management Program, 2002*. Department of Agriculture, Harrisburg, Pennsylvania.

Were, H.K., Kabira, J., Kinyua, Z.M., Olubayo, F.M., Imbuaga, B., Karinga, J., Aura, J., Lees, A.K., Cowan, G.H. and Torrance, L. (2014) Occurrence and distribution of potato pests and diseases in Kenya. *Potato Research* 56, 325–342.

39 Maize–Orange-fleshed Sweetpotato Intercropping: Potential for Use to Enhance Food Security and Scaling-up the Nutrition Effort in Malawi

P.E. Abidin,[1]* F. Chipungu,[2] T. Nyekanyeka,[1] T. Chilanga,[2] O. Mwenye,[2] J. Kazembe,[1] B. Botha[3] and E.E. Carey[1]

[1]*International Potato Center (CIP)-Malawi, Lilongwe, Malawi;*
[2]*Department of Agricultural Research Services (DARS), Blantyre, Malawi;* [3]*Irish Aid, Lilongwe, Malawi*

Abstract

Maize is the staple food in Malawi where it is mainly cultivated by smallholder farmers under rainfed conditions. Due to increasing climatic uncertainty and declining soil fertility, crop failures often result in shortages affecting up to 1.6 million people every year. An estimated 47% of the children in Malawi suffer from stunted growth because of under-nutrition, and are vulnerable to illnesses and learning difficulties. Crop diversification is at the core of Malawi's agriculture policy. The Rooting out Hunger in Malawi with Nutritious Orange-fleshed Sweetpotato (OFSP) project, funded by Irish Aid, has been working with partners since 2009 in Malawi to popularize this type of sweetpotato. The potential of the OFSP cultivar Zondeni to contribute to sustainable intensification of a maize-based cropping system as an intercrop was investigated. Maize–sweetpotato intercropping trials were conducted in two consecutive rainy seasons/years at Bvumbwe Agriculture Research Station in southern Malawi. Four spatial arrangements were compared (two rows OFSP: one row maize; one row OFSP: one row maize; one row OFSP: two rows maize; and three OFSP: one maize plant within the same row) in large plots. Trials, each with three replications, were repeated each year, with maize harvested at 4 months and OFSP harvested at 5, 6 and 7 months after planting. In year 2, land equivalent ratios (LER) were determined, while in both years farmers'assessment of trial performance were obtained, and gross margin analysis were determined. Gross margin analysis indicated that OFSP–maize intercropping was highly profitable in both years, with sweetpotato accounting for greater than 90% of the income. Yields were much lower in the second year compared with the first, and considered to be more representative of on-farm yields in Malawi. In this trial, LER was 1.79 over all spatial arrangements, indicating the high potential of OFSP–maize intercropping to contribute towards sustainable intensification in Malawi. Farmers' rank preferences for OFSP–maize spatial arrangements changed between year 1 and year 2, with the higher OFSP-density (two rows OFSP: one row maize) arrangement preferred in year 2. The marked absence of sweetpotato weevil damage on sweetpotato in the OFSP–maize intercrops was noted in both years. Results are discussed and recommendations made for future testing and dissemination of OFSP–maize intercropping in Malawi.

Keywords: food security, intercropping, intra-cropping, land equivalent ratio (LER), maize, orange-fleshed sweetpotato (OFSP), strip intercropping, sweetpotato, sweetpotato weevils

*p.abidin@cgiar.org

39.1 Introduction

Malawi is a land-locked nation of 16 million people in southern Africa with much of its population engaged in smallholder agriculture under rainfed conditions. Maize is the main staple crop grown by almost all farmers in Malawi, and its production is supported by government policies, including fertilizer subsidies that been have hailed as contributing to a new Green Revolution in Africa (Denning *et al.*, 2009). Maize is, however, notoriously vulnerable to dry spells, and these are increasing in southern Africa as a result of climate change (Tadross *et al.*, 2009; Taylor, 2012). Food shortages in the rural countryside are frequent, particularly during the 'lean season' from January to March (Sassi, 2012; FEWSNET, 2013). An estimated 47% of children under 5 years of age in Malawi have stunted growth because of under-nutrition, and micronutrient deficiencies are widespread placing a large economic burden on the country (World Bank, 2009; NSO and ICF Macro, 2011). Furthermore, soil degradation is a concern due to continuous cropping and erosion of ever-smaller landholdings (Barungi and Maonga, 2011).

Crop diversification is at the core of Malawi's agriculture sector-wide approach to develop a more resilient and sustainable agriculture, and to ensure economic growth and food security (Government of Malawi, 2010; Malawi Government, 2010). Malawi is also an active participant in the Scaling Up Nutrition (SUN) movement, which is working to eliminate under-nutrition, particularly during the critical stages for child development from conception to 24 months (the first 1000 days) by breaking down barriers separating agriculture, health, social protection and finance (SUN, 2012). Sweetpotato is one of Malawi's most widely grown crops (FAOSTAT, 2013), and has the potential to play an increasingly important role in the sustainable intensification of Malawi's farming and food systems. Sweetpotato roots are rich in energy (in contrast to most vegetables) and both roots and leaves are good sources of vitamins and minerals, notably pro-vitamin A in the case of orange-fleshed sweetpotato (OFSP) varieties (Low *et al.*, 2007). Studies in Mozambique and Uganda have shown that OFSP can be an effective means of combatting vitamin A deficiency in vulnerable populations at the community level (Low *et al.*, 2007; Hotz *et al.*, 2011, 2012). The Rooting out Hunger in Malawi with Nutritious OFSP project is an integrated multi-partner effort emphasizing seed systems, integrated crop management and value chain development that is contributing to both agriculture and nutrition objectives in Malawi (International Potato Center, 2012).

Intercropping and rotations with crop and tree legume species have been investigated and show promise for the sustainable intensification of the maize-based systems in Malawi (Snapp *et al.*, 2010; Ajayi *et al.*, 2011; Ngwira *et al.*, 2012). In contrast to monocropped maize, and its associated risks of soil degradation and crop failure, these cultural practices can contribute to greater productivity and stability of cropping systems through a range of ecosystem services, including: (i) fertilizer efficiency; (ii) soil cover; (iii) grain and nutrient yield and quality; (iv) profitability; and (v) reduced pest and disease pressure (Zuo and Zhang, 2009; Snapp *et al.*, 2010). However, adoption of intercropping and other cultural practices which appear to have merit to researchers, is often limited at the farm level by diverse socio-economic factors including access to labour, input supplies, gender, economic and educational status, though participatory approaches can do much to facilitate development of appropriate technologies (Horton *et al.*, 2010; Bezner-Kerr *et al.*, 2011; Ngirwa *et al.*, 2012).

There are relatively few reports of maize–sweetpotato intercropping, and none to date from Malawi. The productivity of maize and sweetpotato when intercropped has been reported to vary considerably depending on: (i) cultivars; (ii) time of planting; (iii) spatial arrangement of intercrops; and (iv) agroecological conditions (Oswald *et al.*, 1996; Amede and Nigatu, 2001; Ossam, 2010; Nedunchezhiyan *et al.*, 2011). Given the importance of both sweetpotato and maize in Malawi, we undertook studies to evaluate suitability of maize–sweetpotato intercropping systems to help in combatting food insecurity, under-nutrition and poverty. This research work was conducted at the Bvumbwe Research Station rather than on-farm, since intercropping maize and sweetpotato is not yet a recommended practice promoted by the extension service in Malawi, though it is practised by some farmers, typically as a relay crop, intra-crop or on border rows around the maize field. Farmer opinions were included in the assessment at open field days. We used a medium-maturing (120 days) maize cultivar, which we hypothesized would yield well, but not compete strongly with the later-maturing sweetpotato. We also evaluated early, medium and late harvests (5, 6 and 7 months) of the OFSP cultivar Zondeni, in both strip-cropping and intra-row arrangements.

39.2 Materials and Methods

OFSP cv. Zondeni, and maize cv. Mkango (SC627) were used in the study. The trial was conducted in the two successive rainy seasons of 2010/2011 and 2011/2012 at the Bvumbwe Research Station at an elevation of 1228 m above sea level in Thyolo District, Southern Region of Malawi. Soil types vary from sandy to clay loam in different parts of the station. Experimental factors were:

- spatial arrangement (strip crops of OFSP: maize at: (i) two: one rows; (ii) one: one rows; (iii) one: two rows; and (iv) an intra-crop of one maize plant: three sweetpotato plants within the row);
- OFSP harvest at 5, 6 and 7 months after planting (AP), maize was harvested at 4 months AP; and
- trial season.

In the second season, OFSP and maize monoculture plots were included to permit the calculation of the land equivalent ratio (LER).

The trial site was prepared by disc harrow, followed by hand hoeing to raise ridges 30 cm high by 30 cm wide, in accordance with farmers' practices in Malawi. The planting distance for both crops was 30 cm between plants and 75 cm between rows/ridges. Individual plots were comprised of 22 ridges 6.6 m long, each planted with 22 plants. One sweetpotato vine cutting or one maize seed was planted per planting point. Plots were bordered, so net plots consisting of 20 ridges with 20 plants per ridge were harvested for data collection. Gap filling was done in the second season. Fertilizers were applied only on maize, at recommended rates as farmers typically do not fertilize sweetpotato. Trials were laid out as a randomized block design with three replicates. The entire trial was repeated three times each season to allow for assessment of sweetpotato harvest at 5, 6 or 7 months AP.

In the first season, the trial was conducted in a well-managed area of the experiment station with good fertility levels, and where crop rotation and fallow were practised. In the second season, the trial was in a less well-managed area of the experiment station where the neighbouring community has used the land for growing maize on an annual basis. The soil was less fertile and more comparable to on-farm conditions. The annual rainfall for 2010/2011 season was 850.1 mm while it was 1146.6 mm for the 2011/2012 season with maximum temperature of 25.2°C for both seasons.

The rains commenced in December in both years. In the first season, a dry spell occurred leading to staggered planting dates of the different blocks. In the second season, temperatures were quite chilly for a few days in early January shortly after trial establishment and the rains stopped early. The assessments conducted in the two seasons had distinctly different environmental conditions, representing the variation in conditions faced by smallholder farmers in southern Malawi.

Trials for 5, 6 and 7 month harvests of sweetpotato were planted on 30 December 2010, 6 and 11 January 2011, respectively, in the first season, and on 17, 19 and 20 December 2011 in the second season. Maize plants were fertilized by individual plant application at 92 kg/ha N and 21 kg/ha P using 23:21:0+4S and urea, with 50% of N and 100% of P applied at or soon after planting, and the remainder of N applied when maize was 'knee high'. Maize in all plots was harvested, shelled and weighed when the maize was mature and dry at roughly 4 months AP and OFSP was harvested at 5, 6 or 7 months AP. Sweetpotato roots >100 g were considered to be of marketable size. Yields from each plot were converted into metric tonne per hectare (t/ha). General ANOVA (analysis of variance) was used for calculating yield and testing the significance for each season and across seasons using Genstat (Anonymous, 2005).

Efficiency of intercropping was measured by using the LER (Mead and Willey, 1980) whereby:

$$\text{Land equivalent ratio (LER)} = \text{L-sweetpotato} + \text{L-maize} \quad (39.1)$$

Where:
L-sweetpotato = yield of sweetpotato under intercropping/yield of sole-cropped sweetpotato
L-maize = yield of maize under intercropping/yield of sole-cropped maize.

In the first year, a total of 52 participants (26 female) including farmers, representatives of relevant government agencies and NGOs (non-governmental organizations) from Dedza, Zomba, Phalombe, Mulanje and Chikhwawa districts participated in two open field days on 9 May and 22 August 2011. In the second year, 75 participants (50 male) came from Dowa, Lilongwe and Salima districts in addition to the five original districts to an open field day held on 17 May 2012. Participants at the field days toured the trials and discussed results with the researchers. A simple questionnaire was given to each participant to elicit thier preference. The survey was conducted once each season, when maize was still present in the field.

Gross margin analysis for maize and sweetpotato was based on costs of inputs and actual yields for each of the spatial arrangements. Calculations were made in the local currency, Malawi kwacha (Mk), and then converted to US dollars (US$) using exchange rates of Mk200 = US$1 in year 1 and Mk300 = US$1 in year 2. Income calculations were based on prevailing maize prices of Mk35/kg and Mk50/kg in years 1 and 2, respectively, while two sweetpotato prices were used: Mk100/kg (actual market price) and the same price as maize.

39.3 Results

No serious damage by pests and diseases was noted for either maize or OFSP during the two trial seasons. However, theft of maize from trial plots did occur. In the first rainy season, the maize from the 7 months AP OFSP harvest trial was severely affected by theft and was dropped from the analysis.

Maize and OFSP yields from year 1 and year 2 trials, and mean maize yields over seasons are presented in Table 39.1. Mean yields of marketable and total sweetpotato roots, and number of roots per plant over seasons are presented in Table 39.2. Farmer opinions about the trials before harvest based on observation are presented in Table 39.3. Results of gross margin analysis of the systems are presented in Table 39.4, with sweetpotato accounting for greater than 90% of income in each of the spatial arrangements.

The average LER of the intercrop systems over all three sweetpotato harvests was 1.79 in year 2, with L-sweetpotato and L-maize values of 1.12 and 0.67, respectively. The highest LER was for the two rows OFSP: one row maize strip crop (LER of 2.2, with L-sweetpotato and L-maize values of 1.42 and 0.78), while the least productive was the intra-crop (LER of 1.52 with L-sweetpotato of 1 and L-maize of 0.52). Only one of the strip-crop arrangements, one row OFSP: two rows maize, had an L-sweetpotato below one (0.92).

39.4 Discussion

Based on experiences during 2 years at the Bvumbwe Research Station, sweetpotato–maize intercropping may have the potential to improve incomes, household nutrition and increase system productivity. Both crops yielded very well under favourable conditions, and when the LER was determined in the second year under considerably less favourable conditions, the intercropping systems were found to provide a very high LER of 1.79 over all trial types. Given that maize is the staple crop in Malawi, cropping arrangements that least affect maize yield are likely to be of interest. Thus, the strip-cropping arrangements appear to be promising.

Table 39.1. Yields of maize (4 months after planting (AP)) and orange-fleshed sweetpotato (OFSP) cv. Zondeni (5, 6 and 7 months AP) in intercropping arrangements evaluation during two seasons/years at Bvumbwe Research Station, Malawi.

	2010/2011 season		2011/2012 season		Mean of seasons
	Maize yield[a] (t/ha)	OFSP yield[b] (t/ha)	Maize yield[a] (t/ha)	OFSP yield[b] (t/ha)	Maize yield[a] (t/ha)
Treatment (OFSP plants: maize plants per plot)					
T1: two rows OFSP: one row maize (260: 140)	5.1	28.8	1.5	11	3.30
T2: one row OFSP: one row maize (200: 200)	5.3	21.2	1.7	12	3.56
T3: one row OFSP: two rows maize (120: 280)	4.6	19.5	1.8	17	3.22
T4: intra-crop three OFSP plants and one maize plant (300: 100)	4.1	26.5	1.2	12	2.71
T5: OFSP monoculture (400: 0)	–	–	–	12	
T6: Maize monoculture (0: 400)	–	–	2.3	–	
P value[c]	**	*	*	*	(*)
LSD	0.8	6.66	0.6	3.5	(0.529)
Block (maize) or harvest (OFSP)					
B1 or H1: 5 months AP	4.7	10.8	1.6	10	3.09
B2 or H2: 6 months AP	4.8	15.9	2.0	13	3.30
B3 or H3: 7 months AP	–	45.2	1.4	15	–
P value[c]	ns	**	(*)	**	ns
LSD 5%		5.74	(0.4)	2.7	
Grand mean	4.8	24	1.7	13	3.20
CV (%)[d]	10	28.3	25.2	28.6	24

[a]Maize yield expressed as shelled weight after drying.
[b]OFSP yield expressed as weight of marketable roots.
[c] **, Highly significant at $P < 0.01$, least significant difference (LSD) 5%; *, significant at $P < 0.05$, LSD 5%; (*), $P < 0.1$, LSD 10%; ns, non-significant.
[d]CV, coefficient of variation.

Nedunchezhiyan *et al.* (2011) reported on the advantages of sweetpotato strip-intercropping systems with a number of crops in India, particularly related to the border rows where crops meet. These authors reported an LER of 1.08 for sweetpotato: maize (three: three rows) using a 90-day maize variety and a 120-day sweetpotato variety. The LER of sweetpotato–maize intercrops has been reported to vary significantly with cultivar and agroecological conditions (Roberts-Nkrumah *et al.*, 1995; Amede and Nigatu, 2001). LERs were highest when both crops were planted at the same time, as opposed to later planting of sweetpotato, under conditions of intermittent and terminal drought in Ethiopia (Amede and Nigatu, 2001), which supports our decision to plant maize and sweetpotato simultaneously in our trials. Furthermore, our choice of the medium maturity maize cultivar, Mkango, to intercrop with cv. Zondeni seems to have been warranted, though it may be that other maize–sweetpotato cultivar combinations might reduce maize yields less.

Sweetpotato clearly benefitted from the maize fertilization, and its yield advantage increased enormously at later harvests. Delaying harvest of sweetpotato in Malawi would not be a problem, since fields lie fallow during the long dry season, and having sweetpotato in the field would provide crop cover and continued supply of sweetpotato.

Table 39.2. Mean yields of OFSP cv. Zondeni harvested at 5, 6 and 7 months after planting (AP), intercropping with maize in four spatial arrangements over 2010/2011 and 2011/2012.

	No. of marketable roots per plot (per plant)	Weight of marketable roots (t/ha)	Total yield storage roots (t/ha)
Treatment (OFSP plants per plot)			
T1: two rows OFSP: one row maize (260)	359 (1.38)	19.9	23.1
T2: one row OFSP: one row maize (200)	345 (1.73)	16.7	19.2
T3: one row OFSP: two rows maize (120)	278 (2.32)	18.0	20.4
T4: intra-crop three OFSP plants and one maize plant (300)	387 (1.29)	19.3	21.5
P value[a]	**	ns	ns
LSD	59.6	–	–
Environment (harvest)			
H1: 5 months AP	304	10.6	12.3
H2: 6 months AP	399	14.5	16.7
H3: 7 months AP	323	30.3	34.1
P value[a]	(*)	**	**
LSD	(73.6)	7.16	7.88
Season			
2010/2011 rainy season	323	24.0	27.3
2011/2012 rainy season	361	13.0	14.7
P value[a]	ns	**	**
LSD	–	2.52	5.56
Grand mean	342	18.5	21.0
CV (%)[b]	41.4	42.1	42.2

[a]**, Highly significant at $P < 0.01$, least significant difference (LSD) 5%; *, significant at $P < 0.05$, LSD 5%; (*), $P < 0.1$, LSD 10%; ns, non-significant.
[b]CV, coefficient of variation.

Table 39.3. Opinions of respondents collected during the open days conducted in rainy season of 2010/2011 ($n = 52$) and 2011/2012 ($n = 36$).

	2010/2011 Season		2011/2012 Season	
Treatment	No. of respondents	Rank	No. of respondents	Rank
T1: two rows OFSP: one row maize	13	2	19	1
T2: one row OFSP : one row maize	12	3	6	2
T3: one row OFPS : two rows maize	24	1	4	4
T4: intra-crop three OFSP plants and one maize plant	3	4	0	6
T5: OFSP monoculture	–	–	5	3
T6: Maize monoculture	–	–	2	5

A further observation from our work was that roots and stems of sweetpotato were noticeably 'clean' at 7 months AP. Typically, populations of sweetpotato weevils (*Cylas* spp.) build up when sweetpotato crops are left in the ground for an extended season, and can result in extensive destruction of the crop (Chalfant *et al.*, 1990). Our observations are consistent with the findings by Nampeera *et al.* (2011) that the non-host crops (i.e. maize

Table 39.4. Income and profit per intercropped (US$) from intercropping maize and OFSP cv. Zondeni during two rainy seasons/years in Malawi.

	Two rows OFSP: one row maize	One row OFSP: one row maize	One row OFSP: two rows maize	Intra-crop three OFSP plants and one maize plant	Maize monoculture	OFSP monoculture
Year 1: 2010/2011 rainy season						
Income from maize sale ($0.175/kg)[a]	312	464	564	179		
Income from OFSP sale ($0.5/kg)[a]	9360	5300	2925	9938		
Income from OFSP sale ($0.175/kg)[a]	3276	1855	1024	3478		
Total variable cost	1743	1756	1770	1736		
Profit (OFSP is $0.5/kg)	7929	4007	1718	8381		
Profit (OFSP is $0.175/kg)	1845	562	−183	1921		
Year 2: 2011/2012 rainy season						
Income from maize sale ($0.167/kg)[a]	88	142	210	50	383	
Income from OFSP sale ($0.33/kg)[a]	2383	2033	1650	3100		4000
Income from OFSP sale ($0.167/kg)[a]	1192	1017	825	1550		2000
Total variable cost	1171	1183	1197	1163	1221	1144
Profit (OFSP is $0.33/kg)	1300	992	663	1987	−838	2856
Profit (OFSP is $0.167/kg)	109	−24	−162	437		856

[a]Exchange rate: US$1 = Mk200 (year 1), US$1 = Mk300 (year 2).

and onion) could be useful as crop barriers to reduce weevil infestations.

The positive gross margins indicate that sweetpotato–maize intercropping is a profitable enterprise. It was observed that higher productivity and price of sweetpotato were the main contributing factors to profit. The variable costs largely came from inorganic fertilizer and labour. The profit in the first year was higher than in the second year due to the devaluation of the Malawi kwacha. The cost of sweetpotato vines was not considered in the analysis of profitability, even though vines can be sold as planting material and a leafy green vegetable. From this initial assessment, sweetpotato–maize intercropping has the potential to improve food and nutrition security for smallholder farmers in Malawi.

39.5 Conclusions

Strip intercropping of maize and sweetpotato appears to have considerable potential to contribute to crop diversification, cropping system intensification and sustainability, farm profitability and household food and nutrition security in increasingly risk-prone maize production systems in Malawi. Subsequent intercropping studies with OFSP, maize and onion using participatory approaches under on-farm conditions are recommended.

Acknowledgements

We sincerely acknowledge the Government of Malawi and Irish Aid.

References

Ajayi, O.C., Place, F., Akinnifesi, F.K. and Sileshi, G.W. (2011) Agricultural success from Africa: the case of fertilizer tree systems in southern Africa (Malawi, Tanzania, Mozambique, Zambia and Zimbabwe). *International Journal of Agricultural Sustainability* 9, 129–136.

Amede, T. and Nigatu, Y. (2001) Interaction of components of sweetpotato–maize intercropping under the semi-arid conditions of the Rift Valley, Ethiopia. *Tropical Agriculture* 78, 1–7.

Anonymous (2005) *Genstat Release 8*. Lawes Agricultural Trust, Rothamsted, Harpenden, 92 pp.

Barungi, M. and Maonga, B.B. (2011) Adoption of soil management technologies by smallholder farmers in central and southern Malawi. *Journal of Sustainable Development in Africa* 12, 28–38.

Bezner-Kerr, R., Berti, P.R. and Shumba, L. (2011) Effects of a participatory agriculture and nutrition education project on child growth in northern Malawi. *Public Health Nutrition* 14, 1466–1472.

Chalfant, R.B., Jansson, R.K., Scal, D.R. and Schalk, J.M. (1990) Ecology and management of sweet potato insects. *Annual Review of Entomology* 35, 157–180.

Denning, G., Kabambe, P., Sanchez, P., Malik, A., Flor, R., Harawa, R., Nkhoma, P., Zamba, C., Banda, C., Magombo, C., Keating, M., Wangila, J. and Sachs, J. (2009) Input subsidies to improve smallholder maize productivity in Malawi: toward an African Green Revolution. *PLOS Biology* 7, 2–10.

FAOSTAT (2013) Agricultural Production in Malawi: Preliminary 2011 Data Available Now. Available at: http://faostat.fao.org/site/339/default.aspx (accessed 2 April 2013).

FEWSNET (2013) Famine Early Warning Systems Network. Available at: http://www.fews.net (accessed 2 April 2013).

Government of Malawi (2010) Malawi CAADP Compact. Available at: http://www.fanrpan.org/documents/d00929/Signed_Malawi_CAADP_Compact.pdf (accessed 2 April 2013).

Horton, D., Akello, B., Aliguma, L., Bernet, T., Devaux, A., Lemaga, B., Magala, D., Mayanja, S., Sekitto, I., Thiele, G. and Velasco, C. (2010) Developing capacity for agricultural market chain innovation: experience with the 'PMCA' in Uganda. *Journal of International Development* 22, 367–389.

Hotz, C., Loechl, C., de Brauw, A., Eozenou, P., Gilligan, D., Moursi, M., Munhaua, B., van Jaarsveld, P., Carriquiry, A. and Meenakshi, J.V. (2011) A large-scale intervention to introduce orange sweet potato in rural Mozambique increases vitamin A intakes among children and women. *British Journal of Nutrition* 108, 163–176.

Hotz, C., Loechl, C., Lubowa, A., Tumwine, J.K., Ndeezi, G., Masawi, A.N., Baingana, R., Carriquiry, A., de Brauw, A., Meenakshi, J.V. and Gilligan, D.O. (2012) Introduction of beta-carotene-rich orange sweet potato in rural Uganda resulted in increased vitamin A intakes among children and women and improved vitamin A status among children. *Journal of Nutrition* 142, 1871–1880.

International Potato Center (CIP) (2012) Rooting Out Hunger in Malawi with Nutritious Orange-Fleshed Sweetpotato: Year 3 Annual Report, 1 November 2011–31 October 2012. Available at: http://sweetpotatoknowledge.org/projects-initiatives/sweetpotato-for-profit-and-health-initiative-sphi/rooting-out-hunger-in-malawi-with-ofsp (accessed 2 April 2013).

Low, J.W., Arimond, M., Osman, N., Cunguara, B., Zano, F. and Tschirley, D. (2007) A food-based approach introducing orange-fleshed sweet potatoes increased vitamin A intake and serum retinol concentrations in young children in rural Mozambique. *Journal of Nutrition* 137, 1320–1327.

Malawi Government (2010) *Agriculture Sector Wide Approach (ASWAp): Malawi's Prioritised and Harmonised Agricultural Development Agenda*. Ministry of Agriculture and Food Security, Lilongwe, Malawi.

Mead, R. and Willey, R. (1980) The concept of 'Land Equivalent Ratio' and advantages in yields from intercropping. *Experimental Agriculture* 16, 217–228.

Nampeera, E.L., Talwana, H. and Potts, M. (2011) Effects of nonhost crop barriers on *Cylas* spp. sweetpotato infestation and damage (abstract). In: Anon. (ed.) *From Soil to Soul: Crop Production for Improved African Livelihoods and a Better Environment for Future Generations*, Entomology Session D3, 10th African Crop Science Society Conference, 10–13 October 2011, Maputo, Mozambique, p. 112.

National Statistical Office (NSO) and ICF Macro (2011) *Malawi Demographic and Health Survey 2010*. NSO and ICF Macro, Zomba, Malawi and Calverton, Maryland.

Nedunchezhiyan, M., Laxminarayana, K., Rao, K., Rajasekhara, K. and Satapathy, B.S. (2011) Sweet potato (*Ipomoea batatas* L.)-based strip intercropping: I. Interspecific interactions and yield advantage. *Acta Agronomica Hungarica* 59, 137–147.

Ngwira, A.R., Kabambe, V.H., Kambauwa, G., Mhango, W.G., Mwale, C.D., Chimphero, L., Chimbizi, A. and Mapfumo, P. (2012) Scaling out best-fit legume technologies for soil fertility enhancement among smallholder farmers in Malawi. *African Journal of Agricultural Research* 7, 918–928.

Ossam, E.M. (2010) Influence of sweet potato/maize association on ecological properties and crop yields in Swaziland. *International Journal of Agriculture and Biology* 12, 481–488.

Oswald, A., Alkamper, J. and Midmore, D.J. (1996) The response of sweet potato (*Ipomoea batatas* Lam.) to inter- and relay-cropping with maize (*Zea mays* L.). *Journal of Agronomy and Crop Science* 176, 275–287.

Roberts-Nkrumah, L.B., Ferguson, T.U. and Wilson, L.A. (1995) Performance of sweetpotato cultivars intercropped with maize 1. Total dry matter production and tuber yield responses to two maize populations. *Tropical Agriculture (Trinidad)* 73, 202–209.

Sassi, M. (2012) Short-term determinants of malnutrition among children in Malawi. *Food Security* 4, 593–606.

Scaling Up Nutrition (SUN) (2012) Sun Movement Progress Report 2011–2012. Available at: http://scalingupnutrition.org/wp-content/uploads/2012/10/SUN-MP-REPORT_EN.pdf (accessed 13 January 2013).

Snapp, S.S., Blackie, M.J., Gilbert, R.A., Bezner-Kerr, R. and Kanyama-Phiri, G.Y. (2010) Biodiversity can support a greener revolution in Africa. *Proceedings of the National Academy of Sciences of the United States of America* 107, 20840–20845.

Tadross, M., Suarez, P., Lotsch, A., Hachigonta, S., Mdoka, M., Unganai, L., Lucio, F., Kamdonyo, D. and Muchinda, M. (2009) Growing-season rainfall and scenarios of future change in southeast Africa: implications for cultivating maize. *Climate Research* 40, 147–161.

Taylor, D. (2012) Risk and abandonment, and the meta-narrative of climate change. In: Castro, A.P., Taylor, D. and Brokensha, D.W. (eds) *Climate Change and Threatened Communities: Vulnerability, Capacity and Action*. Practical Action Publishing, Rugby, UK, pp. 57–68.

World Bank (2009) Nutrition at a Glance – Malawi. Available at: http://siteresources.worldbank.org/NUTRITION/Resources/281846-1271963823772/Malawi.pdf (accessed 2 April 2013).

Zuo, Y. and Zhang, F. (2009) Iron and zinc biofortification strategies in dicot plants by intercropping with gramineous species: a review. *Agronomy Sustainable Development* 29, 63–71.

40 Effects of Staking on Flower Induction, Pollination and Cross-compatibility Among Sweetpotato

M.S. Afolabi[1]* and M.O. Akoroda[2]

[1]*Landmark University, Omu Aran, Nigeria;*

[2]*Department of Agronomy, University of Ibadan, Ibadan, Nigeria*

Abstract

Glasshouses have been used to induce flowering in sweetpotatoes for the purpose of breeding. However, these structures are expensive to construct and/or maintain. This study investigated the effects of staking the sweetpotatoes in the field as a way of flower induction for purposes of pollination and testing cross-compatibility in sweetpotato. Two field experiments (staking and non-staking) were carried out at Ibadan from May 2010 and April 2011 using a randomized complete block design with three replicates. In the first experiment (non-staked), vine cuttings of 39 parent clones were evaluated for flowering while staking was done for the second experiment, which was also used as a crossing block. Vine cuttings of the 39 parent clones were planted on heaps at 1 m × 1 m with two 25 cm cuttings per heap, staked with 2 m wooden poles. The main vines were tied and trained up the stakes to induce flowering. Clones that flowered were crossed in a diallel. Dried fruits were harvested between 30 and 50 days after pollination. Seeds from these fruits were soaked in water and planted in polythene bags filled with top soil to test their viability. Results from these studies showed that 18 out of 39 clones readily produced flowers and that staking increased the number of flowers produced. Blesbok and 440168 clones had longer anthers than their stigma while it was the opposite in the other clones. In clone 440034 the anther and stigma were at the same height. Of the 324 cross-combinations conducted on the 18 clones that flowered, only 109 produced seeds. The mean percentage cross-compatibility for the 18 clones was 44.3%. Compatibility varied considerably among the clones ranging between 5.4% and 68.7%, with clone W-151 showing the highest level (68.7%). All flowering clones were observed to be self-incompatible except for clones Sauti and TIS 8441. The hybrid seeds obtained from the flowering clones had a mean percentage germination of 50.6%. These data suggest that stakes induced flowering in 18 parent stocks and therefore can be used to facilitate hybridization in sweetpotato for improved hybrid seed production. However, further work is needed to enhance success in flowering of clones.

40.1 Introduction

Sweetpotato is ranked seventh in energy production after wheat, rice, maize, potato, barley and cassava in developing countries (Thottapilly, 2009). It is almost always self-incompatible and this, combined with other seed-limiting processes, has impaired the understanding

*afolabimike97@yahoo.com

of its breeding. The practical consequences of self-incompatibility and sterility have been recognized by different researchers in different countries (Wang, 1982; Martin, 1988; Wilson *et al.*, 1989; Mok *et al.*, 1998; Mihovilovich *et al.*, 2000). It is difficult to produce seeds by self-pollination. Hand pollination cannot produce more than four seeds, and often only one or two. Because of its polyploidy, the sweetpotato is not a suitable species for Mendelian genetics and the segregation ratios are quite complex. Despite these practical constraints, numerous breeding programmes were implemented in the 1970s but mostly in temperate climates, in the USA or Japan, and subtropical Taiwan. Self-incompatibility in sweetpotato has been shown to be pollen dependent. Seed-setting failure is often associated with complete pollen-germination failure. Sweetpotato is heterogeneous in nature and exhibits a wide variability in morphological traits such as flowering habits and nutritional qualities. Under normal conditions, some cultivars do not flower, or have scanty flowering and others flower profusely. Several techniques have been developed to promote not only sweetpotato flowering but also fruit and seed production. These are: (i) short photoperiod; (ii) moderate temperature; (iii) limited water supply; (iv) grafting; (v) trellises; (vi) growth regulators; (vii) vine girdling; (viii) pesticide sprays; (ix) soil fertilization; (x) application of boric acid and magnesium sulfate; and (xi) the bouquet method with genetic selection (Wilson *et al.*, 1989). In practice, a combination of these methods is used that is commonly complemented with the determination of sexual fertility and compatibility of the accession to be included in seed production. This study was aimed at achieving the following objectives:

- to examine the effects of staking on flowering induction in sweetpotato; and
- to determine the fertility and compatibility of sweetpotato varieties.

40.2 Materials and Methods

Study area

The experiments were conducted at the experimental field of the Department of Agronomy, University of Ibadan, Ibadan, Oyo State, Nigeria. The University of Ibadan is located on latitude 7° 30′ N and longitude 3° 54′ E. Altitude is 210 m above sea level. Soil from the experimental plot in the field was sampled randomly from 0 cm to 15 cm depths. It was later bulked and a representative sample was taken to the laboratory for analysis. Laboratory results indicated that the soil was a sandy loam type.

Description of the genetic material

The genetic material used for this experiment was from the germplasm collection of the Department of Agronomy, University of Ibadan, Nigeria. This material is comprised of 39 varieties of sweetpotato: 14 orange fleshed, 14 white fleshed and 11 yellow fleshed (Table 40.1). The experimental plot consisted of two rows of a 5 m long ridge (first row for staking, second row for non-staking) with 1 m between the ridges in three replications arranged in a randomized complete block design.

Planting

Planting of 25 cm long vine cuttings was done on the crest of each ridge after light rains on 31 May 2010. Each vine cutting was inserted at a slant, with two-thirds buried below the soil surface at a spacing of 1 m along the ridges and 1 m between the crest of one ridge and another. The plot size used was 5 m × 1 m in two rows. This gave a total of ten plants per clone per replication.

Field cultural practices

Weeding was done 4, 6 and 8 weeks after planting, using a small hoe, to keep the plots and walk-ways between plots free of weeds. No fertilizer or pesticide was used during this experiment. Wooden poles 2 m long were used for the plants that were staked. The main vines were tied and trained up the stakes to induce flowering.

Table 40.1. Sweetpotato varieties used for the experiment.

Variety	Source[a]	Flesh colour
199062.1	CIP Kenya	Yellow
Ejumula	CIP Kenya	Orange
199024.2	CIP Kenya	Orange
440034	CIP Kenya	Orange
440141	CIP Kenya	Orange
440215	CIP Kenya	Orange
Blesbok	CIP Kenya	Orange
Ex-Oyunga	Uganda	Orange
96-117	South Africa	Orange
1999-5-21	South Africa	Orange
Resisto	CIP Kenya	Orange
Excel	South Africa	Orange
W-151	CIP Kenya	Yellow
Egbeda	Oyo State	Yellow
2008-12-8	South Africa	Orange
Hernandez	South Africa	Orange
Shaba	Oyo State	Yellow
BBD	Oyo State	Yellow
Ex-igbariam	Abia State	Yellow
Tanzania	CIP Kenya	Yellow
Naspot 5	Tanzania	Yellow
MGO	CIP Kenya	Yellow
Kaduna	Kaduna State	Yellow
Barth	Oyo State	Yellow
440170	CIP Kenya	White
Sauti	Tanzania	White
Iddo	Iddo	White
Akoroda	Professor Akoroda	White
Famgbe 1	Bayelsa State	White
440168	CIP Kenya	White
TIS 8250	IITA Ibadan	White
Arrow tip	Oyo State	White
TIS 86/0356	IITA Ibadan	White
Nasarawa	Nasarawa	White
199034.1	CIP Kenya	Orange
Ak-Wide	Professor Akoroda	White
TIS 8441	IITA Ibadan	White
TIS 87/0087	IITA Ibadan	White
Benue	Benue State	White

[a]CIP, International Potato Center; IITA, International Institute for Tropical Agriculture.

40.3 Data Collection

Floral characteristics

All the flowers from staked and non-staked plants were examined separately on a plot basis early in the morning for 4 weeks. A total of four flower characteristics were assessed: (i) flower colour (where ten flowers of the same clone were observed and the predominant colour was recorded); (ii) limb shape; (iii) colour of stigma; and (iv) stigma exsertion (i.e. the relative position of the stigma as compared with the highest anther).

Hybridization of sweetpotato clones

On 22 January 2011 hand pollination began on staked and non-staked plants separately. When hand pollinating, the flowers are well protected from pollination by insects. The buds and flowers due to open the following morning were prevented from opening by clipping the tip of the corolla with a drinking straw during the evening hours prior to hand pollination. The flower from the male parent was carried to the female parent and the clip was removed gently without destroying the corolla. The petals were then spread and the anthers of the male parent were rubbed gently over the stigma of the female parent. In order to prevent pollen contamination, the corolla of the pollinated flower were tied together to prevent insects entering the flower. The pollinated flowers were tagged.

Harvesting of sweetpotato seeds

Dried fruits were harvested between 30 and 50 days after pollination between 7.00 a.m. and 8.00 a.m. to prevent seed scattering. Harvesting was done separately for staked and non-staked plants. The fruits were further air dried, shelled and kept in desiccators.

Scarification of sweetpotato seeds

Sweetpotato seeds were soaked overnight in water to break the dormancy in order to improve germination. Immediately after soaking, the seeds were placed in polythene bags filled with top soil and watered. The number of seeds that germinated was recorded and the percentage compatibility and percentage germination were calculated by:

$$\text{Percentage compatibility} = \frac{\text{Number of seeds collected}}{\text{Expected number of seeds}} \times 100 \quad (40.1)$$

$$\text{Percentage germination} = \frac{\text{Number of seeds germinated}}{\text{Number of seeds planted}} \times 100 \quad (40.2)$$

40.4 Results and Discussion

Floral characteristics

Of the 40 clones 18 readily produced flowers. The floral characteristics of these 18 clones were not significantly affected by staking (Table 40.2). Clones 199034.1 and 199024.2 had a pentagonal limb shape while clones Excel and 440215 were semi-stellate. The rest of the clones had a rounded limb shape. Most clones studied had a longer stigma than the highest anther. The exceptions were Blesbok and 440168 in which the anthers were longer than the stigma whereas only 440034 had both anthers and stigma of equal length. This has implications regarding the ease of hand pollination.

Table 40.2. Floral characteristics of flowered sweetpotato clones.

Clone	Limb shape	Stigma exsertion[a]
199024.2	Pentagonal	Exserted
440034	Rounded	Same height
Blesbok	Rounded	Inserted
96-117	Rounded	Exserted
5/21/1995	Rounded	Exserted
Resisto	Rounded	Exserted
Excel	Semi-stellate	Exserted
199034.1	Pentagonal	Exserted
W-151	Rounded	Exserted
Barth	Rounded	Exserted
Famgbe 1	Rounded	Exserted
440215	Semi-stellate	Exserted
Ak-wide	Rounded	Exserted
Sauti	Rounded	Exserted
440168	Rounded	Inserted
TIS 8250	Rounded	Exserted
TIS 8441	Rounded	Exserted
TIS 87/0087	Rounded	Exserted

[a]Exserted, Stigma protrudes beyond the highest anther; inserted, highest anther is longer than the stigma.

Genetic compatibility of 18 sweetpotato clones

Eighteen clones of sweetpotato from both staked and non-staked treatments produced flowers. The flower production of staked clones was significantly higher ($P < 0.5$) than non-staked (Table 40.3). The flowers produced were used for a complete diallel cross for staked and non-staked separately. A total of 324 cross-combinations (18 clones × 18 clones) were expected from the staked and non-staked but only 109 cross-combinations were compatible (i.e. produced seeds) which is equivalent to 33.6% success (Table 40.4).

Table 40.3. Average daily flower production (number of flowers) of staked and non-staked sweetpotato clones.

Clone	No. of flowers: Staked	No. of flowers: Non-staked	Difference (%)
Resisto	123.3	17.1	86.1
Sauti	125.7	45.5	63.8
Ak-wide	167.1	70.8	57.6
TIS 8250	175.1	19.5	88.9
199024.2	97.3	67.0	31.1
96-117	124.6	44.1	64.6
TIS 8441	98.4	83.8	14.8
Excel	44.7	31.8	28.8
440215	12.3	67.8	18.0
1995-5-21	64.1	19.2	70.0
440168	98.3	24.7	74.9
Famgbe 1	49.5	26.6	46.3
Blesbok	123.4	50.0	59.5
Barth	102.0	40.6	60.2
199034.1	54.1	64.8	37.3
440034	67.3	55.6	17.4
TIS 87/0087	67.5	72.2	25.6
W-151	68.7	83.3	14.4
Total	1663.4	884.4	859.5
Mean	92.4	49.1	47.7
Standard deviation	15.3	19.7	
CV (%)[a]	34.5	38.9	

[a]CV, coefficient of variation.

Table 40.4. Genetic compatibility[a] of flowered sweetpotato clones.

	Clone																	
Clone	199024.2	440034	Blesbok	96-117	1995-5-21	Resisto	Excel	W-151	440215	Barth	Sauti	Famgbe 1	440168	TIS 8250	199034.1	Ak-Wide	TIS 8441	TIS 87/0087
199024.2	N	C	C	C	C	C	C	C	N	N	N	N	N	N	C	N	N	N
440034	C	N	C	C	C	C	C	C	N	N	N	N	N	N	C	N	N	N
Blesbok	C	C	N	C	C	C	C	C	N	N	N	N	N	N	C	N	N	N
96-117	C	C	C	N	C	C	C	N	N	N	N	N	N	N	C	N	N	N
5/21/1999	C	C	N	N	N	N	N	N	N	N	N	N	N	N	N	N	N	N
Resisto	C	C	C	N	N	N	C	N	N	N	N	N	N	N	C	N	N	N
Excel	C	C	C	C	C	C	N	C	N	N	N	N	N	N	C	N	N	N
W-151	N	N	N	N	N	N	N	N	C	C	C	N	C	C	N	N	C	C
440215	N	N	N	N	N	N	N	C	N	C	C	C	N	N	N	C	C	C
Barth	N	N	N	N	N	N	N	C	C	N	N	C	N	N	N	N	C	C
Sauti	N	N	N	N	N	N	N	C	C	C	C	N	N	N	N	C	C	C
Famgbe 1	N	N	N	N	N	N	N	C	C	C	C	N	N	N	N	C	C	C
440168	N	N	N	N	N	N	N	C	N	N	N	N	N	N	N	N	N	N
TIS 8250	N	N	N	N	N	N	N	C	N	N	N	N	N	N	N	N	N	C
199034.1	C	C	C	C	C	N	C	N	N	N	N	N	N	N	N	N	N	N
Ak-Wide	N	N	N	N	N	N	N	C	C	C	C	C	N	N	N	N	C	C
TIS 8441	N	N	N	N	N	N	N	C	C	C	N	N	N	C	N	N	C	C
TIS 87/0087	N	N	N	N	N	N	N	C	C	C	C	C	C	C	N	N	C	N

[a]C, compatible; N, not compatible.

Table 40.5. Percentage compatibility and germination of flowered sweetpotato clones.

Clones	Compatibility (%)	Germination (%)
Resisto	16.7	43.3
Sauti	25.0	45.0
Ak-wide	26.7	70.8
TIS 8250	27.5	19.5
199024.2	31.3	67
96-117	34.0	44.1
TIS 8441	39.8	83.8
Excel	44.0	31.8
440215	46.3	67.8
1995-5-21	46.4	19.2
440168	47.5	24.7
Famgbe 1	49.5	26.6
Blesbok	52.9	50.0
Barth	53.3	40.6
199034.1	54.0	64.8
440034	67.1	55.6
TIS 87/0087	67.5	72.2
W-151	68.7	83.3
Mean	44.3	50.6
Standard deviation	15.3	19.7
CV	34.5	38.9

Table 40.6. Genetic compatibility of flowered sweetpotato clones.

Clone	Genetic compatibility (%)	Rank
Resisto	7.2	17
Sauti	11.3	15
Ak-wide	18.9	10
TIS 8250	5.4	18
199024.2	21.0	9
96-117	15.0	11
TIS 8441	33.4	5
Excel	14.0	12
440215	31.4	6
1995-5-21	8.9	16
440168	11.7	14
Famgbe 1	13.2	13
Blesbok	26.5	7
Barth	21.6	6
199034.1	35.0	4
440034	37.3	3
TIS 87/0087	67.5	2
W-151	68.7	1

Details of genetic compatibility between the flowered sweetpotato clones are shown in Table 40.4. For example, clone 199024.2 was reciprocally compatible with 440034, Blesbok, 96-117, 1995-5-21, Resisto, Excel W-151, and 199034.1 while Barth was genetically compatible with W-151, 440215, Famgbe 1, TIS 8441 and TIS 87/0087 but not with other varieties. All the flowering clones were observed to be self-incompatible except for clones Sauti and TIS 8441.

Percentage germination and genetic compatibility of sweetpotato clones

There was variability in terms of percentage compatibility of sweetpotato clones (Table 40.5 and Table 40.6). Clone W-151 had the highest percentage compatibility (68.7%) and Resisto had the lowest percentage variability of 16.7%. Percentage germination for all cross-combinations varied from 19.2% to 83.8% for 1995-5-21 and TIS 8441, respectively, for staked clones (Table 40.5). Clone W-151 had the highest genetic compatibility (68.7%) with other clones, while TIS 8250 has the least (5.4%) (Table 40.6).

40.5 Conclusion

It is important to cross-breed and release improved varieties of sweetpotato that will meet consumer preferences. The study has demonstrated that staking significantly promoted flower production. Although not complete substitutes for glasshouses for inducing flowering in sweetpotatoes, staking could greatly improve breeding activities in programmes, with minimal capital input. Genetic incompatibility was encountered among the clones that produced flowers leading to a reduction in the number of crossings performed. And as expected, a lot of variability in germination was recorded.

References

Martin, F.W. (1988) Genetic and physiological basis for breeding and improving the sweetpotato. In: Degras, L. (ed.) *Proceedings of the 7th Symposium of the International Society for Tropical Root Crops*, Guadeloupe, French West Indies. Institut National de la Recherche Agronomique (INRA) editions, Paris, pp. 749–762.

Mihovilovich, E., Mendozam, H.A. and Salazar, L.F. (2000) Combining ability for resistance to sweet potato feathery mottle virus. *HortScience* 35(7), 1319–1320.

Mok, I.-G., Zhang, D.P. and Carey, E. (1998) Sweetpotato breeding strategy of CIP. In: *Proceedings of International Workshop on Sweetpotato Systems: Toward the 21st Century*, KNAES Miyazaki, Japan, pp. 9–27.

Thottapilly, G. (2009) Introductory remarks. In: Loebenstein, G. and Thottapilly, G. (eds) *The Sweetpotato*. Springer Science + Business Media BV, Houten, The Netherlands, pp. 3–8.

Wang, H. (1982) The breeding of sweet potatoes for human consumption. In: Villareal, R.L. and Griggs, T.D. (eds) *Sweet Potato. Proceedings of the First International Symposium*, Asian Vegetable Research and Development Center (AVRDC) Publication No. 82-172. AVRDC, Taiwan, pp. 297–311.

Wilson, J.E., Pole, F.S., Smith, N.E.J.M. and Taufatofua, P. (1989) *Sweet Potato Breeding*. Institute for Research, Extension and Training in Agriculture (IRETA) University of the South Pacific (USP) Agro-Facts Publication. IRETA-USP, Western Samoa, 39 pp.

41 Sweetpotato and Garden Egg Intercrop Compatibility Studies in Umudike, Nigeria

C.N. Ebeniro,* A. Udealor, A.O. Ano and C.O. Amadi
National Root Crops Research Institute, Umudike, Nigeria

Abstract
This experiment was conducted at Umudike (latitude 05° 29′ N, longitude 07° 33′ E) in the tropical rainforest agroecological zone of Nigeria to determine the compatibility of sweetpotato (*Ipomoea batatas*) and garden egg (*Solanum aethiopicum*) intercrop. Two sweetpotato varieties (NRSP 05/022 and TIS 87/0087) each planted at 1 m × 0.30 m were intercropped with garden egg at three different plant spacings of 1 m × 0.5 m, 1 m × 1 m and 1 m × 1.5 m. The experiment was laid out in a randomized complete block design. Sweetpotato vine length and leaf number were measured at 8 weeks and 12 weeks after planting (AP). Garden egg height, leaf number and number of garden egg fruits per plant were also measured at the two periods. Sweetpotato was harvested and the root yields measured at 16 weeks AP while garden egg fruits were harvested from 60 to 90 days AP and the yield summed up. Intercropping sweetpotato and garden egg had no significant effect ($P = 0.05$) on sweetpotato vine length or on the number of leaves produced. Intercropping with TIS 87/0087 reduced significantly the plant height of garden egg. NRSP 05/022 gave a higher root yield than TIS 87/0087. Intercropping at the three populations of garden egg did not affect garden egg fruit yield ($P > 0.05$). Intercropping sweetpotato with garden egg resulted in a higher land equivalent ratio (LER) over sole cropping of either sweetpotato or garden egg alone. An LER greater than 1 was obtained for sweetpotato variety NRSP 05/022 intercropped with garden egg at 1 m × 0.5 m, 1 m × 1 m and 1 m × 1.5 m thus suggesting that both were compatible.

Keywords: compatibility, garden egg, growth, intercropping, sweetpotato, yield

41.1 Introduction

Intercropping is a common agricultural practice among small-scale farmers in West Africa through which food and income are availed at different periods of the year (Emede and Adegoke, 2011). In Nigeria in particular, between 60% and 70% of cropped land is devoted to growing crop mixtures of two or three crops but also up to eight crops simultaneously (Elemo *et al.*, 1990). In intercropping studies (Palaniappan and Sivaraman, 1996; Muoneke and Asiegbu, 1997) the yield advantage is attributed to better use of land and other resources such as light, nutrients and water (Trenbath, 1976) while ensuring better control of weeds (Anuebunwa, 2000). Sweetpotato (*Ipomoea batatas* (L.) Lam) from the *Convolvulaceae* family is one of the world's most widely grown crops.

*christyebeniro@yahoo.com

In sub-Saharan Africa, the high productivity of sweetpotato per unit area and time is the basis for its choice as a principal crop in agricultural campaigns towards food security and poverty alleviation (Horton *et al.*, 1989; NRCRI, 2005). With production of over 3.5 million t annually (FAO, 2008) from 1.03 million ha (FAO, 2007), sweetpotato is rapidly becoming a staple food in Nigeria.

Garden egg (*Solanum aethiopicum* (L.) family *Solanaceae*) is a fruit vegetable that is widely eaten as a dessert fruit type or as part of a main meal when boiled or fried or as a popular salad eaten fresh. Garden egg is rich in vitamins (vitamins A, B and C) and minerals (Fe) (Udo *et al.*, 2005). According to Schippers (2000), its economic importance equals that of tomatoes.

Intercropping sweetpotato with different crops, for example legumes, cereals, root and tuber crops, has been reported (Onwueme and Charles, 1994; Njoku *et al.*, 2010; Udealor and Ezulike, 2011). Few reports exist on intercropping sweetpotato with garden egg. These crops are planted in Southeastern Nigeria. Based on crop architecture, garden egg is an erectophile while sweetpotato is a planophile, and can be intercropped. Intercropping sweetpotato with garden egg could increase income and productivity of the farming system if the two crops are compatible. The objective of this study was to determine compatibility of sweetpotato and garden egg in an intercropping system.

41.2 Materials and Methods

The study in the 2011 cropping season was conducted at the National Root Crop Research Institute (NRCRI), Umudike, latitude 05° 29′ N and longitude 07° 33′ E and an elevation of 122 m above sea level. The annual rainfall is 2162.7 mm while the relative humidity and minimum and maximum temperature ranges are 54–87%, 20–23°C and 28–33°C, respectively. The soil is acidic and a well-drained sandy clay. The experiment was conducted on land that had been fallow for 2 years (2008/2010) after a cassava crop. Land was cleared, disc ploughed, harrowed and ridged. The experiment was laid out in a randomized complete block design. The plot size was 6 m × 5 m (30 m^2) and this was replicated three times. Treatments consisted of two sweetpotato varieties (NRSP 05/022 light orange flesh and TIS 87/0087 white flesh) and three garden egg plants spaced at 1 m × 0.5 m, 1 m × 1 m and 1 m × 1.5 m with plant populations of 20,000, 10,000 and 6666 plants/ha respectively. Sole crops of sweetpotato and garden egg for all spacings were included for computation of the productivity parameter of intercropping systems. Sweetpotato was planted at 1 m × 0.30 m to give a plant population of 33,333 plants/ha. Garden egg seeds were raised in nursery beds and transplanted into the field at 4 weeks after planting (AP). Garden egg seedlings (var. Ngwa_large) were planted midway between the crest and base of ridges. Weeding was carried out at 4 weeks AP, 200 kg/ha of NPK 15: 15: 15 fertilizer and 3 t/ha of poultry manure were applied to all the plots at 4 weeks AP. Insecticide (Karate) was applied at 4 weeks AP before flowering of garden egg.

Data on sweetpotato were collected on: (i) vine length (in centimetres) at 8 weeks and 12 weeks AP; (ii) number of leaves per plant at 8 weeks and 12 weeks AP; and (iii) yield (in tonnes per hectare) at 16 weeks AP. The land equivalent ratio (LER) was also computed for sweetpotato and garden egg. Data on garden egg were collected on: (i) plant height (in centimetres) at 8 and 12 weeks AP; (ii) number of leaves per plant at 8 weeks and 12 weeks AP; and (iii) fruit yield (in tonnes per hectare). Piecemeal harvest of garden egg fruits was done between 8 weeks and 13 weeks AP. The data were subjected to analysis of variance using GENSTAT DISCOVERY edition 1 (Lawes Agricultural Trust, 2003) and means were compared using Fisher's least significance difference (LSD) at 5% level of probability.

41.3 Results

Data showed sweetpotato varieties differed significantly (P = 0.05) in vine length irrespective of the type of cropping (i.e. whether it was grown as a sole crop or intercropped). Variety TIS 87/0087 had longer vines than NRSP 05/022 at both 8 weeks and 12 weeks

AP (Table 41.1). Intercropping sweetpotato and garden egg at different garden egg spacing had no significant effect ($P = 0.05$) on sweetpotato vine length at 8 weeks and 12 weeks AP. Similarly, the number of leaves of sweetpotato was not significantly affected ($P = 0.05$) by intercropping sweetpotato and garden egg at both sampling dates. However, intercropping sweetpotato with garden egg significantly reduced the plant height of garden egg. Sweetpotato NRSP 05/022 produced a higher root yield than TIS 87/0087 when grown as both a sole crop and as an intercrop.

Intercropping sweetpotato variety NRSP 05/022 with garden egg significantly reduced garden egg plant height at 8 weeks AP but increased it at 12 weeks AP (Table 41.2). The number of leaves produced of garden egg was not affected by intercropping and plant spacing at 8 weeks AP. Intercropping at all spacings studied did not affect the garden egg fruit yield.

The LER as defined Mead and Willey (1980) was used to determine the effect of intercropping on productivity. This was based on the formula:

$$\text{LER} = \left(\frac{\text{Intercrop yield of sweetpotato}}{\text{Sole crop yield of sweetpotato}}\right) + \left(\frac{\text{Intercrop yield of garden egg}}{\text{Sole crop yield of garden egg}}\right) \quad (41.1)$$

The highest LER of 1.58 was obtained for sweetpotato variety NRSP 05/022 intercropped with garden egg at 1 m × 1 m spacing (Table 41.2). Intercropping sweetpotato with garden egg gave LER values higher than 1 except TIS 87/0087 intercropped with garden egg at 1 m × 0.5 m which had a value 0.89.

41.4 Discussion and Conclusion

There was a significant variation in root yield of the sweeetpotato varieties with NRSP 05/022 yielding more than TIS 87/0087. This variation could be attributed to differences in the potential of two varieties. Varietal differences in yield have been

Table 41.1. Effect of intercropping on the growth, yield and land equivalent ratio (LER) of sweetpotato in sweetpotato–garden egg intercrop.

	Vine lengths (cm)		Number of leaves per plant			
Treatments[a]	8 weeks AP[b]	12 weeks AP	8 weeks AP	12 weeks AP	Yield (t/ha) 16 weeks AP	LER
Sole NRSP 05/022	72.2	95	23.6	50	15.23	1.0
Sole TIS 87/0087	226.5	438	17.67	34	9.26	1.0
NRSP 05/022 + GE @ 1 m × 0.5 m	56.5	86	22.67	45.3	4.74	1.12
NRSP 05/022 + GE @ 1 m × 1 m	63.3	83	26.67	49.3	6.42	1.58
NRSP 05/022 + GE @ 1 m × 1.5 m	67.7	80	26.67	53.7	8.69	1.34
TIS 87/0087 + GE @ 1 m × 0.5 m	188.4	443	16.67	33.7	3.19	0.89
TIS 87/0087 + GE @ 1 m × 1 m	210.4	398	18.67	36.3	3.39	1.36
TIS 87/0087 + GE @ 1 m × 1.5 m	226.14	404	18.3	37.7	4.59	1.07
LSD (0.05)[c]	77.4	117.4	NS	NS	3.66	

[a]Sweetpotato varieties NRSP 05/022 and TIS 87/0087; GE, garden egg.
[b]AP, after planting.
[c]LSD (0.05), least significance difference at 5% level of probability; NS, not significant.

Table 41.2. Effect of intercropping on the growth, yield and land equivalent ratio (LER) of garden egg (GE) in sweetpotato–garden egg intercrop.

	Plant height (cm)		Number of leaves per plant			
Treatments[a]	8 weeks AP[b]	12 weeks AP	8 weeks AP	12 weeks AP	Yield (t/ha) 8–13 weeks AP	LER
Sole GE @ 1 m × 0.5 m	53.3	143	24.4	209	8.0	1.0
Sole GE @ 1 m × 1 m	50.3	129	25	272	5.29	1.0
Sole GE @ 1 m × 1.5 m	49.1	115.3	32	248	6.81	1.0
NRSP 05/022 + GE @ 1 m × 0.5 m	41.3	173.3	17.7	119	4.45	1.2
NRSP 05/022 + GE@ 1 m × 1 m	44	116	20.3	204	6.11	1.58
NRSP 05/022 + GE @ 1 m × 1.5 m	55.8	133	31.3	269	5.29	1.34
TIS 87/0087 + GE @ 1 m × 0.5 m	39.9	127.7	24.3	125	5.45	0.89
TIS 87/0087 + GE@ 1 m × 1 m	50.6	121.3	20.3	16.5	5.94	1.36
TIS 87/0087 + GE @ 1 m × 1.5 m	39.2	114	22.7	267	5.29	1.07
LSD (0.05)[c]	2.92	8.75	NS	59.5	NS	

[a]Sweetpotato varieties NRSP 05/022 and TIS 87/0087.
[b]AP, after planting.
[c]LSD (0.05), least significance difference at 5% level of probability; NS, not significant.

reported by many authors including Udealor and Asiegbu (2005). Higher yields obtained in sole crops than when grown as an intercrop in this study is similar to results reported in other studies (Muoneke and Mbah, 2007; Njoku *et al.*, 2010). Intercropping sweetpotato and garden egg did not affect garden egg fruit yield probably because of the difference in architecture of the plants. Sweetpotato is a planophile while garden egg is an erectrophile so interaction in space of the two crops is limited. Intercropping sweetpotato and garden egg resulted in a higher LER over the sole crops showing that it is more productive than sole crops. This result is in agreement with the findings of Udealor and Ezulike (2011) and Muoneke and Mbah (2007). This study suggests that sweetpotato variety NRSP 05/022 is compatible with garden egg at each of the plant spacings studied. Intercropping the sweetpotato variety NRSP 05/022 with garden egg planted at a spacing of 1 m × 1 m gave the optimum productivity and a higher return on investment. A LER greater than 1 was obtained for sweetpotato variety NRSP 05/022 intercropped with garden egg at all the spacings evaluated suggesting that both were compatible.

References

Anuebunwa, F.O. (2000) A bio economical on-farm evaluation of the use of sweetpotato for complementary weed control in yam/maize, egusi/cassava intercrop in pigeon pea hedge rows in rain forest belt of Nigeria. *Biological Agriculture and Horticulture* 18(2), 95–102.

Elemo, K.A., Kumar, V., Olukosi, J.O. and Ogungbile, A.O. (1990) Review of research work on mixed cropping. Miscellaneous paper 127. Institute for Agricultural Research (IAR), Ahmadu Bello University, Zaria, Nigeria.

Emede, T.O. and Adegoke, D.E. (2011) Response of three cultivars of white guinea yam (*Dioscorea rotundata* Poir) to yam/fluted pumpkin (*Telferia occidentalis* Hook F.) intercrop. *Nigerian Journal of Horticultural Science* 16, 19–26.

Food and Agriculture Organization of the United Nations (FAO) (2007) *Production Year Book*. FAO, Rome.

Food and Agriculture Organization of the United Nations (FAO) (2008) *Production Year Book*. FAO, Rome.

Horton, D., Prain, G. and Gregory, P. (1989) High level investment returns for global sweetpotato research and development. *CIP Circular, Lima* 17(3), 1–11.

Lawes Agricultural Trust (2003) GENSTAT DISCOVERY edition 1. Lawes Agricultural Trust (Rothamsted Experimental Station), Rothamsted, UK.

Mead, R. and Willey, R.H. (1980) The concept of land equivalent ratio and advantages in intercropping. *Experimental Agriculture* 16, 217–218.

Muoneke, C.O. and Asiegbu, J.E. (1997) Effects of okra plant density and spatial arrangement in intercrop with maize on the growth and yield of the component species. *Journal of Agronomy and Crop Science* 179, 202–207.

Muoneke, C.O. and Mbah, E.U. (2007) Productivity of cassava/okra intercropping systems as influenced by okra density. *African Journal of Agricultural Research* 2(5), 205–221.

National Root Crops Research Institute (NRCRI) (2005) *NRCRI Annual Report*. NRCRI, Umudike, Nigeria.

Njoku. S.C., Ano, A.O., Amangbo, L.E.F., Akinpelu. A.O. and Ebeniro, C.N. (2010) Effect of cropping system on yield of some sweetpotato and okra cultivars in an intercropping system. *Journal of Agriculture and Social Research* 10(2), 40–47.

Onwueme, I.C. and Charles, W.B. (1994) *Tropical Root and Tuber Crops Production: Perspectives and Future Projects*. Food and Agriculture Organization of the United Nations (FAO), Rome.

Palaniappan, S.P. and Sivaraman, K. (1996) *Cropping System in the Tropics: Principles and Management*. New Age International, New Delhi, 215 pp.

Schippers, R.R. (2000) *African Indigenous Vegetables. An Overview of the Cultivated Species*. Natural Resources Institute/ACP-EU Technical Centre for Agricultural and Rural Cooperation, Chatham, UK.

Trenbath, B.R. (1976) Resource use by intercrops. In: Francis, C.A. (ed.) *Multiple Cropping Systems*. Macmillan, New York, pp. 57–81.

Udealor, A. and Asiegbu, J.E. (2005) Effects of cassava genotype and planting pattern of vegetable cowpea on growth, yield and productivity of cassava/vegetable cowpea intercropping system. *Nigerian Agricultural Journal* 36, 88–96.

Udealor, A. and Ezulike, T.U. (2011) Evaluation of different crop species for compatibility with cocoyam minisett. *Nigerian Agricultural Journal* 42, 122–129.

Udo, D.J., Asuquo, P.E., Ndom, B.A. and Ndaeyo, N.U. (2005) *Crop Production Techniques for the Tropics*. Concept Publications, Lagos, Nigeria, pp. 232–237.

42 Nutrient Uptake and Yield Efficiency of Exotic Sweetpotato Cultivars Under an Organic Soil Management System in Abeokuta, Southwestern Nigeria

O.I. Lawal,[1]* M.O. Atayese,[1] A.A. Oyekanmi,[1] S.O. Afuwape,[2] O.S. Sakariyawo,[1] A.O. Olaiya,[1] O.T.H. Idowu[1] and I.O.O. Aiyelaagbe[1]

[1]Federal University of Agriculture, Abeokuta, Nigeria; [2]National Root Crops Research Institute (NRCRI), Umudike, Nigeria

Abstract

The need to evaluate agronomic performance of crops under organic production systems is necessary due to the high cost of maintaining degraded soil fertility coupled with scarcity and health implications of continuous use of inorganic fertilizers. Field experiments were conducted at the Federal University of Agriculture, Abeokuta, Nigeria in the rainy seasons of 2010 and 2011 to determine the optimum rate of composted cow dung (CCD) and NPK fertilizer on nutrient uptake and yield of exotic sweetpotato cultivars. Vines (20 cm length) of four cultivars of sweetpotato (cv. TIS 87/0087, 199000.1, Ex-egbariam and TIS 86/0356) were planted at 0.3 m distance on ridges spaced 0.75 m apart. Three weeks after planting, CCD was applied at 0 t/ha, 2.5 t/ha, 5.0 t/ha, 7.5 t/ha and 10 t/ha and NPK 15:15:15 at 400 kg/ha. The treatments were arranged factorially in a randomized complete block design with four replicates. Data on precropping soil analyses, leaf fresh and dry weights, leaf nutrient contents and tuber yield of sweetpotato plants were collected at harvest. Sweetpotato plants that received 10 t/ha of CCD had significantly higher ($P < 0.05$) leaf fresh and dry weights, leaf nutrient contents and higher tuber weight than those obtained from sweetpotato plants grown without fertilizer (control) and those that received other fertilizer treatments. Tuber yield and yield components were in the order TIS 87/0087 > TIS 86/0356 > 199000.1 > Ex-egbariam. Since yield and its components are a product of nutrient uptake and leaf dry matter, leaf N content and tuber weight, CCD at 10 t/ha in combination with any of the four sweepotato cultivars can therefore be recommended for adoption by farmers.

Keywords: dry matter, exotic sweetpotato, nutrient uptake, tuber yield

42.1 Introduction

Sweetpotato (*Ipomoea batatas* L. [Lam.]) belongs to the family *Convolvulaceae*. It is one of the most important food crops in the developing countries of the world (Edmond, 1971). It ranks thirteenth on the basis of agricultural commodities and fifth on the list of most valuable food crops of developing countries. It is rich in iron and zinc and

*lawalisiaq@gmail.com

contains starch of about 65–80% of the dry weight (Woolfe, 1992). Compared with grains, sweetpotato is more tolerant of poor soils and resistant to drought, pests and diseases. Furthermore, its roots are storable in the ground for months after they mature. Sweetpotato is a promising raw material for producing both biodegradable plastics and hydrogen gas for use as an energy source in cell batteries (Kozai *et al.*, 1999). The annual starch yield of sweetpotato in tropical and subtropical countries is 1.5 times higher than those of rice (*Oryza sativa* L.) or maize (*Zea mays* L.) and two times higher than that of Irish potato (*Solanum tuberosum* L.) (Kozai *et al.*, 1999), when the plants are protected from pests and diseases.

One of the most limiting factors for sweetpotato production is the inadequate source of planting materials for desirable varieties. The choice of varieties depends on adaptation to prevailing climatic conditions, yield performance, nutritive value, eating quality, horticultural characters and resistance to pest and diseases. Most tropical soils are highly degraded and infertile due to human activities. Thus, continuous cropping on these soils reduces organic matter during the first few years following land cleaning (Brams, 1991; Juo *et al.*, 1995), leading to poor crop yield. External nutrient inputs from either organic or inorganic sources are essential to improve and sustain crop production on these soils (Juo and Wilding, 1996). Although crops grow faster with inorganic fertilizer, the plants may lack useful vitamins and minerals compared with those grown with organic fertilizer. Residual effects in the soil, ecological damage and increased risk of human diseases such as cancer, stomach ulcers and other metabolic disorders may be expected with continous consumption of crops grown with mineral fertilizer when compared with organically grown crops. The use of chemicals (pesticides, herbicides, etc.) in food production has been identified as a major source of health risk and cause of extensive environmental damage (FAO, 2001).

Organic farming is a production system that avoids or largely excludes the use of synthetically produced fertilizers, pesticides, growth regulators and animal additives but relies as much as possible on crop rotation, crop residues, animal manures, legume and green manure. Organic fertilizers improve both the physical and the chemical properties of the soil such as soil structure, soil tilths, cation exchange capacity (CEC), water-holding capacity and crumb formation. They also promote infiltration, protect against erosion and facilitate the spread and penetration of plant roots. The slow release nature of organic fertilizer prevents loss of nutrients through leaching and enhances synchronization of nutrient release with uptake by the crop with a long-term effect on soil fertility. This calls for the need to develop standard organic sweetpotato production practices for farmers in Nigeria. There is limited information on the nutrient uptake and tuber yield of organically grown sweetpotato in Nigeria. The objective of this study is to determine the effect of organic and inorganic fertilizer on nutrient uptake and yield components of exotic sweetpotato varieties in Abeokuta, Southwestern Nigeria.

42.2 Materials and Methods

A field trial was conducted on the Teaching and Research Farm of the Federal University of Agriculture, Abeokuta (FUNAAB), Southwestern Nigeria during the cropping season of 2010 and 2011 to determine the effect of varying rates of composted cow dung (CCD) on nutrient uptake and tuber yield of sweetpotato. This is a complex and multidisciplinary research project involving an agronomist, a plant physiologist, an entomologist, a pathologist and food and animal scientists. The present chapter reflects on the agronomic/plant nutrition aspect. FUNAAB is located in the forest/savannah transitional zone (latitude 7° 23^1, longitude 3° 23^1). The area is characterized by bimodal rainfall with peaks in July and September. Fifteen composite samples of the top soil of the experimental sites were collected each year, bulked together and analysed for the physico-chemical properties (Table 42.1). The land was manually cleared and ridges were made at a spacing

Table 42.1. Pre-planting soil physical and chemical properties of the experimental site (2010 and 2011) and the composted cow dung (CCD).

Soil properties	2010	2011	CCD properties	
Soil pH (H_2O)	5.70	6.50	N (%)	2.58
Organic matter (%)	0.40	0.45	P (%)	1.10
Total N (%)	0.07	0.06	K (%)	0.68
Available P (mg/kg)	4.50	3.43	Ca (%)	3.62
Exchangeable K (cmol/kg)	0.26	0.27	Mg (%)	0.18
Exchangeable Na (cmol/kg)	0.32	0.36	C:N ratio (g/kg)	8.97
Exchangeable Mg(cmol/kg)	2.00	2.14	Na (cmol/kg)	1.75
Exchangeable Ca (cmol/kg)	2.30	2.27	Fe (mg/kg)	4.42
Exchangeable acidity	0.40	0.38	Mn (mg/kg)	3.70
Exchangeable Zn (ppm)	6.20	5.56	Zn (mg/kg)	3.60
Exchangeable Fe (ppm)	10.00	9.68	Cu (mg/kg)	11.45
Cation exchange capacity (CEC) (cmol/kg)	0.02	0.03	Mo (mg/kg)	0.60
Sand (g/kg)	867	765		
Silt (g/kg)	100	102		
Clay (g/kg)	033	133		

of 0.75 m apart. The entire field was demarcated into 48 plots with each individual plot measuring 3.0 m × 3.0 m and separated by a 1.0 m corridor. The treatments were combined following a split-split plot arrangement fitted into a randomized complete block design with four replicates. The main plot comprised of CCD and NPK 15-15-15 fertilizer. The subplot was four sweetpotato cultivars: cv. TIS 86/0356, TIS 87/0087, 199000.1 and Ex-egbariam. The sub-sub plot was six fertilizer rates: CCD at 0 t/ha, 2.5 t/ha, 5.0 t/ha, 7.5 t/ha and 10 t/ha and 400 kg/ha NPK15-15-15. The sweetpotato varieties were planted 30 cm apart on ridges 0.75 cm apart. Three weeks after planting, the two fertilizer types (with proximate analysis of CCD as shown in Table 42.1) were spot applied to the sweetpotato plantlets 10 cm away in ring form.

Data collection and analyses

Data were taken on leaf fresh and dry weights, total fresh weight and tuber yield at harvest. Pre-cropping soil physico-chemical properties and leaf nutrient contents were determined using standard laboratory procedures. Data generated were subjected to analysis of variance at 5% level of probability and the separation of treatment means was done (SAS, version 9.1) using the least significance difference (LSD) test.

42.3 Results and Discussion

The pre-treatment soil nutrient analyses in 2010 and 2011 (Table 42.1) revealed that the soil pH was slightly acidic (5.70 and 6.50), the soil N (0.07% and 0.06%) was below the critical level of 0.15% recommended for sweetpotato crop production. P (4.50 mg/kg and 3.43 mg/kg) and K (0.26 cmol/kg and 0.27 cmol/kg) were lower than 13 mg/kg and 0.34 cmol/kg P and K recommended for crop production (Adeoye and Agboola, 1985), and hence justified the need for additional fertilizer treatment. The soils were sandy loam in texture and were optimal to retain adequate water for normal crop growth.

The CCD contained 2.58%, 1.10%, 0.68%, 3.62% and 0.18% of N, P, K, Ca and Mg, respectively (Table 42.1). The organic materials are rich in plant nutrients with C:N ratios of 8.97g/kg being adequate for quick nutrient release (Palm *et al.*, 1997).

Leaf nutrient contents of sweetpotato cultivars in 2010 and 2011 (Table 42.2) indicated

Table 42.2. Leaf nutrient contents of sweetpotato cultivars in 2010 and 2011.

Variety	N (%)	P (%)	K (%)
		2010	
TIS 87/0087	2.67	0.21	3.43
TIS 86/0356	2.36	0.20	2.57
199000.1	2.25	1.19	2.54
Ex-egbariam	2.32	0.27	2.68
LSD ($P < 0.05$)[a]	ns	0.16	0.14
		2011	
TIS 87/0087	2.45	0.23	3.48
TIS 86/0356	2.41	0.21	2.29
199000.1	2.28	1.17	2.56
Ex-egbariam	2.35	0.28	2.63
LSD ($P < 0.05$)[a]	ns	0.13	ns

[a]LSD, least significance difference; ns, not significant.

Table 42.3. Leaf NPK uptake of sweetpotato varieties in 2010 and 2011.

Variety	N uptake (%)	P uptake (%)	K uptake (%)
		2010	
TIS 87/0087	0.14	0.01	0.17
TIS 86/0356	0.18	0.02	0.19
199000.1	0.28	0.17	0.28
Ex-egbariam	0.25	0.16	0.24
LSD ($P < 0.05$)[a]	0.06	0.09	0.04
		2011	
TIS 87/0087	0.13	0.01	0.18
TIS 86/0356	0.19	0.02	0.21
199000.1	0.29	0.18	0.29
Ex-egbariam	0.25	0.17	0.25
LSD ($P < 0.05$)[a]	0.05	0.08	ns

[a]LSD, least significance difference; ns, not significant.

that significant differences ($P < 0.05$) were observed only in P and K but not in N, with sweetpotato variety 199000.1 showing significant P content compared with other varieties while TIS 87/0087 had significantly more K than other varieties.

Leaf NPK uptake followed a similar trend to that observed for leaf nutrient contents albeit variety 199000.1 projected higher significant ($P < 0.05$) differences for both P and K uptakes (Table 42.3). These observations may be due to the innate ability of the cultivars to source/compete for more of nutrients P and K from the environment and consequently utilized them for growth and development more than other varieties. These observations were in accord with the findings of Lawal (2009; Lawal *et al.*, 2010) on yams and Lawal *et al.* (2008) on citrus where they generally observed that the leaf nutrient contents and uptakes in these crops to be dependent on variety, environmental factors, crop phenology and fertilizer materials used as soil amendments.

Influence of fertilizer treatments on tuber yield and yield components (Table 42.4) indicated that sweetpotato plants that received 10 t/ha CCD were significantly ($P < 0.05$) superior to those plants treated with other fertilizer materials in root length, root weight, stem weight, leaf weight and total fresh weight with values of 78.1 cm, 99.8/12 m^2 plot, 20.7 kg/12 m^2 plot and 38.9 kg/12 m^2 plot, respectively. The lowest values of 50.8 cm, 54.0 kg/plot, 6.7 kg/plot, 7.7 kg/plot and 15.2 kg/plot for root length, root weight, stem weight, leaf weight and total fresh weight, respectively, were obtained in plants that received no fertilizer treatments (Table 42.4).

Sweetpotato plants that received 10 t/ha CCD produced significantly ($P < 0.05$) higher tuber weights than those that received other fertilizer treatments (Table 42.4). Sweetpotato variety TIS 87/0087 was found to be superior to other cultivars in terms of tuber yield and other yield components in both 2010 and 2011 (Table 42.5).

Generally, sweetpotato planted in 2011 performed better than those of 2010 in terms of yield parameters (Table 42.5). This may be attributed to the slight differences in the soil fertility status of the experimental soils and more favourable weather conditions associated with 2011 cropping season relative to that of 2010.

For most of the parameters measured the fertilized plots, most especially at 10 t/ha, were superior to the control. The superiority of sweetpotato plants grown on fertilized plots over the control in performance, especially on yield parameters, indicated that the control plants must have been in short supply of essential nutrients over time

Table 42.4. Effect of fertilizer treatments on tuber yield and yield components of sweetpotato varieties as affected by composted cow dung (CCD) application; each plot is 12 m^2.

Fertilizer rate	Root length (cm)	Root weight (kg/plot)	Stem weight (kg/plot)	Leaf weight (kg/plot)	Total fresh weight (kg/plot)	Tuber yield (t/ha)
	Yield and yield components of sweetpotato 2010					
10.0 (t/ha CCD)	78.1	99.8	17.4	20.7	38.9	9.90
7.5(t/ha CCD)	61.3	82.1	12.7	14.7	22.2	7.50
5.0 (t/ha CCD)	69.5	81.5	14.8	17.6	33.5	8.40
2.5 (t/ha CCD)	58.4	66.5	10.7	8.9	18.2	6.15
400 (kg/ha NPK)	48.0	59.3	7.8	8.7	18.0	3.60
No fertilizer	50.8	54.0	6.7	7.7	15.2	3.10
LSD ($P < 0.05$)[a]	2.43	3.45	1.18	1.60	1.91	0.58
	2011					
10.0 (t/ha CCD)	83.8	97.7	18.5	22.4	43.8	10.12
7.5 (t/ha CCD)	66.7	88.3	13.4	16.8	33.5	8.25
5.0 (t/ha CCD)	67.4	87.4	15.7	19.6	43.2	8.65
2.5 (t/ha CCD)	63.5	68.7	11.8	10.3	20.1	7.22
400 (kg/ha NPK)	52.8	61.4	9.6	10.6	19.4	5.76
No fertilizer	51.2	58.4	7.5	8.4	17.3	4.23
LSD ($P < 0.05$)[a]	2.64	2.85	2.15	2.23	2.19	0.65

[a]LSD, least significance difference.

Table 42.5. Yield and yield components of sweetpotato cultivars at harvest in 2010 and 2011.

Cultivars	Yield (t/ha)	Root length (cm)	Root weight (kg/plot)	Shoot weight (kg/plot)	Leaf weight (kg/plot)	Total fresh weight (kg/plot)	Root dry weight (g/kg/plot)	Leaf dry weight (g/kg/plot)	Shoot dry weight (g/kg/plot)	Total dry weight (g/kg/plot)
	2010									
TIS 87/0087	10.95	67.34	8.97	11.64	13.68	34.29	48.36	45.32	47.41	141.09
TIS 86/0356	9.93	65.08	6.26	11.06	12.04	29.36	43.02	46.67	45.67	135.36
Ex-egbariam	8.64	54.76	7.68	10.35	12.54	31.53	42.37	41.44	43.23	132.54
199000.1	8.73	56.92	8.51	10.35	13.75	32.61	47.38	40.78	43.34	131.50
LSD ($P < 0.05$)	0.71	3.91	2.48	1.24	3.29	2.75	4.08	4.23	3.22	ns
	2011									
TIS 87/0087	11.19	68.13	9.23	12.26	12.76	34.25	45.43	46.53	47.31	143.92
TIS 86/0356	10.23	66.52	7.26	11.23	11.54	30.03	44.32	45.58	46.16	138.24
Ex-egbariam	8.95	56.68	8.16	10.15	11.65	29.75	43.22	40.74	42.63	133.65
199000.1	9.17	58.24	8.15	10.85	12.67	31.67	48.15	43.54	46.23	135.58
LSD ($P < 0.05$)	0.65	2.39	2.48	1.24	2.12	3.23	3.23	14.8	3.72	ns

[a]LSD, least significance difference; ns, not significant.

resulting in low tuber yields in terms of both quantity and quality. The application of CCD at 10 t/ha was in most cases better than NPK in enhancing growth, nutrient uptake and yield of the sweetpotato crop. Scientists are unanimous regarding the contribution of organic fertilizer to soil fertility on different cropping systems. Particularly, organic fertilizers lead to improvements in the physical and chemical properties of the soil, the soil structure, soil tilths, CEC, water-holding capacity and crumb formation, with long-term residual effects and consequently these enhance crop performance and yield. It also promotes infiltration, protects against erosion and facilitates the spread and penetration of plant roots. The relatively poor performance of the control and NPK fertilizer at 400 kg/ha is due the declining availability of the available nutrients over time, inadequate nutrient content of the untreated soil and the quick nutrient-release pattern of NPK fertilizer material triggered by leaching and erosion which does not favour the performance of a long-season crop like sweetpotato. This is in accord with the observation of Giwa and Ojeniyi (2005) who reported that organic manure improves the physical condition of soil and provides adequate amounts of the necessary nutrients for the soil productivity and that continuous use of mineral fertilizer alone leads to soil acidity, nutrient imbalance and declining crop yield (Tendon, 1992).

42.4 Conclusion

Nutrient uptake, yield and yield components of exotic sweetpotato varieties were evaluated under organic soil amendments. Results obtained indicated the superiority of CCD at 10 t/ha over all other fertilizer treatments for most of the uptake and yield parameters measured in this study. Also sweetpotato variety TIS 87/0087 was found to be superior to other cultivars in terms of tuber yield and other yield components. Since nutrient uptake and tuber yield are functions of physiological parameters such as leaf fresh and dry weights, leaf nutrient contents, etc., CCD at 10 t/ha was therefore recommended for adoption in this study. Other sweetpotato cultivars and fertilizer rates need to be evaluated under different soil amendments and agroecologies.

Acknowledgement

The project team acknowledge the Federal University of Agriculture Abeokuta (FUNAAB), Nigeria for providing a conducive research-friendly environment for this work and for the technical support of colleagues; and the National Root Crops Research Institute, Umudike and the Agronomy Department, University of Ibadan for providing the planting materials. Full sponsorship support for this chapter, funded by Bayer East Africa Ltd through the African Potato Association based in Kenya, is highly appreciated.

References

Adeoye, G.O. and Agboola, A.A. (1985) Critical levels for soil PH, available P, K, Zn and Mn and maize ear leaf content of P, Cu and Mn in sedimentary soil of southwest Nigeria. *Fertilizer Research* 6, 65–71.

Brams, E.A. (1991) Continuous cultivation of West African soils. Organic diminution and effects of applied lime and phosphorus. *Plant and Soils* 35, 401–414.

Edmond, J.B. (1971) Sweetpotato production, processing and marketing. Major feed and food series. In: Frank, S.S. and Herman, R. (eds) *Proceedings of the Sixth Triennial Sympossium of the International Society for Tropical Root and Tuber Crops*. International Potato Center (CIP), Lima, pp. 51–59.

Food and Agriculture Organization of the United Nations (FAO) (2001) *World Market for Organic Fruit and Vegetables: Opportunities for Developing Countries in the Production and Export of Organic Horticultural Products*. Technical Trade Centre, Technical Centre for Agricultural and Rural Cooperation, FAO, Rome.

Giwa, D.D. and Ojeniyi, S.O. (2005) Effect of integrated application of pig manure and NPK on soil nutrient content and yield of tomato (*Lycopersicum esculentus* Mill.). In: Salako, F.K., Adetunji, M.J., Ojanuga, A.G., Arowolo, T.A. and Ojeniyi, S.O. (eds) *Proceedings of the 29th Annual Conference of the Soil Science Society of Nigeria,* 6–10 December 2004, University of Agriculture Abeokuta, Nigeria. Soil Science Society of Nigeria, Abeokuta, Nigeria, pp. 164–169.

Juo, A.S.R. and Wilding, L.P. (1996) Soils of the lowland forest of West and Central Africa. In: *Essays on the Ecology of the Guinea-Congo Rainforest. Proceedings of the Royal Society of Edinburgh* Vol. 1043. Royal Society of Edinburgh, Edinburgh, UK, pp. 15–26.

Juo, A.S.R., Franzluebbers, K., Dabiri, A. and Ikhile, B. (1995) Changes in soil properties during long term fallow and continuous cultivation after forest clearing in Nigeria. *Agriculture Ecosystem and Environment* 56, 9–18.

Kozai, T., Kubota, C., Zobayed, S.M.A., Nguyen, Q.T., Afreen-Zobayed, F. and Heo, J. (1999) Developing a mass-propagation system of woody plants. In: Watanabe, K. and Komamine, A. (eds) *Challenge of Plant and Agriculture Sciences to the Crisis of Biosphere on the Earth in the 21st Century*, Proceedings of the 12th Toyota Conference. Landes Bioscience, Austin, Texas, pp. 293–306.

Lawal, O.I. (2009) Response of white yam to different levels of fertilizers in two soil-types in southwestern Nigeria. PhD thesis, Department of Agronomy, University of Ibadan, Nigeria, 210 pp.

Lawal, O.I., Olaniyan, A.A. and Okafor, B.N. (2008) Nutrient uptake and fruit yield of sweet orange as affected by poultry manure and NPK fertilizers application in Ibadan, Nigeria. *Journal of Applied Agricultural Research* (publication of the Agricultural Research Council of Nigeria (ARCN)) 1, 113–117.

Lawal, O.I., Adeoye, G.O., Asiedu, R. and Ojeniyi, S.O. (2010) Organo-mineral fertilizer effects on yield and nutrient uptake of white yam (*Dioscorea rotundata* Poir.) in Ibadan, southwest Nigeria. *Nigerian Journal of Soil Science* 20(2), 35–46.

Palm, C.A., Myers, R.J. and Nandwa, S.M. (1997) Organic–inorganic nutrient interaction in soil fertility replenishment. In: Buresh, R.J., Sanchez, P.A. and Calhoun, F. (eds) *Replenishing Soil Fertility in Africa*. Soil Science Society of America special publication 51. Soil Science Society of America, Madison, Wisconsin, pp. 193–218.

Tendon, H.L.S. (ed.) (1992) *Organic Manures, Recyclable Wastes and Biofertilizer Development*. Fertiliser Development and Consultation Organisation, New Delhi, 148 pp.

Woolfe, J.A. (1992) *Sweet Potato: an Untapped Food Resource*. Cambridge University Press, Cambridge.

Part V

Innovations in Marketing, Processing Technologies and Value Chain Management for Potato and Sweetpotato

43 On the Road to Potato Processing in African Tropical Highlands

A.J. Haverkort,[1]* G. Woldegiorgis,[2] M.J. van Koesveld,[3] S. Ntizo,[4] R. Wustman[3] and X. Zhang[5]

[1]*Plant Research International (PRI), Wageningen University and Research Centre, The Netherlands;* [2]*Ethiopian Institute of Agricultural Research (EIAR), Addis Ababba, Ethiopia;* [3]*Applied Plant Research (PPO), Wageningen University and Research Centre, The Netherlands;* [4]*Rwanda Agriculture Board (RAB), Kigali, Rwanda;* [5]*Institute of Agricultural Economics (LEI), Wageningen University and Research Centre, The Netherlands*

Abstract

Potato processing at the industrial level is relatively insignificant in the tropical highlands of Africa in Ethiopia, Kenya, Uganda and Rwanda. No large international companies are present in this area whereas they have successfully established themselves in North Africa (Egypt) and South Africa. The reason for the presence of large French fries (chips) and crisps factories in northern and southern Africa is the availability of growers who have relatively large areas of irrigated land assuring a year-round flow of raw material with the required specification and consistency for the finished product. The objectives of recently carried out investigations in Rwanda and Ethiopia funded by The Netherlands government were to investigate the challenges and opportunities for the establishment of (private) industrial processing facilities. Such factories and installations would create added value through employment in the whole supply chain from breeding new varieties, creating a supply chain of seed and ware (raw material) potatoes, postharvest handling, processing and trade. The study in Rwanda and Ethiopia consisted of the following elements: (i) a quick scan of potato production related to the potential supply of raw material to the industry; (ii) an inventory of current processing at household and restaurant level such as boiling, mashing, and the preparation of French fries and crisps; and (iii) an inventory of processed potato products such as frozen French fries, crisps and mashed potato powder and packed fresh tubers in markets and supermarkets. An estimate was then made of the current and future market for these products. The type of small- and medium-sized equipment needed and their costs was assessed. Historical meteorological data and a crop growth model that calculated attainable yields and water use in the two countries were used for yield gap analyses. From this and from farm surveys the costs and competitiveness of the raw material were derived and from it conclusions regarding the costs of locally produced versus imported finished products. We concluded that the farm-gate potato price, which is related to the cost of production, is quite competitive – in Ethiopia it is about half that of northern Europe and in Rwanda it is about equal to north-western Europe – which is most important in order to be competitive with imports in the future. An investor planned a frozen French fry (chips) factory in Ethiopia to open in 2014. In Rwanda a similar action is planned with government involvement.

Keywords: chips, crisps, French fries, potato processing quality, sub-Saharan Africa

*anton.haverkort@wur.nl

43.1 Introduction

Quick scans of potato production in Ethiopia and Rwanda

In Ethiopia and Rwanda about 200,000 ha of potatoes are grown with average yields of about 10 t/ha so total production in each country is around 2 million t. Statistics vary somewhat depending on source. The population of Rwanda is close to 12 million of which 1 million live in the capital Kigali. Ethiopia's population is 94 million with well over 4 million living in the capital Addis Ababa. Average annual potato consumption in Rwanda is 125 kg per inhabitant, in Ethiopia this is 17 kg per inhabitant.

Situation in Rwanda

Potatoes were introduced in Rwanda around 1900 by German soldiers and later by Belgian missionaries. Since national independence, potatoes have played an increasingly important role in the Rwandan economy and in the diets of the Rwandese. In 1962, the Rwandan Institute for Agricultural Sciences (ISAR), operating under the auspices of the Ministry of Agriculture and Livestock, assumed responsibility for agricultural research and development. In 1979 the Programme National pour l'Amélioration de la Pomme de terre (PNAP; National Programme for Potato Improvement) was established within ISAR in cooperation with the International Potato Center (CIP). Later it became ISAR-Musanze and now it is Rwandan Agricultural Board (RAB)-Musanze and still mainly works on potato. The programme operates a 4 ha research station at Musanze and a 45 ha seed farm at Kinigi at 2300 m above sea level (masl). The programme's mandate includes: (i) selection and multiplication of improved varieties for Rwandan conditions; (ii) production and distribution of improved seed; (iii) development and transfer of technologies to increase potato yields and reduce storage losses; and (iv) enhancement of research and extension capabilities through training.

In most areas at least two major crops are planted each year. In the volcanic zone and along the Nile–Congo Divide the crop is planted at the onset of the dry season – May in the volcanic highlands of the north, April to June in central Rwanda. Harvest takes place 3–4 months later. A second crop is planted in October/November and harvested from January to March. Along the Nile–Congo Divide, and especially in the volcanic zones, some farmers grow a slightly later crop in marshy valleys, planting in July and harvesting in October. Another minor crop can be grown from January to March where local conditions are favourable and growers accept lower yields in return for higher prices.

Although some commercial producers cultivate up to 10 ha, the most important players of potato production consist of small, traditional, family farms growing less than 0.5 ha each year. An important feature of Rwandese agriculture is that many crops are intercropped, and potato varieties are often selected for their ability to yield well in an intercropping system. The use of purchased inputs such as fertilizers (NPK, 17:17:17), insecticides and fungicides (mainly Ridomil) is gradually increasing and substantially contributes to the significant yield increase from about 6 t/ha some 30 years ago to over 11 t/ha per season presently, with many growers in Musanze achieving yields well over 20 t/ha. All land preparation is done by hand. Along the Nile–Congo Divide and in Byumba potatoes are planted on relatively flat land. In the volcanic areas of Gisenyi and Musanze prefectures large ridges or beds are prepared that are 1 m or more in width and 25–50 cm high. These permit sufficient drainage in these areas of high rainfall. Potatoes are planted with an irregular spacing of 20–40 cm between plants. Some ridging may be performed after planting, followed by two weedings.

Varieties

In broad terms the following waves of variety introduction have taken place since the introduction of the crop into the country. First was the introduction of European varieties from Germany and Belgium. After independence this was halted and the available varieties degenerated due to bacterial wilt and virus diseases. Moreover they were mixed and some components of the mixture were probably derived from self-pollinated berries which produced seedlings and later tubers that subsequently were multiplied clonally. In the late 1970s, ISAR introduced some varieties from mountainous humid Mexico – Toluca – that has similar humid conditions to Rwanda, hence these varieties (Sangema, Montsama and Atzimba – acronyms of the first letters of the names of the team that bred them) were resistant to late blight. As of 1979 PNAP started to rapidly multiply and distribute these varieties countrywide. When PNAP was established more varieties were introduced from elsewhere such as Cruza 148, also from Mexico, that also showed good tolerance to bacterial wilt. Advanced clonal material was also screened and some varieties were named and finally crosses were made between newly released varieties and promising advanced clones yielding varieties such as Kinigi and Gasore. Table 43.1 shows most current varieties and some of their characteristics.

It is obvious that only a handful of varieties dominate the market, Kinigi being the most important one. Kinigi is appreciated by consumers because of its taste and flouriness but it has a long dormancy period, is sensitive to degeneration due to viruses and is susceptible to insects.

Economic and food security importance of the potato in Rwanda

Since 1961, Rwanda's potato output has risen from less than 100,000 t to 1.3 million t in 2005 and currently is over 2 million t/year. The potato underpins Rwanda's food security. Annual consumption is high at 125 kg per person, making potato the country's second most important source of calorie intake after cassava. In recent years, potato production has expanded rapidly in northwestern Rwanda, especially in the zone of the volcanic highlands, where environmental conditions are very favourable for potato production. Along the Musanze–Gisenyi road a large number of traders engage in potato marketing; they not only handle the interregional trade in potatoes by supplying all potato-deficit regions of Rwanda (and exporting to Congo and Burundi as well) but have also set up an efficient input-provision network to supply farmers with knapsack-sprayers, pesticides and improved seeds.

Over the last 5 years, potato production has increased by 70%, reaching 2,171,000 t in 2011. In 2010, potato ranked third among the commodities produced in Rwanda after plantain and cassava (Ministry of Agriculture and Animal Resources, 2011). According to FAOSTAT (2013), the farm-gate price for potato has increased steadily from US$94.5/t in 2002 to US$204.2/t in 2009 with a fall in price between 2004 and 2005.

Rwanda's population increased from 4 to 11 million inhabitants (a 2.75-fold increase) over the period 1979–2014 whereas the potato area increased from 44,000 ha with 250,000 t annual production (5.7 t/ha) to 150,000 ha with 2.2 million t production (14 t/ha) a 2.5-fold yield increase per hectare and an 8.4-fold increase of total potato production: potato output increased three times faster than population growth over a 35 year period. This success is mainly due to variety introduction, breeding programmes and seed distribution of the best performing material by the predecessor of the RAB potato programme (PNAP) in the 1980s, and more recently the introduction and utilization of chemical fertilizers (N:P:K, 17:17:17) and chemical fungicides to control the most important disease, late blight caused by *Phytophthora infestans*. A conservative estimate shows that about half of the national potato production increase (1 million t/year) has been made possible by the new varieties that replaced the ill-adapted European varieties and mixtures of landraces from their spontaneous progeny. With a current farm-gate price of RWF140 (€0.175) this annually represents a value of €175,000,000 (€175 million/year). This is from an estimated investment in the period 1978–1995 of probably less than €3 million in 2013 euro (€) value equivalents; a remarkable return on investment.

Table 43.1. Potato varieties in Rwanda and some of their characteristics (prepared by the farmers' organization Imbaraga).

Variety[a]	Origin[b]	Release	Dormancy (days)	Length of cycle (days)	Colour	Positive and negative points[c]		Importance[d]
						+++	---	
Peko	RAB/Uganda	2010	120	120	White	Y>, LBR, taste	BWS	X
Rwashaki	RAB/Uganda	2010	120	105	Clear red	Y>, DT	LBS, sugars	X
Petero	PNAP	1986	60	90	Vague red	Y>	DE, LBS, MA	X
Victoria	RAB Uganda	2003	70	90	Clear red	Y>, French fries	MA, too watery	X
Kigega	PNAP	1999	90	120	White	Y>, LBR	DS, local MA	XX
Mabondo	PNAP	1989	70	105	White	Y>, taste	BWS	XX
Gikungu	RAB	2006	120	120	Red	Y>, taste	Very BWS	X
Kinigi	PNAP	1985	120	120	Dark red	Taste!!, MA, DMC	Dormancy, poor VR and TMR	XXX
Cruza 148	PNAP	1983	30	120	White	Y>, BWR	MA	XX
Cruza 58	RAB	2014	90	120	White	Y>, LBR	Taste	Not yet
Kirundo	RAB	2006	60	100	White	Y>, MA	Rotting	XX
Sangema	ISAR	1978	90	120	White, red eyes	Good taste!	Seed health (variety easily degenerates)	X

[a]Other very minor varieties include: Montsama, Atzimba, Gahinga, Mizero, Muhabura, Condea, Ngunda, Rwamkume, Marirahinda (Maryland), Nyabizi, Kuruseke, Mokoroni, Gasore (organic potato in Belgium), Makarar and Gashora.

[b]ISAR, Rwandan Institute for Agricultural Sciences; PNAP, Programme National pour l'Amélioration de la Pomme de terre; RAB, Rwandan Agricultural Board.

[c]BWR, bacterial wilt resistant; BWS, bacterial wilt susceptible; DE, deep eyes; DMC, dry matter concentration; DS, drought susceptible; DT, drought tolerant; LBR, late blight resistant; LBS, late blight susceptible; MA, market appreciation (may be positive or negative depending on whether consumers like the variety or not); PE, pink eye; RE, red eye; sugars, brown when fried; TMR, tuber moth resistance; VR, virus resistance; Y>, yield larger than average.

[d]XXX, economically important; XX, some importance; X, unimportant.

The added value of the 1980s varieties is currently in a position of stagnation as new releases are sporadic, whereas there is a strong need to give a new boost to variety introduction and creation for the following reasons:

- Late blight has developed new strains against which new resistances need to be introduced.
- Late blight resistance research has led to the detection of many resistance genes from which Rwanda could benefit.
- The rapidly growing urban population desires potatoes with a more appealing skin appearance and tuber shape than the currently most popular irregularly shaped, deep-eyed variety Kinigi.
- The upcoming processing industry requires new varieties to meet specifications regarding appearance, shape, size, dry matter and reducing sugar concentrations.

It is estimated that up to 30% of Rwandan potato production is marketed. Most farmers reserve a portion of the crop for home consumption, keep about 20% for seed, and sell the rest. A few potato farms are exclusively market-oriented operations. No potatoes are imported except some from Kenya for Kigali supermarkets and some cross-border trade with Uganda in the northwest of the country. About 75% of marketed potatoes are traded through local, rural marketing channels. These include informal exchange among relatives and neighbours and sale to local merchants or itinerant rural traders. Local trading of potatoes appears particularly important in the volcanic zone where consumption by non-producers reportedly exceeds consumption by growers in other areas. Rural-to-urban marketing of potatoes also appears dominated by the volcanic region, with major collection points at Gisenyi and Musanze. Potatoes collected there are shipped to urban markets in Kigali, Butare and Gitamara. Some trade also occurs from Byumba to Kigali. Trucker/traders buy potatoes from rural traders or in prefectural markets and sell them to urban wholesalers or directly to urban or suburban retailers. Also, consumer cooperatives buy and distribute potatoes and producer cooperatives transport and market them. Some potatoes are also shipped from Kigali to other urban areas. Minor amounts of potatoes are marketed as well in non-producing rural areas, brought in directly from the northern producing areas or by traders based in the capital.

The 2000 urban consumption levels were 140 kg of potato (Irish potato), 24 kg rice, 18 kg bread, 37 kg cooking banana, 36 kg beans and 29 kg sweetpotato per capita per year.

The following factors will contribute to a sharply increased demand:

- The total population is expected to double between 2000 and 2020 (from 8.3 million in 2000 to 16.7 million by 2020).
- The Rwandan urbanization rate is still low, but the urban population is expected to grow more rapidly than the total population. An urbanization rate of 30% in 2020, compared with 10% in 2000, will lead to a demand increase for Irish potato of 18% (national average consumption of 90 kg/capita in 2020 compared with 76 kg/capita in 2000).
- Income growth leads to higher potato consumption, both in rural and in urban areas (short-term income elasticity of demand: rural 1.45 and urban 1.25).

Situation in Ethiopia

In Ethiopia, there are two main production seasons for potato. At altitudes of approximately 2500 masl potatoes grow from June to September (*Meher* in Amharic) and a season from April to August (*Belg* in Amharic). There is a gradual shift from *Meher* to *Belg*. The fact is that the late blight pressure is increasing and farmers experience less risk with cultivation during the 'small' rains combined with irrigation during *Belg*. Relatively low temperatures without rainfall in winter allows growers to postpone their harvest – storing the potatoes in the soil – for up to 5 months. This means potatoes are supplied to the market year round, especially with increasing irrigation for a December–March crop under furrow irrigation at mid-elevation altitudes. During the rest of the year it is too hot and humid to grow potatoes there. The fresh potatoes harvested here fetch double the price of those brought to the market from the soil storage. The regions of Amhara, Oromia and the Southern Nations Nationalities and Peoples' Region (SNNPR) in the south of the country are the main potato-producing areas in the country but the Gayint region, 500 km north of Addis Ababa at altitudes above 3000 masl, also qualifies. Known varieties introduced by CIP and selected by the Ethiopian Institute of Agricultural Research (EIAR) are Jalene, Menegasha and Gudene. Fertilization is low, about 100 kg diammonium phosphate (DAP)/ha and crops are sprayed about three times during the growing season with Ridomil or Dithane (Mancozeb).

In October 2011 reports of two studies, the *Potato Value Chain Analysis and Development in Ethiopia, Case of Tigray and SNNP Regions* (Bezabi Emana and Mengistu Negussi, 2011) and the *Roadmap for Investment in the Seed Potato Value Chain in Eastern Africa* (CIP, 2011), were both commissioned by the United States Agency for International Development (USAID) and carried out under the auspices of CIP.

On experimental fields using improved seed, varieties and management, yields varying between 19 t/ha and 46 t/ha have been reported. The main potato season is during the short rainy season, planting in February and harvesting in June; a second planting season is during the intense rainy season (July–August) and marketed as of September when part of the crop may be 'stored' in the soil for many months. To avoid the glut, some growers plant and/or harvest earlier or later, spreading out the flow to the market. Farm-gate, wholesale and retail prices fluctuate accordingly and typically are 2, 3 and 4 Birr in August and 3, 4.5 and 6 Birr during the low season. In supermarkets, a 1 kg plastic bag of washed potatoes may go as high as 10 Birr/kg. Distinct off-season production periods exist: (i) a winter crop in frost-free, low-lying areas in the Rift Valley (1600 masl) from December through to March; and (ii) a minor rainy season crop in the south from January through to April. Over 1 million Ethiopian holdings are producing potatoes: about 10% of all crop producers. Over 80% of the production area can be found in the Amhara and Oromia regions and around 15% in SNNPR. The average size of holding is low, around 500 m^2, and the production is around 400 kg per holding/year. In contrast to the Food and Agriculture Organization of the United Nations (FAO) statistics indicating yields above 10 t/ha in 2009 and 2010 (FAOSTAT, 2013), the yields derived from the Ethiopian Central Statistical Agency are below 10 t/ha (CSA, 2011). The latter is also in line with the findings of Abay and Tesfaye Dejene (2011): yield is below 10 t/ha, 47.5% is used for household consumption and 15.6% as cash crop.

The introduction of new high quality multi- and single-purpose varieties grown outside the rainy season is currently taking place and the establishment of a supply chain of high-quality table potatoes to high-end supermarkets in urban areas is being considered, as well as the construction of a frozen French fries (chip) factory. Solagrow – a commercial potato company established and run by a Dutch company – developed a new farm of a few hundred hectares of ploughed grassland at Wenchi. The land is flat and is intended for irrigated dry-season farming (i.e. now planting the February–May crop with Dutch varieties Red Scarlet, Cesar and Markies). The crop is fertilized using 165 kg DAP and 195 kg urea, which for seed potatoes is applied once in the planting furrow, but for ware potatoes the application is split (45 kg of urea being applied during growth). There are four main problems: (i) diseases spread by seed potatoes; (ii) soil preparation; (iii) N fertilization; and (iv) availability of varieties. In the rainy season yields of 10 kg/ha are normal for Ethiopian varieties, but the HZPC varieties do not perform well due to lack of late blight resistance (they probably mature too early). The Dutch varieties are too sensitive to blight and chemical control is not effective as the heavy rains rinse away the fungicides. Solagrow also grows varieties Gudene, Jalene and Balete seed potatoes for the rainy season crop for other growers. These varieties originate from Holeta and represent 20% of the Solagrow crop.

43.2 Calculation of Attainable Yields

We calculated the attainable yields for both countries as the gap between attainable and actual yields inform us about the future competitiveness of the crop and scope for yield increase through improved management practices. The crop growth model used (LINTUL-POTATO; Kooman and Haverkort, 1994) uses temperature as a development parameter (a higher temperature leads to earlier emergence and leaf area development) and solar radiation for growth (2.5 g/MJ intercepted radiation) with growth slowing down when the daily maximum temperature (T_{max}) exceeds 27°C, and it reaches zero at 33°C or is below 18°C and is zero at 4°C. The input data are daily maximum and minimum temperatures (T_{max} and T_{min}), solar radiation and monthly precipitation and evapotranspiration (ETP). Tables 43.2, 43.3 and 43.4 show these weather data for potato-growing areas at two altitudes in Rwanda and for Addis Ababa. Rwanda being closer to the equator has two rainy seasons, the one around April being somewhat longer and more intense than the one around October, whereas Ethiopia has a major rainy season around August and a minor one around April. For the two potato-growing areas in Rwanda we have not been

Table 43.2. Three year (1982–1985) average weather data[a] at Kinigi, Rwanda (2300 masl).

Month	T_{max} (°C)	T_{min} (°C)	Daily radiation (MJ/m²)	Precipitation (mm/month)	ETP (mm/month)
January	20.6	10.6	12.1	50.0	81.0
February	20.2	10.5	12.2	120.0	83.0
March	19.9	10.4	12.3	210.0	62.0
April	20.6	10.3	10.1	280.0	59.0
May	20.2	10.6	11.0	205.0	60.0
June	19.9	10.4	11.9	78.0	64.0
July	20.0	10.2	12.0	61.0	67.0
August	19.6	11.0	13.3	70.0	65.0
September	20.2	10.6	13.3	91.0	73.0
October	20.2	10.1	12.8	185.0	77.0
November	20.7	10.0	12.1	230.0	76.0
December	20.8	9.0	12.6	142.0	79.0

[a]T_{max}, mean daily maximum temperature; T_{min}, mean daily minimum temperature; ETP, evapotranspiration.

Table 43.3. Three year (1982–1985) average data weather data[a] at Musanze (Ruhengeri town), Rwanda (1900 masl).

Month	T_{max} (°C)	T_{min} (°C)	Daily radiation (MJ/m²)	Precipitation (mm/month)	ETP (mm/month)
January	24.5	10.5	17.1	48.0	85
February	24.1	11.2	17.5	60.0	82.0
March	24.2	11.1	17.8	130.0	75
April	24	10	16.2	171.0	78
May	24	11.3	16.3	80.0	66
June	23.9	11.1	16.6	57.0	88
July	23.8	10.9	16.0	42.0	85
August	24.9	12	17.1	81.0	78
September	25	12.2	15.9	108.0	99
October	25	10.9	15.3	155.0	78
November	23.8	10.8	17.0	150.0	70
December	24.5	10.6	16.6	81.0	82

[a]T_{max}, mean daily maximum temperature; T_{min}, mean daily minimum temperature; ETP, evapotranspiration.

Table 43.4. Long-term average weather data[a] Addis Ababa 2500 masl (national meteorological institute data).

Month	T_{max} (°C)	T_{min} (°C)	Solar radiation (MJ/m²/day)	Rainfall (mm/month)	ETP (mm/month)
January	23	7	21.20	20	78
February	24	8	22.60	30	79
March	25	10	22.70	80	84
April	24	11	23.60	95	83
May	23	11	23.10	55	80
June	23	11	20.40	100	70
July	20	10	18.60	245	60
August	20	10	18.40	265	50
September	21	10	20.70	120	150
October	22	8	22.80	45	30
November	23	7	22.60	10	15.0
December	23	7	21.50	12	14.0

[a]T_{max}, mean daily maximum temperature; T_{min}, mean daily minimum temperature; ETP, evapotranspiration.

able to acquire more recent weather data as no recording takes place. Climate change may have influenced the weather to some degree but we do not expect that it has influenced potato yields significantly. In both countries these rainy seasons are associated with two cropping seasons and in both countries there is a shift of moving the crop from the major towards the minor rainy season because of increasing late blight pressure. Both countries also grow potato in the dry season: in Rwanda in marshy valley bottoms and in Ethiopia in the Rift Valley in winter with irrigation. For the model calculations we added 3°C to the Addis Ababa T_{max} and T_{min} to compensate for the lower and warmer environment.

The results of the model runs are shown in Table 43.5 (Rwanda) and Table 43.6 (Ethiopia). The warmer and more sunny Musanze site has higher attainable yields – even in a 100 day season – than cooler and more cloudy Kinigi where it takes some 122 days to grow a crop to maturity. For the three identified seasons *Belg* (pre-rain), *Meher* (rainy season) and the dry winter season in Ethiopia, the highest attainable yields of 74 t/ha were calculated for the pre-rain season, followed by 61 t/ha for the rainy season and 55 t/ha for the winter season. The latter has the greatest need of irrigation water (3 mm/t), whereas the rainy season crop has an excess of rain so then no irrigation is needed. It is obvious that attainable yields in Ethiopia are substantially higher than in Rwanda.

It should be noted that developed production systems' actual yields are usually 50–70% of attainable yields and the actual irrigation need is about twice the calculated amount. So for Ethiopia *Belg* yields of 37 t/ha, *Meher* yields of 41 t/ha and winter yields of 36 t/ha seem quite economically feasible. These suggest that usually 35 t/ha can be attained two times per year on a farm (preferably not on the same field to avoid soil-borne diseases and pests).

With actual yields around 12 t/ha in Rwanda and 10 t/ha in Ethiopia, theoretically yields can still increase about fourfold in Rwanda and about sixfold in Ethiopia; assuming that economically optimal yields are 60% of calculated attainable yields then yields in Rwanda are expected to increase by 2.5-fold and yields in Ethiopia by 3.5-fold.

43.3 Current Potato Marketing and Value Addition

The Rwandan potato supply chain

Seed potatoes were partly stored in diffused light stores for 3–4 months. This practice has

Table 43.5. Model calculations for two Rwandan sites.

Characteristic modelled	Kinigi 2300 masl 1 April–1 August	Musanze 1900 masl 1 April–10 July
Growing period (days)	122	100
Days between planting and emergence	13	12
Days between emergence and 100% ground cover	42	37
Days between 100% ground cover and harvest	67	51
Dry matter (DM) tuber yield (t DM/ha)	9.7	10.7
Attainable fresh tuber yield at 20% DM (t/ha)	**48.5**	**53.5**
Precipitation between planting and harvest (mm)	610	311
ETP between planting and harvest (mm)	184	181
Accumulated precipitation deficit (mm)	3	130
Irrigation need (mm)	13	50
Irrigation need per tonne fresh potato (mm)	0.27	0.98

Table 43.6. Results of model calculations for the three distinctive seasons in Ethiopia.

Characteristic modelled	*Belg* short rains 15 March–5 July	*Meher* long rains 1 June–20 September	Winter dry 1 December–20 March
Growing period (days)	112	110	109
Days between planting and emergence	14	15	14
Days between emergence and 100% ground cover	37	41	36
Days between 100% ground cover and harvest	61	54	59
Dry matter (DM) tuber yield (t DM/ha)	16.4	13.5	12.3
Attainable fresh tuber yield (t/ha)	**73.8**	**60.9**	**55.5**
Soil field capacity (mm water/m soil)	120	120	120
Irrigation point (mm water/m soil)	100	100	100
Precipitation between planting and harvest (mm)	337	676	112
ETP between planting and harvest (mm)	198	194	197
Accumulated precipitation deficit (mm)	13	5	94
Soil water reserve (mm)	8	8	8
Irrigation need (mm per season)	5	-3	86
Irrigation need per tonne fresh potato (mm)	0.14	0	3.0

been practised successfully since its introduction in the early 1980s. Most ware potatoes are put on the market from November to March, leading to low farm-gate prices. Average farm-gate prices (2006, 2007 and 2008) are highest during August–October. Farm-gate prices (period 2006, 2007 and 2008) show a limited variation during a calendar year; the maximum variation is 35% (RWF120–150/kg). Ware potatoes are not stored in purpose-built constructions but mainly sold and (home) consumed. Some part of the yield is stored in farmhouses where the individual farmer's volume is low. The potato volume harvested at the average 0.7 ha farm may be around 10 t. The amount stored may be a few tonnes per farm. So far individual farmers have not invested in purpose-built ware potato storage. Table 43.7 shows the anticipated losses during a 3 month storage period (January–April) versus an increase in price: 15.5% versus 16.8%. The positive difference is too small to justify the investment costs of a ware potato store.

The main potato actors in the value chain in Rwanda are potato producers, collectors, transporters, agents/commissioners, traders, retailers, institutional customers and individual consumers. There are tens of thousands of potato growers. Some farmers are also part-time collectors. The transporting

Table 43.7. Losses of stored ware potatoes during a 3 month storage period (January–March). (From interviews with ten traders during an April 2012 field mission by the authors.)

Effect	Loss (%)
Moisture loss, shrinkage	6
Rotting	5
Lost interest from bank	4.5
Total loss	15.5
Maximum increased value	16.8

fee from Musanze to Kigali is RWF2/kg (€1 = RWF800 = 23 Birr). The potato agents receive a commission fee of about RWF2/kg for their contacts. Traders send about 80% of potatoes to the Kigali markets while small quantities are also exported to neighbouring countries. Besides traditional marketing channels with multiple stages of transactions in between, shorter channels with direct linkages with supermarkets in Kigali are also found. Farmers wash and package potatoes especially for supermarkets. One kilogram of well-presented potatoes can be sold to the supermarkets at twice the price of the normal potatoes in the markets. In April 2012, the value distribution of a typical potato chain can be formulated as shown in Table 43.8.

During the PNAP years in the 1980s many new potato varieties were introduced from other countries and were bred in a crossing and selection programme. All current varieties stem from that period and no new varieties are being developed or evaluated currently. There is a need to establish a breeding and variety development programme to obtain varieties: (i) with resistance to late blight; (ii) that are well adapted to local conditions; and (iii) that meet consumer specifications such as appearance, taste and suitability for processing into French fries and crisps (so with high dry matter concentrations). Ideally this would be through public–private partnerships.

Clean healthy material is maintained and multiplied as small plantlets *in vitro*. *In vitro* plantlets are used to produce disease-free minitubers (in sterilized soil or in hydro- or aeroponics systems). These are then multiplied in the field to produce pre-basic

Table 43.8. Value addition of the fresh potato chain Musanze–Kigali (€1 = RWF800 = 23 Birr). (From interviews with ten traders during an April 2012 field mission by the authors.)

Actors	Marketing functions	Price level (RWF/kg)
Producers	Production	140
Collectors	Collecting	145
Wholesalers	Wholesaling	170
Retailers	Retailing	180
Consumers	Final consumption	200

seed that when multiplied produces basic seed which enters the market, and from which growers usually keep some seed for a number of years for subsequent production. Currently RAB-Ruhengeri is not meeting the demand for *in vitro* plantlets and there is a need for private *in vitro* laboratories that cooperate with RAB and for example INES (Institut d'Enseignement Supérieur de Ruhengeri). Most of the *in vitro* plantlets are used by RAB itself but currently four screen-houses have been established by private growers that produce minitubers and (pre-) basic seed.

Growers increasingly use chemical fertilizers and fungicides and increasing numbers can afford to use them, especially fungicides. There is a need for decision-support systems such as forecasts of weather and blight pressure for growers to control late blight caused by *P. infestans* more efficiently.

Trade and transport is rapidly increasing, to Kigali and other cities, with 5–10 t trucks being used. More and more potatoes are transported to Burundi-Bujumbura. Trucks even go as far as South Sudan. Upon arrival at Kigali other sellers (wholesalers) and retailers take over the produce and take it to market, some through contracts with supermarkets.

In every shop and supermarket crisps are sold. Most of the branded bags come from Kenya and Uganda and just a few unbranded bags from local artisanal production, supplied with a telephone number. Frozen French fries (chips) – Farm Frites, a 750 g package costing RWF3900 (more than €2) – are available occasionally and mashed potato powder is not found at all. There is a need (or an economic opportunity) for a processing factory producing local chips. Most fresh potatoes are sold unwashed at vegetable markets but increasingly also at supermarkets. One supermarket even sold washed smooth-skinned (Désirée?) potatoes imported from Kenya. Local washed potato had skin shears because they were washed when the skin had not yet hardened. There is a need (or an economic opportunity) for a branded washed-and-graded fresh potato chain.

Indirect actors that are equally important in the potato supply chain are the following: (i) the state; (ii) the national seed service (certification); (iii) research and extension; (iv) non-governmental organizations; (v) traders of inputs; (vi) banks and micro-credit suppliers; and (vii) donor agencies. In various workshops and missions, needs for R&D have been articulated such as breeding new varieties, testing soils and making weather data available to create decision-support programmes for fertilizer applications and late blight control.

The Ethiopian potato supply chain (from a September 2012 survey)

Wholesale markets

Most of the potatoes produced outside Addis Ababa arrive at the wholesale market in the Piazza district of the city. Here potatoes arrive in the early morning in trucks loaded with bags (wide plastic/nylon-meshed bags). The bags are opened, the potatoes placed in heaps and sold to consumers and small-scale retailers, who do their selling at the side of the street or from stalls. At the time of the survey the potatoes were of the 'variety' Shashemene (meaning this is where they were sourced from) and were available in two types: (i) pink eyed (actually variety Cara); and (ii) white (various varieties possible). One was referred to as 'Bokoji' but it was not clear if this is the name of a site of origin (i.e. source) or the name of a variety. The potatoes had lots of insect damage and pressure bruising. The retail price was 6 Birr/kg while the farm-gate price was close to 3 Birr/kg. A labourer's daily wage was about 35–45 Birr (€1 = RWF800 = 23 Birr). Processed potatoes are made in restaurants and a look in the kitchen of small roadside restaurant showed two ladies peeling and slicing; they were neatly dressed in white clothes and the slicer/cook wore a cook's hat. The frying equipment consisted of two oil baths each with a metal net. It all appeared to be quite neat and the fries were well presented.

At the Piazza wholesale market, the largest wholesale market, there were around 70 potato wholesalers. Each sold between 40 and 70 quintal fresh potatoes/day (1 quintal is 100 kg). Potatoes were mainly sourced from Shashemene and Asella regions south of the capital. Their main customers were retailers and individual household consumers. Supermarkets are special segments who demand large-size tubers and good quality products. Wholesalers' purchase prices from farmers were around 18,000 Birr for 70 quintal (one truck), including 4200 Birr of other costs for labour and middlemen (€1 = RWF800 = 23 Birr). For top quality Grade 1 potatoes, the price was 3 Birr/kg while 2.5 Birr/kg was charged for grade 2 and 2 Birr for grade 3. In general, wholesalers felt there had been an increase in the number of potato transactions over the years although it was very much dependent on the production and harvest. The major challenge was the poor organization along the potato supply chain. Farmers often sold their potatoes to a middleman without knowing the market prices. Middlemen often take advantage of their close contacts with farmers and push prices high to wholesalers thus increasing their margin. Unpredictable weather conditions are a great threat to the stability of potato supply.

At the Marcatto market there were only two or three potato wholesalers. They coexisted with retailers in the same outlet. One interviewed wholesaler purchased potatoes directly from farmers at Nazraeth at a price of 1.90 Birr/kg (€1 = 23 Birr). It was cheaper because the quality of potatoes from Nazraeth is not as good as from Shashemene. The transport cost was 4200 Birr for a truck which could hold 55 quintal. It is equivalent to €0.75/kg. He could sell 1 quintal to local retailers at 3.25–3.50 Birr/kg. He also sold a small proportion to consumers. One retailer who bought potatoes from this wholesaler sold the potatoes to individual consumers at 4–4.5 Birr/kg. Another retailer who bought Shashemene potatoes from Piazza sold products to individual consumers at prices of 4.5–5 Birr/kg. Wholesalers have a reasonably optimistic attitude towards the future of their business. They believe that there is a general trend of increasing demand for potato products. However, the unstable supply of potatoes and unpredictable weather conditions are major threats to their business.

Shola market is one of the big markets next to Piazza. At the Shola market there were a total of 15 wholesalers selling both potatoes and vegetables. The volume they sell per year is on average 6000 quintal of potatoes: they sell 15–20 quintals/day in the rainy seasons and 9–10 quintals/day in the dry season. They buy the potatoes directly from the farmers and the place where they buy the potatoes varies from season to season. From June to October they buy from the south of Ethiopia (Shashemene). Then in the dry months from October to February they buy from Asella and northern Ethiopia which is mainly Menagesha and Gojam. Their customers vary from season to season. In the dry season supermarkets take 45%, individual consumers take 35% and hotels some 20%. But in rainy seasons, supermarkets take 5%, individual consumers take 70% and hotels take 25%. The wholesalers got potatoes from farmers for 4.5 Birr/kg in the dry season and from 1.5 to 3.00 Birr/kg in the rainy season. The collecting price was 30 Birr/quintal. The grading price was 10 Birr/quintal. The retailer price was 5.50 Birr/kg. The information outlined below is based on the trade situation in September 2012.

The transportation cost was 41.60 Birr/quintal and 24.00 Birr per truck which holds around 60 quintals.

Farmer__________4.5 Birr/kg
⇓
Agent__________increases 0.20 cents/kg so 4.70 Birr/kg
⇓
Wholesaler__________5.00 Birr/kg
⇓
Retailer __________5.50 Birr/kg
⇓
Customer__________6 Birr/kg

In the Piazza Atkilt Tera market there were a total of around 17 large wholesalers. Their trade volume is around 11,000 quintal/year, which is around 30 quintals/day. The product comes from the southern region and Oromia region, mainly Shahemene. Their customers are hotels (10%), retailers (50%), consumers (5%) and supermarkets (35%). The farm-gate price was 2–2.50 Birr/kg. The sorting/grading price was 10 Birr/ quintal. The wholesale price was 260–300 Birr/quintal depending on the quality. The retail price was from 3.50 Birr/kg which is 350 Birr/quintal. The information below is also based on the September 2012 trade situation:

Farmer__________2.50 Birr/kg
⇓
Agent__________increases 0.20 cents/kg so 2.70 Birr/kg
⇓
Wholesaler__________3.00 Birr/kg
⇓
Retailer__________3.50 Birr/kg
⇓
Customer__________5.50 Birr/kg

Production costs of potato

Production costs of ware potato in Rwanda

The production cost of potatoes in the Musanze region is RWF1,132,300/ha or RWF56,615 (€71)/t at a yield level of 20 t/ha. Land rent (18%), seed (22%) and organic manure (18%) lead to 58% of the total cost (Table 43.9). This price level is competitive compared with north-western European and North American potato producers. The low production cost supports the options to grow fresh potatoes for high-end consumers and

Table 43.9. Costs of raw material for potato processing Rwanda (€1 = RWF800 1 January 2013) May 2012 survey by the authors.

Item	Cost (RWF)	Percentage of total		
Land rent	200,000	18		
Seed	250,000	22		
Transport of seed	25,000	2		
Organic manure	200,000	18		
Transport organic manure	100,000	9		
NPK fertilizers	102,000	9		
Dithane	50,000	4		
Supmetrine insecticide	8,000	1		
Ridomil	14,000	1		
Labour land preparation	35,000	3		
Labour planting and crop	21,000	2		
Labour application NPK	21,000	2		
Weeding	17,500	2		
Hilling, ridging	17,500	2		
Fungicide application	33,600	3		
Watchmen	20,000	2		
Rental cost of spraying pesticides	7,200	1		
Harvest	10,500	1		
Total cost	1,132,300	100		
Variation yields in t/ha	**15**		**20**	**25**
Cost per tonne (RWF)	75,487		56,615	45,292
Cost per tonne (€)	94		71	57

for processing into high-grade products. Note that these costs are calculated based on a survey in May 2012 and based on 20 t/ha yield. Where actual yields often are less than 15 t/ha production costs are close to €100 and the farm-gate price close to €150 which is comparable to potato prices in northern Europe.

Production costs of raw material in Ethiopia

Four groups of potato growers and one individual grower were interviewed in August 2012 to arrive at the cost of production of potatoes per hectare, where data were gathered similar to those in Table 43.9. Table 43.10 provides the summary of the costs, financial return and profit margin (Birr per hectare, €1 = 23 Birr December 2012).

The maximum production cost per kilogram varies from 1.90 Birr (20 t yield) to 0.95 Birr (40 t yield). The minimum production cost per kilogram varies from 0.87 Birr (20 t yield) to 0.44 Birr (40 t yield) (see Table 43.11).

The low production costs provide great scope for processing and will have a rather sound competition with imported processed potato products. Processing needs more attention as it will lead to more financial turnover in the Ethiopian potato industry. The constraints to production mentioned before are interdependent, meaning that the use of high quality seed is only profitable if planted in a fertile soil, adequately provided with fertilizers, water and crop protection.

Production and costs of frozen French fries

In East Africa potato processing and imports are relatively insignificant (Tesfaye *et al.*, 2010). Each of the three chips processors in Kenya on average processed 117 t of chips/year. The infant chip-processing enterprise in Uganda processed only 2 t of chilled chips/year. With increasing demand, the quantity of chips processed in retail outlets and processors is expected to increase. In general, the small quantities processed indicate that trade in chips in Eastern and Central Africa (ECA) countries is not yet strong, even though there is a high potential for chips

Table 43.10. Costs and revenues of potato production per hectare (Birr/ha) of four farmers' groups and one individual grower in Ethiopia.

Costs and revenues	Group 1	Group 2	Group 3	Group 4	Grower
Total cost	20,750	24,298	33,352	17,418.5	37,975
Total estimated revenue	30,000	40,000	74,000	160,000	93,750
Revenue - cost	9,250	15,702	40,648	142,582	55,775

Table 43.11. Production costs at three presumed yield levels of the same interviewees as in Table 43.10 (€1 = 23 Birr).

	Production costs (Birr/kg)				
Yield levels (t/ha)	Group 1	Group 2	Group 3	Group 4	Individual grower
20	1.04	1.21	1.67	0.87	1.90
30	0.69	0.81	1.11	0.58	1.27
40	0.52	0.61	0.83	0.44	0.95

demand and quality potato production suitable for chip making. In Ethiopia, very limited amounts of frozen chips, not exceeding 25 t, are imported from Europe (mainly from Belgium, The Netherlands and Germany) each year to be used in just two five-star hotels. However, no imported frozen chips are available in supermarkets or other retail shops. Burundi and Rwanda did not import frozen chips at the time of this study (2005/06) although Burundi imports potatoes from Rwanda for chip processing. One hotel in Rwanda (Hotel des Milles Collines) used to import frozen chips from Belgium via Sabena Airways. However, Rwanda was not importing at the time of the study because of the lack of an importer. Njoro Canning, Kenya, in addition to depots it has in Nairobi and Mombasa, has one cold store at the factory which is large enough to store 200 t and one refrigerated truck for delivery. Nairobi Airport Service has cold stores and refrigerated trucks to handle customers' goods at a fee. The Horticultural Crop Development Authority has about half a dozen refrigerated trucks which are available for hire. Also big supermarkets that are involved in trade of imported frozen chips in all the importing ECA countries have cold-storage chambers. Incomes of the urban population are increasing from year to year in Africa. The phenomenon of increasing urbanization and the tourism industry is closely linked to increasing incomes of urban households. This contributes to the increasing demand of value added products, such as chips.

French fries (chips) production requires the highest amounts of raw material as their volume is often much greater than that of crisps. There is a difference between 'fresh fries' (chilled) and 'frozen fries' (frozen to −20°C). Fresh fries are fried for a few minutes longer and contain less water (65%) and more oil (5.5%) than frozen fries (72% water and 4.5% oil). Note when calculating the amount of oil needed, always add 0.5% because of losses. Recovery (kilograms of chips produced per tonne of raw material) is roughly 50% (1 t of potatoes yields 0.5 t of French fries). A higher recovery is attained when: (i) the dry matter concentration of the raw material is higher; (ii) the length:width ratio is greater; (iii) steam rather than carborundum peeling is applied; and (iv) the tubers have fewer defects and less reducing sugars (leading to fewer brown fries). Assuming that tubers arrive at the factory washed and graded (40–75 mm) they are then peeled in a carborundum drum. Subsequently they are cut into strips with an Usher (USA), FAM (Belgium) or Fines (The Netherlands) cutter. An optional step is that the strips are blanched for a few minutes; this results in a more homogeneously coloured pre-cooked product. After blanching, the strips need to be dried (shaken).

The next step is frying at 160°C (1 min) or 140°C (1.5 min) in oil without *trans* fats (*trans* fat is the common name for unsaturated fat with *trans*-isomer (E-isomer) fatty acid(s)). Because the term refers to the configuration of a double carbon–carbon bond, *trans* fats are sometimes monounsaturated or polyunsaturated, but are never saturated. *Trans* fats are rare in nature, but can occur in food production processes. So hardened fat is not suitable, nor is linseed oil or soy oil unless it contains less than 1% linolenic acid. Suitable oils include palm oil (refined), canola oil and sunflower oil. A shaker is needed to remove excess oil. An apparatus to measure the water content of the cooled-down product is needed (infrared equipment exists).

The next step is optional and involves using ambient (night) air to cool the product, assisted by refrigeration. However, the oil will harden causing the potato strips to stick together, so moving them from one conveyer to the next may be needed to avoid this. Locating the factory at a high cool site may be advantageous.

The product is packaged in boxes (fusts), plastic bags or plastic bags in a carton box on pallets. Weighing equipment is needed and a sealer (this could potentially be a vacuum sealer). At this stage hygiene in the factory is of utmost importance (i.e. gloves, hair netting) to avoid contamination with for example *Pseudomonas* bacteria.

The next step is distribution (by order only) in an air-conditioned van. Chilled fries for supermarkets with a shelf life of about 1 week need to be stored at 2–4°C in a controlled atmosphere bag (70% CO_2 and 30% N_2). Potatoes stored for a few days need to be chilled and the receiving end needs a fridge as well. (Note: those fried at the consumer end within 12–24 h after leaving the factory may need no refrigeration at all along the chain.)

A cheaper alternative may be: just peel, cut into strips, add ascorbic acid, vacuum seal and distribute (end users deep fry the product before consumption).

Cost of chips production and investments

Tesfaye *et al.* (2010) calculated the economic profitability of processing 100 kg of potatoes into chips as shown in Table 43.12 (Rwanda) and Table 43.13 (Ethiopia).

Table 43.12. Economic profitability of processing 100 kg of potatoes into chips in Rwanda. Designation unit value (RWF). (From Tesfaye *et al.*, 2010.)

Costs	
Purchase of potato	RWF8,000
Labour cost	RWF1,000
Transportation	RWF200
Additives (e.g. cooking oil)	RWF11,200
Total cost	RWF20,400
Benefits	
Sale of 1 kg of fried chips on average	RWF1,500
Quantity of chips obtained from 100 kg of potatoes	60 kg
Gross benefits	RWF90,000
Net benefit in local currency	RWF69,600
Net benefit in hard currency	US$126.09[a]

[a]US$1 = RWF552 (2006).

Table 43.13. Economic profitability of processing 100 kg of potatoes into chips in Ethiopia. (From Tesfaye *et al.*, 2010.)

Costs	
Purchase of potatoes	127.55 Birr
Transportation of potatoes	8.31 Birr
Labour for processing chips	30.00 Birr
Additives (e.g. cooking oil)	51.52 Birr
Depreciation value of processing utensils	2.50 Birr
Frying fuel	22.5 Birr
Total cost	242.38 Birr
Benefits	
Sale of 1 kg of fried chips on average	18 Birr
Quantity of fresh chips that could be produced from 100 kg of potatoes	56.00 kg
Quantity of fried chips obtained from 100 kg of potatoes	46.00 kg
Gross benefits	828.00 Birr
Net benefit in local currency	585.62 Birr
Net benefit in hard currency	US$68.90

[a]US$1 = 8.5 Birr (2004).

Purchasing 100 kg of raw material in Rwanda leads to a profit of US$126 whereas the work clearly shows that the costs of 214 Birr per 100 kg of potatoes are well below (586 Birr) those of the gross benefits (828 Birr) in Ethiopia. Based on such figures a business plan for investments (e.g. in a plant consisting of construction of a building and purchase of machinery) could be made.

The elements of such a business model for Ethiopia (and very similarly for Rwanda) are the following:

- recovery 50%: 1 kg potatoes = 0.5 kg French fries;
- 100 ha × 15 t/ha = 1500 t processing potatoes (raw material) and they yield 750,000 kg French fries for Ethiopian market in the first year;
- vegetable oil content 5–7%;
- estimated investment in the processing line: €1 million (modularly built: initially a single line with one shift, later two shifts, later still three shifts of 8 h/day, later doubling the line);
- 100 ha of production requires 250 kg of seed potatoes/year;
- complex organization of seed distribution, raw material collection, including contract growing;
- farmer knowledge and skills need attention; and
- business group explores investment with, for example, private sector investment support from The Netherlands.

Potato production outgrowers' scheme for French fries production includes the following:

- French fries production requires suitable varieties;
- providing new opportunity for farmers;
- outgrowers' scheme for seed potatoes;
- distribution of high quality seed potatoes;
- outgrowers' scheme for French fries potatoes;
- collection system for potatoes for processing;
- planning, farmer advice and production supervision; and
- potatoes grown as a rotation crop with vegetables and field crops.

The value addition in the supply chain to the high-end consumer products can best be exemplified by following the sequence of crops from minituber to table. Starting with *in vitro* plantlets (variety A for the table and variety B for processing into French fries) the sequence is as follows (where e.g. G1 is the first generation).

Minituber	3 g	e.g. at Wenchi costs €0.25
Seed tuber G1	300 g	e.g. at Wenchi
Seed tuber G2	3000 g	e.g. at Wenchi
Outgrowers G3	30,000 g	e.g. at Meki (town in Ethiopia)
Producers for market G4	300,000 g	e.g. at Meki

So every minituber produces 300 kg of marketable potatoes in the fourth generation (G4) with an estimated value of €30. This shows that the initial costs of a minituber are relatively insignificant. Those G4 tubers may go to either:

- sorting, grading and packing at a packing station followed by distribution to (super)markets; or
- sorting, grading, washing, peeling, cutting, blanching, frying, cooling followed by distribution to supermarkets, hotels and restaurants.

43.4 Concluding Remarks

From literature and our own findings of surveys in 2012 we can conclude that the consumers in the urban areas of the two countries Ethiopia and Rwanda are familiar with processed potatoes such as crisps and chips. They are generally produced at home or in small roadside stalls, shops or restaurants. Crisps are also consumed bought in packages from small-scale cottage industry or imported from Kenya or Uganda (in Rwanda) or from industrialized countries (Ethiopia). In both countries, following Kenya that already has some industrial production of chips and crisps, plans exist to build a frozen chips factory. In Ethiopia an investor was identified and there were plans to have a plant operating in 2014, whereas in Rwanda the government has similar plans and is still identifying commercial parties to join. In Uganda a French fries factory just across the border with Rwanda was scheduled to open in 2013 and this factory is now open and in production and is likely to draw raw material from Rwandan and Ugandan growers. The success of such enterprises will depend largely on a number of factors:

- Production should be demand driven. Following the demand, processing and supply lines should be adjusted with such flexibility that production is hardly in excess of demand. This means that storage capacity of the produce should be relatively modest (e.g. should not exceed a month's demand).
- Production lines in the factory should be aimed at growth. Starting with a single line that has a limited capacity and that only operates for a limited number of hours per day. Extension when the demand increases is then realized by first extending the number of shifts per day and subsequently by extending the capacity of the factory by adding production lines.
- The raw material supply to the factory needs to meet two demands: (i) it should be at an internationally competitive level; and (ii) its quality should be such that it assures the highest possible recovery. Potato prices therefore need to be no higher than those in Europe otherwise the pressure from that market may prove too strong for survival. Comparing modelled and calculated attainable yield per season per field and per year per field of those obtained in Africa with those obtained in Europe will assist in assessing relative competitiveness.
- Therefore a keen eye on national and international market prices is needed and the costs of potato production should go down. This is best assured by achieving high yields by optimizing all inputs: expensive high quality seed is only brought to value when the crop is adequately supplied with water and fertilizers and protected from late blight by fungicide treatments.
- Recovery increase is determined by larger tubers with greater length:width ratio and a higher dry matter concentration with fewer defects such as greening and black spot. New variety introduction and breeding efforts should aim at a reduced number of tubers, but larger tubers that are longer in shape (for chips, for crisps round tubers are preferred) with a high dry matter concentration.
- Supply chain management should assure year-round supply to the factory of high quality raw material. This preferably is arranged by sourcing from growers and areas with varying planting and harvest times and reducing storage to as little as possible (i.e. to what is economically necessary). When storage is necessary to cover a slot in the market when potato prices are prohibitively high, then this is preferably arranged by the growers (associations) (i.e. decentralized) rather than by the processing plant. This allows smaller lots with their own quality characteristics.
- Organizing growers into dedicated (basic) seed growers and processing potato growers and transferring knowledge is probably best facilitated when they form associations or cooperatives. This type of work requires tailor-made activities geared towards the specific local situation.
- When deciding on processing at a large scale the impact on the use of resources such as land, soil fertility and water need to be taken into consideration. For example, what is the carrying capacity of the area and how can its sustainability be assured? How robust is the system of producing seed, potatoes and products vis-à-vis climate change? Crop growth modelling and scenario studies may be most helpful here.
- When making business plans and calculations of returns on investment, use can be made of the various possibilities to obtain funding from international cooperation agencies. The authors of this chapter can be contacted for this purpose.

Potato consumers in Europe and North America pay at least five times as much for their potato and its processed products as the farm-gate price growers receive. This means that the potato industry there adds much value to the crop. In the tropical highlands of Africa this is just beginning but the prospects are great.

References

Abay, A. and Tesfaye Dejene (2011) Integrated application of compost and inorganic fertilizers for production of potato (*Solanum tuberosum* L.) at Angacha and Kokate in southern Ethiopia. *Journal of Biology, Agriculture and Healthcare* 1, 15–24.

Bezabi Emana and Mengistu Negussi (2011) *Potato Value Chain Analysis and Development in Ethiopia, Case of Tigray and SNNP Regions*. International Potato Center (CIP), Addis Ababa, 71 pp.

Central Statistical Agency (CSA) (2011) *Report on Area and Production of Crops*. CSA, Addis Ababa. Available at: http://ethiopinvest.org/investment_regime.php (accessed 14 February 2012).

FAOSTAT (2013) Available at: http://faostat.fao.org/site/384/default.aspx (accessed 1 October 2013).

International Potato Center (CIP) (2011) *Roadmap for Investment in the Seed Potato Value Chain in Eastern Africa*. CIP, Lima, 27 pp.

Kooman, P.L. and Haverkort, A.J. (1994) Modelling development and growth of the potato crop influenced by temperature and day length: LINTUL-POTATO. In: Haverkort, A.J. and MacKerron, D.K.L. (eds) *Ecology and Modelling of Potato Crops Under Conditions Limiting Growth*. Kluwer Academic Publishers, Dordrecht, The Netherlands, pp. 41–60.

Ministry of Agriculture and Animal Resources (2011) *National Post Harvest Staple Crop Strategy. Ministry of Agriculture and Animal* Resources, Kigali, 56 pp.

Tesfaye, A., Lemaga, B., Mwakasendo, J.A., Nzohabonayoz, Z., Mutware, J., Wanda, K.Y., Kinyae, P.M., Ortiz, O., Crissman, C. and Thiele, G. (2010) Markets for fresh and frozen potato chips in the ASARECA region and the potential for regional trade: Ethiopia, Tanzania, Rwanda, Kenya, Burundi and Uganda. *Social Sciences Working Paper*. International Potato Center (CIP), Nairobi.

44 Unlocking the Potential of the Potato Subsector in Kenya – A Roadmap for Revitalizing the Subsector

C. Lung'aho,[1]* D. Kipkoech,[2] N. Ng'ang'a,[2] W. Kaguongo,[3] M. Nyongesa[2] and E. Schulte-Geldermann[3]

[1]*International Potato Center Sub-Saharan Africa (CIP-SSA), Chimoio, Mozambique;* [2]*Kenya Agricultural and Livestock Research Organization (KALRO), Limuru, Kenya;* [3]*National Potato Council of Kenya (NPCK), Nairobi, Kenya;* [4]*International Potato Center Sub-Saharan Africa (CIP-SSA), Nairobi, Kenya*

Abstract

The potato subsector in Kenya has an unexploited potential which, if properly harnessed, can make significant contributions to Kenya's development aspirations – particularly those related to food and nutrition security, and economic development – as articulated in the country's two main policy documents: *Kenya Vision 2030* and the *Agriculture Sector Development Strategy (ASDS) 2010–2020*. Despite the investments made in the potato subsector in the last several years, its growth has been unsatisfactory. The subsector is currently characterized by: (i) low commercialization; (ii) low average yields; (iii) poor competitiveness; (iv) high wastage; and (v) low value addition with limited processing and limited agribusiness activities. There are very few contractual arrangements between producers and end users of potatoes (e.g. processors, fast food restaurants or supermarkets). In order to transform the potato subsector from subsistence to a vibrant, commercially oriented subsector, a new roadmap – one which leverages available resources to deliver economic growth and opportunity, improved food security and nutrition, and environmental sustainability is required. This chapter presents a roadmap that provides a guide on the core investment areas necessary to revitalize the potato subsector. The roadmap is anchored on three investment areas, namely: (i) the seed potato value chain; (ii) the ware potato value chain; and (iii) the processed potato value chain. The targets of the roadmap are to increase yields by at least 20% in the next 10 years to achieve three overarching objectives, namely: (i) a 15% increase in farm incomes; (ii) improved food security; and (iii) more business opportunities for at least 10% of potato-growing households. These growth targets will be achieved through implementation of several flagship projects in each of the priority value chains focusing on several areas including value chain financing, improvement of markets, formulation of favourable policies, adoption of improved farming technologies (e.g. irrigated potato farming) and promotion of value addition.

Keywords: *Agriculture Sector Development Strategy, Kenya Vision 2030*, potato subsector, processing potato value chain, Roadmap, seed potato value chain, ware potato value chain

*c.lungaho@cgiar.org

44.1 Introduction

A key challenge facing Kenya is to ensure food security for present and future generations, while protecting the natural resource base. Kenya's population currently stands at about 40 million and is projected to reach approximately 50 million by the year 2020. Overall, vulnerability to food insecurity in the country is exacerbated by the absence of substantive diversification in food production and consumption. Despite the enormous potential of the potato subsector to make significant contributions to Kenya's development aspirations – particularly those related to food and nutrition security and economic development – as articulated in the country's two main policy documents: *Kenya Vision 2030* (GoK, 2007) and the *Agriculture Sector Development Strategy (ASDS) 2010–2020* (GoK, 2010a), the subsector has lagged behind and its growth has not matched expectations (GoK, 2009). Yields continue to be low, with farmers getting very low returns on their investment. The subsector is not sufficiently commercialized and sufficiently competitive to contribute towards moving the country to the next level of development. Actors and players are fragmented, and activities uncoordinated. There are high postharvest losses, low value addition and limited agribusiness activities in the potato subsector. There are also very few contractual arrangements between producers and end users of potatoes (e.g. processors, fast food restaurants or supermarkets).

Under-investment in infrastructure, research innovation and market development has hampered efforts to profitably cultivate the crop and develop a vibrant and self-sustaining potato subsector. In order to transform the potato subsector from subsistence to a vibrant, commercially oriented subsector, a new roadmap which leverages available resources to deliver economic growth and opportunity, improved food security and nutrition, and environmental sustainability is required.

44.2 Development of the Roadmap

The roadmap (CIP, 2013) was arrived at following a rigorous consultative process. During an inception workshop for a study entitled 'A Policymakers' Guide to Crop Diversification: the Case of Potato in Kenya' (Kaguongo *et al.*, 2013) held on 4 April 2012, problems affecting the potato subsector were outlined and discussed with stakeholders. Following this workshop, a comprehensive study was undertaken by a team of potato experts drawn from the National Potato Council of Kenya (NPCK), Kenya Agricultural Research Institute (KARI), the Ministry of Agriculture and the International Potato Center (CIP). At a potato Round Table meeting held on 25 June 2012, policy makers and major stakeholders discussed the recommendations of the study and deliberated on important issues of the subsector that could contribute to mapping out the way forward for the potato industry. Interventions and action plans necessary to unlock the potential of the subsector were discussed and agreed upon with stakeholders (NPCK, 2012). Based on the two workshops, desktop research and additional discussions with experts and stakeholders in the potato sector, a consensus document 'A roadmap for revitalization of the potato sector in Kenya' (CIP, 2013) was developed.

44.3 Targets to Revitalize the Subsector

The overall objective of the roadmap is to transform the potato subsector from subsistence to a vibrant and commercially oriented subsector. The roadmap aims at stimulating an annual production growth rate of the potato subsector of about 10% in the next 5–10 years to achieve three overarching objectives, namely: (i) a 15% increase in farm incomes; (ii) improved food security; and (iii) more business opportunities for at least 10% of potato-growing households. The focus areas and approaches proposed in this roadmap are consistent with the major policy documents and recent reports on the potato subsector. These include:

- *Kenya Vision 2030* (GoK, 2007) – with the country's current development blue print;

- the *Agriculture Sector Development Strategy (ASDS) 2010–2020* (GoK, 2010a) – the overall agricultural sector policy document;
- the draft root and tuber crops policy (GoK, 2010b);
- the draft potato strategy (GoK, 2010c);
- the seed potato master plan (Kaguongo *et al.*, 2010);
- the *National Potato Taskforce Report* (Gok, 2009);
- review and analysis of staple product value chains (KARI, 2012);
- *A Policymakers' Guide to Crop Diversification: the Case of Potato in Kenya* (Kaguongo *et al.*, 2013); and
- KARI strategic plan (2009–2014) (KARI, 2009).

44.4 Core Investment Areas and Strategies to Turn Around the Subsector

The poor performance of the potato subsector can be turned around through good policies, sustained public and private investments, and promotion of public–private partnerships backed by open, transparent procedures and processes along the potato value chains. Three chains have been prioritized as being key investment areas that will spur the growth and development of the potato subsector: (i) the seed potato value chain; (ii) the fresh potato value chain; and (iii) the processed potato value chain. Some of the strategies identified to strengthen each of the three value chains are presented below.

Strengthening the seed potato value chain

Seed production plans

A description of the actors in the quality seed value chain, both existing and potential, and the status of seed production technology should be undertaken. This should be followed by an analysis of the profitability of different seed enterprises and strengths, weaknesses, opportunities and threats analysis of the seed potato value chain. The national seed potato business plan should then be developed and it is expected to validate the targets for production of different seed categories along the seed potato value chain, analyse the existing capacity of seed potato businesses, and provide estimates of the capacity that is needed to achieve the targets. The types and size of investments that are needed to close the gap should be identified taking cognisance of lessons learned in previous projects.

Infrastructure and decentralized seed multiplication

Improvements in infrastructure and capacity for high-quality seed production need to be undertaken. Part of the investment cost should be covered as an incentive for existing and upcoming entrepreneurs to participate in rapid multiplication techniques (RMTs) and related activities. In addition to strengthening existing tissue culture labs already engaged in *in vitro* production (public and private), new private tissue culture facilities should be identified to deliver on the projected targets.

To make seed readily available in the immediate localities of growers, it is necessary to establish networks of decentralized seed producers. Such networks require strong linkages and coordination with the producers of early generation seed to source their seed where the logistics of seed movement should be facilitated. Agro-input dealers should also be involved in the seed potato trade. Innovative approaches include the use of smaller seed bags and having larger agro-input companies distribute seed regularly through their network of smaller local agro-input dealers. Approaches such as voucher schemes, seed banks (based on commodity loan systems), seed fairs, input loans and input insurance systems should be explored as mechanisms to diffuse clean seed to small private multipliers and ware potato growers. Linking seed growers to ware potato and processing potato growers could also help in making the seed enterprises more profitable.

Skills, quality control and systems for production

Improving entrepreneurs and farmers' skills and knowledge in seed storage, seed quality management and accessing new varieties will facilitate increased uptake and spread of new varieties and improved practices. Seed production should be viewed as a business rather than a technical or development activity if it is to succeed. Consequently, the need for business and entrepreneurial skills, and not just technical skills must be emphasized from the outset of any seed initiative. Private seed potato producers (individuals, groups, associations) will need to be identified and trained in seed multiplication. It will be necessary to complement efforts to increase the capacity of the national potato programme and plant health inspectorate services to enable them to monitor and backstop seed production through:

- advocacy and piloting of farmer/community-based quality standards, practising self-certification and labelling (such as Quality Declared Seed – QDS), with limited regulatory oversight;
- training of community-based or grower association-based inspectors to support QDS; and
- improved and easy-to-use low-cost disease detection methods.

Improvement of *in vitro* and minituber production systems is needed so as to provide more efficient methods and reduce the risk that could be associated with using the technologies. Research should concentrate on development of best agronomic methods such as: (i) improvement in plant nutrition and disease control; (ii) development and shifts in the occurrence and severity of the major seed-borne diseases; and (iii) new technologies for multiplication, well suited to conditions in the respective growing regions (e.g. by fine-tuning aeroponics and sand hydroponics to specific varieties, and managing diseases in aeroponics and hydroponics systems). Decentralized seed potato multipliers can be further supported by testing alternative quality control systems, standards, and easy, low-cost disease detection techniques and effective approaches such as the branding of quality seed. This could be combined with testing options for on-farm seed maintenance. The advantages of high-quality seed should be further developed through approaches such as: (i) establishing plots comparing quality seed with farmer-saved seed; (ii) distribution and/or sale of quality seed to farmers in small quantities (5–25 kg); and (iii) the use of different media channels.

Information flow and delivery of source of seed

Communication and feedback mechanisms need to be developed and maintained to ensure that relevant value chain actors are aware of what the market is responding to and its needs. The registration and cleaning of farmer- and market-preferred unregistered, introduced varieties (e.g. escapes from the variety testing and release process) should be supported to increase the range of varieties available through the regulated seed system. The development of sustainable strategies for managing and delivering source seed – including contracting for sale of pre-basic seed – by the National Agricultural Research System (NARS) can stimulate commercial seed production and farmer-to-farmer diffusion of new improved varieties with additional benefits to farmers.

Farmer training and linkages

Farmer training will support farmers to maximize their investment in clean seed. For example, if farmers understand that viral diseases can be transmitted by aphids, and that aphids acquire viruses from other sources, they can readily see the value of removing virus-infected *Solanum* weeds from hedgerows. Farmers can also avoid planting clean seed in soil infested with bacterial wilt if they understand the contamination routes of this disease. Linkages should be established with initiatives to promote access to and better use of fertilizer, as clean seed is highly responsive to increased input use. New approaches to extension such as, farmer-to-farmer, private local extension services and contract farming with extension components should also be explored.

Existing farmer-training networks need to be strengthened in order to enhance farmers' capacity to extend the quality of their seed. Practices that should be promoted include: (i) positive selection; (ii) on-farm sanitation; (iii) rotation; and (iv) proper seed storage. Appropriate seed storage such as the diffused light store (DLS) is an important pre-condition for the profitability of seed enterprises and should be promoted. When professionally designed, seed storage warehouses can add value to certified and QDS planting material. Farmer-owned seed cooperatives can also increase seed storage capacity, promote trust among seed value chain actors, and provide a good 'incubator' for learning essential business management skills on a small scale.

To develop a viable seed potato value chain, linkages between service providers, basic-seed producers, decentralized seed multipliers and ware producers need to be improved. The Participatory Market Chain Approach (Bernet *et al.*, 2006) brings together small farmers, market actors and service providers for an intense process of facilitated interaction. To overcome the lack of information flow about seed business, seed potato stakeholder platforms should be formed with appropriate linkages to the wider potato value chain at each location. Functional platforms will facilitate the flow of technical and market-related information for the seed business. Skills, knowledge and replicable knowledge need to be scaled up so that widespread benefits and economies of scale can be achieved for the country's emerging seed potato sector. Specialized capacity-building modules, incorporating new curricula and learning tools, need to be developed for trainers, producers and businesses. Capacity-building modules could include 'seed-entrepreneurs' business schools', exchange visits, short training events and on-the-job training.

Diseases, pests, trade in quality seed and storage

Surveys should be conducted to evaluate the status of seed-based pests and diseases as a precondition for the safe movement of seed, leading to the identification of 'clean' areas suitable for seed production and monitoring systems. To manage quarantine risks, public and private sector partners should commit to improving the knowledge base on plant health and develop functioning integrated quality assurance systems for the seed potato sector. An integrated strategy to prevent the spread and introduction of plant pests and pathogens includes: (i) research on plant health; (ii) promotion of appropriate technologies, best practices, capacities and policies for avoidance and detection; (iii) communication; and (iv) institution strengthening.

Market opportunity and relative cost of production studies need to be conducted to identify the possibilities for expanded trade in high-quality seed potato nationally and within the eastern African region. The studies should focus on: (i) relative costs of production and favourable conditions (e.g. low virus vector and soil-borne disease pressure, sufficient land for multiplication, and capacities for seed production across the countries where trade could occur); and (ii) transport and transaction costs to identify market niches in the countries that could be served by a lower cost and more efficient producer of clean seed than another country.

There is a need to explore the possibility of establishment of 'strategic reserves' of seed potatoes by promoting use of cold storage and/or improved storage facilities particularly DLSs made from locally available materials. *In vitro* conservation of plantlets will also ensure that there are sufficient stocks of 'nuclear seed stocks' of priority varieties readily available for further multiplication whenever the need arises.

Strengthening the fresh potato value chain

Enabling environment and business plans

A favourable environment within which the fresh potato value chain can operate efficiently is required. Policy makers need to be sensitized on the nature and functioning of potato value chains so that they can realize the importance of potato, vis-à-vis other crops, for food security and poverty alleviation. Inputs such as seed and fertilizer, and services such as credit, extension, research and information need to be available and accessible if efficient performance of the fresh potato value chain is to be realized.

An approach similar to the one described for the seed potato value chain should be adopted to develop plans for sustainable fresh/ware potato business development. Improving the infrastructure and capacity for high-quality ware potato will also require investments in infrastructure, technical inputs and other factors. To expand ware potato production it will be necessary to identify suitable locations and operators for specialized fresh/ware multiplication with sufficient land to conduct a proper rotation scheme. In some cases this may require development of irrigated production. Promotion of private and public sector investments in collection centres and warehouse receipting systems will be necessary so as to: (i) even out the supply of ware potatoes over the year; (ii) encourage farmers to store (receipting system); (iii) add value through storage (time utility); and (iv) reduce volatility of ware potato prices over the seasons.

Network of ware potato producers and Good Agricultural Practices (GAPs)

A network of decentralized ware producers will need to be established in order to make ware potato available locally in all consumption regions. GAPs and traceability systems for potatoes should be developed. GAP is a basic system to assure food safety on farm while traceability is an indispensable system to communicate information on food safety to both traders and consumers. The traceability system, therefore, serves as a pipeline of information. Combining these two systems is the only way to establish a food safety chain that can supply safe foods with confidence. There is need to reduce risks associated with the use of pesticides, tools and machinery, and to ensure that potatoes for the fresh potato value chain are produced and handled in a manner that does not harm the environment and the health and safety of farmers and consumers.

Research and technology transfer

Research should focus on:

- investing in laboratories for the diagnosis of potato diseases, for measuring mineral concentrations in soils, and manure and fertilizers, and for determining the residue levels of pesticides;
- development of 'climate smart' varieties possessing other key traits such as late blight resistance, virus resistance, earliness and culinary qualities (use of modern biotechnologies would speed up the development of such varieties);
- resistance breeding and integrated management strategies for late blight;
- integrated management to control bacterial wilt with an emphasis on designing improved detection technologies, developing recommendations that farmers are willing and able to adopt, and exploring options for suppressing the disease through improved soil fertility and health;
- consumer surveys to identify growing market segments and types of fresh potato products likely to be in demand in the near future; and
- surveys to determine the most critical problems affecting players along the fresh potato value chain.

A more enabling environment for potato production in the country can be created through further uptake of affordable technologies. There is therefore, need for farmers to adopt technologies to bolster yields from the current national average yields of <10 t/ha. Programmes to inform potato producers about existing technologies and how their use can contribute to enabling environments and improve performance need to be established. It is imperative that existing technologies be scaled up and out to ensure better accessibility and application for farmers and stakeholders. Embracing of e-extension may help relieve some of the constraints faced by extension officers in service delivery.

Risk reduction and marketing

Development of risk reduction strategies will provide significant scope to create a more favourable enabling environment for potato production. Production risk can be mitigated through crop insurance, the adoption of technologies (e.g. use of robust varieties and disease-resistant varieties), the promotion of irrigated production in drought-prone areas, etc. Postharvest losses can be mitigated through the construction of storage facilities and processing capacities, while price and market risk can be mitigated through vertical integration and forward contracts.

Marketability can be improved by pooling production, enabling farmers to create larger potato batches for sale that have been graded and selected specifically for the target market. Establishment of collection centres and warehouse receipting systems will also go a long way in addressing market challenges such as fluctuating supply and prices throughout the year. In order to remove the market cartels operating at major wholesale markets, registration of marketing agents/agencies directly linked to the farmers should be undertaken.

Linking producers and producer groups directly with other value chain participants, such as wholesalers and retailers, will enable smallholders to meet the specific demands of the value chain, thereby adding value and boosting incomes. Promoting an understanding of the market and the demands of the value chain to smallholders will ensure the right product, in terms of variety, grading and packaging, is delivered to the right place at the right time.

Modern information and communication technologies (ICT)

Application of modern ICT applications (m-agriculture,[1] e-extension, etc.), guided by business logic, can offer several interventions in the supply chain including:

- reduction of costs of coordination (collection of production, distribution of inputs, etc.);
- increased transparency in decision making between partners;
- reduction of transaction costs;
- dissemination of market demand and price information;
- dissemination of weather, pest and risk-management information;
- dissemination of best practices to meet quality and certification standards;
- collection management data from the field; and
- ensuring traceability.

ICTs can also be used in extension in what is commonly known as e-extension or m-agriculture.

Demand increase, financial instruments and flow of information

Opportunities exist to increase demand for fresh potatoes in promotion, packaging, niche varieties, taste, education, alternative uses and consistency in product quality. Increases in per capita consumption of potatoes can be in achieved through various marketing processes, such as blanket generic marketing. There is potential for increasing exports of fresh potato beyond the very modest volumes that have been shipped to date. Export efforts geared towards ware potatoes should focus first on stabilizing supplies and improving the quality of the tubers available within the country as part of a more long-term process to make them more competitive in potential export markets. There is potential to expand the packaging industry and increase the value of fresh potato.

There is also a need to build a package of instruments – grants, guarantees, patient capital and commercial money combined – to finance the capital costs associated with transformation of the potato subsector. The various instruments identified should be mutually enabling, for example using crop insurance or financial guarantees as collateral to encourage banks to lend money, or using donor funding and patient capital to co-invest with private investors to improve a project's access to capital through public–private partnerships.

Organizations such as the NPCK and Kenya National Potato Farmers' Association (KENAPOFA) can help to improve the flow of information among players in the fresh potato value chain. There is a need to continuously review and update existing potato standards for both local and export markets. There is also a need to review and enforce legal and regulatory frameworks that affect the fresh potato value chain (e.g. legal notices No. 44 and 113). Markets are in a pathetic state and need to be improved.

Strengthening the processed potato value chain

Good manufacturing practices (GMP) and research

GMP should be developed for the potato processing industry. GMP compliance assures sanitary and processing requirements applicable to all food processing establishments are met. The GMP is a set of regulations, codes and guidelines that control the operational conditions within food establishment allowing for the production of safe food. Research is required to: (i) develop best practices for growing potatoes for the processing industry; (ii) conduct consumer surveys to identify growing market segments and types of processed potato products likely to be in demand in the near future; (iii) conduct surveys to determine the most critical problems affecting players along the processed potato value chain; and (iv) reduce polyacrylamide levels in fried potato products.

Risk reduction and markets

Public policies and private sector initiatives that reduce risk in the processed potato value chain should be promoted. Considerable scope exists to further the development of potato varieties that are suitable for the processing industry. Producers and producer groups in the processed potato value chain should be linked as this will enable smallholders to meet the specific demands of the chain, thereby adding value and boosting incomes. Promoting an understanding of the market and the demands of the processed potato value chain to smallholders will ensure the right product, in terms of variety, grading, quality and packaging is delivered. Effective business administration along the chain should also be promoted.

Opportunities exist to increase demand for processed potatoes in promotion, packaging, niche varieties, taste, education, alternative uses and consistency in product quality. There is potential for increasing exports. Strategies need to be put in place to aggressively find markets for processed potato products beyond the domestic front. There is potential to expand the packaging industry and increase the value of processed potato.

Farmer coordination and standards for products

Just like in the fresh potato value chain, coordination among farmers to facilitate activities along the processed potato value chain is still not sufficiently developed and needs strengthening. Finalization of local and regional standards (e.g. the East African Community (EAC) standards) for crisps and other processed products will go a long way to strengthening the value chain. A contract farming approach could assist farmers to access high-value markets linked to seed (and other inputs) provided by the processor on a loan basis to ensure a regular supply of the desired processing characteristics (e.g. the right variety, potatoes of appropriate age and qualities).

44.5 Managing the Revitalization Process

Revitalization of the potato subsector will require an appropriate approach to design, manage and monitor implementation of activities to drive change at scale. It will require the institutionalization of the partnerships through a formal organization and governance set-up to orchestrate change and monitor progress, either through explicit empowerment of existing actors or establishment of a delivery unit. Adequate resources will be required to drive operational development and deployment of initiatives until they achieve a self-sustaining momentum. The envisaged transformation is unlikely to be achieved without the right mix of people, skills and funding, and a structure that gives and clarifies the authority to manage. Any of the following three options may be adopted to manage the revitalization process:

1. A stand-alone 'delivery' or 'transformation' unit reporting directly to a change leader who has strong amounts of authority and staff to quickly troubleshoot and execute on all aspects of the transformation could be created.

2. A 'lean-and-mean' coordinating steering committee that provides a regular forum for all stakeholders to meet and track roll out could be created.

3. A revitalization/transformational unit or an overarching programme management office could be set up to monitor progress and adjust plans without taking a direct role in implementation.

Note

[1] m-Agriculture refers to agricultural services, technology dissemination and communication using mobile devices such as mobile phones.

References

Bernet, T., Thiele, G. and Zschocke, T. (2006) *Participatory Market Chain Approach (PMCA) – User Guide*. International Potato Center (CIP), Lima.

Government of Kenya (GoK) (2007) *Kenya Vision 2030. A Globally Competitive and Prosperous Kenya*. GoK, Nairobi.

Government of Kenya (GoK) (2009) *National Potato Taskforce Report, Final Report*. Promotion of Private Sector Development in Agriculture Programme. GoK, Nairobi.

Government of Kenya (GoK) (2010a) *Agricultural Sector Development Strategy (ASDS) 2010–2020*. GoK, Nairobi.

Government of Kenya (GoK) (2010b) *National Roots and Tuber Crops Policy*. Ministry of Agriculture (Draft). GoK, Nairobi.

Government of Kenya (GoK) (2010c) *Strategy for Developing Potato Industry in Kenya*. Ministry of Agriculture (Draft). GoK, Nairobi.

International Potato Center (CIP) (2013) A roadmap for revitalization of the potato sector in Kenya. In: *Wealth Creation through Integrated Development of the Potato Production and Marketing Sector in Kenya, Uganda, and Ethiopia,* Final Report. CIP, Lima, (Annex 2) pp. 1–55.

Kaguongo, W., Ng'ang'a, N., Muthoka, N., Muthami, F. and Maingi, G. (2010) *Seed Potato Subsector Master Plan for Kenya (2009–2014)*. National Potato Council of Kenya, Nairobi.

Kaguongo, W., Nyangweso, A., Nderitu, J., Lung'aho, C., Ng'ang'a, N., Kipkoech, D. Kabira, J., Gathumbi, M., Njane, P., Irungu, J., Onyango, A., Borus, D., Schulte-Geldermann, E. and Lutaladio, N. (2013) *A Policymakers' Guide to Crop Diversification: the Case of Potato in Kenya.* Food and Agriculture Organization of the United Nations, Rome.

Kenya Agricultural Research Institute (KARI) (2009) *Kenya Agricultural Research Institute Strategic Plan (2009–2014)*. KARI, Nairobi.

Kenya Agricultural Research Institute (KARI) (2012) *Report on the Review and Analysis of Staple Product Value Chains*. Volume II. KARI, Nairobi.

National Potato Council of Kenya (NPCK) (2012) Proceedings of potato Round Table meeting, 25 June, held at Silver Springs Hotel Nairobi, Kenya. NPCK, Nairobi.

45 Potential of Processing Potato Flakes from Popular Kenyan Potato Varieties

G.O. Abong',[1]* J. Kabira,[2] M.W. Okoth,[1] J.A. Ogolla[1] and J. Ouma[1]
[1]University of Nairobi Nairobi, Kenya; [2]Kenya Agricultural and Livestock Research Organization (KALRO), Limuru, Kenya

Abstract

Losses of potatoes are incurred during seasons of glut and farmers are forced to sell their produce to middlemen at low prices due to the short storage life of the raw tubers. Processing potatoes after harvesting into dehydrated products such as potato flakes is one of the ways to overcome the problem of inadequate and inappropriate bulk storage of raw potatoes. Potato flakes are known to be shelf stable and hence assure users of the availability of the produce all year round. The current study was instituted to establish the potential of processing potato flakes from selected Kenyan potato varieties. Five potato varieties (Tigoni, Dutch Robyjn, Désirée, Kenya Mpya and Sherekea) which are popular commercially were used in this study. The varieties were harvested after maturity and processed into potato flakes, then evaluated for retained vitamin C, colour, oil and moisture content, and sensory attributes. The varieties used in this study had dry matter contents ranging from 18.23% to 23.66%. The oil content of the flakes was significantly ($P \leq 0.05$) lower in Dutch Robyjn (0.23%) compared with Désirée (0.76%). The oil content of flakes was significantly lower ($P = 0.011$) but negatively correlated ($r = -0.956$) with the dry matter content of the raw potatoes. Reduced ascorbic acid (vitamin C) in raw tubers differed significantly ($P \leq 0.05$) among the varieties ranging from 90.37 mg/100 g in Kenya Mpya to 127.56 mg/100 g in Sherekea. The vitamin decreased in flakes, the reduction being higher in Kenya Mpya (60.34%) and Sherekea (71.71%) compared with Tigoni (18.15%) and Désirée (23.26%). Overall, Dutch Robyjn was the best variety for flakes followed by Désirée and Tigoni with regard to overall acceptability; Kenya Mpya was barely acceptable; while Sherekea was unacceptable. Promotion of these varieties for flakes processing will not only diversify the range of potato products, but also add value to local potatoes.

Keywords: drum drying, flakes, potato varieties, reduced ascorbic acid

45.1 Introduction

Potatoes produced in Kenya are used in homemade recipes and industrial processing of French fries, potato crisps, potato sticks and other related products (Abong' *et al.*, 2010a). There are, however, considerable losses during seasons of glut and farmers are forced to sell their produce to middlemen at low prices due to the short storage life of the raw tubers (Walingo *et al.*, 2004). Coupled with lack of appropriate cold storage facilities, many varieties of potatoes grown in Kenya are not suitable for production

*georkoyo@yahoo.com

of fried products even after reconditioning from cold storage. They accumulate high amounts of reducing sugars leading to dark brown fries which are unacceptable to many consumers (Abong' *et al.*, 2009a).

The position of potato in the world as an important food crop is well documented, being one of the staple foods providing food security to millions of people (CIP, 2012). In terms of utilization, potatoes are ranked second after maize in Kenya, and are a major food staple (MoA, 2009). Processing potatoes after harvesting into dehydrated products such as potato flakes is one of the ways to overcome the problem of inadequate and inappropriate on-farm storage of raw potatoes. Potato flakes are known to be shelf stable and hence assures users of the availability of the produce all year round (Neilson *et al.*, 2006; Nayak *et al.*, 2011; Boesch, 2012).

The process of converting potato into flakes involves many stages culminating in drum drying of potato slurry or mash, a process which can be compared to that used for preparing precooked starchy baby foods, casein milk, maltodextrins and fruit pulps. Under normal circumstances, potato slurry is spread on the surface of a heated drum. Heat is transferred through the metal of the drum and water is evaporated resulting in the dried slurry adhering to the drum which is removed usually with an appropriate scraping device (Kakade *et al.*, 2011).

Potato flakes have a wide range of uses including as: (i) a thickening agent (Van Hecke *et al.*, 2012); (ii) extruded products (Dushkova *et al.*, 2011); (iii) fabricated potato chips; (iv) snack pellets; (v) in battered breaded products; and (vi) potato powder (Kakade *et al.*, 2011). There are only two brands of potato flakes in Kenya, an imported brand from the USA and a locally packaged brand whose source is unclear (Abong' and Kabira, 2012). It has been difficult to establish if any of the local varieties are used in processing. This study was instituted to establish the potential of processing potato flakes from selected popular Kenyan potato varieties.

45.2 Materials and Methods

Raw potatoes for processing

Five potato varieties which have previously been found to have good processing characteristics (Abong' *et al.*, 2009b, 2010b) were used in this study. Three established varieties (Tigoni, Dutch Robyjn and Désirée) and two newly released varieties (Kenya Mpya and Sherekea) were grown under standard cultural conditions at the National Potato Research Centre at Tigoni, Kenya between April and August 2012. The crop was dehaulmed after maturity and harvested after 2 weeks. Tubers were stored for 2 weeks more in a dark store under ambient air conditions before processing. Dry matter and reduced ascorbic acid (RAA) contents were analysed in raw tubers and processed flakes.

Processing of potato flakes

About 5 kg of fresh potatoes from each of the five varieties were hand-peeled using a knife, trimmed and washed with clean water to remove surface dirt before cutting into pieces 20–30 mm in size. The potatoes were cooked by boiling until ready (just when a fork could penetrate with minimal force) before being converted into a mash using a cooking stick. A single drum dryer (IOT7578 SERIES OCD, Renold Limited, Germany) comprising a hollow drum allowing for the introduction of steam was used for dehydration. The drum was made of stainless steel and had a length of 300 mm and an outer diameter of 320 mm; the effective surface area of the drum available for drying being 0.302 m^2. A scraper blade was fitted to the drum. The drum speed was set at 10 rpm so that the residence time of feed on the drum surface was 10 s. As the drum rotated towards the scraper blade, the dried potato mash was scraped out as a thin layer of dry material from the surface of the drum. The dried mash was broken into smaller particles, packaged and sealed in flexible polyethylene bags (300 gauge) after cooling to 20–25°C.

Dry matter and moisture content

Dry matter and moisture contents of raw potatoes and processed flakes were determined on triplicate samples by standard analytical methods (KEBS, 2007).

Determination of reducing sugars of raw tubers

Levels of reducing sugars were determined in the raw potato tubers as per the method described by Abong' *et al.* (2009a).

Oil content of potato flakes

The oil content was determined by extraction of 2.5 g of finely ground samples of flakes in a Soxhlet apparatus for 8 h using analytical grade petroleum ether (boiling point 40–60°C) according to the method described by KEBS (2007) and the oil content was calculated as a percentage.

Potato flakes colour

The colour of the flakes was determined using a colour spectrophotometer (NF 333, Nippon Denshoku, Japan) in the Commission International de l'Eclairage (CIE) L*a*b* scale described by Abong' *et al.* (2011c).

Determination of vitamin C

Vitamin C was determined as RAA in raw and processed flakes by indophenol dye as described by Masamba and Nguyan (2008).

Sensory analysis

Potato flakes from each variety were coded and samples presented to ten panellists familiar with processed potato products. Panel members scored for colour, taste, mouth feel, aroma and overall acceptability on a seven-point hedonic rating scale where 1 = dislike very much and 7 = like very much (Larmond, 1977). A score of 4 was the lower limit of acceptability.

Data analysis

Analytical data obtained for each parameter were subjected to analysis of variance (ANOVA) for comparison of means and significance difference tests at 5% level of significance using the Statistical Analysis System (SAS version 9). Significant differences were considered at $P \leq 0.05$. The Pearson correlation was determined between dry matter of raw tubers and oil content of flakes, sensory colour scores and laboratory colour analysis.

45.3 Results and Discussion

Variation in raw potato tuber dry matter content

The varieties used in this study had dry matter contents ranging from 18.23% in Désirée to 23.66% in Dutch Robyjn. Dry matter contents of Désirée and Kenya Mpya were significantly ($P < 0.0001$) lower compared with Sherekea, Tigoni and Dutch Robyjn. The dry matter content of raw potato is an important parameter to be considered, especially if the tubers are meant for processing. Potato dry matter content is influenced by among other factors the variety. Dry matter influences the starch matrix and content and hence affects the functional properties of processed products (Frančáková *et al.*, 2011). Higher dry matter content means that the processor is assured of good textural properties and higher yield (Burton, 1989). Fat content correlates negatively with dry matter content especially in cases where deep-oil frying is involved (Abong' *et al.*, 2010b; Pedreschi *et al.*, 2010). In the case of processing flakes, it means that there will be less energy consumption, less time and higher yields when dehydrating the variety Dutch Robyjn compared with Désirée.

Attributes of potato flakes

Oil and moisture content of dehydrated flakes

The moisture content and oil content (dry weight basis) of flakes differed significantly ($P \leq 0.05$) among the five varieties used in processing (Table 45.1). The oil content of the flakes was significantly ($P \leq 0.05$) lower in Dutch Robyjn (0.23%) compared with Désirée (0.76%). By contrast, moisture content of flakes from Kenya Mpya, Sherekea and Tigoni did not significantly ($P > 0.05$) differ from each other. This was the same for Désirée and Dutch Robyjn; flakes' moisture content did not have any particular varietal pattern. The oil content was significantly lower ($P = 0.011$) but negatively correlated ($r = -0.956$) with the dry matter content of the raw potatoes.

The oil content in potato that has not undergone any type of frying is always low as confirmed with these dehydrated flakes in the current study. It is, however, important to note that the oil present in a potato product, however small it may be, is a critical determinant of flavour and shelf life (Woolfe, 1987; Ziaiifar *et al.*, 2008). Potato flakes therefore provide low-oil products that can be used in a variety of food preparations. The low moisture content of the flakes, ranging from 10.21% in Désirée to 11.65% in Kenya Mpya, makes it possible to store them for a significant period of time. The lower the moisture content, the longer the product can be stored since the water activity will not support oxidation or microbial activities so long as the packaging is suitable (Marsh and Bugusu, 2007; Abong' *et al.*, 2011b).

Effect of processing on RAA (vitamin C) content

The RAA content in raw tubers differed significantly ($P \leq 0.05$) among the varieties, ranging from 90.37 mg/100 g in Kenya Mpya to 127.56 mg/100 g in Sherekea (Table 45.2). The levels, however, reduced significantly ($P \leq 0.05$) when the tubers were processed into flakes. The reductions were higher in Kenya Mpya (60.34%) and Sherekea (71.71%) compared with Tigoni (18.15%) and Désirée (23.26%). Potatoes are known to have high levels of vitamin C, matching or surpassing some fruits. The vitamin C in potato tubers and potato products is readily available and easily absorbed by the human body (Yoshitaka

Table 45.1. Oil and moisture contents of flakes from five Kenyan varieties.[a]

Variety	Oil content (%)[b]	Moisture content (%)	Tuber dry matter content (%)
Désirée	0.76 ± 0.08a	10.21 ± 0.30b	18.23 ± 0.23e
Dutch Robyjn	0.23 ± 0.02d	10.85 ± 0.21b	23.66 ± 0.11a
Kenya Mpya	0.50 ± 0.14b	11.65 ± 0.78a	20.07 ± 0.94d
Sherekea	0.32 ± 0.03c	11.35 ± 0.91a	22.69 ± 0.03b
Tigoni	0.53 ± 0.04b	11.40 ± 0.57a	21.66 ± 0.54c

[a]Values are means of duplicate determinations ± standard deviation. Values with similar letters in the same column are not significantly different at 5% level of significance.
[b]Oil content values are expressed on dry weight basis.

Table 45.2. Reduced ascorbic acid (RAA) (mg/100 g, dry weight basis) in raw potato tubers and dehydrated flakes.[a]

Variety	Raw tubers	Potato flakes	Reduction (%)
Désirée	126.60 ± 0.85a	97.15 ± 4.55a	23.26d
Dutch Robyjn	117.73 ± 4.90b	72.43 ± 1.05c	38.48c
Kenya Mpya	90.37 ± 0.48d	35.85 ± 3.15d	60.34b
Sherekea	127.56 ± 2.42a	36.09 ± 1.40d	71.71a
Tigoni	106.91 ± 2.70c	87.51 ± 3.5b	18.15e

[a]Values are means of duplicate determinations ± standard deviation. Values with similar letters in the same column are not significantly different at 5% level of significance.

et al., 2012). The Irish poor for instance relied on the potato almost exclusively as the main source of vitamin C to such an extent that when potatoes were wiped out in Ireland, leading to the great famine, there was widespread scurvy along with other infections (Geber and Murphy, 2012). The variation in vitamin C content is largely due to the cultivar (Murniece et al., 2011; Abong' et al., 2011a; Lombardo et al., 2012). However, the levels of the vitamin can be influenced by the cultivation system indirectly through control of photosynthesis and metabolism of precursors such as sugars, environmental conditions and cultural practices such as pest control (Lee and Kader, 2000; Marek and Krystyna, 2012). The levels of vitamin C in the current study compare with those reported by Lombardo et al. (2012), which ranged from 49 to 74 mg/100 g on a fresh weight basis, but are higher than those reported by Murniece et al. (2011) of 10–15 mg/100 g fresh weight. Vitamin C is among the most heat-sensitive vitamins during thermal processing or extrusion (Riaz et al., 2009). The loss of the vitamin during dehydration is attributed to irreversible oxidation and thermal destruction that occurs during drying (Vega-Gálvez et al., 2008; Mrad et al., 2012). The effect of thermal dehydration is variety specific, probably due to differences in the individual cell matrix that makes up a given potato variety.

Colour parameters of the flakes

Flake colour differed significantly ($P \leq 0.05$) among the varieties (Table 45.3). The lightness parameter was least in Désirée (78.58) and highest in Dutch Robyjn (84.78). In all cases the lightness parameter was relatively high and therefore all the varieties gave good colour. Values for the redness parameter were all negative indicating less browning among the varieties. Values for the yellowness parameter ranged from 14.7 in Sherekea to 27.80 in Désirée. The colour of flakes and any food product is of utmost importance since it determines the acceptability of the product (Wang-Pruski and Nowak, 2004; Pedreschi et al., 2006). Colour of processed potato products is determined before harvest by the kind of cultural practices such as fertilization, harvesting time and maturity. These activities influence the reducing sugar content which reacts at high temperatures with free amino acids to cause browning (Abong' et al., 2009a, 2011a; Halford et al., 2012). Extensive browning did not occur in flakes as shown by the high lightness and low redness parameters. Since light flakes are preferred by consumers (Kakade et al., 2011), it is expected that flakes in the current study will be acceptable as far as perceived colour is concerned.

Sensory attributes of flakes from five Kenyan varieties

Table 45.4 indicates variations in sensory attributes of flakes from five potato varieties. Sensory colour scores significantly differed ($P \leq 0.05$) among the varieties being lowest in Sherekea (3.4) and highest in Dutch Robyjn (5.5). Scores for Désirée, Kenya Mpya and Tigoni did not, however, differ significantly indicating that panelists did not notice any apparent differences. On the other hand, flakes

Table 45.3. Colour parameters for flakes from five selected potato varieties.[a]

	Parameters[b]		
Variety	L*	a*	b*
Désirée	78.58 ± 1.51c	−0.80 ± 0.00a	27.80 ± 1.23a
Dutch Robyjn	84.78 ± 0.67a	−4.08 ± 0.17d	21.88 ± 1.63b
Kenya Mpya	83.50 ± 3.73a	−2.00 ± 0.22c	15.83 ± 1.12cd
Sherekea	82.03 ± 0.58b	−1.25 ± 0.31b	14.70 ± 1.97d
Tigoni	81.70 ± 0.88b	−2.15 ± 0.19c	17.55 ± 2.37c

[a]Values are means of four determinations ± standard deviation. Values with similar letters in the same column are not significantly different at 5% level of significance.

[b]L*, Lightness parameter ranging from black (0) to white (100); a*, chromatic redness parameter with negative values tending towards green with no browning while high positive values indicate dark browning; b*, yellowness parameter corresponding to yellow when positive and blue when negative.

Table 45.4. Sensory attributes of flakes from five selected potato varieties.[a]

Variety	Sensory attributes[b]				
	Colour	Taste	Mouth feel	Aroma	Overall acceptability
Désirée	4.8 ± 0.8b	4.0 ± 0.9a	4.4 ± 0.8ab	4.4 ± 0.6b	4.7 ± 0.8b
Dutch Robyjn	5.5 ± 1.1a	4.2 ± 1.1a	4.6 ± 1.1a	5.0 ± 1.2a	5.1 ± 1.2a
Kenya Mpya	4.7 ± 1.4b	3.3 ± 1.6cb	3.9 ± 1.2cd	4.1 ± 1.3b	3.9 ± 1.1c
Sherekea	3.4 ± 0.9c	3.2 ± 0.9c	3.7 ± 0.6d	4.0 ± 1.4b	3.4 ± 1.0d
Tigoni	4.6 ± 1.4b	3.6 ± 0.9b	4.2 ± 1.2bc	4.1 ± 1.0b	4.0 ± 1.1c

[a]Values with similar letters in the same column are not significantly different at 5% level of significance.
[b]Evaluation was done on a seven-point hedonic scale where 1 = dislike very much and 7 = like very much. A score of 4 was the acceptable lower limit.

from Sherekea were rejected on the basis of colour, scoring below 4 which was the minimum lower limit, meaning that flakes from Sherekea are unlikely to be acceptable at the first instance when displayed in the markets. There was no significant ($P > 0.05$) correlation between sensory colour score and any of the colour parameters (L*, a* and b*) implying that it is not possible to simply use the objective measurements to predict consumer choice of the flakes by use of colour. Scores for taste were significantly ($P \leq 0.05$) higher in flakes processed from Désirée and Dutch Robyjn and lower in Sherekea and Kenya Mpya. The taste of the flakes from Sherekea, Kenya Mpya and Tigoni was not acceptable.

Sensory scores for the mouth feel and aroma of flakes significantly ($P \leq 0.05$) differed with the variety, being highest in Dutch Robyjn and lowest in Sherekea for both attributes. Whereas all the flakes were acceptable in terms of flavour, Sherekea (3.7) was unacceptable and Kenya Mpya (3.9) was barely acceptable. Regarding overall acceptability, Dutch Robyjn with a score of 5.1 was the best variety for flakes followed by Désirée (4.7) and Tigoni (4.0), respectively; Kenya Mpya (3.9) was barely acceptable and Sherekea (3.7) was unacceptable. It therefore means that varieties Dutch Robyjn, Désirée and Tigoni will be suitable for processing into flakes while Kenya Mpya may either be accepted or rejected; Sherekea does not make acceptable flakes.

45.4 Conclusion

Generally, acceptable potato flakes were processed from the five Kenyan potato varieties selected for this study. Dutch Robyjn, Désirée and Tigoni produced excellent flakes compared with Kenya Mpya which was barely acceptable while Sherekea was unacceptable. Promotion of these varieties for flakes will not only diversify the range of potato products, but will also add value to local potatoes.

Acknowledgements

The authors acknowledge financial support from Kenya Agricultural Productivity Programme and Agribusiness Project (KAPAP) and the Kenya Agricultural and Livestock Research Organization (KALRO). Technical support from J. M'thika, R. Kamau and B. Muroki from the Department of Food Science, Nutrition and Technology, University of Nairobi, and C. Muchoki, B. Kawili and N. Ngone from KALRO-Tigoni is appreciated.

References

Abong', G.O. and Kabira, J.N. (2012) Diversity and characteristics of potato flakes in Nairobi and Nakuru, Kenya. *Global Journal of Science Frontier Research Agriculture and Veterinary Sciences* 12(10), 35–39.

Abong', G.O., Okoth, M.W., Karuri, E.G., Kabira, J.N. and Mathooko, F.M. (2009a) Levels of reducing sugars in eight Kenyan potato cultivars as influenced by stage of maturity and storage conditions. *Journal of Animal and Plant Sciences* 2(2), 76–84.
Abong', G.O., Okoth, M.W., Karuri, E.G., Kabira, J.N. and Mathooko, F.M. (2009b) Evaluation of selected Kenyan potato cultivars for processing into French fries. *Journal of Animal and Plant Sciences* 2(3), 141–147.
Abong', G.O., Okoth, M.W., Imungi, J.K. and Kabira, J.N. (2010a) Consumption patterns, diversity and characteristics of potato crisps in Nairobi, Kenya. *Journal of Applied Biosciences* 32, 1942–1955.
Abong', G.O., Okoth, M.W., Imungi, J.K. and Kabira, J.N. (2010b) Evaluation of selected Kenyan potato cultivars for processing into potato crisps. *Agriculture and Biology Journal of North America* 1(5), 886–893.
Abong', G.O., Okoth, M.W., Imungi, J.K. and Kabira, J.N. (2011a) Losses of ascorbic acid during storage of fresh tubers, frying, packaging and storage of potato crisps from four Kenyan potato cultivars. *American Journal of Food Technology* 6(9), 753–762.
Abong', G.O., Okoth, M.W., Imungi, J.K. and Kabira, J.N. (2011b) Effect of packaging and storage temperature on the shelf life of crisps from four Kenyan potato cultivars. *American Journal of Food Technology* 6(10), 870–881.
Abong', G.O., Okoth, M.W., Imungi, J.K. and Kabira, J.N. (2011c) Effects of cultivar, frying temperature and slice thickness on oil uptake and sensory quality of potato crisps processed from four Kenyan potato cultivars. *Journal of Agricultural Science and Technology A* 1, 156–163.
Boesch, I. (2012) Preferences for attributes of industrial potatoes: an empirical investigation of the Swiss market. *American Journal of Potato Research* 89,199–206.
Burton, W.G. (1989) *The Potato*, 3rd edn. Longman Group, London.
Dushkova, M.A., Menkov, N.D. and Toshkov, N.G. (2011) Functional characteristics of extruded blends of potato flakes and whey protein isolate. *International Journal of Food Engineering* 7(6), Article 12.
Frančáková, H., Líšková, M., Bojňanská, T. and Mareček, J. (2011) Changes of carbohydrates complex influenced by the storage time. *Journal of Medical and Biological Sciences* 1(3), 446–454.
Geber, J. and Murphy, E. (2012) Scurvy in the great Irish famine: evidence of the vitamin C deficiency from the mid-19th century skeletal population. *American Journal of Physical Anthropology* 148(4), 512–524.
Halford, N.G., Curtis, T.Y., Muttucumaru, N., Postles, J., Elmore, J.S. and Mottram, D.S. (2012) The acrylamide problem: a plant and agronomic science issue. *Journal of Experimental Botany* 63(8), 2841–2851.
International Potato Center (CIP) (2012) Potato. Available at: www.cipotato.org/potato (accessed 3 August 2013).
Kakade, R.H., Das, H. and Ali, S. (2011) Performance evaluation of a double drum dryer for potato flake production. *Journal of Food Science and Technology* 48(4), 432–439.
Kenya Bureau of Standards (KEBS) (2007) *Potato Crisps Specifications. Kenya Standards 1094-1: 2007.* KEBS, Nairobi, Kenya.
Larmond, E. (1977) *Methods for Sensory Evaluation of Food*. Food Research Institute, Central Experiment Farm, Canada Department of Agriculture, Ottawa.
Lee, S.K. and Kader, A.A. (2000) Preharvest and postharvest factors influencing vitamin C content of horticultural crops. *Postharvest Biology and Technology* 20(3), 207–220.
Lombardo, S., Pandino, G. and Mauromicale, G. (2012) Nutritional and sensory characteristics of 'early' potato cultivars under organic and conventional cultivation systems. *Food Chemistry* 133, 1249–1254.
Marek, G. and Krystyna, Z. (2012) Vitamin C content in tubers as influenced by insecticide application. *Polish Journal of Environmental Studies* 2(4), 1101–1105.
Marsh, K. and Bugusu, B. (2007) Food packaging – roles, materials, and environmental issues. *Journal of Food Science* 72(3), 39–55.
Masamba, K.G. and Nguyen, M. (2008) Determination and comparison of vitamin C, calcium and potassium in four selected conventionally and organically grown fruits and vegetables. *African Journal of Biotechnology* 7(16), 2915–2919.
Ministry of Agriculture (MoA) (2009) Mainstreaming the potato crop from orphan crop status. Paper presented at Round Table Africa (RTA) potato stakeholders' workshop, 8 May 2009, MoA, Nairobi, Kenya.
Mrad, N.D., Boudhrioua, N., Kechaou, N., Courtois, F. and Bonazzi, C. (2012) Influence of air drying temperature on kinetics, physicochemical properties, total phenolic content and ascorbic acid of pears. *Food and Bioproducts Processing* 9, 433–441.
Murniece, I., Karklina, D., Galoburda, R., Santare, D., Skrabule, I. and Costa, H.S. (2011) Nutritional composition of freshly harvested and stored Latvian potato (*Solanum tuberosum* L.) varieties depending on traditional cooking methods. *Journal of Food Composition and Analysis* 24, 699–710.

Nayak, B., Berrios, J.J., Powers, J.R., Tang, J. and Ji, Y. (2011) Coloured potatoes (*Solanum tuberosum* L.) dried for antioxidant-rich value-added foods. *Journal of Food Processing and Preservation* 35, 571–580.

Neilson, A.P., Pahulu, H.F., Ogden, L.V. and Pike, O.A. (2006) Sensory and nutritional quality of dehydrated potato flakes in long-term storage. *Journal of Food Science* 71(6), S461–S466.

Pedreschi, F., León, J., Mery, D. and Moyano, P. (2006) Implementation of a computer vision system to measure the colour of potato chips. *Food Research International* 39, 1092–1098.

Pedreschi, F., Segtnan, S.V.H. and Knutsen, H. (2010) On-line monitoring of fat, dry matter and acrylamide contents in potato chips using near infrared interactance and visual reflectance imaging. *Food Chemistry* 121, 616–620.

Riaz, M.N., Asif, M. and Ali, R. (2009) Stability of vitamins during extrusion. *Critical Reviews in Food Science and Nutrition* 49(4), 361–368.

Van Hecke, E., Nguyen, P.-U., Clausse, D. and Lanoiselle, J.-L. (2012) Flow behaviour of carrot puree: modeling the influence of time, temperature and potato flakes addition. *International Journal of Food Science and Technology* 47, 177–185.

Vega-Gálvez, A., Lemus-Mondaca, R., Bilbao-Sáinz, C., Fito, P. and Andrés, A. (2008) Effect of air drying temperature on the quality of rehydrated dried red bell pepper (var. Lamuyo). *Journal of Food Engineering* 85, 42–50.

Walingo, A., Lung'aho, C., N'gang'a, N., Kinyae, P.M. and Kabira, J.N. (2004) Potato marketing, storage, processing and utilization in Kenya. Paper presented at the 6th Triennial Congress of the African Potato Association, 5–10 April 2004, African Potato Association, Agadir, Morocco.

Wang-Pruski, G. and Nowak, J. (2004) Potato after-cooking darkening. *American Journal of Potato Research* 81, 7–16.

Woolfe, J.A. (1987) *The Potato in the Human Diet*. Cambridge University Press, Cambridge, pp. 19–54.

Yoshitaka, K., Chihana, H., Mizuki, I., Handa, S., Mugita, H., Maruyama, N., Koga, H. and Ishigami, A. (2012) Bioavailability of vitamin C from mashed potatoes and potato chips after oral administration. *British Journal of Nutrition* 107(6), 885–892.

Ziaiifar, A.M., Achir, N., Courtois, F., Trezzani, I. and Trystram, G. (2008) Review of mechanisms, conditions, and factors involved in the oil uptake phenomenon during the deep-fat frying process. *International Journal of Food Science and Technology* 43, 1410–1423.

46 Assessing Potato Production Efficiency through Contract Farming in Kenya: The Case of Bomet and Molo Farmers

D. Kipkoech,[1*] D. Borus,[2] B. Lemaga,[2] J. Kering,[3] G. Muriithi,[4] N. Ng'ang'a[1] and J. Kabira[1]

[1]Kenya Agricultural and Livestock Research Organization (KALRO), Limuru, Kenya; [2]International Potato Center Sub-Saharan Africa (CIP-SSA), Nairobi, Kenya; [3]Ministry of Agriculture, Bomet Sub County, Kenya; [4]Deepa Industries Limited, Nairobi, Kenya

Abstract

A study on contract farming among 137 randomly selected potato farmers in the Bomet and Molo districts was done to evaluate its impact on potato production and productivity using cross-sectional data. Descriptive and inferential statistics on farmers' socio-economic, institutional and geographical aspects were used. Stochastic frontier model analysis was used to evaluate farmers' attributes that affect their production efficiency under contract farming. Contract farming enhanced potato productivity and incomes while also becoming an incentive for farmers to supply a niche market and it increased their demand for farm inputs. Due to the lack of a sound organized market for potatoes, there was inefficiency in potato production which caused a decline in inputs demanded. To overcome these problems farmers formed an organized production and marketing system under a contractual arrangement with the buyers. This led to increased demand for quality inputs and enhanced access to credit. In response to the challenges in contract farming, a tri-partied system of farmers, service providers and processors was constituted for arbitration. Contract farming in the districts has shown that farmers could produce sufficient quantities of high quality potatoes that fetch premium price if they are organized to do so. However, rules on contracts have to be agreed on and followed by all participating players. Contract farming involved extra costs but the accrued benefits were higher due to improved productivity, efficiency and the premium price paid under the contract. There was also a reduction in processing losses due to delivery of quality tubers (losses due to rotting, browning and peeling were reduced by 0.78%, 0.3% and 0.75%, respectively) compared with the pre-project period. Attributes of respondents such as being the head of the household, highest level of training received, large family size and contract farming participation significantly enhanced production efficiency while credit access and age of respondent had significant negative effects on production efficiency.

Keywords: contract farming, potato, production efficiency, productivity, quality input

*davidnyamasia@gmail.com

46.1 Introduction

In Kenya, the average yearly growth in potato production was 5.5% from the early 1990s to 2003 (FAOSTAT, 2005) and 4.3% afterwards (Scott *et al.*, 2000). This growth has been driven by an increase in demand as a result of urbanization, a general population increase and a shift in food consumption patterns particularly in urban centres (Scott *et al.*, 2000). However, the potato sector continues to be constrained by the following factors: (i) poor quality seed potato (Kinyua *et al.*, 2001); (ii) high incidences of diseases (Schulte-Geldermann *et al.*, 2012); (iii) expensive certified seed (N.M. Ng'ang'a, P. Kinyae, A. Walingo and M. Wakahiu, 2003, unpublished data); (iv) imperfect marketing systems and polices (Gildermacher *et al.*, 2007; GoK, 2009; Nderitu, 2010); (v) a limited varietal pool for processing (Kabira and Lemaga, 2003); and (vi) farmers' production inefficiency (Kipkoech *et al.*, 2008). In many potato-producing areas, including Bomet and Molo, producers who are not empowered are faced by unregulated potato markets where middlemen exploit them. Growth in the sector may require an effective and efficient supply chain arrangement, reduction in the costs of doing potato business and better returns for the various players. This would also encourage investment in potato technologies required for profitable business in the potato industry (Saenz-Segura, 2006).

Another contribution to growth in the sector may come from enforcing contract farming. Contract farming can be defined as an 'agreement between farmers and processing and/or marketing firms for the production and supply of agricultural products under forward agreements, frequently at predetermined prices' (Eaton and Shepherd, 2001). The decision for agribusinesses and farmers to adopt contract farming can be explained by the use of transaction cost and risk theories (Williamson, 1979). Contract farming can be used by both the principal (agribusiness) and the agent (farmer) to mitigate risks (Williamson, 1979). Contracts that provide credit, technology, inputs, information, extension services and risk mitigation: (i) help producers improve production efficiency; (ii) develop a commercial culture; and (iii) augment income and employment (Key and Runsten, 1999; Holloway *et al.*, 2000).

In 2008, the Kenya Agricultural Research Institute (KARI) in collaboration with the International Potato Center (CIP), Ministry of Agriculture (MoA) and Deepa Industries Limited piloted contract farming in Kenya through funding from the Common Funds for Commodities (CFC). One objective of this pilot project was to improve market linkages and communication among potato value chain stakeholders particularly potato farmers and processors of crisps. In Kenya, contract farming in potato, especially in Bomet, was motivated by the processor's efforts to obtain a reliable supply of quality tubers for processing (Key and Runsten, 1999). Several factors have been attributed to the emergence of contract farming and these factors seem to vary between developed and developing countries (Escobal and Cavero, 2007). In order to design an effective and efficient marketing arrangement, interventions will involve a strategy to understand these factors.

46.2 Materials and Methods

One hundred and thirty seven (137) potato farmers were randomly selected in Bomet and Molo districts for interview using a structured questionnaire. Stratified random sampling was done to capture relevant variables both from contracted farmers and from non-contracted farmers. The survey focused on transactions and socio-economic attributes influencing individual producer participation in various market channels and the effects of the choice of market on potato productivity. These markets were mainly contracts and the open market. Quantitative results were supplemented with group discussions and key informants from both contract and non-contract groups.

To determine the difference in productivity of contracted farmers compared with non-contracted, production efficiency was estimated using a stochastic frontier production approach pioneered simultaneously by Aigner *et al.* (1977), Meeusen and van den

Broeck (1977), and Battese and Corra (1977). This approach assumes that the stochastic frontier production function contains an error term that consists of two elements: (i) a symmetric component (common to all farmers: disasters both natural and economic); and (ii) a one-sided component (capturing inefficiency).

To determine efficiency, the basic null hypothesis test is that the farmer is fully efficient, H_0: $\gamma = 0$ and that the explanatory variables have no effect, H_0: $\delta_0 = \delta_1 = \delta_2 =, \ldots, \delta_n = 0$ (where γ has a value between 0 and 1 and represents the total variation in output from the frontier attributable to technical inefficiency and δ is the variance). However, the alternative hypothesis is that the farmer is not efficient, H_1: $\gamma \neq 0$.

The test of hypotheses for the parameters of the frontier model is conducted using generalized likelihood ratio (LR) stipulated in Kodde and Palm (1986) statistic defined by:

$$\lambda = -2\ln\left[L(H_0)/L(H_1)\right] \quad (46.1)$$

Where:

$L(H_0)$ is the value of the likelihood function for the frontier model in which parameters restricted are specified by the null hypothesis. $L(H_1)$ is the value of the likelihood function for the general model, in this case ordinary least squares (OLS).

If the null hypothesis is true, then λ has approximately a chi-square (or mixed square) distribution with the degree of freedom equal to the difference between the parameter estimated under H_1 and H_0, respectively. The parameter λ = σ_u/σ_v(LR) characterizes the distribution and it tests the inefficiency component of the model (λ (lambda) is the generalized likelihood ratio and shows the presence of inefficiency; σ (sigma) represents the explanatory variables; v is the error term associated with natural phenomena that affect farmers' production; μ is the error term specific to each farmer and captures inefficiency; thus total variation due to both natural effects and inefficiency can be represented as: $\varepsilon = v + \mu$). A simple regression model (OLS) results if $\lambda = 0$ and the implication is that every potato producer operates on his/her frontier. The LR test is a specification test and provides a superior explanation of relationships present in the data in comparison with the traditional OLS estimation based on the average response function (Greene, 1990).

The basic idea underlying the LR test is that if the potato farmer is efficient, the values of log-likelihood function generated in the unrestricted model (allowing μ_i (the ordinary two-sided error term assumed to be normally, identically and independently distributed with mean zero and constant variance and takes care of all other omitted inputs on potato output by i farmer) to assume random values) and the restricted average model (forcing $\mu_i = 0$), should be equal. That is LR is zero and there is no difference in OLS and maximum likelihood estimate (MLE). Otherwise, inefficiency effects are present whenever the difference exists. Data were subjected to analysis using the stochastic frontier model as implemented in the Statistical Program for Social Scientists (SPSS).

46.3 Results and Discussion

The comparative statistics for contracted and non-contracted farmers are as shown in Table 46.1. The average age of respondent in both contracted and non-contracted was almost the same at 44 and 45 years, respectively. Also their spouses' age averaged at 34 and 35 years, respectively. There was a notable difference in access to training between contracted and non-contracted farmers whereby 96% (contracted) compared with 52% (non-contracted) had been trained.

Also 46% of the contracted farmers received credit compared with 14% non-contracted. Contracted farmers incurred extra cost of KSh7,214 since the procedure of coming up with a contract agreement involved some costs in terms of time and transport to the contract negation meeting mediated by the project partners. In effect, farmers paid an average of up to KSh2,958 ± 1,377 towards transport and meals costs to attend negotiation meetings. The processor on the other hand hired a field officer to monitor and supervise planting schedules, harvesting and delivery of potatoes to the processor. The supervisor monitored the contract farmers, making up to three visits per production cycle. Depending on the frequency and distance, the cost of

Table 46.1. Descriptive statistics comparing contracted and non-contracted farmers.

Descriptive variable	Measure	Non-contracted, $N = 111$	Contracted, $N = 26$
Age of respondent	Years	45.09	44.27
Age of spouse	Years	35.06	34.42
Total number of family members	Number	6.41	7.77
Total farm size	Acreage	5.54	6.62
Years growing potatoes	Years	8.95	11.73
Total cultivated area	Acreage	3.45	3.57
Potatoes storage (seed or ware or both)	Yes = 1 (No = 0)[a]	0.81 (0.19)	0.85 (0.15)
Period that potatoes were stored	Months	0.97	1.19
Received potato training in the last 12 months	Yes = 1 (No = 0)	0.52 (0.48)	0.96 (0.04)
Received credit	Yes = 1 (No = 0)	0.14 (0.86)	0.46 (0.54)
Costs involved in contract negotiation	KSh[b]	12.61	7,214
Years farmer has known buyer	Years	4.70	2.37
Household head	Yes = 1 (No = 0)	0. 86 (0.144)	0. 77 (0. 231)
Own title to the land	Yes = 1 (No = 0)	0.423 (0.58)	0.39 (0.62)
Times potatoes grown per year	Once (twice) [three times]	0.11 (0.58) [0.33]	0. 3.8 (0.73) [0.23]
Member of any market organization	Yes = 1 (No = 0)	0. 33 (0.67)	0.96 (0.38)
Trust in the trader	Yes = 1 (No = 0)	0.57 (0. 42)	0.89 (0.12)

[a]The dummy variable (yes and no) measure gives the proportion of the descriptive statistic.
[b]US$1 = KSh84.

visits by the supervisor averaged at KSh72 ± 218 (exchange rate KSh84 = 1US$).

Although non-contracted farmers had known the buyer for more years (4.7 years) than the contracted farmers (2.3 years), 57% had trust in the buyer compared with 89% of contracted farmers who reported to have trust in the buyer. Both contracted (85%) and non-contracted (81%) farmers reported to store potatoes in almost the same proportions but the duration for storing differed slightly at over 1 month for contracted and under 1 month for non-contracted farmers. This difference is mainly due to varietal difference. Most (91%) of contracted farmers grew the variety Dutch Robyjn which has long dormancy and non-contracted farmers (61%) grew the farmers' variety Shangi which had no dormancy. Most (96%) contracted farmers were members of a marketing organization or group as compared with only 33% of non-contracted farmers. This was because before farmers entered into a contract they were supposed to form groups to facilitate effective monitoring and collection of potatoes. But farmers joined groups for various reasons, namely: (i) financial gains (12.4%); (ii) for advice on social issues (4.4%); (iii) for advice on farming (13.1%); (iv) to share profits (1.5%); (v) for training (7.3%); and (vi) for marketing (6.4%).

46.4 Challenges and Benefits of Potato Contract Farming

Lack of a contract culture and appropriate trading policies made the coordination and communication among parties in contract farming very difficult. There was side selling to intermediate buyers by farmers when open market prices were high and side buying by the processor. The focus on the terms of agreement for the contract was payment regulation, supply regulation, quality and quantity prescription in a given period and penalties on defaulters.

In Bomet, promotion of high quality tubers for processing, through cooperation between processing firms (Deepa, Norda, Chirag and Pioneer), farmers and financial institutions (Equity Bank, Cooperative Bank) was initiated in 2008. The efforts of the partners in co-ordinating contracts between processors and farmers have made initial establishment possible.

The initiative has spurred progress and built the right conditions of trust between firms and farmers despite there being issues that still need to be addressed. Despite all this problems, 62.8% of the farmers in Bomet trusted the trader to whom they were selling potato compared with 36.5% who doubted them. Most of farmers have known the potato buyer for more than 5 years and the majority (74.5%) of them got paid cash for potatoes deliveries. In case of non-payment on delivery, payment was usually made within a fortnight. There was also reduction in processing losses due to delivery of quality tubers (losses due to rotting, browning and peeling were reduced by 0.78%, 0.3% and 0.75%, respectively) compared with pre-project period.

46.5 Technical Efficiency Among Potato Growers

Farmers producing under contract are exposed to a different set up of technology that is not available to non-contracted farmers therefore there are differences in their production efficiencies as shown in Table 46.2.

The estimated total variation from the frontier (sigma squared) $\delta^2 = \delta_u^2 + \delta_v^2$ and variations due to inefficiency (gamma) $\gamma = \delta^2 / \delta_u^2$ terms (refer to Table 46.2) are positive and significantly different from zero at 5% significant level with farmers' data. This is an indication that the observed output significantly differed from the frontier output and factors which are within the control of the farmer are the cause of these variations. It also indicates a good fit and correctness of the specified half-normal distribution assumption of the composite error term. The values of γ are 80.4% and highly significant at 1% levels. This implies that 80.4% variations in the outputs of potatoes were due to inefficiency. The results depict the systemic influences that are unexplained by the production function are the dominant source of errors. It also shows that variations in inputs productivity among the potato farmers are also due to differences in efficiency (Ajibefun *et al.*, 2006).

The tests for presence of technical inefficiency were done using generalized LR

Table 46.2. Socio-economic factors affecting farmers' technical efficiency.

Inefficiency model[a]	Coefficient	Standard error	*t*-ratio[b]
Constant	3.46	0.35	10.02***
Ln(seed)	−0.14	0.11	−5.31***
Ln (fertilizer)	0.03	0.07	0.46
Ln(chemicals)	−0.14	0.10	−1.50
Ln(labour)	0.10	0.08	1.25
Constant	−0.400	0.502	−0.798
Household head (M = 1, F = 2)	−0.601	0.249	−2.420**
Age (years)	0.027	0.010	2.653**
Educational level (years)	−0.026	0.023	−1.136
Family size (number)	−0.086	0.042	−2.034**
Farm size (acres)	−0.004	0.011	−0.375
Training (Yes = 1, No = 0)	−0.312	0.193	−1.622*
Credit (Access = 1, No access = 0)	0.468	0.223	2.101**
Contract (Yes = 1, No = 0)	−3.468	1.631	−2.127**
Sigma-squared δ^2	0.344	0.084	4.085**
Gamma γ	0.804	0.062	13.014**
Log likelihood function (MLE)	−51.904		
log likelihood function (OLS)	−74.768		
LR test of the one-sided error	45.729		

[a]Ln, natural logarithm of the explanatory variables; LR, likelihood ratio; MLE, maximum likelihood estimate; OLS, ordinary least squares.

[b]Values marked with the symbols ***, ** and * are statistically significant at less than 1%, 5% and 10% level of significance, respectively.

test statistics according to Kodde and Palm (1986). The generalized LR test for γ= 16.50. The corresponding critical values at 5% significant level and seven numbers of restrictions (degrees of freedom) were 45.73 for the potato data (Table 46.2). The null hypotheses are rejected by the data and the alternative hypothesis is accepted. The mean technical efficiency was 75.8% for all farmers. The implication is that given the input bundles and farmers' use of existing technologies, they could improve on production levels by 24.2% (overall), 23.7% (for non-contracted farmers) and 26.5% (for contracted farmers). Additionally, potato productivity could be improved through the existing farmers' technology.

46.6 Socio-economic Factors Influencing Potato Production Efficiency

Table 46.2 also shows the socio-economic effects on farmers' production efficiency. Household head, family size, training and contract farming variables had negative effects on production inefficiency. As the majority of the respondents were household heads who managed production, their direct involvement improved production efficiency through prompt decision making on potato production. A higher number of dependents in the farming household had a negative effect on inefficiency contrary to the finding of Parikh *et al.* (1995). This could be because a large household contributes much in terms of labour in potato farming and an urge to operate efficiently to feed more people motivates farmers to produce more per unit of inputs.

The age of the farmer was positively correlated with inefficiency. This is in agreement with the findings of Ajibefun and Daramola (2003), Ogunyika and Ajjbefun (2004) and Alabi *et al.* (2004). Farmers' age affects efficiency in two ways: (i) a young farmer becomes relatively more efficient over time through learning by doing. This relationship levels off and declines as farmers approach retirement age (Tauer, 1995); and (ii) as Wang Hun-Jen (2002), using a quadratics specification model, found as a farmer ages they become less active and less physically able to be efficient. Access to credit had positive effect on inefficiency contrary to the expectation that credit access enables farmers to make timely purchases of inputs. This could mean farmers did not use the credit they reportedly accessed in potato production. This is so because only a few farmers accessed credit and they were mainly farmers producing dairy products and tea.

46.7 Conclusions and Recommendations

Market access remains the most important driving force for potato production to be revolutionized. The project has demonstrated that improved market access and interactions among the value chain players is a prerequisite for improved potato productivity. Contract farming improves productivity and efficiency of ware potato farmers. Farmers get a premium price under contract and thus more income. Contract farmers demanded more farm inputs to boost their investment in potato, given the assured market for their produce. A policy strategy aimed at improving technical efficiency in the short run should emphasize effective and efficient use of technology transfer instruments which enhance the capacity of the farmer to efficiently use the physical inputs. What is needed is promotion of farmer organizations and corporate responsibility among processing companies. This could support fair trade between agribusiness and small farmers, and additionally improve quality and consistency of products under contract farming.

Service providers should facilitate market access through contracting to improve potato production efficiency. Potato farmers need to utilize the available technology to reduce losses of output or alternatively gain from it by minimizing inputs use while maintaining output levels, *ceteris paribus*. The disparity in efficiency levels showed that productivity could be enhanced through bridging production efficiency levels among the farmers.

Acknowledgements

The study was supported by the Common Fund for Commodities (CFC). The International Potato Center (CIP) provided technical support and guidance to project partners during project implementation. We thank Dr Berga Lemaga, the Regional Project Coordinator, whose coordination role was very instrumental in the achievement of project objectives. Thanks are also due to Kenya Agricultural Research Institute (KARI) Tigoni, MoA (Bomet and Molo), Deepa Industries Ltd and the farmers for successfully implementing the project. Finally, we acknowledge African Lead for sponsorship for the African Potato Association (APA) conference.

References

Aigner, D., Knox Lovell, C.A. and Schmidt, P. (1977) Formulation and estimation of stochastic frontier production function models. *Journal of Econometrics* 6, 21–37.

Ajibefun, I.A. and Daramola, A.G. (2003) *Efficiency of Micro-Enterprise in the Nigerian Economy*. African Economic Research Consortium, Nairobi.

Ajibefun, I.A., Daramola, A.G. and Falusi, A.N.O. (2006) Technical efficiency of small scale farmers, an application of the stochastic frontier production function to rural and urban farmers in Ondo State, Nigeria. *International Economic Journal* 20(1), 87–107.

Alabi, R.A., Daramola, A.G. and Ajibefun, I.A. (2004) The relative advantage of agroforestry system over arable crop farming: empirical evidence from cocoa based agroforestry and arable crop farming system in Oyoo State, Nigeria. *Journal for Food, Agriculture and Environment* 2(2), 169–177.

Battese, G.E. and Corra, G.S. (1977) Estimation of a production frontier model: with application to the pastoral zone of Eastern Australia. *Australian Journal of Agricultural Economics* 21, 169–179.

Eaton, C. and Shepherd, A.W. (2001) *Contract Farming: Partnership for Growth*. Agricultural Services Bulletin No. 145. Food and Agriculture Organization of the United Nations, Rome.

Escobal, J. and Cavero, D. (2007) Transaction costs and institutional arrangements in potato marketing by small producers in rural Peru. *Improving Institutions Pro-Poor Growth (IPPG) Discussion Paper Series* No. 12. Available at: www.ippg.org.uk (accessed 29 May 2015).

FAOSTAT (2005) FAO statistical yearbook. World food and agriculture. Available at: http://www.fao.org/economic/ess/ess-publications/ess-yearbook/en/ (accessed 12 July 2012).

Gildermacher, P., Demo, P., Kinyae, P., Wakahiu, M., Nyongesa, M. and Zchocke, T. (2007) *'Select the Best' Positive Selection to Improve Farmers Saved Seed Potatoes*. Trainers Manual. International Potato Center (CIP), Lima.

Government of Kenya (GoK) (2009) *National Potato Taskforce Report, Final Report. Promotion of Private Sector Development in Agriculture Programme*. GoK, Nairobi, 33 pp.

Greene, W.H. (1990) A gamma-distributed stochastic frontier model. *Journal of Econometrics* 46, 141–164.

Holloway, G., Nicholson, C., Delgado, C., Staal, S. and Ehui, S. (2000) Agro-industrialization through institutional innovations: transaction costs, cooperatives and milk market development in the East African highlands. *Agricultural Economics* 23, 279–288.

Kabira, J.N. and Lemaga, B. (2003) *Potato Processing Quality Evaluation Procedures for Research and Food Industry Applications in East and Central Africa*. KARI, Kenya, p. 24.

Key, N. and Runsten, D. (1999) Contract farming, smallholders, and rural development in Latin America: the organization of agro-processing firms and scale of out-grower production. *World Development* 27, 381–401.

Kinyua, Z.M., Smith, J.J., Lungaho, C., Olanya, M. and Priou, S. (2001) On-farm success and challenges of producing bacterial wilt-free tubers in seed plots in Kenya. *African Crop Science Journal* 9(1), 279–285.

Kipkoech, D.N., Mburu, J. and Ritho, C.W. (2008) Production efficiency in farmers' seed enterprises, case of seed potato in Nyandarua and Molo districts in Kenya. *East African Agricultural and Forestry Journal* 74(2), 113–115.

Kodde, D.A. and Palm, F.A. (1986) Wald criteria for jointly testing equality and inequality restriction. *Econometrica* 54, 1243–1248.

Meeusen, W. and van den Broeck, J. (1977) Efficiency estimation from Cobb–Douglas production functions with composed error. *International Economic Review* 18, 435–444.

Nderitu, J.H. (2010) *Monitoring and Evaluation of Implementation of Legal Notice no. 113 of 5th September 2008* (the local government (adoptive by-laws) (agricultural produce) (standard weight of packages) order, 2008). Kenya National Potato Farmers Association, Nairobi.

Ogunyika, O.E. and Ajjbefun, A.I. (2004) Determinants of technical inefficiency on farm production, Tobit analysis approach to the NDE farmers in Ondao state, Nigeria. *International Journal of Agriculture and Biology* 6(2). doi:1560-8530/2004/06-2-355-358

Parikh, A., Ali, F. and Shah, M.K. (1995) Measurement of economic efficiency in Pakistani agriculture. *American Journal of Agricultural Economics* 77, 675–685.

Saenz-Segura, J.F. (2006) Contract farming in Costa Rica: Opportunities for smallholders. Thesis, Wageningen University, Wageningen, The Netherlands.

Scott, G.J., Rosegrant, M.W. and Ringler, C. (2000) Global projections for root and tuber crops to the year 2020. *Food Policy* 25, 561–597.

Shulte-Gildermann, E., Gildemacher, P.R. and Struik, P.C. (2012) Improving seed health and seed performance by positive selection in three Kenyan potato varieties. *American Journal of Potato Research* 89, 429–437.

Tauer, L.W. (1995) Age and farmers' productivity. *Reviews in Agricultural Economics* 17, 63–69.

Wang Hun-Jen (2002) Heteroscedasticity and non-monotonic efficiency effects of a stochastic frontier model. *Journal of Productivity Analysis* 18(3), 241–253.

Williamson, O. (1979) Transaction cost economics: the governance of contractual relations. *Journal of Law and Economics* 22, 236–261.

47 Improvement of Processing Technology Research and Utilization of Sweetpotato and its Derived Foods in China and Rwanda

K. Sindi,[1]* J. Xie,[2] K. Xie[3] and Y. Zhu[2]
[1]International Potato Center Sub-Saharan Africa (CIP-SSA), Nairobi, Kenya; [2]Institute of Agro-products Processing Science and Technology of Sichuan Academy of Agricultural Sciences, Chengdu, PR China; [3]CIP China Center for Asia Pacific (CCCAP), Beijing, PR China

Abstract
In the past decade, several new technologies in sweetpotato processing based on sweetpotato flours and purées rich in provitamin A and anthocyanin have been developed in China and Rwanda. First, this chapter describes these new technologies in use in China, such as accurate gelatinization with microwave treatment, energy-saving drying and highly hygienic packaging for sweetpotato flour and purée. The chapter also provides a step-by-step explanation on how to make products utilizing these technologies to get nutritious and healthy sweetpotato-based foods. Secondly, recent work done in Rwanda is described whereby sweetpotato purée and flour were used to make various products and the associated results from sensory testing are shown. The results demonstrate that Rwandese consumers prefer products made with sweetpotato purée compared with those made using sweetpotato flour and preferred it or rated it equally to using 100% wheat flour. Exposing consumers to additional knowledge concerning vitamin A content in a product did not make a difference in consumer preferences. Overall in Rwanda, consumers preferred orange-fleshed sweetpotato (OFSP)-based products compared with the competing products without any sweetpotato. Utilizing OFSP purée lowers the per unit cost of all the products in the study. The much more extensive use of sweetpotato by industry in China compared with Rwanda is in part driven by the much higher average sweetpotato yields (22 t/ha) found in China compared with Rwanda (8.1 t/ha).

Keywords: beta-carotene, energy-saving drying, flour, microwave, orange-fleshed sweetpotato (OFSP), purée, starch, sweetpotato

47.1 Background

Sweetpotato is one of the most important agroindustrial crops in Sichuan, PR China, with an enormous planting area of 8.7 million ha. Sweetpotato grows in soils of different conditions from relatively poor to fertile soils rich in organic manure. The crop also adapts

*k.sindi@cgiar.org

well to various geographical conditions from coastal lowlands to highlands. During its growth, it requires less than half and one-fifth of the nitrogen and irrigation water, respectively, required for growing rice, but its output (converted into calories) is 50% higher than that of rice and twice as much as that of potato (Liu, 2002). Sweetpotato starch is easily digestible and therefore a useful ingredient in the preparation of good weaning meals. This chapter explores advances in the processing technologies that have been in development over the last few years, contrasting the Chinese and Rwandan experience. In China, sweetpotato use by agroindustry is highly advanced, whereas it is only starting in a few African countries like Rwanda.

In China, the major postharvest processing technologies concerning sweetpotato in the 1990s were to transform the roots into instant noodles, straight vermicelli and spaghetti. The current technology is much more sophisticated and complete sets of equipment have been developed that permit the industry to explore new markets, that previously were just utilizing wheat.

In the past 25 years, the International Potato Center (CIP) and the Institute of Agro-products Processing Science and Technology of Sichuan Academy of Agricultural Sciences have collaborated on processing technology research. Outputs in China include the use of improved medium- to small-scale machines and energy-saving drying technologies that includes microwaving then a double-screw extrusion method for flour and sweetpotato purée. Sweetpotato is processed into refined starch, modified starch, instant noodles, snack foods, whole nutritious flour and other derived foods. The large-scale agroindustry in Sichuan province processes sweetpotato into crude starch, refined starch, instant noodles and nutritious flours. In the last 3 years, sweetpotato flour development has undergone rapid changes due to the development of a new energy-saving drying technology that has been integrated into the production line. This has led to the development of a series of products such as noodles, biscuits and various confectionary products. Due to lowering of the cost the proportion of sweetpotato flour in processed products has gone up to 24%. In Sichuan province in 2013, 57% of sweetpotato processed products were starch and noodles, 24% sweetpotato flour-based foods, 14% other sweetpotato snack foods and 5% foods made with sweetpotato purée or other processed sweetpotato products.

Figure 47.1 summarizes the processing options for sweetpotato and its derived products in China that can be a template for development for other parts of the world where the technology is under development.

47.2 Processing Sweetpotato Starch and its Derived Products

In Sichuan, China, sweetpotato starch is an important ingredient used for food and non-food industries. The starch agroindustry in China has become very successful over the past two decades, with great social and economic benefits. The traditional crude starch processing that was mainly done by farmers on-farm has transformed to being increasingly done by large agroindustries that process refined starch, instant noodles and snack food products. Starch technologies and machines (e.g. the 'sour liquid' technology) and improved machines for washing, smashing, separating and refining centrifuges have been combined into one processing system between the crude starch and the refined starches. These machines are suitable for small-scale to medium-size processors and are affordable to most farmers and medium- to small-scale producers in China. The combined processing guarantees the quality and quantity of sweetpotato root starch produced as per the industry requirements (Fig. 47.2). Sweetpotato starch is widely used to make: (i) instant noodles with a special tolerance/chewing mouth feel; (ii) baked foods; and (iii) noodles that are mixed with grain crops that can be used as table food.

The technologies and equipment can also be used with cassava, canna (*Canna edulis*)

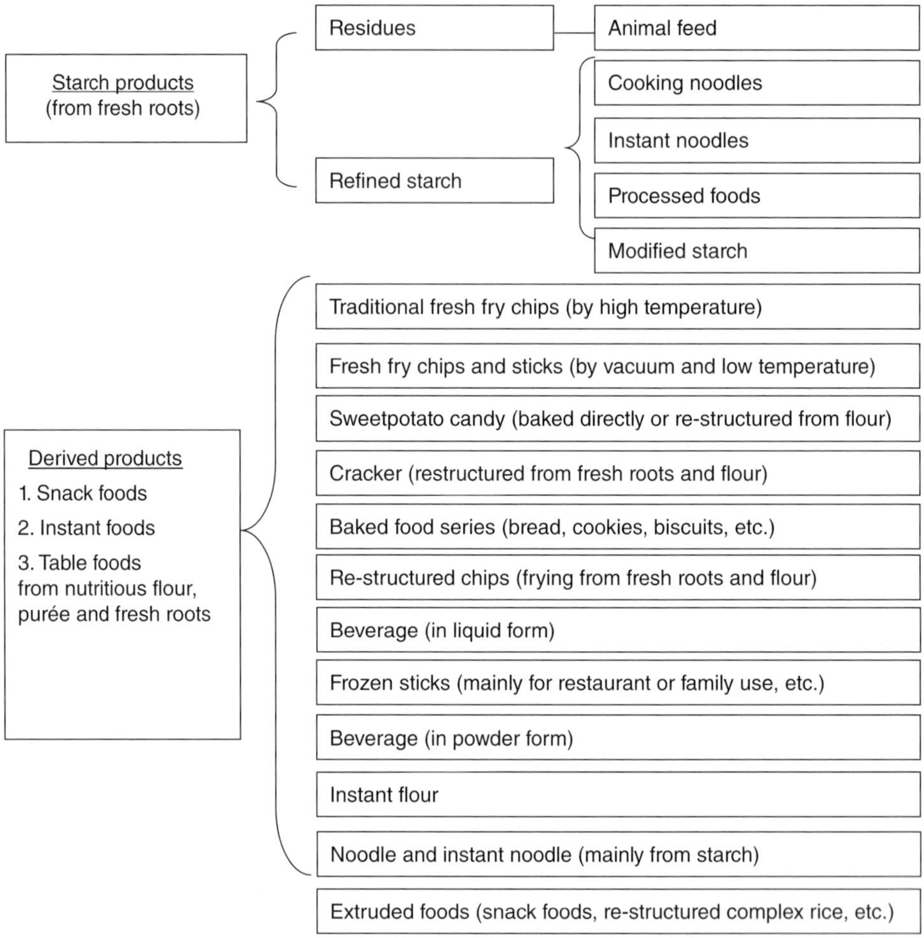

Fig. 47.1. Sweetpotato processing technology in China and its derived products.

and other root and tuber crops. Processed instant noodles utilizing potato, sweetpotato, wheat, maize (corn), buckwheat, oat, bean, broomcorn and black rice use a special double extruder technology under high pressure and high temperature treatment. This process enables nutrient preservation as the original raw material is transformed. A new vacuum package technology is currently being used to process wet purple sweetpotato noodles made from sweetpotato flour, purée or fresh roots. This type of packing makes the noodles convenient to carry, store and cook (Fig. 47.3). By having improved ingredients and accurate sterilization, the shelf life of these products has been extended to over a year.

47.3 Sweetpotato Flour Research and Production in Sichuan, China

Sweetpotato flour is a new product that is developing very fast because of its nutritional benefits. The technology involves microwave treatment for gelatinization (Fig. 47.4) followed by either drying using a hot air house (process one) or drying using a machine

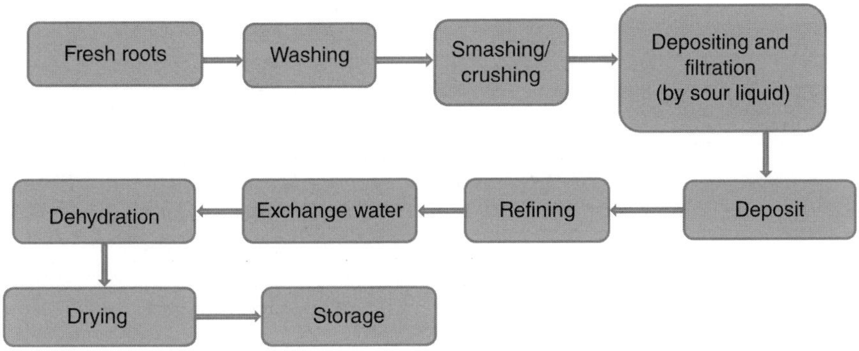

Fig. 47.2. Flow diagram of the improved method of processing sweetpotato starch.

Fig. 47.3. Vacuum package of wet purple sweetpotato noodles.

(Fig. 47.5) that generates a current of hot air (process two) that is energy efficient. The process chosen out of the two methods depends on the scale of production and the usage of the flour in the food industry.

Drying process one – hot air house

This processing technology is based on hot air house drying (Fig. 47.6) with energy saving technology and is suitable for medium- and small-scale use. The flour generated is sold for home use.

Drying process two – hot current drying machine

This process is based on a drying machine that generates a current of hot air (Fig. 47.7). The quality of flour is better than that produced by drying process number one, and is suitable for medium- and large-scale use.

In the two processing methods described above, a microwave can be used instead of steam and hot water in the heat treatment stage. By doing so, production costs can be reduced by 30% because the water content of sweetpotato products is 15–20% lower than that from the hot water and steam processing method. In addition, nutrient retention is higher. The derived sweetpotato flour can be used to manufacture many products such as: (i) baked foods (e.g. baked bread, biscuits, cookies and candy); (ii) instant sweetpotato flour; (iii) combinations of sweetpotato and wheat flour for steamed bread; (iv) beverages from reconstituted flour; (v) noodles; (vi) ice creams; and (vii) crackers.

It is possible to process sweetpotato purée and flour simultaneously. Utilizing sweetpotato purée to make products lowers the cost of production and produces better quality products. The purée can be treated and packaged using: (i) plastic pouches (utilizing canning technology); or (ii) by freezing and packing hygienically in reusable containers such as stainless steel tanks that can be stored for more than 1 month under normal storage conditions.

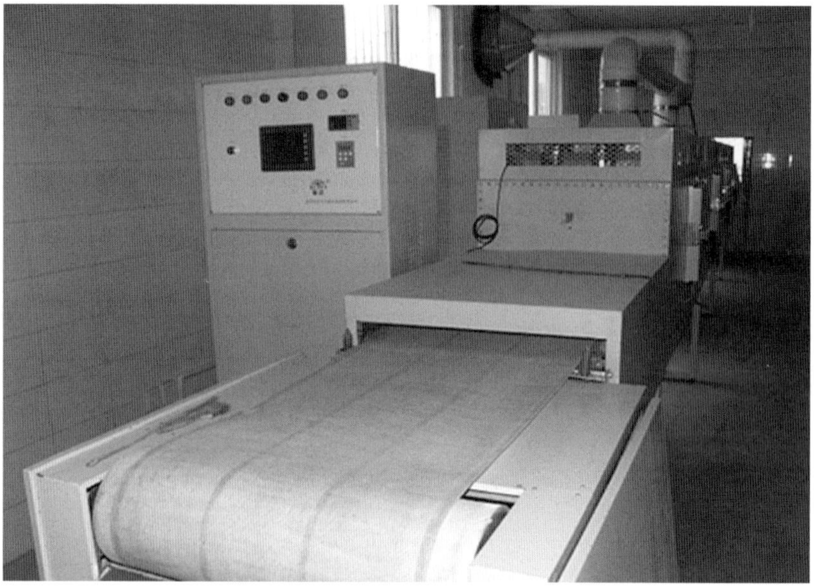

Fig. 47.4. Microwave treatment machine used in flour production.

Fig. 47.5. Hot air current drying machine used in sweetpotato flour production.

47.4 Testing Sweetpotato Flour and Purée-based Products in Rwanda

Sweetpotato bakery products were unknown in Rwanda prior to the launching of the Super Foods project in 2011. As a first step, assessing consumers' perceptions of common bakery products using a mix of wheat flour and either sweetpotato purée or sweetpotato flour, and others that are made purely

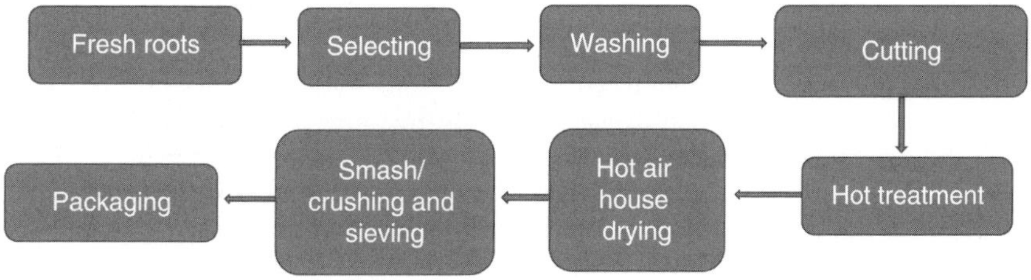

Fig. 47.6. Schematic diagram of the drying process for flour using a hot air house.

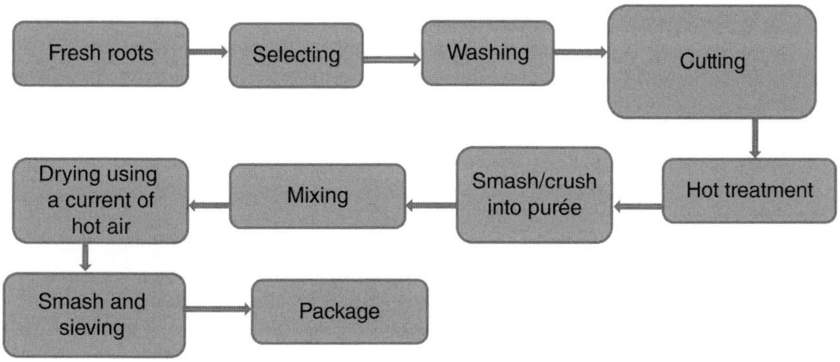

Fig. 47.7. Schematic diagram of the drying process using a hot current drying machine.

(100%) from wheat flour was key. Orange-fleshed sweetpotato (OFSP) varieties were used to replace a percentage of wheat flour in various bakery products because they are rich in beta-carotene (provitamin A). If products derived from OFSP supply a significant amount of vitamin A, this could prove to be a cost-effective way of delivering needed nutrition to the urban population.

For OFSP to be incorporated in the bakery products, processors have to be sure that consumers will either prefer the new products or be indifferent between them and the regular ones made purely from wheat flour. It is also important to ensure that sweetpotato–wheat-based products have appropriate characteristics of appearance, aroma, taste and texture, which are key determinants of consumers' sensory acceptability of bakery products. Other sensory tests of baked products have found that replacing some basic components changes the structure and volume of the crumb (Gujral and Rosell, 2004), and also the structure and texture (Mezaize *et al.*, 2009). This study conducted sensory testing of industrial wheat–sweetpotato blend processed products that have been formulated for markets in Rwanda. This aspect of the project is critical in ensuring that sufficient demand exists to help pull sweetpotato along the value chain and into the market place.

Apart from the purchase aspects, this study also examined the various attributes likely to influence demand patterns. Indeed, Lancaster (1966) and many later studies (Lusk and Hudson, 2004; Lusk and Briggeman, 2009; Takatsuka *et al.*, 2009; Lippe *et al.*, 2010) indicate that consumers usually demand the attributes that are embodied in the product rather than the product per se. They posit that consumers derive satisfaction from the attributes or characteristics of the good. This study, therefore, examined the consumers' perceptions of the size, shape, colour, smell, hardness/crunchiness of the Akarabo Golden Power Biscuit (GPB) against those of the most popular biscuit. The information was collected through both open and blind testing.

Methodology

For the comparative product study we used roots of the OFSP variety Cacearpedo to make purée and flour. In preparing the flour, roots were harvested, washed and the skin removed using a kitchen knife. The roots were cut into very thin pieces to facilitate quick drying. The cut pieces were then soaked in water containing sodium meta-bisulfite, a disinfectant and antioxidant to prevent oxidation of the OFSP and minimize the loss of beta-carotene. The pieces were dried in a curing room using firewood. The dried chips were then milled into flour, which was used by a commercial baker in combination with wheat flour to make the processed products. To make the purée, Cacearpedo roots were harvested, washed and then boiled. After boiling, the skin was removed and purée prepared using an electric mixer. The processor prepared the products following a protocol with scientists from the Postharvest Unit of the Rwanda Agricultural Board. All processed products were distributed to different outlets for sale. We report the testing that was conducted with students at an urban girls' school and at the university in Kigali, Rwanda. These 86 taste testers were selected because they represent key target groups for the proposed products to be marketed. We then tested the mean differences of the scores given for the products using pairwise Student's *t*-tests.

The second study used data collected from a total of 1085 consumers in November 2012. This was testing the consumer acceptability of the GPB 1 week after it was officially launched by the agro-processing company. Consumers were selected randomly in different main markets of Rwanda. The markets represented different socio-economic groups: (i) low-income markets (namely, Musanze, Muhanga, Kimironko/Kigali); (ii) low-/middle-income markets (namely, SINA Nyirangarama, SINA Nyabugogo (Kigali), SINA Kigali town, SINA Musanze, SINA Muhanga); (iii) high-/middle-end markets (represented by Ndoli Supermarket in Kigali); and (iv) high-end markets (represented by Union Trade Centre market in Kigali).

In each market, data were collected through personal interviews by trained enumerators using pre-designed, pre-tested questionnaires. Data collected included: (i) personal characteristics (e.g. age); (ii) quantity of biscuits purchased; (iii) attribute perceptions; and (iv) willingness to pay. To compare consumer perceptions about the various attributes, consumers were further stratified into two categories, namely blind and open. In the blind category, consumers were asked to taste, feel and observe the unlabelled samples of Akarabo GPB and a very popular known biscuit on the local market and state their perception about the specific attributes including taste, colour, hardness, sweetness (sugar content) and packaging. The experiment was then repeated for consumers using labelled samples of the Akarabo GPB and the ordinary biscuit. Both categories of consumers were also asked to state how much they would be willing to pay for Akarabo GPB and the ordinary biscuit.

Results

In this section, the results on overall acceptability from the consumer tests for the various bakery products are presented. The first analysis compared the mean score for the overall acceptability among the three samples of each of the products in the first consumer test.

Descriptive results

Figure 47.8 shows the mean acceptability of bread made with different combinations of wheat flour and sweetpotato flour or purée. Results show that bread made with a mix of 80% wheat flour and 20% sweetpotato purée was preferred to the other two types of bread (the 100% wheat flour bread and the bread made of 80% wheat flour and 20% sweetpotato flour). Differences in means were all statistically significant. Therefore, the panellists clearly preferred bread made with 80% wheat and 20% sweetpotato purée over the other two types of bread. The bread made with a mixture of 80% wheat flour and 20% sweetpotato flour was clearly the least preferred.

The consumer acceptability of cupcakes is as shown in Fig. 47.9. Mean acceptability of cupcakes made out of 100% wheat appeared to be slightly higher than that of cakes made with 60% wheat flour and 40% sweetpotato purée, but there is no significant difference in acceptability between the two. However, testers clearly preferred these two types of cupcakes compared with those made with 60% wheat flour and 40% sweetpotato flour.

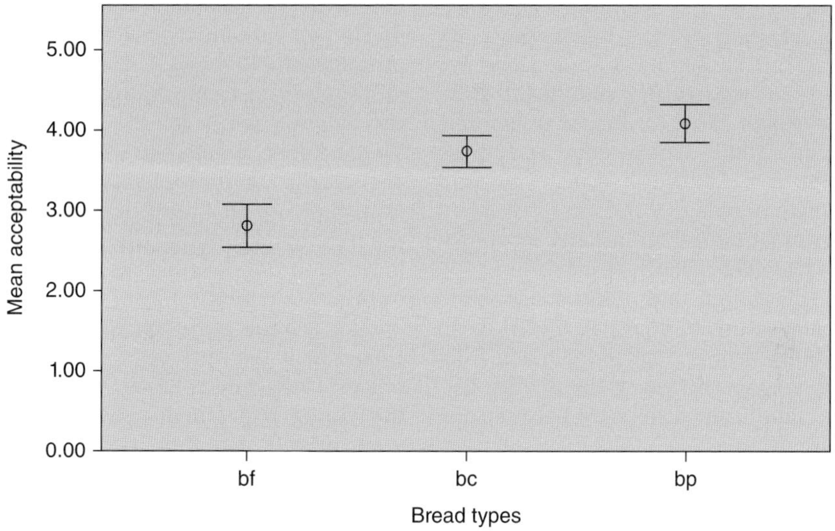

Fig. 47.8. Consumer assessment of normal 100% wheat bread and bread made with a mixture of wheat and sweetpotato flour or purée. bf, bread made with 80% wheat flour, 20% sweetpotato flour; bc, bread made with 100% wheat flour; bp, bread made with 80% wheat flour and 20% sweetpotato purée. Error bars indicate 95% confidence interval (CI).

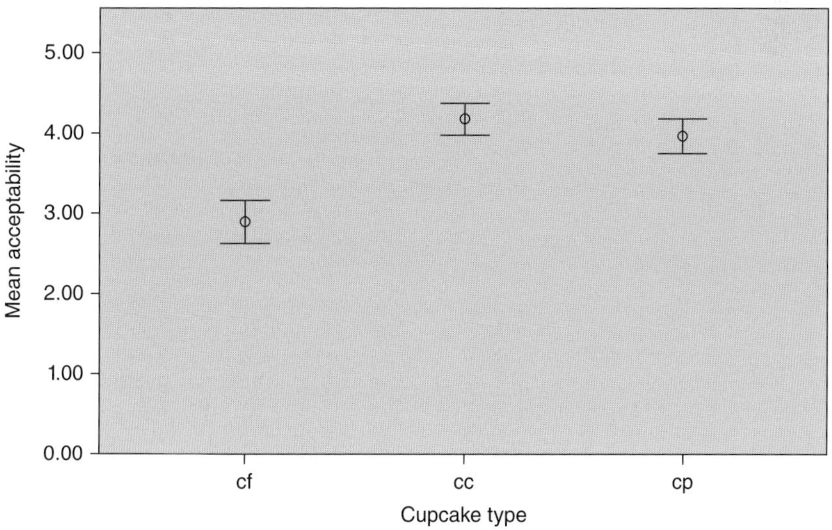

Fig. 47.9. Consumer assessment of normal cupcakes and cupcakes made with a mixture of wheat and sweetpotato flour or purée. cf, cupcakes made with 60% wheat flour and 40% sweetpotato flour; cc, cupcakes made with 100% wheat flour; cp, cupcakes made with 60% wheat flour and 40% sweetpotato purée. Error bars indicate 95% CI.

The ratings for doughnuts are presented in Fig. 47.10. The panellists preferred doughnuts made with 100% wheat flour compared with those doughnuts made with a mix of wheat and sweetpotato flour. Also, there was greater preference for doughnuts made with a mix of wheat flour and sweetpotato purée than for those made with a mix of wheat and sweetpotato flour. However, there was no statistically significant difference in acceptability scores for doughnuts made with 100% wheat flour and those made out of a mixture of wheat flour and sweetpotato purée.

Consumer ratings for biscuits are shown in Fig. 47.11. The biscuits made from a mixture of wheat flour and either sweetpotato purée or sweetpotato flour had higher mean acceptability scores than those made from wheat flour only. However, there is no statistically significant difference between the three biscuit types.

Econometric results

The preceding descriptive results indicate that in some cases, panellists preferred sweetpotato-based products, particularly the purée compared with the control products (100% wheat flour). To determine what the panellists considered as the most important product characteristics that informed the overall acceptability ordered logit regressions were run on each of the purée products with the acceptability score being the dependent variable in each case. Results are presented in Table 47.1. The only statistically significant factor in bread acceptability was flavour, while for doughnuts none of the product characteristics was found to be statistically significant (at 5% or less). However, increase in age of panellists lowered the likelihood of liking the doughnuts. The location of the panellist was important and in this case panellists from Lycee Girls secondary school were compared with those from Kigali Institute of Science and Technology, University of Kigali. Lycee represents the young and female part of the demographic and that increased acceptability of doughnuts with sweetpotato purée. For cupcakes, colour, structure and flavour of the product were significant factors that influenced overall acceptability. In the case of biscuits, flavour was the only significant product characteristic.

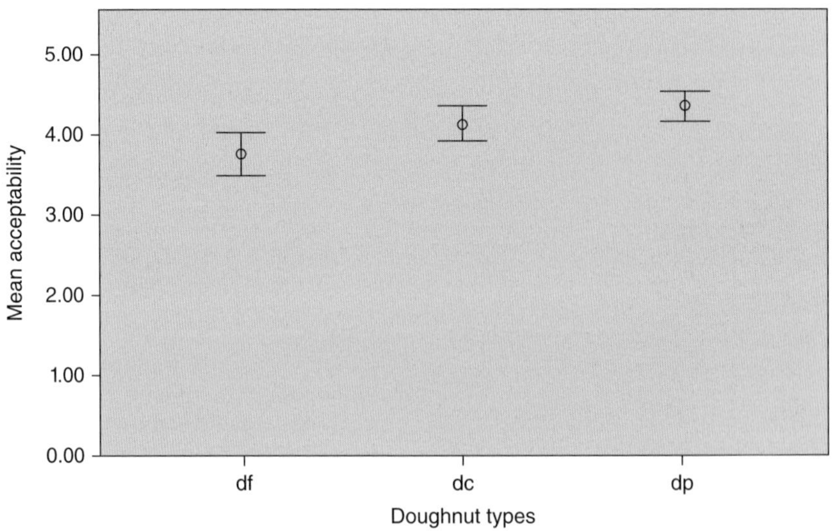

Fig. 47.10. Consumer assessment of normal doughnuts and two types of doughnuts made with a mixture of wheat and sweetpotato flour or purée. df, doughnuts made with 60% wheat flour and 40% sweetpotato flour; dc, doughnuts made with 100% wheat flour; dp, doughnuts made with 60% wheat flour and 40% sweetpotato purée. Error bars indicate 95% CI.

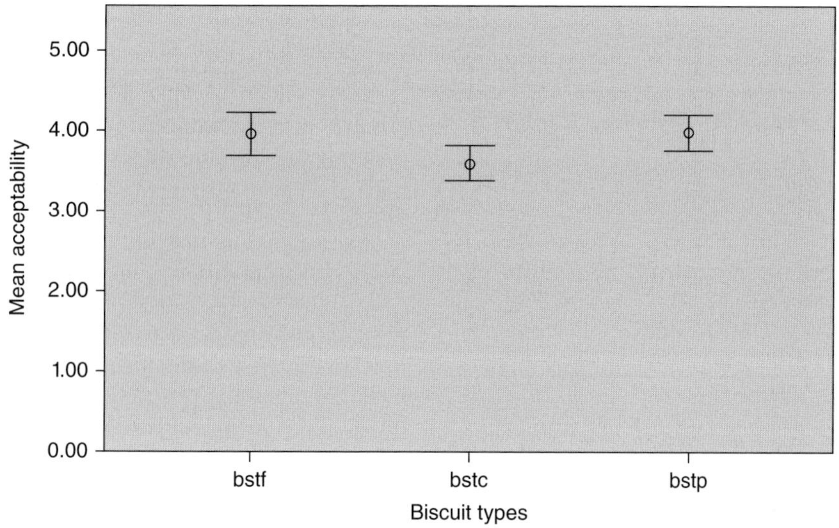

Fig. 47.11. Consumer assessment of normal biscuits and two types of biscuits made with a mixture of wheat and sweetpotato flour or purée. bstf, biscuits made with 60% wheat flour and 40% sweetpotato flour; bstc, biscuits made with 100% wheat flour; bstp, biscuits made with 60% wheat flour and 40% sweetpotato purée. Error bars indicate 95% CI.

Table 47.1. Results of ordered logit analysis of factors affecting consumer acceptability of purée products (absolute value of *z* statistics in parentheses).[a]

	Bread	Doughnuts	Cakes	Biscuits
Colour	0.062 (0.15)	–0.182 (0.69)	0.886 (2.88)***	0.539 (1.55)
Shape	0.002 (1.11)	–	–0.002 (1.28)	0.075 (0.28)
Structure	0.204 (0.58)	–0.001 (0.94)	0.695 (2.30)**	0.001 (0.38)
Flavour	2.772 (5.85)***	0.191 (0.72)	1.082 (3.33)***	0.933 (2.86)***
Texture	–0.001 (0.34)	–0.001 (0.39)	0.005 (0.12)	–0.007 (0.79)
Age	0.001 (0.84)	–0.001 (1.86)*	0.001 (1.18)	–0.001 (1.44)
Lycee Girls School (= 1)	0.961 (1.69)	0.907 (1.87)*	–0.740 (1.51)	–0.441 (0.85)
Tenderness	–	–	–0.002 (1.24)	–
Number of observations	84	85	86	74

[a]Absolute value of *z* statistics in parentheses. *, Significant at 10%; **, significant at 5%; ***, significant at 1%.

Cost analysis of substituting sweetpotato purée for wheat flour

An analysis of cost savings associated with substitution of varying percentages of wheat flour with sweetpotato purée in four different commercially produced and marketed bakery products was conducted. Results showed significant cost savings per unit of product (Table 47.2). This suggests that sweetpotato purée can be a good substitute for wheat flour in the analysed products. Such a substitution confers a number of benefits, namely reduced production costs and addition of more vitamin A into the product. Given that sweetpotato flour is more expensive than wheat flour in Rwanda, substitution with sweetpotato flour does not make economic sense.

Demand for attributes

Following the first study, the private sector agro-processor proceeded to make all four

Table 47.2. Costs savings associated with substitution of varying percentages of wheat flour with sweetpotato purée.

Product	Percentage of wheat flour replaced with sweetpotato purée	Costs saved per unit of product (%)
Doughnuts	30	14
Queen cakes	40	10
Biscuits	39	9
Bread	34	20

products using sweetpotato purée. However, improved packaging was developed for the Akarabo GPB and it was officially launched at a public event in November 2012. This section presents the results from consumer testing of the launched Akarabo biscuits against the established brands in the market.

From the open testing several observations emerge. First, more consumers rated the Akarabo GPB more favourably in most of the non-price attributes when compared with the other dominant biscuit in the market. Secondly, more male than female consumers, in absolute terms, indicated that Akarabo GPB had better looks, colour, taste, hardness and packaging than the dominant competing brand and more sugar. Thirdly, hardness is one striking difference between Akarabo GPB and the competing dominant brand. A significantly higher number of consumers found the dominant competing brand of biscuits harder than the Akarabo GPB.

The results of the blind test showed no differences in the way the male and female consumers perceive the attributes of Akarabo GPB from the perceptions of the open-tested respondents reported above. As in the case of open testing, more male and female respondents found other brands of biscuits harder than Akarabo GPB.

In all the considered attributes, other than hardness of the biscuit, female farmers in the lower income category reported that Akarabo GPB was better than the other brands. They also reported that Akarabo GPB was more sugary (hence sweeter) than the dominant competing brand.

This study also assessed the effect of providing the consumer with information concerning the superior qualities of Akarabo biscuits due to the use of biofortified OFSP in their production. Some of the study respondents were informed about the presence of vitamin A in the Akarabo GPB while others were not. A comparison of the consumer responses regarding the non-price attributes between the two consumer categories (those with and without vitamin A information) shows surprising results. There is not much difference in the number of respondents that rate the Akarabo GPB as being better between those with information about vitamin A content of the OFSP-based biscuit and those that don't. Indeed, the results of a Student's *t*-test of differences in means reveal no statistically significant difference between consumers with and those without vitamin A information. However, slightly more consumers with information about vitamin A indicated that Akarabo GPB has better looks than the other biscuits. Just as in the earlier case of the analysis of preferences, more consumers (both with and without vitamin A information) found the dominant competing biscuit brand harder than the Akarabo GPB.

In the analysis of the consumer ranking of the taste of Akarabo GPB, 79% ($N = 574$) of the consumers rated it as either good or very good with more male consumers rating the GPB biscuit as 'good'. Slightly more female consumers rated Akarabo GPB as tasting 'very good'. Analysis by knowledge of information regarding vitamin A also revealed that most consumers (79%, $N = 580$) rated Akarabo GPB as 'good' or 'very good'. While more consumers with information about vitamin A content in Akarabo GPB rated it as 'good', the converse is true for the 'average' and 'very good' ratings. Overall, the Akarabo GPB is rated very favourably by consumers in general across all market-type categories.

Factors contributing to the rating of the Akarabo biscuits

The results from the ordered logit show that being female increased the odds ratio of choosing the highest rating of the Akarabo biscuits by 0.4 (Table 47.3). When comparing

Table 47.3. Predictions of the log odds of reporting the highest rating of Akarabo biscuits (with 5 being highest and 1 the lowest rating compared with the main competing biscuits brand).

Category	Predictors	Proportional odds ratios[a]
Individual characteristics		
	Gender (being female)	0.40 (3.35)**
	Age	–0.01
Competing brand		
	Riham	0.76 (2.59)**
Purchasing characteristics		
	Frequency of buying biscuits	0.02 (0.35)
	Cost of the competing brand per unit (packet)	–0.00 (0.27)
	Number of packets bought at once	–0.01
Product characteristics		
	Smell	–0.51 (3.67)**
	Taste	–0.64 (4.08)**
	How sugary the biscuit tastes	0.22 (1.78)
	Hardness	0.10 (0.64)
Prior knowledge of vitamin A		
	Vitamin A knowledge	0.11 (0.93)
Informed about vitamin A		
	No vitamin A information	0.60 (4.09)**
	Vitamin A information	0.68 (4.66)**
Market class location		
	High end	0.54 (2.14)*
	High end to mid-end	1.01 (4.52)**
	Middle end	0.71 (3.25)**
N		1085

[a]*, Significant at $P < 0.05$; **, significant at $P < 0.01$.

the Akarabo biscuits with the competing low-end Riham biscuit, the odds ratio of selecting the Akarabo GPB increases by 0.76. However, the purchasing pattern characteristics of the consumer do not seem to impact the odds ratios of the Akarabo biscuits. The smell and the tastes of the biscuit seem to reduce the odd ratios of rating the Akarabo biscuits highly.

Some members of the panel were given more knowledge about the importance of vitamin A to human health. However, irrespective of whether the individual was informed about vitamin A and told the biscuit contains vitamin A, both groups had an increased odds ratio of rating the Akarabo highly compared with the individuals who had a blind test with no information. High-end to mid-end market customers had an increased odds ratio of rating the Akarabo highly compared with the low-end market customers.

47.5 Discussion and Conclusions

The chapter has shown that in China considerable strides have been taken in sweetpotato processing and the accompanying development of technology. Products made out of the various varieties of sweetpotato are also marketed successfully in China. Technologies improving the efficiency and quality of sweetpotato flours are particularly of note. From the work done in Rwanda using OFSP it has been demonstrated that consumers will accept sweetpotato-based products. Further, it has been shown that products processed utilizing OFSP purée make products that consumers like better than those that are made utilizing OFSP flour.

China's ability to produce a broad range of food products using sweetpotato, from purée to noodles, is due to the high mean productivity of sweetpotato – 22 t/ha in 2011 (FAOSTAT, 2011). In comparison in Rwanda, average yields are much lower (estimated to be 8.1 t/ha), making sweetpotato flour uneconomic at this point in time.

Fortunately, purée is economically competitive and, moreover, produces bakery products that are preferred by the consumer to those made with sweetpotato flour and in some cases (bread, queen cakes), those made with 100% wheat flour. Until sweetpotato productivity is raised significantly in sub- Saharan Africa (SSA), substitution with OFSP purée not flour should be promoted. Fortunately, China has also made considerable advances in processing, packaging and storage technologies for sweetpotato purée. Thus, the potential for joint research and further technological exchange between China and selected countries in SSA is quite promising as sweetpotato product development is poised for commercial take-off in several sweetpotato-producing countries in SSA.

References

FAOSTAT (2011) Available at: http://faostat.fao.org/ (accessed 1 May 2013).

Gujral, H. and Rosell, C. (2004) Improvement of the breadmaking quality of rice flour by glucose oxidase. *Food Research Institute* 37(1), 75–81.

Lancaster, K.A. (1966) New approach to consumer theory. *Journal of Political Economy* 74, 134–157.

Lippe, R.S., Mergenthaler, M. and Isvilanonda, S. (2010) Consumer Willingness to Pay for Pesticide Safe Produce: the Case of Cabbage and Yellow Mango in Thailand. Available at: www.globalresearch.com/main/.../PAPER_238_ConsumerWillingness.pdf (accessed 24 April 2013).

Liu, H. (2002) *Post-Harvest Processing Technology of Sweetpotato in Sichuan, China*. Sichuan Research and Design Institute of Agricultural Machinery, Chengdu, Sichuan, China. Available at: http://unapcaem.org/Activities%20Files/A20/4%20China.pdf (accessed 24 July 2014).

Lusk, J.L. and Briggeman, B. (2009) Food values. *American Journal of Agricultural Economics* 91(1), 184–196.

Lusk, J.L. and Hudson, D. (2004) Willingness-to-pay estimates and their relevance to agribusiness decision making. *Review of Agricultural Economics* 26(2), 152–169.

Mezaize, S., Chevallier, S., Le Bail, A. and De Lamballerie, M. (2009) Optimization of gluten-free formulations for French-style breads. *Journal of Food Science* 74(3), 140–146.

Takatsuka, Y., Cullen, R., Wilson, M. and Wratten, S. (2009) Using stated preference techniques to value four key ecosystem services on New Zealand arable land. *International Journal of Agricultural Sustainability* 7(4), 279–291.

48 Building a Sustainable Sweetpotato Value Chain: Experience from the Rwanda Sweetpotato Super Foods Project

J. Ndirigwe,[1*] K. Sindi,[4] J. Low,[2] D. Shumbusha,[3] J.B. Shingiro,[3] J.C. Nshimiyimana,[4] S.Hakizimana[5] and A. Angsten[6]

[1]Rwanda Agriculture Board (RAB), Kigali, Rwanda; [2]International Potato Center Sub-Saharan Africa (CIP-SSA), Nairobi, Kenya; [3]RAB, Huye, Rwanda; [4]International Potato Center (CIP), Kigali, Rwanda; [5]Catholic Relief Services, Kigali, Rwanda; [6]Johns Hopkins University, Baltimore, USA

Abstract

Sweetpotato is widely grown in almost all agroecological zones of Rwanda, where it is prized by most resource-poor farmers as a reliable, low-input, food security crop but with limited commercial potential. The bulkiness, lack of processing technologies and lack of market at the peak of sweetpotato production are reported as major constraints by producers and policy makers for scaling up production. A strategy to promote and make available to farmers disease-free planting material of two elite selected orange-fleshed sweetpotato varieties (Gihingamukungu and Cacearpedo) that yield roots of acceptable consumer quality was initiated and has been adopted by organized farmers' groups in three districts. Processing sweetpotato into products offers the opportunity to increase demand for the crop and create added value. Since 2010, the Rwanda Sweetpotato Super Foods action research project, a collaboration between different stakeholders has been investigating how to connect smallholder sweetpotato producers to agro-processors through effective value chains. Tasty and acceptable sweetpotato products have been developed, processed and promoted among different rural, semi-urban and urban consumers. Farmers in different groups and individual contracted farmers are producing and selling fresh roots. Akarabo Golden Power Biscuits and Mandazi doughnuts are two sweetpotato-based processed products profitably produced and marketed by Urubwitso (SINA) Enterprises.

Keywords: orange-fleshed sweetpotato (OFSP), propagation, Rwanda, super foods, sweetpotato, value addition

48.1 Introduction

The sweetpotato (*Ipomea batatas*) is an important source of calories in the cool humid climates of the highlands of the western branch of the Rift Valley of Africa, of which Rwanda is typical (Tardif, 1991; FAO, 2008). Since its introduction in the early 18th century, the sweetpotato has become one of the most important staple crops to be cultivated

*ndrick3@gmail.com

by nearly all peasant farm households in this densely populated, high elevation 'thousand hills' country (Ferris *et al.*, 2002; Ndirigwe, 2006). It is regarded as a flexible source of food as it can be grown on soils of limited fertility and is relatively drought tolerant. Planting and harvest periods are more flexible than those of maize and other cereals. Sweetpotato is mostly consumed fresh after boiling and also as a snack (Tardif, 1991; Ndirigwe, 2006). In caloric terms, the Rwandan diet consists of beans (22.3%), sweetpotatoes (21.6%), manioc (14.2%), bananas (14.1%), Irish potato (11.9%) and maize (8.6%). Secondary caloric sources include sorghum, peas, yams, rice and vegetables, where the consumption varies by region and season (PASAR, 2002; MINAGRI, 2010).

Sweetpotato is bulky and perishable and the production areas capable of generating surpluses tend to be relatively localized but dispersed, contributing to inadequate market access and distinct price variations. This important food security crop in Rwanda is grown mainly by women for consumption and family income. Per capita sweetpotato production in Rwanda is among the highest in sub-Saharan Africa – over 80 kg per capita – with white-fleshed varieties, containing negligible amounts of beta-carotene, being the most consumed (Tumwegamire *et al.*, 2004). Orange-fleshed sweetpotato (OFSP) varieties were recently introduced in the country to combat vitamin A deficiency (VAD). Children below 5 years of age, children 7–10 years of age and adults need to eat about 1000 g, 140 g and 250 g, respectively, of the vitamin A-rich OFSP to meet the recommended daily intakes. OFSP varieties are believed to represent the least expensive, year-round source of dietary vitamin A to poor families in the SSA region (Kimura *et al.*, 2007). It is estimated that 39% of the children under 5 years of age in Rwanda suffer from VAD (WHO, 2009).

The popularity of sweetpotato in the countryside, among semi-subsistence farmers, does not extend to the rapidly growing urban population (MINECOFIN, 2002). Thiele *et al.* (2009) reported that the dispersed and seasonal nature of sweetpotato production, high costs of marketing, lack of processing opportunities, and its limited consumption in cities lead to low volume urban markets. Similarities have been observed in Rwanda where low prices occur during peak production periods, resulting in many households being reluctant to produce a significant surplus to be sold when there is no guaranteed market (Njeru *et al.*, 2007). Farmers in Rwanda complain of lack of sufficient profitable markets during peak harvest times.

Rwanda produces a significant amount of its own wheat but because conditions are not very suitable for wheat production cost of production is high. Rwanda is a landlocked country and additional wheat is imported through the port in Mombasa, Kenya and then is transported through both Kenya and Uganda to mills and bakeries in Rwanda, most of which are concentrated in the capital, Kigali. Although there is an expanding demand for wheat-based products in urban areas, wheat flour is relatively expensive and cheaper ingredients that could be substituted for wheat flour offer the potential of significant cost savings. Some idea of the large potential savings is that the producer price of wheat at US$195/metric t was about six times as high as that for sweetpotatoes at US$34/million t in 2003 (FAO, 2008).

Given the ability to produce OFSP roots year round in Rwanda, it has been logical to test using sweetpotato as an ingredient for food processing with the potential for increased income generation. Research done by the Rwandan Institute for Agricultural Sciences (ISAR, 2006) demonstrated the technical feasibility of substituting sweetpotato flour for imported wheat flour at 20% for bread and 40% for other bakery products. However, between 2006 and 2010, the willingness of processors to adopt the technology was still limited. Currently, there are some farmer's groups and cooperatives interested in the promotion of white sweetpotato and OFSP by substituting 20% of wheat flour with sweetpotato either as flour or as mashed, fresh sweetpotato in biscuits and potentially in bread.

In addition, Rwanda citizens have little knowledge concerning sweetpotato processing or the potential for year-round use of sweetpotato in commercial products such as chips (i.e. French fries), flour, breads and cakes. Urban households consume about less than half as much sweetpotato as their rural counterparts (Tardif, 1991; Njeru *et al.*, 2007). A principal reason for this difference is related to marketing problems. Sweetpotato production is faced with several production and marketing constraints, which are limiting its benefits to producers and consumers alike. The sector is still largely farmer-driven from a food security perspective and not consumer-driven with a commercial perspective, which is linked to the inadequate and unsustainable system of multiplication and distribution of planting material. As of 2010, the Rwandan sweetpotato market had not developed the linkages and economies necessary for commercializing the crop.

This proof-of-concept study was aimed at addressing this challenge by improving the sweetpotato value chain by including all actors along the chain from growers to processors using Cacearpedo and Gihinga-mukungu, two high-yielding OFSP varieties. These varieties have a comparative advantage over other common staple foods, such as cereals and legumes that provide zero to minor traces of beta-carotene (Bovell-Benjamin, 2007). Processing sweetpotato into products offers the opportunity to increase demand for the crop and create value addition. The Rwanda Super Foods action research project sought to explore how to effectively link smallholder sweetpotato producers to agro-processors and build gender-equitable value chains.

48.2 Materials and Methods

Site selection

The study was implemented in three districts in Rwanda: Rulindo, Muhanga and Kamonyi, with estimated populations of 307,501, 349,094 and 265,365 inhabitants, respectively. These districts are major sweetpotato production areas with high population densities. The average landholding size per household is only 0.7 ha. Processing of sweetpotato products offers the opportunity to increase demand for the crop, create value addition and thereby expand the incomes of smallholder producers.

Key stakeholders' involvement and value chains development

Key stakeholders (the research side of the government's Rwanda Agriculture Board (RAB), the non-governmental organization (NGO) Catholic Relief Services (CRS), the private-sector agro-processor Urubwitso (SINA) Enterprises and local authorities at district level) were consulted and agreed on project objectives, responsibilities and gained consensus on the benefits of strengthening links between all partners. RAB selected two new OFSP varieties (Cacearpedo and Gihinga-mukungu) based on beta-carotene content and consumer preference (e.g. high dry matter content and skin and flesh colour). The sweetpotato research programme of RAB is providing high quality OFSP of disease-free ('clean') selected OFSP varieties which were rapidly multiplied in tissue culture in collaboration with the Kenya Plant Health Inspectorate Service (KEPHIS) to ensure adequate quantities of clean planting material for secondary multiplication sites in collaboration with CRS and Urubwitso (SINA) Enterprises. Clean planting material was then distributed to trained secondary vine multipliers (farmers) at the village level through CRS. Urubwitso (SINA) Enterprises, a large-scale juice and bakery product processor, collaborates with RAB, the International Potato Center (CIP) and CRS. CRS and its implementing partners (the Young Women's Christian Association (YWCA) and the Imbaraga farmers' syndicate) organized farmers into groups that produce roots for sale to the factory. Imbaraga and YWCA extension personnel assisted these groups to form cooperatives, ensuring strong participation of women, so that they could increase their income from growing and supplying sweetpotato fresh roots to a guaranteed market offered by Urubwitso (SINA) Enterprises through contracts. In addition, Urwibutso technicians worked directly with 45 individually contracted farmers who already had relations with Urubwitso supplying other produce, such as strawberries and passion fruits. The introduction of the new OFSP varieties was combined with appropriate management and postharvest grading practices. Moreover, extension personnel assisted groups to develop a staggered production plan to smooth the supply of roots to the factory, with an initial goal of 1 t of roots/day.

Urubwitso (SINA) assists farmer groups in transporting the fresh roots to its factory, where they are processed into various products. Urubwitso (SINA) Enterprises (with support from RAB and Kigali Institute of Food Science (KIST) food scientists) conducted OFSP product testing and consumer evaluation. Two of these products, were selected for fuller line development and promotion: Akarabo Golden Power Biscuits (GPB) and doughnuts (Mandazi). These were developed and tested for consumer acceptability. From the beginning of 2011 to mid-2013, development of the two value chains had evolved to a sufficient scale to warrant an assessment of the steps in the development of the chain as well as the establishment of the market. During the establishment of the two value chains, all stakeholders regularly met to define rules to guarantee both supply and quality of planting material and fresh roots and ensure products met the food safety and regulatory requirements of the Rwanda Bureau of Standards (RBS) and International Organization for Standardization (ISO) regulations to ease national and international market access. Comparing the individual contracted growers versus the NGO-supported group approach (Fig. 48.1) permits an examination of the relative efficiency of the sweetpotato product value chains and their potential to increase farmer income and equity.

48.3 Results and Discussion

In vitro and field-based rapid multiplication of clean planting material was done at Rubona and Karama RAB research stations, producing 4000–5000 high quality cuttings per month. From early 2011 through to mid-2013, approximately 28,000 plantlets of the two OFSP varieties were multiplied in the tissue culture laboratory. 1,212,000 cuttings of Cacearpedo and Gihingamukungu were produced in primary multiplication fields at Rubona and disseminated to farmer groups (model 2) and individually contracted (model 1) farmers in different sites.

The number of model 2 farmer groups expanded to 20 (80% women). In model 1, Urwibutso (SINA) Enterprises directly contracted 45 (72% women) farmers and provided them with clean planting material to produce and supply fresh roots to the factory.

After 18 months, participant farmers were able to meet 20% of their planting material needs from quality planting materials that they had conserved themselves. Provision of disease-free or 'clean' planting material and training farmers on proper vine maintenance increased the production of sweetpotato roots by project farmers from 4 t/ha in early 2011 to an average of 12 t/ha, a 200% increase, by mid-2013.

A feasibility study comparing sweetpotato purée (boiled and mashed sweetpotato) and sweetpotato flour (chipped, dried and milled) showed that the purée makes better products and was more profitable to incorporate and needed less technical knowledge to produce. The following sweetpotato-based products are being produced at the Urubwitso factory and by some farmer groups for sale in urban areas: (i) doughnuts; (ii) bread; and (iii) queen cakes. In these products, 30–45% of wheat flour is substituted with sweetpotato purée. Urubwitso (SINO) Enterprises sold products worth US$33,359 between July 2011 and June 2012, an increase of 178% compared with the sales of the previous year for the same period. The projected savings for the processor obtained by replacing 30% of wheat flour with OFSP purée was calculated to be up to US$525/day (Table 48.1).

A new cookie/biscuit processing machine was imported and installed to boost the demand for fresh roots at the farmer level and develop a new shelf-stable product for the urban market. The production cost per product at Urubwitso has been reduced

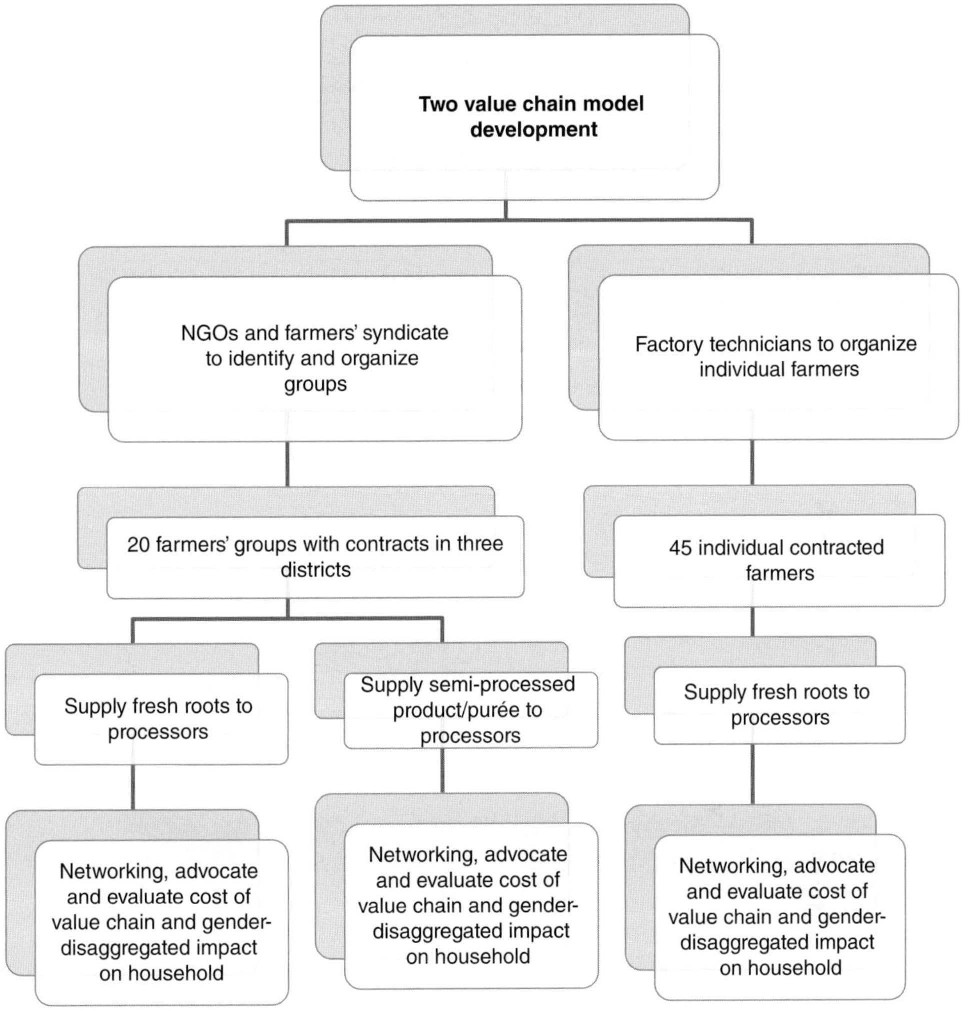

Fig. 48.1. The two value chains in the Rwanda Super Foods project. The value chains are: (i) non-governmental organization (NGO)-supported farmer groups; and (ii) individual farmers who already had contracts with the agro-processor to supply other produce.

significantly by using sweetpotato purée to replace 43% of the wheat flour in a biscuit. The savings in costs per unit product are 14% for biscuits, 15% for doughnuts (Mandazi) and 7% for bread. Frequency of sweetpotato consumption by gender showed that women consume sweetpotato more frequently than men (Fig. 48.2).

The consumer acceptability test showed that bread made out of a mix of 30% OFSP purée and 70% wheat flour was preferred to bread made of 100% wheat flour. No differences in preference were found between a 100% wheat product and a mix of 60% wheat flour and 40% OFSP purée for the doughnut. Eight farmers' groups are producing doughnuts using OFSP purée substituting 30–45% of the wheat flour and selling the doughnuts at local markets and consuming them at home. For biscuits a product with 40% OFSP purée mixed with 60% wheat flour is preferred to one made with 100% wheat flour and for cake there was no difference in consumer preference between products made out of

Table 48.1. Projected mean and maximum savings (RWF and US$)[a] to be gained by incorporating orange-fleshed sweetpotato (OFSP) into the bakery products of 11 surveyed agro-processors.

		Median		Mean (n = 11)		Maximum	
Wheat flour needed (kg/day)		171		320		1400	
OFSP purée needed (kg/day) if 30% wheat replaced		51		96		420	
	Purée price/kg	RWF	US$	RWF	US$	RWF	US$
Daily savings	150	38,475	64	72,000	120	315,000	525
Monthly savings		230,850	385	432,000	720	1,890,000	3,150
Daily savings	200	35,910	60	67,200	112	294,000	490
Monthly savings		215,460	359	403,200	672	1,764,000	2,940
Daily savings	300	30,780	51	57,600	96	252,000	420
Monthly savings		184,680	308	345,600	576	1,512,000	2,520
Daily savings	400	25,650	43	48,000	80	210,000	350
Monthly savings		153,900	257	288,000	480	1,260,000	2,100

[a]RWF, Rwandan francs; US$, US dollars.

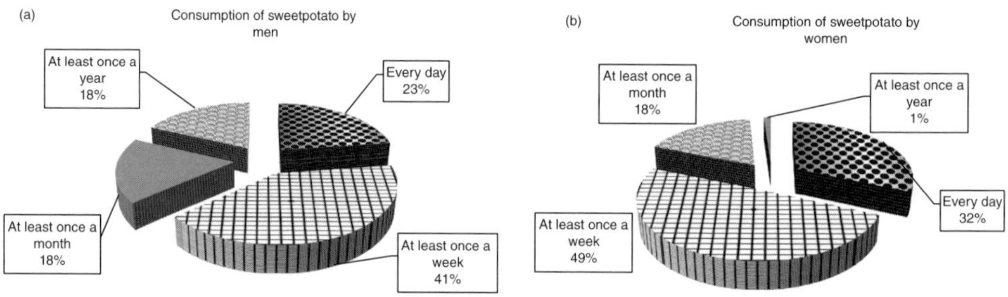

Fig. 48.2. Frequency of sweetpotato consumption by gender.

100% wheat flour and the ones made from a mix of 40% OFSP purée and 60% wheat flour.

Test products have been pilot tested with consumers at more than ten agricultural and trade fairs with high levels of acceptance. Local authorities and other policy makers are beginning to recognize sweetpotato's potential and authorize access to marshland plots for sweetpotato cultivation and conservation.

48.4 Conclusion

From early 2011 through to mid-2013, the Sweetpotato Super Foods project in Rwanda has built sweetpotato value chains that work for women and the poor and also benefit the agro-processor. The Rwanda Super Foods project is showing that sweetpotato purée can be incorporated into various commercially viable bakery products. The study showed that farmers can participate in a sweetpotato value chain either as a group or as an individual through contracts which result in greater participation of and high revenues for women farmers. Use of quality planting material coupled with proper training can significantly increase yield per unit area on farmers' fields. Using evidence-based advocacy can positively change policy makers' opinions about the value of sweetpotato and improve its image. As of mid-2013, clear evidence exists that two sweetpotato-based products, the Akarabo GPB and Mandazi doughnut are profitable and acceptable to rural and urban Rwandan consumers.

References

Bovell-Benjamin, A.C. (2007) Sweetpotato: a review of its past, present and future role in human nutrition. *Advances in Food and Nutrition Research* 52, 1–59.

Ferris, R.S.B., Wanda, K., Tuyisenge, J., Munkankubana, D., Kagiraneza, B., Ngaboyisonga, C., Gatarayiha, C., Uwantege, C., Kayange, C. and Butare, L. (2002) *Sweet Potato Sub-sector Market Survey Rwanda.* Agriculture Technology Development Transfer – International Center for Tropical Agriculture (ATDT-CIAT)/Rwandan Institute for Agricultural Sciences (ISAR)/International Institute of Tropical Agriculture (IITA)-FOODNET and Partnership to Enhance Agriculture in Rwanda through Linkages (PEARL) Project, Kigali.

Food and Agriculture Organization of the United Nations (FAO) (2008) Annual Report. Available at: http://www.fao.org/docrep/012/al047e/al047e00.pdf (accessed 23 May 2013).

Institut des Sciences Agronomiques du Rwanda (ISAR) (2006) *Rapport Annuel (Annual Report).* ISAR, Rwanda.

Kimura, M., Kobori, C.N., Rodriguez-Amaya, D.B. and Nestel, P. (2007) Screening and HPLC methods for carotenoids in sweet potato, cassava and maize for plant breeding trials. *Food Chemistry* 100, 1734–1746.

Ministry of Agriculture (MINAGRI) (2010) Annual Report. Available at: http://www.minagri.gov.rw/fileadmin/user_upload/documents/Reports/Final_AnnualReport_FY2010_2011updated.pdf (accessed 23 May 2013).

Ministry of Economics and Finance (MINECOFIN) (2002) Rwanda development indicators. In: MINECOFIN (ed.) *Annual Report 2002.* MINECOFIN, Kigali.

Ndirigwe, J. (2006) Adaptability and acceptability of orange and yellow-fleshed sweetpotato genotypes in Rwanda. MSc thesis, Makerere University, Kampala, Uganda.

Njeru, R.W., Bagabe, M.C., Kayiranga, D., Nsabimana, J., Ndirigue, J., Kajuga, J. and Nshimiyimana, J.C. (2007) Management of biotic constraints to sweetpotato production for enhanced food security in Rwanda. In: *Proceedings of National Conference on Agricultural Research Outputs,* Kigali, Rwanda. Rwanda Agricultural Board, Kigali, p. 233.

Programme d'Appui à la Sécurité Alimentaire du Rwanda (PASAR) (2002) *Annual Report, 2002.* PASAR, Kigali.

Tardif, D. (1991) The marketing of sweet potatoes in Rwanda: commercializing a perishable crop under adverse conditions. PhD thesis, Cornell University, Ithaca, New York.

Thiele, G., Lynam, J., Lemaga, B. and Low, J. (2009) Sweetpotato value chains. In: Andrade, M., Barker, I., Cole, D., Dapaah, H., Elliott, H., Fuentes, S., Grüneberg, W., Kapinga, R., Kroschel, J., Labarta, R., Lemaga, B., Loechl, C., Low, J., Lynam, J., Mwanga, R., Ortiz, O., Oswald, A. and Thiele, G. (eds) *Unleashing the Potential of Sweetpotato in Sub-Saharan Africa: Current Challenges and Way Forward.* Working Paper 2009-1. International Potato Center (CIP), Lima, pp. 106–129.

Tumwegamire, S., Kapinga, R., Zhang, D., Crissman, C. and Agili, S. (2004) Opportunities for promotion of orange fleshed sweet potato as a mechanism for combating vitamin A deficiency in Sub-Sahara Africa. *African Crop Science Journal* 12(3), 241–247.

World Health Organization (WHO) (2009) Global Prevalence of Vitamin A Deficiency in Populations at Risk 1995–2005: WHO Global Database on Vitamin A Deficiency. WHO, Geneva, Switzerland. Available at: http://whqlibdoc.who.int/publications/2009/9789241598019_eng.pdf (accessed 18 March 2013).

49 Sweetpotato Value Chain Development in West Africa: Matching Products with Farmer Typology

D. Peters*
Baltimore, Maryland, USA

Abstract

Sweetpotato value chain studies conducted in 2012 in three West African countries – Ghana, Nigeria and Burkina Faso – indicated three types of sweetpotato producers. Type I farmers specialize in sweetpotato production, making it the most important cash crop for their farm. Type II farmers grow sweetpotato as one of the cash crops and sweetpotato may rank second or third among these cash crops. Type III farmers are those who grow it mainly for home consumption though still sell a part of the roots due to perishability. The marketing assessment indicated three potential product value chains worth developing, each appropriate for different types of producers.

The fresh root value chain is well suited for Types I and II farmers for obvious reasons. Potential interventions to improve the system include: (i) breeding/selection of high-yielding varieties with the characteristics acceptable to the markets; (ii) best practices for production including ridging and weeding technologies to reduce labour inputs, appropriate fertilizer application, identifying best intercropping practices; and (iii) organizing farmers to connect to the national collectors directly to reduce costs and time spent on individual marketing efforts.

Sweetpotato as livestock feed is best suited for Type I farmers, though it may be appropriate for some Type II farmers. These farmers typically allocate large amounts of land to sweetpotato production, thus producing an abundance of vines at the time of harvest once, twice or even three times a year. They also produce lots of low-value and no-value small roots. These vines and roots would yield much better value if converted into feed that translates to meat or milk production. A better developed livestock system contributes in turn to improved soil fertility important to sweetpotato production.

Orange-fleshed sweetpotato (OFSP) is suited for developing as a home consumption product for all types, particularly Type III, farmers, who grow sweetpotato mainly for consumption. Promoting a product with more nutritional value for the family, particularly children, would be a good substitute for the local varieties. As OFSP lacks the characteristics sought by buyers in the market, until a serious and long-term awareness campaign has increased the awareness of OFSP in the markets, it should be treated principally as a product for home consumption, not commercialization.

Keywords: farmer typology, orange-fleshed sweetpotato, value chain, West Africa

*daipeters@gmail.com

49.1 Introduction

Sweetpotatoes have the potential to provide a source of income, food security, nutrition and feed to smallholder farmers. However, it is generally viewed as a home consumption crop that many smallholder farmers grow, but only enough for consumption. Thus, it has not received adequate attention from West African governments. Both the production and the markets for sweetpotatoes are not well covered by major sources of agricultural data, therefore there is lack of knowledge of the current production status and the extent of regional, country and local markets for this product.

Some literature was found on Nigeria and Ghana, while there was little available from Burkina Faso. Drawing on secondary data, mainly Food and Agriculture Organization of the United Nations (FAO) statistics, Akoroda (2009) summarized sweetpotato production areas in Ghana, Nigeria, Burkina Faso and other West African countries in terms of production area, varieties, economics and production practices. This review contains a section on harvesting, marketing and profitability showing sweetpotato economics in relation to cassava, yam and maize. In 2008, Akoroda also prepared a background paper on agroecological zones and their relations to sweetpotato production systems in sub-Saharan Africa, considering soil and climate, which have direct effects on plants through root interactions (M.O. Akoroda, 2008, unpublished data).

In Nigeria, Fawole (2007) indicated that yields of 4–7 t/ha in Kwara State, were obtained by smallholder farmers. He attributed the low productivity to: (i) limited cultivated land; (ii) lack of improved practices; (iii) poor storage facilities; (iv) lack of credit facilities and lack of extension training; (v) poor transportation; (vi) high cost of inputs; and (vii) farmers' disincentive to improve because marketing of the crop is dominated by middlemen, to whom most of the profits accrue. Due to the common practices of intercropping, researchers in Nigeria examined the effects of this practices on yield (A.B. Adeniji, 2004, unpublished data; I.T. Okunola, 2004, unpublished data). S. David and T. Madu (2012, unpublished data) investigated the roles and responsibilities of women and men in producing and marketing the crop and making decisions about resource allocation, consumption and marketing. On the subject of sweetpotato utilization, literature focuses on processing sweetpotato into flour or *gari* for food security purposes, such as Meludu *et al.* (2003) and one similar study 5 years later by Odebode (2008). Ukpabi *et al.* (2012), on the other hand, examined the potential use of orange-fleshed sweetpotato (OFSP) roots to produce β-carotene-rich chips. Onumah *et al.* (2012), while acknowledging the growing market for sweetpotato fresh roots, also placed emphasis on processed products such as crisps and flour. On fresh root consumption, Aikpokpodion *et al.* (2001) screened various genotypes of sweetpotato on dry matter content (DMC), sweetness, colour and texture of the cooked roots in order to classify these accessions based on consumption characteristics, rather than production traits. Sokoto and Ibrahim (2007) found little consumption of sweetpotato in Sokoto State but concluded that sweetpotato should be further explored considering the population increase in the state, particularly the metropolis.

Doku's (1969) review of root crops in Ghana, including cassava, yam, cocoyam and sweetpotato, concluded that these crops to Ghanaians are as important as Irish potato is to European diets, as they are relatively cheap to produce while still yielding heavily. The other advantage identified by Doku is their resistance to pests and diseases and storability, especially relative to Irish potato. Sam and Dapaah (2009) stated that sweetpotato was the most important crop on a fresh weight basis in developing countries next to rice, wheat, maize and cassava. The author's only knowledge of any sweetpotato marketing system studies is the one conducted by Birago (2005, unpublished) at Moree Junction. Some workers are concerned with production issues (e.g. Ennin *et al.*, 2007) while A.A. Dankyi (2004 and 2005, unpublished data) examined sweetpotato marketing chains and the process of disseminating improved varieties from their neighbours. The World Bank flyer 'Nutrition at a Glance: Ghana' showed that three-quarters of the pre-school-aged children and one-fifth of pregnant woman suffer from vitamin A deficiency (World Bank, 2013). The FAO report (2009), while pointing out Ghanaians' heavy reliance on root and tuber crops for energy consumption, did not make a reference to the potential role of sweetpotato in addressing this deficiency. In an effort to incorporate and expand sweetpotato use in Ghana, a number of studies screened the various clones to examine the suitability of their characteristics for processing *fufu*, *gari*, flour and starch (Adu-Kwarteng *et al.*, 2001; Oduro *et al.*, 2001; Ellis *et al.*, 2001, 2003). Attempting to address dietary micronutrient deficiencies, Amagloh *et al.* (2011) proposed an alternative blend based on β-carotene-rich sweetpotato to complement cereal-based foods; while Bonsi *et al.* (2001) developed dishes and recipes for both sweetpotato roots and leaves which showed moderate to high levels of acceptance.

This chapter is based on a study which centres on integrating investments along the entire agricultural value chain from 'molecule' (e.g. investments into plant research and development) to 'mouth' (investments that connect farmers to consumers). This value chain study aims to inform the development endeavours to make better decisions on how to prioritize research, increase farmers' market access, and address policy constraints through a better understanding of the dynamics of the sweetpotato sub-sector in West Africa.

The chapter contains a comparison of the status regarding sweetpotato production and marketing in selected areas of Nigeria, Ghana and Burkina Faso in late 2012. Detailed country reports, complete with recommendations for policies, research and development interventions to develop sweetpotato for each country, are available as separate reports from the author.

49.2 Methodology

The process of the study consisted of the following:

- review of available literature;
- planning meetings to design fieldwork, held in conjunction with the International Potato Center (CIP) in Kumasi, Ghana in November 2012;
- fieldwork in Ghana, Nigeria and Burkina Faso, including visits to key informants in relevant national and regional institutions, larger businesses and service providers to collect additional secondary information and expert opinions (Table 49.1);
- field surveys of specific value chains of interest, comprising market chain actors, service providers and other stakeholders;
- focus group discussions for consumers and farmers on relevant production, market and consumption issues (including on-farm consumption);
- development of survey instruments for primary data collection by the Ghanaian local consultants; and
- a final workshop with stakeholders, including farmers, private-sector actors, National Agriculture Research System (NARS) partners, non-governmental organizations (NGOs) and other relevant

Table 49.1. Number of farmers and other stakeholders interviewed by country.

	Dates of fieldwork	Farmer households	Collectors/traders/ wholesalers/retailers
Ghana	24–31 October 2012, 17–20 February 2013[a]	184	14
Nigeria	2–15 November 2012	162	15
Burkina Faso	9–17 February 2013	200	10[b]

[a]Additional interviews and field data in Ghana were conducted and collected by local consultants in November 2012 and February/March 2013.
[b]The Burkina Faso study was conducted during the off season, thus it was difficult to find traders and collectors.

programmes (e.g. school feeding programmes, Farm Radio International) from all three countries to discuss and finalize the report.

The field surveys of value chain actors used semi-structured interviews, based on topic guides customized for each type of actor.

49.3 Results and Discussion

The sweetpotato production and marketing chains in Nigeria, Ghana and Burkina Faso share some similarities on the macro level while they also diverge in many aspects at the micro level. The macro-level similarities indicate the common intervention areas that are appropriate for all three countries, while the differences dictate that different specific intervention activities are also needed. To avoid redundant efforts with a concurrent gender study being conducted by CIP, this study did not delve into the gender aspects of production or marketing.

The similarities of the sweetpotato value chain in the three countries

The similar characteristics of the sweetpotato value chain in the three countries are summarized in Table 49.2. These characteristics have clear implications for the areas where interventions are most appropriate, most relevant, and most immediate.

Different intervention strategies for different farmers

While the fresh root chain should be the major focus of interventions for all three countries, the specific strategies, from breeding to seed system to production management

Table 49.2. Similarities among the sweetpotato value chains in Ghana, Nigeria and Burkina Faso, and the implications of the findings.

Similarities	Implications for intervention areas for all three countries
The fresh roots have evolved into an important cash crop in recent years. Even among farmers who grow sweetpotato mainly for home consumption, most of the roots are sold because there is market demand and the roots are perishable. Furthermore, all three countries show that demand outstrips supply, indicating a growing market.	Sweetpotato fresh roots as a cash crop should be the most important intervention area to help farmers generate greater income from sweetpotato through increased productivity and/or decreased costs, as well as access to markets. Any improvement in this regard will have widespread benefits for a large number of producers.
Large amounts of vines during harvest are dried and fed to cattle and goats. Better management of vines, in combination with the low-valued, or non-marketable small roots, can make far more efficient animal feed systems than currently in use.	Sweetpotato as animal feed is an intervention area that can help farmers generate income by improving their animal feed system. This intervention may be most relevant to farmers with large amounts of vines.
There is low awareness of the nutritional value of sweetpotato for family health (particularly for children). Instead, some people in all three countries incorrectly believe in the alleged harmful effects of sweetpotato on health. Meanwhile, efforts promoting the use of orange-fleshed sweetpotato (OFSP) have been misdirected to persuading farmers to grow OFSP as a market product, which has, in some cases, caused farmers to suffer loss.	Sweetpotato for nutrition needs to be an intervention area that is clearly introduced as a family health and home consumption option (except, to some extent, in Burkina Faso). In other words, OFSP should be introduced and promoted specifically and clearly for such a purpose. Until a market awareness campaign has taken effect, or firm linkage has been made to selling to a specific programme such as 'School Feeding', farmers should not be led to believe that OFSP is a marketable product.

and marketing, may differ depending on the particular situation in each country.

Before discussion of the differences between the three countries, it is important to acknowledge that the sweetpotato farmers within each country are also differentiated by their sizes, land allocation for sweetpotato, productivity and prices obtained (Table 49.3). The larger farmer types (e.g. Type I farmers in Nigeria, Southern farmers in Ghana and large farmers in Burkina Faso) allocate higher percentages of land to sweetpotato, obtain higher yields and higher prices, and achieve higher net income from sweetpotato. Hence farmers must be viewed as differentiated types of producers (instead of treating them as one undifferentiated group of sweetpotato producers) with differentiated intervention strategies for specific types of farmers. The farmer typologies were assigned with regard to sweetpotato production and marketing but also refer to the farmers' overall economic conditions. Type III farmers, though not making much money from sweetpotato, could be wealthier farmers who grow other more profitable crops.

Varieties

The number of years of growing sweetpotato as a cash crop, ways that sweetpotatoes are consumed, market access, and market requirements of the varieties all lead to the current varieties grown in each country (Table 49.4). In Nigeria, the need to be transported 2–3 weeks

Table 49.3. Various types of sweetpotato farmers in each country and their productivity and prices.

Country	Farmer typology[a]	Farmgate prices (US$/kg)	Mean yield (t/ha)[b]	Proportion of land cultivated with sweetpotato (%)[b]
Nigeria	**Type I** – specializes in sweetpotato production as a cash crop, targets national markets with one or two specialized varieties	0.10–0.15 (for national markets)	6.8 (3.5–10)	57 (35–70)
	Type II – produces sweetpotato as one of the cash crops, various varieties targeted for local markets		3.7 (2.5–7.1)	34 (20–47)
	Type III – produces sweetpotato as a consumption crop but still sells a big portion of it		3.6 (1–6.3)	12 (12)
Ghana	**Southern farmers** – produce sweetpotato but it is not considered an important cash crop (little fertilizer used) but it is still high yielding and there are high profits due to high prices	0.30 (in peak season in south)	14.1 (1.3–17.4)	36 (30–61)
	Northern farmers – produce sweetpotato as an important cash crop, apply fertilizer, but low profits due to low prices	0.1 (in peak season in north)	7.2 (2.5–12)	22 (10–43) (Upper East Region) 7 (4–9) (Upper West Region/ North Region)
Burkina Faso	**Large sweetpotato farmers** – produce up to 5 ha per household **Small sweetpotato farmers** – produce 0.2 ha per household	0.05–0.06 (peak season), 0.09 (in dry season)	19.5 (14.7–30)	21 (10–40)

[a]Types I, II and III refer to the three types of sweetpotato farmers in the Nigerian country report.
[b]Ranges are shown in parentheses.

Table 49.4. Characteristics of sweetpotato that influence variety requirements.

	Grown as cash crop (years)	Ways consumed	Market access	Variety requirements	Current varieties
Nigeria	< 10	Fried (60%), boiled (40%)	Four major cities plus the local markets	Long shelf life (2–3 weeks) due to long chain, sweet taste	Two varieties targeting national markets
Ghana	5–10	Fried (80%), boiled (20%)	Spread around the country	Medium shelf life (1–2 weeks), fry well, bright colour	Diverse varieties with no champion
Burkina Faso	15	Boiled (70%), fried (30%)	A few particular provinces	High dry matter content (DMC), sweet taste	Two varieties targeting national markets

For details of farmer typology see Table 49.3.

before reaching consumers, coupled with fried chip consumption, led to the predominant requirements of long shelf life, frying quality and taste.

In Ghana, sweetpotato has been grown specifically as a cash crop for the shortest time, and it has evolved into a staple crop that is consumed in snack form (e.g. fried chips or slices). Thus, frying quality has become a predominant consideration. Due to the lack of centralized markets as in Nigeria, and the generally low frequency of movement between the north and south, no major varieties have emerged as a national variety. It is not clear how much the yield discrepancy between the north and south can be attributed to varietal difference and how much to production management.

Burkina Faso has the longest history of growing sweetpotato as a cash crop and the roots are consumed as a staple crop in boiled form in a few provinces. Thus, the traditional characteristics of staple crops with widespread appeal, such as high DMC and good taste are preferred.

Due to these differences, the varietal requirements recommended for each country's breeding programme differ.

Seed

Most of the farmers try to maintain a nursery during the dry season in order to avoid paying high prices for seed, but the majority still have to purchase some seed (Table 49.5). In areas where the dry season is not very long, farmers are able to rely on a combination of ratooning and staggered planting to supplement seed deficiencies, thus reducing the seed cost. Therefore, in Nigeria and southern Ghana where farmers are able to plant two seasons per year, ratooning is more likely than in Burkina Faso where the sweetpotato harvest period is followed by a long dry season.

One of the reasons that Type I farmers in Nigeria and Southern farmers in Ghana need to buy seed, despite the ratoon option, is that these farmers may cultivate large tracts of sweetpotato and their need for seed is too great to simply rely on ratooning.

Since sweetpotato is cultivated for cash, it is not unusual for farmers to pay for planting material, or seed. The implications are that: (i) it is possible to establish a commercial seed system for market-oriented varieties; (ii) the existing seed producers (i.e. those who have access to suitable land for seed maintenance during the dry season) can be considered as seed multipliers; (iii) assisting farmers at large to improve the seed system would help them decrease their production cost considerably; and (iv) some areas may rely on improving the ratoon system while other areas, such as Burkina Faso, do not have such options and must rely on other means to produce seed.

Table 49.5. Seed systems of the different types of farmers[a] in the three countries.

	Number of seasons	Farmers who maintain nurseries in dry season (%)	Farmers who buy seed (%)	Mean seed cost (US$/ha)[b]
Nigeria	Two seasons are common, occasionally three seasons	7 (Type I), 45 (Type II), 50 (Type III)	93 (Type I), 35 (Type II), 60 (Type III)	25.8 (6.5–45.2) (from neighbour), 45.2 (20–200) (from market)
Ghana	One season in north, two to three seasons in south	41 (South), 35 (North)	48 (South), 35 (North)	55.5 (30–200)
Burkina Faso	Mainly one season	83	63	53.5 (0–400)

[a]For details of farmer typology see Table 49.3.
[b]Ranges are shown in parentheses.

Production and marketing

Fertilizer, ridging or mounding labour, seed and transportation comprise the majority of the expenses in sweetpotato production (Table 49.6). Fertilizer application seems confusing as it does not always correlate with productivity, which of course is also affected by native soil fertility, variety grown and production practices. Nevertheless, there should not be such inconsistency between fertilizer application and productivity.

Transportation is a major cost for smaller farmers who have to transport their products to the market, while larger farmers do not have to pay for transport as collectors purchase directly from the farm. Establishing market linkages and market centres where farmers can stockpile products to entice collectors to come to collect would eliminate, or reduce, transportation costs for smaller farmers.

The net incomes are also in direct correlation with the expenses; in other words, the higher the investment the higher the net income. This of course does not mean that as long as farmers invest more in fertilizer, ridging, weeding and so forth that this will automatically lead to higher net income, because the net incomes are also highly related to prices, which vary greatly. However, it does indicate that farmers are in a better position to increase profits by moving into higher scales of production.

49.4 Recommendations

Table 49.7 outlines interventions recommended in the country reports, taking into consideration the different characteristics of farmers. OFSP is well accepted in fresh root markets in Burkina Faso, but not known in Nigeria and Ghana. Unless the consumers' preference and transport issues can be resolved by breeding OFSP varieties with market characteristics, it is not likely that farmers will want to grow them as a cash crop in the short to medium term. Consumer and farmers' acceptance of OFSP for their nutritional value will depend on how much, how extensive and how successful the OFSP nutrition promotion campaign is. Until then, it is most reasonable to start introducing it as a nutritional product to the sweetpotato producers for home consumption to improve their family health.

One needs to keep in mind there are two types of processing. The 'industrial' processing is to process sweetpotato into purée, flour or starch to use as an ingredient in other food and industrial products. This is the type of processing suitable for farmers to add value to sweetpotato. However, in order to make these products, sweetpotato must directly compete with other flour or starch sources, most commonly cassava. Sweetpotato flour can be used as a wheat-flour substitute to reduce costs in noodles or baked goods. This has been tested in countries where sweetpotato is abundant, and has rarely been shown to be a success at any

Table 49.6. Production and marketing costs of the different types of farmers in the three countries.

Farmer type (net income)	Costs	Fertilizer	Ridge	Weed	Planting	Harvest	Seed	Transport	Total
Nigeria Type I (US$136 per person/year)	Cost (US$/ha)	84.1	181.1	38.6	121.9	40.4	47.7	14.6	528.4
	Proportion of total costs (%)	15.9	34.3	7.3	23.1	7.6	9.0	2.8	100
Nigeria Type II (US$41 per person/year)	Cost (US$/ha)	108.8	146.7	17.4	60.2	43	12.7	42.9	431.7
	Proportion of total costs (%)	25.2	34.0	4.0	13.9	10.0	2.9	9.9	100
Nigeria Type III (US$14 per person/year)	Cost (US$/ha)	56.1	50	0	54.8	0	10.3	60.3	231.5
	Proportion of total costs (%)	24.2	21.6	0.0	23.7	0.0	4.4	26.0	100
Ghana Southern (US$677 per person/year)	Cost (US$/ha)	26	275.1	107.1	12	65.2	76.5	160.2	722.1
	Proportion of total costs (%)	3.6	38.1	14.8	1.7	9.0	10.6	22.2	100
Ghana Northern (US$17 per person/year)	Cost (US$/ha)	121.6	77.5	0	0	0	123.1	177.2	499.4
	Proportion of total costs (%)	24.3	15.5	0.0	0.0	0.0	24.6	35.5	100
Burkina Faso (large farmers: US$366 per person/year, small farmers US$99 per person/year)	Cost (US$/ha)	127.8	81.2	18.2	16.3	29.2	58.9	111.6	443.2
	Proportion of total costs (%)	28.8	18.3	4.1	3.7	6.6	13.3	25.2	100

Table 49.7. Suggested sweetpotato products and interventions for the different farmer types.

	Fresh roots for market	Sweetpotato as livestock feed	OFSP as nutrition product
Appropriate farmers	Type I and Type II	Type I (all) and Type II (some)	Type III
Why the product is suited for certain types of farmers	Fresh roots are already important products for these two types of farmers and proper interventions may make sweetpotato more profitable to farmers while also affordable to consumers	This product chain is suited for farmers who have large amounts of vines at harvest, and small roots which command low prices, both would yield higher value, with proper processing and feeding technology, as livestock feed	OFSP does not have the characteristics desired by the fresh market, but has potential to improve nutritional status as a home consumption product. Type III farmers mainly grow sweetpotato for consumption and OFSP is appropriate for them.

significant scale, which is also likely to be the case in Ghana, Nigeria and Burkina Faso. The other type of processing is to process snacks from various forms of sweetpotato (roots directly or flour or mush, etc.) and this already is the way sweetpotato is consumed in Ghana, and to a lesser extent in Nigeria. The final actor in the chain is a fryer who fries sweetpotato into chips, or some make them into crisps. The target population of this processing are the small urban and peri-urban processors, not farmers. Assuming the development for sweetpotato focuses on farmers, processing is thus not included as one of the recommended areas of intervention.

References

Adu-Kwarteng, E., Otoo, J.A. and Oduro, I. (2001) Screening of sweetpotato for poundability into fufu. In: *Proceedings of the 8th International Society for Tropical Root Crops – African Branch (ISTRC-AB) Symposium,* 12–16 November, Ibadan, Nigeria. ISTRC-AB, Ibadan, Nigeria.

Aikpokpodion, P.O., Ng, Q. and Akoroda, M.O. (2001) Utilization potentials of some selected sweetpotato, *Ipomoea batatas* (L.) Lam. Accessions for staple consumption in African diet and industrial processing. In: *Proceedings of the 8th International Society for Tropical Root Crops – African Branch (ISTRC-AB) Symposium, 12–16 November, Ibadan, Nigeria*. ISTRC-AB, Ibadan, Nigeria.

Akoroda, M.O. (2009) Sweetpotato in West Africa. In: Loebenstein, G. and Thottappilly, G. (eds) *The Sweetpotato*. Springer Science + Business Media BV, Houten, The Netherlands, pp. 441–468.

Amagloh, F.K., Weber, J.L., Brough, L., Hardacre, A., Mutukumira, A.N. and Corad, J. (2011) Complementary food blends and malnutrition among infants in Ghana: a review and proposed solution. *Scientific Research and Essays* 7(9), 972–989.

Bonsi, E., Zabawa, R. and David, P. (2001) Enhanced production, recipe development, and consumption of sweetpotato in Ghana, West Africa. Poster presented at the 8th International Society for Tropical Root Crops – African Branch (ISTRC-AB) Symposium, 12–16 November, Ibadan, Nigeria.

Doku, E.V. (1969) Root crops in Ghana. In: *Proceedings of the 1st International Symposium on Tropical Root Crops* Vol. 1 (section III), pp. 39–68.

Ellis, W.O., Oduro, I., Fianko, K. and Otoo, J.A. (2001) Quality of gari from eighteen sweetpotato varieties. In: *Proceedings of the 8th International Society for Tropical Root Crops – African Branch (ISTRC-AB) Symposium,* 12–16 November, Ibadan, Nigeria. ISTRC-AB, Ibadan, Nigeria.

Ellis, W.O., Oduro, I., Barimah, J. and Otoo, J.A. (2003) Quality of starch from six Japanese sweetpotato varieties in Ghana. *African Journal of Root and Tuber Crops* 5(2), 38–41.

Ennin, S.A., Dapaah, H.K. and Asafu-Agyei, J.N. (2007) Land preparation for increased sweetpotato production in Ghana. In: *Opportunities for Poverty Alleviation and Sustainable Livelihoods in the Developing World,* Proceedings of 13th International Society for Tropical Root Crops (ISTRC) Symposium, 2003, Arusha, Tanzania. ISTRC, Arusha, Tanzania.

Fawole, O.P. (2007) Constraints to production, processing and marketing of sweet-potato in selected communities in Offa local government area, Kwara State Nigeria. *Journal of Human Ecology* 22(1), 23–25.

Food and Agriculture Organization of the United Nations (FAO) (2009) *Nutrition Country Profile: Republic of Ghana.* Nutrition and Consumer Protection Division, FAO, Rome.

Meludu, N.T., Ajala, C.G. and Akoroda, M.O. (2003) Poverty alleviation through the processing of sweetpotato tubers into toasted granules and consumer preferences in Nigeria. *African Journal of Root and Tuber Crops* 5(2), 56–59.

Odebode, S.O. (2008) Promotion of sweetpotato spari among rural women for food security and economic empowerment in south-west Nigeria. *Research on Crops* 9(2), 418–422.

Oduro, I., Ellis, W.O., Nyarko, L., Koomson, G. and Otoo, J.A. (2001) Physicochemical and pasting properties of flour from four sweetpotato varieties in Ghana. In: *Proceedings of the 8th International Society for Tropical Root Crops – African Branch (ISTRC-AB) Symposium,* 12–16 November, Ibadan, Nigeria. ISTRC-AB, Ibadan, Nigeria.

Onumah, G., Dipeolou, A. and Ganiyat Feguga (2012) Report on sweetpotato value chain study: exploring opportunities to promote greater exploitation of the benefits of sweetpotato in representative states of Nigeria. In: *Sweetpotato Action for Security and Health (SASHA) Project Report.* SASHA, Nigeria.

Sam, J. and Dapaah, H. (2009) *Baseline Survey Report – Ghana*. West African Agricultural Productivity Programme (WAAPP), Accra.

Sokoto, M.B. and Ibrahim, Z. (2007) Analysis of sweetpotato consumption in Sokoto Metropolis of Sokoto State of Nigeria. *Journal of Biological Sciences* 7(2), 445–447.

Ukpabi, U.J., Ekeledo, E.N. and Ezigbo, V.U. (2012) Potential use of roots of orange-fleshed sweet potato genotypes in the production of β-carotene rich chips in Nigeria. *African Journal of Food Science* 6(2), 29–33.

World Bank (2013) Nutrition at a Glance: Ghana. Flyer produced with support from the Japan Trust Fund for Scaling Up Nutrition. World Bank and Japan Trust Fund, Accra.

50 Vacuum-frying Processing Technology Improves Quality Attributes of Fried Sweetpotato Chips

O.P. Sobukola,* T.A. Esan, H.A. Bakare and L.O. Sanni
Federal University of Agriculture, Abeokuta, Nigeria

Abstract

Sweetpotato (*Ipomea batatas*) is a very important crop in developing countries. Orange-fleshed varieties provide beta-carotene, a key antioxidant and source of provitamin A. However, sweetpotato does not store for a long period unless properly cured; hence the need for processing into shelf-stable forms such as fried snacks. This study investigated the effect of sweetpotato cultivars (EX-OYUNGA, 440216 and SPK 004), type of frying oil (groundnut oil, refined, bleached and deodorized oil (RBDO) and palm oil) and vacuum frying conditions (temperature – 108°C, 122°C and 136°C, vacuum pressure – 6.54 kPa, 13.21 kPa and 19.88 kPa and time – 3 min, 6 min and 9 min) on some quality attributes of fried sweetpotato chips. The response surface methodology technique based on the Box-Behnken design was used to optimize vacuum frying processes and study the effects of the variables on quality of fried chips. Twenty-eight combinations including five replicates of centre points were performed in random order. Seven responses such as oil and moisture content, beta-carotene, texture, lightness, redness and yellowness were considered to evaluate effects of independent variables on sweetpotato chips. Fried sweetpotato chips from optimized vacuum frying conditions were compared with atmospheric fried samples using the concept of equivalent thermal driving force (ETDF). Significant differences ($P < 0.05$) were observed among sweetpotato varieties and frying oil type. EX-OYUNGA and groundnut oil produced chips with acceptable quality attributes. Frying temperature and time as well as vacuum pressure have significant effect on beta-carotene, texture, lightness, redness and yellowness with coefficient of determination (R^2) of models for responses investigated varying between 0.7179 and 0.951. The optimized coded vacuum frying conditions for the adopted cultivar (EX-OYUNGA) and frying oil (groundnut oil) were –0.64, 1 and –1 for frying temperature, vacuum pressure and frying time, respectively, based on the desirability concept of 0838. This choice was based on fried chips of lower oil and moisture contents, higher level of beta-carotene retention, lower breaking force, lighter and intense yellow colouration. When the optimized vacuum fried samples were compared with atmospherically fried ones, the former retained about 35% beta-carotene, had improved texture, were lighter and had a more intense golden yellow colour. Hence, vacuum frying could be an alternative for producing more nutritious deep-fat fried sweetpotato chips compared with atmospheric fried samples, the dominant current practice.

Keywords: beta-carotene, colour, oil content, sweetpotato, vacuum frying

*olajidephilip@yahoo.com

50.1 Introduction

Sweetpotato (*Ipomoea batatas*) is a large, starchy, sweet-tasting, tuberous root (Woolfe, 1992). Over 95% of the global sweetpotato crop is produced in developing countries, where it is the fifth most important food crop in terms of fresh weight.

The sweetpotato is high in carbohydrates and can produce more edible energy per hectare per day than wheat, rice or cassava. The orange-fleshed varieties also provide provitamin A. In some studies, sweetpotato has been shown to be a better source of bioavailable beta-carotene than green leafy vegetables (Bengtsson *et al.*, 2010). Consequently, sweetpotatoes are now being used in Africa to combat a widespread vitamin A deficiency that results in blindness and even death for 250,000–500,000 African children a year (Low *et al.*, 2007).

Frying, used widely both in domestic and in industrial processes, consists of cooking food in oil at a temperature above the boiling point of water, usually between 150°C and 200°C (Mir-Bel *et al.*, 2009). This quick and easy process results in products with organoleptic qualities (colour, texture and flavour) that are in demand by consumers. It involves simultaneous heat and mass transfers resulting in significant microstructural changes in the product. Heat transfer causes protein denaturation, starch gelatinization, water vaporization, crust formation and colour development (Singh, 1995). Mass transfer is characterized by volatiles and soluble materials escaping from the food and oil penetrating it. However, these products have a high fat content that in some cases reaches one-third of their total weight (USDA, 2008). The popularity of potato chips suggests that suitable substitutes might be found among tropical roots, tubers and plantain (Vitrac *et al.*, 2000). Sweetpotato is a relatively underutilized tuber crop whose commercial outlook could be improved by diversifying its use. With a growing middle class in urban areas in many developing countries, demand for convenient snack foods is increasing, thus creating an opportunity for increased sweetpotato utilization and consumption.

Consumers of fried food prefer healthy products that taste good. Fried products are produced using non-hydrogenated oil and contain no saturated or trans fats. Previous studies on oil uptake during frying using diverse cooking techniques point to the need for alternatives to the common processes (Mellama, 2003; Ziaiifer *et al.*, 2008). Vacuum frying is a potential option that could yield high quality dried fruits and vegetables in a shorter time than conventional frying (Troncoso *et al.*, 2009). In this method, the samples are heated under a negative pressure that lowers the boiling points of the frying oil and water and improves the quality of the processed product. Absence of air during vacuum frying may inhibit lipid oxidation and enzymatic browning (Fan *et al.*, 2005; Shyu *et al.*, 2005). Many studies have evaluated vacuum frying and its adaptations with different vegetables, comparing it with atmospheric pressure frying (Troncoso and Pedreschi, 2009; Dueik *et al.*, 2010). Apart from reducing the fat content (Therdthai *et al.*, 2007; Da Silva and Moreira, 2008; Nunes and Moreira, 2009), vacuum frying lowers the acrylamide content and improves the organoleptic and nutritional qualities (Troncoso *et al.*, 2009).

Products such as sweetpotato chips can be fried under vacuum yielding less oil absorption with higher retention of natural colours (provitamin A carotenoid) and flavours. It is a technology that can be used to produce fruits and vegetables with the necessary degree of dehydration without excessive darkening or scorching. This technology could improve nutrition and health by producing products that are tasty, but when compared with conventionally fried snacks: (i) retain more nutritional value; (ii) have a lower oil content; (iii) are safer due to little or no acrylamide formation; and (iv) keep longer.

Cultivar effects, type of frying oil and processing conditons during vacuum frying of sweetpotato in Nigeria has not been investigated. Because of compositional differences, cultivars of sweetpotato and types of frying oil influence mass and heat transfers during vacuum frying and define processing conditions. Characteristics of vacuum-fried sweetpotato chips, particularly those that influence storage and acceptability have also received limited attention. This research aims to address these issues by understanding the mechanisms of vacuum frying of Nigerian cultivars of sweetpotato, examining processing conditions and identifying the process–quality–composition interactions. The objectives are to study the effects of sweetpotato cultivars and types of frying oil on the quality attributes of vacuum-fried sweetpotato chips and to optimize the appropriate vacuum frying conditions using the response surface methodology.

50.2 Materials and Methods

Three different cultivars of yellow and orange-fleshed sweetpotatoes (EX-OYUNGA, 440216 and SPK 004) were purchased from a market in Abeokuta, south-west Nigeria. For each set of experiments, sweetpotato roots were stored until use at 10°C and a relative humidity of 80–90%. Three types of frying oil (groundnut oil, refined, bleached and deodorized oil (RBDO) and palm oil) commonly used for frying in the study area were purchased from the local market in Abeokuta. The effect of sweetpotato cultivar, type of frying oil, type of frying and frying temperature on moisture content, oil content, shrinkage, provitamin A retention, texture and colour (redness, yellowness and lightness) parameters of vacuum-fried sweetpotato chips was investigated. Fresh sweetpotato were washed, peeled and sliced (3 mm thick). A vegetable slicer (model ART: NO: SF-923-1, Texas, USA) was used to provide slices with a diameter of 40 mm. Surface starch of sweetpotato slices was washed off by placing the slices in a water bath at 30°C for 10 min. Excess surface water was removed by placing the slices between moistened towels. The five factors considered include: (i) sweetpotato cultivar (EX-OYUNGA, 440216 and SPK 004); (ii) type of frying oil (groundnut oil, RBDO and palm oil); (iii) frying temperature (108°C, 122°C and 136°C); (iv) vacuum pressure (6.54 kPa, 13.21 kPa and 19.88 kPa); and (v) frying time (3 min, 6 min and 9 min) (Table 50.1).

Table 50.1. Coded values for the independent variables in the sweetpotato frying trial.

Variable	Coded value –1	0	+1
Temperature (°C)	108	122	136
Vacuum pressure (kPa)	6.54	13.21	19.88
Frying time (min)	3	6	9
Cultivars	EX-OYUNGA	440216	SPK 004
Frying oils	RBDO[a]	Palm oil	Groundnut oil

[a]RBDO, refined, bleached and deodorized oil.

Frying

The equipment for both atmospheric and vacuum frying methods was filled with oil and pre-heated for 1 h prior to frying. Slices of sweetpotato were fried at different times according to the type of frying, cultivar, oil temperature and frying oil, until reaching a final moisture content of about 2 kg/100 kg (on a wet basis). The 28 different runs are summarized in Table 50.2.

Vacuum-frying set up and frying conditions

A laboratory-size vacuum fryer equipped with a centrifuge was used for the vacuum frying. It has a maximum temperature and vacuum pressure of 190°C and 0.335 MPa for an oil capacity of 70 l. The vacuum vessel was set at the target temperature and vacuum for about 1 h before the start of frying. Once the oil temperature reaches the target value, the sweetpotato slices were placed into the basket, the lid closed, and the vessel evacuated. The samples were lowered into the oil and frying began for the desired frying time. Fried sweetpotato chips were removed from the basket, allowed to cool to room temperature on a paper towel prior to storing in polyethylene bags for further analysis.

Atmospheric deep-fat frying experiments

For the atmospheric frying experiments, an electric deep-fat fryer (model S-576, Hong

Table 50.2. Description of the 28 experimental runs.[a]

Run	Cultivar	Frying oil	Frying temperature	Vacuum pressure	Frying time
1	0	0	−1	0	0
2	1	1	−1	−1	1
3	0	0	0	0	0
4	0	0	0	−1	0
5	−1	−1	−1	−1	1
6	1	1	1	1	1
7	0	0	0	1	0
8	−1	1	−1	1	1
9	−1	1	1	1	−1
10	1	1	1	−1	−1
11	−1	0	0	0	0
12	1	−1	−1	−1	−1
13	0	1	0	0	0
14	0	−1	0	0	0
15	1	−1	−1	1	1
16	0	0	1	0	0
17	0	0	0	0	−1
18	1	1	−1	1	−1
19	−1	1	−1	−1	−1
20	0	0	0	0	0
21	−1	−1	1	1	1
22	−1	1	1	−1	1
23	1	0	0	0	0
24	−1	−1	−1	1	−1
25	1	−1	1	1	−1
26	−1	−1	1	−1	−1
27	1	−1	1	−1	1
28	0	0	0	0	1

[a]The coded values −1, 0 and 1 are defined in Table 50.1.

Kong, China) was used. The fryer was set to the required frying temperature and left for 30 min for equilibration. Samples were fried to about 2 kg/100 kg moisture content based on the equivalent thermal driving force (ETDF) to compare with the optimized vacuum fried samples.

Analytical methods

Oil and moisture content determination

Oil and moisture contents of fried sweetpotato chips were determined using Association of Official Analytical Chemists (AOAC) (1995) methods.

Determination of chips' colour parameters

To determine colour parameters, 1 g of well-ground sample was weighed and 25 ml of HPLC methanol was added to extract the colour by shaking and stirring with a glass rod for 30 min. The mixture was then allowed to stand for 10 min and filtered through hardened Whatman No. 42 filter paper into another 100 ml conical flask. The organic filtrate obtained was used to determine the colour by taking the absorbance or optical density readings at wavelengths of 420 nm, 520 nm and 620 nm on a Cecil 2483 spectrophotometer. Colour intensity is the sum of the absorbance of each sample at the following wavelengths: 420 nm, 520 nm and 620 nm. Colour parameters were calculated as shown in Eqns 50.1–50.4.

$$\text{Lightness}(\%) = \frac{\text{Absorbance at 420 nm} - \text{Absorbance at 520 nm}}{\text{Colour intensity}} \times \frac{100}{1} \quad (50.1)$$

$$\text{Yellow}(\%) = \frac{\text{Absorbance at 420 nm}}{\text{Colour intensity}} \times \frac{100}{1} \quad (50.2)$$

$$\text{Red}(\%) = \frac{\text{Absorbance at 520 nm}}{\text{Colour intensity}} \times \frac{100}{1} \quad (50.3)$$

$$\text{Blue}(\%) = \frac{\text{Absorbance at 620 nm}}{\text{Colour intensity}} \times \frac{100}{1} \quad (50.4)$$

Determination of texture

Texture of the fried samples was analysed using a three-point bending test where a sample is supported at two parallel edges and the load is applied centrally. The system was mounted in a TA.XT2 Texture Analyser (Texture Technologies Corp., USA) using a support span of 16 mm. A steel blade of 2.5 mm thickness with a flat edge was used to crush the sample at a constant speed rate of 10 mm/s. The force (N) at the fracture point (the highest value in the plot) was used as resistance to breakage (Garayo and Moreira, 2002).

Determination of beta-carotene of sweetpotato chips

Carotenoid content of raw and fried sweetpotato chips was determined spectrophotometrically using the UV-1601 spectrophotometer at 450 nm. Acetone (50 ml) was added to about 1 g of each ground sample. The flask was closed and allowed to rest in the dark at 22°C for 24 h. The solution was filtered and the extracted solution placed in decantation balloons and 50 ml of petroleum ether added. The residue was washed five to seven times with 100 ml of distilled water. Sodium sulfate (0.5 ml) was added to the flask to bind any remaining water. The carotene extract was recovered by concentrating the previous extract for 10 min to evaporate all the petroleum ether and acetone. Spectrophotometric determination of samples was done by diluting previous concentrates with hexane. Total carotenoid was determined using a standard curve, with beta-carotene used as a standard (Da Silva and Moreira, 2008).

Statistical analysis

Randomization of experiments and statistical analysis were done using Design Expert version 7.0. The statistical significance of factor effects and their interactions were analysed using analysis of variance (ANOVA) tests ($P < 0.05$). For the optimization processes, data were analysed using Design Expert version 7.0 and multiple regression used the least square method to fit the second order model to all variables to obtain the coefficients of determination, probability values and F ratios.

50.3 Results and Discussion

Response surface analysis (RSA) was used to optimize the process conditions (frying temperature, vacuum pressure and frying time) and sample varieties and frying oils based on the Box-Behnken design (Table 50.3). The RSA results for different experimental runs are presented in Table 50.3 and regression coefficients and ANOVA for the dependent variables in Table 50.4. Oil content varied between 5.36% and 18.60% (dry weight basis, dwb) and moisture content between 1.76% and 7.25% (Table 50.3). Beta-carotene of the fried samples varied between 1.89 ppm and 109.90 ppm while the breaking force (texture) varied between 3.15 N and 20.55 N. For the colour parameters, lightness, redness and yellowness varied between 1.04 and 62.60, 0.64 and 74.25, and 21.95 and 72.97, respectively. As shown in Table 50.4, all main, linear, quadratic and interactive effects were calculated for each model and the adequacy of the models was tested using coefficient of determination (R^2) and F value. The high R^2 results indicate that the variables were adequately fitted to the regression model. Response surface and contour plots for responses as affected by frying temperature and vacuum pressure at constant frying time using a particular variety of sweetpotato and frying oil were obtained and are available upon request from the authors.

The vacuum frying process involves heating of food under reduced pressure in a closed system which lowers the boiling points of both water and frying oil. Under these conditions, unbound water is removed quickly when the oil temperature attains the boiling point of water. First, RSA results found that moisture content of fried sweetpotato chips decreases as frying temperature and vacuum pressure increases at constant frying time. These results support those of Garayo and Moreira (2002) and Sobukola *et al.* (2013b) who report a similar trend. Various researchers have reported that vacuum frying reduces oil content (Dueik *et al.*, 2010; Tarzi *et al.*, 2011; Sobukola *et al.*, 2013a) and preserves natural colours and flavours of fried foods (Shyu *et al.*, 1998; Dueik *et al.*, 2010). Secondly, RSA results show that the oil content of fried chips increased at the same rate as frying temperature and vacuum pressure. These results are similar to the work of Shyu *et al.* (1998). This could be due to structural changes that occur at higher temperatures and pressure triggering degradation and increasing infiltration of adhered surface oil (Bouchon *et al.*, 2003).

Table 50.3. Response surface analysis (RSA) results as a function of the independent variables.[a]

Experimental run	Oil (%)	Moisture (%)	Beta-carotene (ppm)	Texture (N)	Lightness	Redness	Yellowness
1	16.38	2.10	53.37	6.9	26.58	0.84	71.88
2	13.72	1.76	4.89	5.75	1.04	74.25	23.88
3	17.07	2.12	82.74	9.50	32.53	10.87	56.33
4	17.10	4.27	60.79	14.00	2.02	69.57	27.96
5	11.33	2.45	48.32	8.65	27.96	1.37	69.86
6	17.83	1.85	62.21	5.05	26.58	31.23	41.96
7	9.93	3.07	54.79	3.25	26.33	1.26	71.54
8	16.99	1.66	1.89	3.15	1.96	71.37	25.97
9	7.51	2.85	68.21	20.55	31.24	42.37	25.65
10	19.32	1.87	80.69	4.90	19.93	22.71	56.78
11	10.85	2.14	75.48	9.45	28.12	1.45	70.03
12	17.34	1.94	86.06	7.10	25.91	1.16	72.62
13	6.93	2.94	71.85	8.60	41.27	35.96	21.95
14	7.10	1.97	87.95	4.90	62.60	7.19	29.87
15	10.92	2.07	75.32	15.00	34.95	3.02	61.66
16	18.30	2.17	104.69	5.60	38.26	2.96	57.97
17	16.16	2.07	69.00	5.90	39.64	1.87	57.94
18	17.00	1.83	72.16	7.80	4.52	67.84	27.48
19	17.29	1.77	94.58	12.20	1.76	71.27	25.56
20	18.60	2.13	103.27	10.40	1.84	38.74	58.63
21	17.58	1.68	96.64	6.35	26.30	0.64	72.83
22	5.36	7.25	66.16	5.15	58.31	9.86	31.14
23	10.41	2.05	82.89	6.60	26.05	1.04	72.74
24	15.20	1.94	107.85	9.45	1.78	39.86	58.33
25	17.83	2.38	109.90	4.45	23.86	11.26	63.96
26	8.17	5.84	71.37	5.32	1.87	72.40	24.96
27	15.00	1.93	8.53	3.65	1.74	73.84	23.96
28	15.80	2.25	72.48	12.85	26.09	0.91	72.97

[a]Values are means of duplicates.

Researchers on beta-carotene (a precursor of vitamin A) have reported that it is deficient in diets of most people in developing countries. Carotenoids are subject to destruction by oxidation, light, minerals, heat, moisture and length of storage. In this study, beta-carotene is significantly ($P < 0.05$) affected by: (i) linear effects of temperature and time; (ii) the quadratic effect of temperature; and (iii) interaction effects of temperature and time, temperature and variety, time and vacuum pressure, and time and variety. RSA results found that increasing temperature decreases the level of beta-carotene retention; at higher vacuum pressures, lower levels of retention were obtained. Similar results were reported by Yang *et al.* (2012). Clearly, high frying temperature easily degrades heat-labile functional compounds in foods.

Crispness is the most important textural attribute of fried food products and an indicator of freshness and quality (Tarzi *et al.*, 2011). Breaking force (N) was reported as the crispness of fried sweetpotato chips. Fried samples with a lower breaking force value are referred to as being crispy. Texture is significantly different ($P < 0.05$) and is affected by: (i) time, variety, oils, temperature and interaction effects of frying temperature and oils; (ii) frying time and vacuum pressure, frying time and variety; and (iii) frying time and frying oil. RSA findings indicate increasing frying temperature and vacuum pressure, increases the breaking force of the fried chip.

Colour of the fried product is also a quality driving consumer acceptability. The changes in colour during frying are the

Table 50.4. Regression coefficients for the quadratic models for all the responses.[a]

Parameters	Oil	Moisture	Beta-carotene	Texture	Lightness	Redness	Yellowness
Intercept	9.85	1.78	81.68	7.15	23.23	14.69	61.66
A	−0.02	−0.50	−15.68*	1.37	−11.38*	20.04*	−9.63
B	0.23	−0.42	−2.90	1.19	0.66	−11.25	10.56
C	0.85	−1.08	−13.74*	2.29*	−11.57*	14.61	−2.90
D	0.77	−0.23	−4.47	−1.73*	5.58	−7.91	2.43
E	1.07	0.01	3.58	−2.08*	−4.65	2.06	2.48
A^2	−1.34	0.45	−44.31*	−4.70*	−2.07	34.28*	−32.28*
B^2	0.57	0.46	2.68	−0.85	2.23	−3.58	1.29
C^2	0.13	−0.33	11.62	0.04	−5.01	12.35	−7.56
D^2	0.15	0.27	17.99	2.02	−6.65	0.83	5.69
E^2	−2.89	0.17	−3.68	4.22*	6.83	−17.17	10.55
AB	0.16	0.29	11.57*	0.69	9.26	−2.60	−6.84
AC	−0.57	0.08	−4.30	−1.66	−0.61	4.56	−4.09
AD	1.01	0.40	−13.46*	−0.30	−2.56	−0.65	3.15
AE	−0.46	0.54	3.84	−3.27*	8.11	−9.91	1.69
BC	0.51	0.71	16.45*	−2.96*	0.27	3.20	−3.60
BD	−0.14	0.18	−14.71*	−0.56	−1.08	−1.79	2.79
BE	−0.68	0.06	1.83	0.07	9.16*	−11.99	2.79
CD	−1.34	0.50	−11.28*	−3.55*	11.19*	−9.20	−2.13
CE	1.14	0.25	−8.32	2.45*	1.05	−0.72	−0.56
DE	−0.44	−0.09	8.37	−0.03	6.06	5.97	0.0012
R^2	0.76	0.72	0.95	0.95	0.89	0.91	0.875
Press	2,209.46	1,380.80	37,000.23	3,415.56	48,307.68	200,000	95,030.57
F value	1.75	0.89	6.80	5.51	2.89	3.34	2.46

[a]Values marked with * are significant at $P < 0.05$.

result of Maillard reaction between reducing sugars and amino acids at the surface coupled with temperature and frying time (Marquez and Anon, 1986). Lightness of fried sweetpotato chips was significantly different ($P < 0.05$) and was affected by temperature, frying time and interaction of frying time and variety. This quality decreases as temperature and vacuum pressure increases due to caramelization and chemical oxidation of phenolic compounds. Similarly, the redness parameter was greatly influenced by frying temperature and vacuum pressure. This parameter decreases as vacuum pressure increases but increases as frying temperature increases. The redness parameter (a*) was significantly ($P < 0.05$) affected by linear and quadratic effects of temperature (Table 50.4). A golden yellow colour is usually preferred by consumers in fried products. The yellowness parameter of fried chips was significantly ($P < 0.05$) affected by temperature (Table 50.4). Increasing frying temperature and vacuum pressure encourages the Maillard reaction and caramelization which increases yellowness.

Table 50.5. Comparison of findings between atmospheric and optimized vacuum-fried sweetpotato chips.

	Atmospheric	Optimized vacuum
Oil (% dwb)	4.22	9.33
Moisture (%)	3.49	2.68
Beta-carotene (ppm)	83.39	112.04
Texture (N)	10.34	7.75
Lightness	42.17	28.06
Redness	1.08	−15.96
Yellowness	70.91	72.97

Based on desirability concept (0.835) in which samples of lower oil content, golden yellow and high retention of beta-carotene were preferred, the coded optimized vacuum frying conditions and materials are -0.64, -1, 1, 1 and -1 for frying temperature, vacuum pressure, frying time, cultivar and frying oil. The optimized vacuum fried samples were then compared with atmospheric fried samples based on ETDF where the same difference was obtained between the frying oil temperature and boiling point of water. Table 50.5 shows that the optimized vacuum fried samples have better quality attributes than atmospheric fried ones despite having higher oil contents. This can be due to the inability of vacuum frying conditions to encourage surface starch gelatinization that can prevent infiltration of adhered oil on the surface of the fried chips.

Acknowledgements

The authors are grateful to the International Foundation for Science for providing the corresponding author, O.P. Sobukola, with funds (E/4891/1) for this research.

References

Association of Official Analytical Chemists (AOAC) (1995) *Official Methods of Analysis*, 16th edn. AOAC, Washington DC.

Bengtsson, A., Brackmann, C. and Enejder, A. (2010) Effects of thermal processsing on the *in vitro* bioaccessibility and microstructure of β-carotene in orange fleshed sweet potato. *Journal of Agricultural and Food Chemistry* 58(20), 11090–11096.

Bouchon, P., Aguilera, J.M. and Pyle, D.L. (2003) Structure–oil relationship during deep fat frying. *Journal of Food Science* 68, 2711–2716.

Da Silva, P.F. and Moreira, R.G. (2008) Vacuum frying of higher quality fruit and vegetable-based snacks. *Lebensmittel-Wissenschaft und Technologie – Food Science and Technology* 41, 1758–1767.

Dueik, V., Robert, P. and Bouchon, P. (2010) Vacuum frying reduces oil uptake and improves the quality parameters of carrot crisps. *Food Chemistry* 119, 1143–1149.

Fan, L.P., Zhang, M., Xiao, G.N., Sun, J.C. and Tao, Q. (2005) The optimization of vacuum frying to dehydrate carrot chips. *International Journal of Food Science and Technology* 40, 911–919.

Garayo, J. and Moreira, R. (2002) Vacuum frying potato chips. *Journal of Food Engineering* 55, 181–191.

Low, J.W., Arimond, M. Osman, N., Cunguara, B., Zano, F. and Tschirley, D. (2007) A food-based approach introducing orange-fleshed sweet potatoes increased vitamin A intake and serum retinol concentrations in young children in rural Mozambique. *Journal of Nutrition* 137(5), 1320–1327.

Marquez, G. and Anon, M. (1986) Influence of reducing sugars and amino acids in the colour development of fried potatoes. *Journal of Food Science* 51, 157–160.

Mellama, M. (2003) Mechanism and reduction of fat uptake in deep fat fried foods. *Trends in Food Science and Technology* 14, 364–373.

Mir-Bel, J., Oria, R. and Salavdor, M.L. (2009) Influence of the vacuum break conditions on oil uptake during potato post-frying cooling. *Journal of Food Engineering* 95, 416–422.

Nunes, Y. and Moreira, R.G. (2009) Effect of osmotic dehydration and vacuum-frying parameters to produce high quality mango chips. *Journal of Food Science* 74(7), 355–362.

Shyu, S., Hau, L. and Hwang, L. (1998) Effect of vacuum frying on the oxidative stability of oils. *Journal of the American Oils Chemist's Society* 75, 1393–1398.

Shyu, S., Hau, L. and Hwang, L. (2005) Effects of processing conditions on the quality of vacuum fried carrot chips. *Journal of the Science of Food and Agriculture* 85, 1903–1908.

Singh, R.P. (1995) Heat and mass transfer in food during deep fat frying. *Food Technology* 49(4), 134–137.

Sobukola, O.P., Dueik, V. and Bouchon, P. (2013a) Understanding the effect of vacuum level in structure development and oil absorption in vacuum-fried wheat starch and gluten based snacks. *Food and Bioprocess Technology* 6(8), 2010–2017.

Sobukola, O.P., Dueik, V., Mounoz, L. and Bouchon, P. (2013b) Comparison of vacuum and atmospheric deep fat frying of wheat starch and gluten based snacks. *Food Science and Biotechnology* 22(S), 177–182.

Tarzi, B.G., Bassiri, A., Ghavami, M. and Bamenimoghadam, M. (2011) Process optimization in vacuum frying of mushroom using response surface methodology. *World Applied Sciences Journal* 14(7), 960–966.

Therdthai, N., Wuttijumnong, P., Jangchud, A. and Kusucharid, C. (2007) Optmization of vacuum frying condition of shallot. *Kasetsart Journal* 41, 338–342.

Troncoso, E. and Pedreschi, F. (2009) Modeling water loss and oil uptake during vacuum frying of pre-treated potato slices. *Lebensmittel-Wissenschaft und Technologie – Food Science and Technology* 42(6), 1164–1173.

Troncoso, E., Pedreschi, F. and Zuniga, R.N. (2009) Comparative study of physical and sensory properties of pre-treated potato slices during vacuum and atmospheric frying. *Lebensmittel-Wissenschaft und Technologie – Food Science and Technology* 42, 187–195.

United States Department of Agriculture (USDA) (2008) USDA National Nutrient Database for Standard Reference. Release 21. Available at: http://ndb.nal.usda.gov/ (accessed 15 September 2015).

Virtrac, O., Dufour, D., Trystram, G. and Raoult-Wack, A. (2000) Deep-fat frying of cassava: influence of raw material properties on chip quality. *Journal of the Science of Food and Agriculture* 81, 227–236.

Woolfe, J.A. (1992) *Sweet Potato: an Untapped Food Resource*. Cambridge University Press, Cambridge.

Yang, J.H., Park, H.Y., Kim, Y.S., Choi, I.W., Kim, S. and Choi, H.D. (2012) Quality characteristics of vacuum-fried snacks prepared from various sweetpotato cultivars. *Food Science and Biotechnology* 21, 525–530.

Ziaiifar, A.M., Achir, N., Courtois, F., Trezzani, I. and Trystram, G. (2008) Review of mechanisms, conditions and factors involved in the oil uptake phenomenon during deep fat frying process. *International Journal of Food Science and Technology* 43(8), 1410–1423.

51 Price Integration of Sweetpotato Marketing: Implications for an Efficient Marketing System in Nigeria

H.N. Anyaegbunam,[1]* A.C. Nwosu[2] and J.A. Mbanasor[2]
[1]*National Root Crops Research Institute, Umudike, Nigeria;*
[2]*Michael Okpara University of Agriculture, Umudike, Nigeria*

Abstract
This 2010 study was carried out in the south-east agroecological zone of Nigeria to determine the degree of market integration and infrastructural and social problems associated with sweetpotato marketing. In a food market economy like Nigeria, an efficient, integrated and spatial pricing system should lead to economic growth. This will be true if the price differential in traded product due to costs of transportation, storage and processing is effectively minimized over time and space. A multistage random sampling technique was used in selecting states, markets and respondents. A total of four states with 360 respondents (comprising 240 retailers, 120 wholesalers) in 24 markets were randomly chosen. The list of wholesalers and retailers who pay stall fees formed the sampling frame. Primary data were collected using a structured questionnaire. Analyses of data provided descriptive statistics and bivariate correlation coefficients. Results revealed that the markets were spatially separated but not fully integrated. The major problems of sweetpotato marketing identified include, in order of severity (most to least): (i) the lack of credit facilities; (ii) perishability of sweetpotato; (iii) lack of good processing machines; (iv) inadequate market stalls; (v) high transport cost; (vi) bulkiness of the commodity; (vii) lack of storage facilities; and (viii) the negative attitude of law enforcement agents. The study concludes that in general, the markets in the zone are integrated but not yet perfect. Consequently, we recommend that advocacy on structural lending reforms be conducted and facilities provided to enhance marketing activities and efficiency in the zone. Market actors should be encouraged to form cooperatives to capture economies of scale, especially for capital build-up and to establish bulking centres.

Keywords: correlation, marketing, Nigeria, price integration, sweetpotato

51.1 Introduction

Sweetpotato (*Ipomoea batata* (L.) Lam.) is an important tropical root crop, second after cassava. The crop can be considered very important in promoting nutritional security, particularly in agriculturally semi-arid areas. Besides carbohydrates, it is a rich source of protein, lipid, calcium and carotene. It is an ideal crop for areas with poor soils and poor agricultural facilities (Low *et al.*, 2009).

*helenkol8@yahoo.com

In annual production, sweetpotato ranks as the fifth most important food crop on a fresh-weight basis in developing countries after rice, wheat, maize and cassava. Sweetpotato is cultivated in 114 countries and ranks among the five most important food crops in over 50 countries. Asia has the world's major production area for sweetpotato. In Asia, the greatest share of production is in China, which accounts for 85% of global production (Srinivas, 2009). According to FAOSTAT (2008), 23 countries produce 90% of all sweetpotato in sub-Saharan Africa (SSA). Uganda and Nigeria dominate in terms of overall sweetpotato production in SSA, accounting for 33% of total production.

In Nigeria, sweetpotato has diversified uses. It is eaten boiled, fried or roasted. Sometimes, the roots are eaten in pounded form (Chukwu, 2001). Omodamiro *et al.* (2009) highlighted that orange-fleshed sweetpotato (OFSP)-based drinks containing lemon as preservative competed favourably with similar drinks in the market. Similarly, sweetpotato confectionary recipes have been formulated by the National Root Crops Research Institute, Umudike, Nigeria. Sweetpotato flour can be used to make cake, strips, chin-chin, salad cream and bread balls (Aniedu and Oti, 2008). OFSP varieties are currently being used in Africa to combat widespread vitamin A deficiency that is linked with impaired immune systems, blindness and even death among young children (Low *et al.*, 2009).

Sweetpotato production in Nigeria rose from 2.52 million t in 2006 to 3.4 million t in 2007 (FAOSTAT, 2008; Akoroda, 2009; Srinivas 2009). This increase has been attributed to improved technological inputs and international and national research efforts. Tewe *et al.* (2003) asserted that increased sweetpotato production is not being matched by adequate promotion and marketing activities to absorb surpluses from production, which is detrimental to the sustainability of sweetpotato production in Nigeria.

Sweetpotato is perishable and seasonal, resulting in variation in the quantity and quality of roots in the market, with associated price swings (Low *et al.*, 2009). Rising consumer prices may indicate market inefficiency. Spatial analysis or market integration studies seek to depict the relationship between price movements in two markets. They measure the extent to which markets are associated in terms of price behaviour and inter-market commodity movements (Timmer *et al.*, 1993). The ability of a marketing system to effectively and efficiently perform its functions depends on the ease with which price changes and responses are transmitted spatially and temporally (Onyeka *et al.*, 2006). In a food market economy, an efficient spatial pricing system usually leads to economic growth, *if* it effectively minimizes the price differential in traded product due to costs of transportation, storage and processing over time and space. Transportation of commodities from a surplus area to a deficit area takes place due to price differentials and consumption needs over time and space. Price differentials are also due to storage of the commodity over time, and processing costs lead to price differences at a particular point in time and space (Reddy *et al.*, 2006). Spatial pricing efficiency is usually affected by the rigidity of the marketing costs and the nature and degree of competition in marketing.

In Nigeria, sweetpotatoes are transported from major areas of production to non-producing or less-producing areas. Thus, it is of interest to determine how integrated prices are in the different markets. Nigeria operates an open market economy implying that market integration can be used to test the economic efficiency of the market (Chikwendu, 2003). Therefore, the objectives of this study were to:

- determine the extent of market integration for sweetpotato markets; and
- identify infrastructural and social problems associated with sweetpotato marketing.

51.2 Methodology

The study was carried out in the south-east agroecological zone of Nigeria in 2010. This zone is known for high levels of production and marketing of sweetpotato. The sampling frame used was the list of wholesalers and retailers who pay stall fees. A multistage random sampling technique was used in the selection of markets and respondents. In the first stage Cross River, Anambra, Ebonyi and Enugu were selected randomly out of the nine states that make up the south-east agroecological zone. In the second stage, two agricultural zones were selected in each of the chosen states, for a total of eight agricultural zones. Three markets were chosen from the selected agricultural zone in stage three while in stage four, ten retailers and five wholesalers were randomly selected in each of the selected 24 markets from the list of marketers who pay stall fees. A total of 360 respondents comprised of 240 retailers and 120 wholesalers were chosen. A structured questionnaire guided data collection. Time series as well as cross-sectional data on prices from the various markets were also collected on a weekly basis for 1 year. This ensured that data across all seasons were collected since sweetpotato is seasonal. Data collected were on: (i) prices of sweetpotato per kilogram (kg); (ii) transportation costs in the local currency (Naira); (iii) use of infrastructural facilities; and (iv) other problems faced by the marketers.

Analysis of data

Data collected were analysed, generating correlation coefficients and simple descriptive statistics such as percentages, frequencies, means and tables. In testing pricing efficiency, the bivariate correlation coefficients between sweetpotato prices for the spatially separated markets were computed to determine how spatially integrated the sweetpotato markets were, following methods used by Diavosavvas (1995), Mendoza and Rosegrant (1995) and Asumugha *et al.* (2003). The bivariate correlation coefficient, a statistical measure of association, is the most common methodological approach used for determining spatial integration (Trotter, 1992). It is practical for applications in developing countries because of its minimal data requirements (Mendoza and Rosegrant, 1995). Trotter (1992) described it as a scale-free measure of association.

The model is specified generally as Eqn 51.1:

$$P_{ij} = bo + bi\,P_{ik} + e \qquad (51.1)$$

Where:
P_{ij} = price series of market j
P_{ik} = price series of market k
bo, *bi* = coefficients
e = error term

The equation was estimated for pairs of sweetpotato markets. The closer *bi* is to unity, the more spatially integrated the sweetpotato markets are. The size of this coefficient and its significance shows the level of inter-market dependence. A correlation coefficient (*r*) greater than 0.8 signifies strong correlation, when '*r*' falls between 0.6 and 0.8, it is regarded as a moderate correlation, while a correlation coefficient (*r*) of less than 0.6 is termed weak correlation (Asumugha *et al.*, 2003).

Theoretical framework

The procedure for analysing the spatial price differentials utilized the competitive market model and pricing efficiency to test the extent of sweetpotato integration among the 24 selected markets in different parts of the zone. In this respect, the bivariate correlation coefficients between pairs of sweetpotato markets at different locations were computed (Asumugha, 2003).

Price correlation analyses have been described as means of assessing economic efficiency and have been used to indicate how well and how far and fast price changes are transmitted through the market system (Scaborough and Kydd, 1992). This assumes the law of one price. This is the basis on which correlation coefficient analyses are used in determining the size and degree of integration of markets in various time, space and dimensions.

51.3 Results and Discussion

Market price integration in the four states

Table 51.1 shows the price movement of sweetpotato for the wholesale sector in the

Table 51.1. Wholesale price correlation results[a] in four south-east Nigerian states. (From field data, 2010.)

States	Cross River	Enugu	Ebonyi	Anambra
Cross River	1.00			
Enugu	0.372	1.00		
Ebonyi	0.908	0.932	1.00	
Anambra	0.146	0.734	0.42	1.00

[a]Correlation is significant at the 0.05 level (two tailed) $r > 0.8$ = strong correlation; $r = 0.6–0.8$ = moderate correlation; $r < 0.6$ = weak correlation.

Table 51.2. Retail price correlation results[a] in four south-east Nigerian states. (From field data, 2010.)

States	Cross River	Enugu	Ebonyi	Anambra
Cross River	1.00			
Enugu	0.68	1.00		
Ebonyi	0.036	0.710	1.00	
Anambra	0.759	0.546	0.546	1.00

[a]Correlation is significant at the 0.05 level (two tailed) $r > 0.8$ = strong correlation; $r = 0.6–0.8$ = moderate correlation; $r < 0.6$ = weak correlation.

four states under study. The results revealed strong positive correlations between Cross River and Ebonyi states ($r = 0.9$), and Enugu and Ebonyi states ($r = 0.9$). A moderate positive correlation was recorded between Enugu and Anambra states ($r = 0.7$). The correlation between Cross River and Enugu states, Cross River and Anambra states, then Ebonyi and Anambra states were weak with $r = 0.4$, 0.1 and 0.4, respectively. The implication is that for states that were strongly and moderately correlated, there were positive price movements across the states, tending towards identical price movement where $r = 1.00$. Markets in these states can be considered spatially integrated though not perfectly integrated.

According to Scarborough and Kydd (1992), a coefficient of 1.00 indicates identical or perfect price movements in two markets but this does not occur in reality due to imperfect resource mobility resulting from transport, storage and processing costs, imperfect information and in some cases product differentiation. States with weak correlation coefficients have markets that were not spatially integrated. This may be as a result of market information constraints and transport problems which is typical of less developed countries (Parmod and Sharma, 2003).

The correlation results for price movement in the four states for retailers are shown in Table 51.2. The analysis revealed that moderate correlations existed between Cross River and Enugu states ($r = 0.7$), Cross River and Anambra states ($r = 0.8$) and then Enugu and Ebonyi states ($r = 0.7$). The correlations between Cross River and Ebonyi states, Enugu and Anambra states and then Ebonyi and Anambra states were weak with $r = 0.03$, 0.5 and 0.5, respectively. No strong correlations between states were recorded in the retail sector. This may be attributed to the low capital base associated with retailing activity and high marketing costs. The wholesale markets often showed better price integration than retail markets, as reflected in stronger correlation coefficients.

Problems of sweetpotato marketing

There are numerous problems facing sweetpotato marketers (wholesalers and retailers) (Table 51.3). Data collected on the constraints were only on the sale of sweetpotato roots, not vines. Lack of credit facilities was the major problem facing sweetpotato marketers because of bureaucratic processes involved in securing loans from formal institutions. About 96% of the wholesalers and 99% of the retailers indicated that lack of a credit facility was a major problem. This made the marketers rely on their own limited resources or on informal credit sources such as *isusu* (monetary contribution by a group of people of common interest from which members borrow money at a minimal interest rate or collect the whole amount of money depending on one's turn) and money lenders (charging high interest rates).

Table 51.3 Problems of sweetpotato marketing as indicated by wholesalers and retailers. (From field data, 2010.)

	Wholesalers		Retailers	
Problem	Frequency[a]	%	Frequency[a]	%
Lack of credit facilities	115	95.83	239	99.58
Perishability of sweetpotato	110	91.67	237	98.75
Lack of good processing technology	105	87.50	230	95.83
Seasonality of sweetpotato	100	83.33	220	91.67
Poor access roads	100	83.33	225	95.75
Inadequate market stalls	90	75.00	200	83.33
High transport cost	85	70.83	200	83.33
Bulkiness of the commodity	83	66.67	150	62.50
Lack of storage facilities	70	58.33	100	41.67
Negative attitude of law enforcement agents	50	41.67	90	37.50

[a]Some respondents indicated more than one problem so multiple responses were recorded.

Sweetpotato is a perishable crop, it does not store for a long time once harvested. Because of this, it has a lot of risks, such as deterioration while in transit or during storage, and uncertainties such as price fluctuations. About 92% of the wholesalers and 99% of the retailers indicated perishability as a major problem. The retailers were more prone to this risk since they bought and stored the sweetpotatoes, selling them piecemeal while the wholesalers sell in bulk.

Close to 88% of the wholesalers and 96% of the retailers indicated that lack of a good processing technology was one of the major problems militating against an efficient sweetpotato marketing system. This problem restricted the marketing of sweetpotato due to lack of diversification in its use commercially.

Sweetpotato is a seasonal crop, with its availability tied to a particular period. Because of this, some middlemen find it difficult to finance its marketing activities. About 83% of the wholesalers and 92% of the retailers indicated that this is a serious problem. Retailers demonstrated more concern than wholesalers, because the latter had more financial strength and were able to buy sweetpotatoes at higher prices during the off-season, whereas the retailers are restricted by their very low capital base.

Poor access roads were another major constraint militating against an efficient marketing system, highlighted as a problem by 83% of the wholesalers and 96% of the retailers. Some roads in the zone were not accessible especially during the rainy season. This resulted in increased prices of sweetpotato, which were eventually transferred to the consumers.

Inadequacy of market stalls was one of the problems confronting sweetpotato marketers (75% of wholesalers; 83% of retailers) in both rural and urban markets. Inadequate market stalls resulted in traders displaying their products in open spaces in the open air. Such products are prone to rapid deterioration. Those traders that managed to secure market stalls in the rural markets paid significant fees, between (₦)100 and ₦200 every market day, which occurred every 4 days. Traders in the urban markets paid between ₦500 and ₦1000 monthly. The deplorable state of the rural–urban roads resulted in high transport fares and price fluctuations.

About 67% of the wholesalers and 63% of the retailers indicated that the bulkiness of sweetpotato roots was another problem facing them. This problem resulted in high transportation and handling charges, especially for wholesalers (58% reported this problem specifically compared with only 42% of retailers, who deal in smaller quantities). These problems need to be addressed properly to ensure efficient sweetpotato marketing in the zone.

51.4 Conclusions and Recommendations

The findings of this study revealed that the prices of sweetpotato in the study region were spatially integrated but not perfectly, though prices between some markets are tending towards identical/perfect integration. Marketers faced many problems and these problems hinged primarily on lack of credit and inadequate infrastructural facilities.

Therefore, it is recommended that advocacy on structural lending reforms be undertaken and infrastructural facilities provided in order to enhance marketing activities and efficiency in the zone. Market actors should be encouraged to form cooperatives to capture economies of scale, especially for capital build-up and to establish bulking centres.

Acknowledgements

The first author is grateful to the Director Dr J.C. Okonkwo and the management of the National Root Crops Research Institute, Umudike, Abia State, Nigeria for sponsoring the field work and also to the International Potato Center (CIP), Lima, Peru for the scholarship award to present this work at the 2013 African Potato Association (APA) conference in Kenya.

References

Akoroda, M.O. (2009) Sweetpotato in West Africa. In: Loebenstein, G. and Thottappilly, G. (eds) *The Sweetpotato*. Springer Science + Business Media BV, Houten, The Netherlands, pp. 441–468.

Aniedu, C. and Oti, E. (2008) *Development of Sweetpotato Based Recipes*. Extension Guide No. 24. National Root Crops Research Institute, Umudike, Nigeria, pp. 10–11.

Asumugha, G.N., Njoku, J.E. and Nweke, F.I. (2003) Marketing system and spatial differentiation of ginger in Nigeria. *African Journal of Root and Tuber Crops* 3(2), 1–6.

Chikwendu, D.O. (2003) Price communication and market integration: the case of banana and plantain in Nigeria. *Samaru Journal of Agricultural Research* (Institute of Agricultural Research, Zaria, Nigeria) 19, 29–35.

Chukwu, G.O. (2001) Seasonality and climate period effects on crop evapotranspiration of sweet potato. In: Akoroda, M.O. and Ngeve, J.M. (compilers) *Root Crops in the 21st Century*. Proceedings of the 7th Triennial Symposium of the International Society for Tropical Root Crops – African Branch (ISTRC-AB), 11–17 October 1998, Centre International des Conférence, Cotonou, Benin Republic. ISTRC-AB, Cotonou, Benin Republic, pp. 280–284.

Diakosavvas, D. (1995) How integrated are world beef markets? The case of Australian and US beef markets. *Agricultural Economics* 12, 37–53.

FAOSTAT (2008) *Production Statistics*. Food and Agricultural Organization of the United Nations (FAO), Rome. Available at: http://faostat.fao.org/site/339/default.aspx (accessed 15 September 2015).

Low, J., Lynam, J., Lemaga, B., Crissman, C., Barker, I., Thiele, G., Namanda, S., Wheatley, C. and Andrade, M. (2009) Sweetpotato in sub-Saharan Africa. In: Loebenstein, G. and Thottappilly, G. (eds) *The Sweetpotato*. Springer Science + Business Media BV, Houten, The Netherlands, pp. 355–386.

Mendoza, M.S. and Rosegrant, M.W. (1995) Pricing conduct of spatially differentiated markets. In: Scott, G.J. (ed.) *Prices, Products and People: Analyzing Agricultural Markets in Developing Countries*, Reprint No. 318. International Potato Center (CIP) and Lynne Rienner Publishers, Boulder, Colorado.

Omodamiro, R.M., Afuaekpe, S. and Etundaye, H.A. (2009) Consumer acceptance of orange fleshed-sweetpotato based drinks containing lemon as a preservative. In: *National Root Crops Research Institute, Umudike, Nigeria, Annual Report*. National Root Crops Research Institute, Umudike, Nigeria, pp. 219–220.

Onyeka, U.P., Olayemi, J.K., Kormawa, P. and Mifimisebi, T.E. (2006) Testing market integration of staple foods in Borno state, Nigeria. *Nigerian Agricultural Journal* 37, 11–17.

Parmod, K. and Sharma, R.K. (2003) Spatial price integration and price efficiency: a study of paddy in Haryana. *Indian Journal of Agricultural Economics* 58(2), 201–217.

Reddy, S.S., Ram, P.R., Sastry, T.V.N. and Devi, I.B. (2006) *Agricultural Economics*. Oxford and IBH Publishing Co., New Delhi, 652 pp.

Scarborough, V. and Kydd, J. (1992) *Economic Analysis of Agricultural Markets: a Manual.* Natural Resources Institute, Chatham Maritime, UK, 66 pp.

Srinivas, T. (2009) Economics of sweetpotato production and marketing. In: Loebenstein, G. and Thottappilly, G. (eds) *The Sweetpotato.* Springer Science + Business Media BV, Houten, The Netherlands, pp. 235–268.

Tewe, O.O., Ojeniyi, F.E. and Abu, O.A. (2003) *Sweetpotato Production, Utilization and Marketing in Nigeria.* International Potato Center (CIP), Lima and the University of Ibadan, Oyo State, Nigeria, 54 pp.

Timmer, C.P., Falcon, W.P. and Pearson, S.R. (1983) *Food Policy Analysis.* Johns Hopkins University Press for the World Bank, Baltimore, Maryland.

Trotter, B.W. (1992) *Applying Price Analysis to Marketing Systems: Methods and Examples from Indonesian Rice Markets.* Marketing Series 3. Natural Resources Institute, Chatham Maritime, UK, p. 60.

52 Transaction Costs and Agricultural Household Supply: Response of Sweetpotato Farmers in Kwara State of Nigeria

C.O. Farayola,* S.C. Nwachukwu and B.I. Alao
Agricultural and Rural Management Training Institute (ARMTI), Ilorin, Nigeria

Abstract
Transaction costs arise in the course of market exchange and involve the cost of information, search, negotiation, screening, monitoring, coordination and enforcement. These are costs related exclusively to coordination of exchange among market actors that are distinct from the physical costs of transferring goods, such as transport, handling and storage costs. Transaction costs vary by individual, leading to heterogeneous market behaviours. This chapter looks into the relationship between transaction costs and household supply response among sweetpotato farmers in Kwara State. A multistage random sampling technique was employed in selecting 120 respondents for the study. A structured interview schedule was used to collect data from the respondents in September–October 2012. Data were analysed using descriptive statistics and an estimation Cobb-Douglas regression model. The descriptive analysis revealed the mean age of respondents was 43 years, of which 81% were male. Only 18% of the sample had no formal education and more than half of the respondents (65%) had farming as their main occupation. Thirty-nine per cent of the respondents cultivated less than 1 ha of land, 53% cultivated 1–2 ha while 8% cultivated above 2 ha. Farmers in the study area grow sweetpotato on a small-scale level. The majority (71%) of sweetpotato farming households used their personal savings to finance cultivation. The elasticity of supply response showed that with respect to price, area of land cultivated, negotiation cost, agent's fee, harvesting cost, assemblage cost, storage cost, transportation cost and land rent transactions cost, a 10% change in each of the variables resulted in 7.2%, 10.5%, 1.0%, 13.4%, 4.8%, 0.8%, 1.2%, 1.5% and 9.3% variation in the quantity of sweetpotatoes supplied to households, respectively. The adjusted *R*-squared (R^2) was 0.734, showing that 73.4% of the variations in quantity of sweetpotato supplied were explained by the estimated variables. Results showed that significant relationships exist between transaction costs and agricultural household supply response in the study area. Hence, reducing transaction costs will complement price policies in affecting the supply response. Policy recommendations based on these results are made.

Keywords: household, Nigeria, supply response, sweetpotato, transaction cost

*walecharless@yahoo.com

52.1 Introduction

Transaction costs refer to the cost of measuring the valuable attributes of the commodity exchanged and the costs of providing and ensuring the desired attributes. Transaction costs of sales are composed of information, negotiation, screening, and enforcement and supervision costs. Information costs represent the costs of searching for a partner with whom to exchange, screening the potential trading partners' trustworthiness and searching for the best price (Key *et al.*, 2000). Bargaining costs are costs incurred when negotiating with potential trading partners in order to reach an agreement. After the transaction has occurred, the agreement needs to be monitored and enforced (Hobbs, 1997; Staal *et al.*, 1997; Key *et al.*, 2000). The transfer of the product between the parties involves costs as well (Staal *et al.*, 1997; Key *et al.*, 2000). Also, in transforming agriculture into agribusiness, valuing compensation for farmers and their produce is important. A high level of trust and strong civic and social norms can lower transactions costs. On the other hand, uncertainty can increase transaction costs. Another factor that can affect transaction costs is the frequency of any decision-making process. Hence, research into transaction costs suggests that they can be quite varied depending on the sector and the country.

More than 700 million people in the developing world do not have access to sufficient food to live a healthy and productive life (Pinstrup-Andersen, 1994). The annual demand for food keeps growing at 3.3% and may not be matched by the growth in agricultural production (IITA, 2005). Not surprisingly, per capita calorie intake remains at low levels in sub-Saharan Africa (SSA), much below the developing world average. If current trends continue, there will be approximately 370 million persons that would be undernourished in 2050 (FAO, 2009), with SSA emerging as the region with the highest number of inhabitants who are chronically malnourished. Sweetpotato has a high yield potential realizable within a relatively short growing season. It is grown in a wide ecological range from 0 to 2000 m above sea level. According to the Food and Agriculture Organization of the United Nations (FAO, 2000) production of sweetpotato in Nigeria increased from 149,000 t in 1961 to 2,468,000 t in 2000. The area of cultivation increased from 13,000 ha to 381,000 ha while yield decreased from 11 t/ha to 6.8 t/ha over the same period.

Idachaba (2004) observed that food insecurity can be caused by supply-side factors and demand-side factors. One of the supply-side causes is the food marketing problem. He argued that the dwindling agricultural production in Nigeria is a confirmation of low returns to farmers, which tends to discourage increased production.

Most previous research focuses on price and effect on the agricultural supply response. Ajetomobi *et al.* (2006) carried out a supply analysis for food crops in Oyo State but only considered the own price factor. Abebe (2005) measured the supply response with respect to own price and cross-prices of cereals in Ethiopia. Krishna (1967) looked at agricultural price policy and economic development. Askari and Cummings (1977) looked into the agricultural supply response to price. Ahmed and Rustagi (1987) looked at marketing and price incentives in African and Asian countries while Mamingi (1996, 1997) measured the impact of prices and macroeconomic policies on agricultural supply. Farayola (2006) estimated the supply responses of Nigerian agricultural export crops at both aggregate and individual levels, and found that all crops considered are price responsive. S.O. Odunuga (1988, unpublished data) looked at the acreage response to prices in small-scale food-crop agriculture in Oyo State. Murova *et al.* (2001) and Leaver (2003) measured the responsiveness of agricultural output for Ukrainian and Zimbabwean farmers, respectively, to price but did not consider any market factors. Chibber (1988) worked on raising agricultural output through price and non-price factors but never took into account any market factor. Goetz (1992), however, used a selectivity model in which marketed surplus is estimated to be conditional on market participation. Key *et al.* (2000) also carried out a similar study on Mexican farmers and suggested that the issue of transaction costs creates a situation where some producers buy, others sell, and others do not participate in the markets.

The main objective of this study was to investigate the role of transaction costs in determining the sweetpotato supply response of farmers in Kwara State. The specific objectives were to: (i) analyse the socio-economic characteristics of sweetpotato farmers in the study area; (ii) identify variables associated with transaction costs in the study area; (iii) determine the magnitude and the direction to which the level of transaction costs influence changes in sweetpotato supply in the area; and (iv) estimate the elasticity of sweetpotato supply in the study area. The hypothesis that there is no significant relationship between transactions costs and the quantity of sweetpotato supplied by farmer households was tested.

52.2 Materials and Methods

The study area

The study was carried out in Kwara State in the North Central region of Nigeria. The state covers an area of 74,256 km^2, one-twelfth of the total area of Nigeria. In the state, there are 247,975 farm families with 254,242 ha of cropped area. The annual rainfall ranges from 600 mm to 1500 mm, between the months of April and October with peak rains in May–June and September–October. The months of November–February are virtually without rainfall and the mean temperatures ranges from 20°C to 22°C.

Agriculture is the mainstay of the economy and the main crops are: sweetpotato, cassava, yam, cowpea, groundnut, maize, sorghum, wheat, melon, kola nut, shea nut, tobacco, benseed, palm produce, okra, melon, pepper and some leafy vegetables (KWADP, 1996). The prevailing agricultural system combines bush fallow and mixed cropping with emphasis on subsistent farming, while some farmers engage in craft activities such as weaving, blacksmithing, bricklaying, carpentry and welding.

Sampling procedure and sample size

A multi-stage random sampling technique was employed to select 120 sweetpotato farmers. First, a purposive sampling technique was used to select Offa and Oyun local government areas (LGAs) where production and marketing of sweetpotato are predominant among LGAs of Kwara State (KWADP, 1996; MANR, 2010). Secondly, half the villages in Oyun (19 out of 38) were randomly selected. Half of the villages (one out of two) were also randomly selected in Offa LGA (a total of 20 villages).

Research instrument and methods of data analysis

A structured interview schedule was used to collect information on personal and socio-economic characteristics from sampled sweetpotato farmers beginning in mid-September 2012.

Descriptive and inferential statistics were used for data analysis. The descriptive statistics used includes mean, standard deviation, percentage and frequency counts while multiple regression models were used to test for significant relationships.

The regression model

Among commonly used production function forms, economists prefer using the log-linear Cobb-Douglas (C-D) model as well as constant elasticity of substitution (CES) models (Shang, 1990). The C-D model has unity elasticity of substitution whereas the CES permits the empirical data to determine the degree of substitutability among inputs. CES is, however, difficult to apply when more than two inputs are used; therefore the C-D model is mostly preferred by economists (Smith, 1982) and was used in this study.

Regression analysis was used to estimate the relationship between the dependent and the independent variables. This is statistical tool for predicting the value of a dependent variable given the values of the independent variables and measures the degree of association between two or more variables; hence, the coefficient of determination (R^2) shows the level of variation in quantity of sweetpotato supplied by the respondents (Q), which is explained by variation in X(s). The C-D production function used has evolved since its development early in the 1900s and has been widely used in both theoretical and empirical production analyses. There are, however, some criticisms to the use of this model. These are as follows: (i) it cannot handle a large number of inputs; (ii) the function is based on restrictive assumptions of perfect competition in the factor and product markets; (iii) it assumes constant returns to scale; (iv) serial correlation and heteroscedasticity are common problems that beset this function too; (v) labour and capital are correlated and the estimates are bound to be biased; (vi) unitary elasticity of substitution is unrealistic; (vii) it is inflexible in form; (viii) single equation estimates are bound to be inconsistent; and (ix) it cannot measure technical efficiency levels and growth very effectively (Bhanumurthy, 2002).

The C-D model was chosen for the following advantages (Shang, 1990; Bhanumurthy, 2002; Munzir and Heidhues, 2002):

- The partial elasticities of production, which measures the responsiveness of output to unit increase of input, are identical to the production coefficient (β_{ij}). Therefore, a percentage change in output that results from a given percentage change in input use can be easily identified.
- The sum of partial elasticities of production ($\Sigma\beta_i$) can be interpreted as a measure of economies of scale (i.e. the percentage change of output relative to the percentage change in all inputs used). If $\Sigma\beta_i > 1$, for example, positive economies of scale exist. This implies that a doubling of the use of all inputs will result in more than a doubling of output.
- Estimation is simple because input and output data can readily be used without aggregation as they are in the CES function.
- The unconstrained C-D function can describe a production surface that demonstrates increasing, unitary or decreasing returns to scale depending upon the data.
- Unlike the quadratic function that requires more degrees of freedom because of interaction terms, the C-D function requires only one degree of freedom per explanatory variable.
- It facilitates computations and has the properties of explicitly representing ability, uniformity, parsimony and flexibility.

The relationship between the dependent and all independent variables was analysed using Eqn 52.1 (C-D function):

$$\ln Q = \beta_o + \sum_{i=1}^{9} \beta_i \ln X_i + \sum_{i=1}^{9}\sum_{j=1}^{9} \beta_{ij} \ln X_i \ln X_j \quad (52.1)$$

Where:
Q – denotes output
X_i – denotes inputs
β_o – denotes a constant
β_i – denotes model coefficients (the elasticities of production)
β_{ij} – denotes production coefficient.

This is a linearized exponential equation using log transformation for the linearization:

Log $Q = b_0 + b_1$ Log $X_1 + b_2$ Log $X_2 + b_3$ Log $X_3 + b_4$ Log $X_4 + b_5$ Log $X_5 + b_6$ Log $X_6 + b_7$ Log $X_7 + b_8$ Log $X_8 + b_9$ Log $X_9 + b_{10}$
While $b_1 > 0, b_2 > 0, b_3 < 0, b_4 < 0, b_5 < 0, b_6 < 0, b_7 < 0, b_8 < 0, b_9 < 0$
Where:
Q = Quantity of sweetpotato supplied (kg)
X_1 = Natural log of area of land cultivated to sweetpotato (ha)
X_2 = Natural log of market price for sweetpotato in Naira (₦), the local currency
X_3 = Natural log of harvest cost (₦)
X_4 = Natural log of storage cost (₦)
X_5 = Natural log of cost of transport (₦)
X_6 = Natural log of assemblage cost (₦)
X_7 = Natural log of negotiation/bargaining cost (₦)
X_8 = Agent's fee (₦)
X_9 = Natural log of transaction for land rent (₦)

52.3 Results and Discussion

Demographic characteristics of the respondents

Table 52.1 reveals that the majority of household heads interviewed were males (81%) with the rest (19%) being female divorcees and widows. Family members were usually engaged in activities such as fetching water for spraying, fertilizer applications and transportation of the harvested sweetpotatoes from the farm to the roadside. The men were the farm owners and the heads of households and were more involved in land preparation and planting than women. The sweetpotato farmers had a mean age of 43 years, ranging from 19 to 71 years. Only 37% were below 40 years, supporting Adewumi and Fatimoh's (2008) claim that the young adults are not extensively involved in sweetpotato farming. Ninety-four per cent of the respondents were married, with 73% of the households having between six and 15 house members. Only 27% had fewer than five family members. Thus, most households had significant family labour available for cultivating sweetpotato.

Attained formal educational levels of the household varied widely. Eighteen percent had no formal education, 33% had primary education, 17% had secondary education, 24% had Arabic education and only 8% had post-secondary education. More than half of the respondents (65%) had farming as their main occupation; others that engaged in it on a part-time basis were mainly artisans and civil servants. The average number of years of farming experience was 25 years; 81% had been planting sweetpotato for the past 11–40 years.

Access to land appears not to be a constraint in the study area, as the respondents claimed not to have a problem in land acquisition for production. The majority (53%) acquired land through inheritance, only 12% purchased land, 17% acquired land through lease and 18% acquired land by freehold and other sources. The average amount of land under sweetpotato cultivation in the study area was 1.5 ha. Most of the respondents

Table 52.1. Frequency and percentage distribution of respondents according to their socio-economic characteristics (sweetpotato growers in Kwara State). (From field survey, 2012.)

Socio-economic characteristics	Frequency	Percentage
Age group of the household head		
< 20 years	6	5
21–40 years	38	32
41–60 years	65	54
> 60 years	11	9
Sex		
Male	97	81
Female	23	19
Education level		
No formal education	22	18
Primary	39	33
Secondary	20	17
Arabic school	29	24
Post-secondary education	10	8
Marital status		
Single	2	2
Married	113	94
Divorcee/widow/widower	5	4
Primary occupation of the household head		
Farming only	78	65
Formal	25	21
Non-formal	17	14
Family size of respondents		
1–5	32	27
6–10	71	59
11–15	17	14
Household head's farming experience		
< 10 years	23	19
11–20 years	20	17
21–30 years	58	48
31–40 years	19	16
Membership of association		
Cooperative	22	18
Farmer association	52	44
Nil	46	38
Sources of land		
Inherited	64	53
Leasehold	20	17
Freehold	8	6
Purchased	14	12
Other source	14	12

Continued

Table 52.1. Continued.

Socio-economic characteristics	Frequency	Percentage
Farm size (ha)		
< 1 ha	47	39
1–2 ha	63	53
3–4 ha	10	8
> 4 ha	0	0
Source of planting materials		
Own farm	65	54
Friends and relatives/own farm	47	39
Purchased	8	7
Usage of fertilizer		
Used	21	18
Not used	99	82
Source of capital		
Personal saving	85	71
Friends and relatives	17	14
Cooperative society	18	15
Total	120	100.0

were small-scale farmers, defined by Aliyu *et al.* (1997) as having up to 5.99 ha. Thirty-nine per cent of the respondents cultivated 1 ha or less, 53% cultivated 1–2 ha while 8% cultivated above 2 ha. Results indicate that the majority (71%) of the sweetpotato farming households used their personal savings to finance their activities, while the rest (29%) sourced money from friends/relatives or a cooperative society. Lastly, the results reveal that about 93% of the farmers' sourced their vine cuttings planted from their own previous harvest, friends or relatives while only 7% purchased vines. Only 18% used fertilizer for their production. Those that did not use it claimed they had fertile land; some said its procurement procedure was too tedious. Sixty-two per cent of the sweetpotato farming households belonged to farmer or cooperative associations.

Transaction costs

Table 52.2 shows the descriptive statistics of transaction costs incurred by the respondents per annum. Variables found to be associated with transaction costs in the study area include: (i) harvesting; (ii) storage; (iii) assemblage; (iv) negotiation and/or bargaining; (v) the agent's fee; (vi) transaction land rent; and (vii) transportation to points of sale. Four variables out of the nine variables (Table 52.3) were found to be statistically significant in relation to supply decisions made by agricultural households: (i) price of sweetpotato; (ii) area of land cultivated to sweetpotato; (iii) the agent's fee; and (iv) the transaction land rent. The first three of these affect the quantity of sweetpotato supplied positively, however, the fourth item (land rent) has a significant inverse relationship with the quantity of sweetpotato supplied. Contrary to expectation, the agent's fee was found to be positively related to quantity supplied. According to the respondents, this could be attributed to the fact that qualified agents usually charge higher fees than the quacks based on their unique value addition and their broad contacts to the end users of the produce. The farmers prefer the service of professional agents, not minding the higher fee because such agents have positive effects on their sales.

Adjusted R-squared (R^2) from the regression analysis was 0.734, showing that 73.4% of the variation in the quantity of sweetpotato supplied by the respondents was explained by the estimated variables.

Elasticity of supply response sweetpotato transaction costs

Results showed that with respect to price, area, negotiation cost, agent's fee, harvesting cost, assemblage cost, storage cost, transportation cost and transaction land cost, a 10% change in each of the variables resulted in elasticity of supply response percentages ranging from 0.8% to 13.4% for different transaction cost components (Table 52.4). Supply response was highly significant with respect to area of land cultivated and agent's fee and moderately elastic with respect to price and transaction land rent.

Table 52.2 Distribution of sweetpotato transaction costs per annum. (From field survey, Kwara State, Nigeria 2012.)

Transaction costs variables	Minimum (₦)	Maximum (₦)	Mean (₦)	Standard deviation	Variance
Harvest cost	720	51,480	5,143.94	334.746	11,120,529
Storage cost	360	27,440	2,798.06	1,857.502	3,450,313
Cost of transport	960	68,540	7,035.38	4,604.020	21,196,667
Assemblage cost	120	8,580	929.20	573.199	328,557
Negotiation/bargaining cost	230	6,220	761.66	434.677	188,944
Agent's fee	300	7,780	956.78	546.069	298,190
Transaction land rent	300	10,360	1,242.38	729.800	532,607

Table 52.3. Regression results (dependent variable: quantity of sweetpotato supplied; $n = 120$). (From field survey, Kwara State, Nigeria 2012.)

Independent variable	Coefficient	*t*-value[a]
Constant term	3.753	2.766
Log (AREA)	1.051	16.526***
Log (PRICE)	0.717	2.071**
Log (HARVEST)	–0.482	1.039
Log (STORAGE)	0.120	0.826
Log (TRANSPORT)	0.146	0.402
Log (ASSEMBLAGE)	–0.079	–0.570
Log (NEGOTATION)	0.100	0.310
Log (AGENT)	1.340	3.664***
Log (RENT)	–0.927	–2.417**

[a]$R^2 = 0.734$; ***, significant at 1%; **, significant at 5%.

Table 52.4. Elasticity of sweetpotato supply response to transaction costs. (From field survey, Kwara State, Nigeria 2012.)

Independent variable[a]	Elasticity of supply response (%)
Area of land cultivated	10.5
Market price	7.2
Harvest cost	4.8
Storage cost	1.2
Transportation cost	1.5
Assemblage cost	0.8
Negotiation cost	1.0
Agent's fee	13.4
Transaction land rent	9.3

[a]A 10% change in each of the variables resulted in elasticity of the sweetpotato supply response shown.

52.4 Conclusion, Recommendations and Policy Implications

In conclusion, sweetpotato supply responses to transaction costs in the study area were statistically significant, responding positively to market price and area of land cultivated. Market factors as well as non-market factors significantly affected agricultural household supply response. Contrary to expectations, the marketing agent's role and services are important and positive factors.

These results call for policy recommendations that will reduce transactions costs that can complement price policies in affecting the supply response. The effects of institutional deficiencies on the functioning of markets should also be addressed. Since proper market institutions promote competition and induce a more efficient market organization these should be invested in to further reduce transactions costs. The quality of road infrastructure should be improved to reduce transport costs.

Agricultural households should consider strengthening themselves financially by forming cooperative groups that enable members to have access to loans at accessible interest rates and farm inputs at lower prices due to bulk purchase. The produce could also be sold in bulk, lowering the average transaction costs. Local farmers' cooperative groups could act as catalysts to correct any market failure. Team action could enhance trade through decreasing uncertainties and creating benefits from reduced transaction costs.

References

Abebe, H.G. (2005) The supply responsiveness of peasant agriculture in Ethiopia: some macro econometric results from cereal production. Center Discussion Paper.

Adewumi, M.O. and Fatimoh, A.A. (2008) Profitability and technical efficiency of sweet potato production in Nigeria. *Journal of Rural Development* 31(5), 105–120.

Ahmed, R. and Rustagi, N. (1987) Marketing and price incentives in African and Asian countries: a comparison. In: Dieter, E. (ed.) *Agricultural Marketing Strategy and Pricing Policy*. World Bank, Washington, DC, pp. 104–118.

Ajetomobi, J.O., Ajao, A.O. and Olarinde, L.O. (2006) Supply analysis for food crops in an oil deregulated economy. *International Journal of Applied Agricultural and Apicultural Research* 3(2), 144–150.

Aliyu, J., Shaib, B. and Bakshi, J.S. (1997) *Nigeria: National Agricultural Research Strategy Plan: 1996–2010*. Federal Ministry of Agriculture and Natural Resources, Abuja, Nigeria, pp. 22–35.

Askari, H. and Cummings, J.T. (1977) Estimating agricultural supply response with the Nerlove Model: a survey. *International Economic Review* 18(2), 257–292.

Bhanumurthy, K.V. (2002) Arguing a case for the Cobb-Douglas production function. *Review of Commerce Studies* 20–21(1), 75–91. Commerce Department, School of Economics, University of Delhi, India. Available at: http://ssrn.com/abstract=598082 (accessed 17 March 2015).

Chibber, A. (1988) Raising agricultural output: price and non-price factors. *Finance and Development* June, 44–47.

Farayola, C.O. (2006) Supply response of Nigerian agricultural export crops: 1970–2003. MTech thesis in agricultural economics, Ladoke Akintola University of Technology, Ogbomoso, Nigeria.

Food and Agriculture Organization of the United Nations (FAO) (2000) *FAO Production Year Book*. FAO, Rome.

Food and Agriculture Organization of the United Nations (FAO) (2009) High level expert forum – how to feed the world in 2050. In: Economic and Social Development Department (ed.) *Global Agriculture Towards 2050*. FAO, Rome, pp. 1–4.

Goetz, S.J. (1992) A selectivity model of household food marketing behaviour in sub-Saharan Africa. *American Journal of Agricultural Economics* 74, 444–452.

Hobbs, J.E. (1997) Measuring the importance of transaction costs in cattle marketing. *American Journal of Agricultural Economics* 79(4), 1083–1095.

Idachaba, F.S. (2004) Food security in Nigeria: challenges under democratic dispensation. Ninth Agricultural and Rural Management Training Institute (ARMTI) Annual Lecture, ARMTI, Ilorin, Nigeria.

International Institute of Tropical Agriculture (IITA) (2005) *Research to Nourish Africa: Overview of the 2006–2008 Medium-term Plan*. IITA, Ibadan, Nigeria.

Key, N., Sadoulet, E. and Janvry, A. (2000) Transaction costs and agricultural household supply response. *American Journal of Agricultural Economics* 82, 245–259.

Krishna, R. (1967) Agricultural price policy and economic development. In: Southworth, H.M. and Johnson, B.F. (eds) *Agricultural Development and Economic Growth*. Cornell University Press, Ithaca, New York, Chapter 13.

Kwara State Agricultural Development Project (KWADP) (1996) *Annual Agronomic Survey Report*. Kwara State Ministry of Agriculture, Ilorin, Kwara State, Nigeria.

Leaver, R. (2003) Measuring the supply response function of tobacco in Zimbabwe. Paper presented at the Agricultural Economic Association of South Africa, 2003 Annual Conference, 2–3 October, Pretoria, South Africa.

Mamingi, N. (1996) *The Impact of Price and Macroeconomic Policies on Agricultural Supply and the Environment: a Synthesis*. World Bank, Washington DC.

Mamingi, N. (1997) The impact of prices and macroeconomic politics on agricultural supply: a synthesis of available results. *Agricultural Economics* 16, 17–34.

Ministry of Agriculture and Natural Resources (MANR) (2010) *Report of Kwara State Farmers' Census*. MANR, Ilorin, Kwara State, Nigeria.

Munzir, A. and Heidhues, F. (2002) Toward sustainable development of floating net cage culture for income security in rural Indonesia: a case study of common carp production at Lake Maninjau, Indonesia. In: Edwards, P., Little, D.C. and Demaine, H. (eds) *Rural Aquaculture*. CAB International, Wallingford, UK, pp. 143–153.

Murova, O., Cobel, K. and Trueblood, M. (2001) Supply response of Ukrainian agriculture economics and agribusiness. Selected Paper, University of Arkansas, Faye Hevile, Arkansas.

Pinstrup-Andersen, P. (1994) *World Food Trends and Future Food Security*. A food policy report. International Food Policy Research Institute, Washington, DC, pp. 1–3.

Shang, C.Y. (1990) *Aquaculture Economic Analysis: an Introduction*. World Aquaculture Society, Baton Rouge, Louisiana.

Smith, I.R. (1982) Micro-economics of existing aquaculture production systems: basic concepts and definitions. In: *Aquaculture Economics Research in Asia*, Proceedings of a workshop, 2–5 June 1981, Singapore. International Development Research Centre, Ottawa, Canada and International Center for Living Aquatic Resources Management (ICLARM), Manila, The Philippines.

Staal, S., Delgado, C. and Nicholson, C. (1997) Smallholder dairying under transaction costs in East Africa. *World Development* 25(5), 779–794.

53 Promotion of Vitamin A-enriched Sweetpotato for Production by Small-scale Commercial Farmers in South Africa

S.M. Laurie,* M.M. Mtileni, W.M. Mphela, A.A. Van den Berg, T. Ramathavhana, L. Sediane, T. Maraganedzha and C.P. du Plooy
Agricultural Research Council-Roodeplaat Vegetable and Ornamental Plant Institute (ARC-VOPI), Pretoria, South Africa

Abstract
The National Department of Science and Technology funded a project on demonstration agronomy of vitamin A-enriched sweetpotato in South Africa for job creation, enterprise development and ultimately addressing food security and malnutrition, particularly vitamin A deficiency. This chapter reports on a competitive system for small-scale commercial farmers for the informal market, based on experiences of the well-developed commercial sweetpotato industry in South Africa. Capacity was developed in nursery management and handling practices for good quality cuttings, and agronomic and postharvest practices for fresh produce. Improved cultivars (high yield, dry texture) were promoted, focusing on orange-fleshed cultivars, but not excluding cream-fleshed cultivars. A gross margin analysis of cutting production was performed and market information collected through informal interviews. The project resulted in setting up of infrastructure at four nursery sites in four provinces (the Agricultural Research Council-Roodeplaat Vegetable and Ornamental Plant Institute (ARC-VOPI) and at the Universities of Fort Hare, Venda and Zululand) using disease-indexed stock plants of seven new cultivars propagated by the ARC sweetpotato scheme. Training was provided for 85 farmers in three provinces on: (i) vitamin A nutrition; (ii) cultivation practices; and (iii) handling, grading and packaging of sweetpotatoes for informal markets. Over 600,000 cuttings were issued to outreach activities of various government departments, and directly to farmers. The gross margin for cuttings production at ARC-VOPI was R36,110/ha (US$4248/ha). Small-scale commercial cultivation of vitamin A-enriched cultivars, ranging from 0.3 ha to 1 ha, in different climate and socio-economic settings, achieved a mean farm-level yield of 21.3 t/ha. During the 2012/13 growing season, an average of 25 ha of the vitamin A-rich sweetpotato was produced. The key attributes of a competitive system for small-scale commercial farmers were improved vine management procedures, and quality care at harvest time leading to quality fresh produce. In January 2013, roll-out of the programme to provincial level commenced, focusing on nursery enterprises as well as sweetpotato growers.

Keywords: disease-indexed cuttings, orange-fleshed sweetpotato, vine production

*slaurie@arc.agric.za

53.1 Introduction

A low intake of fruits and vegetables was identified as one of the main deficiencies in the South African diet (Vorster *et al.*, 1997). Vegetables and fruits are important sources of many vitamins, minerals, fibre and other substances such as antioxidants, and therefore a diet rich in fruits and vegetables is protective against several diseases. It has been estimated that, for example in the European Union, an increased intake of fruits and vegetables could result in a considerable improvement in public health (Joffe and Robertson, 2001).

Vitamin A deficiency (VAD) is particularly important in South Africa as a national survey in 2005 revealed that 64% of 1–9-year-old children and 27% of women at child-bearing have low vitamin A status (Labadarios *et al.*, 2007). Production and consumption of foods rich in vitamin A is considered a sustainable long-term strategy to address VAD and is used in many parts of the developing world (Agble, 2004). In this regard, use of vitamin A-enriched (orange-fleshed) sweetpotato, a rich source of β-carotene (the major precursor of vitamin A), in programmes to provide vitamin A to the human diet is escalating (Burri, 2011). This is also the case in several African countries as part of the Vitamin A for Africa Initiative (Kapinga *et al.*, 2007) (e.g. in Tanzania, Uganda and Kenya) and the Sweetpotato for Profit and Health Initiative targeting 17 African countries (Low, 2011). In South Africa, the crop-based programme is managed by the Agricultural Research Council (ARC), the Medical Research Council, government departments and non-governmental organizations (NGOs) (Magasana *et al.*, 2004; Faber and Laurie, 2011; Faber *et al.*, 2013).

In South Africa, sweetpotato production for the fresh market is dominated by less than 100 large-scale commercial farmers who specialize in one or two cream-fleshed cultivars which are not addressing VAD of the broader population. The crop-based initiatives create opportunities for an expanded scale of production of vitamin A-enriched sweetpotato. In addition, government policy is focusing strongly on increasing, expanding and empowering sustainable rural enterprise development, particularly in agriculture (Department of Agriculture, Forestry and Fisheries, 2011). The experiences of the well-developed commercial sweetpotato industry can be valuable in developing an efficient system for small-scale farmers.

At the ARC, sweetpotato research, for more than a decade, focused on developing vitamin A-enriched cultivars with high yield, good root quality, sweet taste and dry texture according to needs of the informal market (Laurie *et al.*, 2009). Conventional breeding is employed involving the polycross system, followed by field evaluation and selection of progenies. A number of promising cultivars have been identified through multi-location trials where farmers participated in selection, conducting consumer acceptability trials, as well as determination of nutrient content (Laurie *et al.*, 2012a, b).

Availability of quality planting material of sweetpotato has been accentuated widely (Nsibande and McGeoch, 1999; Fuglie, 2007; Kapinga *et al.*, 2007). Low production volumes of vitamin A-enriched cultivars in South Africa can be related to the lack of availability of cuttings of improved cultivars, especially in rural farming settlements, and knowledge about planting, as well as where and what cultivars to buy. Traditionally white- and cream-fleshed landraces have been grown, with growers retaining their planting material for future use, while none of the commercial nurseries offered cuttings of the new ARC sweetpotato varieties for sale.

The National Department of Science and Technology, South Africa, over the period 2010–2013, funded promotion of vitamin A-enriched varieties and capacity development in agronomic practices of sweetpotato. This chapter reports on a competitive system for small-scale commercial farmers for the informal market, based on experiences of the well-developed commercial sweetpotato industry in South Africa. The project contributed to job creation, enterprise development and ultimately combatting food security and VAD.

53.2 Materials and Methods

Infrastructure was established for four nursery sites in four provinces of South Africa. In conjunction with the main site at the Agricultural Research Council-Roodeplaat Vegetable and Ornamental Plant Institute (ARC-VOPI) (Gauteng Province), the project targeted rural-based universities, namely the Universities of Fort Hare (Eastern Cape Province), Venda (Limpopo Province) and Zululand (KwaZulu-Natal Province). In order to produce top quality propagation material the major infrastructure upgrading included: (i) purchasing vehicles; and (ii) upgrading glasshouse infrastructure, soil pasteurization equipment, vine handling and storage facilities, and irrigation systems.

Seven improved cultivars were included, based on recommendations flowing from multi-location trials where farmers participated in selection, consumer acceptability, as well as nutrient content (Laurie *et al.*, 2012a, b). The major emphasis was on three orange-fleshed cultivars, namely Bophelo and Impilo, bred at the ARC-VOPI, and W-119, imported from the USA, and one yellow-orange cultivar 199062.1 (an import from the International Potato Center (CIP), Peru). ARC cream-fleshed cultivars Ndou and Monate and, to limited extent, Mvuvhelo were also included. Disease-indexed stock plants of the selected cultivars were propagated in glasshouses at the ARC-VOPI and distributed at the beginning of each season to the four nurseries. Frequent visits by ARC research technicians ensured transfer of skills in nursery management, record keeping, labelling, multiplication and handling of cuttings. Where necessary, diseases and pests were controlled by spraying chemicals.

A nursery production plan was developed for each site according to climatic conditions, varying from frost-free in Zululand and Venda, to light frost at ARC-VOPI to very cold winters in Fort Hare. In the case of ARC-VOPI and Fort Hare, the nurseries had an 18-month cycle. Foundation blocks were established yearly during October–December, enlarged into base blocks yearly during January–February and to nursery blocks in February–March. Following sprouting, cuttings were sold from October to March and all blocks were phased out in April–May. Meanwhile, new planting material was obtained in October–December for a new multiplication cycle. In the case of Zululand and Venda, foundation blocks were established in February, and multiplication to base blocks and nursery blocks were done in March–May. Cuttings were issued all-year round. All blocks were phased out in June of the following year; therefore a 16-month cycle was followed.

Top cuttings of 30 cm in length (with bottom leaves removed) were made according to orders from clients, tied in bundles of 200 and labelled. The price of cuttings was 50 cents (US$0.06) per cutting for orders less than 3000, and 30 cents (US$0.04) per cutting for orders exceeding 3000 cuttings. The first 20 cuttings were given out free to clients. For short-distance transportation, bundles with 200 cuttings were packed in the back of trucks/light-duty vehicles. Packaging of the cuttings for sending by courier entailed covering the bases of bundles of stem cuttings with moist paper towel and wrapping that part in a plastic bag and packing the bundles in cardboard boxes – these were able to keep for 7 days at room temperature. A gross margin analysis was performed for vine production at ARC-VOPI (agricultural supplies, labour, tractor running cost) based on the estimated total number of cuttings which can be produced on 1 ha of land.

Farmers' days or information days were held annually at each site. During these events nutrition education was provided, demonstrations on planting practices were conducted, and processed products of orange-fleshed sweetpotato were exhibited.

The ARC further established projects producing vitamin A-enriched cultivars at small commercial farms in three areas that varied in climate and socio-economic setting. The first was at eManguzi (Kangwanase, in the KwaZulu-Natal–Maputo Corridor), with three communal production sites each having access to 7 ha of arable land for production; each group consisting of between eight and ten members. The soil was sandy, with a high water table. Supplementary irrigation was limited to a hand pump used for watering during planting. The second site was at Tshiombo (Thohoyandou, Limpopo Province), an area where sweetpotato is a very popular traditional crop. This is a high rainfall area with a fertile red-clay soil and ample water supply through furrow irrigation. The Tshiombo Sweet Potato Cooperative consisted of 80–100 farmers each with 1.5 ha of land. The third project was conducted in collaboration with the Department of Social Development's Women Empowerment Initiative at five communal groups (with between eight and 15 members per group; land size 1–15 ha) in the Uitenhage and King Williamstown peri-urban areas, Eastern Cape Province. Irrigation systems had been provided earlier by another funder. Project activities included site assessment, training and monthly visits to monitor production. Training included information on: (i) the nutritional value of vitamin A-enriched cultivars; (ii) cultivation practices; and (iii) handling, grading and packaging of roots for informal markets. Market information was collected through informal discussions with growers and retailers during project visits.

53.3 Results and Discussion

Dissemination of cuttings

The sizes of nurseries varied according to availability of land: 0.5 ha at Zululand, 1 ha at Venda, 1 ha at Fort Hare and 2 ha at ARC-VOPI – the latter having the potential to produce 2 million cuttings. From ARC-VOPI, between 160,815 and 372,350 cuttings were issued over the 2011/12 and 2012/13 seasons, including both orange-fleshed and cream-fleshed cultivars (Table 53.1). Large numbers of cuttings were issued to outreach activities of various government departments which were purchasing the cuttings to supply small-scale commercial farmers and community projects. The major customers were: (i) the Gauteng Province Department of Agriculture and Rural Development (130,000); (ii) Mogwase Prison (38,000); (iii) NGOs (41,100); (iv) the Limpopo Province Department of Agriculture (13,600); and (v) ARC-managed projects (25,590). Each season, five to ten individual farmers purchased 8000–30,000 cuttings for growing 1–2 ha of sweetpotato.

The off-station nurseries (taken together) disseminated between 30,000 and 44,000 cuttings per season, almost exclusively of orange-fleshed varieties (Table 53.1). These cuttings were mostly issued for small-scale commercial production. The four nurseries disseminated 205,514 cuttings in 2011/12 and 402,124 cuttings in 2012/13. Of these, 48% were cuttings of vitamin A-enriched cultivars.

The gross margin analysis for vine sales at ARC-VOPI (Table 53.2) was R36,110/ha (US$4248/ha). The total production cost was R323,890/ha (US$38,105/ha). Labour cost constituted 66% of the cost. Published studies related to production of sweetpotato plant

Table 53.1. Numbers for sweetpotato cuttings disseminated from four nursery sites during 2011/12 and 2012/13.

	Numbers of cuttings				
Nursery[a]	2011/12	2012/13	Total for two seasons	OFSP[b]	OFSP (%)
Rural universities	44,699	29,774	74,473	64,414	86%
UNIZUL	8,849	9,214	18,063	16,141	89%
UNIVEN	25,000	11,560	36,560	28,820	79%
UFH	10,850	9,000	19,850	19,453	98%
Roodeplaat	160,815	372,350	533,165	229,762	43%
Total	205,514	402,124	607,638	294,176	48%

[a]UFH, University of Fort Hare; UNIVEN, University of Venda; UNIZUL, University of Zululand.
[b]OFSP, orange-fleshed sweetpotato.

material are scarce (as opposed to production of sweetpotatoes). Hotbed production of sweetpotato slips in the San Joaquin Valley, USA, had a total operating cost of US$34,313/acre (US$85,782/ha) (University of California, 2006).

Small-scale commercial production

Three communal sites (Kwasa, Owethusonke and Isambane) in eManguzi in the Northern KwaZulu-Natal each planted between 0.75 ha and 1 ha of vitamin A-enriched sweetpotatoes, delivering sweetpotatoes in bags (Fig. 53.1) to the informal market in Umhlabuyalingana local municipality and local community (Table 53.3). The yield varied greatly from 16.4 t/ha in season 1, to 3.8 t/ha in season 2, due to lower rainfall. Consequently, the ARC planned to upgrade the irrigation systems to improve crop management during dry periods. Another constraint was that stipends were still paid from project funds since these could not be replaced by income from sales. However, the harvest was of very good quality and was sold at the same price as poor quality roots (i.e. those that were cracked, small and/or crooked) of local available landraces; also the price was relatively low compared with Tshiombo (Table 53.3). The price should thus be increased, linked to marketing as high quality produce. In addition to sweetpotato, the farmers also planted door-sized plots of conventional vegetables, such as onion, Swiss chard and carrot, for home consumption; only sweetpotato survived in large plantings under rainfed conditions. As for other frost-free areas, eManguzi had the advantage of delivering sweetpotato during August–December, which was the time of peak prices for sweetpotato since the temperate production areas can only deliver from January to July.

Table 53.2. Gross margin analysis of cutting sales at ARC-VOPI.

Item	R/ha[a]
Income (cuttings sales/ha)	360,000
Variable costs (production/ha)	323,890
Disease-indexed stock plants	54,035
Tractor running cost	12,345
Fertilizer	4,110
Chemicals (pesticides, fungicide, herbicides)	8,400
Transportation	18,400
Water and electricity	3,100
Labour	215,000
Packaging materials	8,500
Gross margin (per ha)	36,110

[a]US$1 = R8.50.

A project visit to the Tshiombo site, Limpopo Province, in April 2013, revealed that about 50 farmers of the Tshiombo Sweet Potato Cooperative were growing cultivar Bophelo on a third of their field space (1.5 ha), thus covering about 15 ha. Initially, the farmers were reluctant to buy cuttings due to the tradition of saving and sharing of planting material. However, after seeing the success

Fig. 53.1. Graded sweetpotatoes packed in bags for marketing at eManguzi (a) while at Tshiombo unwashed roots were marketed in crates (b).

Table 53.3. Site and marketing information for vitamin A-enriched sweetpotato in three provinces.

Marketing information	Site			
	eManguzi	Tshiombo	TATI[a]	Ilitha
Province	KwaZulu-Natal	Limpopo	Eastern Cape	Eastern Cape
Land size	7 ha per group	1.5 ha per farmer	15 ha	1 ha
Area planted with OFSP[b]	0.75–1 ha per group	0.5 ha per farmer	0.5 ha	0.25 ha
Yield of OFSP[b]	3.8–16.4 t/ha	~15–30 t/ha	30.3 t/ha	35 t/ha
Wash roots	No, rub off sand lightly	Sold in crates not washed, sold in buckets washed by retailer	Yes	Yes
Market	Hawkers, directly to community	Wholesalers and directly to the community	Grant pay point, local community, schools	Local community
Unit for sales	Bags (2 kg, 4 kg, 10 kg)	Crates (20 kg), buckets (3 kg, 5 kg)	Bags (10 kg)	Bags (10 kg)
Price[c]	R4/kg	R100/crate = R5/kg Makonde: R6–R6.50/kg for buckets Tshakuma: R8–R10/kg for buckets	R3/kg	R2.50/kg
Other crops/ income	Very small vegetable plots, project salary	Beans 0.5 ha per farmer, job, grants	Large-scale vegetable production	Selling chickens

[a]TATI, Tinarha Agri Tourism Initiative.
[b]OFSP, orange-fleshed sweetpotato.
[c]US$1 = R8.50.

of others, more farmers purchased the cuttings. Some were saving Bophelo material for the second season, but this was strongly discouraged due to increased disease build-up and weevil infestation on root tubers. Wholesalers purchased crates (20 kg) of unwashed roots which could easily be stacked during transport (Fig. 53.1). Retailers at two informal markets washed and sold root tubers. In some cases, farmers sold the sweetpotatoes directly to consumers at the road side. The retailers indicated that they experienced a large demand for this cultivar. At Tshakuma market, the selling price was higher than at Makonde market, because of the further distance from Tshiombo (Table 53.3). The prices at both markets were considerably higher than at the other project sites.

In Eastern Cape, the Ilitha project grew 0.25 ha of Bophelo sweetpotato at a site in the community supported by the Women Empowerment Initiative of the Department of Social Development. They produced good quality sweetpotatoes and achieved high yields. However, the selling price was also relatively low. The Thembisa group and Alice groups, except for selling fresh produce to the local community, also produced processed products of vitamin A-enriched sweetpotato for a local pre-school day care centre. The largest site was at the Tinarha Agri Tourism Initiative (TATI) group in Uitenhage, consisting of 13 members who earned their full income from vegetable sales from 15 ha of land.

Bophelo cuttings were issued from the four nurseries to approximately 279 direct beneficiaries during the 2012/13 season. A rough estimation is that during the 2012/13 cropping season, 18 ha of this variety had been grown, 5 ha of 199062.1 and 2.5 ha of Impilo. This implies that 25.5 ha of the vitamin A-rich sweetpotato were grown. At an estimated farm-level yield of 21.3 t/ha, 540 t were produced.

A study is in process to determine: (i) the production efficiency levels of farmers growing vitamin A-enriched sweetpotato; (ii) the socio-economic factors that affects this efficiency level; (iii) the profitability of the promoted cultivars; and (iv) the marketing channels used by farmers. In analysis of productivity efficiency of sweetpotato production in Imo State, Nigeria (Nwaru *et al.*, 2011), membership of cooperatives was found to be significant and positively related to production efficiency. In the current study, growers at all the sites were operating as groups/cooperatives and based on the finding of Nwaru *et al.* (2011) this was an advantage. It is further noted from Nwaru *et al.* (2011) that other significant factors for production efficiency were farming experience and access to credit. Cost returns analysis by Kassali (2011) indicated that sweetpotato production was profitable and further listed planting material, transportation to markets, adoption of new cultivars, fertilizer level, full-time farming and also experience as positive to sweetpotato production efficiency. The critical importance of traders in sourcing of produce and marketing of surplus orange-fleshed sweetpotato was emphasized by Cloote and Massingue (2012). Training was provided for traders, including the health benefits of orange-fleshed sweetpotato, novel ways of preparation and financial benefits of selling it, while farmers were given training in market requirements. This led to increasing the market segment of orange-fleshed sweetpotato from 18% in 2008, to 50% in 2009 in the Zambezia Province of Mozambique. These reported results will be valuable for larger-scale implementation in South Africa.

53.4 Conclusion

The key value of the present results was the evolution of a quality vine and root production system in order for small-scale commercial farmers to be competitive. Experiences with three projects pertaining to small-scale commercial production of vitamin A-enriched sweetpotato transpired into the following lessons:

- To convince farmers of the commercial value of quality cuttings 1–2 years may be needed.
- Small-scale farmers can benefit from a commercial system being adapted for them to be competitive.
- For small-scale commercial farmers in drier areas, vitamin A-enriched sweetpotato that is planted at the appropriate time offers reasonable production and fair income prospects when compared to other conventional field-crops.
- Producers should charge premium prices for quality sweetpotatoes produced, as well as for the health benefit of vitamin A-enriched sweetpotatoes. Nutrition education will be crucial in this regard.

The information and experience obtained was valuable when implementing a new project in January 2013 funded by the Department of Rural Development and Land Reform to establish sweetpotato enterprises in five provinces of South Africa, targeting both small-scale commercial vine growers and sweetpotato producers.

Acknowledgements

The authors gratefully thank the National Department of Science and Technology, South Africa, for funding the project. Collaborative efforts of the Eastern Cape Department of Social Development, Umhlabuyalingana local municipality, Gauteng Department of Agriculture and Rural Development and Limpopo Department of Agriculture are acknowledged. The authors thank Dr Jan Low for valuable advice on the chapter.

References

Agble, R. (2004) Towards an integrated programme for improving nutrition – the role of food-based approaches. In: Brouwer, I.D., Traoré, A.S. and Trèche, S. (eds) *Food-based Approaches for a Healthy Nutrition in West Africa,* Proceedings of the 2nd International Workshop, 23–28 November 2003, Ouagadougou. Presses Universitaires de Ouagadougou, Burkina Faso, pp. 21–31.

Burri, B.J. (2011) Evaluating sweet potato as an intervention food to prevent vitamin A deficiency. *Comprehensive Reviews in Food Science and Food Safety* 10, 118–130.

Cloote, C. and Massingue, J. (2012) Evaluation of a trader-focused marketing strategy for the sustained uptake of orange flesh sweet potato in Zambézia province, Mozambique. In: *Sixteenth Book of Abstracts Triennial Symposium of the International Society for Tropical Root Crops,* 23–28 September 2012, Federal University of Agriculture, Abeokuta, Nigeria. International Society for Tropical Root Crops, Abeokuta, Nigeria.

Department of Agriculture, Forestry and Fisheries (2011) *Strategic Plan for Department of Agriculture, Forestry and Fisheries 2011/12 to 2014/15.* Directorate Agricultural Information Services, Pretoria, South Africa, 29 pp.

Faber, M. and Laurie, S.M. (2011) A home-gardening approach developed in South Africa to address vitamin A deficiency. In: Thompson, B. and Amoroso, L. (eds) *Combating Micronutrient Deficiencies: Food-based Approaches.* CAB International, Wallingford, UK and Food and Agriculture Organization of the United Nations (FAO), Rome, pp. 163–182.

Faber, M., Laurie, S., Ball, A. and Andrade, M. (2013) *A Crop-based Approach to Address Vitamin A Deficiency in South Africa.* Medical Research Council, Cape Town/Agricultural Research Council (ARC)-Roodeplaat, Pretoria, South Africa.

Fuglie, K.O. (2007) Priorities for sweetpotato research in developing countries: results of a survey. *Horticultural Science* 42(5), 1200–1206.

Joffe, M. and Robertson, A. (2001) The potential contribution of increased vegetable and fruit consumption to health gain in the European Union. *Public Health Nutrition* 4, 893–901.

Kapinga, R., Tumwegamire, S. and Ndunguru, J. (2007) *Status Report of VITAA (Vitamin A for Africa).* A partnership program combating vitamin A deficiency through increased utilization of orange-fleshed sweetpotato in sub-Saharan Africa. International Potato Center (CIP)-VITAA, Kampala.

Kassali, R. (2011) Economics of sweet potato production. *International Journal of Vegetable Production* 17(4), 313–321.

Labadarios, D., Moodie, I.M. and Van Rensburg, A. (2007) Selected micronutrient status: vitamin A. In: Labadarios, D. (ed.) *National Food Consumption Survey – Fortification Baseline (NFCS-FB): South Africa, 2005.* Directorate Nutrition, Department of Health, Stellenbosch, South Africa, pp. 409–446.

Laurie, S.M., Van den Berg, A.A., Tjale, S.S., Mulandana, N.S. and Mtileni, M.M. (2009) Initiation and first results of a biofortification program for sweetpotato in South Africa. *Journal of Crop Improvement* 23(3), 235–251.

Laurie, S.M., Faber, M., Calitz, F.J., Moelich, E.I., Muller, N. and Labuschagne, M.T. (2012a) The use of sensory attributes, sugar content, instrumental data and consumer acceptability in selection of sweet potato varieties. *Journal of the Science of Food and Agriculture* 27, 151–159.

Laurie, S.M., van Jaarsveld, P.J., Faber, M., Philpott, M.F. and Labuschagne, M.T. (2012b) *Trans*-β-carotene, selected mineral content and potential nutritional contribution of 12 sweetpotato cultivars. *Journal of Food Composition and Analysis* 27, 151–159.

Low, J.W. (2011) Unleashing the potential of sweet potato to combat poverty and malnutrition in sub-Saharan Africa through a comprehensive initiative. *Acta Horticulturae* 921, 171–179.

Magasana, V., Witten, C.B., Romano, R.M. and Beeforth, M. (2004) Mdantsane orange-fleshed sweetpotato gardens improve community participation in the vitamin A supplementation programme in the Eastern Cape. *South African Journal of Clinical Nutrition* 17, S16–S17.

Nsibande, M.L. and McGeoch, M.A. (1999) Sweet potato, *Ipomoea batatas* (L.), cropping practices and perceived production constraints in Swaziland: implications for pest management. *International Journal of Pest Management* 45(1), 29–33.

Nwaru, J.C., Okoye, B.C. and Ndukwa, P.C. (2011) Measurement and determinants of production efficiency among small-holder sweet potato (*Ipomoea batatas*) farmers in Imo State, Nigeria. *European Journal of Scientific Research* 59(3), 307–317.

University of California (2006) *Sample Cost to Produce Sweetpotatoes in California San Joaquin Valley.* PO-SJ-06. University of California Cooperative Extension, Berkeley, California.

Vorster, H.H., Oosthuizen, W., Jerling, J.C., Veldman, F.J. and Burger, H.M. (1997) *The Nutritional Status of South Africans. A Review of the Literature from 1975 to 1996.* The Health Systems Trust, Durban, South Africa.

54 Getting the Equation Right: Engendering Sweetpotato Value Chains in East Africa

S. Mayanja[1]* and M. McEwan[2]
[1]*International Potato Center Sub-Saharan Africa (CIP-SSA), Kampala, Uganda;* [2]*CIP-SSA, Nairobi, Kenya*

Abstract

Upgrading sweetpotato value chains is constrained by its bulkiness, its short shelf life and it being predominantly seen as a woman's and poor person's crop. Commercialization efforts require increased involvement of men and young people, without jeopardizing benefits accruing to women.

The International Potato Center (CIP) is supporting value chain development of orange-fleshed sweetpotato (OFSP) using the Participatory Market Chain Approach (PMCA) in Kenya, Uganda and Tanzania. However, while the PMCA helps generate commercial, technological and institutional innovations, it does not explicitly address equitable participation and utilization of benefits by the different gender categories. To address this gap, the authors undertook a 9 month gender coaching trajectory organized by AgriProFocus with the aim of engendering the PMCA. During the coaching sessions, a critical analysis of the PMCA was done which resulted in development of a gender action plan. The plan emphasized gender-specific interventions using three strategies: (i) **mitigating resistance by building on tradition** (i.e. transforming women's roles from reproductive to productive – subsistence to commercial farming); (ii) **capacity building** (business literacy, sensitizing men, gender-responsive financial services); and (iii) **gender-responsive businesses** (better business options for women, men and young people). Gender-responsive tools were used in the three PMCA phases, and in tracking the resultant changes. CIP staff then coached the core country teams to employ the tools and strategies.

The chapter presents findings from the application of gender tools to identify, analyse and prioritize gender-based constraints and use the resulting strategies to upgrade value chains. It is too early to assess whether the resultant innovations and benefits will be accessed equitably by women, men and young people. We conclude with a discussion around: (i) what tools are most appropriate; (ii) how the PMCA methodology might be strengthened through their use; and (iii) operational issues which might arise through using an engendered-value-chain approach.

Keywords: East Africa, gender responsive, orange-fleshed sweetpotato, value chains

*s.mayanja@cgiar.org

54.1 Introduction

Sweetpotato is a versatile root crop that is grown widely in East and Central Africa (ECA) mainly for food and income. It is easy to grow and fits into many farming systems. Sweetpotato is important in people's diet and food security, with a high per capita consumption in Uganda (i.e. 85 kg per capita/year, ranking third after bananas and cassava (Mwanga and Ssemakula, 2011). The orange-fleshed sweetpotato (OFSP) varieties are an important source of beta-carotene, a vitamin A precursor, and can be used to combat vitamin A deficiency, which if not addressed can lead to increased child mortality (Low *et al.*, 2007). However, sweetpotato is considered a subsistence or 'orphan' crop, in many areas in ECA. The crop is also predominantly grown by women and rarely has priority in the crop planting cycle (Andrade *et al.*, 2009). This limits its popularity and commercial potential, especially for OFSP varieties. Marketing of sweetpotatoes is hampered by its bulkiness and high perishability. There have been several efforts to counter these challenges in ECA through the Agricultural Innovation Systems (AIS) approach. The approach recognizes that a broad participation and interaction of actors from public and private sectors is required for agricultural innovation to take place (McEwan, 2009). The International Potato Center (CIP) with a consortium of partners is implementing the 'Scaling up technologies in OFSP using AIS to address food and nutrition security in Eastern and Central Africa' (OFSP-AIS) project. Its purpose is to enhance utilization and scale up innovation and knowledge in provitamin A-rich OFSP in Kenya, Uganda and Tanzania. The project is being implemented in Gem and Ugunja districts, Western Kenya by two partners, the Ugunja Community Resource Center and Kenya Agricultural Research Institute – Kakamega. The area largely depends on maize for food and income. In Tanzania, the project is being implemented in Missenyi and Muleba districts, by the Kolping Society of Tanzania and the Lake Zone Agricultural Research and Development Institute. The major crops grown include bananas, coffee and sugarcane. In Northern Uganda, the project is being implemented in Kitgum and Pader, a former war zone where resettlement initiatives are ongoing; Building Resources Across Comunities (a non-governmental organization) and Gulu University are the implementing partners in Uganda. OFSP is a recent crop in all the project areas. The project uses the PMCA as the conceptual framework for value chain development. The approach centres on generating technological, commercial and institutional innovations along market chains by increasing trust, confidence and linkages among market chain actors and improving market access for small-scale farmers (Bernet *et al.*, 2006, 2008). The PMCA is structured in three phases each with clear objectives. In Phase one, exploratory, diagnostic market research is conducted to understand the chain and chain actors; and to motivate them to participate in chain-upgrading activities. Phase two centres on analysing and selecting joint market opportunities in thematic groups, for which business plans are developed. In Phase three, chain actors collaborate in practical innovation processes and focus on activities needed to launch specific innovations. Each phase closes with a final event at which results are show cased.

We (the CIP team) identified methodological gaps in the PMCA during implementation. The PMCA is 'silent' on the gender equity perspective and assumes that men and women will benefit equally from market interventions. It has no explicit strategy for promoting women's participation and benefit from marketing activities. Yet women farmers face more gender-specific barriers and other general barriers to engaging in markets. This is in addition to production constraints such as restricted access to land, credit and inputs (Baden, 2013). Furthermore, underperformance of the agricultural sector in ECA is linked to women's unequal access to productive resources and opportunities (FAO, 2011). But, as argued by Apotheker *et al.* (2012), focusing on gender equity in agricultural value chains is good for women, for society and for business. They present three arguments: (i) social justice: fair distribution of advantages, assets and benefits among all members of a society; (ii) fighting poverty is difficult when gender blind; and (iii) gender inequity is a missed business opportunity. This prompted the CIP team to undertake a gender coaching trajectory, through which a gender action plan (GAP) was developed, and guided engendering the PMCA.

The process included: (i) identifying and selecting tools for engendering the PMCA; (ii) hands-on coaching on utilization of the tools by the CIP team; (iii) training country teams on the tools; and (iv) application of the tools with the market chain actors. Three tools were identified and were used by country teams to identify, analyse and select 'best bets' for gender-responsive businesses. This enabled development of engendered business plans that guided development of gender-responsive OFSP businesses. This chapter outlines our experiences in using the tools, methods and skills obtained from the gender coaching in engendering the PMCA.

54.2 Materials and Methods

We joined the coaching trajectory after an AgriProFocus (APF) workshop on gender in value chains, in which some gender tools and a book titled *Challenging Chains to Change* were introduced. We then developed a GAP based on three strategies. The first, 'mitigating resistance by building on tradition' was ideal since OFSP is a woman's crop. The strategy focuses on capacity building for women actors, links to better markets, transformation of petty businesses and improving women's position in the chains. The second, 'organizing for change', focuses on breaking down structural constraints and building human agency: confidence, self-skills and capacities; while the third, 'gender-responsive businesses' focused on promoting better business options for women, men and young people (KIT, APF and IIRR, 2012). Activities involved collective action, sensitizing men and access to finance. Coaching commenced immediately after rolling out the GAP. It involved: (i) face-to-face and Skype interactions; (ii) hands-on sessions with the tools; (iii) provision of resource materials; (iv) linkages to resource persons; (v) sharing at learning events; and (vi) formal workshops.

The CIP team was coached on how to use three tools. The first, the gender-sensitive mapping tool (Mayoux and Mackie, 2008) is used to map chains with a gender lens and assists in identifying the chain actors (disaggregated by sex), the major blockages to the chain, power and value in the chain, as well as identifying invisible actors. The second, the gender-based constraints analysis tool (Terrilon *et al.*, 2013) enables profiling of the chain actors and identifying constraints that could limit participation in a business opportunity. The third, the risk–benefit tool (Vanderschaeghe and Lindo, 2012) is used for *ex ante* assessment of potential risks and benefits of business opportunities to women and men, and aids in selecting 'best bets'.

The CIP team trained the country teams, which were comprised of 20 facilitators (12 women, eight men) and also provided support during application of the tools with 86 thematic group members (25 in Kitgum (16 men, nine women); 21 in Ugunja (11 men, ten women) and 40 in Bukoba (15 men, 25 women)). The tools were further improved during a regional workshop on gender in value chains and business planning where the APF coach supported 14 project team members (nine women, five men) to integrate them in the PMCA.

54.3 Results

Gender-responsive value chains mapping

The gender in value chains mapping tool was used by country teams in Phase one to develop gender sensitive maps. The results of the diagnostic market survey guided the preliminary mapping, after which the actors applied a 'gender' lens using their experience. In Kenya, the tool showed that very few women engaged in the vines chain while the root chain was dominated by women; apart from the wholesale function. The blockage hindering transformation of the root chain was identified to be lack of quality planting materials, while power was concentrated in the male-controlled wholesale activities. The gender map for processed products revealed that women were 'invisible actors', and were engaged in time-consuming activities like peeling, chipping and drying, while men dominated in nodes that required larger capital outlay such as processing and distribution. The blockage was inconsistent supply of roots, and power lay with the processor. The gender maps for vines, roots and processed products in Tanzania were similar to the Kenyan ones. The difference was children were also actors (retailers) in Tanzania. The gender mapping exercise was done in mixed gender groups for both countries, however, in Tanzania this restrained women's participation after they realized men were opposing their ideas.

The maps were shared with the coach, who advised that analysis of constraints and opportunities should be done immediately after mapping. The coach noted that chain support functions should also be included in the map, and advised the use of single-sex groups. In Uganda, these two adaptations were included. The major blockage for upgrading the vines and roots chain was identified to be drought. Further analysis in single-sex groups revealed that men advocated for irrigation equipment to address the challenge, while women selected access to credit for petty trading. Men were surprised to learn that gender-based violence and divorce had strong links to drought, as revealed by the women (Pyburn and Terrilon, 2013).

The teams noted that the tool was very detailed (six steps) and took long to apply. It also required adequate gender-sensitive information prior to application.

Gender-based constraint analysis tool

The information obtained from the gender mapping exercise was further analysed using the gender-based constraint analysis tool. The Kenya team used this to analyse the constraints that could curtail the opportunity to supply OFSP vines to institutions and large buyers. For women, access to land and quality starter vines were the challenges. The consequence would be production of poor quality vines, resulting in poor sales. Proposed interventions included the establishment of primary multiplication sites within the community, and sensitizing men to allow women access to productive land. For men, constraints identified were limited capital to irrigate their plots during the dry season. The consequence would be high costs of production, and the remedy was linking vine producers to microfinance institutions to procure irrigation equipment.

The Tanzanian team identified a market opportunity to supply fresh OFSP roots to markets in Mwanza, Sengerema and Geita. Analysis revealed that women farmers faced challenges in accessing arable land to produce sweetpotatoes, since men allocate prime land for 'male' cash crops. The consequence would be low productivity. The potential remedies were sensitization of men and leaders', women's access to resources, training in soil fertility management and access to low-cost fertilizer. The constraints faced by men included limited access to inputs such as labour and manure, although it was agreed that women would face this constraint, albeit at a greater scale. The consequence to men was low productivity and late harvests, and the potential action was access to loans to procure fertilizer, and training in good agricultural practices.

The Uganda team used the tool to analyse gender-based constraints that would curtail the opportunity to supply OFSP-based processed products to retail markets. While larger (male) and small-scale (female) processors faced challenges in accessing fresh roots due to inconsistencies in supply, the remedies were different. The large male processor could be linked to female producer groups, while the women opted to link with individual producers. The second challenge for women was accessing affordable baking equipment. The remedy was renting equipment from the male baker.

The results were used: (i) to further refine analysis and selection of market opportunities; and (ii) the actions and strategies were included in business plans.

Gender-responsive risk–benefit analysis

The risk–benefit tool can be used for *ex ante* analysis, or as a planning tool. In our case, it was used to assess the potential impact of implementing the business opportunity on women and male actors, and the community at large. This enabled selection of gender-sensitive market opportunities. The criteria included effect on time and workload, income and control of resources, social position (i.e. change in social position and gender relations) and market position (i.e. change in economic power position in the chain) as a result of the upgrading strategy. The Kenya team used the tool to assess the impact of upgrading the vines chain. Analysis revealed that men would benefit from the training and would also provide more family labour. Men and women would benefit from increased income which would elevate women's position in society hence increasing their visibility and potential to vie for leadership positions. However, men would use wetlands for vine production, displacing women's vegetable gardens since the vines would now be generating higher income, thus warranting prime resource allocation. The expected impact on women was increased workload. Women would also have to contend with having small roots for food from male-owned commercial plots, given the land constraints. This analysis enabled the team to identify mitigating strategies to reduce the expected negative impacts.

A number of challenges were experienced in using the tools, especially the time required to apply the tools, ease of use and comprehensiveness of information collected.

54.4 Discussion

There were many similarities in the information generated across the three countries. This could be attributed to the fact that the OFSP chains were similar, though more sophisticated and elaborate in Kenya, due to linkages to earlier interventions. In contrast, the information captured in Uganda on gender constraints was more elaborate, due to working with sex-disaggregated groups, with same-gender facilitators. This is helpful and takes cognisance of cultural and gender barriers that would otherwise impede contributions from women.

The choice of tools was critical and had to be in line with the strategies identified in the GAP, but also complementing the phased nature of the PMCA. The in-depth analysis obtained from applying the tools generated information that allowed market chain actors to get a better understanding of the challenges faced by fellow actors, and how best to address them within their scope of operation. This allowed identification of activities that needed to be conducted to roll out the GAP strategies. In Uganda, the link between gender based-domestic violence and drought, prompted the District Production Officer to caution and implore men to be more supportive and open with their wives, as a prerequisite for development. The strong advocacy message played a crucial role in rolling out the 'organizing for change' strategy, and paved the way for differentiated interventions such as loans for women. The potential actions to address constraints faced by men were noted to be easier and faster (e.g. access to loans) than those for women (e.g. advocacy for increased access to land).

Our experience is in agreement with Baden (2013), who stresses that development and policy interventions are needed to address deeply embedded attitudes and beliefs, as well as gender-based asset gaps. If the constraints were addressed and the upgrading strategy undertaken, the benefits would outweigh the risks. The gender mapping tool also raised consciousness about the distribution of workload, resources and benefits among men and women. These issues were further amplified with the constraints analysis tool. It was clearly observed that women had unequal access to resources compared with men and were hence more constrained to participate in value-chain-upgrading activities supporting findings from the Food and Agriculture Organization of the United Nations (FAO, 2011). What was interesting to note was the 'agency' nature of the potential benefits for women. Examples of this include: (i) elevated status; and (ii) leadership positions and power to control or lead, empower or involve other women, as observed from the information generated from the risk–benefit tool.

The PMCA could benefit from two types of modifications: (i) introduction of new tools (e.g. the gender mapping tool); and (ii) engendering existing tools (e.g. the survey and market assessment tools by including gender disaggregated questions and in the business plan by integrating potential gender-specific actions to address identified constraints). Gender transformative activities embedded in the business plans will guide implementation of the market opportunities in a gender-responsive manner. This will lead to more cohesion of the groups and trust building, hence creating a better space or 'theatre' for innovation involving different actors (Critchley *et al.*, 2006 cited in Thiele *et al.*, 2011).

The tools could be improved to allow for flexibility especially with regards to the differing nature of various agricultural commodity chains. Also, all three tools require adequate time for effective application. In future, graphics could be included and the tools translated into the local language. The other challenge that needs to be addressed lies with translating information and conclusions drawn from analyses into practical activities.

Most research and development (R&D) gender interventions are seen to be the domain of women. While the core teams initially were comprised of women, the CIP team has tried to 'lure' men with increasing success. In Uganda, a man is at the forefront of engendering the PMCA. However, men facilitators still want to deal with 'core' PMCA work, leaving the gender mainstreaming activities to women. Process-wise, facilitators need time to reflect and understand the tools and how best they can be applied. Though most practitioners think they can easily be adapted (likening them to rapid rural appraisal tools), our experience has shown that this may not be the case. At thematic group level, the mapping tool 'broke the ice' and led to more interactions between men and women, especially in Kenya and Tanzania. These interactions have continued through the PMCA activities.

54.5 Conclusion

In conclusion, engendering value chains is vital for holistic agricultural marketing development interventions. Most development practitioners have been grappling with the challenge of identifying comprehensive value-chain-upgrading approaches that address the varying needs of different classes/groups in society, more so the most vulnerable. In the recent past, the focus was on the numbers reached reporting sex-disaggregated data. The tools employed during this study have enabled core teams to conduct deeper analysis and collect more qualitative sex-disaggregated data that has resulted in identifying which strategies could be more effective for women or men, and also the possible impacts if so implemented. Support from a gender expert is invaluable as one learns how to use tacit but important gender-sensitive language (e.g. entrepreneur instead of businessman; or supervisor instead of foreman), and how to use gender analysis for a more in-depth identification of constraints and subsequently a more appropriate intervention.

References

Andrade, M., Barker, I., Cole, D., Dapaah, H., Elliott, H., Fuentes, S., Grüneberg, W.J., Kapinga, K., Kroschel, J., Labarta, R., Lemaga, B., Loechl, C., Low, J., Lynam, J., Mwanga, R., Ortiz, O., Oswald, A. and Thiele, G. (eds) (2009) *Unleashing the Potential of Sweetpotato in Sub-Saharan Africa: Current Challenges and Way Forward*. Working Paper 2009-1. International Potato Center (CIP), Lima.

Apotheker, R., Pyburn, R. and Laven, R. (2012) Why focus on gender equity in value chains? In: *Challenging Chains to Change: Gender Equity in Agricultural Value Chain Development*. Karlsruhe Institute of Technology (KIT) publishers, Royal Tropical Institute, Amsterdam, pp. 13–26.

Baden, S. (2013) *Women's Collective Action: Unlocking the Potential of Agricultural Markets*. Oxfam International, London.

Bernet, T., Thiele, G. and Zschocke, T. (2006) *Participatory Market Chain Approach (PMCA) User Guide*. International Potato Center (CIP), Lima.

Bernet, T., Devaux, A., Thiele, G., López, G., Velasco, C., Manrique, K. and Ordinola, M. (2008) The participatory market chain approach: stimulating pro-poor market-chain innovation. *ILAC Brief* no. 21. Institutional Learning and Change Initiative (ILAC-CGIAR), Washington, DC.

Critchley, W., Verburg, M. and van Veldhuizen, L. (2006) *Facilitating Multi-stakeholder Partnerships: Lessons from PROLINNOVA*. A publication in the series for Promoting Local Innovation. Prolinnova International Secretariat, ETC EcoCulture, Silang, Cavite, The Philippines.

Food and Agriculture Organization of the United Nations (FAO) (2011) *The State of Food and Agriculture: Closing the Gender Gap for Development*. FAO, Rome.

Karlsruhe Institute of Technology (KIT), AgriProFocus (APF) and International Institute of Rural Reconstruction (IIRR) (2012) *Challenging Chains to Change: Gender Equity in Agricultural Value Chain Development*. KIT Publishers, Royal Tropical Institute, Amsterdam.

Low, J.W., Arimond, M., Osman, N., Cunguara, B., Zano, F. and Tschirley, D. (2007) A food-based approach: introducing orange-fleshed sweet potatoes increased vitamin A intake and serum retinol concentrations in young children in rural Mozambique. *Journal of Nutrition* 137, 1320–1327.

Mayoux, L. and Mackie, G. (2008) *Making the Strongest Links: a Practical Guide to Mainstreaming Gender Analysis in Value Chain Development*. International Labor Organization, Geneva, Switzerland.

McEwan, M. (2009) Out-scaling and up-scaling orange fleshed sweetpotato technologies: the potential role of innovation platforms. Paper presented at the 15th Triennial Symposium of the International Society for Tropical Root Crops (ISTRC), 6–9 November, Lima, Peru. Available at: http://www.istrc.org/istrc-symposiums/20-fifthteenth-triennial-symposium/53-fifthteenth-triennial-symposium-session-x (accessed 23 March 2015).

Mwanga, R.O.M. and Ssemakula, G. (2011) Orange-fleshed sweetpotatoes for food, health and wealth in Uganda. *International Journal of Agricultural Sustainability* 9(1), 42–49.

Pyburn, R. and Terrilon, J. (2013) Achieving gender equity in agricultural value chains – sweetpotato, drought and pork: making the link. *New Agriculturist* March 2013. Available at: http://www.new-ag.info/en/focus/focusItem.php?a=2923 (accessed 4 May 2013).

Terrilon, J., McEwan, M. and Mayanja, S. (2013) Tool 3.3.a. Activity mapping and the identification of gender based constraints. In: Senders, A., Lentink, A., Vanderschaeghe, M. and Terrilon, J. (eds) *Gender in Value Chains: a Practical Toolkit to Integrate a Gender Perspective in Agricultural Value Chain Development*. AgriProFocus, Kampala.

Thiele, G., Devaux, A., Reinoso, I., Pico, H., Montesdeoca, F., Purnisacho, M., Andrade-Piedra, J., Velasco, C., Flores, P., Esprella, R., Thomann, A., Manrique, K. and Horton, D. (2011) Multi-stakeholder platforms for linking small farmers to value chains: evidence from the Andes. *International Journal of Agricultural Sustainability* 9(3), 423–433.

Vanderschaeghe, M. and Lindo, P. (2012) *Another Point of View: a Manual on Gender Analysis Training for Grassroots Workers*. Adapted from *Gender Analysis Matrix in Rani Parker (1993)*. UNIFEM, Geneva, Switzerland.

Part VI

Evidence on Nutritional Value, Product Acceptance and Changing Behaviours Concerning Sweetpotato

55 Assessing Nutritional Value and Changing Behaviours Regarding Orange-fleshed Sweetpotato Use in Sub-Saharan Africa

J. Low,[1]* A. Ball,[2] P.J. van Jaarsveld,[3] A. Namutebi,[4] M. Faber[3] and F.K. Grant[1]

[1]International Potato Center Sub-Saharan Africa (CIP-SSA), Nairobi, Kenya; [2]HarvestPlus, Kampala, Uganda; [3]South African Medical Research Council, Tygerberg, South Africa; [4]Makerere University, Kampala, Uganda

Abstract

During the past 15 years, significant investments have been made in the development and promotion of orange-fleshed sweetpotato (OFSP) varieties in sub-Saharan Africa (SSA) due to their high β-carotene contents and thus, their potential to contribute towards reducing vitamin A deficiency. As the dominant varieties in SSA are white-fleshed, lacking in β-carotene, the introduction of OFSP means that producers and consumers need to accept the very visible change in colour along with any other trait differences.

We first review the building of the evidence base for OFSP in SSA, summarizing the key lessons learned to date, focusing on interventions aiming for impact on vitamin A intakes or status. Second, we review the state of knowledge concerning how to maximize the nutritional value of OFSP when processed and identify appropriate entry points to reach rural and urban consumers. Third, we explore how the health sector and private-sector marketing firms are tackling behavioural change and based on this accumulated multi-sector experience develop improved recommendations to guide practitioners on how they should approach reaching consumers. Finally, we pinpoint opportunities to enable the successful scaling out of OFSP adoption and utilization and identify areas of research needed to address remaining knowledge gaps.

In developing OFSP-focused food-based approaches, several key questions are addressed: (i) Are OFSP varieties competitive with existing local varieties?; (ii) Do producers and consumers accept a variety with a distinct colour difference?; (iii) What does it take to get proper utilization of OFSP at the household and the young child level?; (iv) What are the key nutrition messages?; and (v) What do we need to do to break into rural and urban markets?

We also examine the best techniques for assessing micronutrient contents of OFSP and its impact on status, the minimum amount of fat consumption linked to OFSP consumption and the bioaccessibility of OFSP processed in different ways. After reviewing how other sectors are approaching behavioural change, we propose 12 key recommendations on how to approach rural households and present dominant factors likely to influence urban consumer behaviour. The current policy environment is very favourable to integrated agriculture–nutrition interventions, and the Scaling Up Nutrition (SUN) movement emerges as the most promising opportunity for OFSP integration.

Keywords: behavioural change, beta-carotene, bioaccessibility, nutritional value, orange-fleshed sweetpotato (OFSP), scaling out, vitamin A

*j.low@cgiar.org

55.1 Introduction

Overview of orange-fleshed sweetpotato (OFSP) benefits and introduction in sub-Saharan Africa (SSA)

The nutrition community has long recognized the negative impact of micronutrient deficiencies on the growth and development of children; the resultant economic cost of such deficiencies was recently estimated at 2.65% of gross domestic product (GDP) in Tanzania (Scaling Up Nutrition, 2013). Vitamin A deficiency (VAD) is widespread among young children on the African continent, with an estimated 36.4% of children under 5 years of age affected by VAD in 2007 (UN Standing Committee on Nutrition, 2011). VAD can limit growth, weaken immunity, cause xeropthalmia leading to blindness, and increase mortality (Sommer and West, 1996). Partners[1] in the Vitamin A Global Initiative, founded in 1997, are committed to the eradication of VAD as a public health concern.

Food-based approaches to combatting VAD aim to increase access to and intake of vitamin A-rich foods. There are two types of vitamin A available in foods: (i) highly bioavailable preformed retinol (vitamin A itself) typically found in animal foods such as eggs, liver and milk; and (ii) provitamin A carotenoids found in plant foods such as dark-green leafy vegetables and yellow and orange vegetables and fruits (McLaren and Frigg, 2001). Poor households typically cannot afford to consume animal foods on a regular basis. Among plant sources, β-carotene is the major provitamin A carotenoid and its bioavailability varies considerably. OFSP have good to excellent amounts of *trans*-β-carotene, which is highly bioavailable (Haskell *et al.*, 2004; van Jaarsveld *et al.*, 2005) and its true retention after boiling is high (70–92%) (van Jaarsveld *et al.*, 2006). Just 100–125 g of boiled or steamed OFSP available in SSA meet the daily recommended intake levels of vitamin A for children under 5 years of age (Low *et al.*, 2009). Moreover, unlike many vegetables, sweetpotato has significant amounts of energy (e.g. 76 kcal/100 g in sweetpotato versus 26 kcal/100 g in pumpkin) (USDA-ARS, 2011). Hence, OFSP is considered a **biofortified**[2] staple food crop that can tackle the problem of inadequate energy intake as well as VAD.

During the past 15 years, considerable research undertaken in Mozambique, Uganda, South Africa and Kenya has built the evidence base demonstrating that OFSP can be successfully introduced and is an effective tool for combatting VAD among children under 5 years of age (Hagenimana *et al.*, 1999; Faber *et al.*, 2006; Low *et al.*, 2007b; Hotz *et al.*, 2012a, b). Solid evidence for food-based approaches is considered to be limited (Ruel and Levin, 2000; Berti *et al.*, 2004; Masset *et al.*, 2011; Girard *et al.*, 2012). It is therefore important to understand how the evidence base for OFSP has been built to convince the nutrition community and policy makers of its potential, especially since the research on OFSP is being cited as one of the few well-documented food-based strategies demonstrating impact on vitamin A intakes and status (Masset *et al.*, 2011; Girard *et al.*, 2012). OFSP is the first biofortified crop to reach farmers' fields in SSA, and the lessons learned from this experience can potentially guide the design of other dissemination efforts of biofortified crops with a visible trait as well as inform other food-based efforts.

In spite of the practical experiences to date integrating agriculture and nutrition, our understanding of what actually makes individuals change their dietary behaviours is still very limited. In SSA, however, considerable investments in behavioural change research have been made with regards to tackling the human immunodeficiency virus (HIV)/acquired immune deficiency syndrome (AIDS) crisis, dietary practices, and water and sanitation access. Private-sector marketing firms also invest in understanding different consumer segments. Joining the experience with OFSP to date with that of these sectors could lead to an improved set of recommendations on how to maximize the uptake of OFSP and its potential health benefit.

Objectives of the chapter

There are four major objectives of this chapter:

1. Review the building of the evidence base for OFSP and summarize the key lessons learned to date from its introduction in SSA, focusing on interventions aiming for impact on vitamin A intakes or status.

2. Review the current state of knowledge concerning how to maximize the nutritional value of OFSP when it is processed and examine what might be the most appropriate entry points to reach rural and urban consumers.

3. Explore how the health sector and private-sector marketing firms are tackling behavioural change and based on the combined experience to date develop improved recommendations to guide practitioners on how they should approach reaching rural consumers and urban consumers.

4. Pinpoint opportunities to enable the successful scaling out of OFSP adoption and utilization and identify areas of research needed to address remaining knowledge gaps.

55.2 Major Phases in the OFSP Story in SSA

Phase I: Confronting conventional wisdom (1995–2000)

Most dominant sweetpotato varieties in Africa or Asia are white or yellow fleshed. Prior to 1995, the attitude towards OFSP use in SSA by many scientists is captured in the major book on sweetpotato by Woolfe published in 1992: '*Sweet potato cultivars with deep yellow or orange-fleshed roots are unfortunately rejected in many developing countries in favour of white or cream-fleshed types having little or no pro-vitamin A activity*'. Attempts to introduce dark OFSP varieties in Bangladesh and other countries failed in the 1980s, not due to the colour but due to the moist texture (low dry matter content) of the varieties used. However, conventional wisdom reduced the message African and Asian consumers do not like orange-fleshed varieties.

In 1995, recognizing the seriousness of VAD in SSA, the International Potato Center (CIP) and the Kenya Agriculture Research Institute (KARI) began OFSP research as part of a broader effort to develop and test women-based approaches for addressing micronutrient deficiencies. In Kenya, sweetpotato production is largely under the control of women. The research compared OFSP uptake among women's groups receiving agricultural extension advice and the new OFSP varieties to women's groups receiving the same agricultural intervention *plus* nutrition education. The study found that nutrition education was essential for seeing an increase in the frequency of intake of vitamin A-rich foods among young children (Hagenimana *et al.*, 1999). Several OFSP varieties used in the study (selected from introduced varieties and local landraces) were found to yield as well or were superior to dominant local white- or yellow-fleshed varieties. Children were found to like OFSP varieties that were lower in dry matter content (more watery) than adults, who preferred higher dry matter content varieties similar to dominant existing varieties.

Breeding programmes began selecting for OFSP varieties with higher dry matter content. Small-scale OFSP development activities were funded in Uganda, other parts of Kenya and Tanzania, but the decade was notable for declining agricultural investment, and the nutrition community emphasized vitamin A capsule supplementation as the major strategy for combatting VAD.

Phase II: Building the evidence base (2001–2009)

While the agricultural community had begun to recognize the potential contribution of OFSP by the year 2000, the nutrition community to a large extent was not yet supportive, often citing lack of convincing evidence. An interest group spearheaded by CIP, the Vitamin A for Africa (VITAA) Platform, was created in 2001 to raise awareness and serve as a forum for exchange initially among five SSA countries. A 2001 *ex ante* study drawing on available data detailing where sweetpotato was produced and where VAD existed, estimated that switching from white-fleshed sweetpotato to OFSP could significantly contribute to reducing VAD in 50 million African children (Low *et al.*, 2001). Around the same time, a major programme looking at breeding micronutrients across several staple crops, HarvestPlus, received significant funding and the term **biofortification** was coined. OFSP emerged at the forefront of the biofortification effort because very high levels of β-carotene already existed in the germplasm, and the breeding effort could focus on developing OFSP materials adapted to target areas in SSA and Asia. In other crops, the amount of the micronutrient within the crop had to first be increased to biologically significant levels.

An OFSP efficacy study conducted among school children (5–10 years old) in South Africa in 2002 measured vitamin A status using the modified-relative-dose response test (van Jaarsveld *et al.*, 2005). The treatment group ($n = 90$) consumed 125 g of boiled, mashed OFSP, while the control group ($n = 90$) ate white-fleshed sweetpotatoes for 53 school days. The treatment group showed significant improvement in vitamin A liver stores compared with the control group, with the proportion of children in the former group with normal vitamin A status increasing from 78% to 87% after the intervention. There was no significant change in vitamin A liver stores among the control group (86% to 82%). This evidence was bolstered by a community-level intervention in a very resource-poor area of Mozambique (the Towards Sustainable Nutrition Improvement (TSNI) project), where VAD prevalence among the children at the beginning of the study was 71%, using serum retinol concentrations as proxy for vitamin A status (Low *et al.*, 2007b).

The intervention consisted of an integrated approach along three intersecting pathways:

1. Agriculture: *Introduction of a new source of vitamin A and energy, using biofortified OFSP.* Intervention farmers organized in groups received planting material of high-yielding OFSP varieties, combined with lessons on how to improve crop management and storage practices to maximize the availability of OFSP throughout the year.
2. Nutrition: *Demand creation and empowerment through knowledge.* At the village level, principal caregivers, both women and men, were encouraged and enabled to improve infant and young child feeding practices, hygiene practices, and diversify the household diet. A nutrition extentionist conducted monthly group sessions for a year. Demand creation efforts focused on building awareness among the broader community to create: (i) demand for the new OFSP cultivars and derivatives; (ii) demand for other vitamin A-rich foods; and (iii) a supportive environment to accelerate behaviour change at the household level. These included provincial-wide radio programmes, community theatre performances, painted stalls and signs in local markets, and T-shirts, caps and sarongs worn by women decorated with the slogan '*O doce que dá saúde*' (the sweet that gives health).
3. Marketing: *Market development for OFSP roots and processed products.* This component aimed to link farmers to traders and to inform consumers about where they could purchase OFSP. Farmers with identified market outlets were more likely to expand the area under production. Thus, generated demand combined with market development stimulated production, enhanced producer income and spread the health benefits of OFSP to a wider population, all of which contributed to farmers' willingness to retain OFSP and expand production. Demand for OFSP was enhanced if profitable processed products using OFSP as a major ingredient were developed.

The intervention lasted 18 months in two of the poorest districts in rural Zambézia Province, Mozambique. World Vision, an international non-governmental organization (NGO), posted pairs of extensionists at the community level: one for agriculture and marketing, the other for nutrition. Each pair served 14 farmers' groups. In total there were 498 mother–child pairs captured in the study that were compared to 243 mother–child pairs from 'control' areas where no intervention was made.

By the end of the study, 90% of intervention households produced OFSP. Vitamin A intakes among intervention children ($n = 498$) were much higher than those of control children ($n = 243$) (median 426 versus 56 μg retinol activity equivalents, $P < 0.001$). OFSP contributed 35% to the vitamin A intakes of all children in the intervention area and 90% among those who had consumed it the previous day. Controlling for infection/inflammation and other cofounders, a 15% decline in the prevalence of VAD was attributable to the integrated intervention (Low *et al.*, 2007b). OFSP was well accepted and liked by both adults and children.

Concurrently, interest in OFSP interventions began to grow and the VITAA Platform expanded to 11 SSA countries. The next important step was to investigate whether such an OFSP-led food-based integrated approach could be taken to scale at reasonable cost. Drawing on a modified version of the three pathways outlined above, simultaneous 2-year intervention studies conducted in Uganda and Mozambique as part of the Reaching End Users (REU) project tested two different levels of intensity of extension contact using extentionists supported by non-paid promoters recruited from the community (HarvestPlus, 2010). The intervention in Mozambique reached 14,000 households, while the intervention in Uganda reached 10,000 farmer-group-member households. A randomized, controlled effectiveness study evaluated the intervention's impact on the intake of OFSP and dietary vitamin A among children 6–35 months and 3–5 years of age and women in both countries (Hotz *et al.*, 2012a, b), and on the vitamin A status of the 3–5-year-old children and women in Uganda (Hotz *et al.*, 2012b). In Mozambique 77% of households adopted OFSP, compared to 65% in Uganda. In both countries, vitamin A intakes increased significantly among both women and young children, with OFSP contributing 78% of total dietary vitamin A intake among children 6–35 months of age in Mozambique and 53% in Uganda. Average costs per target beneficiary were US$86/household in Mozambique and US$56/household in Uganda. OFSP vines are easy for farmers to share. Hence with time, the initial investment had considerable spillover effects. In terms of just the vitamin A benefit (not considering the food security and other micronutrient benefits), the intervention in Uganda cost US$15–20 per disability-adjusted life year saved (HarvestPlus, 2010).[3] This amount falls within the 'highly cost-effective' category of interventions as defined by the World Health Organization (WHO).

Phase III: Addressing the bottlenecks to exploiting OFSP's full potential (2009 to date)

Building on growing donor support, CIP and over 30 partner organizations launched the 10 year Sweetpotato for Profit and Health Initiative (SPHI) in October 2009. The SPHI emerged from a 7 month consultative process to identify the constraints blocking the full exploitation of sweetpotato and develop interventions in breeding, propagation and dissemination of healthy planting material, crop management, human nutrition and marketing. The initiative's vision is to reposition sweetpotatoes in African food economies, particularly in expanding urban markets, to reduce child malnutrition and improve smallholder incomes. It seeks to positively affect the lives of 10 million African families by 2020 and is establishing support platforms in three sub-regions (East and Central Africa, Southern Africa, West Africa) to enable the creation of a vibrant community of practice. SPHI targets 17 countries and breeding or varietal selection activities are underway in 14 of those countries (Low, 2011). In addition to breeding, research has focused on addressing the problem of timely access by smallholders to adequate quantities of quality planting material and testing delivery systems that lead to enhanced vitamin A intakes and improved value chains for farmers.

The publishing of evidence and release of improved, more-adapted OFSP varieties in Mozambique and Uganda has led to substantial political and donor support in the dissemination of those materials. HarvestPlus continues to study going-to-scale with an integrated agriculture–nutrition strategy to reach 225,000 households by 2016; in Mozambique mass distribution of OFSP led by CIP to 122,000 additional households with a limited awareness campaign occurred in 2011–2013. Pressure is on to go-to-scale as cost-effectively as possible. Understanding what the key components must be so that nutrition goals are still achieved is critical.

55.3 Key Lessons Learned

In developing OFSP-focused food-based approaches, several key questions had to be addressed:

1. Will the OFSP varieties be competitive with existing local varieties?
2. Are producers and consumers willing to accept a sweetpotato variety with a distinct colour difference?
3. What does it take to get proper utilization at the household level and the young child level?
4. What are the key nutrition messages that we must include?
5. What do we need to keep in mind if we want to have OFSP break into rural and urban markets?

Will the OFSP varieties be competitive with existing local varieties?

We learned that it was essential to cross locally adapted materials with orange-fleshed varieties to generate OFSP varieties with enough dry matter to appeal to adult tastes and with agronomic performance that was competitive with dominate local varieties. In drought-prone areas, it is essential to have varieties with vigorous vines, whose roots can re-sprout effectively if left unharvested. As of 2013, nine countries (Mozambique, Uganda, Rwanda, Kenya, Tanzania, South Africa, Zambia, Malawi and Nigeria) have released OFSP materials bred in the relevant country.

Are producers and consumers willing to accept a sweetpotato variety with a distinct colour difference?

The flesh of OFSP has a distinctive visual attribute (i.e. an orange colour because of the β-carotene content). The colour of the sweetpotato is directly related to the β-carotene content, and colour intensity (cream, yellow, yellow-orange, dark orange) may therefore be used as an indicator of provitamin A value (Takahata *et al.*, 1993). The vitamin A value of the sweetpotato is therefore visible to the consumer. In the TSNI project, the colour orange became clearly associated with healthy foods and proved to be an effective tool for demand creation and marketing. Consumers found the attractive golden colour of bread made with OFSP (replacing 38% of wheat flour with OFSP with medium orange intensity) superior to that of white bread (Low and van Jaarsveld, 2008), indicating that the visual trait is also beneficial in postharvest products. In Uganda, OFSP was particularly liked by children, and caregivers cited its orange colour as one of the reasons for the children's preference (Nagujja and Yanggen, 2005).

Associating colour in plant foods and health also occurs in more developed countries. Formal dietary guidelines often use a very simple message – eat a variety of different colours of vegetables and fruit. Australian women, who were not familiar with the β-carotene content of sweetpotato, considered OFSP to be healthy, purely based on the orange colour of the flesh. It was important for them to eat a variety of vegetable colours, of which yellow or orange was specifically referred to, and they valued OFSP as an easy way to add colour to meals (Henderson *et al.*, 2012).

The orange colour is not a barrier to adoption (Hotz *et al.*, 2012b), and building an 'orange brand' to raise awareness of vitamin A and OFSP as part of a marketing campaign can be very effective. For example, the orange colour can be used as a background for T-shirts, caps, sarongs, posters, market stalls, etc. that carry key messages associating OFSP and good health. However, one challenge in many SSA countries is that local languages have no specific word for orange and there is often confusion among users between yellow and orange. Since the yellow colour in sweetpotato is often partially caused by other carotenoids, the amount of β-carotene can be limited. Hence, special attention must be paid to eliminate this common confusion.

What does it take to get proper utilization at the household level and the young child level?

A key concern of planners and practitioners is understanding what kinds of investments are essential for achieving impact and how can the cost of an intervention be as low as possible, yet effective. The TSNI and REU projects described above, along with work in Kenya are the basis of the key lessons learned about what it takes to get adoption and utilization of OFSP at the household level and what it takes to actually make an impact on young child nutritional status. In the REU project, the two selected countries were highly contrasting: Uganda had better growing conditions for sweetpotato, higher population densities (hence, lower service delivery costs), better educated pools of extension personnel, and more highly educated farmers than Mozambique. In the TSNI project, which was only in Mozambique, two extensionists (one for agriculture and marketing, one for nutrition) per site were responsible for all farmer training and follow up; in the follow-up REU project, extensionists worked with village-level non-paid promoters, who were provided with job tools and promotional clothing. The agriculture and health promoters, in turn, taught fellow farmers with backstopping from the extension personnel.

Some key lessons learned from our experience to date are summarized below.

1. *If the goal is to improve young child vitamin A intake, investment in community-level nutrition education is needed.* If farmers like the introduced OFSP variety's agronomic performance and taste, it will be adopted and utilized as part of the family diet. However, to make an impact on young child vitamin A status, particularly the very vulnerable 6–18 month olds, use of OFSP-based complementary foods (e.g. porridges) and increase in feeding frequency are critical and most effectively delivered through community-level approaches.

2. *Designing the project intervention from the beginning with a gender perspective in mind is key.* From the outset, project designers realized: (i) the importance of recruiting both women and men as extentionists and promoters; and (ii) that men influence the land allocated for sweetpotato, what foods are consumed within the household and can enhance or deter the adoption of improved care-giving practices of other members. Data from Mozambique show that female nutrition extension workers were significantly more successful than male nutrition extension workers in teaching messages about child feeding and vitamin A to nutrition promoters. However, in rural settings, male extension personnel typically outnumber female extension personnel. When male nutrition personnel are employed, it is important to give them extra training on how to conduct cooking demonstrations and interact with women effectively about breastfeeding.

3. *In addition to the agriculture intervention, it is necessary to have regular interaction with caregivers with young children in nutrition-focused groups with repetition of consistent messages alongside demonstrations to ensure OFSP uptake and incorporation into the young child's diet as well as the household diet.* Extension personnel working with groups (each having 15–30 members with young children) can effectively achieve uptake and incorporation of OFSP in to the household diet and the young child's diet and overall nutritional knowledge can be improved by holding monthly meetings with caregivers, assuming the extension personnel are properly trained and equipped with appropriate tools and skills. In the TSNI project, findings showed that six additional home visits did not lead to any better use of OFSP and other vitamin A-rich foods, but there was more use of other vegetables and fruits and fat in the diet among groups receiving additional home visits. In the REU project, having a second year of nutritional lessons, often repeating first-year messages, did not result in any greater adoption or better utilization of OFSP by young children. If it is difficult to get fathers to attend nutrition sessions, consider holding separate sessions for men or incorporate key nutrition messages into agricultural/marketing training that they attend.

4. *Extension personnel need quality job aids.* Formative research is needed in each country setting to tailor messages to the education level of the target population and produce written and promotional materials in local languages whenever possible. Clearly, utilization of existing, approved nutrition messaging from government makes sense. Demonstrations (especially for recipe preparation) were the preferred method of transmitting knowledge to farmers. Effective job tools to accompany participatory learning sessions can include: (i) portable-designed flipcharts or laminated A4 cards with key pictures and messages on the front and instructions on the back to remind extentionists about how to best conduct the lesson; (ii) an agricultural calendar describing key crop management tasks to be done each month; and (iii) simple recipe books. Having groups write and sing songs about OFSP and how to use it is an effective way to reinforce key messages.

5. *The total number and depth and frequency of training extension personnel and community promoters/volunteers are dependent on the level of formal education of the pool of candidates.* For example, most extentionists in Uganda were college graduates, whereas most extentionists in Mozambique had completed just 2 years of secondary school. Most community volunteers in Uganda had completed primary school, as had male promoters in Mozambique, but the vast majority of female nutrition promoters in Mozambique were practically illiterate. Personnel with higher levels of formal education are able to absorb a wider range of knowledge and retain and transmit it more effectively than those with substantially less education. In Uganda, single extension agents worked on agriculture, nutrition, and marketing modules. In Mozambique, the project had separate extension personnel for agriculture and nutrition and a university graduate coordinated the marketing work. When educational levels are low, frequent refresher training is required for extension personnel. In Mozambique, these refresher sessions were aligned with monthly staff meetings scheduled when personnel came for their monthly salary payment. In Uganda, extension staff members were trained at the beginning of every season and expected to write weekly progress reports.

6. *In some settings, it will be possible to save costs by using existing farmer or social groups as an entry point.* In the TSNI project, considerable time was spent working with communities to establish farmer groups. In the REU project, to lower costs, existing farmer and church groups were used as entry points. The advantage is that the core members already know and trust each other. The main drawback to working with existing groups is that often an insufficient number of members are households with young children and hence the group size has to expand. Other projects are using antenatal clinics as entry points for recruiting pregnant women to participate in clubs where nutrition education and OFSP utilization are the focus.

Many projects, like the REU, included the use of village resident promoters/volunteers as a way to reach more direct beneficiaries at a lower cost. Criteria are established for the community to select the person that will be trained as an agriculture or nutrition promoter. Some argue that an advantage of using promoters is that people are more likely to adopt something promoted by their peers than by an outsider. Since farmers' groups often meet at weekends, promoters have an advantage over extension services that only work weekdays.

In using community-level promoters/volunteers, the following lessons have been learned:

1. *Use of volunteer promoters is not cost free.* To be effective and remain motivated, incentives for community volunteers are a critical part of any programme and need to be adequately budgeted for. These typically include job tools and attire (badges, T-shirts, hats, decorated sarong) and even bicycles. Promoters require initial training, retraining and monitoring. Staggering the provision of incentives helps to periodically re-motivate promoters. Most promoters are effective for a maximum of 2 years and typically can only be involved a limited number of days per month.

2. *Be aware of government policy and practices of other organizations.* Dissatisfaction among promoters is likely to be high if other organizations are offering better 'incentives'. Some SSA governments are now advocating standardized packages/practices for organizations utilizing community-level workers/promoters. Other projects have found it very effective to pay promoters a small salary each month.

3. *It is critical to monitor the accuracy of the promoters training of caregivers.* While nutrition promoters were able to achieve the desired impact, REU findings clearly demonstrate that for extension personnel accurate knowledge retention was higher than that of the nutrition promoters, and the nutrition promoters in turn showed higher message retention than the mothers they taught. The latter was due to: (i) receiving better training (from extentionists); (ii) receiving regular message 'recaps' from extentionists in their training; (iii) physically possessing job aids (as a memory jog); and (iv) being required to train other people, which thus solidified their own knowledge.

Since the end of the REU project, we have seen in some countries the rapid spread of cell phone uptake in rural areas. This could potentially revolutionize the ability to engage in repeat messaging and monitoring of community-level activities in addition to empowering caregivers to actively seek knowledge on a demand basis. This should further lower the cost of going-to-scale of the integrated approach.

What are the essential messages to include in a nutrition component?

Experience in implementing programmes in Uganda and Mozambique has demonstrated that the topline nutrition message to be communicated to the target population is primarily about the linkage between OFSP and vitamin A. The success of this topline message is, however, partially dependent on the existence of a vitamin A supplementation (VAS) programme being implemented by the Ministry of Health in the country. VAS programmes do exist in most developing countries and the communication messages are often linked to child or family health days that are being promoted by the respective government. Where these communication programmes exist, parents are instructed to bring their children under the age of 5 years for immunization, including VAS, at the nearest health clinic or centre twice per year. As a result, many mothers are familiar with the term 'vitamin A' and they know that their child should take it as part of the immunization package. Consequently the link between vitamin A and OFSP is a natural one and becomes the first message to be delivered to farmers and mothers alike – 'OFSP has vitamin A. Vitamin A is good for your health'. The specifics of how vitamin A improves health become the supporting messages for the topline nutrition message. Thus the supporting messages about vitamin A are that children need it for good health, to fight common infections and to maintain normal vision.

OFSP should not be considered a 'silver bullet' solution to VAD and neither should it be presented as one to the recipient populations. Child health is affected by many more factors than just vitamin A and situating a vitamin A intervention such as OFSP within the larger context should be considered if the child is to benefit from the increased vitamin A being delivered through OFSP. Children with VAD often live in situations where there is a dearth of quality health care services and nutritional information available to parents and/or caregivers. Basic infrastructure including safe water sources and waste disposal are often lacking and these combine to work against good child health practices. As a result, it becomes imperative to consider which other supporting nutrition messages are required by the recipient populations so that the intervention of OFSP is maximally supported. Telling parents to feed their children OFSP can be detrimental if the parents do not understand that exclusive breastfeeding should be practised from birth until 6 months of age, after which foods such as OFSP can be introduced. Mothers must also understand that their own nutrition will affect the nutrition of their babies.

Consequently, there is a set of additional essential messages that must be considered so that OFSP/vitamin A messages are set within an appropriate context. These messages can be considered as supporting messages insofar as they provide the check-and-balance that helps prevent parents and caregivers from taking the OFSP/vitamin A messages out of context. In our experience, we have learned that the specific messages that most caregivers do understand include:

- *Breastfeeding:* Parents are taught that babies should be exclusively breastfed from birth to 6 months and that breastfeeding should continue until the age of 2 years. Caregivers are taught that vitamin A is transmitted from mother to baby via breast milk hence it is important that the breastfeeding mother is consuming enough food which contains vitamin A. The importance of giving the infant the first milk (**colostrum**) in most cases is an easy practice for mothers to adopt; however, preventing the provision of herbal drinks or water as part of traditional ceremonies during the first weeks of life has proved challenging.
- *Young child feeding:* Parents and caregivers are taught about which foods can be introduced to children after 6 months and how to prepare them so that the child will eat enough. Frequency of feeding, food combinations and feeding of sick children is also addressed. Increased feeding frequency is often achievable, with OFSP being used as a breakfast food or snack significantly contributing to its achievement. Cooking demonstrations, where OFSP is combined with other locally available food sources, are more effective than lectures because they are interactive and participatory.
- *Balanced diet and vitamin A-rich foods:* Using/adapting the food guide being promoted by the respective Ministry of Health, mothers and caregivers can be taught about how different foods are necessary for functions within the body. They can learn how to combine them and prepare them to improve family diets. In this case it is imperative to use foods that are available to the population. Particular attention is paid to the sources of vitamin A – both animal and plant sources, including the OFSP and sources of fat. Most nutrition projects do not include much detail about food sources of vitamin A and tend to recommend VAS.
- *Hygiene, sanitation and safe water:* It has been well documented that diarrhoeal diseases are one of the top causes of mortality and morbidity among young children thus any gains that are made with increasing vitamin A intakes through the diet (i.e. OFSP) are diminished if hygiene, sanitation and safe water issues are not addressed in households. Safe water sources are still lacking for millions of households, and basic hygiene and sanitation facilities are beyond the reach of many yet with simple practices it is possible to break the major transmission routes. In our experience, while these messages are important, they have proved very difficult to change without a fairly intensive education strategy and sufficient financial and technical resources to address this component. Hence, some OFSP programmes concentrate their resources on the nutrition/agriculture messages and not on these topics.

In deciding which messages are likely to result in behaviour change, the Trials for Improved Practices (TIPs) approach is highly recommended and a step-by-step guide is available for practitioners (Dickin *et al.*, 1997). It involves conducting consultative research with mothers of different socio-economic status or cultural backgrounds to identify simple and effective actions within the household that will improve child feeding and then test these recommended practices in homes with caregivers to determine which are the most practical and culturally acceptable. Those that work are then promoted at the group level.

OFSP becomes the *entry point* to address family diets and child feeding (Low *et al.*, 2007a). As a result this intervention can be added to existing programmes which are providing additional supportive activities, such as growth monitoring. Mothers, caregivers and communities can be empowered to make changes and an enabling environment will allow for the uptake of the crop and the messages.

What do we need to keep in mind in order to have OFSP break into rural and urban markets?

Having markets for roots drives sweetpotato farmers to expand and sustain OFSP production. General awareness campaigns to reach potential OFSP producers and caregivers, especially those using songs on the radio, concurrently contribute towards building market demand for OFSP roots. The REU project again provided an opportunity to study two very different sets of conditions: (i) breaking into well-established, larger sweetpotato markets in a country (Uganda) where sweetpotato is a primary staple and there are strong consumer preferences for existing non-OFSP varieties; and (ii) breaking into a smaller and seasonal sweetpotato market in a country (Mozambique) where sweetpotato is a secondary staple and poorer consumers are more price conscious than variety conscious. Clearly, marketing is a complex target, requiring knowledgeable personnel for implementation. A project should be at least 3 years and preferably 5 years long to have a marketing component as it takes time to build up sufficient production for a meaningful effort.

Overarching lessons learned about breaking into markets include:

1. *Understand the existing sweetpotato marketing chains and relative price of sweetpotato compared to other food staples and foods rich in vitamin A as an input into the design of a marketing strategy.* Sweetpotato varieties are unique to each country and consumer behaviour varies based on: (i) whether sweetpotato is a staple food or is viewed as a vegetable or snack; (ii) what crops it competes against; (iii) how it is marketed and its relative prices; (iv) whether significant areas already exist with commercialized sweetpotato growers; and (v) whether and how women are engaged in trading sweetpotato.

2. *Health messages linked to the orange colour is the demand-pull for OFSP.* The link to the benefits of vitamin A and the high amount of provitamin A in OFSP is the key message for all areas. In urban zones, upper end consumers are increasingly concerned about diabetes and the good dietary fibre content of sweetpotato peels should also be stressed in this regard (Bovell-Benjamin, 2010). Use of radio and promotional jingles has been very successful in building urban market demand for OFSP.

3. *Every country has a unique set of sweetpotato varieties and it is important to understand the preferences of different sub-groups (young children, women, men, youth and agro-processors).* Invest in consumer surveys and training a sensory panel to evaluate different varieties. Make sure each variety has a market-friendly name and assess its suitability for different uses (boiled, chipped, crisped, bakery products). Unfortunately, there is a tendency to lump OFSP varieties with the same skin colour into a common category. Investing in labelling and creating consumer awareness about the uniqueness of each variety will facilitate monitoring the growth of OFSP commercialization.

4. *Facilitate linking traders to farmers and traders to market opportunities.* It is difficult to break into strong, existing sweetpotato market chains where traders have an understanding of existing consumer preferences and certain traders control access to key wholesale markets in urban centres. However, to facilitate links the following should be considered: (i) hosting a meeting between traders and growers; (ii) encouraging contractual commitments between traders and farmers; and (iii) subsidizing painting of market stalls and providing decorated umbrellas and aprons. Specialized marketing organizations can facilitate linking traders to credit opportunities, as farmers prefer being paid up front for their roots.

5. *Train and treat retail traders as change agents, capable of spreading nutritional messages, and stress separating OFSP from* *other non-OFSP type varieties at point of sale.* Experience has shown that when there are media promotion activities and OFSP is sold separately from other sweetpotato types, either a premium price (up to 15% higher than white-fleshed) for OFSP emerges, or traders note that OFSP sells faster, even when it is selling at the same price.

6. *Even if economically viable opportunities for OFSP-based product development emerge, the major commercialized product during the next 5 years will remain roots.* Investing in appropriate harvest and post-harvest handling methods and fresh root storage to avoid market gluts should be considered. Sweetpotato supply chains tend to be short because of the bulkiness of the crop, so initiate market activities in areas with good access to major trade routes.

7. *Since smallholder farmers typically prioritize home consumption before sales of food crops, including some larger, commercially oriented sweetpotato farmers from the outset helps assure consistent supply.* The most difficult challenge in all marketing projects has been stabilizing root supply throughout the year. This is critical for OFSP processed product development.

In marketing campaigns in Rwanda, efforts to break into the urban market have focused on attracting youth through hiring a young IT specialist to use social media (Twitter, Facebook, etc.) to promote consumption of the OFSP among urban youth. In Bukedea district (Uganda), where sweetpotatoes were not widely consumed at the start of the project, a whole new trading system sprang up based on young, male bicycle traders who assembled OFSP from project growers and took it to more distant markets.

55.4 Reflections on the Current State of Knowledge and Remaining Gaps Concerning Nutritional Value

Clearly to make a difference, OFSP varieties or products being promoted need to have the desired nutritional quality and the amount of provitamin A (β-carotene) is the key micronutrient of interest. Moreover, we need to be able to evaluate impact of OFSP consumption on vitamin A status. In this section, we will identify advances in knowledge concerning nutritional values and assessment and pinpoint remaining gaps.

Best techniques for assessing micronutrient content of OFSP and its impact on status under laboratory versus field conditions

Evaluating OFSP nutritional content

As mentioned above, intensity of orange colouring is highly correlated with β-carotene content in sweetpotato. Detailed field charts have been developed for use by breeders in the field that include approximate β-carotene content.

Due to the inherent difficulties in carotenoid analysis such as incorrect quantification by various analytical methods available, as well as improper sampling and sample preparation, the reliability and/or validity of the values reported for this important micronutrient in the available literature is often questioned (Rodriguez-Amaya, 1997). In fact, errors incurred in sampling can easily surpass those from the analysis itself (Gross, 1991; Rodriguez-Amaya, 1993). It is also important to include in the results the variety of the OFSP, the maturity stage, the portion size analysed as well as the season and geographical origin (Gross, 1991; Rodriguez-Amaya, 1993).

Laboratory assessment of β-carotene contents of OFSP is best achieved by chromatographic separations, with high performance liquid chromatography (HPLC) being the most appropriate and widely used. The main problems in HPLC include obtaining and maintaining carotenoid standards for quantification, and the extremely high capital, maintenance and operational costs required. Thus, most studies using HPLC in developing countries are usually supported by external funding (Rodriguez-Amaya, 1997).

Fortunately, during the past decade, much work has gone into developing the correct calibration curves between HPLC or other wet chemistry standards and readings from near infrared reflectance spectroscopy (NIRS). Zum Felde *et al.* (2009) have developed a fast-throughput system for sweetpotato using NIRS, that can analyse raw root samples for protein, sugars and major micronutrients, including β-carotene, and minerals in a few minutes. This has revolutionized the ability to engage in breeding for quality traits in sweetpotatoes cost-effectively. A remaining gap is to develop similar calibration curves for cooked roots.

Techniques for assessing vitamin A status in the field

Studies seeking to identify VAD populations or evaluate the effectiveness of an intervention need to select appropriate biomarkers. Tanumihardjo (2011) describes how each marker performs and its suitability for assessing vitamin A status across the continuum from deficient to adequate to toxic. In SSA, most studies are concerned with measuring subclinical levels of vitamin A depletion, when liver reserves are marginal. In field studies, serum retinol concentrations are the most common indicator used. In addition, methods have been developed that are less expensive that use surrogate analyses for the carrier protein retinol-binding protein (RBP). However, both serum retinol and RBP concentrations are static measures and do not always change in response to an intervention (Tanumihardjo, 2011). It is important to measure and control for infection and inflammation and detection of change is more likely when working with a population with high initial levels of VAD.

For lactating women, breast-milk retinol concentrations can be used as an indicator of vitamin A status, with potential extrapolation to the breastfeeding infant (Tanumihardjo and Penniston, 2002). However, it may be more indicative of recent dietary intake as opposed to status.

Currently, the two best methods for assessing vitamin A status are the dose–response test and the isotope dilution test, both of which require laboratory settings for data collection. The modified relative-dose–response test (MRDR) is very useful for evaluating a deficient-through-normal vitamin A status and requires one blood sample from each subject. Tanumihardjo cites several examples where MRDR detected significant differences between two populations when the serum retinol test did not. The isotope dilution test is the most sensitive biomarker of vitamin A status of liver reserves, but also the most expensive. Hence, this method is rarely used for programme evaluation, although consideration could be given to applying it in a randomly selected sub-sample as was done for children participating in a sugar-fortification programme in Nicaragua (Ribaya-Mercado *et al.*, 2004).

Clearly, there is a need for more accurate field-suitable methods than serum retinol or RBP concentrations. Until this is available, it is critical to have vitamin A intake indicators to help build the chain of evidence along the impact pathway of the intervention.

What is the minimum amount of fat consumption linked to OFSP root consumption in terms of timing and quantity and the bioaccessibility of OFSP when fat availability is extremely limited?

The bioaccessibility and bioavailability of a nutrient are governed by the physical properties of the food matrix, which affect the efficiency of the physical, enzymatic and chemical digestion processes (Parada and Aguilera, 2007). It is well understood that the co-consumption of dietary fat is required for the best absorption of carotenoids from a meal. The release of nutrients from the food matrix during digestion makes them **bioaccessible** and fat is essential for the process of miscellarization, as well as the presence of bile salts, during which carotenoids are emulsified and solubilized for absorption by the intestine's enterocytes (Olson, 1994) making them **bioavailable**.

Carotenoid absorption from a meal containing β-carotene sources is enhanced with increasing amounts of dietary fat or food with a high fat content (e.g. avocado fruit) (Jalal *et al.*, 1998; Brown *et al.*, 2004; Unlu *et al.*, 2005). The absorption of carotenoids, including β-carotene, is negligible after ingestion of fresh vegetable salads with fat-free salad dressing (Brown *et al.*, 2004). Over and above the importance and the amount of dietary fat, the type of fat also affects postprandial carotenoid concentrations. The absorption of β-carotene is significantly enhanced with co-ingestion of long-chain fatty acyl groups compared with medium-chain fatty acyl groups (Borel *et al.*, 1998), whereas the degree of unsaturation of the fatty acyl groups has no effect on the *in vitro* bioaccessibility of carotenoids (Huo *et al.*, 2007). In Mongolian gerbils, bioconversion of provitamin A carotenoids from sweetpotato was improved as the amount of fat in the diet increased from 3% to 12% (Mills *et al.*, 2009).

Studies using different food matrixes as sources of β-carotene, for example a root crop such as sweetpotato (Jalal *et al.*, 1998), carotenoid-enriched spread (Roodenburg *et al.*, 2000), or green leafy vegetable such as spinach (Jayarajan *et al.*, 1980), suggest that a minimum amount of between 3 g (Jalal *et al.*, 1998; Roodenburg *et al.*, 2000) and 5 g (Jayarajan *et al.*, 1980) fat per meal is required to ensure intestinal carotenoid uptake. Food-based interventions should therefore incorporate recommendations to add this minimum amount of fat to the meal containing OFSP in order to improve vitamin A status of at-risk populations. In very poor rural environments, with few sources of fat in the diet, consideration may need to be given to introducing a fat-rich food source, such as avocado or groundnut, concurrent with the introduction of OFSP.

What are the OFSP processed products that have the best nutritional value and highest potential adoption and impact on those most at risk of VAD in urban versus rural communities?

'Natural' food structures remain at the cornerstone of influencing the bioavailability of carotenoids and other nutrients. Food processing improves bioavailability by disrupting cell walls of plant tissues, dissociation of the nutrient-matrix complexes, or transformation into more active molecular structures. Carotenoids present in plants are partially concentrated in chromoplasts or chloroplasts. Therefore the extent of release from these food matrices is highly variable and will depend on whether carotenoids are non-covalently bound to protein or fibre or dissolved in oil or in crystalline form (Parada and Aguilera, 2007).

During the past decade, considerable research has been done on retention of β-carotene in OFSP when processed and cooked. Retention is influenced by: (i) the cultivar; (ii) the cooking process used (temperature and oxygen exposure, in particular); and (iii) the duration of cooking. Almost all β-carotene in raw OFSP is in the available *trans*-configuration and the degree of isomerization to the less available *cis*-configuration during processing is thus also important. Boy and Miloff (2009) summarized findings from 20 studies that met stated quality standards for seven processing techniques: boiling, steaming, frying, roasting, microwaving, baking and drying. Boiling, the most common and affordable form of preparation in SSA, has one of the highest retention rates (84%, range 50–130%) among the techniques studied. Retention was also high for steaming (77%, range 48–95%) and frying (79%, range 67–95%). Roasting whole sweetpotato on a grill showed higher retention (74%; range 40–100%) than baking (69%). Numerous drying techniques in both dry and wet weather yielded β-carotene retention results above 50%. Oven-dried sweetpotato retained on average 87% β-carotene content (range 79–96%). Overall, *cis*-isomer levels were minimal, except for baking and microwaving.

Heat exposure during processing increases bioaccessibility by disrupting cell walls and breaking up the protein complexes in which the β-carotene is embedded. Tumuhimbise *et al.* (2009) studied the effect of traditional heat processing methods on the microstructure and *in vitro* bioaccessibility of β-carotene among OFSP varieties currently in use in Uganda (Table 55.1). They

Table 55.1 Effect of processing methods on the amount of *trans*-β-carotene, bioaccessible *trans*-β-carotene and the estimated bioaccessible retinol activity equivalents (RAE) in three varieties of orange-fleshed sweetpotato used in Uganda.

	Variety								
	Ejumula			Kabode (Naspot 10/SPK 4/6/6)			Kakamega (SPK004)		
Flesh colour	Dark orange			Medium orange			Light orange		
Columns[a]	a	b	c	d	e	f	h	i	j
Form[b]	All *trans*-β-carotene (μg/g dm)	Bioaccessible *trans*-β-carotene (μg/g dm)	Estimated bioaccessible RAE (μg/100 g fw)[c]	All *trans*-β-carotene (μg/g dm)	Bioaccessible *trans*-β-carotene (μg/g dm)	Estimated bioaccessible RAE (μg/100 g fw)[c]	All *trans*-β-carotene (μg/g dm)	Bioaccessible *trans*-β-carotene (μg/g dm)	Estimated bioaccessible RAE (μg/100 g fw)[c]
Raw	314.5	34.8	174	212.4	41.5	194	100.6	18.2	85
Boiled (20 min at 92°C)	248.5	106.5	530	153.4	70.7	330	64.8	37.2	174
Steamed (30 min at 94°C)	246.9	105.8	529	157.0	64.6	302	59.6	25.3	118
Deep fried slices (10 min at 170°C)	259.8	150.8	750	169.8	101.1	472	64	40.5	189
Baked (15 min in 108°C)	229.7	78.3	390	127.0	48	224	50.7	18.5	87

[a]Results reported in columns a, b, d, e, h and i are adapted from Tumuhimbise *et al*. (2009). For details on standard deviations, consult the article.

[b]The amount of 'all *trans*-β-carotene' and bioaccessible *trans*-β-carotene (measured in μg/g dry matter (dm)) in the micellar fraction was determined by high performance liquid chromatography (HPLC). The values are means based on three samples of each preparation type.

[c]Estimated bioaccessible RAE (measured in μg/100 g fresh weight (fw)) was calculated by multiplying the bioaccessible *trans*-β-carotene value by the dry matter proportion (0.30 for Ejumula, 0.28 for Kabode and Kakamega) and adjusting for the amount in 100 g, and then using a bioconversion factor of 6 units β-carotene to 1 unit retinol because bioaccessibility has already been accounted for.

showed that the bioaccessibility of β-carotene differed due to processing technique. Bioaccessibility of β-carotene increased from raw, to baking, to steaming/boiling, to deep frying. The β-carotene in deep-fried OFSP was more bioaccessible than that in OFSP prepared by other processing procedures, indicating that fat increases bioaccessibility. Therefore, although heat processing reduces β-carotene retention, the loss in retention is compensated for by improved bioaccessibility because of the presence of fat.

It is best to use OFSP varieties of at least medium orange intensity in nutrition programmes, at least 55 μg/g of β-carotene on a fresh weight basis (fwb). Moreover, indications are that for a given OFSP variety, fried products will provide more provitamin A per 100 g than baked products. However, the heavy fat content of fried products means that they should not be promoted as healthy foods. For young children, the best product would be boiled and mashed sweetpotato with at least 3–5 g of fat added and ideally a good protein source.

Bechoff *et al.* (2011) investigated retention and bioaccessibility of three products, fried doughnuts (mandazi), chapatti (flat bread) and porridge, made with 30% sweetpotato flour by local agro-processors. They found that the efficiency of miscellarization of all-*trans*-β-carotene and bioaccessibility was greater in the fried products (chapattis and mandazis) than that in the boiled products (porridge and purée). Taking bioaccessibility into account alters the amount needed of a product to meet the recommended dietary allowances. For example, an individual would need to eat either 220 g of OFSP purée or 100 g of OFSP chapatti to get the equivalent amount of vitamin A (Table 55.2). However, if a little bit of oil was added to the boiled OFSP purée, no doubt the bioaccessibility would increase significantly. As more products emerge on the market, there will be need for bioaccessibility assessments, not just retention studies.

One constraint to product development has been the relatively high price of sweetpotato flour compared with wheat flour. It takes 4–5 kg of fresh sweetpotato to make 1 kg of flour. In most SSA countries sweetpotato flour is as expensive as wheat flour. Hence, there has been more interest in using sweetpotato purée (boiled and mashed) for wheat flour substitution in bakery products as it takes only 1.25 kg of fresh root to make 1 kg of purée. In Rwanda, 45% of wheat flour is replaced with OFSP purée in a 'Golden Power' biscuit.

A major challenge in using sweetpotato purée compared with flour is that it is difficult to store. In the USA, sweetpotato purée is most often marketed in pails or drums that must remain frozen or chilled until use. There is, however, an advanced method for rapid sterilization and aseptic packaging of OFSP purées using a continuous flow microwave system in use in the USA (Coronel *et al.*, 2005). The product is packaged in flexible plastic containers or aluminum-polyethylene laminated bags, demonstrating shelf stability for 12 months and high retention of β-carotene.

There is an urgent need for a more affordable way than the microwave system of storing OFSP purée without refrigeration, but retaining high levels of β-carotene. Scant information is available on this because it is a relatively recent topic. Techniques using vacuum frying with carrots could potentially be applied to sweetpotato (Dueik *et al.*, 2010). It has been shown that vacuum frying can reduce oil content by nearly 50% and preserve approximately 90% of *trans*-α-carotene and 86% of *trans*-β-carotene in carrots. Exclusion of oxygen (e.g. through vacuum or hot filling, oxygen-impermeable packaging, inert atmosphere), protection from light and low temperature diminish carotenoid degeneration during storage (Rodriguez-Amaya and Kimura, 2004). This is a priority for future research.

Should we be investing more in promoting leaf consumption?

To date, OFSP promotion efforts in SSA have focused principally on roots. However, in many African countries (e.g. Zambia, Sierra Leone, Liberia and Angola) sweetpotato leaves are widely consumed and often have

Table 55.2. Amount of all *trans*-β-carotene in different OFSP products (variety Ejumula). (From Bechoff *et al.*, 2011, p. 10378 with permission.)

OFSP product	Fat (%)	All *trans*-β-carotene (μg/g)	All *trans*-β-carotene bioaccessibility (μg/g)	Unit	After *in vitro* digestion: μg retinol equivalent per unit shown in the previous column
Boiled OFSP	0	95.0 ± 2.0	9.9 ± 0.1	Purée portion (100 g)	183.7
Porridge (30% OFSP flour)[a]	0	8.7 ± 0.3	16.3 ± 0.9	One mug (300 g)	80.0
Chapatti (30% OFSP flour)[a]	7.4	31.5 ± 1.4	72.7 ± 5.4	One chapatti (100 g)	401.7
Mandazi (30% OFSP flour)[a]	3.3	32.9 ± 1.7	49.0 ± 3.0	Two mandazi (90 g)	289.6

[a]Porridge, chapatti and mandazi all contained 30% OFSP flour made from the Ejumula variety.

value as a cash crop in peri-urban/urban settings because they can be ratooned for a long period. In East Africa, the higher protein content of sweetpotato leaves (16.0% crude protein) compared with Napier grass (10% crude protein) is appreciated in the dairy industry (Lukuyu *et al.*, 2012).

Sweetpotato leaves have moderate amounts of β-carotene after cooking (550 μg/100 g), but the bioavailability of that β-carotene has still not yet been determined. The leaves are also a rich source of dietary lutein (which can delay macular degeneration), having higher levels than other cruciferous leafy vegetables (Menelaou *et al.*, 2006). Levels of polyphenolics, namely anthocyanins and phenolic acids, are also high compared with major commercial vegetables; polyphenolics protect against diseases linked to oxidation such as cancer, allergies and cardiovascular disease (Islam, 2006). The stems and leaves of sweetpotato also contain chlorogenic acids, which have been shown to improve glucose tolerance in humans (Bovell-Benjamin, 2010). Clearly further research work is needed on the functional properties and application potential of isolated bioactive compounds from sweetpotato leaves. Given the nutritious nature of the leaves, increased promotion of their use for human consumption is warranted.

55.5 Reflections on the Current State of Knowledge and Remaining Gaps Concerning Behavioural Change

What can we learn from other behavioural change efforts in health (HIV/water sanitation) that could help us develop a lens for behaviour change for an integrated agriculture–nutrition project?

The field of behaviour change has developed and expanded substantially in the past 30 years in part because of the global AIDS pandemic which has required us to re-examine our understanding of how much we really know about changing behaviours. Since that time the water and sanitation sector (WATSAN) has faced their own limited success in improving the situation in the world despite massive infrastructural interventions and they too have realized that a deep understanding of human behaviour is called for if the investment in infrastructure is to be properly valued. The valuable lessons learned about behaviour change through AIDS and WATSAN can now be applied to the agriculture and nutrition sectors in particular as these two sectors now strive to integrate their efforts in projects that make a difference. Key to these lessons is the understanding that people interpret and at times create new meanings of information based on their own context; culture, norms and surrounding influencing social networks. People can't always control the issues that create their behaviour and, moreover, people are not always rational in deciding what is best for their health and well-being (Communication-Change, year unknown). Social and behaviour change communication (SBCC) uses the socio-ecological model to find a tipping point for change and uses multiple strategies (advocacy, social mobilization and behaviour change communication) to achieve the goal of bringing about behavioural change.

Know the context

Evidence from the AIDS and WATSAN sectors demonstrate that interventions are more effective if they address the many levels of influence on behaviour. The socio-ecological model illustrates how individuals with their preferences, knowledge, beliefs and practices, social and cultural norms and empowerment are influenced at the interpersonal level with family, friends and social networks. Likewise, the community and organizational level influences through neighbourhoods or villages, cultural practices, schools, clinics, rules and policies. Social structures, policy, media and systems overarch all other levels of influences (Contento, 2008a). Thus, in an agriculture and nutrition project, it would be short sighted to focus a behaviour-change message or intervention only on married women without taking into consideration that their husbands and extended families influence their decisions regarding farming and family nutrition. For example, in areas of Mozambique the mother-in-law is particularly influential in what is acceptable behaviour by her daughter-in-law. In this case, a behaviour change intervention must address several different levels of influence because this is the context in which the woman is operating.

Formative research to understand the 'influencers' of conduct is essential to behaviour change if programming is going to be targeted accurately. This may include barrier analysis (to identify the obstacles which individuals face in changing their behaviour), doer-non-doer studies (what are the differences between these groups and how does this inform us?), and examination of positive deviance (who has already changed behaviours and how did they do it?). In this way, it is possible to explore the determinants of behaviour so that the right messaging and activities can be chosen for programme.

Information is never enough

The process of behaviour change starts with providing information but it is only the first step in a much longer process. The mistake of assuming that because people have information, they will change their behaviours has been repeated in all sectors. In reality having information or knowledge rarely translates into changed behaviour. Literature abounds about this disjuncture between knowledge and behaviour change. Hand washing as a critical interruption of disease transmission has been known since 1846 and has been taught widely in the general population and in the health community (CDC, 2000; Grol and Grimshaw, 2003; Twomey, 2006) and yet the *practice* of hand washing must be taught and retaught in countries around the world. Campaigns on smoking, alcohol consumption and nutrition (balanced diets) have educated the public but until they were reinforced at other levels, the information *alone* did not bring the desired behavioural results. Both the AIDS and the WATSAN sectors have suffered from this problem. Many programmes have focused on telling people how to prevent HIV infection, or the importance of defecating in a toilet or pit latrine, but have not taken the target population beyond the information stage. As a consequence, they have not realized the obstacles that prevent people from putting into practice the information they have received, and developing the skills to enact positive behavioural change.

Gender matters

While it is commonly acknowledged that men and women operate differently and are differently affected by interventions and projects, this knowledge is not taken forward into programming so that adjustments are made to address gender difference. For example, the lack of sanitation facilities at schools affects boys and girls differently, particularly as they mature. It is now widely documented that girls will stay home from school when they are menstruating if pit latrines/toilets and water for hand washing are not provided at school. Likewise in an integrated agriculture–nutrition project, access to resources (land, seed, labour) is different for men than for women. Sweetpotato tends to be a crop of home consumption which is the responsibility of the woman until such time as it is commercialized and it becomes the concern of men. As such, conducting a gender analysis should be prioritized in an integrated project. This will affect programming, interventions, training materials and methods, as well as data.

Tap into aspirations (hook onto the familiar and modify existing behaviours)

Understanding the perceived positive or negative consequences of a person performing a behaviour can provide direction in messaging. There are advantages and disadvantages to changing behaviours and unless we try to gain insight and understand it will be difficult to convince people to make changes. Fear as a motivator to change can work but only for a limited time. Early AIDS campaigns used skull and crossbones to scare people that they would die from the disease. In fact for many who were struggling to feed themselves on a daily basis, the threat of death in the future was less frightening than the imminent hunger of today. Learning from the advertising sector, however, aspirational goals are powerful. Life, health, strength and money appeal to people and advertisers use these aspirational goals liberally in their campaigns. Reducing money spent on medicine and clinics by reducing disease levels is appealing. Showing off healthy smart children appeals to the hearts of most parents. Society approves of these positive aspirations; they fit the societal norms. Programmes in AIDS, WATSAN and health that have tapped into aspirations and hooked onto familiar (rather than unfamiliar) constructs have been more successful than those that did not (Duhigg, 2012).

What can we learn about behavioural change from other food-based approaches to improving diet diversity and nutritional status?

Arimond *et al.* (2011) identified five non-mutually exclusive main pathways through which agricultural interventions may impact on nutrition. These are: (i) increased food availability and access (production for own consumption, filling seasonal gaps); (ii) increased income (through production for sale in markets and hence potential to increase food purchases or higher quality food purchases); (iii) reduction in real food market prices (with increased agricultural production); (iv) shifts in consumer preference (e.g. when programmes include behaviour-change communication); and (v) shifts in control of resources within households and communities (in particular interventions that shift resources, e.g. income, time, etc. towards women). Various factors, for example market access and integration, potential for surplus production within a given setting, pre-existing nutritional knowledge, and consumer preferences and demand will determine which pathway dominates in impacting on nutrition.

Regardless of which of the above pathways is dominant, successful food-based strategies all include nutrition education and behaviour change communication to ensure that the increased access to food translates into improved nutrition and empower women to make educated consumption and nutritionally informed food purchasing choices (HKI, 2010). Demonstration sessions, growth monitoring and community theatre are preferred, rather than lecture sessions (Low *et al.*, 2007a). Cooking demonstrations are used to illustrate the best cooking methods to retain nutrients and enhance absorption of micronutrients from garden produce (Faber *et al.*, 2006; HKI, 2010). Incorporating gender considerations and livelihoods frameworks into the design of food-based strategies strengthens the impact on nutrition (Arimond *et al.*, 2011).

In South Africa, where sweetpotato is eaten as a vegetable, OFSP is introduced in food-based projects together with a variety of other vitamin A-rich vegetables and fruit. In addition, consumption of wild-growing African leafy vegetables is encouraged (Faber *et al.*, 2006). Promoting a variety of β-carotene vegetables and fruit helps to overcome seasonal variability and adds variety to the diet. In the Asia-Pacific region, a more holistic approach is used by focusing on all foods rich in micronutrients (not only vitamin A). Year-round availability of micronutrient-rich foods is achieved by planting a large variety of vegetables and fruit. In addition, small animal husbandry (e.g. chickens) is included because of the higher bioavailability of micronutrients in animal foods. This was shown to be successful in improving dietary intake and decreasing micronutrient deficiencies (HKI, 2010).

Partnerships across disciplines are required, particularly a close collaboration between agriculture and nutrition. Berti *et al.* (2004) argues that agricultural projects that had impact on nutrition also invested in four or five types of 'capital' (physical, natural, financial, human and social) as defined in the Sustainable Livelihoods Framework. When selecting crops to plant, agronomic traits, potential to address nutrient deficiencies in the local diets, and motivations and constraints that determine the household consumption decisions must be considered (Arimond *et al.*, 2011). One of the major challenges to a multi-crop food-based intervention is managing the increased range of seed systems and ensuring sufficient technical expertise for all of the crops and livestock in use.

What can we learn from rural and urban consumers so we can best meet their needs?

People's food choices are influenced by a complex set of interacting factors, namely: (i) biological-determined behavioural predispositions; (ii) experience with food (associative conditioning) leading to people's food preferences; (iii) personal factors, such as perceptions, attitudes and beliefs; and (iv) environmental factors, including food, social, economic and informational environments (Contento, 2008b). An understanding of factors affecting food choices of the target population is therefore needed to develop appropriate behaviour change strategies.

Foods may be viewed differently by urban and rural consumers. In Cameroon, for example, rural adolescents, who had a shortage of food at home, reported that they ate to 'live' and be healthy. Urban poor adolescents had better food availability, although affordability was problematic, and for them food was for health and beauty. Urban rich adolescents had more than enough food, and for them the purpose of the food was for pleasure and beauty (Dapi *et al.*, 2007).

In many African countries, agriculture is the major component of the rural economy and household livelihoods. In rural areas, eating traditional food at home is still common and diet diversity in many settings limited. In urban areas, households rely mostly on purchased food, including processed foods. Poorer households often face high prices of wood and charcoal, leading to reduced frequency of home meal preparation. Urban women who are employed outside the home have less time available to shop for ingredients and prepare meals for the family. In such circumstances, consumption of fast foods outside the house easily increases, and street-food purchases are particularly high in urban areas of Africa (Fanzo, 2012). In Nairobi, one study found 50% of consumers from both a low and a middle-income area of the city consuming street foods 3 days of the week or more (Ogubi, 2007). Fried sweetpotato slices are popular informal foods in some East and West African cities and towns.

Environmental factors affecting food choices are rapidly changing. This is reflected in the global nutrition transition with diets converging from the more traditional diet to a more Westernized diet, characterized by energy dense, often highly refined, foods high in fat, added sugar and salt, and low in micronutrients. In some settings, Westernized diets are considered 'modern' and convenient; foods such as sweetpotato are considered foods of the poor and time-consuming to prepare. The nutrition transition has been documented for low- and middle-income countries in all urban areas and increasingly in rural areas. In African countries, a double burden of malnutrition is emerging (i.e. hunger alongside the health problems associated with over nutrition, such as obesity, diabetes and stroke). Obesity is higher in urban areas in most African countries; but rural obesity is catching up (Popkin *et al.*, 2012). African women, often from poor, food-insecure households are particularly vulnerable to obesity when they experience the nutrition transition (Vorster *et al.*, 2011). Modern food systems, characterized by processing and packaging of food products, and corporate concentration in retailing and distribution (e.g. brand names, large supermarkets) play an important role in the nutrition transition (Ericksen, 2008).

Within the context of the double burden of malnutrition, food-based strategies focusing on diversification of diets should aim for optimal, but prudent diets for all, and under- and over-nutrition should be addressed simultaneously in the communication strategies (Vorster *et al.*, 2011).

What can we learn from private-sector marketing firms about consumers' behaviour and their food choices?

In developed economies, private agro-processing firms spend considerable resources in understanding what consumers want. There is widespread recognition that consumers differ in their preferences. There are consumers who 'eat-to-live' and follow predictable purchasing patterns; others love innovative products and are adventurous in approaching new foods. A market-oriented company seeks to understand and exploit distinct consumer segments in a timely fashion, paying attention to socio-economic, age group, ethnic and gender differences. Any company in the world seeks to build brand loyalty. This is often done through advertising that implies a connection between a product and a desirable goal, evoking certain emotions (Lister, 2013).

Consumer food choice, the way in which consumer's choose to spend their money on food, is driven in all societies by the consumer's income, the food products' prices and quality. The concept of quality, however, has undergone considerable change in the past 20 years, particularly in regards to urban consumers. Grunert (2003) describes four groups of quality attributes:

1. *Sensory attributes* are the classical food quality aspects of taste, appearance and smell, with taste as the dominant aspect.
2. *Health/nutritional attributes* have been increasing in importance, as evidence by the emergence of *functional foods*, food products which have an added positive health benefit.
3. *Process attributes* refer to consumers' interest in the way a food product has been produced. In Europe and the USA, some consumers pay premiums for organic products.
4. *Convenience attributes* are aspects of the product which save time or energy for the household, whether during shopping, storage, preparation, eating and/or disposal.

Cultural differences between and within countries determine the relative weight of the four categories of quality attributes on food choice. Differences in taste preferences are widely recognized. In the case of sweetpotato, adult consumers in the Americas prefer low dry matter varieties (18–23%); whereas adults in SSA prefer moderate to high dry matter types (28% and above). However, the perception of what is healthy, what is convenient, and which types of production are acceptable also can differ significantly (Grunert, 2003).

Nutrition education efforts in SSA focus on developing and transmitting key messages and it is often assumed that if the target group understands those messages, the desired change in behaviour will occur. However, organizational behaviour and marketing specialists have investigated how the way in which an innovative idea is communicated determines whether it will survive or not in the competitive market place, in other words, what makes the idea *stick* (Heath and Health, 2008).

Their six principles should be considered when designing a campaign for a new product such as OFSP roots or OFSP-based processed products, namely:

1. Simplicity. Focus on ideas/messages that are simple, yet profound.
2. Unexpectedness. Generate interest and curiosity by violating people's expectations.
3. Concreteness. Explain ideas in terms of human actions and non-ambiguous language.
4. Credibility. The idea must be credible; preferably suggest how to test this idea oneself.
5. Emotions. Make the client feel something.
6. Stories. Tell stories to get people to act on the idea/message.

The factors influencing urban consumer food choice fall into five broad categories: personal, social, economic, environmental and psychological (Dolceta (European Union online consumer education), 2013). Most lower socio-economic status urban consumers purchase their OFSP as fresh roots or fried slices in informal markets. A few have access to OFSP processed products such as doughnuts. Higher socio-economic status consumers also can use informal markets, but many in major urban centres have shifted to more formal markets, including shops and supermarkets. Higher end consumers are more likely to prioritize their selection based on taste (personal), current trends (social) and style of packaging (psychological) than poorer consumers.

To reach lower socio-economic status consumers, the dominant factors will be price relative to other sources of energy, followed by taste and health-driven preferences and convenience. The price will be determined to a large extent by transport costs from production zones to the urban centre. Consideration should be given to training wholesale and retail traders about the value of OFSP, facilitating linkages between traders and producers, and providing promotional signs, aprons and umbrellas to those agreeing to separate out OFSP from other types of sweetpotato.

55.6 Discussion

By reflecting on the OFSP experience to date and reviewing progress of the health and marketing sectors concerning behavioural change, it is clear that the knowledge of what works at the community level to improve vitamin A intakes among women and young children is much greater than knowledge about the how to cost-effectively reach rural and urban consumers purchasing OFSP roots and processed products. It is clear that nutrition interventions at the community level should discuss the benefits of OFSP and provide key messaging concerning breastfeeding, complementary feeding of young children, combining foods to create a balanced diet and vitamin A rich foods. That is the minimum.

Key recommendations for how to approach rural households can be summarized in a checklist:

1. Define which behaviours must be addressed or developed in order to integrate OFSP into the farming system and rural household. Conduct formative research to understand barriers to behaviour change and how best to tackle them. Consider bringing in a specialist to train designated staff on how to do this kind of qualitative work, so that periodically teams can engage in operations research and make adjustments during the project life span.

2. Target households (i.e. both men and women). Although women are the gatekeepers to household nutrition, they do not act in isolation. Access to family resources (land, seed, labour) are frequently controlled by men. Family (mothers, mothers-in-law) and friends influence the women – either by positive or negative reinforcement of behaviours based on social norms and the information they have. In this respect it is more beneficial to address the household unit even if it means providing certain messages to women and other messages to the influential groups such as men and grandmothers (mothers-in-law).

3. Segment the population and design messages and interventions that are targeted to their specific behaviours, aspirations and obstacles. While the overarching key message may be for everyone, the information, skills and obstacles being faced are not the same. Mothers need to learn how to incorporate OFSP into the family diets while men may be more concerned with marketing OFSP. Use methods such as TIPs to help understand the challenges caregivers face in making the proposed change(s).

4. Understand the cultural beliefs and practices around certain foods and design messages appropriately.

5. Behaviour change messages should be clear, comprehensive and actionable to the group for which they are intended. Keep it simple with one action per message or people will become confused.

6. Repeat behaviour change messages frequently, and reinforce the messages with practice so that change is long lasting.

7. Identify and promote the use of lowest-cost fat source in OFSP-containing porridges and other recipes, given the importance of a small amount of fat in increasing bioefficacy of β-carotene in OFSP.

8. Raise awareness and create a supportive environment for change through cost-effective community radio programmes, jingles and talk shows in local languages.

9. Focus on the skills that rural consumers need to develop in order to sustain the behaviour change which has been identified.

Organize practical cooking demonstrations using locally available foods and involving the mothers so that they gain the skills to enable them to feed their children dishes enriched with OFSP.

10. Use pre-existing social networks to encourage the sharing of information to friends, neighbours and peers. This creates positive peer pressure and encourages a change at the community level which supports everyone. It is easier to make a change if others around you are doing the same. If resources permit, get a key public figure to endorse the effort in the local media.

11. Involve structures within the community (e.g. schools, health centre and clinics, radios, extension/community workers) to reinforce messages from all directions. Align messages to be consistent with existing messages being promoted by government to build a synergy that can take things forward.

12. Build in an operations research component as part of the intervention strategy to periodically reassess which techniques are succeeding in getting adoption of OFSP and its appropriate use.

On the marketing side, it is clear that there is much to be learned from the private sector and the health community. All seem to agree that *aspirational* messaging, using images of healthy children and productive families, is better than showing the negative consequences of VAD.

Reaching higher socio-economic status consumers will require a more substantial financial investment, preferably done in collaboration with a private-sector agro-processor with existing marketing channels, funds to support promotion campaigns and a commitment to support local farmers in the development of reliable supply chains. Extensive advertising and other promotional efforts will be needed as the dominant image of sweetpotato as a food for the poor must be shifted, with OFSP being promoted as a health food for all. Investing in communication specialists will help to design promotion messages that will 'stick' in the selected environment. In Rwanda, the launch of the Golden Power biscuit has begun to change the opinion of policy makers about sweetpotato's potential commercial value. Facebook, Twitter and the radio are used to make consuming Golden Power biscuits 'cool' among young adults. In areas with internet services, investments in social media can be an effective way to create demand and interest, particularly among younger consumers. If funds for reaching urban consumers are limited, emphasis should be placed on reaching consumers in informal markets, as the risk of VAD is higher in this group.

Evidence is strong that consuming OFSP can significantly contribute to lowering the prevalence of VAD. Research is still needed, however, to devise more 'field friendly' and precise biological indicators than serum retinol.

Considerable progress has been made in understanding the bioaccessibility of β-carotene in OFSP roots and processed OFSP-based products. Although heat processing reduces β-carotene retention, the loss in retention is compensated for by improved bioaccessibility. Fat in the meal increases the bioaccessibility of β-carotene in OFSP, but only small amounts (3–5 g) are needed. Emphasis should be placed on using medium intensity or dark orange OFSP varieties whenever possible, particularly for baked products. Since sweetpotato purée is more economically viable than sweetpotato flour, research is needed to identify cost-effective ways to store purée without significant loss of quality. Also, greater investment in postharvest storage is needed.

55.7 Way Forward

To build the evidence base, most interventions used OFSP as the key intervention. Building on that solid base, going-to-scale at reasonable cost with OFSP also will need to focus on how to successfully integrate it into existing initiatives and frameworks seeking to improve nutrition and/or agricultural productivity and/or market opportunities. The options will be broad, recalling the multiple roles that sweetpotato can play in the food system: (i) as a staple food; (ii) as a vegetable (both the roots and the leaves); (iii) as a breakfast or snack food; (iv) as a resilient, food security crop in maize-based systems subject to the impact of climate change; (v) as an animal feed (dual purpose for food and feed); and (vi) as an ingredient in processed products.

In the past 5 years, there has been increasing interest among policy makers in SSA about integrating nutritional concerns into agricultural interventions. African leaders committed themselves in 2003 to developing detailed agriculture plans at the country level that align with the principles of the Comprehensive African Agriculture Development Programme (CAADP) and ensuring that 10% of the GDP is used for agriculture. In 2012 and 2013, efforts were being made to help mainstream nutrition into existing and developing agriculture investment plans through a series of subregional workshops. CAADP documentation explicitly recognizes investment in nutrient-dense crops such as OFSP as quality investment (NEPAD and CAADP, 2008). The CAADP framework also heavily supports the use of home-grown school feeding programmes as well as vegetable gardens to address chronic undernutrition. OFSP could be built into both of these programmes.

The Scaling Up Nutrition (SUN) movement is perhaps the best opportunity for OFSP integration. It emphasizes combatting chronic undernutrition in children under 2 years of age through establishing multi-dimensional community-based nutrition programmes. As of March 2013, 22 African countries had joined the SUN movement, with 21 of them already having developed nutrition investment plans. Integrating OFSP into SUN-financed community nutrition programmes will provide an opportunity to significantly lower the cost of the OFSP intervention because they are focused on the key target groups at risk of VAD and they recognize the need for staff working at the community level. Drawing on behavioural change and social marketing techniques outlined above will enable researchers and practitioners to improve the techniques and tools needed to get adoption and proper utilization of OFSP.

There is no doubt that OFSP can also play a significant role in a range of food and nutrition security, value chain and climate change mitigation interventions. Additional research is needed on integrated crop management, postharvest management, marketing systems and product development for its full potential to be realized in these areas. Attention should also be paid to how to best promote sweetpotato leaves, to maximize their potential nutrition contribution. To monitor uptake, it is critical for national surveys and other data collection efforts to distinguish between OFSP and non-OFSP sweetpotato types in their survey instruments. With a combined multi-disciplinary effort, prospects are excellent for OFSP to significantly contribute towards reducing food insecurity and VAD in the coming decade.

Notes

[1] United Nations Children's Fund (UNICEF), the World Health Organization (WHO), the Canadian International Development Agency (CIDA), the UK's Department for International Development (DFID), the United States Agency for International Development (USAID), and the Micronutrient Initiative (MI).

[2] Biofortification is the process of breeding staple food crops with high contents of key micronutrients. Currently the emphasis is on breeding crops with higher provitamin A (β-carotene), iron and zinc.

[3] The disability-adjusted life year is a measure of overall disease burden, reflecting the number of years lost due to ill health, disability or early death.

References

Arimond, M., Hawkes, C., Ruel, M.T., Sifri, Z., Berti, P.R., Leroy, J.L., Low, J.W., Brown, L.R. and Frongillo, E.A. (2011) Agricultural interventions and nutrition: lessons from the past and new evidence. In: Thompson, B.

and Amoroso, L. (eds) *Combatting Micronutrient Deficiencies: Food-Based Approaches*. Food and Agriculture Organization of the United Nations (FAO), Rome and CAB International, Wallingford, UK, pp. 41–75.

Bechoff, A., Poulaert, M., Tomlins, K.L., Westby, A., Menya, G., Young, S. and Dhuique Mayer, C. (2011) Retention and bioaccessibility of β-carotene in blended foods containing orange-fleshed sweet potato flour. *Journal of Agricuture and Food Chemistry* 59, 10373–10380.

Berti, P.R., Krasevec, J. and Fitzgerald, S. (2004) A review of the effectiveness of agriculture interventions in improving nutrition outcomes. *Public Health Nutrition* 7, 599–609.

Borel, P., Tyssandier, V., Mekki, N., Grolier, P., Rochette, Y., Alexandre-Gouabau, M.C., Lairon, D. and Azaïs-Braesco, V. (1998) Chylomicron β-carotene and retinyl plamitate responses are dramatically diminished when men ingest β-carotene with medium-chain rather than long-chain triglycerides. *The Journal of Nutrition* 128, 1361–1367.

Bovell-Benjamin, A.C. (2010) Sweet potato utilization in human health, industry and animal feed systems. In: Ray, R.C. and Tomlins, K.L. (eds) *Sweetpotato: Post Harvest Aspects in Food, Feed, and Industry*. Nova Science Publishers, New York, pp. 193–224.

Boy, E. and Miloff, A. (2009) Provitamin A carotenoid retention in orange sweet potato. *Site and Life* 3, 27–33.

Brown, M.J., Ferruzzi, M.G., Nguyen, M.L., Cooper, D.A., Eldridge, A.L., Schwarts, S.J. and White, W.S. (2004) Carotenoid bioavailability is higher from salads ingested with full-fat than with fat-reduced salad dressings as measured with electrochemical detection. *The American Journal of Clinical Nutrition* 80, 396–403.

Centers for Disease Control (CDC) (2000) CDC Media Relations: Why is Handwashing Important? Available at: http://www.cdc.gov/media/pressrel/r2k0306c.htm (accessed 6 March 2000).

Communication-Change (year unknown) A Short Guide to Social and Behaviour Change (sbcc) Theory and Models. Available at: http://www.slideshare.net/CChangeProgram/communication-for-change-a-short-guide-to-social-and-behavior-change-sbcc-theory-and-models (accessed 1 June 2012).

Contento, I. (2008a) *Nutrition Education: Linking Research, Theory, and Practice*. Jones and Bartlett Publishers, Sudbury, Massachusetts.

Contento, I.R. (2008b) Nutrition education: linking research, theory, and practice. *Asia Pacific Journal of Clinical Nutrition* 17, 176–179.

Coronel, P., Truong, V.D., Dimunovic, J., Sandeep, K.P. and Cartwright, G.D. (2005) Aseptic processing of sweet potato purées using a continuous flow microwave system. *Journal of Food Science* 70, 531–536.

Dapi, L.N., Omoloko, C., Janlert, U., Dahlgren, L. and Håglin, L. (2007) 'I eat to be happy, to be strong, and to live.' Perceptions of rural and urban adolescents in Cameroon, Africa. *Journal of Nutrition Education and Behaviour* 39, 320–326.

Dickin, K., Griffiths, M. and Piwoz, E. (1997) *Designing by Dialogue: a Program Planners' Guide to Consultative Research for Improving Young Child Feeding*. Prepared for the Health and Human Resources Research and Analysis for Africa (HHRAA) project. Academy for Educational Development, Washington, DC, 318 pp.

Dolceta (European Union online consumer education) (2013) Factors Which Influence Consumer Choice. Available at: http://www.dolceta.eu (accessed 5 May 2013).

Dueik, V., Robert, P. and Bouchon, P. (2010) Vacuum frying reduces oil uptake and improves the quality parameters of carrot crisps. *Food Chemistry* 119, 1143–1149.

Duhigg, C. (2012) *The Power of Habit: Why We Do What We Do in Life and Business*. Random House, New York.

Ericksen, P.J. (2008) Conceptualizing food systems for global environmental change research. *Global Environmental Change* 18, 234–245.

Faber, M., Laurie, S. and Venter, S. (2006) *Home-gardens to Address Vitamin A Deficiency in South Africa: a Food-based Approach*. Agricultural Research Council-Roodeplaat Vegetable and Ornamental Plant Institute and the Nutritional Intervention Research Unit of the Medical Research Council, South Africa.

Fanzo, J. (2012) *The Nutrition Challenge in Sub-Saharan Africa*. United Nations Development Programme, New York. Available at: http://www.undp.org/content/dam/rba/docs/Working Papers/Nutrition Challenge.pdf (accessed 21 April 2013).

Girard, A.W., Self, J.L., Mcauliffe, C. and Oludea, O. (2012) The effects of household food production strategies on the health and nutrition outcomes of women and young children: a systematic review. *Paediatric and Perinatal Epidemiology* 26(suppl. 1), 205–222.

Grol, R. and Grimshaw, J. (2003) From best evidence to best practice: effective implementation of change in patients' care. *Lancet* 362, 1225–1230.

Gross, J. (1991) *Pigments in Vegetables: Chlorophylls and Carotenoids*. Van Nostrand Reinhold, New York.

Grunert, K.G. (2003) How changes in consumer behavior and retailing affect competence requirements for food producers and processors. Paper presented at the Economic Research Service (ERS)/United States Department of Agriculture (USDA) workshop on Global Markets for High Value Food, 14 February 2003, Washington, DC.

Hagenimana, V., Oyunga, M.A., Low, J.W., Njoroge, S.M., Gichuki, S. and Kabira, J. (1999) The effects of women farmers' adoption of orange-fleshed sweet potatoes: raising vitamin A intake in Kenya. *International Center for Research on Women (ICRW)/Obstetrics Maternal Newborn Investigation (OMNI) Research Program, Research Report* No. 3. ICRW/OMNI, Washington, DC.

HarvestPlus (2010) *Disseminating Orange-fleshed Sweet Potato: Findings from a HarvestPlus Project in Mozambique and Uganda*. HarvestPlus, Washington, DC.

Haskell, M.J., Jamil, K.M., Hassan, F., Peerson, J.M., Hossain, M.L., Fuchs, G.J. and Brown, K.H. (2004) Daily consumption of Indian spinach (*Basella alba*) or sweet potatoes has a positive effect on total body vitamin A stores in Bangadeshi men. *The American Journal of Clinical Nutrition* 80, 705–714.

Heath, C. and Health, D. (2008) *Made to Stick*. Random House, New York.

Helen Keller International (HKI) (2010) Homestead food production model contributes to improved household food security, nutrition and female empowerment – experience from scaling-up programs in Asia (Bangladesh, Cambodia, Nepal and Philippines). *Helen Keller International Nutrition Bulletin* volume 8 (issue 1), March 2010.

Henderson, G., Gething, K., Dennien, S. and McIntyre, A. (2012) Sweetpotato Consumer Research. Insights to Increasing Consumption. Queensland Government. Available at: era.deedi.qld.gov.au/3572/1/sweetpotato_consumer_insights.pdf (accessed 3 May 2013).

Hotz, C., Loechl, C., de Brauw, A., Eozenou, P., Gilligan, D., Moursi, M., Munhaua, B., van Jaarsveld, P., Carriquiry, A. and Meenakshi, J.V. (2012a) A large-scale intervention to introduce orange sweet potato in rural Mozambique increases vitamin A intakes among children and women. *British Journal of Nutrition* 108(1), 163–176.

Hotz, C., Loechl, C., Lubowa, A., Tumwine, J.K., Ndeezi, G., Masawi, A.N., Baingana, R., Carriquiry, A., Brauw, A.D., Meenakshi, J.V. and Gilligan, D.O. (2012b) Introduction of β-carotene-rich orange sweet potato in rural Uganda results in increased vitamin A intakes among children and women and improved vitamin A status among children. *The Journal of Nutrition* 142, 1871–1880.

Huo, T., Ferruzzi, M.G., Schwartz, S.J. and Failla, M.L. (2007) Impact of fatty acyl composition and quantity of triglycerides on bioaccessibility of dietary carotenoids. *Journal of Agricultural and Food Chemistry* 55, 8950–8957.

Islam, S. (2006) Sweetpotato (*Ipomoea batatas* L.) leaf: its potential effect on human health and nutrition. *Journal of Food Science* 71, r13–r21.

Jalal, F., Nesheim, M.C., Agus, Z., Sanjur, D. and Habicht, J.P. (1998) Serum retinol concentrations in children are affected by food sources of β-carotene, fat intake, and anthelmintic drug treatment. *The American Journal of Clinical Nutrition* 68, 623–629.

Jayarajan, P., Reddy, V. and Mohanram, M. (1980) Effect of dietary fat on absorption of β-carotene from green leafy vegetables in children. *Indian Journal of Medical Research* 71, 53–56.

Lister, J. (2013) The Influence of Product Marketing on Food Choices. Available at: http://smallbusiness.chron.com/influence-product-marketing-food-choices-25763.html (accessed 28 April 2013).

Low, J.W. (2011) Unleashing the potential of sweetpotato to combat poverty and malnutrition in sub-Saharan Africa through a comprehensive initiative. *Acta Horticulturae* 921, 171–179.

Low, J.W. and van Jaarsveld, P.J. (2008) The potential contribution of bread buns fortified with β-carotene-rich sweetpotato in Central Mozambique. *Food and Nutrition Bulletin* 29, 98–107.

Low, J.W., Walker, T. and Hijmans, R. (2001) *The Potential Impact of Orange Flesh Sweetpotatoes on Vitamin A Intake in Sub-Saharan Africa*. International Potato Center (CIP), Lima.

Low, J.W., Arimond, M., Osman, N., Cunguara, B., Zano, F. and Tschirley, D. (2007a) Ensuring supply and creating demand for a biofortified crop with a visible trait: lessons learned from the introduction of orange-fleshed sweet potato in drought-prone areas of Mozambique. *Food and Nutrition Bulletin* 28, s258–s270.

Low, J.W., Arimond, M., Osman, N., Cunguara, B., Zano, F. and Tschirley, D. (2007b) A food-based approach introducing orange-fleshed sweet potatoes increased vitamin A intake and serum retinol concentrations in young children in rural Mozambique. *The Journal of Nutrition* 137, 1320–1327.

Low, J.W., Lynam, J., Lemaga, B., Crissman, C., Barker, I., Thiele, G., Namanda, S., Wheatley, C. and Andrade, M. (2009) Sweetpotato in sub-Saharan Africa. In: Loebenstein, G. and Thottappilly, G. (eds) *The Sweetpotato*. Springer Science + Business Media BV, Houten, The Netherlands, pp. 359–390.

Lukuyu, B., Gachuiri, C.K., Agili, S., Leon-Velarde, C. and Kirui, J. (2012) *Making High Quality Sweetpotato Silage: an Improved Tube Silage Making Method*. Sweetpotato Action for Security and Health in Africa project, Nairobi.

Masset, E., Haddad, L., Cornelius, A. and Isaza-Castro, J. (2011) A systematic review of agricultural interventions that aim to improve nutritional status of children. Institute of Education, University of London, London.

McLaren, D.S. and Frigg, M. (2001) *Sight and Life Guidebook on Vitamin A in Health and Disease*. Task Force Sight and Life, Basel, Switzerland.

Menelaou, E., Kachatryan, A., Losso, J.N., Cavalier, M. and Bonte, D.L. (2006) Lutein content in sweet potato leaves. *HortScience* 41, 1269–1271.

Mills, J.P., Tumuhimbise, G.A., Jamil, K.M., Thakkar, S.K., Failla, M.L. and Tanumihardjo, S.A. (2009) Sweet potato β-carotene bioefficacy is enhanced by dietary fat and not reduced by soluble fiber intake in Mongolian gerbils. *The Journal of Nutrition* 139(1), 44–50.

Nagujja, S. and Yanggen, D. (2005) Post harvest utilization of sweetpotato and implications for reducing incidence of vitamin A deficiency in Uganda. In: *Proceedings of African Crop Science Conference*, Vol. 7, 5–9 December, Entebbe, Uganda. African Crop Science Society, Kampala, pp. 663–668.

New Partnership for Africa's Development (NEPAD) and Comprehensive African Agriculture Development Programme (CAADP) (2008) *The Pan African Nutrition Initiative*. NEPAD, Johannesburg, South Africa, 31 pp.

Ogubi, M.A. (2007) The role of street foods as a source of micronutrients in Nairobi. MSc thesis, Kenyatta University, Nairobi, Kenya.

Olson, J.A. (1994) Absorption, transport and metabolism of carotenoids in humans. *Pure and Applied Chemistry* 66, 1011–1016.

Parada, J. and Aguilera, J.M. (2007) Food microstructure affects the bioavailability of several nutrients. *Journal of Food Science* 72, r21–r32.

Popkin, B.M., Adair, L.S. and Ng, S.W. (2012) Now and then: the global nutrition transition: the pandemic of obesity in developing countries. *Nutrition Reviews* 70, 3–21.

Ribaya-Mercado, J.D., Solomons, N.W., Medrano, Y. and Al, E. (2004) Use of the deuterated-retinol-dilution technique to monitor the vitamin A status of Nicaraguan schoolchildren 1 year after initiation of the Nicaraguan national program of sugar fortification with vitamin A. *The American Journal of Clinical Nutrition* 80, 1291–1298.

Rodriguez-Amaya, D.B. (1993) Nature and distribution of carotenoids in foods. In: Charalambous, G. (ed.) *Shelf-life Studies of Foods and Beverages: Chemical, Biological, Physical and Nutritional Aspects*. Elsevier Science Publishers, Amsterdam, pp. 547–589.

Rodriguez-Amaya, D.B. (1997) *Carotenoids and Food Preparation: the Retention of Provitamin A Carotenoids in Prepared, Processed, and Stored Foods*. John Snow Inc./Obstetrics Maternal Newborn Investigation (OMNI) project, Washington, DC.

Rodriguez-Amaya, D.B. and Kimura, M. (2004) HarvestPlus handbook for carotenoid analysis. *HarvestPlus Technical Monograph 2*. International Food Policy Research Institute (IFPRI) and International Center for Tropical Agriculture (CIAT), Washington, DC.

Roodenburg, A.J.C., Leenen, R., Van Het Hof, K.H., Weststrate, J.A. and Tijburg, L.B.M. (2000) Amount of fat in the diet affects bioavailability of lutein esters but not of α-carotene, β-carotene, and vitamin E in humans. *The American Journal of Clinical Nutrition* 71, 1187–1193.

Ruel, M.T. and Levin, C. (2000) Assessing the potential for food based strategies to reduce vitamin A and iron deficiencies: a review of recent evidence. *Food Consumption and Nutrition Division discussion paper.* International Food Policy Research Institute, Washington, DC.

Scaling Up Nutrition (2013) Tanzania: 2.7% of GDP Lost Due to Micronutrient Deficiencies. Available at: http://scalingupnutrition.org/news/tanzania-2-7-of-gdp-lost-due-to-micronutrient-deficiencies#.uanao0bmiso (accessed 4 May 2013).

Sommer, A. and West, K.P. (1996) *Vitamin A Deficiency: Health, Survival and Vision*. Oxford University Press, New York.

Takahata, Y., Noda, T. and Nagata, T. (1993) HPLC determination of β-carotene of sweet potato cultivars and its relationship with colour values. *Japanese Journal of Breeding* 43, 421–427.

Tanumihardjo, S.A. (2011) Vitamin A: biomarkers of nutrition for development. *The American Journal of Clinical Nutrition*, 94(suppl), 658s–665s.

Tanumihardjo, S.A. and Penniston, K.L. (2002) Simplified methodology to determine breast milk retinol concentrations. *Journal of Lipid Research* 43, 350–355.

Tumuhimbise, G.A., Namutebi, A. and Muyonga, J.H. (2009) Microstructure and *in vitro* beta-carotene bioaccessibility of heat processed orange fleshed sweetpotato. *Plant Foods for Human Nutrition* 64, 312–318.

Twomey, C. (2006) Hand Hygiene: Best Practices for 2006. Available at: http://www.infectioncontroltoday.com/articles/2006/02/hand-hygiene.aspx (accessed 2 May 2013).

United Nations (UN) Standing Committee on Nutrition (2011) *Sixth Report on the World Nutrition Situation: Progress in Nutrition*. UN Standing Committee on Nutrition, Geneva, Switzerland.

United States Department of Agriculture Agricultural Research Service (USDA-ARS) (2011) USDA National Nutrient Database for Standard Reference. Available at: http://ndb.nal.usda.gov/ (accessed 15 March 2015).

Unlu, N.Z., Bohn, T., Clinton, S.K. and Schwartz, S.J. (2005) Carotenoid absorption from salad and salsa by humans is enhanced by the addition of avocado or avocado oil. *The Journal of Nutrition* 135, 431–436.

van Jaarsveld, P.J., Faber, M., Tanumihardjo, S.A., Nestel, P., Lombard, C.J. and Benade, A.J.S. (2005) β-carotene-rich orange-fleshed sweetpotato improves the vitamin A status of primary school children assessed with the modified-relative-dose-response test. *The American Journal of Clinical Nutrition* 81, 1080–1087.

van Jaarsveld, P.J., Marais, D.W., Harmse, E., Nestle, P. and Rodriguez-Amaya, D.B. (2006) Retention of β-carotene in boiled, mashed orange-fleshed sweet potato. *Journal of Food Composition Analysis* 19, 321–329.

Vorster, H.H., Kruger, A. and Margetts, B.M. (2011) The nutrition transition in Africa: can it be steered into a more positive direction? *Nutrients* 3, 429–441.

zum Felde, T., Burgos, G., Espinoza, J., Eyzaguirre, R., Porras, E. and Grüneberg, W. (2009) Screening for protein, β-carotene, iron, zinc, starch and individual sugars in sweetpotato germplasm accessions by near-infrared reflectance spectroscopy (NIRS). Paper presented at the 15th Triennial International Society of Root and Tuber Crops Symposium, 2–6 November, Lima, Peru. Available at: http://r4d.dfid.gov.uk/PDF/Outputs/Misc_Crop/s8_zumfelde.pdf (accessed 15 September 2015).

56 Effect of Sweetpotato Meal and Composite Sweetpotato-meal-based Diets on Performance of Weaner Rabbits

C.C. Ekwe,* J.I. Obi and K.C. Ekwe
National Root Crops Research Institute, Umudike, Nigeria

Abstract
An experiment was conducted to evaluate the effect of sweetpotato and its composite meal-based diets on the biological performance and bio-economics of weaner rabbit production. Thirty weaner rabbits averaging 0.35 kg each were randomly allotted to five treatment groups in a completely randomized design. Each treatment group was replicated three times. Sweetpotato meal and composite sweetpotato meal replaced maize at a 20% inclusion level. Treatment numbers 1, 2, 3, 4 and 5 contained the control, sun-dried sweetpotato meal (SSPM), boiled sweetpotato meal (BSPM), composite sun-dried sweetpotato meal (CSSPM) and composite boiled sweetpotato meal (CBSPM), respectively, at 20% levels of inclusion. The weaner rabbits were fed *ad libitum* throughout the experimental period of 10 weeks. Significant differences were observed for growth performance of the rabbits in average daily weight gain, average daily feed intake and feed conversion ratio (FCR) in the weaner rabbits using sweetpotato and composite sweetpotato-meal-based diets. The values of mean daily weight gain ranged from 16.09 g in treatment 3 (BSPM) to 18.47 g in treatment 1 (control). The values of average daily feed intake ranged from 42 g in treatment 3 (BSPM) to 50 g in treatment 1 (control). Moreover, the values of the FCR ranged from 2.47 in treatment 5 (CBSPM) to 2.70 in treatment 1 (control). Significant differences ($P < 0.05$) were also observed in the bio-economics of production of weaner rabbits using composite sweetpotato meals. The cost per kilogram of feed, cost of total feed consumed per rabbit and cost per kilogram weight gain of the rabbits fed the test diets were significantly lower than the control diet. For revenue, no significant difference was observed among the treatments. Gross margins of CSSPM and CBSPM use were the same but both significantly higher than the control. We conclude that using sweetpotato and composite sweetpotato meal can be a cost-effective feeding method for weaner rabbit production.

Keywords: feed, gross margin, Nigeria, sweetpotato, sweetpotato meal, weaner rabbits

56.1 Introduction

Sweetpotato (*Ipomoea batatas* (L.) Lam.) is a tropical crop with a relatively short vegetative cycle, the storage roots being used for both human and animal consumption (Woolfe, 1992). Both foliage and storage roots have been used widely as an alternative feed for livestock in tropical countries. The roots have low protein, fat and fibre content, but high nitrogen-free extract, which thus indicates their potential value as an energy source. The vines

*chiomaekwe@yahoo.com

have low carbohydrate content but are higher in fibre and protein, and their principal value is as a source of vitamins and protein.

The chemical composition of the leaves, stems and tubers varies depending on the time of harvesting as well as on genotypic differences. The leaves have superior content of dry matter (DM) and crude protein (CP) compared with stems (Le Van An *et al.*, 2003). CP content of sweetpotato vines on a dry weight basis ranges from 16% to 29% (Dung, 2001).

There is competition between man and livestock for maize in Nigeria. Maize serves as a staple food for Nigerians and also as a good energy source for livestock. Due to this competition and its scarcity as a result of low yields in Nigeria (Durunna *et al.*, 2000) maize is costly, resulting in raising the cost of livestock feed. Maize is the major source of energy in monogastric animal feeds because of its high starch content. It provides about 3400 kcal/kg of metabolizable energy in feed (Olomu, 1995).

If the problem of low protein intake of Nigerians (less than the recommended 35 g per head/day; FAO, 1982) is to be solved through accelerated animal production, there is the need to replace maize with alternative, less costly feedstuffs such as sweetpotato meal processed into different forms, such as sun-dried sweetpotato meal (SSPM), boiled sweetpotato meal (BSPM), composite sun-dried sweetpotato meal (CSSPM) and composite boiled sweetpotato meal (CBSPM). Rabbit, on the other hand, is a highly desired meat in Nigeria, because it is a white meat with low cholesterol content.

Rabbits are a good choice to help meet the protein needs of Nigerians. They are highly prolific, have short gestation period and grow very fast. Their high feed conversion efficiency means that there is a quick return on investment, compared with other animals (Aduku and Olukosi, 1990). For these reasons, we selected the rabbit as the experimental animal for this trial.

The objective of this study was to determine the effect of sweetpotato meal and composite sweetpotato-meal-based diets on growth performance of rabbits and the economics of production of rabbits fed such diets.

56.2 Materials and Methods

Preparation of sweetpotato meals

The experiment was carried out at the rabbit production unit of the National Root Crops Research Institute in Umudike, Nigeria. The sweetpotato used in this study was also obtained from this same institute. The sweetpotato was harvested, washed and divided into four parts. One part was chipped, sun-dried and then milled (sun-dried sweetpotato meal or SSPM). The second part was boiled, chipped, sun-dried and then milled (boiled sweetpotato meal or BSPM). The third part was chipped, sun-dried, milled and then mixed with milled sun-dried sweetpotato leaves at the ratio of 5:3 (composite sun-dried sweetpotato meal or CSSPM). The fourth part was boiled, chipped, sun-dried, milled and then mixed with milled sun-dried sweetpotato leaves at the ratio of 5:3 (composite boiled sweetpotato meal or CBSPM). The resulting samples which partially replaced maize in the diets at 20% levels in treatments 2, 3, 4 and 5 were then incorporated into the diets as a mash. Treatment 1 (control) is a maize-based diet containing no sweetpotato.

Feeding of experimental rabbits

Thirty weaner rabbits averaging 0.35 kg each were randomly allotted into five treatment groups of six rabbits each. Each treatment group was replicated three times with two rabbits (male and female) per replicate. The groups were randomly assigned to five isocaloric and isonitrogenous diets in a completely randomized design. Five experimental diets were formulated such that:

- treatment 1 (T1 – control) contained no sweetpotato meal;
- treatment 2 (T2) contained sun-dried sweetpotato meal (SSPM);
- treatment 3 (T3) contained boiled sweetpotato meal (BSPM);
- treatment 4 (T4) contained composite sun-dried sweetpotato meal (CSSPM); and
- treatment 5 (T5) contained composite boiled sweetpotato meal (CBSPM).

Other fixed ingredients were added to meet the needs of weaner rabbits for vitamins and minerals as indicated in Table 56.1.

All experimental rabbits were given feed and water *ad libitum*. Records of mean final weight gain, average daily weight gain, average daily feed intake and feed conversion ratio (FCR) were taken. The experimental diets were fed for 10 weeks.

Economics of the diets

Economics of the diets were determined as described by Ukachukwu and Anugwa (1995), and modified by Ekwu (2008). It involves:

Cost per kilogram of feed = Total cost of producing 100 kg of feed divided by 100

Cost of feed consumed in Naira (₦) = Cost per kilogram of feed × Total feed consumed

Cost per kilogram of weight gain = Cost per kilogram of feed × FCR

Revenue = Price of 1 kg meat × Mean weight gain in kilograms

Gross margin = Revenue − Cost of production

Determination of proximate composition

The feeds were analysed for their proximate composition according to AOAC (1990) methods.

Data analysis

The data collected on growth performance and parameters of economics of production were subjected to analysis of variance (ANOVA) and means evaluated for their significant differences using the Duncan multiple range-test as outlined by Steel and Torrie (1980). SPSS was the statistical package used for data analysis.

56.3 Results and Discussion

Table 56.2 presents the proximate composition of the sweetpotato meal and the composite

Table 56.1. Composition of experimental diets (in percentage terms).

	Experimental diets (%)[a]				
Ingredients	T1 Control	T2 SSPM	T3 BSPM	T4 CSSPM	T5 CBSPM
Maize	32.00	12.00	12.00	12.00	12.00
SSPM	–	20.00	–	–	–
BSPM	–	–	20.00	–	–
CSSPM	–	–	–	20.00	–
CBSPM	–	–	–	–	20.00
Wheat offal	50.00	50.00	50.00	50.00	50.00
Groundnut meal	14.20	14.20	14.20	14.20	14.20
Bone meal	3.00	3.00	3.00	3.00	3.00
Salt	0.50	0.50	0.50	0.50	0.50
Vitamin premix	0.30	0.30	0.30	0.30	0.30
Total	100	100	100	100	100
Calculated composition					
Crude protein (%)	18.09	17.83	17.80	17.93	17.84
Crude fibre (%)	3.35	5.75	6.75	6.55	6.35
Energy (ME/Kcal)	2409	2206	2204	2202	2205

[a]T1, treatment 1 – control; T2, treatment 2 – sun-dried sweetpotato meal (SSPM); T3, treatment 3 – boiled sweetpotato meal (BSPM); T4, treatment 4 – composite sun-dried sweetpotato meal (CSSPM); T5, treatment 5 – composite boiled sweetpotato meal (CBSPM).

Table 56.2. Proximate composition of sweetpotato meal and composite meals.[a]

	Nutrients (%)			
	SSPM	BSPM	CSSPM	CBSSM
Crude protein (CP)	8.68	8.53	9.19	8.75
Crude fibre (CF)	8.16	6.75	9.90	9.10
Ether extract (EE)	9.19	8.00	9.00	10.55
Nitrogen free extract (NFE)	64.78	60.50	69.00	69.25

[a]SSPM, sun-dried sweetpotato meal; BSPM, boiled sweetpotato meal; CSSPM, composite sun-dried sweetpotato meal; CBSSM, composite boiled sweetpotato meal.

sweetpotato meal treatments. The values of protein observed ranged from 8.53% in BSPM to 9.19% in CSSPM, which are higher than the value (4%) observed by Dominguez (1992). The values of crude fibre observed ranged from 6.75% in BSPM to 9.90% in CSSPM are also higher than the value (4%) reported by Dominguez (1992).

Growth performance of the weaner rabbits

The growth performance showed significant differences ($P < 0.05$) among the treatments for average daily weight gain, average daily feed intake and the FCR (Table 56.3). There was a significant difference among the treatments for average daily weight gain. Treatment 5 (CBSPM) (18.13 g) was similar ($P > 0.05$) to treatment 1 (control) (18.47 g) but significantly ($P < 0.05$) higher than treatment 4 (CSSPM) (17.89 g). Treatment 4 was also higher than treatments 3 (BSPM) (16.09 g) and 2 (SSPM) (16.33 g); the latter two did not differ significantly ($P > 0.05$). The average daily weight gain of the rabbits fed composite boiled sweetpotato-meal-based diet (CBSPM) compared very well with that of the control diet. Raw sweetpotato storage roots contain medium levels of trypsin inhibitors that are known to decrease protein digestibility (Bradbury *et al.*, 1992), which could consequently affect the weight gain of the rabbits. Moist heat treatments are effective in eliminating trypsin inhibitor activity in sweetpotato (Zhang and Corke, 2001). Therefore, the significant value observed in treatment 5 could be as a result of better detoxification occasioned by the processing method and also due to incorporating leaves into the diet that contributed to better nutrient availability.

Significant differences ($P < 0.05$) were observed for average daily feed intake. Treatment 1 (control) (50 g) was significantly higher than treatments 4 (CSSPM) (45 g) and 5 (CBSPM) (45 g) that were similar ($P > 0.05$) but higher ($P < 0.05$) than treatments 2 (SSPM) (43 g) and 3 (BSPM) (42 g), that were themselves different ($P < 0.05$). The higher feed intake values observed in treatments 4 (CSSPM) and 5 (CBSPM) could be as a result of leaves incorporated in the diets. This finding agrees with those of Olorunnisomo (2007), who found that mixing sweetpotato foliage and roots resulted in better intake and weight gain than roots alone. The FCR shows significant differences ($P < 0.05$) in means across treatments, ranging from 2.70 to 2.47 with the control diet (treatment 1) having the poorest value (2.70) and the CBSPM (treatment 5) the best value 2.47. The best value observed for CBSPM could be as a result of the method of processing and leaf meal included in the diet. This finding agrees with results found by Giang *et al.* (2004) that sweetpotato roots mixed with vine meal at ratio 1:1 improved feed efficiency.

Bio-economics of production of the weaner rabbits

Significant differences ($P < 0.05$) were observed for all the bio-economic parameters

Table 56.3. Growth performance of weaner rabbits fed different treatment meals.[a]

	Treatments[b]				
Parameters	T1 Control	T2 SSPM	T3 BSPM	T4 CSSPM	T5 CBSPM
Mean initial live weight (kg)	0.35	0.35	0.34	0.36	0.36
Mean final live weight (kg)	1.65a	1.50b	1.46c	1.61a	1.63a
Average daily weight gain (g)	18.47a	16.33c	16.09c	17.89b	18.13a
Average daily feed intake (g)	50.00a	43.00c	42.00d	45.00b	45.00b
Feed conversion ratio (FCR)	2.70a	2.62b	2.67a	2.54c	2.47d

[a]Values in the same row with the different letters alongside are significantly different from each other.
[b]For details of treatments see Table 56.1.

Table 56.4. Bio-economics of production of weaner rabbits fed sweetpotato meal and composite sweetpotato-meal-based diets in Naira (local currency).[a]

	Treatments[b]				
Parameters	T1 Control	T2 SSPM	T3 BSPM	T4 CSSPM	T5 CBSPM
Cost per kilogram of feed (₦)	66.06a	42.06b	42.20b	42.16b	42.36b
Cost of total feed consumed per rabbit (₦)	218.87a	116.60b	124.06b	132.80b	133.43b
Cost per kilogram of weight gain (₦)	178.36a	110.47b	112.81b	107.22b	104.91b
Revenue (₦)	336.26	297.26	292.93	325.86	330.20
Gross margin (₦)	117.39b	180.66ab	168.86ab	193.06a	196.76a

[a]Values in the same row with the different letters alongside are significantly different from each other.
[b]For details of treatments see Table 56.1.

of production among the treatments means (Table 56.4). The values of cost per kilogram of feed, cost of total feed consumed per rabbit and cost per kilogram of weight gain followed the same pattern, treatment 1 (control) was significantly higher ($P < 0.05$) than treatments 2, 3, 4 and 5 which were themselves similar ($P > 0.05$). The costs observed in treatments 2, 3, 4 and 5 were lower than that of the control diet. This is in line with findings of Ukachukwu and Anugwa (1995) that the least cost feed formulation seeks to achieve cost input minimization and output maximization.

No significant difference ($P > 0.05$) was observed for revenue among the treatment means (Table 56.4). Treatments 2 (₦)297.26), 3 (₦292.93), 4 (₦325.86) and 5 (₦330.20) were comparable to the control diet (₦336.26). In terms of gross margin, treatments 4 (₦193.06) and 5 (₦196.76) were similar ($P > 0.05$) but significantly higher ($P < 0.05$) than treatment 1 (₦117.39) but slightly similar to treatments 2 (₦180.66) and 3 (₦168.86). Treatments 4 (CSSPM) and 5 (CBSPM) yielded better economic gains.

56.4 Conclusions

Using sweetpotato and composite sweetpotato meal can be a good feeding method in rabbit production. Therefore, use of sweetpotato meal and a composite sweetpota-to-meal-based diet in lieu of maize should be encouraged in rabbit production in Nigeria to maximize profits.

References

Aduku, A.O. and Olukosi, J.O. (1990) *Rabbit Management in the Tropics. Production, Processing, Utilization, Marketing, Economic Practical Training, Research and Prospects*. Living Books Series. G.U. Publications, Abuja, Nigeria, pp. 43–56.

Association of Official Analytical Chemists (AOAC) (1990) *Official Methods of Analysis*, 5th edn. AOAC, Washington DC.

Bradbury, J.H., Hammer, B.C. and Sugani, I. (1992) Heat stability of trypsin inhibitors in tropical root crops and rice and its significance for nutrition. *Journal of Science, Food and Agriculture* 58(1), 95–100. Available at: http://dx.doi.org/DOI:10.1002/jsfa.2740580116 (accessed 29 May 2015).

Dominguez, P.L. (1992) Feeding of sweetpotato to monogastrics. In: Machin, D. and Nyvold, S. (eds) *Roots, Tubers, Plantains, and Bananas in Animal Feeding*. Food and Agriculture Organization of the United Nations (FAO) Animal Production and Health Paper No. 95. FAO, Rome. Available at: http://www.fao.org/docrep/003/t0554e/t0554e00.htm (accessed 15 September 2015).

Dung, N.N.X. (2001) Evaluation of green plants and by-products from the Mekong Delta with emphasis on fibre utilization by pigs. PhD thesis, Swedish University of Agricultural Sciences, Uppsala, Sweden.

Durunna, C.S., Udedibie, A.B.I. and Anyanwu, G.A. (2000) Combination of maize/sorghum and cassava tuber meal as a substitute for maize in diet of laying hens. In: *Proceedings of the 25th Annual Conference of the Nigerian Society of Animal Production*, March 2000, pp. 169–173.

Ekwu, U. (2008) The effect of forage to concentrate ratios on the performance of weaner rabbits. A project report. Michael Okpara University of Agriculture, Umudike, Nigeria.

Food and Agriculture Organization of the United Nations (FAO) (1982) *Production Year Book*, Vol. 36. FAO, Rome.

Giang, H.H., Ly, L.V. and Ogle, B. (2004) Evaluation of ensiling methods to preserve sweetpotato roots and vines as pig feed. *Livestock Research for Rural Development* 16(7), Article #45. Available at: http://www.lrrd.org/lrrd16/7/gian16045.htm (accessed 15 September 2015).

Le Van An, Lindberg, B.E.F. and Lindberg, J.E. (2003) Effect of harvesting interval and defoliation on yield and chemical composition of leaves, stems, and tubers of sweet potato (*Ipomoea batatas* (L.) Lam) plant parts. *Field Crops Research* 82, 49–58.

Olomu, J.M. (1995) *Monogastric Animal Nutrition*. A Jachem Publication, Benin City, Nigeria, pp. 141–145.

Olorunnisomo, O.A. (2007) A cost–benefit analysis of sweetpotato production for fish feeding in the southwest of Nigeria. *Livestock Research for Rural Development* 19(6). Available at: http://www.lrrd.org/lrrd19/6/olor19080.htm (accessed 15 September 2015).

Steel, R.G.D. and Torrie, J.H. (1980) *Principles and Procedures of Statistics: a Biometric Approach*, 3rd edn. McGraw Hill, London.

Ukachukwu, S.N. and Anugwa, F.O.I. (1995) Bioeconomics of feeding raw or heat treated soyabeans to broilers. *Nigerian Journal of Animal Production* 22(2), 137–140.

Woolfe, J.A. (1992) *Sweet Potato: an Untapped Food Resource*. Cambridge University Press, Cambridge.

Zhang, Z. and Corke, H. (2001) Trypsin inhibitor activity in vegetative tissue of sweetpotato plants and its response to heat treatment. *Journal of Science, Food, and Agriculture* 81(14), 1358–1363. Available at: http://onlinelibrary.wiley.com/doi/10.1002/jsfa.945/abstract (accessed 15 September 2015).

57 Determining Availability of Nutrients in Sweetpotato Grown in the Kenyan Highlands Using *In Sacco* Degradability and *In Vitro* Digestability

J. Kinyua,[1*] L. Musalia,[2] P. Migwi,[3] C. Gachurii,[4] B. Lukuyu,[5] S. Agili[6] and G. Muriithi[1]

[1]Kenya Agricultural and Livestock Research Organization (KALRO), Lanet, Kenya; [2]Chuka University, Chuka, Kenya; [3]Egerton University, Njoro, Kenya; [4]University of Nairobi, Nairobi, Kenya; [5]International Livestock Research Institute (ILRI), Nairobi, Kenya; [6]International Potato Center Sub-Saharan Africa (CIP-SSA), Nairobi, Kenya

Abstract

Six sweetpotatoes cultivars, specifically Naspot1, 103001, Kemb23, Gweri, Kemb36 and Wagabolige, were grown in three agroecological locations within the highlands of Kenya: (i) Nyeri County (high altitude); (ii) Nandi County (medium altitude); and (iii) in Bomet County (low altitude). The principal factors related to sweetpotato's protein and energy digestibility are protein and dry matter content, trypsin inhibitor activity and starch digestibility. Considerable variation exists in intestinal digestion of protein among and within different protein sources, and this variation can possibly be influenced by the area where the forage was grown, the nutrient content and various other factors. The high degradability of sweetpotato suggests that it can be good protein forage and accumulation of this type of information will ensure that protein supplements are allocated on the basis of their flow rate from the rumen with a given type of diet. The digestibility and rumen *in vivo* (*in sacco*) degradation are some of the major tools used in assessing the value of a feed because they give an indication of the proportion of the feed that is digested and therefore, available for absorption. Samples of storage roots and vines of the six sweetpotato cultivars were oven dried, milled to pass through a 3 mm screen and pooled. A subsample of 3.0 g was taken for digestibility analysis and *in vivo* degradation determination. A general linear model (GLM) in a three way factorial arrangement of the statistical software SAS version 9.1 was applied to compute analysis of variance (ANOVA) to test for any significant differences in *in vitro* digestibility. A GLM in a complete randomized design of SAS version 9.1 was used to compute ANOVA to test for any significant differences on *in vivo* degradability. There was a significant difference in vine and storage roots' *in vitro* organic matter digestibility (IVOMD) due to variation in locations ($P < 0.05$), cultivars ($P < 0.05$) and management ($P < 0.05$). There was also significant interaction between management and cultivars ($P < 0.05$) both in vines and storage roots. Gweri overall had the ratoon with the highest IVOMD among all the varieties, while medium-altitude Nandi was the location with the highest average IVOMD. *In vivo* degradation characteristics of vines and storage roots of various

*jimmy.kinyua@yahoo.com

sweetpotato cultivars indicated that there was a difference in solubility or washing loss ($P < 0.05$) and degradation potential ($P < 0.05$) of both vines and storage roots in various cultivars. The high degradability and degradation of sweetpotato suggests that it can be good protein and energy forage.

Keywords: feed, forage, general linear model (GLM), *in sacco* degradability, *in vivo* digestibility, Kenya, protein, sweetpotato

57.1 Introduction

One of the major challenges facing smallholder dairy farmers in the highlands of Kenya is low milk production from their dairy cows, due to their poor quality basal diet that is not able to support high milk production. To solve the problem farmers can supplement their cows with concentrates, however, use of concentrates is prohibitive due to their high cost. Feeding sweetpotato vines and storage roots can be considered as an additional way to supplement. Sweetpotato is affordable and nutritious and farmers grow them in different agroecological locations within highlands of Kenya for both food security and feed for animals. Sweetpotato storage roots are a source of energy and the vines provide protein, while both can be used in fresh and dried form or fermented into silage (Woolfe, 1992; Hoang *et al.*, 2004).

The usefulness of a feed to a ruminant animal such as a dairy cow is determined at two stages: (i) the percentage of degradable nutrients in the foregut (rumen); and (ii) the percentage of metabolizable nutrients in the lower gut (small intestine). In the rumen (foregut), different parts of sweetpotato are degraded and fermented by ruminal microorganisms (MOs), resulting in the generation of volatile fatty acids (VFAs) among other metabolites. The VFAs are used as a source of energy for both the MOs and the host animal. The quantity of a feed that is ruminally degraded is determined by the percentage that is *in sacco* degraded. That is, a feed is put in a nylon bag and is incubated in the rumen of a fistulated animal for a determined period and later the degradability characteristics of a feed are determined and the rate of passage of the digesta (Wilkerson *et al.*, 1993). Little is known about the relative degradabilities of the wide range of tropical forages and browses available or potentially available (Ørskov *et al.*, 1980). The quantity of a feed that is digested in the abomasum and absorbed in the duodenum and ileum (lower gut) into the bloodstream is determined by *in vitro* organic matter digestibility (IVOMD) percentage. It gives an indication of the proportion of the feed that the animal is able to metabolize after by-passing the rumen undegraded, commonly referred to as a by-pass feed.

These two factors determine value of a feed when fed to a ruminant animal such as a dairy cow. The nutritional value of feed for livestock is determined in part by digestibility of nutrients (Abate *et al.*, 1984; Claessens *et al.*, 2009). Ruminant animals risk being underfed when they feed on rations based on chemical content without necessarily considering the two factors of *in sacco* degradation and IVOMD in such a ration.

In the Kenyan highlands, the sweetpotato production system is characterized by a range of different cultivars being grown in varied agroecological locations under different management regimes. The effect of variations in the production system on *in sacco* degradation and IVOMD percentages of the sweetpotato have not yet been established. Once established, these factors can form the basis of selecting preferred sweetpotato cultivars that can be grown in different locations under different management regimes for feeding cattle, in particular dairy cattle. The actual potential of the different sweetpotato cultivars as a feed supplement for dairy cows can be based on information on the degradabilities potential of varieties (Ørskov *et al.*, 1980). This study was conducted to determine the quantity of organic matter of sweetpotato vines and storage roots that is degradable by ruminal MOs in the rumen and the by-pass feed that is metabolized at the lower gut of six sweetpotato cultivars when grown under two different crop management regimes, in three different agroecological locations in the highlands of Kenya.

57.2 Materials and Methods

Six sweetpotatoes cultivars, namely Naspot1, 103001, Kemb23, Gweri, Kemb36 and Wagabolige, were grown in three agroecological locations within the highlands of Kenya: (i) Nyeri County (high altitude); (ii) Nandi County (medium altitude); and (iii) in Bomet County (low altitude). The cultivars were managed under two cropping regimes: (i) defoliating-ratooning at 75 days post-planting at 30 cm staple height; and (ii) no defoliating-ratooning ('unratooned') during the growth period. Samples of storage roots and vines of the six sweetpotato cultivars were oven dried, milled to pass through a 3 mm screen and pooled. A subsample weighing 3.0 g was taken for IVOMD percentage analysis using near-infrared reflectance spectroscopy (NIRS) at the International Livestock Research Institute in Addis Ababa, Ethiopia (Ohlsson *et al.*, 2007). *In sacco* analysis was done in the nutrition laboratory at Egerton University using procedures described below.

Subsamples of vines from six sweetpotato cultivars weighing 3.0 g were put in nylon bags (porosity 26%, size 80 mm × 60 mm, pore size 40 μm; Nybold, Switzerland). In duplicate, the bags were tied to a plastic loop, making a total of six loops. Using a complete randomized design, five of the six loops were incubated in the rumen of fistulated mature Friesian steers (weighing 400 kg) fitted with rumen cannulae. The sixth loop acted as a control and therefore, it was not incubated. The loops were later removed from the rumen, one at a time after 3, 6, 12, 24 and 48 h. Once removed,the bags were immediately immersed in a bucket of water. The bags were later untied from the loop and washed with running water, including the control loop. All bags were later oven dried to a constant weight (for approx. 24 h). The dried bags were later removed from the oven and weighed to determine the loss in weight. The lost weight was then used to determine various degradation attributes of vines according to Ørskov *et al.* (1980). This was later repeated with the storage root subsamples.

Three way factorial arrangements, general linear model (GLM) of the statistical software SAS version 9.1 (SAS Institute, Cary, North Carolina) was used to perform analysis of variance (ANOVA) to test for any significant differences ($P < 0.05$) on IVOMD. One-way GLM of the statistical software SAS version 9.1 was used to perform ANOVA to test for any significant difference $P < 0.05$ on *in sacco* degradation. Mean separation was done using least significant difference (LSD) procedures (Steel and Torrie, 1980) at a probability level of 5%.

57.3 Results and Discussion

In sacco degradation characteristics of various cultivars of sweetpotato vines indicated that there were significant differences ($P < 0.05$) in the soluble fraction between varieties. Kemb36 had the highest percentage of soluble fraction ($P < 0.05$) (Table 57.1).

The percentage of sweetpotato vine that was potentially degradable was significantly different ($P < 0.05$) in some cases (Table 57.1). Kemb36 and Naspot1 had the highest percentages; these two varieties were completely degraded. Gweri had the lowest value, indicating that most of its nutrients by-passed the rumen, hence forming the major feed for lower gut digestion.

In sacco degradation characteristics of various cultivars of sweetpotato storage roots indicated that there was a significant difference ($P < 0.05$) in the soluble fraction in different cultivars. Naspot1 had the highest percentage ($P < 0.05$) of soluble fraction (Table 57.2) among the different root types, meaning that it was the first to be subjected to degradation. The potential degradable percentage was significantly different ($P < 0.05$) among cultivars, with Kemb23 and Wagabolige having the highest ($P < 0.05$) values (Table 57.2). Given that they have the lowest degradation potential, most of the nutrients in 103001 vines and Naspot1 roots would by-pass the rumen for the lower gut digestion.

Table 57.1. *In sacco* parameters of various cultivars of sweetpotato vines.[a]

	Cultivars					
Degradation parameters	103001	Gweri	Kemb23	Kemb36	Naspot1	Wagabolige
Soluble fraction (%)	6.3c	11.4b	5.1c	13.3a	8.7b	9.8b
Potential degradable (%)	94.0a	91.4c	98.3a	100.0a	100.0a	94.6a
Degradable rate/h	0.05a	0.05a	0.05a	0.05a	0.05a	0.05a

[a]Means followed by the same letter in t he same row are not significantly different ($P > 0.05$).

Table 57.2. *In sacco* parameters of various cultivars of sweetpotato roots.[a]

	Cultivars					
Degradation parameters	103001	Gweri	Kemb23	Kemb36	Naspot1	Wagabolige
Soluble fraction (%)	10.7b	9.8b	8.6c	8.7c	13.4a	9.7b
Potential degradable (%)	91.8c	95.8b	100.0a	95.4b	79.5d	100.0a
Degradable rate/h	0.05a	0.05a	0.05a	0.05a	0.05a	0.05a

[a]Means followed by the same letter in the same row are not significantly different ($P > 0.05$).

Cultivars with high soluble fractions were the first to be subjected to degradation as they were broken down into smaller particle sizes than other cultivars at this particular time. This is in agreement with Walker *et al.* (2006), who found that many species of ruminal bacteria actively degrade starch and utilize the intermediate products of starch degradation. Cultivars that were 100% potentially degradable were completely degraded in the rumen, attributable to these cultivars attracting a high number of diverse rumen MOs.

There were significant differences ($P < 0.05$) in the percentage of IVOMD of vines of the various cultivars of sweetpotato due to location, cultivar type and management regime (Table 57.3).

There was also significant interaction ($P < 0.05$) among the three variables. Gweri vines had the highest IVOMD percentage in all the locations except one (the 150 day, unratooned, low-altitude location). This was due to the low fibre content of this particular cultivar. For cultivars 103001, Kemb23, Kemb36 and Wagabolige, vines grown at medium altitudes that were harvested 75 days after planting had the highest IVOMD (Table 57.3). This may be due to vines at an early growth stage having lower fibre content than more mature vines.

There were significant differences ($P < 0.05$) in the IVOMD percentages of storage roots of various sweetpotato cultivars due to location, cultivar type and management regime. There was also significant interaction ($P < 0.05$) among the three variables (Table 57.4).

The storage roots of Gweri had low IVOMD percentages in all the locations, under both unratooned and ratooned management regimes (Table 57.4). This was due to the high fibre content in the storage roots of Gweri, a distinct attribute when the roots are cooked. Clearly, there is a high level of interaction by fibre content with digestibility of the vines and storage roots of Gweri.

Storage roots of Naspot1 had the highest ($P < 0.05$) IVOMD percentages across all locations. However, the percentage was reduced when the cultivar was defoliated at 75 days, which increased the fibre content in the cultivar. Storage roots had higher average IVOMD percentages (below 80%) than vines (below 65%). Digestibility of energy in sweetpotato diets was high as expected (Dominguez, 1992; Hoang *et al.*, 2004). However, levels were not as high as reported by Maphosa *et al.* (2002) who found the digestibility of the sweetpotato carbohydrate fraction in storage roots to be above 90%.

Table 57.3. *In vitro* organic matter digestibility (IVOMD) percentages[a] for vines of various sweetpotato cultivars in different locations under different management regimes (75, 150R and 150U)[b].

	Location and management regime								
	High altitude			Medium altitude			Low altitude		
Cultivar	75	150R	150U	75	150R	150U	75	150R	150U
103001	57.2b	56.1c	55.3c	61.0a	55.1b	56.5b	58.0c	51.4c	50.2d
Gweri	60.8a	62.2a	61.9a	62.5a	60.7a	59.6a	62.9a	59.3a	54.3b
Kemb23	58.3b	58.9b	57.0b	59.4a	56.1b	56.4b	58.5c	53.1b	53.7b
Kemb36	55.3d	57.4a	57.3b	62.6a	56.1b	56.6b	57.7c	55.0b	54.4b
Naspot1	56.1c	54.2d	57.6b	60.0a	57.3b	58.1b	60.4b	51.4c	52.4c
Wagabolige	58.9b	56.2c	59.4b	60.5a	58.3b	57.7b	60.3b	55.7b	57.3a
Sed[c]	1.31	1.31	1.31	1.31	1.31	1.31	1.31	1.31	1.31

[a]Means followed by the same letter in the same column are not significantly different ($P > 0.05$); number of observations per mean = 6.
[b]Management regimes were: 75, crop harvested 75 days after planting; 150R, crop harvested 150 days after planting and the crop was first cut (ratooned) at 75 days; 150U, crop harvested 150 days after planting, and the crop was never cut during the growth period.
[c]Sed, standard error of deviation.

Table 57.4. *In vitro* organic matter digestibility (IVOMD) percentages[a] for storage roots of various sweetpotato cultivars in different locations under different management regimes (R, rationed; U, unratooned).

	Locations and management regime					
	High altitude		Medium altitude		Low altitude	
Cultivar	R	U	R	U	R	U
103001	74.5a	73.8a	75.2a	74.9a	69.8b	74.9b
Gweri	73.9a	64.5b	72.3a	71.0b	69.8b	70.6a
Kemb23	60.2b	77.2a	72.4a	72.9b	59.4d	68.9c
Kemb36	77.1a	76.8a	76.0a	72.0b	62.0b	63.0c
Naspot1	78.2a	76.2a	74.6a	76.4a	72.3a	75.0a
Wagabolige	65.6c	76.9a	73.5a	74.7b	67.7b	67.3c
Sed[a]	1.31	1.31	1.31	1.31	1.31	1.31

[a]Means followed by the same letter in the same column are not significantly different ($P > 0.05$); number of observations per mean = 6.

57.4 Conclusion and Recommendations

Gweri and Naspot1 vines and storage roots had the lowest potential degradable percentages. Gweri vines had the highest IVOMD percentage in all the locations except at the 150 day unratooned low-altitude location, while vines cut at 75 days grown at medium altitudes had the highest IVOMD percentage. Naspot1 storage roots had high IVOMD percentages across all the locations. The percentage reduced when the cultivar was defoliated at 75 days. Gweri cut at 75 days and undefoliated Naspot1 harvested at 150 days would be the preferred cultivars to be grown as a source of vines and storage roots, respectively, for feeding the cattle at most locations in the Kenyan Highlands.

Acknowledgements

The authors acknowledge the International Potato Center (CIP), Kenya Agricultural and Livestock Research Organization, Egerton University, Dairy Management Groups and the African Potato Association for their support.

References

Abate, A., Kayongo-Male, H., Abate, A.N. and Wachira, J.D. (1984) Chemical composition, digestibility and intake of Kenya feedstuffs by ruminants. *A Review: Nutrition Abstracts and Reviews* 54, 1–13.

Claessens, L., Stoorvogel, J.J. and Antle, J.M. (2009) *Ex-ante* assessment of dual-purpose sweetpotato in the crop–livestock system of western Kenya: a minimum-data approach. *Journal of Agricultural Systems* 99, 13–22.

Dominguez, P.L. (1992) *Feeding Sweetpotato Vines to Monogastrics*. FAO Animal Production and Health Paper No. 95. Food and Agriculture Organization of United Nations (FAO), Rome, pp. 217–234.

Hoang, H.G., Viet, L.L. and Ogle, B. (2004) Digestibility of dried and ensiled sweetpotato storage roots and vines and their effect on the performance and economic efficiency of F1 crossbred fattening pigs. *Livestock Research for Rural Development* 16, 7–8.

Maphosa, T., Gunduza, K.T., Kusina, J. and Mutungamiri, A. (2002) Evaluation of sweetpotato storage root as a feed ingredient in broiler chicken diets. *Livestock Research for Rural Development* 15, 1.

Ohlsson, C. H., Weisbjerg, M.R., Lund, P. and Hvelplund, T. (2007) Effective rumen degradation of dry matter, crude protein and neutral detergent fibre in forage determined by near infrared reflectance spectroscopy. *Physiology and Nutrition (Belgium) Journal* 91, 498–507.

Ørskov, E.R., Hovell, F.D.D. and Mould, F. (1980) The use of the nylon bag technique for the assessment of feedstuffs. *Tropical Animal Production* 5(3), 195–213.

Steel, R.G.D. and Torrie, J.H. (1980) *Principles and Procedures in Statistics*, 1st edn. McGraw Hill, New York.

Walker, N.D., Newbold, C.J. and Wallace, R.J. (2006) Nitrogen metabolism in the rumen. In: Pfeffer, E. and Hristov, A. (eds) *Proceedings of Nitrogen and Phosphorus Nutrition in Cattle*. CAB International, Wallingford, UK, pp. 71–115.

Wilkerson, V.A., Klopfenstein, T.J., Britton, R.A., Stock, R.A. and Miller, P.S. (1993) Metabolizable protein and amino acid requirements for growing cattle. *Journal of Animal Science* 71, 2777–2784.

Woolfe, J.A. (1992) *Sweet Potato: an Untapped Food Resource*. Cambridge University Press, Cambridge.

58 Sweetpotato-based Formulation: An Alternative Food Blend for Complementary Feeding

F.K. Amagloh[1]* and J. Coad[2]
[1]*University for Development Studies, Tamale, Ghana;*
[2]*Massey University, Palmerston North, New Zealand*

Abstract

In an effort to reduce the prevalence of protein-energy malnutrition among older infants, cereal-legume blends are being promoted as suitable complementary foods instead of cereal-only porridges. An example of such blend is Weanimix, which is a blend of non-dehulled maize, groundnut and non-dehulled soybean/cowpea. Although Weanimix has adequate protein and energy densities, it contains a significant level of phytate (an antinutrient that reduces iron, zinc and calcium biovailability) and virtually no β-carotene (vitamin A precursor) when white maize is used for processing. Additionally, cereal-based infant foods form a very thick porridge that requires dilution with water to get the desirable viscosity, leading to 'energy and nutrient thinning' (i.e. reduction of energy and nutrient densities). This chapter summarizes the potential of an alternative complementary food formulated from sweetpotato (denoted as ComFa) using both household- and industrial-level processing methods. Both ComFa formulations (home-based and industrial-level) and Weanimix (prepared for comparison) met the stipulated energy and fat values specified in the Codex Standards. However, the protein content of the industrial ComFa formulations (extrusion-cooked ComFa and roller-dried ComFa) was lower by 17%, but the oven-toasted ComFa (household-level formulation) and Weanimix met the protein specification. The phytate content in all the ComFa formulations was approximately a quarter of the level of 0.80 g/100 g in the Weanimix. Only the sweetpotato-based infant foods contained measurable levels of β-carotene, resulting in significantly higher vitamin A content of the oven-toasted ComFa compared with enriched Weanimix (28.73 versus 1.20 µg retinol equivalents/100 kcal). On average, sweetpotato-based formulations were higher in maltose (26 times), sucrose (five times), free glucose (19 times) and fructose (seven times) than levels in enriched Weanimix, but the ComFa formulations contained significantly less starch (10–13 versus 47 g/100 g). The high simple sugar and low starch levels in the ComFa formulations could explain the lower apparent viscosity (nine-, 13- and 20-times, for peak, 'consume' and final viscosities, respectively), higher water solubility index (seven times), and higher consumer acceptance compared with the maize-based formulation. On the basis of the compositional, functional and sensory findings of this study, the sweetpotato-based formulations have significant advantages as a complementary food compared with Weanimix due to the low level of phytate, the high levels of endogenous β-carotene and low viscosity.

Keywords: carotene, complementary/infant food, Ghana, maize, phytate, sweetpotato, vitamin A

*fkamagloh@uds.edu.gh

58.1 Introduction

Chronic vitamin A, zinc and iron deficiencies during early childhood may increase susceptibility to infections, cause stunted growth or blindness and may have irreversible effects on cognitive development (Grantham-McGregor *et al.*, 2007; Black *et al.*, 2008). Nutritional deficiencies during childhood limit an individual from achieving his/her full potential during adulthood, and have been associated with the intergenerational transmission of poverty in low-income countries (Grantham-McGregor *et al.*, 2007; Victora *et al.*, 2008; Atinmo *et al.*, 2009).

Vitamin A deficiency (VAD) among children under 5 years in sub-Saharan Africa (SSA) remains unacceptably high (44%), second to South-east Asia (50%), compared with the worldwide occurrence of 33% (WHO, 2009). Anaemia is similarly a worldwide public health problem. None the less, prevalence of anaemia (defined as haemoglobin concentration below 110 g/L) among infants and young children from 6 months old to 5 years old, is highest in Africa (68%) and South-east Asia (66%) compared with a global occurrence of 47% (de Benoist *et al.*, 2008). The main caustive factors for iron deficiency anaemia include low iron intake and poor bioavailability due to diets high in phytate or phenolic compounds (Hurrell and Egli, 2010).

In SSA, complementary foods for infants and young children are mainly prepared from cereals or legumes (Gibbs *et al.*, 2011), and have been associated with VAD because the widely used white maize, and legumes are devoid or low in β-carotene/provitamin A (Lartey *et al.*, 1998; Dewey and Brown, 2003; Nuss and Tanumihardjo, 2010). Compounding the low provitamin A levels, is the high concentration of phytate in cereal-legume blends, limiting the bioavailability of essential micronutrients including iron (Gibson *et al.*, 2010). The limitations of some of the suggestions to reduce the level of phytate in complementary foods have been discussed elsewhere (Amagloh *et al.*, 2012c).

As well as the low vitamin A and high phytate contents of cereal-based blends, another issue of nutritional concern is the high viscous porridge formed when prepared from unmalted cereals (Mosha and Svanberg, 1983, 1990). Suitable viscosity of cereal-based porridge for infants is obtained by dilution with water, leading to 'energy and nutrient thinning' (Amagloh *et al.*, 2013b), that is the reduction of energy and nutrient densities (Mosha and Svanberg, 1983, 1990; Kikafunda *et al.*, 1997).

Considering the nutritional issues mentioned above with regards to cereal-based complementary foods, we proposed sweetpotato-based (ComFa) formulations that could be prepared as a household- or an industrial-level product (Amagloh *et al.*, 2012c). Sweetpotato is: (i) suited to the tropical conditions prevailing in SSA (Padmaja, 2009); (ii) presumably low in phytate (Gibson *et al.*, 2010); and (iii) depending on the variety, could be a dietary source of β-carotene (Hagenimana *et al.*, 2001). Low (2013) and Greiner (2013) have reported that the orange-fleshed sweetpotato (OFSP) could significantly contribute to the reduction of VAD based on research findings from some African countries.

Data based on compositional analyses, viscosity and sensory attributes between household- and industrial-level ComFa formulations and Weanimix, a maize-soybean-groundnut blend, are evaluated in this chapter. Weanimix, was formulated through collaboration between United Nations Children's Fund (UNICEF) and the Nutrition Unit of the Ministry of Health, Ghana, to address protein-energy malnutrition among Ghanaian infants and young children (Agble, 1997; Lartey *et al.*, 1999). It contains 75–80% maize, 10–15% soybean/cowpea and 10% groundnut.

58.2 Food Formulation

With the aid of a nutrition calculator computer program, which was developed by Global Alliance for Improved Nutrition (GAIN), household- and industrial-level ComFa products were formulated (Table 58.1). As previously discussed by Amagloh *et al.* (2012c), the anchovy fish powder was included in the household-level ComFa formulation to improve the protein quantity and quality; skimmed milk powder was included for the industrial-level formulations as it is a common ingredient in industrial-manufactured infant cereals.

The proposed method for the household-level formulation was oven-toasting; thus, the product was referred to as oven-toasted ComFa. Extrusion-cooked ComFa and roller-dried ComFa were used to denote products processed using an extruder and roller-drier, respectively, as industrial-manufactured formulations. A detailed description of the processing of the ComFa formulations is available elsewhere (Amagloh *et al.*, 2012b).

58.3 Nutrient Composition of the Complementary Foods

All the formulations satisfied the energy (400 kcal/100 g) and fat (10–25 g/100 g) stipulated values for complementary food in the Codex Standard (Codex Alimentarius Commission, 1991). For protein, the extrusion-cooked ComFa and roller-dried ComFa were both 17% short of the protein specification of 15 g/100 g (Codex Alimentarius Commission, 1991) for complementary foods. It was suggested for the ComFa industrial-level formulations, the amount of skimmed milk used could be adjusted to satisfy the protein requirement (Amagloh *et al.*, 2012b).

However, data from Table 58.2 show that all the formulations met the estimated daily protein and fat intakes for 6–8-month-old breastfeeding infants from complementary food (Dewey and Adu-Afarwuah, 2008). None the less, for the energy requirements the ComFa formulations were lower by approximately 23% compared with the reference value of 200 kcal/day (Table 58.2). Weanimix was about 19% lower than the estimated requirement of energy per day. As it has been previously suggested, a serving size of 43 g (dry weight) of the ComFa formulations, instead of 33 g, would meet the suggested energy requirement (Amagloh *et al.*, 2012b).

Due to the lower starch:simple sugars ratio and the starch content of the sweetpotato-based formulations compared with that of Weanimix (Table 58.3), the ComFa formulations

Table 58.1. Ingredients and estimated levels of macronutrients of proposed sweetpotato-based complementary food (denoted as ComFa). (Adapted from Amagloh *et al.*, 2012c.)

Ingredient	Household-level product (g/100 g)	Industrial-level product (g/100 g)[a]
Cream-fleshed sweetpotato flour	66	72
Full-fat soybean flour	10	15
Soybean oil	6	6
Iodized salt	0.5	0.5
Sugar	0.5	0.5
Skimmed milk powder	–	6
Fish powder (anchovies)	17	–

[a]Formulation for the industrial-level ComFa products: extrusion-cooked ComFa and roller-dried ComFa.

Table 58.2. Estimated daily nutrient intake from sweetpotato- and maize-based complementary foods against recommended daily requirements for 6–8-month-old breastfeeding infants in developing countries.[a] (Adapted from Amagloh *et al.*, 2012b; Copyright 2012 © The United Nations University.)

Nutrient composition	ComFa			Weanimix	Reference value
	Extrusion-cooked	Roller-dried	Oven-toasted		
Energy (kcal/day)	153.20 ± 0.0c	152.93 ± 0.1c	154.98 ± 0.0b	162.14 ± 0.1a	200
Protein (g/day)	4.09 ± 0.03c	4.12 ± 0.03c	6.64 ± 0.05b	7.33 ± 0.04a	2
Fat (g/day)	3.32 ± 0.07c	3.43 ± 0.06b,c	3.52 ± 0.02b	3.71 ± 0.05a	0

[a]Values shown are means of triplicate estimation ± standard error of the mean for ComFa formulations and Weanimix, based on estimated average daily intake of 33 g (dry weight) of complementary food from findings by Lartey *et al.* (1999); means in a row with the same letters are not significantly different ($P > 0.05$). Reference value: Dewey and Adu-Afarwuah (2008) for infants 6–8 months old with an average daily breastmilk intake of 600–650 ml/day.

Table 58.3. Carbohydrate composition[a] (g/100 g dry matter basis) of sweetpotato- and maize-based complementary foods. (Adapted from Amagloh *et al.*, 2013b.)

Complementary food	Maltose	Sucrose	Free glucose	Free fructose	Starch[b]	Total dietary fibre	Total available carbohydrate
Extrusion-cooked ComFa	27.50 ± 2.54a	10.20 ± 0.31a	1.24 ± 0.04b	3.07 ± 0.37a	11.32 ± 1.87b	10.25 ± 2.29a	56.07 ± 1.97a
Roller-dried ComFa	30.85 ± 3.84a	10.53 ± 1.06a	1.34 ± 0.03a	2.94 ± 1.04a	10.53 ± 3.70b	10.57 ± 1.21a	58.92 ± 0.83a
Oven-toasted ComFa	25.43 ± 1.17a	10.08 ± 0.43a	1.40 ± 0.05a	2.61 ± 0.46a	13.75 ± 0.72b	8.16 ± 0.77a,b	53.28 ± 0.94a,b
Weanimix	1.06 ± 0.18b	2.01 ± 0.12b	0.07 ± 0.02c	0.39 ± 0.13b	46.72 ± 0.64a	6.08 ± 0.44b	50.25 ± 0.75b

[a]Values shown are means ± standard deviation of triplicate estimates; values with the same letter in a column are not significantly different ($P > 0.05$).

[b]Starch = total available carbohydrate minus the sum of maltose, sucrose, free glucose and free fructose; the starch content for extrusion-cooked ComFa and roller-dried ComFa was corrected for approximately 3 g/100 g of lactose from skimmed milk powder (FoodWorks, 2009) used as an ingredient.

formed a porridge with lower viscosity than Weanimix when cooked (Fig. 58.1). The naturally low viscosity of the ComFa products indicates that they would require less water during porridge preparation than the Weanimix.

Thus, there will be less 'energy and nutrient thinning' for the ComFa formulation. Also, it is likely that more of the composite flour of the sweetpotato-based formulation will be utilized during porridge preparation, and this may lead to increase in total solids per serving size as suggested above, contrary to dilution of porridge to reduce the viscosity in the case of Weanimix. Although the high viscosity of cereal porridge could be reduced by either malting or addition of amylase-rich flour (Mosha and Svanberg, 1990; Afoakwa *et al.*, 2010), this treatment is not required for the sweetpotato-based products.

It has been suggested that infants efficiently digest starch when it is present in food in small quantities (Weaver, 2000; Lentze, 2008). As a result of the lower starch:simple sugars ratio and starch content, the ComFa product would be easier to digest by infants compared with Weanimix. The ease of digestibility could make nutrients more bioavailable. Human studies are required to investigate the true digestibility of sweetpotato- and cereal-based complementary foods.

Twenty female caregivers from SSA who had experience of feeding infants were recruited to carry out a consumer preference study of roller-dried ComFa, oven-toasted ComFa and Weanimix (Amagloh *et al.*, 2013b). Results (Fig. 58.2) indicate higher preference for the ComFa formulations than Weanimix. Additionally, the data show that caregivers were more willing to give the sweetpotato-based complementary foods than Weanimix. The levels of simple sugars in the ComFa compared with those in Weanimix (see Table 58.3) could partly explain the caregivers' greater liking for ComFa in the consumer preference test.

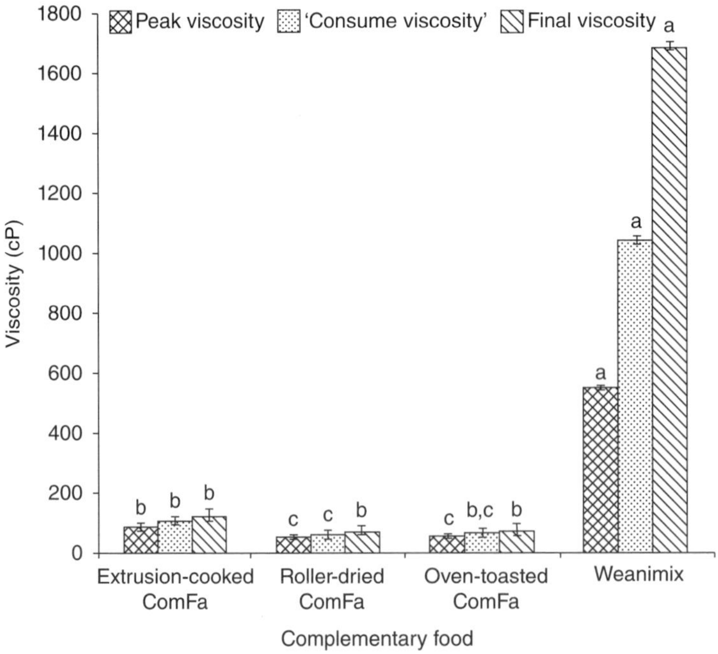

Fig. 58.1. Viscosity during cooking (peak viscosity), at serving ('consume viscosity') and during storage (final viscosity) of porridge from the complementary foods. Values shown are means of triplicate determinations ± standard deviation; bars with a different letter for each variable for the complementary foods were significantly different ($P < 0.0001$). (Adapted from Amagloh *et al.*, 2013b.)

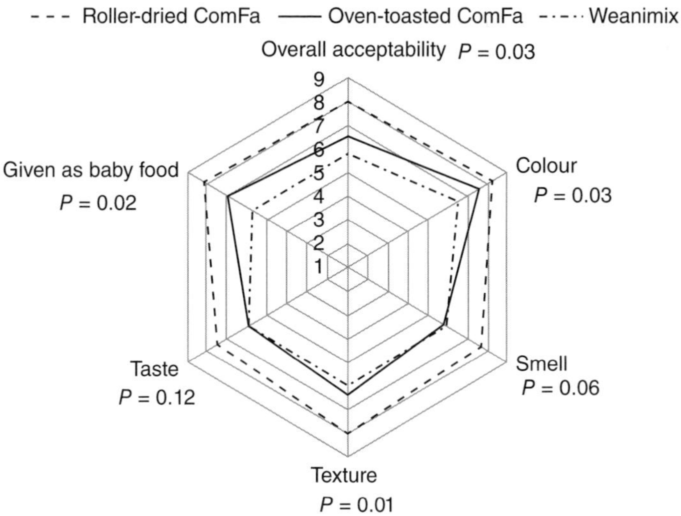

Fig. 58.2. Diagrammatic presentation of results of consumer preference study showing ratings given by 20 female caregivers for sensory attributes of formulations of infant complementary food (roller-dried ComFa, oven-toasted ComFa and Weanimix). A nine-point hedonic scale was used (where for sensory attributes: 1 = least acceptable/extremely dislike, 5 = neutral and 9 = highly acceptable/like very much; and for willingness to give product to babies: 1 = not likely, 5 = neutral, 9 = very likely). A sensory attribute with $P < 0.05$ indicates a significant difference among the complementary foods. (Adapted from Amagloh *et al.*, 2013b.)

58.4 Micronutrient and Antinutrient Levels

Only the ComFa formulations (Table 58.4) contained measurable amounts of vitamin A (β-carotene). Processing infant foods with OFSP could make the products an excellent dietary source of vitamin A (Low, 2013). Of nutritional signifance is that most of the β-carotene in the oven-toasted ComFa was retained for up to 8 weeks when stored in containers with a good moisture barrier under simulated conditions of an average temperature of 32°C and relative humidity of 85% (Amagloh *et al.*, 2013a).

Data from Table 58.4 show that the ComFa formulations may inhibit calcium, iron and zinc absorption to a lesser extent than Weanimix using the phytate:calcium, iron and zinc molar ratios (Amagloh *et al.*, 2012a), indices that has been used by other researchers (Abebe *et al.*, 2007; Chan *et al.*, 2007; Gibbs *et al.*, 2011), to estimate the relative availability of these essential micronutrients. Also, the β-carotene, which was present only in the ComFa products, has been reported to reduce the inhibitory effect of phytate and polyphenols on iron absorption (Garcia-Casal *et al.*, 2000; Layrisse *et al.*, 2000; Garcia-Casal, 2006). However, the high levels of polyphenols in the sweetpotato-based complementary foods (Table 58.4) may offset the benefit of the low levels of phytate on iron absorption (Cercamondi *et al.*, 2013). Petry *et al.* (2010) have shown that total polyphenols extracted from the hull of common bean (> 20 mg gallic acid equivalents per meal) reduced iron absorption, while Abizari *et al.* (2012) found that in cowpea-based flours, high concentrations of polyphenols do not limit iron absorption as there is a high phytate: iron ratio. Therefore, the source of polyphenols, and not polyphenols per se, could predict iron bioavailability. To our knowledge the effects of polyphenols extracted from sweetpotato on iron absorption have not been reported.

Table 58.4. Levels of micronutrients and antinutrients in sweetpotato- and maize-based complementary foods.[a] (Adapted from Amagloh *et al.*, 2012a; Copyright 2012 © Informa UK, Ltd.)

Complementary food	β-carotene (µmol/kg)[b]	Calcium (mg/kg)	Iron (mg/kg)	Zinc (mg/kg)	Phytate (mg/kg)	Total polyphenols (mg gallic acid equivalents/kg)
Oven-toasted ComFa	28.73 ± 0.79a	6008.40 ± 633.34a	69.74 ± 0.37a	21.17 ± 0.75b	2263.70 ± 320.01b	2195.09 ± 162.35a
Roller-dried ComFa	24.84 ± 0.87a	2017.70 ± 18.28b	27.88 ± 1.26b	15.79 ± 0.25c	1949.70 ± 208.40b	2039.85 ± 85.61a
Weanimix	ND[c]	3707.80 ± 251.17b	81.22 ± 4.49a	30.69 ± 0.21a	8032.70 ± 356.81a	1923.95 ± 152.25a

[a]Values shown are means of triplicate estimations ± standard error of the mean reported on a dry matter basis, except for β-carotene which is the mean of duplicate estimations. Mean values within a column with the same letters are not significantly different ($P > 0.05$).

[b]Not detected at the minimum detection limit of 0.09 µmol/kg for β-carotene; 1 µmol β-carotene = 537 µg β-carotene.

[c]ND, not detected.

58.5 Some Nutritional Concerns of the Sweetpotato-based Products

The sweetpotato-based formulations had total dietary fibre (Table 58.3) that was about twice the maximum specification of 5 g/100 g by the Codex Standard (Codex Alimentarius Commission, 1991; FAO and WHO, 2011). This may be a nutritional concern for using the sweetpotato-based product for complementary feeding based on the recommendation in the Codex Standard if the fibre was all insoluble. Preliminary data from subsequent work yet to be published indicate that about 20% of the fibre is soluble so it has the potential to act as a substrate for the growth of beneficial colonic microflora such as lactobacilli and bifidobacteria (Anderson *et al.*, 1994, 2009).

Other limitations of the ComFa formulations are: (i) the cream-fleshed sweetpotato used in our studies has lower β-carotene content compared with OFSP (Burri, 2011); (ii) the drudgery and cost involved in the flour preparation; and (iii) the total degradation of ascorbic acid as a result of the flour processing.

Therefore, there is the need to investigate other product development methods that will eliminate the processing of flour.

58.6 Recommendation for an Alternative Approach for Preparation of Sweetpotato-based Complementary Food

This alternative approach does not involve flour preparation from the sweetpotato roots. The roots, after washing, could be chipped and all the other ingredients listed in Table 58.1 could be added, and the mixture boiled to cook. The chips could be mashed into the broth when soft, and could be served to infants. The proposed formulation is more attractive as: (i) there will be a cost saving (the drying and milling cost is eliminated); (ii) preparation is less time consuming (no need to dry the chipped roots); and (iii) it could possibly result in a formulation with higher levels of essential micronutrients such as ascorbic acid and β-carotene.

58.7 Conclusion

The sweetpotato-based formulations contain less phytate than Weanimix, and thus there is likely to be less inhibition of iron and zinc absorption; and they also contain carotenoids, making such formulations a sustainable source of dietary vitamin A to complement vitamin A supplementation initiatives in low-income countries in SSA. More importantly, the sweetpotato is extensively grown as a secondary food crop in most countries in SSA (Low, 2013).

On the basis of the compositional analyses carried out in this study, the ComFa formulations have the potential to be a valuable complementary food introduced after the period of exclusive breastfeeding, particularly for infants in low-income countries, where prevalence of micronutrient deficiencies, particularly vitamin A, remains high.

Acknowledgements

We acknowledge the Nutricia Research Foundation, The Netherlands, for providing funds for this research (Project number: 2011-30), and the New Zealand International Aid and Development Agency (NZAID) for the award of a Commonwealth Scholarship for a PhD (2009–2012) to F.K. Amagloh, during which this study was done. Also, the initial funding provided by the Institute of Food, Nutrition and Human Health, Massey University, New Zealand during the formulation stage of the sweetpotato-based complementary food project is appreciated. F.K. Amagloh is most grateful to the CGIAR Research Programme on Roots, Tubers and Bananas for the full scholarship provided to attend the 9th Triennial African Potato Association conference, held on 30 June–4 July 2013 at Naivasha, Kenya.

References

Abebe, Y., Bogale, A., Hambidge, K.M., Stoecker, B.J., Bailey, K. and Gibson, R.S. (2007) Phytate, zinc, iron and calcium content of selected raw and prepared foods consumed in rural Sidama, Southern Ethiopia, and implications for bioavailability. *Journal of Food Composition and Analysis* 20(3–4), 161–168.

Abizari, A.-R., Moretti, D., Schuth, S., Zimmermann, M.B., Armar-Klemesu, M. and Brouwer, I.D. (2012) Phytic acid-to-iron molar ratio rather than polyphenol concentration determines iron bioavailability in whole-cowpea meal among young women. *Journal of Nutrition* 142(11), 1950–1955.

Afoakwa, E., Aidoo, P. and Adjonu, R. (2010) Effect of spontaneous fermentation and amylase-rich flour on the nutritive value, functional and viscoelastic properties of cowpea-fortified nixtamalized maize. *International Journal of Food Sciences and Nutrition* 61(3), 256–271.

Agble, R. (1997) Effective programmes in Africa for improving nutrition: effective programmes for improving nutrition in Ghana. *SCN News No.* 15, 9–10.

Amagloh, F.K., Brough, L., Weber, J.L., Mutukumira, A.N., Hardacre, A. and Coad, J. (2012a) Sweetpotato-based complementary food would be less inhibitory on mineral absorption than a maize-based infant food assessed by compositional analysis. *International Journal of Food Sciences and Nutrition* 63(8), 957–963.

Amagloh, F.K., Hardacre, A., Mutukumira, A.N., Weber, J.L., Brough, L. and Coad, J. (2012b) Sweet potato-based complementary food for infants in low-income countries. *Food and Nutrition Bulletin* 33(1), 3–10.

Amagloh, F.K., Weber, J.L., Brough, L., Hardacre, A., Mutukumira, A.N. and Coad, J. (2012c) Complementary food blends and malnutrition among infants in Ghana – a review and a proposed solution. *Scientific Research and Essays* 7(9), 972–988.

Amagloh, F.K., Brough, L., Weber, J.L., Mutukumira, A.N., Hardacre, A. and Coad, J. (2013a) Retention of β-carotene in cream-fleshed sweetpotato-based complementary food stored in different containers under simulated tropical temperature and humidity. *Food and Nutrition Sciences* 4, 23–28.

Amagloh, F.K., Mutukumira, A.N., Brough, L., Weber, J.L., Hardacre, A. and Coad, J. (2013b) Carbohydrate composition, viscosity, solubility, and sensory acceptance of sweetpotato- and maize-based complementary foods. *Food and Nutrition Research* 57, 18717.

Anderson, J.W., Smith, B. and Gustafson, N. (1994) Health benefits and practical aspects of high-fiber diets. *American Journal of Clinical Nutrition* 59(5), 1242S–1247S.

Anderson, J.W., Baird, P., Davis Jr, R.H., Ferreri, S., Knudtson, M., Koraym, A., Waters, V. and Williams, C.L. (2009) Health benefits of dietary fiber. *Nutrition Reviews* 67(4), 188–205.

Atinmo, T., Mirmiran, P., Oyewole, O.E., Belahsen, R. and Serra-Majem, L. (2009) Breaking the poverty/malnutrition cycle in Africa and the Middle East. *Nutrition Reviews* 67(5), S40–S46.

Black, R.E., Allen, L.H., Bhutta, Z.A., Caulfield, L.E., de Onis, M., Ezzati, M., Mathers, C. and Rivera, J. (2008) Maternal and child undernutrition 1 – global and regional exposures and health consequences. *Lancet* 371(9608), 243–260.

Burri, B.J. (2011) Evaluating sweet potato as an intervention food to prevent vitamin A deficiency. *Comprehensive Reviews in Food Science and Food Safety* 10(2), 118–130.

Cercamondi, C.I., Egli, I.M., Zeder, C. and Hurrell, R.F. (2013) Sodium iron EDTA and ascorbic acid, but not polyphenol oxidase treatment, counteract the strong inhibitory effect of polyphenols from brown sorghum on the absorption of fortification iron in young women. *British Journal of Nutrition* 111(3), 481–489.

Chan, S.S.L., Ferguson, E.L., Bailey, K., Fahmida, U., Harper, T.B. and Gibson, R.S. (2007) The concentrations of iron, calcium, zinc and phytate in cereals and legumes habitually consumed by infants living in East Lombok, Indonesia. *Journal of Food Composition and Analysis* 20(7), 609–617.

Codex Alimentarius Commission (1991) *Guidelines for Formulated Supplementary Foods for Older Infants and Young Children. Codex Standards*. Codex Alimentarius Commission, Rome.

de Benoist, B., McLean, E., Egli, I. and Cogswell, M. (eds) (2008) *Worldwide Prevalence of Anaemia 1993–2005: WHO Global Database on Anaemia*. World Health Organization (WHO) Press, Geneva, Switzerland.

Dewey, K.G. and Adu-Afarwuah, S. (2008) Systematic review of the efficacy and effectiveness of complementary feeding interventions in developing countries. *Maternal and Child Nutrition* 4, 24–85.

Dewey, K.G. and Brown, K.H. (2003) Update on technical issues concerning complementary feeding of young children in developing countries and implications for intervention programs. *Food and Nutrition Bulletin* 24 (Special Issue Based on a World Health Organization Expert Consultation on Complementary Feeding), 5–28.

Food and Agriculture Organization of the United Nations (FAO) and World Health Organization (WHO) (2011) Appendix IV – Guidelines on formulated complementary foods for older infants and young children. In: *Report of the Thirty-third Session of the Codex Committee on Nutrition and Foods for Special Dietary Uses: Codex Standards*. FAO and WHO, Bad Soden am Taunus, Germany, pp. 45–54.

FoodWorks (2009) FoodWorks 2009 Version 6. Xyris Software (Australia) Pty Ltd, Brisbane, Australia.

Garcia-Casal, M.N. (2006) Carotenoids increase iron absorption from cereal-based food in the human. *Nutrition Research* 26(7), 340–344.

Garcia-Casal, M.N., Leets, I. and Layrisse, M. (2000) Beta-carotene and inhibitors of iron absorption modify iron uptake by Caco-2 cells. *Journal of Nutrition* 130(1), 5–9.

Gibbs, M., Bailey, K.B., Lander, R.D., Fahmida, U., Perlas, L., Hess, S.Y., Loechl, C.U., Winichagoon, P. and Gibson, R.S. (2011) The adequacy of micronutrient concentrations in manufactured complementary foods from low-income countries. *Journal of Food Composition and Analysis* 24, 418–426.

Gibson, R.S., Bailey, K.B., Gibbs, M. and Ferguson, E.L. (2010) A review of phytate, iron, zinc, and calcium concentrations in plant-based complementary foods used in low-income countries and implications for bioavailability. *Food and Nutrition Bulletin* 31(2), S134–S146.

Grantham-Mcgregor, S., Cheung, Y.B., Cueto, S., Glewwe, P., Richter, L. and Strupp, B. (2007) Child development in developing countries 1: developmental potential in the first 5 years for children in developing countries. *Lancet* 369(9555), 60–70.

Greiner, T. (2013) *Vitamin A: Moving the Food-based Approach Forward*. Food and Agriculture Organization of the United Nations (FAO) and the World Health Organization (WHO), Rome.

Hagenimana, V., Low, J., Anyango, M., Kurz, K., Gichuki, S.T. and Kabira, J. (2001) Enhancing vitamin A intake in young children in Western Kenya: orange-fleshed sweet potatoes and women farmers can serve as key entry points. *Food and Nutrition Bulletin* 22(4), 376–387.

Hurrell, R.F. and Egli, I. (2010) Iron bioavailability and dietary reference values. *American Journal of Clinical Nutrition* 91(5), 1461S–1467S.

Kikafunda, J.K., Walker, A.F. and Abeyasekera, S. (1997) Optimising viscosity and energy density of maize porridges for child weaning in developing countries. *International Journal of Food Sciences and Nutrition* 48(6), 401–409.

Lartey, A., Manu, A., Brown, K.H., Peerson, J.M. and Dewey, K.G. (1998) Vitamin A status of Ghanaian breast-fed infants 6–12 months fed improved complementary foods. *FASEB Journal* 12(5), A648.

Lartey, A., Manu, A., Brown, K.H., Peerson, J.M. and Dewey, K.G. (1999) A randomized, community-based trial of the effects of improved, centrally processed complementary foods on growth and micronutrient status of Ghanaian infants from 6 to 12 mo of age. *American Journal of Clinical Nutrition* 70(3), 391–404.

Layrisse, M., Garcia-Casal, M.N., Solano, L., Baron, M.A., Arguello, F., Llovera, D., Ramirez, J., Leets, I. and Tropper, E. (2000) New property of vitamin A and beta-carotene on human iron absorption: effect on phytate and polyphenols as inhibitors of iron absorption. *Archivos Latinoamericanos De Nutricion* 50(3), 243–248.

Lentze, M.J. (2008) Gastrointestinal development, nutrient digestion and absorption. In: Koletzko, B. (ed.) *Pediatric Nutrition in Practice*. Karger, Basel, pp. 76–79.

Low, J.W. (2013) Biofortified crops with a visible trait: the example of orange-fleshed sweet potato in sub-Saharan Africa. In: Preedy, V.R., Srirajaskanthan, R. and Patel, V.B. (eds) *Handbook of Food Fortification and Health*. Springer, New York, pp. 371–384.

Mosha, A.C. and Svanberg, U. (1983) Preparation of weaning foods with high nutrient density using flour of germinated cereals. *Food and Nutrition Bulletin* 5(2), 10–14.

Mosha, A.C. and Svanberg, U. (1990) The acceptance and intake of bulk-reduced weaning foods: the Luganga village study. *Food and Nutrition Bulletin* 12(1), 69–74.

Nuss, E.T. and Tanumihardjo, S.A. (2010) Maize: a paramount staple crop in the context of global nutrition. *Comprehensive Reviews in Food Science and Food Safety* 9(4), 417–436.

Padmaja, G. (2009) Uses and nutritional data of sweetpotato. In: Loebenstein, G. and Thottappilly, G. (eds) *The Sweetpotato*. Springer Science + Business Media BV, Houten, The Netherlands, pp. 189–234.

Petry, N., Egli, I., Zeder, C., Walczyk, T. and Hurrell, R.F. (2010) Polyphenols and phytic acid contribute to the low iron bioavailability from common beans in young women. *Journal of Nutrition* 140(11), 1977–1982.

Victora, C.G., Adair, L., Fall, C., Hallal, P.C., Martorell, R., Richter, L. and Sachdev, H.S. (2008) Maternal and child undernutrition 2: consequences for adult health and human capital. *Lancet* 371(9609), 340–357.

Weaver, L.T. (2000) Complex carbohydrates and sugars. *Pediatrics* 106(Suppl. 5), 1291.

World Health Organization (WHO) (2009) *Global Prevalence of Vitamin A Deficiency in Populations at Risk 1995–2005: WHO Global Database on Vitamin A Deficiency*. WHO Press, Geneva, Switzerland.

59 Evaluation of Sensory Quality Characteristics of Muffins Developed from Sweetpotato Flours

M.S.E. Williams,[1]* M.K. Sawi,[2] N.M. Anthony[1] and S. Sowe[1]
[1]*Sierra Leone Agricultural Research Institute, Njala, Sierra Leone;*
[2]*Njala University, Njala, Sierra Leone*

Abstract

The need for alternatives to wheat flour is gradually increasing in sub-Saharan Africa. In recent years, wheat import substitution, which also promotes diversification of food products on local and national markets, is often driven by local content policies. Sweetpotato flour is one such local content material that can be used in the production of diverse food products. A study to evaluate the sensory characteristics of muffins developed from sweetpotato flours was conducted in Sierra Leone in 2010. Two sweetpotato varieties, one improved (Slipot III) and one local (Pa-Usman) were processed into flour by peeling, washing, chipping and sun drying within 16 h after harvest. Two processing methods were used to produce the sweetpotato flour: (i) sweetpotato flour from non-limed chips (SFNLCs: untreated); and (ii) sweetpotato flour from limed chips (SFLCs: treated). For the 'lime-treated' chips, sliced tubers were immersed in lime juice (16 parts lime juice to 600 parts water) for 35 min before sun drying, to improve flour whiteness. Sixty percent of each of the treated and untreated sweetpotato flours obtained from the two varieties were mixed with 40% wheat flour. These sweetpotato/wheat composite flours and wheat flour only (as a control) were used to prepare muffins using the creaming method. Fifteen trained panelists evaluated the muffin samples using a five-point hedonic scale. Results indicated that processing methods significantly affected the quality (flavour, colour and texture) of the muffins produced. Sweetpotato/wheat flour composite muffins and 100% wheat-flour muffins were not significantly different in taste. The results of the sensory attributes of the muffins developed indicated that sweetpotato/wheat composite flour can be used to prepare quality flour products such as muffins with no significant variance in consumer acceptability. Promotion of the recipe developed has the potential to enable, improve and sustain livelihoods of farmers in Sierra Leone.

Keywords: consumer acceptability, flour, muffins, processed product, sensory characteristics, Sierra Leone, sweetpotato, wheat flour substitution

59.1 Introduction

Sweetpotato (*Ipomoea batatas* L.) is one of the most important root and tuber crops in Sierra Leone, second only to cassava (*Manihot esculenta*). Sweetpotato is a highly nutritious vegetable, containing high energy, dietary fibre, biologically active phytochemicals, vitamins and minerals, which offers great benefit for use as a functional food ingredient (Brinley

*mrthswilliams@yahoo.com

et al., 2008). It is a food security crop usually grown throughout the year for food and cash. The storage roots are the most economically important part of the crop, although the leaves can also serve as a valuable source of vitamins and minerals. The storage roots are eaten as a vegetable after boiling, baking or frying and are sometimes sliced and sun-dried to produce chips which are pounded into sweetpotato flour (Woolfe, 1992).

After harvest, sweetpotato roots undergo rapid undesirable physiological changes if not properly cured. This problem, therefore, calls for processing to reduce high post-harvest losses. The high nutrient content of sweetpotato also calls for an alternative utilization of the crop other than boiling and frying of whole fresh tubers that is acceptable and cheap for different people in society. The roots can be processed into a variety of products depending on local customs and preferences. Flour is one of the forms in which sweetpotato roots can be processed to add value and increase shelf life. Raised white breads, cakes and other flour-derived products made from wheat flour have become increasingly popular in many tropical and subtropical countries, including Sierra Leone. Wheat is not grown in the country and must be imported, unfavourably altering the import–export balance. Hence considerable interest has developed in the possibility of using locally produced sweetpotato flour as a substitute for a portion of the wheat flour. Sweetpotato flour can also be used as an ingredient in complementary baby foods.

Sensory evaluation is a major determinant in the acceptability, subsequent adoption and use of food products (Otoo and Asiedu, 2009). In order to introduce new products to a community or country there is a need to evaluate the sensory properties (e.g. taste, texture, flavour, colour) of the processed products which form an integral part of the general acceptance of any food product despite their high energy density or nutrient content. This study was carried out with the aim of evaluating the sensory quality characteristics of muffins derived from sweetpotato flour in Sierra Leone.

59.2 Materials and Methods

Two sweetpotato varieties, one improved variety (Slipot III) and one local variety (Pa-Usman), were used for all the investigations undertaken. Freshly harvested roots of the sweetpotato varieties were obtained from the Njala Agricultural Research Centre (NARC) farm at Njala and loaded in sacks and conveyed to the processing facility about 100 m away.

Processing of sweetpotato into flour

Two methods were used to process the sweetpotato varieties (Slipot III and Pa-Usman) into flour. These were: (i) sweetpotato flour from non-limed chips (SFNLCs: untreated); and (ii) sweetpotato flour from limed chips (SFLCs: treated).

For the SFNLCs, 5 kg of each of the freshly harvested sweetpotato varieties were peeled using stainless steel knives, washed, sliced and sun-dried on a tarpaulin for 16–20 h for Pa-Usman and 22–24 h for Slipot III. The dried sweetpotato slices were pounded manually using a wooden pestle and mortar and then sieved using a 180 mm dimension sieve into a smooth flour. The flour was then packaged into airtight dark polythene bags and stored at room temperature.

For the SFLCs the same procedure was followed as for SFNLCs except that the sliced sweetpotato tubers of each of the two varieties were immersed in a solution of lime juice (16 parts lime juice to 600 parts water) for 35 min before sun drying to improve flour whiteness. The lime juice solution was made by squeezing limes to produce the juice, which was then sieved into a 500 ml cylinder and added to the water in a clean bowl.

Preparation of muffins for sensory evaluation

Five types of muffins were made according to the composition of the flour used:

- SFNLCs – variety Slipot III;
- SFNLCs – variety Pa-Usman;

- SFLCs – variety Slipot III;
- SFLCs – variety Pa-Usman; and
- wheat flour (control).

For the control, 100% wheat flour was used whereas for the four samples of sweetpotato muffins the flour was a sweetpotato/wheat composite flour made up of 60% sweetpotato flour and 40% wheat flour. To make this sweetpotato/wheat composite flour for all four samples, three cups (300 g) of sweetpotato flour was thoroughly mixed with two cups (200 g) of wheat flour (i.e. the sweetpotato flour:wheat flour ratio was 3:2). The mixing was done in a large clean bowl.

Ingredients used in the muffin recipe were as shown in Table 59.1 (except for the 100% wheat-flour muffins five cups of wheat flour were used instead of the three cups sweetpotato flour + two cups wheat flour). The ingredients were measured using both local measures and accurate kitchen scales. The same ingredients and measurements were applied to all of the five flour samples (including the 100% wheat flour sample) in baking the muffin cakes.

Steps used in developing the muffins

The following steps were used in making the muffins.

1. Sugar and margarine were creamed until fluffy.
2. Eggs were beaten until they became light and then they were added to the creamed mixture.
3. Vanilla essence was added and mixed.
4. Baking powder and grated nutmeg were added to the flour and mixed.
5. The flour sample was folded into the creamed mixture until it became stiff.
6. The muffin pans were then greased with margarine and the greased pan dusted with some flour. Then the creamed mixture/batter was scooped into the muffin pans to fill two-thirds of the pan.
7. The muffins were baked in a moderately hot oven at 175°C (350°F) for 20–25 min or until they were golden brown.

Table 59.1. Ingredients for composite muffins.

Ingredient	Local measure	Weight (g)
Sweetpotato flour	3 cups	300
Wheat flour	2 cups	200
Margarine	2 cups	400
Eggs	6 medium sized	270
Sugar	1½ cups	300
Vanilla essence	2 teaspoons	10
Baking powder	3 teaspoons	15
Nutmeg (grated)	1 level tablespoon	5

Sensory analysis

The sweetpotato/wheat composite muffins for each sample were served (in three batches of five each) to a panel of 15 trained judges drawn randomly from the staff of the NARC and students of the Institute of Home Sciences, Njala University. The judges were asked to score the muffin samples for colour, texture, flavour, taste and overall rating. A five-point hedonic scale ranging from 1 to 5 was designed to measure the degree of acceptance/preference of the samples with '1' indicating dislike very much, '2' indicating dislike moderately, '3' neither like nor dislike, '4' indicating like moderately, and '5' indicating like very much (Watts *et al.*, 1989). The samples were presented in identical containers, coded with the letters A, B, C, D and E, so that panelists did not know which products were made from which variety or processing method, and they were served simultaneously to ease the possibility of the panelists re-evaluating a sample (Eddy *et al.*, 2007). The evaluation exercise was done in the sensory evaluation room of the Food and Nutrition Laboratory at NARC. Precautions were taken to prevent carry over flavour during the tasting by ensuring that panelists rinsed their mouth with distilled water after each stage of the sensory evaluation.

Experimental design and statistical analysis

A 3 × 2 factorial design with three replicates was used. Data were subjected to analysis of variance (ANOVA) and Duncan's multiple range test was used to compare treatment means ($P < 0.05$). Data analysis was done using Statistical Analysis System SAS version 6.12.

59.3 Results

Colour of muffins

The results showed there was no significant difference ($P < 0.05$) in colour between the control muffins and muffins made with composite flour from the local sweetpotato variety Pa-Usman derived from lime-treated sweetpotato chips (SFLCs) – and the highest scores (4.57 and 4.53, respectively) were recorded for these two types of muffin (Table 59.2). However, there was a significant difference ($P < 0.05$) between these two muffin samples and the other three (muffins made from Slipot III flour from limed-treated sweetpotato chips (SFLCs) (4.32) and composite muffins from untreated sweetpotato flour of both local and improved varieties which scored the lowest values – Pa Usman SFNLCs 4.22 and Slipot III SFNLCs 3.87).

Taste of muffins

Data in Table 59.2 indicated no significant differences in the taste of composite muffins and the 100% wheat-flour muffins regardless of the variety and processing methods used. However, panelists appreciated the taste of muffins from 100% wheat flour more than the composite muffins, although the difference is not statistically significant.

Flavour of muffins

Data in Table 59.2 show there were significant differences ($P < 0.05$) in flavour across the muffins with 100% wheat-flour muffins scoring the highest (4.67), followed by muffins made from Slipot III untreated flour (SFNLCs) (4.28), and Slipot III lime-treated flour (SFLCs) muffins (4.15). Muffins made from Pa-Usman both treated and untreated flours scored the lowest values (4.14). These results indicate that there was a distinct difference in the flavour of the control muffins and the composite muffins.

Texture of muffins

For texture, 100% wheat-flour muffins scored a higher value (4.87) than the composite ones. The difference was significant ($P > 0.05$). Composite muffins developed from SFLCs and SFNLCs of the improved (Slipot III) variety scored lower texture values (3.86 and 4.20, respectively) than those of the local (Pa-Usman) variety (4.24 and 4.26) as shown in Table 59.2. These values indicate that there was no significant difference in the mean scores for texture of the composite muffins irrespective of the variety and processing methods.

Table 59.2. Mean scores[a] for sensory analysis of sweetpotato flour/wheat flour (ratio 3:2) composite muffins prepared from two sweetpotato varieties and two processing methods.

Variety[b]	Processing method[c]	Colour	Taste	Flavour	Texture	Overall rating
Slipot III	SFLCs	4.32b	4.35a	4.15c	3.86b	4.23b
	SFNLCs	3.87b	4.40a	4.28b	4.20b	4.41a
Pa-Usman	SFLCs	4.53a	4.38a	4.14c	4.24b	4.51a
	SFNLCs	4.22b	4.35a	4.14c	4.26b	4.43a
Wheat flour (control)		4.57a	4.44a	4.67a	4.87a	4.54a

[a]Values shown are means of scores from a five-point hedonic scale where: 1 = dislike very much; 2 = dislike moderately; 3 = neither like nor dislike; 4 = like moderately; and 5 = like very much. Means in a column with the same letter are not significantly different at $P > 0.05$ (Duncan's multiple range test).
[b]Pa-Usman is the local sweetpotato variety while Slipot III is the improved sweetpotato variety.
[c]SFLCs, sweetpotato flour from limed chips; SFNLCs, sweetpotato flour from non-limed chips.

Overall rating of sensory characteristics of muffins

With respect to overall rating, data in Table 59.2 demonstrate that there were no significant differences ($P < 0.05$) in the muffins developed from the 100% wheat flour and those of the sweetpotato composite flour muffins. However, there was a significant difference observed between the rest of the muffins and the Slipot III lime-treated (SFLCs) muffins. This implies that there was no difference in the overall acceptability of the composite muffins with the control muffins, except for treated improved sweetpotato muffins.

59.4 Discussion

As far as sensory quality characteristics of muffins developed from sweetpotato flours are concerned, there was no published research found in the literature. Colour serves as a driver in the acceptance of food. People do not eat what is not appealing and colour serves as a signal of the quality of food accepted (Kõster, 2009). The results show that lime-juice treatment significantly improved the colour of the flour and hence, the baked muffins even though improvement in colour observed for muffins from the flour of Slipot III lime-treated chips was not as marked as that for Pa-Usman. This might have been as a result of the 2–4 h difference in drying of the Slipot III chips (22–24 h instead of 16–20 h for the other samples) which might have significantly affected the baked muffins.

Taste is an important parameter when evaluating sensory attributes of food. The product might be appealing and have high energy density but without good taste, such a product is likely to be unacceptable (Muhimbula *et al.*, 2011). The difference in flavour of the composite muffins and the control muffins might be due to the fact that panelists are used to the flavour of 100% wheat-flour muffins and that the proportion of wheat flour substituted was high (3/2). The panelists accepted the taste of the composite muffins. Note that surface texture can give an important clue to the acceptance of food (Darko, 2010). The difference in texture of the 100% wheat-flour muffins and the composite muffins might have been due to the fact that a pestle and mortar was used to pound the sweetpotato chips into flour instead of a hammer mill. Finally, panelists' overall acceptance of the developed muffins was positive which implies that substitution of some amount of wheat flour with sweetpotato flour could be used in muffin production with some degree of sensory acceptance.

59.5 Conclusion and Recommendations

The results of the sensory attributes of the muffins developed from various sources have shown that with the same processing methods, sweetpotato flour mixed with a portion of wheat flour can make a virtually unnoticeable difference with regards to the sensory qualities of muffins and other flour products. Addition of lime juice can also help improve the whiteness of the resulting products.

It was recommended that the use of hammer mills and an appropriate drying facility would improve the quality of chips and flours and hence, the developed products. Also promotion of recipes developed and other stakeholders' involvement, especially marketers, will enable sustainability and improvement in the livelihoods of farmers and the general populace in Sierra Leone.

Future studies should look at substituting sweetpotato flour with different proportions of wheat flour in order to determine the best ratio for attaining high consumer acceptability.

Acknowledgements

The authors would like to thank Njala Agricultural Research Centre (NARC) for letting them use their facilities and for providing the sweetpotato varieties. They would also want to thank Professor Tunde Oguntona of the Federal University of Abeokuta, Prince Norman of NARC and Mr Augustine Mansaray for proofreading and making necessary inputs to this work. Also the authors would like to say a very big thank you to the sensory panelists.

References

Brinley, T.A., Truong, V.D., Coronel, P., Simunovic, J. and Sandeep, K.P. (2008) Dielectric properties of sweet potato puree at 915 MHz as affected by temperature and chemical composition. *International Journal of Food Properties* 11, 158–172.

Darko, S. (2010) Sensory evaluation of multiple fortified stock powder. *Pakistan Journal for Nutrition* 9(8), 742.

Eddy, N.O., Udofia, P.G. and Eyo, D. (2007) Sensory evaluation of wheat/cassava composite bread and effect of label information on acceptance and preference. *Africa Journal of Biotechnology* 6, 2415–2418.

Kõster, E.P. (2009) Diversity in the determinants of food choice. A psychological perspective. *Food Quality and Preference* 20, 70–82.

Muhimbula, H.S., Issa-Zacharia, A. and Kinabo, J. (2011) Formulation and sensory evaluation of complementary foods from local, cheap and readily available cereals and legumes in Iringa, Tanzania. *African Journal of Food Science* 5(1), 26–31.

Otoo, E. and Asiedu, R. (2009) Sensory evaluation: the last hurdle in varietal development of yams (*Dioscorea rotundata* Poir) in Ghana. *African Journal of Biotechnology* 8(21), 5747–5754.

Watts, B.M., Ylimalei, G.L., Jeffery, L.E. and Elias, L.G. (1989) *Basic Sensory Methods for Food Evaluation*. International Development Research Centre, Ottawa, 160 pp.

Woolfe, J.A. (1992) *Sweet Potato: an Untapped Food Resource*. Cambridge University Press, Cambridge.

Index

Page numbers in **bold** refer to figures and tables.